Pietro Fre' received his PhD from the University of Torino in 1974. He is currently Professor of Theoretical Physics at the International School for Advanced Studies (SISSA/ISAS), Trieste. He was Associate Professor of Theoretical Physics at the University of Torino until 1990, and has also worked as Research Associate at the University of Bielefeld, the California Institute of Technology, Torino University, and at CERN. His research activities have been focused on particle physics and statistical mechanics, especially supergravity, superstrings and topological field theories. Prof Fre' has, in collaboration with Leonardo Castellani and Riccardo D'Auria, written for World Scientific the three-volume textbook *Supergravity and Superstrings: A Geometric Perspective.*

Paolo Soriani received his "laurea" degree from the University of Milano in 1987 and his PhD in particle theory from SISSA in 1992. He is currently a post-doctoral fellow at the University of Milano. Dr Soriani has started his scientific career under the supervision of Prof Luciano Girardello and Prof Fre'. His main research interests have been in string theory, conformal field theories, topological models, and applications of complex geometry to those subjects.

# THE N=2 WONDERLAND

## From Calabi–Yau manifolds to topological field theories

**Pietro Fre'**
SISSA–Trieste

**Paolo Soriani**
Università degli Studi di Milano

World Scientific
*Singapore • New Jersey • London • Hong Kong*

*Published by*

World Scientific Publishing Co. Pte. Ltd.
P O Box 128, Farrer Road, Singapore 9128
*USA office:* Suite 1B, 1060 Main Street, River Edge, NJ 07661
*UK office:* 57 Shelton Street, Covent Garden, London WC2H 9HE

**THE N=2 WONDERLAND:
FROM CALABI–YAU MANIFOLDS TO TOPOLOGICAL FIELD THEORIES**



ISBN 981-02-2027-8

Printed in Singapore.

**To Paola and Tiziana**

**and**

**to the memory**

**of our fathers**

*We can forgive a man for making a useful thing as long as he does not admire it. The only excuse for making a useless thing is that one admires it intensely.*

Oscar Wilde

*from the Preface to*
*The Picture of Dorian Gray, 1890*

# PREFACE

*This book is based on a series of lectures given by Pietro Fre' at SISSA, at DESY, at the University of Torino and also at UCLA in the academic years 1991–92 and 1992–93.*

*Lecture notes were taken by Paolo Soriani and later the two authors elaborated and considerably extended to the present book form the material, which includes also results from Soriani's Ph.D. Thesis.*

*The aim was to present in a unitary perspective the logical development that unifies into a single, fascinating subject the topics related with N=2 supersymmetry in two and four space–time dimensions. Beginning with the Kähler structure of low energy supergravity lagrangians, through the analysis of string compactifications on Calabi–Yau manifolds, one reaches the heart of the matter by considering the chiral ring structure of N=2 superconformal models and of their parent N=2 field theories in two dimensions. The concept of topological twist relates in a profound way these theories with d=2 topological field theories, deepening the understanding of mirror symmetry, of the special Kähler geometry of Calabi–Yau moduli spaces and the analysis of the Picard–Fuchs equations associated with the Griffiths period mapping. The relation between the Landau–Ginzburg picture of N=2 superconformal models and the σ-model picture is elucidated by showing, as Witten recently did, that the two kinds of field theories are effective low energy theories of the same spontaneously broken gauge model in two different phases. Our emphasis, which is pedagogical, is on a self consistent presentation of this beautiful subject that blends algebraic geometry with quantum field theory, providing also new techniques to current research in various areas.*

## Suggestions to the Reader

The first chapter, "An Introduction to the Subject", is written in a style substantially different from that of the other chapters, since it is a descriptive essay that covers, in a simplified way, all the main ideas and the main results contained in the rest of the book. The second chapter, "A Bit of Geometry and Topology", provides a summary of the mathematics the reader should be familiar with in order to study the subsequent chapters. From the point of view of physics, the logical development of our subject begins with Chapter 3 "Supergravity and Kähler Geometry". Our suggestion is to read Chapters 3–8 in the given order since they are organized according to a consistent line of thought.

Chapter 1, instead, can be read independently as an introductory primer and it is mainly directed to newcomers to the present field. Chapter 2 is a sort of reference chapter where the reader can refresh his memory about mathematical definitions and theorems utilized elsewhere. However, it is also conceived as a self-contained presentation of the mathematical environment where the physical ideas are rooted.

## Comments on Bibliography

As the reader will realize, we have supplied a list of references that, to the best of our knowledge, should be a sufficient basis for further reading, although it is far from being exhaustive. We apologise to all the authors of papers relevant to the subject that escaped our attention.

## Acknowledgements

We thank, for the many useful comments and illuminating discussions, our friends D. Anselmi, M. Bianchi, M. Billo', L. Bonora, V. Bonservizi, R. D'Auria, B. Dubrovin, S. Ferrara, M. Francaviglia, F. Fucito, L. Girardello, F. Gliozzi, R. Iengo, M. Martellini, A. Van Proeyen, C. Reina, G. Rossi and W. Troost.

# CONTENTS


**1 AN INTRODUCTION TO THE SUBJECT** 1
- 1.1 The Remarkable Interplay . . . 2
  - 1.1.1 Supergravity and Kähler geometry . . . 2
  - 1.1.2 Special Kähler geometry . . . 4
- 1.2 Moduli and Criticality . . . 4
  - 1.2.1 Landau–Ginzburg critical models and the moduli . . . 4
  - 1.2.2 N=2 superconformal field theories . . . 11
- 1.3 Moduli and Algebraic Varieties . . . 12
  - 1.3.1 The chiral ring in N=2 superconformal theories . . . 14
  - 1.3.2 The vanishing locus of the superpotential as a Calabi–Yau manifold . . . 15
  - 1.3.3 The Griffiths residue map and the Hodge ring . . . 15
- 1.4 The Art of Quantizing Zero . . . 17
- 1.5 Mirror Maps . . . 22
- 1.6 Bibliographical Note . . . 23

**2 A BIT OF GEOMETRY AND TOPOLOGY** 25
- 2.1 Introduction . . . 25
- 2.2 Fibre Bundles . . . 25
  - 2.2.1 Definition of a fibre bundle . . . 25
  - 2.2.2 Sheaves and Cech cohomology . . . 27
  - 2.2.3 Sections of a fibre bundle . . . 30
  - 2.2.4 Bundle maps . . . 30
  - 2.2.5 Equivalent bundles . . . 31
  - 2.2.6 Pull-back bundles . . . 31
- 2.3 Vector Bundles, Connections and Curvatures . . . 33
  - 2.3.1 Fibre metrics . . . 33
  - 2.3.2 Product bundle . . . 34
  - 2.3.3 Whitney sum . . . 34
  - 2.3.4 Tensor product bundle . . . 35
  - 2.3.5 Principal fibre bundles . . . 35
  - 2.3.6 Connections on a vector bundle . . . 36

2.4 Complex Structures on 2n-Dimensional Manifolds . . . . . . . . . . . . . . 37
2.5 Metric and Connections on Holomorphic Vector Bundles . . . . . . . . . . 41
2.6 Kähler Metrics . . . . . . . . . . . . . . . . . . . . . . . . . . . . . . . . . 44
2.7 Characteristic Classes and Elliptic Complexes . . . . . . . . . . . . . . . 46
2.8 Hodge Manifolds and Chern Classes . . . . . . . . . . . . . . . . . . . . . 57
2.9 Bibliographical Note . . . . . . . . . . . . . . . . . . . . . . . . . . . . . 64

**3 SUPERGRAVITY AND KÄHLER GEOMETRY 65**
3.1 Introduction . . . . . . . . . . . . . . . . . . . . . . . . . . . . . . . . . 65
3.2 The Geometric Structure of Standard N=1 Supergravity . . . . . . . . . . 66
3.2.1 Holomorphic Killing vectors on the scalar manifold and the momentum map . . . . . . . . . . . . . . . . . . . . . . . . . . . . 71
3.2.2 The momentum map and the complete bosonic lagrangian of N=1 matter-coupled supergravity . . . . . . . . . . . . . . . . . . . . 74
3.2.3 Extrema of the potential and Kähler quotients . . . . . . . . . . 76
3.2.4 Effective N=1 supergravities obtained from Calabi–Yau compactifications . . . . . . . . . . . . . . . . . . . . . . . . . . . . . 79
3.3 Special Kähler Geometry . . . . . . . . . . . . . . . . . . . . . . . . . . 81
3.3.1 Special Kähler manifolds with special Killing vectors . . . . . . . 81
3.3.2 Special geometry and N=2, D=4 supergravity . . . . . . . . . . . 87
3.4 Bibliographical Note . . . . . . . . . . . . . . . . . . . . . . . . . . . . . 91

**4 COMPACTIFICATIONS ON CALABI–YAU MANIFOLDS 93**
4.1 Introduction to Calabi–Yau Compactifications . . . . . . . . . . . . . . . 93
4.1.1 D=10, N=1 matter-coupled supergravity . . . . . . . . . . . . . . 94
4.1.2 Killing spinors and SU(3) holonomy . . . . . . . . . . . . . . . . 96
4.1.3 The plan of this chapter . . . . . . . . . . . . . . . . . . . . . . 99
4.2 D=10 Anomaly-Free Supergravity . . . . . . . . . . . . . . . . . . . . . 101
4.2.1 The role of anomaly-free supergravity in the derivation of Calabi–Yau compactifications . . . . . . . . . . . . . . . . . . . . . . . 101
4.2.2 Strategy to derive anomaly-free supergravity . . . . . . . . . . . 103
4.2.3 The free differential algebra (*step 1*) . . . . . . . . . . . . . . . 107
4.2.4 Parametrization of the super-Poincaré curvatures (*step 2*) . . . . . 108
4.2.5 Cohomology of superforms (*step 3*) . . . . . . . . . . . . . . . . 109
4.2.6 Discussion of the homogeneous $H$-Bianchi (*step 4*) . . . . . . . . 110
4.2.7 The BPT-theorem (*step 5*) . . . . . . . . . . . . . . . . . . . . . 111
4.2.8 Construction of the 3-form $X$ (*step 6*) . . . . . . . . . . . . . . 112
4.2.9 Field equations of MAFS (*step 7*) . . . . . . . . . . . . . . . . . 113
4.2.10 Calabi–Yau compactifications as exact solutions of minimal anomaly-free supergravity . . . . . . . . . . . . . . . . . . . . . . . . 115
4.3 Properties of Calabi–Yau Manifolds . . . . . . . . . . . . . . . . . . . . 115
4.3.1 Ricci-flatness and $SU(n)$ holonomy . . . . . . . . . . . . . . . . 116
4.3.2 Harmonic forms and spinors . . . . . . . . . . . . . . . . . . . . 117

4.3.3 The covariantly constant spinor . . . 119
4.3.4 The holomorphic $n$-form . . . 119
4.3.5 The Hodge diamond of Calabi–Yau 3-folds . . . 122
4.4 Kaluza–Klein zero-modes and Yukawa Couplings . . . 127
4.4.1 Analysis of the gauge sector . . . 128
4.4.2 Analysis of the gravitational sector . . . 131
4.4.3 Yukawa couplings . . . 135
4.5 Complete Intersection Calabi–Yau Manifolds . . . 136
4.6 Bibliographical Note . . . 141

**5 N=2 FIELD THEORIES IN TWO DIMENSIONS 143**
5.1 Introduction . . . 143
5.2 Abstract N=2 Superconformal Theories . . . 146
5.3 N=2 Minimal Models . . . 153
5.4 The Rheonomy Framework for N=2 Field Theories . . . 158
5.4.1 N=2 2D supergravity and the super-Poincaré algebra . . . 158
5.4.2 Chiral multiplets in curved superspace . . . 165
5.5 An N=2 Gauge Theory and Its Two Phases . . . 169
5.5.1 The N=2 abelian gauge multiplet . . . 170
5.5.2 N=2 Landau–Ginzburg models with an abelian gauge symmetry . 172
5.5.3 Structure of the scalar potential . . . 174
5.5.4 Extension to non abelian gauge symmetry . . . 175
5.5.5 R-symmetries and the rigid Landau–Ginzburg model . . . 178
5.5.6 N=2 sigma models . . . 182
5.5.7 Extrema of the N=2 scalar potential, phases of the gauge theory and reconstruction of the effective N=2 $\sigma$-model . . . 186
5.6 N=2 Landau–Ginzburg Models and N=2 Superconformal Theories . . . 196
5.7 Landau–Ginzburg Models and Calabi–Yau Manifolds . . . 203
5.8 Landau–Ginzburg Potentials and Pseudo-Ghost First Order Systems . . 206
5.9 The Griffiths Residue Mapping and the Chiral Ring . . . 215
5.9.1 Rational meromorphic $(n+1)$-forms and the Hodge filtration . . 216
5.9.2 Interpretation of the residue map in N=2 conformal field theory . 220
5.9.3 Explicit construction of the harmonic $(n-k,k)$-forms and the realization of the chiral ring on the Hodge filtration . . . 221
5.10 Bibliographical Note . . . 226

**6 MODULI SPACES AND SPECIAL GEOMETRY 229**
6.1 Introduction . . . 229
6.2 The Special Geometry of $(2,1)$-Forms . . . 232
6.3 The Special Geometry of $(1,1)$-Forms . . . 238
6.4 Special Geometry from N=2 World Sheet Supersymmetry . . . 240
6.5 Concluding Remarks . . . 256
6.6 Bibliographical Note . . . 256

**7 TOPOLOGICAL FIELD THEORIES** **259**

7.1 Introduction . . . . . 259
7.2 The Geometric Formulation of BRST Symmetry . . . . . 265
7.3 Topological Yang–Mills Theories . . . . . 274
7.4 Topological Sigma Models . . . . . 283
7.5 The A and B Topological Twists of an N=2 Field Theory . . . . . 289
7.6 Twists of the Two-Phase N=2 Gauge Theory . . . . . 293
7.6.1 The topological BRST algebra . . . . . 293
7.6.2 Interpretation of the A-model and topological $\sigma$-models . . . . . 297
7.6.3 Interpretation of the B-model and topological Landau–Ginzburg theories . . . . . 301
7.7 Correlators of the Topological Sigma Model . . . . . 305
7.7.1 The topological $\sigma$-model or A-twist case . . . . . 307
7.7.2 Topological $\sigma$-models on Calabi–Yau 3-folds . . . . . 319
7.7.3 The B-twist case and the Hodge structure deformations . . . . . 325
7.8 Topological Conformal Field Theories . . . . . 334
7.9 Correlators of the Topological Landau–Ginzburg Model . . . . . 341
7.9.1 Applications of the residue pairing formula . . . . . 345
7.10 Topological Observables in the Two-Phase Theory . . . . . 352
7.11 Bibliographical Note . . . . . 360

**8 PICARD–FUCHS EQUATIONS AND MIRROR MAPS** **363**

8.1 Introduction to Mirror Symmetry . . . . . 363
8.1.1 The mirror quintic . . . . . 368
8.1.2 The issue of flat coordinates and Picard–Fuchs equation . . . . . 375
8.2 Picard–Fuchs Equations for the Period Matrix . . . . . 377
8.2.1 Picard–Fuchs equations for the cubic torus . . . . . 380
8.2.2 Picard-Fuchs equation for the one-modulus $M\mathbb{CP}_{p-1}(p)$ hypersurfaces, and its singularity structure . . . . . 383
8.2.3 Perspective . . . . . 386
8.3 Picard–Fuchs Equations and Special Geometry . . . . . 387
8.3.1 Introduction and summary . . . . . 387
8.3.2 Differential equations and $W$-generators . . . . . 391
8.3.3 Associated first order linear systems . . . . . 395
8.3.4 The flat holomorphic connection of special Kahler manifolds . . . 398
8.3.5 Holomorphic Picard–Fuchs equations for n-dimensional special manifolds . . . . . 403
8.3.6 The non-holomorphic Picard–Fuchs equations of special manifolds 408
8.4 Monodromy and Duality Groups . . . . . 410
8.4.1 Introduction . . . . . 410
8.4.2 The duality group $\Gamma_W$ of $M\mathbb{CP}_{p-1}(p)$ hypersurfaces . . . . . 411
8.4.3 Monodromy group of the cubic torus . . . . . 416

8.4.4 Barne's integral transform and the calculation of the monodromy matrix $T_0$ . . . 419
8.5 The Mirror Map and the Sum over Instantons . . . 429
8.5.1 Yukawa coupling as the fusion coefficient of the chiral ring . . . 431
8.5.2 General strategy for the evaluation of the Yukawa coupling of the mirror quintic . . . 434
8.5.3 Logarithmic behaviour of the solutions in the neighbourhood of $\psi = \infty$ . . . 438
8.5.4 The instanton expansion of the Yukawa coupling and the prediction of the number of rational curves on the quintic 3-fold . . . 440
8.5.5 The special Kählerian metric of the moduli space of Kähler class deformations for the quintic 3-fold . . . 443
8.5.6 Summary and conclusion . . . 446
8.6 Bibliographical Note . . . 447

**9 FAREWELL** **449**

**BIBLIOGRAPHY** **450**

# Chapter 1

# AN INTRODUCTION TO THE SUBJECT

The subject of this book is the remarkable interplay between *the geometry of complex manifolds* and *quantum field theory* that occurs in the context of N=2 supersymmetry. This interplay has led to the development of a variety of new insights where physical ideas, arising from *statistical field theory and the renormalization group*, have been conjugated with other physical ideas arising in the context of *superstring compactifications*, via the common roots of these two physical problems in the structure of *superconformal field theories* and the geometry that these theories involve. Last but not least, N=2 supersymmetry is related, by means of a map (*the topological twist*), which is well defined both in two and four dimensions, to topological field theories. These are quantum field theories where the functional integral becomes exactly calculable in a non-perturbative way since the space of physical configurations reduces to the finite dimensional *moduli space* of *instantons*. The value of topological field theories lies in many directions. On one hand it provides the way of recasting a classical mathematical problem, namely *intersection theory* on the *moduli space* of certain geometrical structures, like *self-dual connections* or *holomorphic maps*, into the physical language of *field theory* and the *path integral*: this allows a beneficial exchange of results from mathematics to physics and vice-versa. A notable example of this exchange from physics to mathematics is provided by the prediction of the number of *rational curves* of degree $k$ on *quintic 3-folds*, prediction that is achieved by the use of *mirror maps* and that is discussed in the last chapter of this book. On the other hand the relation between the framework of standard field theories and the framework of topological field theories, provided by the topological twist, deepens our non-perturbative understanding of the former, allowing, by means of the latter, a *"topological calculation"* of important classes of *Green functions* in ordinary field theories. The main example considered in the present book of such a use of topological field theories is the evaluation of *Yukawa couplings* in *Calabi–Yau compactifications* of the heterotic superstring.

As an introduction to this subject, whose systematics is postponed to the later chapters, in the following pages we take a bird's eye view of the whole battlefield, making a survey of the main conceptual structures the reader will encounter in the sequel. This survey aims both at familiarizing the newcomer with the content of this research field and at presenting the logical structure that links the various topics treated in this book into a single fabric.

## 1.1 The Remarkable Interplay

The interplay between complex geometry and N=2 supersymmetry occurs at various stages both in two and four dimensions. Let us make a brief introductory survey of the main highlights.

### 1.1.1 Supergravity and Kähler geometry

The first stage where complex geometry mixes with supersymmetry is actually related with the structure of the N=1 theory, the additional request of an N=2 invariance imposing only further constraints. We consider N=1 matter-coupled supergravity in four dimensions [93], [25], [272] (for a review see [76]). In this theory the scalar and pseudoscalar fields sitting in each Wess–Zumino multiplet $\left[0^+, 0^-, \frac{1}{2}\right]$ combine into as many complex scalar fields $z^i$ that can be regarded as local coordinates on a *Hodge–Kähler manifold* $\mathcal{M}_{scalar}$. By definition Hodge–Kähler manifolds are Kähler manifolds $\mathcal{M}$ such that there exists a line bundle $\mathcal{L} \xrightarrow{\pi} \mathcal{M}$ constructed over $\mathcal{M}$, whose first Chern class $c_1(\mathcal{L})$ equals the Kähler class $K(\mathcal{M})$ of the base manifold. Given a holomorphic section $W(z) \in \Gamma(\mathcal{L}, M)$ of this line bundle, the Hodge condition implies that

$$
\begin{aligned}
K \stackrel{\text{def}}{=} \frac{\mathrm{i}}{2\pi}\, g_{ij^\star}\, dz^i \wedge d\overline{z}^{j^\star} &= \frac{\mathrm{i}}{2\pi}\, \partial\overline{\partial} \log \|W(z)\|^2 \\
\|W(z)\|^2 &= |W(z)|^2\, e^{\mathcal{K}(z,\overline{z})}
\end{aligned}
\tag{1.1.1}
$$

where $\mathcal{K}(z,\overline{z})$ is the Kähler potential for the Kähler metric on the scalar manifold $\mathcal{M}_{scalar}$. The physical meaning of the section $W(z)$ is that of superpotential. Together with the Kähler metric $g_{ij^\star}(z,\overline{z})$ it encodes all the self-interactions of the Wess–Zumino multiplets and the interactions of these latter with the gravitational supermultiplet. In particular the potential energy of the scalar fields and the gravitino mass term are given by the following celebrated formulae [93]:

$$
\begin{aligned}
V(z,\overline{z}) &= e^{G(z,\overline{z})} \left(\partial_i G\, \partial_{j^\star} G\, g^{ij^\star} - 3\right) \\
m_{\frac{3}{2}}(z,\overline{z}) &= \exp\left[\frac{1}{2}\, G(z,\overline{z})\right] \\
G(z,\overline{z}) &\stackrel{\text{def}}{=} \log\|W(z)\|^2 = \mathcal{K}(z,\overline{z}) + \log|W(z)|^2
\end{aligned}
\tag{1.1.2}
$$

Also the mass matrix of the fermions and the Yukawa couplings are codified in the structure of $W(z)$. Suppose, for instance, that all the Wess–Zumino multiplets $\left(z^I, \chi^I\right)$

are tree-level massless. Then $W(z)$ has no quadratic term and it begins with a cubic term:

$$W(z) = W_{IJK} z^I z^J z^K + \mathcal{O}(z^4) \tag{1.1.3}$$

where $W_{IJK}$ is a symmetric tensor (actually a section of $\mathcal{L} \otimes^3_{sym} T^{(1,0)}\mathcal{M}_{scalar}$). This tensor is named the *Yukawa coupling* since it appears in the supergravity lagrangian through terms of the following form:

$$\mathcal{L}_{Yukawa} = W_{IJK} \overline{\chi}^I \chi^J z^K \tag{1.1.4}$$

In the present situation, the Yukawa coupling $W_{IJK}$ is just an array of constants since, by definition, it is the coefficient of the cubic term in $W(z)$. However, there is a more interesting possibility that is implicit in the structure of formulae (1.1.2). Let the set of Wess–Zumino multiplets $\{z^I\}$ be the union of two subsets

$$\{z\} = \{M\} \cup \{\mathcal{C}\} \tag{1.1.5}$$

where, by definition, the fields $\{M^a\}$ $(a = 1, \ldots, n)$, named the *moduli*, have the following property. The scalar potential in the first of Eqs. (1.1.2) does not depend on their values when the remaining fields $\{\mathcal{C}^a\}$ are set to zero:

$$\frac{\partial}{\partial M^a} V(M, \mathcal{C})\,|_{\mathcal{C}=0} = 0 \qquad \forall\, M^a \in \mathbb{C} \tag{1.1.6}$$

In other words the moduli are the *flat directions* of the scalar potential. The existence of the moduli fields implies that the *classical vacua* of supergravity, corresponding to the *extrema* of the scalar potential, rather than a discrete set, form a *manifold*, by definition the *moduli space* $\mathcal{M}_{moduli}$. Moduli space has a Kähler geometry, described by the Kähler potential

$$\mathcal{K}_{mod}\left(M, \overline{M}\right) = \mathcal{K}\left(M, \overline{M}, \mathcal{C} = 0, \overline{\mathcal{C}} = 0\right) \tag{1.1.7}$$

which we obtain by restricting the full Kähler potential $\mathcal{K}(z, \overline{z})$ to the subspace of moduli fields. The study of such a Kähler geometry will be one of our main concerns in the sequel. Indeed, important information on the geometry of $\mathcal{M}_{moduli}$ can be obtained when the moduli fields have an interpretation as deformation parameters of geometrical structures constructed on other manifolds. This is what happens in *Calabi–Yau compactifications* of superstrings [70]. In the effective low energy N=1 supergravity lagrangians obtained from these compactifications, the moduli fields are interpreted as the *deformation parameters* of either the *Kähler class* or the *complex structure* of the *internal six-dimensional* manifold. In any case, if there are moduli, irrespectively of their geometrical interpretation, the superpotential (1.1.3) can be replaced by

$$W(M, \mathcal{C}) = W_{\alpha\beta\gamma}(M)\, \mathcal{C}^\alpha \mathcal{C}^\beta \mathcal{C}^\gamma + \mathcal{O}\left(\mathcal{C}^4\right) \tag{1.1.8}$$

where the Yukawa couplings $W_{\alpha\beta\gamma}(M)$ of the non-moduli fields are now dependent on the moduli. As already pointed out, $W_{\alpha\beta\gamma}(M)$ is a section of the bundle $\mathcal{L} \otimes^3_{sym} T^{(1,0)}\mathcal{M}_{moduli}$ and it carries a great deal of information on the geometry of $\mathcal{M}_{moduli}$. Most of the theoretical machinery developed in this book is directed to the understanding and to the calculation of $W_{\alpha\beta\gamma}(M)$.

### 1.1.2 Special Kähler geometry

The second level of interplay between complex geometry and supersymmetry occurs in the coupling of vector multiplets to four-dimensional N=2 supergravity. Here a new geometrical structure has been uncovered by physicists [75] that of *special Kähler geometry.* The complex scalar fields $z^i$ belonging to an N=2 vector multiplet $\left[1, 2 \times \frac{1}{2}, 0^+, 0^-\right]$ span a space $\mathcal{SK}$ that, in addition to being a Hodge–Kähler manifold, is also characterized by the following identity satisfied by the Riemann tensor:

$$\mathcal{R}_{i^\star jl^\star k} = g_{l^\star j} g_{ki^\star} + g_{l^\star k} g_{ji^\star} - e^{2\mathcal{K}} W_{i^\star l^\star s^\star} W_{tkj} g^{s^\star t} \tag{1.1.9}$$

where $W_{tkj} \in \Gamma\left(\mathcal{L} \otimes^3_{sym} T^{(1,0)}\mathcal{SK}, \mathcal{SK}\right)$ is a suitable holomorphic section of the same bundle we considered above when we discussed Yukawa couplings. The $n$-dimensional complex manifolds characterized by Eq. (1.1.9) have an underlying symplectic structure which consists of a flat $(2+2n)$-dimensional complex vector bundle with $Sp(2+2n, \mathbb{R})$ as structural group and $\mathcal{SK}$ as base manifold. The sections of such a bundle are an essential ingredient in the construction of N=2 supergravity, but they are also prearranged for the alternative interpretation of $\mathcal{SK}$ manifolds as moduli spaces of either *Calabi–Yau 3-folds* $\mathcal{M}_3^{CY}$ or N=2, $c = 9$ superconformal field theories. Indeed we shall prove that such moduli spaces have the property (1.1.9) and that the sections of the associated symplectic vector bundle are identified with the *periods*, along a *homology basis of 3-cycles*, of the unique *holomorphic* 3-form $\Omega^3$, whose existence is equivalent to the defining Calabi–Yau property $c_1\left(\mathcal{M}_3^{CY}\right) = 0$. Furthermore, from the point of view of the previous discussion, the sections $W_{tkj} \in \Gamma\left(\mathcal{L} \otimes^3_{sym} T^{(1,0)}\mathcal{SK}, \mathcal{SK}\right)$ are precisely the Yukawa couplings.

The constraints of N=2 supersymmetry lead, therefore, to a remarkable situation where a set of holomorphic data like the Yukawa couplings seem sufficient to determine the hermitean geometry of the underlying manifold.

## 1.2 Moduli and Criticality

Let us now turn to the interpretation of the moduli as deformations of appropriate structures and to the interplay between concepts of geometrical origin with concepts rooted in statistical mechanics.

### 1.2.1 Landau–Ginzburg critical models and the moduli

Landau-Ginzburg theory was originally introduced as a phenomenological theory to discuss the behaviour of a statistical system $\mathcal{S}$ in the vicinity of a critical point (for a review see standard textbooks like [208], [197]). The lagrangian

$$\begin{aligned} \mathcal{L}_{LG} &= \partial_\mu \phi \partial^\mu \phi - V(T, \phi) \\ V(T, \phi) &= \frac{1}{4!}\lambda(T)\phi^4 + \frac{1}{2!}\mu^2(T)\phi^2 \end{aligned} \tag{1.2.1}$$

describes the dynamics of a field $\phi(x)$, that corresponds to some order parameter of the system $\mathcal{S}$, for instance the magnetization $M(x)$, at site $x \in \mathcal{S}$ in the physical sample. The coupling constants in the lagrangian (1.2.1), namely $\lambda(T)$ and $\mu^2(T)$, are functions of the temperature $T$, in such a way that we actually have a family of euclidean field theories parametrized by $T$. The criticality of the system manifests itself in the qualitatively different analytical structure of the scalar potential $V(T,\phi)$ for values of the temperature above and below the critical temperature $T_c$. The energetically most favourable state is given by a constant field

$$\phi(x) = \phi_0 = \langle\phi\rangle \tag{1.2.2}$$

whose value is an extremum of the potential:

$$\frac{\partial}{\partial\phi} V(T,\phi)|_{\phi=\phi_0} = 0 \tag{1.2.3}$$

and represents the average value of the order parameter, for instance the average magnetization $\langle M\rangle$. Now, as is well known from the standard analysis of the Higgs phenomenon, the physical situation is very different if the potential $V(T,\phi)$ is such that $\lambda(T)$ and $\mu^2(T)$ have the same sign or opposite signs. Indeed, in the first case, the extremum equation:

$$\frac{1}{3!}\lambda(T)\,\phi_0^3 + \mu^2(T)\phi_0 = 0 \tag{1.2.4}$$

has the only solution $\phi_0 = 0$, while in the second case we have

$$\phi_0 = \pm\sqrt{-\frac{6\,\mu^2(T)}{\lambda(T)}} \tag{1.2.5}$$

A smooth transition from one regime to the other is realized if, as a function of the temperature, the parameter $\mu(T)^2$ behaves as follows:

$$\begin{cases} \mu(T)^2 < 0 & \text{for } T < T_c \\ \mu(T)^2 = 0 & \text{for } T = T_c \\ \mu(T)^2 > 0 & \text{for } T > T_c \end{cases} \tag{1.2.6}$$

Above the critical point the euclidean field $\phi(x)$ is massive and the inverse of its mass provides an estimate of the correlation length:

$$r(T) \sim m^{-1} = \sqrt{\frac{1}{\mu^2(T)}} \tag{1.2.7}$$

Below the critical point $\phi(x)$ has a negative mass signalling the instability due to the spontaneous breaking of the $\mathbb{Z}_2$ symmetry:

$$\phi(x) \longrightarrow -\phi(x) \tag{1.2.8}$$

At the critical point $\phi(x)$ is a massless field and, as expected from general arguments, the correlation length in the statistical system becomes infinite:

$$r(T_c) = \infty \tag{1.2.9}$$

In this well known analysis, the point of interest for our purposes is the following: *the main features of the possible physics below and above the critical point are encoded in the structure of the critical scalar potential*:

$$V(T_c, \phi) = \frac{1}{4!}\lambda(T_c)\phi^4 \tag{1.2.10}$$

Which is the property that makes the function $V(T_c, \phi)$ in Eq. (1.2.10) be critical? It is an elementary fact: the extremum equation

$$\frac{\partial}{\partial \phi} V(T_c, \phi)|_{\phi=\phi_0} = 0 \qquad \frac{1}{3!}\lambda(T)\,\phi_0^3 = 0 \tag{1.2.11}$$

has a solution:

$$\phi_0 = 0 \tag{1.2.12}$$

which is three times degenerate. This is just the signal that we are sitting on the top of a hill such that any small move away from it will change the situation in a qualitative way. Indeed, it suffices to perturb $V(T_c, \phi)$ with a small term of lower polynomial order

$$V(T_c, \phi) \longrightarrow V(T_c, \phi) + \delta\mu^2(T)\phi^2 \tag{1.2.13}$$

and the degeneracy of the extrema (1.2.12) is immediately removed. These rather trivial observations are the starting point of a structured and rich chapter of modern mathematics, mostly developed by the Russian school of Arnold and collaborators [14], that goes under the name of *singularity theory*. We define *singularity* a function $f(\phi)$ whose derivative $\frac{\partial}{\partial\phi}f(\phi)$, like that of $V(T_c, \phi)$, has degenate zeros in a singular critical point $\phi = \phi_0$: note that we are now referring to critical points in order parameter space, rather than in temperature space. Then one names *resolution of the singularity* a perturbation of the original function:

$$f'(\phi) = f(\phi) + \sum_{i=1}^{N} t_i\,\delta f_i(\phi) \tag{1.2.14}$$

such that the zeros of $\frac{\partial}{\partial\phi}f'(\phi)$, while being in the same number as those of the unperturbed one, are no longer degenerate: in other words the perturbation achieves a *splitting of the critical points* that for the *singular* function $f(\phi)$ were coalescing. In Eq. (1.2.14) $t_i$ are arbitrary numerical parameters and $\delta f_i(\phi)$ is a basis of perturbations that is essentially determined by the very singular function $f(\phi)$. Indeed the basic question is: *What are the possible perturbations $\delta f_i(\phi)$, given $f(\phi)$?* The answer provided by the mathematical theory is remarkably tuned with the physical way of arguing based on the *renormalization group analysis*. From the physical point of view one considers all the possible local operators $\mathcal{O}_n(x)$ by means of which the potential $V(T_c, \phi)$ can be deformed: these are all the powers $\mathcal{O}_n(x) = \phi^n$ of the field $\phi(x)$. Considering the anomalous dimensions developed by these operators, one classifies them as *relevant*, *marginal* or *irrelevant* according to the way they scale under a renormalization group transformation:

$$\mathcal{O}_n(\Lambda x) \sim \Lambda^{\delta_n}\mathcal{O}_n(x) \Longrightarrow \begin{cases} \delta_n + D > 0 & \text{relevant operator} \\ \delta_n + D = 0 & \text{marginal operator} \\ \delta_n + D < 0 & \text{irrelevant operator} \end{cases} \tag{1.2.15}$$

by $D$ denoting the euclidean space–time dimensions. The irrelevant operators are washed away at the *infrared critical point* of the euclidean field theory, while the relevant and the marginal ones have to be taken into account. The presence of $N$ marginal operators is a signal that the *singularity* $V(T_c, \phi) = f(\phi)$ is not isolated, rather it is just an element of a family of functions $f(\mu_1, \mu_2, \ldots, \mu_N, \phi)$, depending on $N$ continuous parameters $\mu_i$, which, for all values of these parameters, share the property of admitting a singular critical point. In other words, for all values of $\mu_i$ the zero $\phi_0(\mu_i)$ of $\frac{\partial}{\partial\phi}f(\mu_1, \mu_2, \ldots, \mu_N, \phi)$ is degenerate. The relevant operators, instead, are those responsible for the *splitting of the critical points*, namely, in physical language, for the spontaneous symmetry breaking that characterizes the phase transition. Now in the mathematical theory of singularities the admissible perturbations correspond precisely to the physical notion of marginal and relevant operators and are distinguished among themselves by the effect they have or not have on the splitting of critical points. Loosely speaking, for a power like singularity $\phi^n$, there are no marginal operators and the relevant ones are the powers $\phi^l$ with $l \leq N - 2$.

To make these observations slightly more precise it is convenient to consider the following lagrangian density in two space–time dimensions:

$$\begin{aligned} \mathcal{L}^{(LG)}_{N=2} &= -(\partial_+ X^{i^*}\partial_- X^i + \partial_- X^{i^*}\partial_+ X^i) \\ &\quad + 2i(\psi^i\partial_-\psi^{i^*} + \widetilde{\psi}^i\partial_+\widetilde{\psi}^{i^*}) + 2i(\psi^{i^*}\partial_-\psi^i + \widetilde{\psi}^{i^*}\partial_+\widetilde{\psi}^i) \\ &\quad + 8\Big\{(\psi^i\widetilde{\psi}^j\partial_i\partial_j\mathcal{W} + \text{c.c.}) + \partial_i\mathcal{W}\partial_{i^*}\mathcal{W}^*\Big\} \end{aligned} \tag{1.2.16}$$

where $X^i$ $(i = 1, \ldots, M)$ is a collection of $M$ complex scalar fields, $\psi^i$ and $\widetilde{\psi}^i$ are the components of as many two-dimensional Dirac spinors:

$$\Psi^i = e^{-i\pi/4}\begin{pmatrix}\psi^i \\ \widetilde{\psi}^i\end{pmatrix} \tag{1.2.17}$$

the analogous starred objects are the complex conjugates of the above spinors and $\mathcal{W}(X)$, named the *superpotential*, denotes a holomorphic function (usually a polynomial):

$$\mathcal{W} : \mathbb{C}^M \longrightarrow \mathbb{C} \tag{1.2.18}$$

Finally $\partial_\pm = \partial_0 \pm \partial_1$ denotes the derivatives in light-cone coordinates. The lagrangian (1.2.16) defines the N=2 supersymmetric Landau–Ginzburg theory in two dimensions.

Such a field theory will be one of the focuses of our attention in the sequel. All of its quantum structure is codified in the properties of the superpotential, regarded from the point of view of singularity theory. To appreciate the reason why we name (1.2.16) the N=2 supersymmetric version of the Landau–Ginzburg model, consider the case with only one field $X$, where the potential is a simple power:

$$\mathcal{W}(X) = \frac{1}{(n+1)!} X^{n+1} \tag{1.2.19}$$

In this case the bosonic part of the lagrangian (1.2.16) becomes very similar to the form of the Landau–Ginzburg lagrangian (1.2.1) considered at the beginning, when the values of the coupling constants are those corresponding to the critical temperature:

$$\mathcal{L}^{(n)}_{N=2} = -(\partial_+ X^* \partial_- X + \partial_- X^* \partial_+ X) + \frac{1}{(n!)^2}(X^* X)^n \tag{1.2.20}$$

For any choice of the superpotential $\mathcal{W}(X)$ the potential energy

$$V(X) = \sum_{i=1}^{N} |\partial_i \mathcal{W}(X)|^2 \tag{1.2.21}$$

is a sum of squares, so that its extrema are at $V(X) = 0$ and are given by the solution to the holomorphic equation:

$$\frac{\partial}{\partial X^i} \mathcal{W}(X)|_{X^i = X^i_0} \stackrel{\text{def}}{=} \partial_i \mathcal{W}(X_0) = 0 \tag{1.2.22}$$

The superpotential $\mathcal{W}(X)$ is a critical potential, namely it corresponds to a Landau–Ginzburg theory at the critical temperature $T_c$ if the its critical points, defined by Eq. (1.2.22), are degenerate. Such a situation is realized if $\mathcal{W}(X)$ is a quasi-homogeneous function, namely if there exists $M$ positive rational numbers $\omega_i \in \mathbf{Q}_+$ and a positive integer $d_\mathcal{W} \in \mathbf{Z}_+$ such that

$$\mathcal{W}(\lambda^{\omega_1} X_1, \ldots, \lambda^{\omega_M} X_M) = \lambda^{d_\mathcal{W}} \mathcal{W}(X_1, \ldots, X_M) \quad \forall \lambda \in \mathbf{C} \tag{1.2.23}$$

A simple example of such a superpotential is given by

$$\mathcal{W}(X) = \sum_{i=0}^{M} X_i^{N_i} \qquad N_i \in \mathbf{Z}_+ \tag{1.2.24}$$

In this case the property (1.2.23) is verified if one sets

$$d_\mathcal{W} = \text{m.c.m.}\{N_i\} \qquad \omega_i = \frac{d_\mathcal{W}}{N_i} \tag{1.2.25}$$

where m.c.m. $\{N_i\}$ denotes the minimum common multiple of the exponents $N_i$. The case $M = 1$ is the already considered example of a simple power superpotential. Singularity theory gives a very simple answer for the *possible deformations* of the critical superpotential. Consider the infinite dimensional ring

$$\mathbb{C}\,[X_1,\ldots,X_M] \stackrel{\text{def}}{=} \mathbb{C}\,[X] \tag{1.2.26}$$

of polynomials in the $M$ complex variables $X_1,\ldots,X_M$ and in $\mathbb{C}\,[X]$ consider the ideal

$$\mathcal{I}_W\,[X] \stackrel{\text{def}}{=} \partial\mathcal{W}(X) \subset \mathbb{C}\,[X] \tag{1.2.27}$$

of all those polynomials that are of the form

$$P(X) = P^i(X)\frac{\partial\mathcal{W}(X)}{\partial X_i} \tag{1.2.28}$$

the coefficients $P^i(X)$ being arbitrary polynomials in the same variables. Consider then the quotient

$$\mathcal{R}_W \stackrel{\text{def}}{=} \frac{\mathbb{C}\,[X]}{\partial\mathcal{W}(X)} \tag{1.2.29}$$

As defined above $\mathcal{R}_W$ is a finite-ring, named the *chiral ring* of the superpotential $\mathcal{W}(X)$, and it contains elements of a degree ranging from zero to a maximum value

$$d_{\max} = \deg H(X) \qquad H(X) \stackrel{\text{def}}{=} \det\{\partial_i\partial_j\mathcal{W}(X)\} \tag{1.2.30}$$

The polynomial $H(X)$ defined in Eq. (1.2.30) is named the *hessian* of the superpotential and it can be shown that its equivalence class, modulo the *vanishing ideal* $\partial\mathcal{W}(X)$, is the unique element of the chiral ring with the maximum degree. It is named the *top element.*

Then according to singularity theory the deformed superpotential is given by

$$\mathcal{W}_{def}\,(X) = \mathcal{W}\,(X) + \sum_{P\in\mathcal{R}_W} t_P\,P(X) \tag{1.2.31}$$

where the sum is extended to all the elements of the chiral ring, the complex parameters $t_P$ playing the same role as in our previous elementary examples was played by the parameter $\mu^2(T)$. By inserting the deformed superpotential (1.2.31) into the lagrangian (1.2.16) we perform an N=2 supersymmetry preserving perturbation of the original critical Landau–Ginzburg model. From the point of view of *statistical field theory,* the criticality of the original quasi-homogeneous superpotential manifests itself into the fact that the theory (1.2.16), although not conformal invariant at the classical level, becomes *conformal invariant at the infrared fixed point.* This, combined with N=2 supersymmetry, ensures that the theory (1.2.16) with a quasi-homogeneous superpotential $\mathcal{W}(X)$ is equivalent, at the infrared fixed point, to an *N=2 superconformal model.* This is no longer true if the critical superpotential is replaced by the deformed one $\mathcal{W}_{def}\,(X)$. Conformal invariance is broken although N=2 supersymmetry is preserved. Indeed each element

$P(X)$ of the chiral ring corresponds to a deformation of the field-theory by means of an operator that can be relevant, marginal or irrelevant depending on its homogeneity degree with respect to the weights $\omega_i$:

$$P\left(\lambda^{\omega_1}X^1,\ldots,\lambda^{\omega_M}X^M\right) = \lambda^{d_P}P\left(X^1,\ldots,X^M\right) \quad \begin{cases} d_P < d_{\mathcal{W}} & \text{relevant} \\ d_P = d_{\mathcal{W}} & \text{marginal} \\ d_P > d_{\mathcal{W}} & \text{irrelevant} \end{cases} \tag{1.2.32}$$

For instance if we consider the following cubic superpotential in three variables

$$\mathcal{W}(X,Y,Z) = \frac{1}{3}\left(X^3 + Y^3 + Z^3\right) \tag{1.2.33}$$

the *chiral ring* is composed of the following eight elements:

$$\begin{array}{c} \mathbf{1} \\ X \;,\; Y \;,\; Z \\ XY \;,\; YZ \;,\; ZX \\ XYZ \end{array} \tag{1.2.34}$$

the cubic polynomial $XYZ$ being the hessian. Correspondingly the perturbed superpotential is

$$\begin{aligned} \mathcal{W}_{def}(X,Y,Z) &= \frac{1}{3}\left(X^3 + Y^3 + Z^3\right) + t_7XYZ \\ &\quad + t_0 + t_1X + t_2Y + t_3Z \\ &\quad + t_4XY + t_5YZ + t_6ZX \end{aligned} \tag{1.2.35}$$

In this case there are seven relevant operators and one marginal operator, $XYZ$, which, in this example, coincides with the top element of the chiral ring. Such an identification is by no means a general rule. Usually the top element has degree higher than $d_{\mathcal{W}}$ and it is an irrelevant operator. Giving a non-vanishing value to the seven parameters $t_0, t_1, t_2, t_3, t_4, t_5, t_6$ that correspond to relevant operators, the original potential is deformed in such a way as to split its critical points. The superpotential, instead, remains critical if $t_0 = t_1 = t_2 = t_3 = t_4 = t_5 = t_6 = 0$ and a non-vanishing value is assigned only to the marginal parameter $t_7$. This state of affairs is understood also in terms of symmetry breaking. The critical superpotential (1.2.33) possesses a manifest $\mathbb{Z}_3$-symmetry acting as follows:

$$(X\,,Y\,,Z) \longrightarrow (\alpha X\,,\alpha Y\,,\alpha Z) \qquad \alpha = \exp\left[\frac{2\pi k i}{3}\right] \quad k = 0,1,2 \bmod 3 \tag{1.2.36}$$

Such a symmetry is broken by all the terms with coefficients $t_i$ $(i = 0,\ldots,6)$ but it is preserved by the marginal perturbation with coefficient $t_7$.

The coefficients of the *marginal perturbations*, like the $t_7$ of the previous example, are the *moduli* of critical Landau–Ginzburg models. Indeed, as already hinted, the presence of marginal operators signals the fact that the considered critical superpotential $\mathcal{W}(X)$ is not isolated, rather it is just an element of a continuous family of such superpotentials, parametrized by a number of moduli equal to the number of marginal operators. In the above example, the critical family of superpotentials is given by

$$\mathcal{W}(X,Y,Z;\psi) = \frac{1}{3}\left(X^3 + Y^3 + Z^3\right) - \psi XYZ \tag{1.2.37}$$

where for consistency with later notations $t_7$ has been renamed $-\psi \in \mathbb{C}$.

This situation is completely parallel to that occurring in the analysis of N=1 supergravity lagrangians. There the presence of moduli fields was the signal of a degeneracy of the vacuum states. Classical supergravity vacua, namely the extrema of the scalar potential, fill a manifold, the *moduli space* $\mathcal{M}^{SUGRA}_{moduli}$, parametrized by the vacuum expectation values of the moduli fields. In the same way critical superpotentials admitting marginal perturbations fill another *moduli space*, $\mathcal{M}^{LG}_{moduli}$.

*Is there a relation*

$$\mathcal{M}^{SUGRA}_{moduli} \stackrel{?}{\leftrightarrow} \mathcal{M}^{LG}_{moduli} \tag{1.2.38}$$

*between these two kinds of moduli spaces?* The answer is of course yes and such a relation is one of the main goals pursued in this book. It is sometimes described as the relation between the *macroscopic* and the *microscopic* viewpoint on string theory. Under suitable circumstances the relation (1.2.38) is in fact an equality. To tell this story we still have some way to go.

### 1.2.2 N=2 superconformal field theories

As we already stated, if the superpotential is critical, at the infrared fixed point the N=2 Landau–Ginzburg model flows to an *N=2 superconformal theory*, namely to a field theory where the physical states are left-right factorized, being constructed out of the tensor product of two representations of the *N=2 superconformal algebra* [1, 2, 3] (see the later Eqs. (5.2.1) and (5.2.3) for a description of this algebra). In general we have

$$\begin{aligned} |\text{state}\rangle &= \lim_{z,\bar{z}\leftarrow 0} \Psi_{\alpha\beta}(z,\bar{z})\,|0\rangle_L \otimes |0\rangle_R \\ \Psi(z,\bar{z}) &= \psi_\alpha(z)\,\widetilde{\psi}_\beta(\bar{z}) \end{aligned} \tag{1.2.39}$$

where $\psi_\alpha(z)$, $\widetilde{\psi}_\beta(\bar{z})$ are *primary operators* of the left (respectively right) moving superconformal algebra and $|0\rangle_L$, $|0\rangle_R$ are the corresponding vacua; $z, \bar{z}$ are complex coordinates on the two-dimensional world-sheet. Which tensor products of which primary operators are allowed is an intrinsic property of the N=2 superconformal theory one considers and it is determined by the parent N=2 Landau–Ginzburg model. To begin with the most basic things let us recall that any *conformal field theory* is characterized by the value of its central charge $c$. If we start from a Landau–Ginzburg model with a quasi-homogeneous superpotential as defined by Eq. (1.2.23), then the central charge of the corresponding

N=2 superconformal model is given by the formula

$$c = \sum_i^M 3\left(1 - 2\frac{\omega_i}{\cdot d_W}\right) \tag{1.2.40}$$

In the case of the quasi-homogeneous potential (1.2.24), setting

$$N_i = k_i + 2 \qquad k_i \in \mathbb{Z}_+ \tag{1.2.41}$$

Eq. (1.2.40), upon use of Eq. (1.2.25), yields

$$c = \sum_i^M \frac{3\,k_i}{k_i + 2} \tag{1.2.42}$$

This formula leads to one question and to one important conclusion. The question is easily posed and easily answered. Why, by means of the position (1.2.41), have we excluded quadratic terms in the potential? Just because we want massless scalar fields at the critical point. The conclusion, instead, has far-reaching consequences. In the representation theory of the N=2 superconformal algebra, the values

$$c = \frac{3\,k}{k + 2} \qquad \forall k \in \mathbb{Z}_+ \tag{1.2.43}$$

correspond to the so-called *minimal models*, namely to the *unitary representations* with a finite number of *primary fields.* Hence the N=2 Landau–Ginzburg model, with superpotential (1.2.24), flows, at the infrared critical point, to a *tensor product of N=2 minimal models.* These latter are solvable theories and, as a consequence, also the Landau–Ginzburg model is solvable at the critical point. *Of this fact one takes advantage [161] in order to construct the solvable $N = 2$, $c = 9$ conformal field theory that describes superstring propagation on the six compactified extra dimensions, when these latter are chosen to correspond to a Calabi–Yau 3-fold. This is the way the moduli space of the Landau-Ginzburg model is eventually identified with the moduli space of the low-energy effective lagrangian.*

## 1.3 Moduli and Algebraic Varieties

We have concluded the previous section with a quite strong and, we suspect, rather unexpected statement. Clearly this statement advocates a *three-sided relation* between $N = 2$ Landau-Ginzburg models, $N = 2$ superconformal field theories and $N = 2$ sigma models on a particular type of internal manifolds, namely Calabi-Yau $n$-folds. To be definite we just mention that Calabi–Yau $n$-folds are complex manifolds with vanishing first Chern class and $SU(n)$-holonomy [61, 268]. In the case $n = 3$, they are proposed as good candidates to fill the additional six compact dimensions predicted by superstrings, since they lead to a chiral N=1 supersymmetric theory in the physical four dimensions.

This being clarified the reason why the quoted statement on string propagation brings into the game the N=2 sigma model is that at loop order $g$, string propagation on a manifold $\mathcal{M}$ is precisely described by a sigma model action:

$$S_{sigma\ model} = -\frac{1}{4\pi\alpha'}\int_{\Sigma_g} d^2\xi\left[\sqrt{-h}h^{\alpha\beta}\partial_\alpha X^\mu \partial_\beta X^\nu g_{\mu\nu}(X)\right], \tag{1.3.1}$$

where $\alpha'$ is the string tension, $h_{\alpha\beta}$ is the two-dimensional metric and

$$X^\mu(\xi) : \Sigma_g \to \mathcal{M} \tag{1.3.2}$$

is a map from a genus $g$ Riemann surface $\Sigma_g$ to the target space $\mathcal{M}$ under consideration, whose background geometry is encoded into the metric $g_{\mu\nu}(X)$. Strictly speaking one should also consider the coupling to a background axion field $B_{\mu\nu}$ and to a dilaton, but these are complicacies that, for the sake of the present argument, we can just disregard. If superstring rather than string propagation is the problem under consideration, then it is the supersymmetric version of the sigma-model (1.3.1) that one should consider. It was shown that the request of N=1 supersymmetry in target four-space implies an N=2 supersymmetry on the world-sheet [239, 240, 30, 29], so that appropriate to our case is precisely the N=2 sigma model. In order to be defined, this two-dimensional field theory requires the target space $\mathcal{M}_K$ to be a complex Kählerian manifold. Let $n = \dim_{\mathbb{C}} \mathcal{M}_K$ and denote by $X^i$ $(i = 1,\ldots,n)$ the complex coordinates. Then the field content of the N=2 sigma model is identical with that of the N=2 Landau–Ginzburg model but the action, rather than being the integral of the lagrangian (1.2.16), is the following one:

$$\begin{aligned} S^{N=2}_{sigma\ model} \;=\; \int \Big[ &- g_{ij^*}\left(\partial_+ X^i \partial_- X^{j^*} + \partial_- X^i \partial_+ X^{j^*}\right) \\ &+ 2i\, g_{ij^*}\left(\psi^i \nabla_- \psi^{j^*} + \psi^{j^*}\nabla_- \psi^i\right) \\ &+ 2i\, g_{ij^*}\left(\tilde{\psi}^i \nabla_+ \tilde{\psi}^{j^*} + \tilde{\psi}^{j^*}\nabla_+ \tilde{\psi}^i\right) \\ &+ 8\, R_{ij^*kl^*}\psi^i \psi^{j^*} \tilde{\psi}^k \tilde{\psi}^{l^*}\Big]\, d^2 z \end{aligned} \tag{1.3.3}$$

where we have denoted by

$$\begin{aligned} \nabla_\pm \psi^i &= \partial_\pm \psi^i - \Gamma^i_{jk}\,\partial_\pm X^j\, \psi^k \\ \nabla_\pm \psi^{i^*} &= \partial_\pm \psi^{i^*} - \Gamma^{i^*}_{j^*k^*}\,\partial_\pm X^{j^*}\, \psi^{k^*} \end{aligned} \tag{1.3.4}$$

the world-sheet components of the target space covariant derivatives, identical equations holding for the tilded fermions and $R_{ij^*kl^*}$ being the curvature tensor of the target manifold while $g_{ij^*}$ is its Kähler metric.

Of the *three-sided relation* we mentioned above, one side was already surveyed, namely the relation between N=2 Landau–Ginzburg models and N=2 superconformal field theories. The second side can now be suspected in view of the fact that the N=2 Landau–Ginzburg and the N=2 sigma-model are two field theories constructed out of the same supermultiplets and with just two apparently different interaction structures, the former codified by the superpotential $\mathcal{W}_K$, the latter codified by the Kählerian metric $g_{ij^\star}$ and by the global topology of the manifold $\mathcal{M}$. The question is: *What is the relation between these two types of analytical and geometrical data and what is their common relation with the algebraic structure of N=2 superconformal field theories, namely what is the third side of the relation triangle?* The answer to this question is centred on the algebraic datum shared by these three theories, namely the *chiral ring.*

### 1.3.1 The chiral ring in N=2 superconformal theories

The chiral ring $\mathcal{R}_\mathcal{W}$ of polynomial deformations defined above has a clear-cut interpretation within the N=2 superconformal theories one obtains at the infrared critical point [214]. Among all the possible primary fields of a given N=2 superconformal model one considers the two subsets of the *chiral* and the *anti-chiral* ones. The elements of these two subsets $\psi^\pm(z)$ are characterized by the following two properties that imply each other:

i) the conformal weight $h$ and the $U(1)$ charge $q$ of $\psi^\pm(z)$ satisfy the relation

$$q = \pm 2h \qquad h > 0 \tag{1.3.5}$$

ii) the operator product expansion of $\psi^\pm(z)$ with one of the two supercurrents is regular:

$$G^\pm(z)\,\psi^\pm(w) \;=\; \text{reg.} \tag{1.3.6}$$

It is now a simple matter to verify that:

i) The primary chiral or (anti-chiral) fields form a ring since their operator product expansion is regular:

$$\lim_{z\leftarrow w} \psi_1^\pm \begin{bmatrix} 2|q_1| \\ q_1 \end{bmatrix}(z)\; \psi_2^\pm \begin{bmatrix} 2|q_2| \\ q_2 \end{bmatrix}(w) \;=\; \psi_3^\pm \begin{bmatrix} 2|q_1+q_2| \\ q_1+q_2 \end{bmatrix}(w) \tag{1.3.7}$$

ii) This ring is finite, and it possesses a unique *top chiral element*

$$\psi_{top}\begin{bmatrix} c/6 \\ \pm c/3 \end{bmatrix}(z) \tag{1.3.8}$$

whose U(1)-charge is one third of the central charge of the superconformal field theory. Since by means of an operation named *"spectral flow"* the chiral ring $\mathcal{C}$ can be mapped into the anti-chiral one $\mathcal{A}$, there are essentially two rings associated with a given N=2 superconformal field theory, once we pair the *left-moving* sector with the *right-moving* one, namely the ring $(\mathcal{C},\mathcal{C})$ and the ring $(\mathcal{C},\mathcal{A})$. The first of these rings is isomorphic with the chiral polynomial ring $\mathcal{R}_\mathcal{W}$ any time the considered N=2 superconformal field theory originates as the infrared fixed point of a Landau–Ginzburg model:

$$(\mathcal{C},\mathcal{C}) \sim \mathcal{R}_\mathcal{W} \tag{1.3.9}$$

The interpretation of the second ring $(\mathcal{C},\mathcal{A})$ requires further insight and it is postponed.

### 1.3.2 The vanishing locus of the superpotential as a Calabi–Yau manifold

Let us go back and consider a family of quasi-homogeneous critical potentials with scaling weights $\omega_i$ and degree $d_{\mathcal{W}}$. Let the family $\mathcal{W}(X_1, \ldots, X_{n+2}; \psi^{\alpha_1}, \ldots, \psi^{\alpha_r})$ be parametrized by $r$ moduli parameters. The vanishing loci:

$$\mathcal{M}_n(\psi) \stackrel{\text{def}}{=} \{X_\Lambda \in W\mathbb{CP}_{n+1;\omega_1,\ldots,\omega_n} \,|\, \mathcal{W}(X_1, \ldots, X_{n+2}; \psi_{\alpha_1}, \ldots, \psi_{\alpha_r}) = 0\} \quad (1.3.10)$$

are a family of $n$ complex dimensional hypersurfaces algebraically embedded in a *weighted projective space* $W\mathbb{CP}_{n-1;\omega_1,\ldots,\omega_n}$. As topological spaces the manifolds $\mathcal{M}_{n-2}(\psi)$ are all identical for each choice of the moduli parameters $\psi^\alpha$, yet, as analytical varieties their complex structure varies while the $\psi^\alpha$ change value. It can be shown that, provided the degree $d_{\mathcal{W}}$ satisfies certain conditions, the manifolds $\mathcal{M}_n(\psi)$ are Calabi–Yau $n$-folds, namely their first Chern class vanishes. At the same time and under the same conditions it can be shown that the functional integral involved by the quantum version of the Landau-Ginzburg model with potential $\mathcal{W}(X,\psi)$ is actually concentrated on the configurations lying in the vanishing locus (1.3.10). Hence the Landau-Ginzburg model is quantum-mechanically equivalent to the N=2 sigma model with target space identical to the vanishing locus (1.3.10). This is just one way of revealing the advocated relationship between N=2 sigma models on Calabi–Yau $n$-folds and Landau–Ginzburg models. There are other more refined ways and they are analysed in this book. For the time being it suffices to note that we have discovered the geometrical interpretation of the *Landau-Ginzburg moduli.* They are the *moduli of the complex structures* of the corresponding *Calabi–Yau* $n$-fold. What about the ring structure in which these moduli are naturally fitted, namely the chiral ring $\mathcal{R}_{\mathcal{W}}$? From the point of view of N=2 superconformal theories it has already been identified with the chiral–chiral ring $(\mathcal{C},\mathcal{C})$. What is the corresponding ring structure in the N=2 sigma model picture? It is the *Hodge ring.*

### 1.3.3 The Griffiths residue map and the Hodge ring

Given a complex manifold $\mathcal{M}$ its relevant cohomology is the Dolbeault cohomology. The Dolbeault cohomology groups $H^{(p,q)}(\mathcal{M})$ are composed of the equivalence classes of closed $(p,q)$-forms

$$\omega_{i_1\ldots i_p j_1^\star \ldots j_q^\star}\, dz^{i_1} \wedge \ldots \wedge dz^{i_p} \wedge d\bar{z}^{j_1^\star} \wedge \ldots \wedge d\bar{z}^{j_q^\star} \quad (1.3.11)$$

modulo exact forms. The direct sum

$$H^\star_{Dbt}(\mathcal{M}) = \oplus_{p+q\le 2n} H^{(p,q)}(\mathcal{M}) \quad (1.3.12)$$

constitutes a ring under the ordinary *wedge product operation.* We name it the *classical Dolbeault ring.* It is also customary to arrange the dimensions of the Dolbeault cohomology groups:

$$h^{(p,q)} \stackrel{\text{def}}{=} \dim_{\mathbb{C}} H^{(p,q)}(\mathcal{M}) \quad (1.3.13)$$

named the *Hodge numbers*, into a diamond-shaped array named the *Hodge diamond*. The general rule is easily understood by looking at the example of 3-folds for which the aforementioned array is the following one:

$$\begin{array}{ccccccc} & & & h^{(0,0)} & & & \\ & & h^{(1,0)} & & h^{(0,1)} & & \\ & h^{(2,0)} & & h^{(1,1)} & & h^{(0,2)} & \\ h^{(3,0)} & & h^{(2,1)} & & h^{(1,2)} & & h^{(0,3)} \\ & h^{(3,1)} & & h^{(2,2)} & & h^{(1,3)} & \\ & & h^{(3,2)} & & h^{(2,3)} & & \\ & & & h^{(3,3)} & & & \end{array} \tag{1.3.14}$$

In the case of Calabi–Yau $n$-folds the Hodge diamond has the peculiarity that $h^{(n,0)} = 1$ and the cohomology groups $H^{(n-k,k)}$ $(k = 0, \ldots, n)$, whose dimensions are arranged on the middle line of the Hodge diamond, constitute a ring under a binary operation, named the $\star$-product, that operates as follows:

$$H^{(n-k_1,k_1)} \star H^{(n-k_2,k_2)} \subset H^{(n-k_1-k_2,k_1+k_2)} \tag{1.3.15}$$

Equipped with the $\star$-product, which is clearly distinct from the ordinary wedge product, the direct sum

$$\text{Hod}\,(\mathcal{M}) \stackrel{\text{def}}{=} \oplus_{k=0}^{n} H^{(n-k,k)}\,(\mathcal{M}) \tag{1.3.16}$$

becomes a ring, the *Hodge ring*, which, whenever the Calabi–Yau $n$-fold $\mathcal{M}$ is the vanishing locus of $\mathcal{W}(X,\psi)$, turns out to be isomorphic to the polynomial subring

$$\mathcal{R}_{\mathcal{W}}^{integer} \subset \mathcal{R}_{\mathcal{W}} \tag{1.3.17}$$

obtained by restricting $\mathcal{R}_{\mathcal{W}}$ to the polynomials of order

$$k\,\nu \quad (k = 0, 1, \ldots, n) \qquad \nu \;=\; \deg \mathcal{W} \tag{1.3.18}$$

This isomorphism is revealed by a crucial geometric construction, known as the *Griffiths residue map* [182]. It represents the harmonic $(n-k,k)$-forms on $\mathcal{W}(X,\psi) = 0$ as the residue of meromorphic $(n+1)$-forms that live in the ambient weighted projective space, and have a pole of order $(k+1)\,\nu$ at $\mathcal{W}(X,\psi) = 0$. In this representation the non-trivial polynomials $P_{k|\nu}(X) \in \mathcal{R}_{\mathcal{W}}^{integer}$ are the numerators of a fraction that has $(\mathcal{W}(X,\psi))^{k+1}$ as denominator. In particular the polynomials of order $P_{1|\nu}(X)$ of the same order as the defining polynomial $\mathcal{W}(X)$ are in one-to-one correspondence with the harmonic $(n-1,1)$-forms and describe the complex structure deformations.

Summarizing we come to the conclusion that the three-sided relation between N=2 sigma models on Calabi–Yau $n$-folds $\mathcal{M}_n^{CY}$, N=2 Landau–Ginzburg models with a superpotential $\mathcal{W}(X,\psi)$ and $c = 3n$ N=2 superconformal theories is based on the following chain of ring isomorphisms:

$$\text{Hod}\,(\mathcal{M}) \;\sim\; \mathcal{R}_{\mathcal{W}}^{integer} \;\sim\; (\mathcal{C},\mathcal{C}) \tag{1.3.19}$$

The basic goal of our investigation is the calculation of the *structure constants* of these isomorphic rings as a function of the moduli parameters. In the case of Calabi–Yau 3-folds these structure constants are essentially identified with the *Yukawa couplings* and this fact provides the link between the just analysed conceptual structures and the *special Kähler geometry* of moduli spaces outlined above. To appreciate this better one turns to topological field theories.

## 1.4 The Art of Quantizing Zero

Topological field theories [259] have been described as the quantization of zero. One starts from a classical lagrangian that either vanishes or is a characteristic class $c_n(V)$ of some vector bundle $V \xrightarrow{\pi} M$ constructed over space–time, so that the value of the action:

$$S_{class} = \int_M \mathcal{L}_{class} = \int_M c_n(V) = k \in \mathbf{Z} \tag{1.4.1}$$

is either zero or a topological integer number ( a winding number of some sort). A typical example is the A-type topological sigma model [260, 32], where the classical action is the integral on a Riemann surface $\Sigma_g$ of the pull-back, through an embedding map

$$X : \Sigma_g \longrightarrow \mathcal{M}_K \tag{1.4.2}$$

of the Kähler 2-form pertaining to the target space $\mathcal{M}_K$:

$$S_{class} = -\pi i \int_{\Sigma_g} X^\star K = \int d^2\xi \, g_{ij^\star}\left(X,\overline{X}\right) \partial_\alpha X^i \partial_\beta X^{j^\star} \varepsilon^{\alpha\beta} \tag{1.4.3}$$

This classical action is invariant under a gigantic group of symmetries: the reparametrizations of the Riemann surface:

$$\xi^\alpha \longrightarrow \xi^{\alpha'} \tag{1.4.4}$$

times the arbitrary continuous deformations of the embedding map:

$$X^i(\xi) \longrightarrow X^i(\xi) + \delta X^i(\xi) \tag{1.4.5}$$

Continuous deformations that are contractible to the identity do not change the homoptopy class of the embedding map $X$. The main idea of topological field theory, namely the *art of extracting something out of nothing* is to apply the standard techniques of BRST quantization to systems possessing groups of classical symmetries as huge as the group we described above. Because of their nature, such symmetry groups that correspond to continuous deformations of the classical fields are named *topological symmetries*. The heart of the matter is the BRST quantization of topological symmetries. Indeed, in the BRST approach, what really matters is the structure of the algebra one declares to be one's algebra of classical symmetries, the classical action being little more than a way of providing such a definition. As in any BRST framework, also in topological field theories one needs a suitable *gauge-fixing condition* and the main course of the day is the analysis of the relevant *BRST cohomology.* There is a hierarchy of symmetries that one respects while implementing the gauge-fixings. So the topological symmmetry is gauge-fixed in such a way as to preserve the other ordinary gauge symmetries. These are subsequently broken by more conventional gauge functions. The key point of interest in the whole game is that one can choose the classical *instanton equations* of the corresponding ordinary field theory as the gauge-fixing functions of its topological version. In this way the space of *physical states* is reduced to the *moduli space* of instantons. The classical action usually functions as a counter that evaluates the *instanton number* and the whole Hilbert space of the theory reduces to an infinite direct sum of finite dimensional spaces:

$$\text{Hilbert space} \;=\; \sum_k \mathcal{M}^{(k)}_{moduli} \tag{1.4.6}$$

Indeed, for each chosen *instanton number* $k$, the instantons come into families parametrized by a set continuous parameters, $\mu_1\,,\,\mu_2\,,\,\dots\,\mu_n$, filling, by definition, the moduli space $\mathcal{M}^{(k)}_{moduli}$ of level $k$. To be definite, in the considered example of the A-type topological sigma model, the instantons are, among all the maps (1.4.2), those that are holomorphic, namely those satisfying

$$\partial_- X^i \;=\; \partial_+ \overline{X}^{i} \;=\; 0 \tag{1.4.7}$$

where $\partial_+ = \frac{\partial}{\partial z}$ ($\partial_- = \frac{\partial}{\partial \bar z}$) denotes the derivative with respect to the holomorphic (respectively anti-holomorphic) coordinate on the world-sheet. To be even more specific, if both the target space and the world-sheet are taken to be Riemann two spheres $S_2 \sim \mathbb{CP}_1$, then the general form of a degree $k$ holomorphic instanton is a rational function:

$$X \;=\; \alpha \frac{\prod_{i=1}^{k}(\,z\,-\,\beta_i)}{\prod_{i=1}^{k}(\,z\,-\,\gamma_i)} \qquad X\,\in\,\mathbb{CP}_1^{target} \quad z\,\in\,\mathbb{CP}_1^{world\text{-}sheet} \tag{1.4.8}$$

and the $2k+1$ complex numbers $\alpha\,,\,\beta_i\,,\,\gamma_i$ fill the degree $k$ moduli space:

$$\mathcal{M}^{(k)}_{moduli} \;\sim\; \otimes^{2k+1}\mathbb{CP}_1 \tag{1.4.9}$$

As we see, topological field theories introduce once more a notion of *moduli* and the natural question that arises is: *What is the relation of these moduli with those previously introduced?* We postpone for a moment the answer to this question and we just observe that the very set up of topological field theories translates into a correspondence between the *BRST cohomology ring* and the *cohomology ring* of *instanton moduli spaces.* We have on one side the BRST cohomology groups:

$$H^{(g)}_{BRST} \;=\; \frac{\ker Q^{(g)}_{BRST}}{\operatorname{im} Q^{(g-1)}_{BRST}} \tag{1.4.10}$$

where $Q^{(g)}_{BRST}$ is the BRST charge acting on operators of ghost number $g$, and on the other side the de Rham cohomology groups:

$$H^{(i)}\left(\mathcal{M}^{(k)}_{moduli}\right) \;=\; \frac{\ker d^{(i)}}{\operatorname{im} d^{(i-1)}} \tag{1.4.11}$$

where $d^{(i)}$ is the ordinary exterior derivative acting on differential forms of degree $i$ that live on $\mathcal{M}^{(k)}_{moduli}$. The topological field theory realizes an injection of rings:

$$H^{\star}_{BRST} = \otimes_{g=0}^{\Delta U_k} H^{(g)}_{BRST} \longrightarrow H^{\star}\left(\mathcal{M}^{(k)}_{moduli}\right) = \otimes_{i=0}^{d_k} H^{(i)}\left(\mathcal{M}^{(k)}_{moduli}\right) \tag{1.4.12}$$

that under suitable circumstances becomes a ring isomorphism. In Eq. (1.4.12),

$$\Delta U_k = \int \partial^{\mu} J^{(ghost)}_{\mu} \, d^D x \tag{1.4.13}$$

is the integrated anomaly of the ghost current in the instanton sector of level $k$, while

$$d_k = \dim_{\mathbb{R}} \mathcal{M}^{(k)}_{moduli} \tag{1.4.14}$$

is the dimension of moduli space at the same level $k$. As we shall argue, using appropriate versions of the Atiyah–Singer index theorem, the two numbers $\Delta U_k$ and $d_k$ are essentially equal and this is the first prerequisite for the isomorphism (1.4.12) to be realized. What actually happens is that after all the necessary formal manipulations have been duely performed, the correlation function of a set of local physical operators

$$G(x_1, \ldots, x_n) = \langle \mathcal{O}(x_1)\, \mathcal{O}(x_2) \, \ldots \, \mathcal{O}(x_n) \rangle \tag{1.4.15}$$

reduces to an integral over moduli spaces of an equal number of cohomology classes of that space. For instance in the A-type topological sigma model, the possible local observables $\Theta^{(0)}_A(x)$ are in one-to-one correspondence with the elements $A \in H^{\star}_{Dbt}(\mathcal{M}_K)$ of the target manifold Dolbeault cohomology ring (the degree of $A$ as a form being equal to the ghost number of the operator $\Theta^{(0)}_A(x)$) and we have

$$\begin{aligned} &\langle \textstyle\prod_{r=1}^{N} \Theta^{(0)}_{A_r}(x_r) \rangle \\ &= \textstyle\sum_{\{k_A\}} \exp\left[2\pi i \sum_{a=1}^{h^{(1,1)}} t^a \int_{\Sigma_g} X^{\star}(\omega_a)\right] \int_{\mathcal{M}^{(k)}_{moduli}} \mathcal{O}_{A_1}(x_1) \wedge \ldots \wedge \mathcal{O}_{A_N}(x_N) \\ &= \textstyle\sum_{\{k\}} \exp\left[2\pi i \sum_{a=1}^{h^{(1,1)}} t^a k_a\right] \int_{\mathcal{M}^{(k)}_{moduli}} \mathcal{O}_{A_1}(x_1) \wedge \ldots \wedge \mathcal{O}_{A_N}(x_N) \end{aligned} \tag{1.4.16}$$

where the complex coefficients $t^a$ parametrize the Kähler class of the target manifold

$$K = \sum_{a=1}^{h^{(1,1)}} t^a \, \omega^{(1,1)}_a , \tag{1.4.17}$$

where $\omega^{(1,1)}_a$ denote a basis of the Dolbeault group $H^{(1,1)}(M_K)$, and where $\mathcal{O}_{A_i}(x_i)$ are suitable closed differential forms on instanton moduli space, whose complete intersection integral $\int_{\mathcal{M}^{(k)}_{moduli}} \mathcal{O}_{A_1}(x_1) \wedge \ldots \wedge \mathcal{O}_{A_N}(x_N)$ just produces an *integer number.*

We can now answer the question we left open some lines above. The instanton moduli spaces that pop out in the evaluation of topological correlation functions are not the moduli spaces we have encountered before either as flat directions of supergravity lagrangians or as deformation parameters of the target manifold in the underlying N=2 sigma model. Nevertheless also these moduli appear in topological field theories and play an essential role: that of *topological coupling constants.*

Indeed the topological correlators (1.4.15) are actually *correlation constants*, in the sense that they do not depend on the locations $x_i$ of the operators $\mathcal{O}(x_i)$:

$$\frac{\partial}{\partial x_i} G(x_1, \ldots, x_n) = 0 \qquad \forall x_i \tag{1.4.18}$$

the only dependence being on the *topological coupling constants* or *topological deformation parameters* of the action. To be once more quite short (the reader will find all the details in Chapter 7), in a generic two-dimensional topological field theory, the correlators are of the following form:

$$c_{A_1,A_2,\ldots,A_N}(t) = \langle\, \Theta^{(0)}_{A_1}, \ldots, \Theta^{(0)}_{A_N} \exp\left[\sum_A t_A \int \Theta^{(2)}_A\right] \rangle \tag{1.4.19}$$

where $\Theta^{(0)}_{A_1}$ constitute a complete set of BRST-closed local observables and are related to the corresponding integrated observables by means of the *descent equations*:

$$\begin{aligned} s\Theta^{(2)} &= d\Theta^{(1)} \\ s\Theta^{(1)} &= d\Theta^{(0)} \\ s\Theta^{(0)} &= 0 \end{aligned} \tag{1.4.20}$$

In Eq. (1.4.19), $\langle \ldots \rangle$ denotes the average performed with respect to the measure - $\exp\left[-S_{quantum}\right]$, where $S_{quantum}$ is the BRST quantum action. Hence, inserting the exponential factor

$$\exp\left[\sum_A t_A \int \Theta^{(2)}_A\right] \tag{1.4.21}$$

corresponds to a deformation of the quantum action

$$S_{quantum} \longrightarrow S_{quantum} + \sum_A t_A \int \Theta^{(2)}_A \tag{1.4.22}$$

The basic goal is to calculate the dependence of $c_{A_1,A_2,\ldots,A_N}(t)$ on the parameters $t$. These are of two types. The $t_A$ that multiple integrated operators $\int \Theta^{(2)}_A$ with a non-vanishing ghost number are the analogues of the *relevant parameters* in the Landau–Ginzburg model. On the other hand let $\mu_a$ be those $t_A$ parameters that multiply integrated operators with a zero ghost number. These are the analogues of the *marginal parameters* and deserve the name of *moduli.* Indeed their presence is a signal that the classical action by means of which we defined the topological field theory was actually not isolated, rather it was an element of a continuous family of such actions parametrized by the moduli $\mu_a$. This is quite evident in the example of the A-type topological sigma model outlined above. Here the classical action is completely defined by the choice of a Kähler class and, as a consequence, the topological correlators are functions of the Kähler class moduli.

In the case of a Calabi–Yau 3-fold, the *Yukawa couplings* $W_{abc}(t)$ of the **27** chiral families, which enter the basic special geometry relation (1.1.9) relative to Kähler class moduli space, can be interpreted as a three-point correlator of the A-type topological sigma model. As such it admits the following expression:

$$\begin{aligned} W_{abc}(t) &= \langle \Theta^{(0)}_a(x_1)\, \Theta^{(0)}_b(x_2)\, \Theta^{(0)}_c(x_3) \rangle \\ &= d_{abc} + \sum_{\{k_a\}} \sum_{m=1}^{\infty} n_{\{k_a\}}\, k_a\, k_b\, k_c \left( \prod_{a=1}^{h^{(1,1)}} q_a^{k_a} \right)^m \\ &= d_{abc} + \sum_{\{k_a\}} n_{\{k_a\}}\, k_a\, k_b\, k_c \frac{\prod_{a=1}^{h^{(1,1)}} q_a^{k_a}}{1 - \prod_{a=1}^{h^{(1,1)}} q_a^{k_a}} \end{aligned} \tag{1.4.23}$$

where

i)

$$d_{abc} = \int_{M_3^{CY}} \omega_a^{(1,1)} \wedge \omega_b^{(1,1)} \wedge \omega_c^{(1,1)} \tag{1.4.24}$$

is the intersection integral of three $(1,1)$-forms in some basis of $H^{(1,1)}(\mathcal{M})$

ii) $\{k_a\}$ are the integer numbers specifying the *degree* of the holomorphic instanton $U : S_2 \longrightarrow M_3^{CY}$, namely the decomposition along an integer homology basis of 2-cycles $C_2^a$ of the rational curve $\mathcal{C}^U = U(S_2) \subset M_3^{CY}$:

$$\left[\mathcal{C}^U\right] = \sum_a^{h^{(1,1)}} k_a\, [C_2^a] \tag{1.4.25}$$

iii) the integer $n_{\{k_a\}}$ denotes the number of such rational curves

iv)

$$q_a \stackrel{\text{def}}{=} \exp\left[2\pi i t^a\right] \qquad [K] = \sum_{a=1}^{h^{(1,1)}} t^a\, \omega_a^{(1,1)} \tag{1.4.26}$$

are exponential factors containing all the dependence on the Kähler class moduli. (In the above equation $K$ denotes the Kähler 2-form.) Equation (1.4.23) is one of the master formulae in the present book. It shows all the difference from a naive Kaluza–Klein approach to string compactifications and the true stringy result. If one takes first the field theory limit of superstring theory and then compactifies on a Calabi–Yau 3-fold $M_3^{CY}$, the result for the Yukawa couplings is merely given by the first constant term $d_{abc}$. The true result receives an infinite number of corrections from all the world-sheet instantons [123, 124, 125] and it has a clear-cut meaning within *topological field theory.* The only obstacle to its effective calculation is that one should know the number $n_{\{k_a\}}$ of rational curves of a certain degree and this information is usually not provided by the mathematical literature.

Actually it is precisely in this respect that one of the most fruitful instances of feedback on mathematics of these types of physical ideas has occurred. Such a story is shortly told in the next section. We just close the present section with the specialization of Eq.

(1.4.23) to the case of a one-dimensional Kähler class moduli space ( $\dim_{\mathbb{C}} H^{(1,1)}(\mathcal{M}_3) = 1$):

$$W(t) = d_{111} + \sum_{k=1}^{\infty} \frac{n_k\, k^3\, q^k}{1 - q^k} \tag{1.4.27}$$

The prediction of the numbers $n_k$ for the case of a quintic three-fold has been one of the successes of

$$\text{Physics} \bigcap \text{Mathematics} \tag{1.4.28}$$

## 1.5 Mirror Maps

Why did we insist on qualifying the topological sigma model discussed in the previous section as being of the A-type? Quite obviously because there is also a B-type model. The reader has certainly guessed as much as that. What is then this *dichotomy*? It is a general phenomenon intrinsic to N=2 supersymmetry in two dimensions. Indeed in D=2, any N=2 supersymmetric field theory can be *topologically twisted* in two alternative ways, in want of better ideas named the *A-twist* and the *B-twist*, that give rise to two different topological field theories, enucleating complementary topological properties of the parent N=2 model. The twists consist of a systematic reassignment of spin and ghost numbers, together with a reinterpretation of one of the supersymmetry charges as the topological BRST-charge. This essentially amounts to a projection of the original theory onto a sector where the Green functions satisfy all the properties of *topological correlators* as described in the previous section. The fact that there are two independent twists implies that there are two such topological sectors in any N=2 theory. They are discriminated by the *type of instantons* that are selected as *physical configurations* and by the *type of moduli* that are promoted to the role of *topological coupling constants*. In the A-twisted sigma model the instantons are the *holomorphic maps* $X : \Sigma_g \longrightarrow \mathcal{M}_n^{CY}$ and the topological coupling constants are the *moduli of the Kähler class*. In the B-twisted sigma model the instantons are the *constant maps* $X : \Sigma_g \longrightarrow \mathcal{M}_n^{CY}$ and the topological coupling constants are the *moduli of the complex structures*. The calculation of correlators in the B-model is infinitely simpler than the analogous calculation in the A-model, since it corresponds to the solution of an algebraic rather than a trascendental problem. While in the A-model the topological observables $\Theta_A^{(0)}(x)$ are in one-to-one correspondence with the elements of the Dolbeault cohomology ring $H^{\star}_{Dbt}(\mathcal{M}_K)$, in the B-model the BRST-closed local observables are in correspondence with the elements $Z^{(0,p_i)}_{(q_i,0)} \in H^{(p_i)}_{\bar{\partial}}\left(\mathcal{M}_n, \Lambda^{q_i} T^{(1,0)}\mathcal{M}_n\right)$ namely with the $\bar{\partial}$-closed multi-holomorphic vector valued $(0,q)$-forms:

$$Z^{(0,p_i)}_{(q_i,0)} = d\bar{z}^{i_1^\star} \wedge \ldots \wedge d\bar{z}^{i_p^\star}\, Z^{j_1 \ldots j_q}_{i_1^\star \ldots i_p^\star} \frac{\partial}{\partial z^{j_1}} \wedge \ldots \wedge \frac{\partial}{\partial z^{j_q}} \tag{1.5.1}$$

After all the necessary functional integral manipulations the B-model correlator of a string of such observables turns out be a classical intersection integral:

$$\langle \prod_{i=1}^{N} \Theta^{(0)} \left[ Z^{(0,p_i)}_{(q_i,0)} \right] (x_i) \rangle = \int_{\mathcal{M}_n} \Omega^{(n,0)} \wedge Z^{(0,p_1)}_{(q_1,0)} \wedge \ldots \wedge Z^{(0,p_N)}_{(q_N,0)} \tag{1.5.2}$$

the wedge product utilized in the above formula being the obvious one and $\Omega^{(n,0)}$ being the unique $(n,0)$-form defined on the Calabi–Yau $n$-fold $\mathcal{M}_n^{CY}$.

In the case of a 3-fold, the Yukawa couplings $W_{\mu\nu\rho}$ of $\overline{\mathbf{27}}$ families, which enter the basic special geometry relation (1.1.9), relative to the complex structure moduli space, can be interpreted as a three-point correlator of the B-type topological sigma model. Every complex structure deformation is represented by a $\overline{\partial}$-closed vector valued 1-form:

$$V_\mu = V^i_{\mu;j^\star} \, d\overline{z}^{j^\star} \frac{\partial}{\partial z^i} \tag{1.5.3}$$

and we have

$$W_{\mu\nu\rho}(\psi) = \int_{\mathcal{M}_3} \Omega^{(3,0)}(\psi) \wedge V_\mu \wedge V_\nu \wedge V_\rho \tag{1.5.4}$$

which is the other master formula of this book. Differently from the case of Kähler classes, the exact string result (1.5.4) coincides with the naive Kaluza–Klein result.

It would just be great if we could calculate A-twisted correlators of a theory defined on a certain manifold as B-twisted correlators of another manifold. The hypothesis of *mirror symmetry* states that this is possible. Indeed it appears that Calabi–Yau manifolds can be arranged into *mirror pairs*

$$\mathcal{M}_n^{CY} \leftrightarrow M\mathcal{M}_n^{CY} \tag{1.5.5}$$

such that the A-model of one element of the pair is equivalent to the B-model of the other. In this way the *complicated* Yukawa coupling (1.4.27) can be evaluated by means of the easy intersection integral (1.5.4). There is just a problem in relating the two coordinate systems on moduli space, namely the $t$-coordinate system utilized by the A-model formula and the $\psi$ coordinate system utilized by the B-model. The searchedfor relation is established by solving a set of differential equations, the Picard–Fuchs equations for the periods of the holomorphic $n$-form. The theory, interpretation and uses of such a system of differential equations are the subjects of the last chapter in this book. It is the conclusive episode of this story and the patient reader is requested to wait until that point in order to rejoice at the happy ending of this novel.

## 1.6 Bibliographical Note

*In this as in all the following chapters we have reduced the number of references spread in the text to a bare minimum. We have done this in order to avoid interrupting our exposition with long strings of quotations.*

*As a reader's guide to the literature, we concentrate the bibliographical information on the various topics treated in each chapter in short notes placed at the end of the corresponding chapters. The present is the first of such notes.*

*In this introduction we briefly mentioned almost all the topics we shall deal with. Correspondingly almost all of our sources are quoted here. In the spirit of the chapter, however, they are grouped by large classes. In subsequent notes the same references will reappear with a few new ones, in relation with a finer analysis of the involved topics.*

- *Supergravity and Kähler geometry. See: [272, 25, 26, 93], [76], [256], [249]*
- *Special Kähler geometry. See: [99, 96, 108, 75, 74], [146, 153, 110, 78, 77], [112, 97, 82, 81, 147], [244, 66], [126], [143, 88, 90, 89]*
- *Calabi–Yau compactifications. See: [70, 221, 61, 268], [64, 65], [179], [258]*
- *N=2 Landau–Ginzburg theories. See: [181, 84, 85, 254, 219], [248, 196, 245, 246, 211], [118, 250, 151, 129, 48], [86, 247, 218, 203], [45, 262]*
- *N=2 superconformal theories. See: [1, 2, 3, 134], [135, 136, 133, 55, 223], [115, 176, 177, 200, 165], [156, 72], [239, 240, 30, 29], [138, 210, 142]*
- *Moduli spaces of Calabi–Yau n-folds and Yukawa couplings. See: [63, 173, 243], [82, 81, 78, 146], [153, 147], [244, 66], [143, 88, 90, 89, 59]*
- *Griffiths residue map. See: [182], [213]*
- *Topological field theories. See: [234, 47, 261, 119, 127, 33], [260, 259, 21], [32], [118, 250, 151, 129, 48], [86, 247, 218, 203], [45, 262], [215, 117, 204, 54], [12, 11, 36], [10, 266, 267, 264], [8, 7]*
- *BRST cohomology. See: [52, 31, 34, 35], [32], [215, 117, 204, 54], [12, 11, 36]*
- *Mirror maps. See: [263, 58, 180, 71, 41], [42, 40, 16, 18, 17], [114, 149, 229, 230, 199], [20, 27, 28], [128, 232], [69, 68] [67, 39], [193], [148, 213], [172, 120, 207, 130], [170, 171, 169], [144, 220], [235, 251, 145]*

# Chapter 2

# A BIT OF GEOMETRY AND TOPOLOGY

## 2.1 Introduction

In the present Chapter we summarize the mathematical lore needed in our subsequent discussion of the Kähler geometry of supergravity theory, of Calabi–Yau compactifications and of topological field theories. The material that follows can be found in many mathematical textbooks [222], [255], [183], [253], [225, 224] and we have no claim to originality: we just collect it here in order to make the present book self-contained and to focus the reader's attention on the concepts and results we are going to utilize.

## 2.2 Fibre Bundles

In the present section we start our survey of the needed mathematical lore from the basic notions of fibre bundle theory.

### 2.2.1 Definition of a fibre bundle

We begin with

**Definition 2.2.1** *A fibre bundle* $(E, \pi, M, F, G)$ *is a geometrical structure that consists of the following list of elements:*

*i) A differentiable manifold E named the* **total space**.

*ii) A differentiable manifold M named the* **base space**.

*iii) A differentiable manifold F named the* **standard fibre**.

*iv) A Lie group G, named the* **structure group**, *which acts as a transformation group on the standard fibre:*

$$\forall g \in G \quad ; \quad g : F \longrightarrow F \quad \{\text{i.e.} \forall f \in F \;\; g.f \in F\} \tag{2.2.1}$$

*v) A surjection map* $\pi : E \longrightarrow M$, *named the* **projection**. *If* $n = \dim M$, $m = \dim F$, *then we have* $\dim E = n + m$ *and* $\forall p \in E$, $F_p = \pi^{-1}(p)$ *is an* $m$*-dimensional manifold diffeomorphic to the standard fibre* $F$. *The manifold* $F_p$ *is named the* **fibre at the point** $p$.

*vi) A covering of the base space* $\bigcup_{(\alpha \in A)} U_\alpha = M$, *realized by a collection* $\{U_\alpha\}$ *of open subsets* ($\forall \alpha \in A$ $U_\alpha \subset M$), *equipped with a diffeomorphism:*

$$\phi_\alpha : U_\alpha \times F \longrightarrow \pi^{-1}(U_\alpha) \tag{2.2.2}$$

*such that*

$$\forall p \in U_\alpha, \forall f \in F : \pi \cdot \phi_\alpha(p, f) = p \tag{2.2.3}$$

*The map* $\phi_\alpha$ *is named a* **local trivialization** *of the bundle, since its inverse* $\phi_\alpha^{-1}$ *maps the open subset* $\pi^{-1}(U_\alpha) \subset E$ *of the total space into the direct product* $U_\alpha \times F$.

*vii) If we write* $\phi_\alpha(p, f) = \phi_{\alpha,p}(f)$, *the map* $\phi_{\alpha,p} : F \longrightarrow F_p$ *is the diffeomorphism required by point v) of the present definition. For all points* $p \in U_\alpha \cap U_\beta$ *in the intersection of two different local trivialization domains, the composite map* $t_{\alpha\beta}(p) = \phi_{\alpha,p}^{-1} \cdot \phi_{\beta,p} F \longrightarrow F$ *is an element of the structure group* $t_{\alpha\beta} \in G$, *named the* **transition function**. *Furthermore the transition function realizes a smooth map* $t_{\alpha\beta} : U_\alpha \cap U_\beta \longrightarrow G$. *We have*

$$\phi_\beta(p, f) = \phi_\alpha(p, t_{\alpha\beta}(p) . f) \tag{2.2.4}$$

To denote a fibre bundle it is customary to utilize the shorthand notation $E \xrightarrow{\pi} M$, where one makes reference only to the total space, the projection, and the base manifold. From the physicist's viewpoint that privileges the explicit construction in terms of coordinate patches, $E \xrightarrow{\pi} M$ is defined by giving the explicit form of the transition functions. For consistency these latter must satisfy the following conditions:

$$\begin{aligned} t_{\alpha\alpha}(p) &= 1 \\ t_{\alpha\beta}(p) &= t_{\alpha\beta}^{-1}(p) \\ t_{\alpha\beta}(p)\, t_{\beta\gamma}(p)\, t_{\gamma\alpha}(p) &= 1 \end{aligned} \tag{2.2.5}$$

The non-trivial nature of the bundle is encoded in its transition functions. Indeed a fibre bundle is trivial if all of its transition functions can be reduced to the identity element of the structure group $G$. In this case the total space $E$ is really diffeomorphic to the direct product of the base manifold $M$ with the fibre $F$.

Given a fibre bundle $E \xrightarrow{\pi} M$, the possible set of transition functions is obviously not unique. We can utilize different sets of local trivializations for the same fibre bundle and each of them yields an equivalent set of transition functions. It is therefore important to understand which property makes two sets of transition functions equivalent and, in particular when a given set defines a *trivial* bundle, namely all the transition functions can be reduced to the identity element.

Let $\{U_\alpha\}$ be a covering of the base manifold $M$ and $\{\phi^\alpha\}$ and $\{\tilde{\phi}^\alpha\}$ be two sets of local trivializations. The transition functions are

$$\begin{aligned} t_{\alpha\beta}(p) &= \phi_\alpha^{-1}(p)\,\phi_\beta(p) \\ \tilde{t}_{\alpha\beta}(p) &= \tilde{\phi}_\alpha^{-1}(p)\,\tilde{\phi}_\beta(p) \end{aligned} \tag{2.2.6}$$

At each point of the base manifold $p \in M$ we can define a map $g_\alpha(p) : F \longrightarrow F$ of the typical fibre into itself, by setting

$$g_\alpha(p) = \phi_\alpha^{-1}(p)\,\tilde{\phi}_\alpha(p) \tag{2.2.7}$$

In order for $\{\phi^\alpha\}$ and $\{\tilde{\phi}^\alpha\}$ to be local trivializations of the same bundle, the maps $\{g_\alpha\}$ must be homeomorphisms belonging to the structural group $\forall\alpha : g_\alpha \in G$. Indeed, if this happens the two sets of transition functions can be related by:

$$\tilde{t}_{\alpha\beta}(p) = g_\alpha^{-1}(p)\,t_{\alpha\beta}(p)\,g_\beta(p) \tag{2.2.8}$$

Hence in a trivial bundle, where the transition functions can be reduced to the identity, their most general form is the following one:

$$t_{\alpha\beta}(p) = g_\alpha^{-1}(p)\,g_\beta(p) \tag{2.2.9}$$

Reconsidering the construction of fibre bundles in terms of transition functions and looking at Eqs. (2.2.5) together with Eq. (2.2.9), one realizes that the classification of equivalence classes of equivalent bundles $E \xrightarrow{\pi} M$, on a given manifold $M$ and with a given structural group $G$ can be recast in a more abstract set up that utilizes the language of cohomology and captures the profound essence of the bundle, namely its *twisting* or deviation from global triviality. This is the language of *sheaves* and of *sheaf* cohomology.

### 2.2.2 Sheaves and Cech cohomology

We introduce

**Definition 2.2.2** *A sheaf over a manifold $M$ is a family of groups $F(U)$ associated to each open subset $U \subset M$. The elements $\sigma \in F(U)$ of the group $F(U)$ are called the sections of the sheaf over the open chart $U$. Given a subspace $U \subset V \subset M$ of a subspace $V \subset M$ there is a map:*

$$r_{V,U} : F(V) \longrightarrow F(U) \tag{2.2.10}$$

*between the two associated groups. The map $r_{V,U}$ is called the restriction map. It must satisfy the following axioms:*

*a) Given three nested charts $U \subset V \subset W$ we must have*

$$r_{W,U} = r_{V,U} \circ r_{W,V} \tag{2.2.11}$$

*We denote by $\sigma|_U$ the restriction to $U$ of our element $\sigma \in F(V)$.*

*b) Given any two submanifolds $U$ and $V$ and any given two sections $\sigma \in F(U)$ and $\tau \in F(V)$ such that*

$$\sigma|_{U\cap V} = \tau|_{U\cap V} \tag{2.2.12}$$

*there exists a section $\rho \in F(U \cup V)$ such that*

$$\rho|_U = \sigma \quad ; \qquad \rho|_V = \tau \tag{2.2.13}$$

*c) If $\sigma \in F(U\cup V)$ and $\sigma|_U = \sigma|_V = 0$ then $\sigma = 0$.*

The prototype of a sheaf is the sheaf of $\mathbb{C}^\infty$-functions. Indeed $\mathbb{C}^\infty(U)$ is clearly an abelian group under addition.

$$f(x) + g(x) = h(x) \tag{2.2.14}$$

and the restriction of a function to a subdomain obviously satisfies the axiom a). Axiom b) is the principle of analytic continuation that states that if two functions coincide on a open intersection of their domains then they are the same function. Axiom c) is also obvious in this case.

Other important examples of sheaves are, for instance:

$\mathcal{O}(\mathcal{M})$ = the sheaf of holomorphic functions on a complex manifold $\mathcal{M}$

$\mathcal{O}^\star(\mathcal{M})$ = the sheaf of never vanishing holomorphic functions on $\mathcal{M}$

$\Omega^{(p)}(\mathcal{M})$ = the sheaf of holomorphic p-forms on $\mathcal{M}$

In all these cases, for each $U \subset M$ the group $F(U)$ is a continuous infinite-dimensional group (the group of all $\mathbb{C}^\infty$-functions on the open set $U$, for instance), but we can also consider sheaves where the $F(U)$ are discrete groups like $\mathbb{Z}_n$.

In the application of sheaf theory to the classification of fibre bundles, the group $F(U)$ is made by all continuous maps $g : U \longrightarrow G$ from the open set $U$ to the structural group $G$ (which is a finite-dimensional Lie group); in other words by all $G$-valued $\mathbb{C}^\infty$ functions on $U$. The functions $\forall p \in U$ ; $g(p) \in G$ are what physicists call *local gauge transformations.* Hence the relevant sheaf for the classification of fibre bundles $E \xrightarrow{\pi} M$ with base manifold $M$ and structural group $G$ is the *sheaf over $M$ of $G$-gauge transformations.*

Next we discuss the notion of *sheaf cohomology* (Cech cohomology). As in any other cohomology construction we define a sequence of vector spaces $C^p$ of $p$-cochains ($p = 0, 1, 2...$) and a nilpotent coboundary operator $\delta^2 = 0$ that defines a sequence of linear maps:

$$\delta : C^p \longrightarrow C^{p+1} \tag{2.2.15}$$

Let $(F, m)$ be a sheaf of groups over the manifold $M$. Let $\cup_{\alpha\in A} U_\alpha = M$ be a finite atlas covering $M$. Then we have the groups $F(U_1)$, $F(U_2)$, ..., $F(U_n)$. The space of 0-cochains $C^0(U, F)$ is defined by

$$C^0(U, F) = \otimes_{\alpha\in A} F(U_\alpha) \tag{2.2.16}$$

In other words a 0-cochain is a collection of $n$ sections $g_\alpha \in F(U_\alpha)$ of the sheaf over the open charts $U_\alpha$:

$$C^0 = (g_1, g_2 \dots g_n) \tag{2.2.17}$$

On the other hand a 1-cochain $C^1$ is a collection of $\frac{1}{2}n(n-1)$ sections of the sheaf $g_{\alpha\beta} = g_{\beta\alpha}^{-1}$

$$C^1 = \left(g_{12}, g_{13}, \dots g_{(n-1)n}\right) \tag{2.2.18}$$

defined on all possible intersections $U_\alpha \cap U_\beta$. A 3-cochain is a collection of $\frac{1}{6}n(n-1)(n-2)$ sections $g_{\alpha\beta\gamma}$ of the sheaf:

$$C^2 = \left(g_{123}, g_{124}, \dots g_{(n-2)(n-1)n}\right) \tag{2.2.19}$$

defined over all possible intersections $U_\alpha \cap U_\beta \cap U_\gamma$ and so on. The definition of the *coboundary operator* is as follows. Let $C^{(p-1)} = \{g_{\alpha_1\alpha_2\dots\alpha_n}\}$ be a $(p-1)$-cochain. The corresponding $(p)$-cochain $\delta\, C^{(p)}$ that is in the image of $\delta_{(p-1)}$ is given by the following alternating product of sections and inverse sections:

$$\begin{aligned}(\delta g)_{\alpha_1\alpha_2\dots\alpha_n\alpha_{n+1}} &= \\ g_{\alpha_2\alpha_3\dots\alpha_{n+1}}|U_{\alpha_1\dots\alpha_{n+1}} &\circ g^{-1}_{\alpha_1\alpha_3\dots\alpha_{n+1}}|U_{\alpha_1\dots\alpha_{n+1}} \circ \\ g_{\alpha_1\alpha_2\alpha_4\dots\alpha_{n+1}}|U_{\alpha_1\dots\alpha_{n+1}} &\circ \dots \circ g_{\alpha_1\alpha_2\dots\alpha_n}|U_{\alpha_1\dots\alpha_{n+1}}\end{aligned} \tag{2.2.20}$$

where $U_{\alpha_1\dots\alpha_{n+1}} = U_{\alpha_1} \cap U_{\alpha_2} \cap \dots \cap U_{\alpha_{n+1}}$. The nilpotency of the operator $\delta$ is easily verified. As usual we define the $p$-th cohomolgy group $H^p(F, M)$ of the sheaf $(F, M)$ as

$$H^p(F, M) = \frac{\ker \delta_p}{\mathrm{im}\delta_{p-1}} \tag{2.2.21}$$

namely as the space of $p$-cocycles $\delta C^{(p)} = 0$ modulo the $p$-coboundaries $\delta C^{(p-1)}$. At this point, if we reconsider Eqs. (2.2.5) and (2.2.9) we can appreciate the sheaf-theoretic interpretation of the transition functions of a fibre bundle. The set $\mathcal{T} = \{t_{\alpha\beta}\}$ defines a 1-cocycle of the sheaf $(\mathcal{G}, M)$ of local G-gauge transformations over the base manifold $M$. The last of Eqs. (2.2.5) is precisely the statement that $\delta\mathcal{T} = 0$. On the other hand Eq. (2.2.9) states that a fibre bundle is trivial if $\mathcal{T} = \delta(\mathit{something})$ is a coboundary. Hence the space of non-trivial $G$-bundles over a given manifold is isomorphic to the first cohomology group of the sheaf $(\mathcal{G}, M)$:

$$\text{set of } G\text{-bundles on } M \approx H^1(\mathcal{G}, M) \tag{2.2.22}$$

In this identification we have tacitly assumed that fibre bundles form a group. This is quite obvious if as transition functions of the product we take the point-wise product of the transition functions. As an example let us consider a complex manifold $\mathcal{M}$ and the complex line bundles $L \xrightarrow{\pi} \mathcal{M}$ constructed over it. A complex line bundle is a fibre bundle where the structural group is just $\mathbb{C}$, namely the set of complex numbers. According to our previous discussion we have

$$\text{group of line bundles on } \mathcal{M} \;=\; H^1\left(\mathcal{O}^\star\,,\,\mathcal{M}\right) \tag{2.2.23}$$

where $\mathcal{O}^\star$ denotes the sheaf of non-vanishing holomorphic functions. If $\mathcal{M} = \Sigma_g$ is a Riemann surface of genus $g$, then the group of line bundles $H^1\left(\mathcal{O}^\star\,,\,\Sigma_g\right) = \mathrm{Pic}\left(\Sigma_g\right)$ is named the Picard group of the surface and can be shown to be isomorphic to the group of divisor classes, namely to the group of divisors modulo the principal divisors [183].

In the subsequent Chapters we shall be especially concerned with holomorphic bundles on complex manifolds. These are bundles where the transition function have a holomorphic dependence on the coordinates of the base manifold. For the time being, however, we keep our exposition at a general level.

### 2.2.3 Sections of a fibre bundle

Let $E \stackrel{\pi}{\longrightarrow} M$ be a fibre bundle.

**Definition 2.2.3** *A section $s : M \longrightarrow E$ is a smooth map from the base manifold to the total space that composed with the projection yields the identity $\pi \circ s = \mathrm{id}_M$. The set of all sections of a bundle is denoted by $\Gamma\left(M, E\right)$.*

In other words a section is a rule to assign in a smooth way a point in the fibre to each point of the base manifold. All the fields considered in physical field theories are sections of some bundles. For instance a vector field on a manifold $X = X^\mu\,\partial_\mu \in \Gamma\left(M, TM\right)$ is a section of the tangent bundle, a 1-form $\omega = \omega_\mu dx^\mu \in \Gamma\left(M, T^\star M\right)$ is a section of the dual cotangent bundle, the Higgs field is a section of a two-dimensional complex vector bundle with structural group $SU(2) \otimes U(1)$ and four-dimensional space-time as base manifold and so on.

### 2.2.4 Bundle maps

Let $E \stackrel{\pi}{\longrightarrow} M$ and $E' \stackrel{\pi'}{\longrightarrow} M'$ be two fibre bundles. A smooth map

$$\overline{f} \,:\, E \longrightarrow E' \tag{2.2.24}$$

is named a *bundle map* if it preserves the fibres. This means that if $\pi(x) = \pi(y)$ then $\pi'(x) = \pi'(y)$. A bundle map induces a smooth map of the base manifolds:

$$f(p) \;=\; p' \in M' \tag{2.2.25}$$

### 2.2.5 Equivalent bundles

Two bundles $E \xrightarrow{\pi} M$ and and $E' \xrightarrow{\pi'} M$ constructed over the same base space $M$ are equivalent if there exists a bundle map $\overline{f} : E \longrightarrow E'$ such that the induced map $f : M \longrightarrow M$ is the identity, while $\overline{f}$ is a diffeomorphism:

$$\begin{array}{ccc} E' & \xrightarrow{\overline{f}} & E \\ {\scriptstyle \pi'}\downarrow & & \downarrow{\scriptstyle \pi} \\ M & \xrightarrow{\mathrm{id}_M} & M \end{array} \qquad (2.2.26)$$

### 2.2.6 Pull-back bundles

Let $E \xrightarrow{\pi} M$ be a fibre bundle and $N$ some other smooth manifold $N$. Let

$$f : N \longrightarrow M \qquad (2.2.27)$$

be a smooth map from the manifold $N$ to the base space of the fibre bundle $E$. Then we can define a new fibre bundle $f^* E \xrightarrow{f^*\pi} N$ that admits $N$ as base space through the following construction.

Consider the space

$$S = N \otimes E \qquad (2.2.28)$$

The points in $S$ are the pairs $\{p, u\}$, where $p \in N$ is a point in the base space of the new bundle to be constructed and $u \in E$ is a point in the total space of the old bundle. Consider now the subspace $f^* E \subset N \otimes E$:

$$f^*E = \{ (p, u) \in S \mid f(p) = \pi(u)\} \qquad (2.2.29)$$

of those pairs where the image of $p \in N$ through the map $f$ coincides with the image of $u \in E$ through the projection $\pi$ of the old bundle. The space $f^*E$ is the total space of the new bundle. The projection

$$f^*\pi : f^*E \longrightarrow N \qquad (2.2.30)$$

is defined by

$$f^*\pi\,(p, u) = p \qquad (2.2.31)$$

The new bundle $f^* E \xrightarrow{f^*\pi} N$ is named the pull-back of $E \xrightarrow{\pi} M$ through $f$. In this way we have a bundle map:

$$\begin{array}{ccc} f^* E & \xrightarrow{\pi_2} & E \\ {\scriptstyle f^*\pi}\downarrow & & \downarrow{\scriptstyle \pi} \\ N & \xrightarrow{f} & M \end{array} \qquad (2.2.32)$$

where $\pi_2 : f^* E \longrightarrow E$ is defined by $\pi_2\,(p, u) = u$. (Remember that the pair $(p, u) \in f^* E$ satisfies, by definition, $f(p) = \pi(u)$.) In terms of transition functions the pull-back bundle is described as follows. The standard fibre $F$ for $E \xrightarrow{\pi} M$ and

$f^\star E \xrightarrow{f^\star\pi} N$ are the same; the transition functions of the former are the pull-back of the transition functions of the latter. Indeed let $\{U_\alpha\}$ be a covering of $M$ and $\{\phi_\alpha\}$ be a corresponding local trivialization of the original bundle:

$$\phi_\alpha \; : \; U_\alpha \otimes F \longrightarrow \pi^{-1}(U_\alpha) \qquad (2.2.33)$$

The inverse images $f^{-1}(U_\alpha)$ provide a covering of the new base manifold $N$. We have an associated local trivialization of the pull-back bundle:

$$\psi_\alpha \; : \; f^{-1}(U_\alpha) \otimes F \longrightarrow f^\star \pi^{-1}\left(f^{-1}(U_\alpha)\right) \qquad (2.2.34)$$

defined as follows. Take $p \in f^{-1}(U_\alpha)$ and consider any point $u \in E$ such that $\pi(u) = f(p)$. In this case the pair $(p,u) \in f^\star E$ is an element of the total space of the new bundle. Let $\phi^{-1}(u) = (f(p), f_\alpha)$ , where $f_\alpha \in F$ is a point in the standard fibre. We set $\psi_\alpha^{-1}(p,u) = (p, f_\alpha)$ and this defines $\psi_\alpha$. The transition function $t_{\alpha\beta}(f(p))$ at a point $f(p) \in U_\alpha \cap U_\beta$ in the intersection of two charts of $M$ maps a point $f_\alpha \in F$ of the standard fibre to a new point $f_\beta = t_{\alpha\beta}(f(p))\, f_\beta$. By definition the transition function $t^\star_{\alpha\beta}(p)$ at a point $p \in f^{-1}(U_\alpha) \cap f^{-1}(U_\alpha)$ in the intersection of two charts of $N$, does the same, namely we have

$$t^\star_{\alpha\beta}(p) = t_{\alpha\beta}(f(p)) \qquad (2.2.35)$$

Summarizing, the transition functions of the pull-back bundle are the pull-back of the transition functions in the original bundle, as already stated. This is particularly significant from the point of view of sheaf cohomology introduced above.

An important result for the classification of fibre bundles is contained in the following theorem due to Steenrod (1951), whose proof we omit [222]:

**Theorem 2.2.1** *Let $E \xrightarrow{\pi} M$ be a fibre bundle and $N$ be some other manifold. Let*

$$\begin{aligned} f \; &: \; M \longrightarrow N \\ g \; &: \; M \longrightarrow N \end{aligned} \qquad (2.2.36)$$

*be two homotopic maps from $M$ to $N$. The $f^\star E$ and $g^\star E$ are equivalent bundles.*

This theorem sheds light on the fact that the non triviality of a fibre bundle is possible only if the base space has a non-trivial topology, namely a non trivial homotopy. Only in this case can the fibres be aligned in such a way as to wind along non-trivial homotopy cycles. In particular we have:

**Corollary 2.2.1** *All bundles constructed over a base manifold that is contractible to a point are trivial bundles.*

Indeed let $E \xrightarrow{\pi} M$ be a fibre bundle and let $M$ be contractible. This means that there is a homotopy:

$$F \; : \; M \otimes [0,1] \longrightarrow M \qquad (2.2.37)$$

such that

$$\begin{aligned} F[p,0] &= p \; ; \quad \forall p \in M \\ F[p,0] &= p_0 \; ; \quad \forall p \in M \end{aligned} \tag{2.2.38}$$

where $p_0 \in M$ is some fixed point. Consider the continuous family of bundles $f_t^\star E$ where $f_t(p) = F(p,t)$. Due to Steenrod's theorem, all these bundles are equivalent. On the other hand, since $f_0 = \mathrm{id}$, $f_0^\star E = E$, while $f_1^\star E$ is a trivial bundle constructed over a base manifold that is just the single point $p_0$. Hence $E \xrightarrow{\pi} M$ is also trivial.

## 2.3 Vector Bundles, Connections and Curvatures

From the point of view of physical applications the most interesting fibre bundles are the vector bundles and their underlying principal bundles, whose sections correspond to the matter fields and to the gauge fields, respectively.

**Definition 2.3.1** *A real vector bundle $E \xrightarrow{\pi} M$ is a fibre bundle whose typical fibre $F \approx \mathbf{R}^k$ is endowed with a real vector space structure. The transition functions act on $F$ as linear maps. The dimension $k$ of the fibre is named the rank of the vector bundle and the structural group $G \subset GL(k, \mathbb{R})$ is necessarily a subgroup of the general linear group. Alternatively in a complex vector bundle the typical fibre is a complex vector space $F \approx \mathbb{C}^k$ and $G \subset GL(k, \mathbb{C})$. Also in this case $k$ is named the rank of the bundle.*

The vector space structure of the typical fibre $F$ is extended to the sections $s \in \Gamma(M,E)$ in a canonical point-wise fashion. Let $s$ and $s'$ be two sections of a vector bundle $E \xrightarrow{\pi} M$. Then we can define the sum of the two sections by

$$\left(s + s'\right)(p) = s(p) + s'(p) \quad ; \quad \forall p \in M \tag{2.3.1}$$

Similarly if $f \in C^\infty(M)$ is a function defined on $M$, for each section $s \in \Gamma(M,E)$ we can define a new section $f\, s$ by

$$(f\, s)(p) = f(p)\, s(p) \quad ; \quad \forall p \in M \tag{2.3.2}$$

Any vector bundle admits a global **null section** $s_0 = \Gamma(M,E)$ such that $\phi_\alpha^{-1}(s_0(p)) = (p,0)$ in any local trivialization.

### 2.3.1 Fibre metrics

On vector bundles one can introduce *fibre metrics*. A fibre metric on a complex vector bundle

$$\begin{aligned} \left(s, s'\right) &\in C^\infty(M) \quad ; \quad \forall s, s' \in \Gamma(M,E) \\ \left(s + s', s''\right) &= \left(s, s''\right) + \left(s', s''\right) \\ \left(s, s'\right) &= \left(s', s\right)^\star \end{aligned} \tag{2.3.3}$$

is a bilinear hermitean form that is point-wise defined on sections, namely

$$\left(s\,,\,s'\right)(p) \;=\; \overline{s}^i_\alpha(p)\, s^j_\alpha(p)\, h^{(\alpha)}_{ij}(p) \quad ; \quad \forall p \in M \tag{2.3.4}$$

where $s^i(p)(i=1,\ldots k)$ are the components of the section $s$ in some basis of the standard fibre $F$ and in some local trivialization $\phi^{-1}_\alpha\,(s\,(p)) \;=\; (p, s^i_\alpha\,(p))$. The local functions $h^{(\alpha)}_{ij}(p)$, forming a $k\times k$ hermitean matrix $h^{(\alpha)}$, are named the coefficients of the metric in the local trivialization $\phi_\alpha$. The transition functions $t_{(\alpha\beta)}$ are elements of $GL(k,\mathbb{C})$, namely they are $k\times k$ matrices $t^i_{(\alpha\beta)\ j}(p)$ such that

$$s^i_\alpha(p) \;=\; t^i_{(\alpha\beta)\ j}(p)\, s^j_\beta(p) \tag{2.3.5}$$

Correspondingly the relation between the fibre metric coefficients in two different local trivializations is the following one:

$$h^{(\alpha)} \;=\; \left(t^{-1}_{\alpha\beta}\right)^\dagger h^{(\beta)}\, t^{-1}_{\alpha\beta} \tag{2.3.6}$$

In this way the inner product of two sections is invariantly defined independently from the local trivialization chosen to calculate it. If we deal with real rather than complex vector bundles then the fibre metric is just a symmetric bilinear form and all the above formulae are canonically restricted to real numbers.

### 2.3.2 Product bundle

On vector bundles we can define the operations of addition and tensor product.

If $E \xrightarrow{\pi} M$ and $E' \xrightarrow{\pi'} M'$ are two vector bundles, we can define the product vector bundle by

$$E\times E' \xrightarrow{\pi\times\pi'} M\times M' \tag{2.3.7}$$

where the fibre is $F\oplus F'$ and the projection works in the following way. If $(u,u')\in E\otimes E'$ and $\pi(u) \;=\; p\;;\;\pi'(u') \;=\; p'$, then

$$\left(\pi\times\pi'\right)\left(u\,,\,u'\right) \;=\; \left(p\,,\,p'\right)\in M\times M' \tag{2.3.8}$$

### 2.3.3 Whitney sum

Let $E \xrightarrow{\pi} M$ and $E' \xrightarrow{\pi'} M$ be two vector bundles constructed over the same base manifold and let

$$\text{diag}\;:\; M \longrightarrow M\times M \tag{2.3.9}$$

be the map that associates to each point $p\in M$ the pair $(p,p)\in M\times M$. The pull-back of $E\times E'$ by the map diag is what we call $E\oplus E'$, namely the **Whitney sum** of the two vector bundles.

$$E\oplus E' \;=\; \text{diag}^\star\left(E\times E'\right) \tag{2.3.10}$$

By definition

$$E \oplus E' = \left\{ \left(u, u'\right) \in E \oplus E' \mid \left(\pi \times \pi'\right) \left(u, u'\right) = (p, p) \right\} \tag{2.3.11}$$

The fibre of $E \oplus E'$ is $F \oplus F'$. Furthermore let $\{U_\alpha\}$ be a covering of the base manifold. The transition functions of the Whitney sum $E \oplus E'$ are block-diagonal $(k+k') \times (k+k')$ matrices of the form

$$T^{E\oplus E'}_{\alpha\beta} = \begin{pmatrix} t^E_{\alpha\beta} \, , & 0 \\ 0 \, , & t^{E'}_{\alpha\beta} \end{pmatrix} \tag{2.3.12}$$

where the $k \times k$ matrices $t^E_{\alpha\beta}$ are the transition function of the first bundle, while the $k' \times k'$ matrices $t^{E'}_{\alpha\beta}$ are the transition functions of the second bundle.

### 2.3.4 Tensor product bundle

In a similar way, if $E \xrightarrow{\pi} M$ and $E' \xrightarrow{\pi'} M$ are two vector bundles constructed over the same base manifold $M$, their tensor product $E \otimes E' \xrightarrow{\pi\otimes\pi'} M$ is the vector bundle that has $F \otimes F'$ as typical fibre and that admits as transition functions the tensor product of the transition functions of the two bundles $E$ and $E'$.

### 2.3.5 Principal fibre bundles

Let $E \xrightarrow{\pi} M$ be a rank $k$ vector bundle with structural group $G \subset GL(k, \mathbf{R})$ or $G \subset GL(k, \mathbf{C})$, depending on whether it is real or complex. The structure of such a vector bundle is in fact completly determined by the structure of the principal bundle to which it is associated.

**Definition 2.3.2** *A principal bundle $P \xrightarrow{\pi} M$ is a fibre bundle where the typical fibre $F$ coincides with the structural group $G$. The action of the structural group on the fibre is just left-multiplication in the group $G$.*

Quite obviously, in a principal bundle, the transition functions $t_{\alpha\beta}(p) = g_{\alpha\beta}(p)$ are smooth functions $g_{\alpha\beta} : U_\alpha \cap U_\beta \longrightarrow G$ from the intersection of two local charts of the base manifold $M$ to the abstract structural group $G$. Given a principal bundle $P \xrightarrow{\pi} M$ and given a linear representation of $G$, $D : G \longrightarrow \mathrm{Hom}(F, F)$, that to each element $g \in G$ associates a linear map $D(g)$ ; $F \longrightarrow F$ of a $k$-dimensional vector space $F$ into itself, we can uniquely construct a vector bundle of rank $k$ , $E \xrightarrow{\pi} M$, where the standard fibre is $F$ and where the transition functions are $t_{\alpha\beta}(p) = D\left(g_{\alpha\beta}\left(p\right)\right)$. Any vector bundle can be constructed in such a way from underlying principal bundle. In physical field theories the principal bundle is related to the gauge sector of the theory, while the various possible associated bundles contain the matter fields. Indeed, as already remarked, every matter field is a section $s \in \Gamma(M, E_D)$ of an associated vector bundle $E_D \xrightarrow{\pi} M$ in representation $D$ of the structural (=gauge) group.

### 2.3.6 Connections on a vector bundle

The concept of connection is the mathematical counterpart of the physical concept of gauge field. From the mathematical point of view a connection is firstly introduced on the principal fibre bundle $P(G, M)$ which underlies any vector bundle with structural group $G$ and then it is canonically extended to any associated bundle in representation $D$. From a physical view point it is, however, more convenient to introduce the concept of a connection by working directly on the associated bundles.

**Definition 2.3.3** *Let $E \xrightarrow{\pi} M$ be a vector bundle, $TM$ the tangent bundle to the base manifold and $\Gamma(M, E)$ the space of all sections of $E$. A connection $\nabla$ is a rule that to each vector field $X \in TM$ associates a map $\nabla_X : \Gamma(M, E) \longrightarrow \Gamma(M, E)$ satisfying the following properties:*

$$\begin{array}{lrcl} a) & \nabla_X\,(a_1\, s_1 + a_2\, s_2) & = & a_1 \nabla_X\, s_1 + a_2 \nabla_X\, s_2 \\ b) & \nabla_{(a_1\, X_1 + a_2\, X_2)}\, s & = & a_1 \nabla_{X_1}\, s + a_2 \nabla_{X_2}\, s \\ c) & \nabla_X\,(f\, s) & = & X[f]\, s + f\, \nabla_X\, s \\ d) & \nabla_{fX} & = & f\, \nabla_X\, s \end{array} \tag{2.3.13}$$

*where $a_i \in \mathbf{R}$ , $s_i \in \Gamma(M, E)$ and $f \in \mathbf{C}^\infty(M)$*

The operator $\nabla_X$ is what physicists call a covariant derivative. In a more condensed way we can say that a connection $\nabla$ is a map

$$\nabla : \Gamma(M, E) \longrightarrow \Gamma(M, E) \otimes T^*M \tag{2.3.14}$$

that to each section $s$ of the vector bundle associates a section-valued 1-form $\nabla\, s$, in such a way that $\forall X \in TM$, $\nabla_X\, s = \nabla\, s\,(X)$. In this formulation the properties (2.3.13) satisfied by the connection are equivalent to

$$\begin{array}{lrcl} a) & \nabla\,(a_1\, s_1 + a_2\, s_2) & = & a_1 \nabla\, s_1 + a_2 \nabla\, s_2 \\ b) & \nabla(\,f\, s\,) & = & (\,df\,)\, s + f\, \nabla\, s \end{array} \tag{2.3.15}$$

Let us now consider a local trivialization $\phi_\alpha : F \otimes U_\alpha \longrightarrow \pi^{-1}(U_\alpha)$.

**Definition 2.3.4** *A frame over $U_\alpha$ is a set of $k$ sections $\{\, s^1_\alpha\,, \ldots\,, s^k\,\}$ such that $\{\, s^1_\alpha(p)\,, \ldots\,, s^k_\alpha(p)\,\}$ is a basis for $\pi^{-1}(p)\, \forall\, p \in U_\alpha$.*

Given a frame over $U_\alpha$, we can write the connection 1-form in that frame by setting

$$\begin{array}{rcl} \nabla\, s^i_\alpha & = & \mathcal{A}^i_{(\alpha)\; j}\, s^j_\alpha \\ \mathcal{A}^i_{(\alpha)\; j} & = & A^A_{(\alpha)\,\mu}\, dx^\mu\, (T_A)^i{}_j \end{array} \tag{2.3.16}$$

where the $k \times k$ matrices $(T_A)^i{}_j$ are a set of generators of the structural group Lie algebra in the representation $D$ carried by the fibre $F$ and $A^A_{(\alpha)\,\mu}\, dx^\mu$. In other words $\mathcal{A}_{(\alpha)}$ is a $G$

Lie algebra-valued 1-form over $U_\alpha$. If we consider two overlapping local trivializations, over the intersection $U_\alpha \cap U_\beta$ we have two definitions of the connection 1-form $\mathcal{A}_{(\alpha)}$ and $\mathcal{A}_{(\beta)}$. They are related by the formula

$$\mathcal{A}_{(\beta)} = t_{\alpha\beta}^{-1} \mathcal{A}_{(\alpha)} t_{\alpha\beta} + t_{\alpha\beta}^{-1} d t_{\alpha\beta} \tag{2.3.17}$$

where $t_{\alpha\beta}$ is the transition function between the two local trivializations. In each local trivialization the **curvature 2-form** associated with the connection 1-form $\mathcal{A}_{(\alpha)}$ is given by

$$\mathcal{F}_{(\alpha)} = d\mathcal{A}_{(\alpha)} + \mathcal{A}_{(\alpha)} \wedge \mathcal{A}_{(\alpha)} = d\mathcal{A}_{(\alpha)} + \frac{1}{2}\left[\mathcal{A}_{(\alpha)}, \mathcal{A}_{(\alpha)}\right] \tag{2.3.18}$$

and the relation between $\mathcal{F}_{(\alpha)}$ and $\mathcal{F}_{(\beta)}$ in the intersection of two local trivializations is given by:

$$\mathcal{F}_{(\beta)} = t_{\alpha\beta}^{-1} \mathcal{F}_{(\alpha)} t_{\alpha\beta} \tag{2.3.19}$$

## 2.4 Complex Structures on 2n-Dimensional Manifolds

Let $\mathcal{M}$ be a 2n-dimensional manifold, $T\mathcal{M}$ its tangent space and $T^*\mathcal{M}$ its cotangent space. Denoting by $\{\phi^\alpha\}$ $(\alpha = 1, \ldots, 2n)$ the $2n$ coordinates in a patch, a section $\vec{t} \in \Gamma(T\mathcal{M}, \mathcal{M})$ is represented by a linear differential operator:

$$\vec{t} = t^\alpha \vec{\partial}_\alpha \tag{2.4.1}$$

while a section in $T^*\mathcal{M}$ is a differential 1-form

$$\omega = d\phi^\alpha \omega_\alpha(\phi) \tag{2.4.2}$$

The contraction is an operation that to each vector field $\vec{t} \in \Gamma(T\mathcal{M}, \mathcal{M})$ associates a map

$$i_{\vec{t}} : T^*\mathcal{M} \longrightarrow \mathbb{C}^\infty(\mathcal{M}) \tag{2.4.3}$$

of 1-forms into 0-forms locally given by the following expression:

$$i_{\vec{t}}\omega = t^\alpha(\phi)\omega_\alpha(\phi) \tag{2.4.4}$$

In particular, if $\omega = df$ we have

$$i_{\vec{t}} df = t^\alpha \partial_\alpha f = \vec{t} f \tag{2.4.5}$$

The contraction is also canonically extended to higher forms:

$$\forall \vec{t} \in \Gamma(T\mathcal{M}, \mathcal{M}) \begin{cases} i_{\vec{t}} : \Omega^p(\mathcal{M}) \longrightarrow \Omega^{p-1}(\mathcal{M}) \\ i_{\vec{t}}\omega = t^\alpha(\phi)\omega_{\alpha\beta_1\ldots\beta_{p-1}}(\phi)\, d\phi^{\beta_1} \wedge \cdots \wedge d\phi^{\beta_{p-1}} \end{cases} \tag{2.4.6}$$

Now we can consider a linear operator $L$ acting on the tangent bundle $T\mathcal{M}$, or more precisely acting on $\Gamma(T\mathcal{M}, M)$:

$$\begin{array}{l} L \ : \Gamma(T\mathcal{M},\mathcal{M}) \rightarrow \Gamma(T\mathcal{M},\mathcal{M}) \\ \forall \ \vec{t} \in \Gamma(T\mathcal{M},\mathcal{M}): \quad L\vec{t} \in \Gamma(T\mathcal{M},\mathcal{M}) \\ \forall \ \alpha,\beta \in C, \ \ \forall \vec{t}_1,\vec{t}_2 \in \Gamma(T\mathcal{M},\mathcal{M}): \quad L(\alpha\vec{t}_1 + \beta\vec{t}_2) = \alpha L\vec{t}_1 + \beta L\vec{t}_2 \end{array} \tag{2.4.7}$$

In every local chart $L$ is represented by a mixed tensor $L_\alpha^\beta(\phi)$ with one covariant index and one controvariant index such that

$$L\vec{t} = t^\alpha(\phi) L_\alpha^\beta(\phi)\vec{\partial}_\beta \tag{2.4.8}$$

Moreover the action of $L$ is naturally pulled back on the cotangent space:

$$L : \Gamma(T\mathcal{M}^*, \mathcal{M}) \rightarrow \Gamma(T\mathcal{M}^*, \mathcal{M}) \tag{2.4.9}$$

by defining

$$i_{\vec{t}} L\omega = i_{L\vec{t}}\,\omega \tag{2.4.10}$$

which in a local chart yields

$$L\omega = d\phi^\alpha L_\alpha^\beta(\phi)\omega_\beta \tag{2.4.11}$$

**Definition 2.4.1** *A 2n-dimensional manifold $\mathcal{M}$ is called almost complex if it has an almost complex structure. An almost complex structure is a linear operator $J : \Gamma(T\mathcal{M},\mathcal{M}) \rightarrow \Gamma(T\mathcal{M},\mathcal{M})$ which satisfies the following property:*

$$J^2 = -\,\mathbb{1} \tag{2.4.12}$$

In every local chart the operator $J$ is represented by a tensor $J_\beta^\alpha(\phi)$ such that

$$J_\alpha^\beta(\phi) J_\beta^\gamma(\phi) = -\delta_\alpha^\gamma \tag{2.4.13}$$

and by a suitable change of basis at every point $p \in \mathcal{M}$ we can reduce $J_\alpha^\beta$ to the form

$$\begin{pmatrix} 0 & \mathbb{1} \\ -\mathbb{1} & 0 \end{pmatrix} \tag{2.4.14}$$

where $\mathbb{1}$ is the $n \times n$ unity matrix. A local frame where $J$ takes the form (2.4.14) is called a "well-adapted" frame to the almost complex structure. Naming

$$\vec{e}_\alpha = \vec{\partial}_\alpha = \frac{\partial}{\partial \phi_\alpha} \tag{2.4.15}$$

the basis of the well-adapted frame we have

$$\begin{array}{l} J\vec{e}_\alpha = -\vec{e}_{\alpha+n} \quad if \quad \alpha \le n \\ J\vec{e}_\alpha = \vec{e}_{\alpha-n} \quad if \quad \alpha > n \end{array} \tag{2.4.16}$$

At this point, introducing the index $i$ with range $i = 1, \cdots, n$ we can define the complex vectors:

$$\begin{aligned} \vec{E}_i &= \vec{e}_i - i\vec{e}_{i+n} \\ \vec{E}_{i^*} &= \vec{e}_i + i\vec{e}_{i+n} \end{aligned} \qquad (2.4.17)$$

and we obtain the following result:

$$\begin{aligned} J\vec{E}_i &= i\vec{E}_i \\ J\vec{E}_{i^*} &= -i\vec{E}_{i^*} \end{aligned} \qquad (2.4.18)$$

The tangent vectors $\vec{E}_i$ are the partial derivatives along the complex coordinates:

$$z^i = \phi^i + i\phi^{i+n} \qquad (2.4.19)$$

while $\vec{E}_{i^*}$ are the partial derivatives along the complex conjugate coordinates $\overline{z}^{i^*} = \phi^i - i\phi^{i+n}$:

$$\vec{E}_i = \vec{\partial}_i = \frac{\partial}{\partial z^i} \quad \vec{E}_{i^*} = \vec{\partial}_{i^*} = \frac{\partial}{\partial z^{i^*}} \qquad (2.4.20)$$

This construction is the reason why $J$ is called an almost complex structure: the existence of this latter guarantees that at every point $p \in \mathcal{M}$ we can replace the $2n$ real coordinates by $n$ complex coordinates, corresponding to a well-adapted frame. Moreover every two well-adapted frames are related to each other by a coordinate transformation which is a holomorphic function of the corresponding complex coordinates.

Indeed let

$$\phi^\alpha \rightarrow \phi^\alpha + \zeta^\alpha(\phi) \qquad (2.4.21)$$

be an infinitesimal coordinate transformation connecting two well adapted frames. By definition this means

$$\partial_\alpha \zeta^\beta J^\gamma_\beta = J^\beta_\alpha \partial_\beta \zeta^\gamma \qquad (2.4.22)$$

which is nothing but the Cauchy–Riemann equation for the real and imaginary parts of a holomorphic function. Hence Eq. (2.4.21) can be replaced by

$$z^i \rightarrow z^i + \zeta^i(z) \qquad (2.4.23)$$

where $\zeta^i(z)$ is a holomorphic function of $z^j$.

Conversely if $\mathcal{M}$ is a complex analytic manifold[1], in every local chart $\{z^i\}$ we can set

$$\phi^\alpha = \mathrm{Re} z^i \ (\alpha \leq n) \quad \phi^\alpha = \mathrm{Im} z^i \ (\alpha > n) \qquad (2.4.24)$$

and we can define an almost complex structure $J$.

[1] Complex analytic manifold means a manifold whose transition functions in the intersection of two charts are holomorphic functions of the local coordinates.

Now let $J$ act on $T^*(\mathcal{M})$. In a well-adapted frame we have

$$\begin{aligned} Jdz^i &= idz^i \\ Jdz^{i^*} &= -idz^{i^*} \end{aligned} \tag{2.4.25}$$

Equations (2.4.25) characterize the holomorphic coordinates.

More generally let $\{x^\alpha\}$ be a generic coordinate system (not necessarily well-adapted) and let $w(x)$ be a complex-valued function on the manifold $\mathcal{M}$: we say that $w$ is holomorphic if it satisfies the equation:

$$Jdw = idw \tag{2.4.26}$$

which in the generic coordinate system $\{x^\alpha\}$ reads as follows:

$$J_\alpha^\beta \partial_\beta w(x) = i\partial_\alpha w(x) \tag{2.4.27}$$

As we have seen, at every point $p \in \mathcal{M}$, $J$ can be reduced to the canonical form (2.4.14) by a suitable coordinate transformation: what is not guaranteed is whether $J$ can be reduced to this canonical form in a whole open neighbourhood $\mathcal{U}_p$. This amounts to asking the question whether Eq. (2.4.27) admits $n$ $\mathbb{C}$-linearly independent solutions in some open subset $\mathcal{U} \in \mathcal{U}_X$, where $\mathcal{U}_X$ is the domain of the considered local chart $\{x^\alpha\}$.

If these solutions $w^i(x)$ exist we can consider them as the holomorphic coordinates in the neighbourhood $\mathcal{U}$, that is we can set

$$z^i = w^i(z) \tag{2.4.28}$$

In view of what we discussed before, the transition function between any two such coordinate systems is holomorphic. Hence if Eq. (2.4.26) is integrable, then a holomorphic coordinate system exists and any function $\phi$ on the manifold can be viewed as a function of $z^i$ and $z^{i^*}$: $\phi = \phi(z, z^{i^*})$. In this case we have

$$\begin{aligned} d\phi &= \partial_i \phi dz^i + \partial_{i^*} \phi d\overline{z}^{i^*} \\ Jd\phi &= i(\partial_i \phi dz^i - \partial_{i^*} \phi dz^{i^*}) \end{aligned} \tag{2.4.29}$$

By taking the exterior derivative of Eq. (2.4.29) we obtain

$$dJ \wedge d\phi = -2i\partial_i \partial_{i^*} \phi dz^i \wedge dz^{i^*} \tag{2.4.30}$$

and we can verify the equation

$$(1 - J) dJ \wedge d\phi = 0 \tag{2.4.31}$$

which follows from

$$JdJ \wedge d\phi = -2i\partial_i \partial_{j^*} \phi Jdz^i \wedge Jdz^{j^*} = -2i\partial_i \partial_{j^*} \phi dz^i \wedge dz^{j^*} = dJ \wedge d\phi \tag{2.4.32}$$

Equation (2.4.31) is true in a holomorphic coordinate system and, being an exterior algebra statement, must be true in every coordinate system.

In the real coordinate system Eq. (2.4.31) reads

$$T^{\alpha}_{\beta\gamma}\partial_{\alpha}\phi dx^{\beta}\wedge dx^{\gamma}=0 \tag{2.4.33}$$

where the tensor

$$T^{\alpha}_{\beta\gamma}=\partial_{[\beta}J^{\alpha}_{\gamma]}-J^{\mu}_{\beta}J^{\nu}_{\gamma}\partial_{[\mu}J^{\alpha}_{\nu]} \tag{2.4.34}$$

is called the "torsion", or the Nienhuis tensor of the almost complex structure $J^{\alpha}_{\beta}$. The vanishing of $T^{\alpha}_{\beta\gamma}$ is a necessary condition for the integrability of Eq. (2.4.27) and hence for the existence of a complex structure. It can be shown that it is also sufficient provided $T^{\alpha}_{\beta\gamma}$ is real analytic with respect to some real coordinate system.

## 2.5 Metric and Connections on Holomorphic Vector Bundles

In the previous section we considered the structure of complex manifolds. When both the base space and the standard fibre are complex manifolds we can refine the notion of fibre bundle by requiring that the transition function be locally holomorphic functions. In particular a very relevant concept, which plays a major role in our subsequent developments, is that of holomorphic vector bundle. For convenience we recall the complete definition that follows from the general definition 2.3.1.

**Definition 2.5.1** *Let $M$ be a complex manifold and $E$ be another complex manifold. A holomorphic vector bundle with total space $E$ and base manifold $M$ is given by a projection map:*

$$\pi\ :\ E\longrightarrow M \tag{2.5.1}$$

*such that*

*a) $\pi$ is a holomorphic map of $E$ onto $M$*

*b) Let $p\in M$, then the fibre over $p$*

$$E_p\ =\ \pi^{-1}(p) \tag{2.5.2}$$

*is a complex vector space of dimension $r$. (The number $r$ is called the rank of the vector bundle.)*

*c) For each $p\in M$ there is a neighbourhood $U$ of $p$ and a holomorphic homeomorphism*

$$h\ :\ \pi^{-1}(U)\longrightarrow U\times\mathbb{C}^r \tag{2.5.3}$$

*such that*

$$h\left(\pi^{-1}(p)\right)\ =\ \{p\}\times\mathbb{C}^r \tag{2.5.4}$$

*(The pair $(U,h)$ is called a local trivialization.)*

*d) The transition functions between two local trivializations* $(U_\alpha, h_\alpha)$ *and* $(U_\beta, h_\beta)$*:*

$$h_\alpha \circ h_\beta^{-1} : (U_\alpha \cap U_\beta) \otimes \mathbb{C}^r \longrightarrow (U_\alpha \cap U_\beta) \otimes \mathbb{C}^r \tag{2.5.5}$$

*induce holomorphic maps*

$$g_{\alpha\beta} : (U_\alpha \cap U_\beta) \longrightarrow GL(r, \mathbb{C}) \tag{2.5.6}$$

Let $E \longrightarrow M$ be a holomorphic vector bundle of rank $r$ and $U \subset M$ an open subset of the base manifold. A frame over $U$ is a set of $r$ holomorphic sections $\{ s_1, \ldots, s_r \}$ such that $\{ s_1(z), \ldots, s_r(z) \}$ is a basis for $\pi^{-1}(x)$ for any $x \in U$. Let $f \stackrel{\text{def}}{=} \{ e_I(z) \}$ be a frame of holomorphic sections. Any other holomorphic section $\xi$ is described by

$$\xi = \xi^I(z)\, e_I \tag{2.5.7}$$

where

$$\overline{\partial} \xi^I = d\overline{z}^{j^\star} \overline{\partial}_{j^\star} \xi^I = 0 \tag{2.5.8}$$

Given a holomorphic bundle with a frame of sections we can discuss metrics connections and curvatures, as we already did for the general case of bundles.

In general a connection $\theta$ is defined, according to the definition 2.3.3 and Eqs. (2.3.14), (2.3.15), by introducing the covariant derivative of any section $\xi$

$$D\xi = d\xi + \theta\xi \tag{2.5.9}$$

where $\theta = \theta^I{}_J$, the connection coefficient, is an $r \times r$ matrix-valued 1-form. On a complex manifold this 1-form can be decomposed into its parts of holomorphic type (1,0) and (0,1)

$$\begin{aligned} \theta &= \theta^{(1,0)} + \theta^{(0,1)} \\ \theta^{(1,0)} &= dz^i\, \theta_i \\ \theta^{(0,1)} &= d\overline{z}^{i^\star}\, \theta_{i^\star} \end{aligned} \tag{2.5.10}$$

Let now a hermitean metric $h$ be defined on the holomorphic vector bundle, according to (2.3.3). This is a sesquilinear form that yields the scalar product of any two holomorphic sections $\xi$ and $\eta$ at each point of the base manifold:

$$\langle \xi, \eta \rangle_h \stackrel{\text{def}}{=} \overline{\xi}^{I^\star}(\overline{z})\, \eta^J(z)\, h_{I^\star J}(z, \overline{z}) = \xi^\dagger h \eta \tag{2.5.11}$$

As is evident from the above formula, the metric $h$ is defined by means of the point-dependent hermitean matrix $h_{I^\star J}(z, \overline{z})$, which is requested to transform, from one local trivialization to another, with the inverses of the transition functions $g_{\alpha\beta}$ defined in Eq. (2.5.6). This is so because the scalar product $\langle \xi, \eta \rangle_h$ is by definition an invariant (namely a scalar function globally defined on the manifold).

**Definition 2.5.2** *A hermitean metric for a complex manifold $\mathcal{M}$ is a hermitean fibre metric on the canonical tangent bundle $T\mathcal{M}$. In this case the transition functions $g_{\alpha\beta}$ are given by the jacobians of the coordinate transformations.*

In general $h$ is just a metric on the fibres and the transition functions are different objects from the Jacobian of the coordinate transformations.

In any case, given a fibre metric on a holomorphic vector bundle we can introduce a canonical connection $\theta$ associated with it. It is defined by requiring that

$$\begin{array}{lrcl} A) & d\langle \xi , \eta \rangle_h & = & \langle D\xi , \eta \rangle_h + \langle \xi , D\eta \rangle_h \\ B) & D^{(0,1)}\xi \stackrel{\text{def}}{=} \left[ \bar\partial + \theta^{(0,1)} \right] \xi & = & 0 \end{array} \tag{2.5.12}$$

namely by demanding that the scalar product be invariant with respect to the parallel transport defined by $\theta$ and by requiring that the holomorphic sections be transported into holomorphic sections. Let $f$ be a holomorphic frame. In this frame the canonical connection is given by

$$\theta(f) = h(f)^{-1}\,\partial\, h(f) \tag{2.5.13}$$

or, in other words, by

$$\theta^I{}_J = dz^i\, h^{IJ^*}\,\partial_i\, h_{K^*J} \tag{2.5.14}$$

In the particular case of a manifold metric (see definition (2.5.2)), where $h$ is a fibre metric on the tangent bundle $T\mathcal{M}$, the general formula (2.5.14) provides the definition of the Levi–Civita connection:

$$dz^k\,\Gamma^i_{kj} = -g^{il^*}\,\partial\, g_{l^*j} \tag{2.5.15}$$

Given a connection we can compute its curvature by means of the standard formula $\Theta = d\theta + \theta \wedge \theta$. In the case of the above-defined canonical connection we obtain

$$\Theta\,(f) = \partial\theta + \bar\partial\theta + \theta\wedge\theta = \bar\partial\theta \tag{2.5.16}$$

This identity follows from $\partial\,\theta + \theta \wedge \theta = 0$, which is identically true for the canonical connection (2.5.13). Component-wise the curvature 2-form is given by

$$\Theta^I{}_J = \bar\partial_i \left( h^{IK^*}\,\partial_j\, h_{K^*J} \right) d\bar z^i \wedge dz^j \tag{2.5.17}$$

For the case of Levi–Civita connection defined in Eq. (2.5.15) in summary we have

$$\begin{array}{rcl} \Gamma^i_j & = & \Gamma^i_{kj} dz^k \\ \Gamma^i_{kj} & = & -g^{i\ell^*}(\partial_j g_{k\ell^*}) \\ \Gamma^{i^*}_{j^*} & = & \Gamma^{i^*}_{k^*j^*} d\bar z^{k^*} \\ \Gamma^{i^*}_{k^*j^*} & = & -g^{i^*\ell}(\partial_{j^*} g_{k^*\ell}) \end{array} \tag{2.5.18}$$

for the connection coefficients and

$$\begin{aligned}
\mathcal{R}^i_j &= \mathcal{R}^i_{jk^\star\ell} d\overline{z}^{k^\star} \wedge dz^\ell \\
\mathcal{R}^i_{jk^\star\ell} &= \partial_{k^\star}\Gamma^i_{j\ell} \\
\mathcal{R}^{i^\star}_{j^\star} &= \mathcal{R}^{i^\star}_{j^\star k\ell^\star} dz^k \wedge d\overline{z}^{\ell^\star} \\
\mathcal{R}^{i^\star}_{j^\star k\ell^\star} &= \partial_k \Gamma^{i^\star}_{j^\star\ell^\star}
\end{aligned} \tag{2.5.19}$$

for the curvature 2-form. The Ricci tensor has a reamarkable simple expression:

$$R^n_{m^\star} = R^i_{m^\star n\, i} = \partial_{m^\star}\Gamma^i_{ni} = \partial_{m^\star}\partial_n \, ln\,(\sqrt{g}) \tag{2.5.20}$$

where $g = det|g_{\alpha\beta}| = (det|g_{ij^\star}|)^2$.

## 2.6 Kähler Metrics

In the previous section we have discussed the general notion of hermitean fibre metrics on holomorphic vector bundles and in particular of hermitean manifold metrics defined on the tangent bundle. In this section we introduce the more restricted concept of Kählerian metrics that plays a fundamental role in the sequel at two levels: it appears in the N=1 matter-coupled supergravity lagrangian as the scalar field metric and it is the metric of the internal compactified manifolds of heterotic superstrings, namely Calabi–Yau manifolds. The definition of the previous section (2.5.2) can also be restated in the following way: a manifold metric $g$ is a symmetric bilinear scalar valued functional on $\Gamma(T\mathcal{M},\mathcal{M}) \otimes \Gamma(T\mathcal{M},\mathcal{M})$

$$g : \Gamma(T\mathcal{M},\mathcal{M}) \otimes \Gamma(T\mathcal{M},\mathcal{M}) \to C^\infty(\mathcal{M}) \tag{2.6.1}$$

In every coordinate system it is represented by the familiar symmetric tensor $g_{\alpha\beta}(x)$. Inded we have

$$g(\vec{u},\vec{w}) = g_{\alpha\beta}u^\alpha w^\beta \tag{2.6.2}$$

where $u^\alpha, w^\beta$ are the components of the vector fields $\vec{u}$ and $\vec{w}$, respectively. In this language the hermiticity of the manifold metric $g$ can be rephrased in the following way:

**Definition 2.6.1** *Let $\mathcal{M}$ be a 2n-dimensional manifold with an almost complex structure $J$. A metric $g$ on $\mathcal{M}$ is called hermitean with respect to $J$ if*

$$g(J\vec{u},J\vec{w}) = g(\vec{u},\vec{w}) \tag{2.6.3}$$

Given a metric $g$ and an almost complex structure $J$ let us introduce the following differential 2-form $K$:

$$K(\vec{u},\vec{w}) = \frac{1}{2\pi} g(J\vec{u},\vec{w}) \tag{2.6.4}$$

The components $K_{\alpha\beta}$ of $K$ are given by

$$K_{\alpha\beta} = g_{\gamma\beta} J^\gamma_\alpha \tag{2.6.5}$$

and by direct computation we can easily verify that:

**Theorem 2.6.1** *$g$ is hermitean if and only if $K$ is anti-symmetric.*

**Definition 2.6.2** *a hermitean almost complex manifold is an almost complex manifold endowed with a hermitean metric $g$.*

In a well-adapted basis we can write

$$g(u,w) = g_{ij}u^i w^j + g_{i^\star j^\star}u^{i^\star} w^{j^\star} + g_{ij^\star}u^i w^{j^\star} + g_{i^\star j}u^i w^{j^\star} \tag{2.6.6}$$

Reality of $g(u,w)$ implies

$$\begin{aligned} g_{ij} &= (g_{i^\star j^\star})^\star \\ g_{i^\star j} &= (g_{ij^\star})^\star \end{aligned} \tag{2.6.7}$$

symmetry ($g(u,w) = g(w,u)$) yields

$$\begin{aligned} g_{ij} &= g_{ji} \\ g_{j^\star i} &= g_{ij^\star} \end{aligned} \tag{2.6.8}$$

while the hermiticity condition gives

$$g_{ij} = g_{i^\star j^\star} = 0 \tag{2.6.9}$$

Finally in the well-adapted basis the 2-form $K$ associated to the hermitean metric $g$ can be written as

$$K = \frac{i}{2\pi} g_{ij^\star} dz^i \wedge d\overline{z}^{j^\star} \tag{2.6.10}$$

**Definition 2.6.3** *A hermitean metric on a complex manifold $\mathcal{M}$ is called a Kähler metric if the associated 2-form $K$ is closed:*

$$dK = 0 \tag{2.6.11}$$

*A hermitean complex manifold endowed with a Kähler metric is called a Kähler manifold.*

Equation (2.6.11) is a differential equation for $g_{ij^\star}$ whose general solution in any local chart is given by the following expression:

$$g_{ij^\star} = \partial_i \partial_{j^\star} \mathcal{K} \tag{2.6.12}$$

where $\mathcal{K} = \mathcal{K}^\star = \mathcal{K}(z, z^\star)$ is a real function of $z^i, z^{i^\star}$. The function $\mathcal{K}$ is called the Kähler potential and it is defined only up to the real part of a holomorphic function $f(z)$. Indeed one sees that

$$\mathcal{K}'(z, z^{i^\star}) = \mathcal{K}(z, z^{i^\star}) + f(z) + f^\star(z^\star) \tag{2.6.13}$$

give rise to the same metric $g_{ij^\star}$ as $\mathcal{K}$. The transformation (2.6.13) is called a Kähler transformation. The differential geometry of a Kähler manifold is described by Eqs. (2.5.18) and (2.5.19) with $g_{ij^\star}$ given by (2.6.12).

## 2.7 Characteristic Classes and Elliptic Complexes

One of the main issues in the present book are Calabi–Yau manifolds, discussed in Chapter 4. These manifolds, which play a crucial role in the compactification of superstrings, are topologically characterized by the vanishing of the first Chern class. Hence the notion of characteristic classes of fibre bundles is an essential ingredient in the development of our subject. Characteristic classes are maps from the ring $I^*(\mathcal{G})$ of invariant polynomials on the Lie algebra $\mathcal{G}$ of the structural group to the de Rham cohomology ring $H^*(M)$ of the base manifold. They provide an intrinsic way of measuring the twisting, or deviation from triviality, of a fibre bundle. They are also an essential ingredient of the index theorems that express the difference of zero modes of an elliptic operator minus its adjoint precisely in terms of integrals of characteristic classes. Index theorems play a fundamental role in topological field theories and this is another reason why we have to focus on characteristic classes.

We begin by recalling the notion of de Rham cohomology groups. The differential forms of degree $r$ on a $k$-dimensional manifold $M$ are sections of a vector bundle, namely of the completely antisymmetrized tensor product $\Lambda^r(T^*M)$ of the cotangent bundle $T^*M$, $r$ times with itself. We name $\Omega^r = \Gamma(M, \Lambda^r(T^*M))$ the space of sections of this bundle, namely the space of $r$-forms. The exterior derivative $d$ provides a sequence of maps $d_i$:

$$\Omega^0(M) \xrightarrow{d_0} \Omega^1(M) \xrightarrow{d_1} \ldots \xrightarrow{d_{k-2}} \Omega^{k-1}(M) \xrightarrow{d_{k-1}} \Omega^k(M) \xrightarrow{d_k} 0 \tag{2.7.1}$$

where $d_r$ is the exterior derivative acting on $r$-forms and producing $r+1$-forms as a result. The well-known property of the exterior derivative $d^2 = 0$ implies that

$$d_i\, d_{i+1} = 0 \qquad \forall i = 0, \ldots, k \tag{2.7.2}$$

What we have just described is named the **de Rham complex** and provides the first and most prominent example of an elliptic complex. More generally we have

**Definition 2.7.1** *An elliptic complex $(E^*, D)$ is a sequence of vector bundles $E_i \xrightarrow{\pi_i} M$ constructed over the same base manifold and a sequence of Fredholm operators $D_i$ mapping the sections of the i-th bundle into those of the (i+1)-th bundle:*

$$\Gamma(M, E_0) \xrightarrow{D_0} \Gamma(M, E_1) \xrightarrow{D_1} \ldots \xrightarrow{D_{k-2}} \Gamma(M, E_{k-1}) \xrightarrow{D_{k-1}} \Gamma(M, E_k) \xrightarrow{D_k} 0 \tag{2.7.3}$$

*such that*

$$D_i\, D_{i+1} = 0 \qquad \forall i = 0, \ldots, k \tag{2.7.4}$$

A Fredholm operator is a differential operator of elliptic type with finite kernel and cokernel, as we will discuss below.

To each elliptic complex and to the de Rham complex, in particular, we can attach the notion of cohomology groups. The $i$-th cohomology group is defined as follows:

$$H^i\,(E^*\,,M) = \frac{\ker D_i}{\operatorname{im} D_{i-1}} \tag{2.7.5}$$

It is the space of sections of the $i$-th bundle $E_i$ satisfying $D_i\,s = 0$, modulo those of the form $s = D_{i-1}\,s'$. In the de Rham complex $H^r\,(\Omega^*(M))$ is the space of closed $r$-forms modulo exact forms.

For any Fredholm operator $D_i$ appearing in the elliptic complex (2.7.3) we denote $D_i^\dagger$ its adjoint, which is defined by

$$\begin{aligned} D_i^\dagger : \Gamma(M, E_{i+1}) &\rightarrow \Gamma(M, E_i) \\ (s', D_i s)_{E_{i+1}} &= (D_i^\dagger s', s)_{E_i} \end{aligned} \tag{2.7.6}$$

where $s \in \Gamma(M, E_i)$, $s' \in \Gamma(M, E_{i+1})$ and $(\ ,\ )_E$ denotes the fibre metric in the specified fibre. The laplacian operator is defined by

$$\begin{aligned} \Delta_i &: \Gamma(M, E_i) \rightarrow \Gamma(M, E_i) \\ \Delta_i &\equiv D_{i-1} D_{i-1}^\dagger + D_i^\dagger D_i \end{aligned} \tag{2.7.7}$$

The cohomology group $H^i(E^*, M)$ is isomorphic to the kernel of the operator $\Delta_i$, so that we have

$$\dim H^i(E^*, D) = \dim \operatorname{Harm}^i(E^*, D) \tag{2.7.8}$$

where by $\operatorname{Harm}^i(E^*, D)$ we denote the vector space spanned by sections $h_i \in \Gamma(M, E_i)$ which satisfies

$$\Delta_i h_i = 0\,. \tag{2.7.9}$$

Given a section $s_i \in \Gamma(M, E_i)$ we can write the Hodge decomposition:

$$s_i = D_i s_{i-1} + D_i^\dagger s_{i+1} + h_i \tag{2.7.10}$$

where $s_{i\pm1} \in \Gamma(M, E_i)$.

**Definition 2.7.2** *Given an elliptic complex $(E^*, D)$ we define the index of this complex by*

$$\operatorname{ind}(E^*, D) = \sum(-)^i \dim H^i(E^*, D) = \sum(-)^i \dim \ker\Delta_i \tag{2.7.11}$$

Equation (2.7.11), when specialised to the de Rham complex, gives the Euler characteristic

$$\operatorname{ind} d = \sum(-)^i \dim H^i(E^*, d) \equiv \chi(M) = \sum(-)^i b^i \tag{2.7.12}$$

where $b^i$ is the $i$-th Betti number, which is equal to the number of linearly independent harmonic $i$-forms.

For a generic Fredholm operator $D : \Gamma(M, E) \to \Gamma(M, F)$ we can define the *analytical index* of $D$ as

$$\mathrm{ind}D = \dim \ker D - \dim \ker D \tag{2.7.13}$$

To show the relation between Eqs. (2.7.11) and (2.7.13) we have to resume our discussion on Fredholm operators. Let $D : \Gamma(M, E) \to \Gamma(M, F)$ be an elliptic operator. The kernel of $D$ is the following set of sections:

$$\ker D = \{s \in \Gamma(M, E) | Ds = 0\}. \tag{2.7.14}$$

We define the cokernel of $D$ by

$$\mathrm{coker}D = \frac{\Gamma(M, F)}{\mathrm{im}D} \tag{2.7.15}$$

We now state the following theorem:

**Theorem 2.7.1** *Let $D : \Gamma(M, E) \to \Gamma(M, F)$ be a Fredholm operator. Then*

$$\mathrm{coker}D \sim \ker D^{\dagger} \tag{2.7.16}$$

Using Theorem 2.7.1 we immediately rewrite Eq. (2.7.11) as

$$\mathrm{ind}D = \dim \ker D - \dim \ker D^{\dagger} \tag{2.7.17}$$

Consider now the one-operator complex $\Gamma(M, E) \xrightarrow{D} \Gamma(M, F)$, which can also be written as

$$0 \xrightarrow{i} \Gamma(M, E) \xrightarrow{D} \Gamma(M, F) \xrightarrow{\phi} 0 \tag{2.7.18}$$

where $i$ is the inclusion map (defined by $i(0) = 0$), and $\phi$ is a map from a generic section in $\Gamma(M, F)$ into 0. Using Eq. (2.7.11) for the complex (2.7.18) we find

$$\dim \ker D - [\dim \Gamma(M, F) - \dim \mathrm{im}D] = \dim \ker D - \dim \mathrm{coker}\, D \tag{2.7.19}$$

The above equation shows the simple relation between the analytical index (2.7.13) and the index of the elliptic complex (2.7.11). Equation (2.7.13) provides an easy formula that is always recalled in physical literature. Moreover, given an elliptic complex, it is always possible to construct a Fredholm operator whose analytical index coincides with the index of the complex $(E^*, D)$. Indeed if we define

$$E_+ = \oplus_i E_{2i}, \qquad E_- = \oplus_i E_{2i+1} \tag{2.7.20}$$

which are respectively called the even and the odd bundles and we consider the operators

$$D \equiv \oplus_i (D_{2i} + D^{\dagger}_{2i-1}) \quad D^{\dagger} \equiv \oplus_i (D_{2i+1} + D^{\dagger}_{2i}) \tag{2.7.21}$$

we easily verify that

$$\begin{aligned} D : \Gamma(M, E_+) &\to \Gamma(M, E_-) \\ D : \Gamma(M, E_-) &\to \Gamma(M, E_+) \end{aligned} \tag{2.7.22}$$

Next, if we define

$$\Delta_+ \equiv D^\dagger D = \oplus_i \Delta_{2i} \quad \Delta_- \equiv D D^\dagger = \oplus_i \Delta_{2i+1} \tag{2.7.23}$$

then we have

$$\text{ind}(E_\pm, D) = \dim\ker\Delta_+ - \dim\ker\Delta_- = \sum(-)^i \dim\ker\Delta_i = \text{ind}(E^*, D) \tag{2.7.24}$$

In general the index of an elliptic complex can be expressed by an integral over $M$ of suitable characteristic classes. At the beginning of the present section we have defined characteristic classes as maps from the ring of invariant polynomials on the Lie algebra of the structural group to the de Rham cohomology group ring of the base manifold. Let us now go a little deeper on the meaning of this definition. Let $M(k, \mathbb{C})$ be the set of complex $k \times k$ matrices. We denote by $S^r(M(k, \mathbb{C}))$ the vector space of symmetric $r$-linear $\mathbb{C}$-valued functions on $M(k, \mathbb{C})$. A map

$$\widehat{P} : \otimes_r M(k, \mathbb{C}) \to \mathbb{C} \tag{2.7.25}$$

belongs to $S^r(M(k, \mathbb{C}))$ if it satisfies, in addition to linearity in each entry, the symmetry

$$\widehat{P}(a_1, \ldots, a_i, \ldots, a_j, \ldots a_r) = \widehat{P}(a_1, \ldots, a_j, \ldots, a_i, \ldots, a_r) \qquad \forall i, j \le r \tag{2.7.26}$$

Consider now the formal sum

$$S^*(M(k, \mathbb{C})) = \oplus_0^\infty S^r(M(k, \mathbb{C})) \tag{2.7.27}$$

and define a product of $\widehat{P} \in S^p(M(k, \mathbb{C}))$ and $\widehat{Q} \in S^q(M(k, \mathbb{C}))$ by

$$\widehat{P} \cdot \widehat{Q}(a_1, \ldots, a_{p+q}) = \frac{1}{(p+q)!} \sum_P \widehat{P}(a_{P(1)}, \ldots, a_{P(p)}) \widehat{Q}(a_{P(p+1)}, \ldots, a_{P(p+q)}) \tag{2.7.28}$$

where $P$ denotes the permutation of the set $(1, \ldots, p+q)$. $S^*(M(k, \mathbb{C}))$ equipped with the product (2.7.28) is an algebra.

If we now consider a Lie algebra $\mathcal{G} \in M(k, \mathbb{C})$, and the corresponding simply connected Lie group $G = \exp[\mathcal{G}]$, in full analogy with Eqs. (2.7.27) and (2.7.26), we can define the sum $S^*(\mathcal{G}) = \oplus_{r \ge 0} S^r(\mathcal{G})$. An element $\widehat{P}(h_1, \ldots, h_r) \in S^r(\mathcal{G})$ ($h_i \in \mathcal{G}$) is said to be invariant if, for any $g \in G$, it satisfies

$$\widehat{P}(g^{-1} h_1 g, \ldots, g^{-1} h_r g) = \widehat{P}(h_1, \ldots, h_r) \tag{2.7.29}$$

The set of invariant elements of $S^r(\mathcal{G})$ is denoted by $I^r(\mathcal{G})$. The product defined in (2.7.28) induces a natural multiplication

$$\cdot : I^p(\mathcal{G}) \otimes I^q(\mathcal{G}) \to I^{p+q}(\mathcal{G}) \tag{2.7.30}$$

The sum $I^* = \oplus_{r \geq 0} I^r(\mathcal{G})$ equipped with the product (2.7.30) is an algebra. The diagonal combination $P(h) = P(h, \ldots, h)$ containing $r$-times the element $h \in \mathcal{G}$ is a polynomial of degree $r$, which is said to be an *invariant polynomial.*

Let now $P(M, G)$ be a principal bundle that has as structural group a Lie group $G$ with Lie algebra $\mathcal{G}$. We extend the domain of invariant polynomials from $\mathcal{G}$ to $\mathcal{G}$-valued $p$-forms on $M$. We define

$$\widehat{P}(h_1\omega_1, \ldots, h_r\omega_r) \equiv \omega_1 \wedge \cdots \wedge \omega_r \widehat{P}(h_1, \ldots, h_r) \tag{2.7.31}$$

where $h_i \in \mathcal{G}$, $\omega_i \in \Omega^{p_i}(M)$ $(i = 1 \cdots r)$. The diagonal combination is now given by

$$P(h\omega) = \omega \wedge \cdots \wedge \omega P(h) \tag{2.7.32}$$

where the wedge product of $\omega \in \Omega^p(M)$ is repeated r-times in (2.7.32).

Consider now the curvature 2-form $\Theta$ associated with a connection in a complex fibre bundle. In the following we are particularly interested in invariant polynomials of the form $P(\Theta)$. We can state the following theorem (*Chern -Weil theorem*).

**Theorem 2.7.2** *Let $P(\Theta)$ be an invariant polynomial in the curvature 2-form; then*
*i) $dP(\Theta) = 0$*
*ii) Let $\Theta, \Theta'$ be curvature 2-forms corresponding to different connections $\theta, \theta'$ on the fibre bundle. Then the difference $P(\Theta) - P(\Theta')$ is exact.*

This theorem proves that an invariant polynomial $P(\Theta)$ is closed and in general non-trivial. We can then associate to $P(\Theta)$ a cohomology class of $M$. Moreover Theorem 2.7.2 ensures that this cohomology class is independent of the chosen connection. The cohomology class defined by $P(\Theta)$ is called a *characteristic class.* The characteristic class defined by an invariant polynomial $P$ is denoted by $\chi_E(P)$, where $E$ is the fibre bundle on which curvatures and connections are defined.

**Theorem 2.7.3** *Let $P$ be an invariant polynomial in $I^*(\mathcal{G})$ and $E$ be a fibre bundle over $M$, whose structural group $G$ has $\mathcal{G}$ as Lie algebra. The map*

$$\chi_E : I^*(\mathcal{G}) \to H^*(M) \tag{2.7.33}$$

*defined by $P \to \chi_E(P)$ is a homomorphism.*

Theorem 2.7.3 establishes a homomorphism, called the Weil homomorphism, between the ring $I^*(\mathcal{G})$ and the de Rham cohomology ring $H^*(M)$, defined by

$$H^*(M) = \oplus_r H^r(M) \tag{2.7.34}$$

where $H^r$ is the $r$-th cohomolgy group.

The Weil homomorphism is the fundamental instrument that allows one to relate the index of an elliptic complex with the integral of particular characteristic classes, through the so called *index theorem* (stated below in Eq. (2.7.56)). Before giving the statement of this theorem, due to Atiyah and Singer, we list some specific examples of characteristic classes, which will be useful in the following.

**Definition 2.7.3** *Given a complex vector bundle $E$ equipped with a connection $\theta$, whose fibre is $\mathbb{C}^r$, we can define its total Chern class $c(E,\Theta)$ as the following formal determinant:*

$$c(E,\Theta) = det\left(\mathbf{1} + \frac{i}{2\pi}\Theta\right) \tag{2.7.35}$$

*where $\Theta$ is the matrix-valued curvature 2-form.*

The determinant is calculated with respect to the matrix indices. As is well known, the determinant $det(\mathbf{1} + A)$ is a polynomial in the matrix elements of $A$ and can be expanded in powers of A. Such an expansion of the total Chern class yields the definition of the individual Chern classes $c_k(E,\Theta)$. In particular, if we call $x_1, \cdots x_r$ the (formal) eigenvaules[2] 2-forms of the matrix $\frac{i}{2\pi}\Theta$ we easily find

$$\det\,(1+\frac{i}{2\pi}\Theta) = \prod_1^r(1+x_j) = 1+(x_1+\cdots+x_r)+(x_1x_2+\cdots+x_{r-1}x_r)+\cdots+(x_1x_2\cdots x_r) \tag{2.7.36}$$

so that, by writing

$$c(E,\Theta) = \sum_{k=0}^{r} c_k(E,\Theta) \tag{2.7.37}$$

we get

$$\begin{aligned} c_0 &= 1, \\ c_1 &= \frac{i}{2\pi}\,tr\,(\Theta), \\ c_2 &= \frac{1}{8\pi^2}\left[tr\left(\Theta^2\right) - (tr\Theta)^2\right] \\ \vdots &\;\vdots\;\vdots \\ c_r &= \det\frac{i\Theta}{2\pi} \end{aligned} \tag{2.7.38}$$

where, for a generic form $\Omega$, by $\Omega^n$ we mean the $n$-th wedge product $\wedge^n\Omega$. A remarkable property of the Chern class is the following: given two complex vector bundles $E \xrightarrow{\pi} M$, $F \xrightarrow{\pi'} M$ we have

$$c(E\oplus F) = c(E)\wedge c(F) \tag{2.7.39}$$

[2] We stress the word "formal eigenvalues" because the correct framework to understand these eigenvalues is the "splitting principle", which, for convenience, is mentioned after the equation (2.7.59).

**Definition 2.7.4** *Given a rank $r$ vector bundle $E \xrightarrow{\pi} M$ we define the total Chern character by*

$$ch(E,\Theta) = tr\, exp(\frac{i\Theta}{2\pi}) = \sum_{l=1} \frac{1}{l!} tr(\frac{i\Theta}{2\pi})^j \tag{2.7.40}$$

*and the $j$-th Chern character by*

$$ch_j(E,\Theta) = \frac{1}{j!} tr(\frac{i\Theta}{2\pi})^j \tag{2.7.41}$$

From now on, for notational convenience we refer to $\mathrm{ch}(E,\Theta)$ as $\mathrm{ch}\, E$ or $\mathrm{ch}\,\Theta$ indifferently (and similarly for the Chern class $c(E,\Theta)$). In terms of the eigenvectors $x_j$ we get

$$\mathrm{ch}(\Theta) = \sum_{j=1}^{r} (1 + x_j + \frac{1}{2} x_j^2 + \cdots) \tag{2.7.42}$$

so that we can write

$$\begin{aligned} \mathrm{ch}_0(\Theta) &= r \\ \mathrm{ch}_1(\Theta) &= c_1(\Theta) \\ \mathrm{ch}_2(\Theta) &= \frac{1}{2}[c_1^2(\Theta) - 2c_2(\Theta)] \end{aligned} \tag{2.7.43}$$

**Theorem 2.7.4** *Let $E$ and $F$ be two vector bundles over a manifold $M$. The Chern character of $E \otimes F$ and $E \oplus F$ are given by*

$$\begin{aligned} ch(E \otimes F) &= ch(E) \wedge ch(F) \\ ch(E \oplus F) &= ch(E) + ch(F) \end{aligned} \tag{2.7.44}$$

Another useful characteristic class associated with a complex vector bundle is the **Todd class** defined by

$$\mathrm{Td}(\Theta) = \prod_{j=1}^{r} \frac{x_j}{1 - e^{-x_j}} \tag{2.7.45}$$

where $x_j$ are the eigenvalues of the curvature 2-form $\frac{i}{2\pi}\Theta$. We obtain

$$\begin{aligned} \mathrm{Td}(\Theta) &= 1 + \frac{1}{2}\sum_j x_j + \frac{1}{12} x_j^2 + \cdots \\ &= \prod_j (1 + \frac{1}{2} x_j + \sum_{k\geq 1} (-)^{k-1} \frac{B_k}{2k!} x_j^{2k}) \\ &= 1 + \frac{1}{2} c_1(\Theta) + \frac{1}{12}[c_1^2(\Theta) + c_2(\Theta)] + \cdots \end{aligned} \tag{2.7.46}$$

where the numbers $B_k$ appearing in Eq. (2.7.46) are the Bernoulli numbers.

Finally we define the **Euler class.** The characteristic classes previously introduced are naturally defined for complex vector bundles. On the other hand the Euler class can be defined for real vector bundles over an orientable Riemann surface $M$. In particular it is consistently defined for even rank real bundles, while it is zero for odd rank bundles.

Given a rank $k$ real bundle $E$ it is useful to construct a complex vector bundle from $E$ by a *complexification* procedure. The complexification of $E$ is the bundle over $M$ obtained by replacing the fibres $\mathbb{R}^k$ by $\mathbb{C}^k = (\mathbb{R} \oplus i\mathbb{R})^k$. We denote the complexification of $E$ by $E^{\mathbb{C}}$. We can think of $E^{\mathbb{C}}$ as the following product

$$E^{\mathbb{C}} = E \otimes (\mathbb{R} \oplus i\mathbb{R}) \tag{2.7.47}$$

Complex vector bundles can also be complexified by converting them into real vector bundles and then complexifying the result. If the starting complex bundle has rank $r$, its complexification has rank $2r$. Notice that, given a complex vector bundle $E$, and denoting by $E_{\mathbf{R}}$ the underlying real bundle, we have

$$E^{\mathbb{C}}_{\mathbf{R}} = E_{\mathbf{R}} \otimes (\mathbb{R} + i\mathbb{R}) \sim E \oplus \overline{E} \tag{2.7.48}$$

where $\overline{E}$ denotes the conjugate complex bundle, defined by applying complex conjugation to the coordinates of the fibres $\mathbb{C}^r$ of $E$.

Having outlined the complexification procedure for a real vector bundle, we define the Euler class through another typical characteristic class defined in real bundles: the Pontrjagin class. Let $E$ be a real vector bundle of rank $r$ over $M$, the $i$-th Pontrjagin class is defined as

$$p_i(E) = (-)^i c_{2i}(E^{\mathbb{C}}) \tag{2.7.49}$$

where $c_{2i}(E^{\mathbb{C}})$ is the $2i$-th Chern class of the complexified bundle. The total Pontrjagin class is defined as

$$P(E) = 1 + p_1(E) + \cdots + p_{[r/2]} \tag{2.7.50}$$

where $[r/2]$ is the largest integer not greater than $r$.

Consider now real vector bundles $E$ of *even rank* over an orientable manifold $M$. The Euler class is defined by

$$e^2(V) = p_{[r/2]} \tag{2.7.51}$$

The Euler class of a Whitney sum $E \oplus V$ is

$$e(E \oplus V) = e(E)e(V) \tag{2.7.52}$$

where we denote $c(E)c(V) = c(E) \wedge c(V)$.

For a complex vector bundle the Pontrjagin and the Euler class are the Pontrjagin and the Euler class of the underlying real bundle. Since the eigenvalues of the curvature 2-form in the conjugate bundle are given by $-x_i$, we have

$$c(E^{\mathbb{C}}) = c(E \oplus \overline{E}) = c(E)c(\overline{E}) = \prod_{i=1}^{r}(1 + x_i)(1 - x_i) = \prod_{i=1}^{r}(1 - x_i^2) \tag{2.7.53}$$

so that

$$c_r(E^{\mathbb{C}}) = (-)^r x_1^2 \cdots x_r^2 \tag{2.7.54}$$

and (recalling that $E^{\mathbb{C}}$ has rank $2r$)

$$\begin{aligned} p_r(E) &= x_1^2 \cdots x_r^2 \\ e(E) &= x_1 x_2 \cdots x_r = c_r(E) \end{aligned} \tag{2.7.55}$$

We are now able to state the *Atiyah–Singer index theorem* in its full generality:

**Theorem 2.7.5** *Given an elliptic complex $(E^*, D)$ over an $m$-dimensional* ($\dim_{\mathbb{R}} M = m$) *compact manifold $M$ without a boundary, then*

$$ind(E^*, D) = (-)^{\frac{m(m+1)}{2}} \int_M ch\left(\oplus_j (-)^j E_j\right) \frac{Td(TM^{\mathbb{C}})}{e(TM)} \tag{2.7.56}$$

*where $TM$ is the tangent bundle over $M$.*

Let us now consider the application of the index theorem to some particular elliptic complexes. Consider an $m$-dimensional compact orientable manifold without boundaries and the elliptic de Rham complex:

$$\cdots \xrightarrow{d} \Omega^{r-1}(M)^{\mathbb{C}} \xrightarrow{d} \Omega^r(M)^{\mathbb{C}} \xrightarrow{d} \Omega^{r+1}(M)^{\mathbb{C}} \xrightarrow{d} \cdots \tag{2.7.57}$$

with $\Omega^r(M)^{\mathbb{C}} = \Gamma(M, \wedge^r T^* M^{\mathbb{C}})$, where we have complexified the forms to apply the Atiyah–Singer theorem. The analytical index is given by

$$\text{ind}\, d = \sum_{r=0}^{m} (-)^r \dim_{\mathbb{C}} H^r(M, \mathbb{C}) = \sum_{r=0}^{m} (-)^r \dim{}_{\mathbb{R}} H^r(M, \mathbb{R}) = \chi(M) \tag{2.7.58}$$

where $\chi(M)$ is the Euler characteristic of $M$. Suppose $M$ is even dimensional $m = 2l$. Theorem 2.7.56 gives the following result for the de Rham index:

$$\text{ind}\, d = (-)^{l(2l+1)} \int_M \text{ch}\left(\oplus_r^{2l} (-)^r \wedge^r T^* M^{\mathbb{C}}\right) \frac{\text{Td} TM^{\mathbb{C}}}{e(TM)} \tag{2.7.59}$$

To compute $\text{ch}\left(\oplus_r^m (-)^r \wedge^r T^* M^{\mathbb{C}}\right)$ we employ the splitting principle. The splitting principle uses the fact that in order to prove an identity for characteristic classes, it is sufficient to prove it only for bundles which decompose into a sum of line bundles. Suppose that a fibre bundle $F$ is a Whitney sum of $n$ line bundles $L_i$; then

$$\wedge^p F = \oplus_{1 \le i_1 \cdots i_p \le n} \left(L_{i_1} \otimes \cdots \otimes L_{i_p}\right) \tag{2.7.60}$$

This means that

$$\text{ch}(\wedge^p F) = \sum_{1 \le i_1 \cdots i_p \le n} \text{ch}(L_{i_1}) \text{ch}(L_{i_2}) \cdots \text{ch}(L_{i_p}) \tag{2.7.61}$$

Since for any line bundle appearing in the Whitney sum $\mathrm{ch}(L_i) = e^{x_i}$, we finally get

$$\mathrm{ch}(\wedge^p F) = \sum_{1 \le i_1 \cdots i_p \le n} e^{x_{i_1} + \cdots + x_{i_p}} \tag{2.7.62}$$

Applying this result to $\oplus_r^m (-)^r \wedge^r T^* M^{\mathbb{C}}$, and using the fact that taking the dual bundle merely changes the sign of $x_i$ we get

$$\mathrm{ch} \oplus_r^m (-)^r \wedge^r T^* M^{\mathbb{C}} = \prod_{i=1}^m (1 - e^{-x_i})(TM^{\mathbb{C}}) \tag{2.7.63}$$

Moreover we can write

$$\mathrm{Td}(TM^{\mathbb{C}}) = \prod_{i=1}^m \frac{x_i}{1 - e^{-x_i}} (TM^{\mathbb{C}}) \tag{2.7.64}$$

Then the index of the de Rham complex is given by

$$\mathrm{ind}\, d = (-)^l \int_M \frac{\prod_{i=1}^m x_i(TM^{\mathbb{C}})}{e(TM)} = (-)^l \int_M \frac{c_m(TM^{\mathbb{C}})}{e(TM)} = \int_M e(TM) \tag{2.7.65}$$

where we have used $c_m(TM^{\mathbb{C}}) = (-)^{m/2} e(TM \oplus TM) = (-)^l x_1^2 \cdots x_m^2 = (-)^l e^2(TM)$.

By combining the results for the analytical index and for the Atiyah–Singer index (often referred to as the topological index), we get the Gauss–Bonnet theorem

$$\int_M e(TM) = \chi(M) \tag{2.7.66}$$

For $m$ odd, the de Rham index is zero.

Let us consider now the application of the index theorem to the Dolbeault complex, which we are going to define below. Consider a complex manifold $M$ with $dim_{\mathbb{C}} M = m$. We denote by $T^{(1,0)}M$ the tangent bundle spanned by the vectors $\{\partial/\partial z^\mu\}$ and by $T^{(0,1)}M$ its complex conjugate. The space dual to $T^{(1,0)}M$ is spanned by the 1-forms $\{dz^\mu\}$. We denote it by $T^{*(1,0)}M$. The space $\Omega^r(M)^{\mathbb{C}}$ of complexified $r$-forms is decomposed as

$$\Omega^r(M)^{\mathbb{C}} = \oplus_{p+q=r} \Omega^{p,q}(M) \tag{2.7.67}$$

where by $\Omega^{p,q}(M)$ we denote the space of $(p,q)$ forms. The exterior derivative can be written as

$$d = dz^\mu \wedge \frac{\partial}{\partial z^\mu} + d\bar{z}^\mu \wedge \frac{\partial}{\partial \bar{z}^\mu} \tag{2.7.68}$$

It is immediate to verify that $\partial, \bar{\partial}$ satisfy the following relations:

$$\partial\bar{\partial} - \bar{\partial}\partial = \partial^2 = \bar{\partial}^2 = 0 \tag{2.7.69}$$

Moreover $\partial$ maps $(p,q)$-forms into $(p+1,q)$-forms and $\bar{\partial}$ maps $(p,q)$ forms into $(p,q+1)$ forms. Let us consider the sequence

$$\cdots \xrightarrow{\bar{\partial}} \Omega^{(0,q)}(M) \xrightarrow{\bar{\partial}} \Omega^{(0,q+1)}(M) \xrightarrow{\bar{\partial}} \cdots \tag{2.7.70}$$

This sequence is called the **Dolbeault complex**. It can be shown that (2.7.70) defines an elliptic complex. The index theorem in this case gives

$$\mathrm{ind}\,\bar{\partial} = \int_M \mathrm{ch}\left(\oplus_r (-)^r \wedge^r T^{*(0,1)} M\right) \frac{\mathrm{Td} TM^{\mathbb{C}}}{e(TM)} \tag{2.7.71}$$

The left hand side of the above equation can be computed using the Eq. (2.7.13), so that

$$\mathrm{ind}\,\bar{\partial} = \sum_{r=0}^{n} (-)^r h^{(0,r)} \tag{2.7.72}$$

where

$$h^{(0,r)} = \mathrm{dim}_{\mathbb{C}} H^{(0,r)}(M) = \mathrm{dim}_{\mathbb{C}} \frac{\mathrm{ker}\bar{\partial}_r}{\mathrm{im}\bar{\partial}_{r-1}} \tag{2.7.73}$$

is the complex dimension of the the cohomology group $H^{(0,r)}$ . The application of Theorem (2.7.56) to this case is analogous to the one presented for the de Rham complex and gives

$$\sum_{r=0}^{n} (-)^r b^{(0,r)} = \int_M \mathrm{Td}(T^{(1,0)} M) \tag{2.7.74}$$

In the Dolbeault complex the space $\Omega^{(0,r)}$ can be replaced by a tensor product bundle $\Omega^{(0,r)} \otimes V$, where $V$ is a holomorphic vector bundle. In this case we define the following elliptic complex, named the **twisted Dolbeault complex**:

$$\ldots \stackrel{\bar{\partial}_V}{\rightarrow} \Omega^{(0,q)}(M) \otimes V \stackrel{\bar{\partial}_V}{\rightarrow} \Omega^{(0,q+1)}(M) \otimes V \stackrel{\bar{\partial}_V}{\rightarrow} \ldots \tag{2.7.75}$$

The Atiyah–Singer theorem for this particular complex reduces to the Hirzebruch–Riemann–Roch theorem:

$$\mathrm{ind}\,\bar{\partial}_V = \int_M \mathrm{Td}\,(T^{(1,0)} M) \mathrm{ch}(V) \tag{2.7.76}$$

In the case of complex dimension one, namely $dim_{\mathbb{C}} M = 1$, we get

$$\mathrm{ind}\,\bar{\partial}_V = \frac{1}{2} \mathrm{dim} V \int_M c_1(T^{(1,0)} M) + \int_M c_1(M) \tag{2.7.77}$$

Since it can be shown that

$$\int_M c_1(T^{(1,0)} M) = \int_M e(TM) = 2(1-g) \tag{2.7.78}$$

where $g$ is the genus of the base manifold, which in complex dimension one is nothing but a Riemann surface $\Sigma_g$, in this case we get

$$\mathrm{ind}\,\bar{\partial}_V = \mathrm{dim} V(1-g) + \int_{\Sigma_g} \frac{i\Theta}{2\pi} \tag{2.7.79}$$

In the general case of a complex manifold $\mathcal{M}$ of complex dimension $n$, the dimensions

$$h^{(p,q)} \stackrel{\text{def}}{=} \dim_{\mathbb{C}} H^{(p,q)}(\mathcal{M}) \tag{2.7.80}$$

of the Dolbeault cohomology groups are named *Hodge numbers*. They can be arranged into a picture of the following type:

$$\begin{array}{ccccccccc}
 & & & & h^{(0,0)} & & & & \\
 & & & h^{(1,0)} & & h^{(0,1)} & & & \\
 & & h^{(2,0)} & & h^{(1,1)} & & h^{(0,2)} & & \\
 & .. & & \dots & & \dots & & .. & \\
h^{(n,0)} & .. & & \dots & & \dots & & .. & h^{(0,n)} \\
 & .. & & \dots & & \dots & & .. & \\
 & & h^{(n,n-2)} & & h^{(n-1,n-1)} & & h^{(n-2,n)} & & \\
 & & & h^{(n,n-1)} & & h^{(n-1,n)} & & & \\
 & & & & h^{(n,n)} & & & &
\end{array} \tag{2.7.81}$$

which is named the *Hodge diamond*. We shall have a lot to say about Hodge diamonds of Calabi–Yau manifolds in Chapter 4.

## 2.8 Hodge Manifolds and Chern Classes

As already pointed out, the definition of Calabi–Yau manifolds involves the notion of Chern classes. The purpose of the present section is the calculation of the first Chern class of a projective variety obtained as a complete intersection of polynomial constraints. Indeed this is the type of Calabi–Yau manifolds we shall mostly consider and for which there exists, via the Landau–Ginzburg picture, a direct relation with superconformal (2,2) theories (see Chapter 5). Furthermore this is also the type of manifold that will enter our discussion of the Griffith's residue mapping (see Section 5.9) and its associated Picard–Fuchs equations for the periods (see Chapter 8). The notion of line bundle treated in this section is also vital for an appropriate discussion of the special Kähler geometry of moduli spaces (see Chapter 6). This discussion involves also the concept of Hodge Kähler manifold that is the one required for the description of N=1 matter-coupled supergravity.

Consider a *line bundle* $\mathcal{L}$ , namely a holomorphic vector bundle of rank $r = 1$. In this case there is only one Chern class, the first:

$$c_1(\mathcal{L}) = \frac{i}{2\pi} \, tr\, \Theta = \frac{i}{2\pi} \overline{\partial} \left( h^{-1} \partial h \right) = \frac{i}{2\pi} \overline{\partial} \partial \log h \tag{2.8.1}$$

where the 1-component real function $h(z, \overline{z})$ is some metric on $\mathcal{L}$. Let $\xi(z)$ be a holomorphic section of the line bundle $\mathcal{L}$: noting that under the action of the operator $\overline{\partial}\partial$ the term $log \left( \overline{\xi}(\overline{z}) \, \xi(z) \right)$ yields a vanishing contribution, we conclude that formula (2.8.1) for the first Chern class can be re-expressed as follows:

$$c_1(\mathcal{L}) = \frac{i}{2\pi} \overline{\partial} \partial \log || \, \xi(z) \, ||^2 \tag{2.8.2}$$

where $\|\xi(z)\|^2 = h(z,\bar{z})\bar{\xi}(\bar{z})\xi(z)$ denotes the norm of the holomorphic section $\xi(z)$.

Formula (2.8.2) is the starting point for the definition of Hodge Kähler manifolds, an essential notion, as we already pointed out, in supergravity theory.

**Definition 2.8.1** *A Kähler manifold M is a Hodge manifold if and only if there exists a line bundle $\mathcal{L} \longrightarrow M$ such that its first Chern class equals the cohomology class of the Kähler 2-form K:*

$$c_1(\mathcal{L}) = [K] \tag{2.8.3}$$

In local terms this means that there is a holomorphic section $W(z)$ such that we can write

$$K = \frac{i}{2\pi} g_{ij^\star} dz^i \wedge d\bar{z}^{j^\star} = \frac{i}{2\pi} \bar{\partial}\partial \log \| W(z) \|^2 \tag{2.8.4}$$

Recalling the local expression of the Kähler metric in terms of the Kähler potential $g_{ij^\star} = \partial_i \partial_{j^\star} \mathcal{K}(z,\bar{z})$, it follows from (2.8.5) that if the manifold $M$ is a Hodge manifold, then the exponential of the Kähler potential can be interpreted as the metric $h(z,\bar{z}) = \exp(\mathcal{K}(z,\bar{z}))$ on an appropriate line bundle $\mathcal{L}$.

This structure is precisely that advocated by the lagrangian of N=1 matter coupled supergravity. Indeed, as we are going to see in Chapter 3, the holomorphic section $W(z)$ of the line bundle $\mathcal{L}$ is just the superpotential and the logarithm of its norm $\log \| W(z) \|^2 = \mathcal{K}(z,\bar{z}) + \log | W(z) |^2 = G(z,\bar{z})$ is precisely that invariant function in terms of which one writes the potential and Yukawa coupling terms of the supergravity action (see Eqs. (1.1.2)). In other words the Wess–Zumino multiplets of any supergravity theory span a Hodge manifold. In the case of Calabi–Yau compactifications the manifold of the moduli fields is not only Hodge but also a special Kähler manifold, namely it belongs to a subclass of Hodge manifolds with additional distinctive properties. Special Kähler geometry is discussed in Chapter 3.

In the case of compact Kähler manifolds, we must remark that the Hodge condition has profound topological implications. Indeed since it can be shown that the first Chern class of any line bundle $\mathcal{L} \longrightarrow M$ is always an integral cohomology class (in the sense that it can be expressed as a linear combination with integer coefficient of the generators of $H^2(M)$) then the Hodge condition implies that the same integrality property holds also for the Kähler 2-form. A fundamental embedding theorem due to Kodaira states that the integrality of the Kähler form is the necessary and sufficient condition for the manifold $M$ to be projective algebraic, namely to be the vanishing locus of homogeneous polynomial contraints in some complex projective space $\mathbb{CP}_N$.

Our next step deals precisely with the structure of the complex projective space $\mathbb{CP}_N$. Relying on the fact that it is a coset manifold:

$$\mathbb{C}P_N = \frac{SU(N+1)}{SU(N) \otimes U(1)} \tag{2.8.5}$$

we can easily calculate its total Chern class and show that it is given by

$$c_{total}(\mathbb{CP}_N) = (1 + K)^{N+1} \tag{2.8.6}$$

where $K$ denotes the Kähler 2-form. In this way all the Chern classes of $\mathbb{CP}_N$ are known. In addition we can prove that $\mathbb{CP}_N$ is the prototype of a Hodge manifold. The line bundle whose first Chern class equals the Kähler class is named the hyperplane bundle and its sections are the homogeneous coordinates on $\mathbb{CP}_N$. The knowledge of the first Chern class of this line bundle enables us to obtain the first Chern class of the normal bundle $\mathcal{N}(\mathcal{M})$ to any algebraic hypersurface $\mathcal{M} \subset \mathbb{CP}_N$ immersed in $\mathbb{CP}_N$ and, by means of this token, also the total Chern class of the hypersurface $\mathcal{M}$. This will be our main instrument for the explicit construction of Calabi -Yau manifolds. Let us then derive the results we have anticipated.

As usual we define $\mathbb{CP}_N$ as the set of equivalence classes of the $(N+1)$-tuples of complex numbers

$$\left\{X^\Lambda\right\} \qquad (\Lambda = 0,1,\ldots,N) \tag{2.8.7}$$

under the equivalence relation:

$$\left\{X^\Lambda\right\} \approx \left\{X^{\Lambda'}\right\} \quad \text{iff} \quad X^\Lambda = \lambda X^{\Lambda'} \qquad \lambda \in \mathbb{C} \tag{2.8.8}$$

The numbers $X^\Lambda$ are the so-called homogeneous coordinates. Coordinate patches in $\mathbb{CP}_N$ are constructed by considering the open subsets $U_\Lambda \subset \mathbb{CP}_N$ where the $\Lambda$-th homogeneous coordinate does not vanish: $X^\Lambda \neq 0$. In each of these patches we can introduce inhomogeneous coordinates. Consider for instance the patch $U_0$: the points of $U_0$ can be uniquely labelled by the coordinates:

$$z^A = \frac{X^A}{X^0} \qquad (A = 1,\ldots,N) \tag{2.8.9}$$

Using these coordinates, we can introduce a natural Kähler metric on $\mathbb{CP}_N$, derived from the following Kähler potential:

$$\begin{aligned} \mathcal{K}_{FS}\,(z,\overline{z}) &= \log\left(1 + |z|^2\right) \\ \text{where} \quad |z|^2 &= \sum_{A=1}^{N} \overline{z}^{A^\star} z^A \end{aligned} \tag{2.8.10}$$

We have used the subscript $FS$ for the Kähler potential, because the corresponding metric is named the Fubini–Study metric: as one can easily verify, it has the form

$$\begin{aligned} g_{AB^\star} &= \frac{1}{1+|z|^2}\left(\delta^{AB} - \frac{\overline{z}^{A^\star} z^B}{1+|z|^2}\right) \\ g^{AB^\star} &= \left(1+|z|^2\right)\left(\delta^{AB} + \overline{z}^{A^\star} z^B\right) \end{aligned} \tag{2.8.11}$$

and the corresponding connection and curvature are

$$\begin{aligned} \Gamma^A{}_B &= \Gamma^A_{BC}\, dz^B = -\frac{1}{1+|z|^2}\left(\delta^{AC}\,\overline{z}^{B^\star} dz^B + \overline{z}^{C^\star} dz^A\right) \\ R^A{}_C &= -\delta^A_C \frac{1}{(1+|z|^2)^2}\left[(1+|z|^2)\, d\overline{z}^{M^\star} \wedge dz^M - z^M d\overline{z}^{M^\star} \wedge z^{N^\star} dz^N\right] \\ &\quad - \frac{1}{(1+|z|^2)^2}\left[(1+|z|^2)\, d\overline{z}^{C^\star} \wedge dz^A - z^M d\overline{z}^{M^\star} \wedge z^{C^\star} dz^A\right] \end{aligned} \tag{2.8.12}$$

If we regard $\mathbb{CP}_N$ as the coset manifold (2.8.5), then we can describe its geometry by means of the Maurer–Cartan equation

$$d\Omega^\Lambda{}_\Sigma + \Omega^\Lambda{}_\Gamma \wedge \Omega^\Gamma{}_\Sigma = 0 \tag{2.8.13}$$

where

$$\Omega^\Lambda{}_\Sigma = \left(L^\dagger\right)^\Lambda{}_\Gamma \, dL^\Gamma{}_\Sigma \tag{2.8.14}$$

is the left-invariant 1-form $\Omega = L^{-1}dL$ associated with a parametrization

$$L(z) \in G \quad ; \quad z \in G/H \tag{2.8.15}$$

of the coset manifold. A convenient parametrization of the unitary $(N+1) \times (N+1)$ matrix $L \in SU(N+1)$ that uses the inhomogeneous coordinates (2.8.9) as labels of $\mathbb{CP}_N$-points is the following one:

$$L(Z) = \begin{pmatrix} \left(1 + Z Z^\dagger\right)^{-1/2} & Z\left(1 + Z^\dagger Z\right)^{-1/2} \\ -Z^\dagger\left(1 + Z Z^\dagger\right)^{-1/2} & \left(1 + Z^\dagger Z\right)^{-1/2} \end{pmatrix} \tag{2.8.16}$$

where

$$Z = \begin{pmatrix} z^1 \\ z^2 \\ \dots \\ \dots \\ z^N \end{pmatrix} \tag{2.8.17}$$

is the column vector of the inhomogeneous coordinates. The left-invariant 1-form (2.8.14) takes the form

$$\Omega^\Lambda{}_\Sigma = \begin{pmatrix} \Omega^A{}_B & E^A \\ -\overline{E}_B & \Omega^0{}_0 \end{pmatrix} \tag{2.8.18}$$

where we have split the index range as it follows $\Lambda = A\ (= 1, 2, ...., N)\,,\ 0$ and where the explicit expressions for the 1-forms $E^A$, $\Omega^A{}_B$, and $\Omega^0{}_0$ are easily calculated from the definition (2.8.14), upon use of (2.8.16). In particular, for the 1-form vector $E^A$, which has the interpretation of vielbein of our manifold, we obtain

$$E = \left(1 + Z Z^\dagger\right)^{-1/2} dZ \left(1 + Z^\dagger Z\right)^{-1/2} \tag{2.8.19}$$

and using (2.8.19) in the definitions of the metric and of the Kähler 2-form

$$\begin{aligned} ds^2 &= E^A \otimes \overline{E}_A = g_{AB^\star}\, dz^A \otimes d\overline{z}^{B^\star} \\ K &= \frac{i}{2\pi} E^A \wedge \overline{E}_A = \frac{i}{2\pi} g_{AB^\star}\, dz^A \wedge d\overline{z}^{B^\star} \end{aligned} \tag{2.8.20}$$

we find that the results are identical with the Fubini–Study metric (2.8.11) and its associated Kähler 2-form. In this way we have shown that the Fubini–Study metric is the maximally symmetric metric on $\mathbb{CP}_N$ admitting $SU(N+1)$ as isometry group. This identification greatly facilitates the calculation of the Chern classes. Indeed defining the spin connection

$$\omega^A{}_B = -\Omega^A{}_B + \Omega^0{}_0 \delta^A_B \tag{2.8.21}$$

the Maurer–Cartan equation (2.8.13) with indices $\Lambda = A, \Sigma = 0$ and $\Lambda = 0, \Sigma = 0$ become the torsion equation relating the spin connection to the vielbein:

$$dE^A - \omega^A{}_B \wedge E^B = 0 \tag{2.8.22}$$

On the other hand the the Maurer–Cartan equation (2.8.13) with indeces $\Lambda = A$, $\Sigma = B$ becomes the identity

$$\mathcal{R}^A{}_B = \left( E^A \wedge \overline{E}_B + \delta^A_B E^M \wedge \overline{E}_M \right) \tag{2.8.23}$$

satisfied by the Riemann curvature:

$$\mathcal{R}^A{}_B = d\omega^A{}_B - \omega^A{}_C \wedge \omega^C{}_B \tag{2.8.24}$$

Using Eq. (2.8.23), the identification (2.8.20) of the Kähler 2-form and the definitions (2.7.38) of the Chern classes, for the tangent bundle to the manifold $\mathbb{CP}_N$ we obtain

$$\begin{aligned}
c_0 &= 1, \\
c_1 &= (N+1)\, K \\
c_2 &= \frac{(N+1)\, N}{2}\, K \wedge K \\
c_3 &= \frac{(N+1)\, N\, (N-1)}{6}\, K \wedge K \wedge K \\
c_4 &= \ldots
\end{aligned} \tag{2.8.25}$$

which can be condensed in the already stated formula (2.8.6) for the total Chern class.

Coming next to the question of the Hodge condition satisfied by the Kähler structure of $\mathbb{CP}_N$, we define the *hyperplane bundle* as that line bundle

$$U_{1,N+1} \longrightarrow \mathbb{CP}_N \tag{2.8.26}$$

that has the set of $(N+1)$-tuples of homogeneous coordinates $\left\{ X^\Lambda \right\}$ as holomorphic sections. Thus, a frame of holomorphic sections for $U_{1,N+1}$ is given by

$$s(z) = X^\Lambda(z) \tag{2.8.27}$$

where for each $z \in \mathbb{CP}_N$ , $\{X^\Lambda(z)\}$ is some $(N+1)$-tuple of homogeneous coordinates describing the point $z$. Since any other $(N+1)$-tuple with the same property is given by $\{X^{\Lambda'}(z)\} = c(z)\{X^{\Lambda'}(z)\} = c(z)\, s(z) = s'(z)$, the point-dependent multiplication constant $c(z)$ can be taken as a local coordinate for the sections of the line bundle in the frame (2.8.27). A canonical metric on this line bundle can be introduced by defining the norm of a section as follows:

$$\| s(z)' \|^2 = \sum_{\Lambda=0}^{N} \overline{X}^{\Lambda'}(\overline{z})\, X^{\Lambda'}(z) = c^\star(\overline{z})\, c(z)\, \overline{X}^{\Lambda}(\overline{z})\, X^{\Lambda}(z) \stackrel{\mathrm{def}}{=} c^\star(\overline{z})\, c(z)\, h(z,\overline{z}) \tag{2.8.28}$$

Calculating the canonical connection and curvature associated with this metric, according to Eq. (2.5.13) we obtain

$$\begin{aligned} \theta &= h^{-1}\partial h = |X|^{-2}\overline{X}^{\Lambda}\partial X^{\Lambda} \\ \Theta &= -\frac{1}{|X|^4}\left[|X|^2\delta^{\Pi\Delta} - \overline{X}^{\Pi}X^{\Delta}\right]\partial X^{\Pi}\wedge\overline{\partial X}^{\Delta} \\ |X|^2 &= \sum_{\Lambda=0}^{N}\overline{X}^{\Lambda}X^{\Lambda} \end{aligned} \tag{2.8.29}$$

where the Dolbeault external derivatives $\partial$ and $\overline{\partial}$ are taken with respect to the coordinates of the underlying $\mathbb{CP}_N$ manifold. We observe that the curvature 2-form $\Theta$ is a well-defined 2-form on the base manifold. Indeed under a rescaling $X^{\Pi}(z) \longrightarrow c(z)\, X^{\Pi}(z)$, the terms proportional to $\partial c$ or $\overline{\partial c}$ cancel by means of the identity

$$\left[|X|^2\delta^{\Pi\Delta} - \overline{X}^{\Pi}X^{\Delta}\right]X^{\Pi} = 0 \tag{2.8.30}$$

and those without derivatives cancel in the numerator against those in the denominator. Using this fact we can rescale $X^{\Pi}(z) \longrightarrow \frac{X^{\Pi}(z)}{X^0(z)}$ and we get

$$c_1\left(U_{1,N+1}\right) \stackrel{\mathrm{def}}{=} \frac{i}{2\pi}\Theta = K \tag{2.8.31}$$

where $K$ is the Kähler 2-form associated with the Fubini–Study metric calculated in Eqs. (2.8.20) and (2.8.11). Equation (2.8.31) shows that indeed $\mathbb{CP}_N$ is a Hodge–Kähler manifold, the hyperplane bundle being the line bundle whose first Chern class equals the Kähler class.

From our previous analysis it follows that a generic section of the hyperplane bundle is a homogeneous polynomial of degree one in the homogeneous coordinates. This means that a homogeneous polynomial $\mathcal{W}_\alpha(X)$ of degree $deg\,\mathcal{W}_\alpha = \nu_\alpha$ transforms as a section of the $\nu_\alpha$-th power of the hyperplane bundle. As a consequence the total Chern class of the line bundle $\mathcal{N}_\alpha \approx \left(U_{1,N+1}\right)^{\nu_\alpha}$ whose sections are the $\mathcal{W}_\alpha$ polynomials is given by

$$c_{total}(\mathcal{N}_\alpha) = (1 + \nu_\alpha K) \tag{2.8.32}$$

Consider now an algebraic $n$-dimensional hypersurface $\mathcal{M}_n \subset \mathbb{CP}_N$, which is defined as the locus of simultaneous zeros of $r = N - n$ polynomial constraints $\mathcal{W}_\alpha(X)$ of degree

$\nu_\alpha$:

$$X^\Lambda \in \mathcal{M}_n \quad \Longrightarrow \begin{cases} \mathcal{W}_1(X) = 0 \ ; & deg\,\mathcal{W}_1 = \nu_1 \\ \mathcal{W}_2(X) = 0 \ ; & deg\,\mathcal{W}_2 = \nu_2 \\ \dots\dots & \dots \\ \mathcal{W}_\alpha(X) = 0 \ ; & deg\,\mathcal{W}_\alpha = \nu_\alpha \\ \dots\dots & \dots \\ \mathcal{W}_r(X) = 0 \ ; & deg\,\mathcal{W}_r = \nu_r \end{cases} \tag{2.8.33}$$

In this case the tangent bundle $T(\mathbb{CP}_N)$ to the ambient manifold can be decomposed into the tangent bundle to the surface $T(\mathcal{M}_n)$ plus the normal bundle $\mathcal{N}(\mathcal{M}_n)$:

$$T(\mathbb{CP}_N) = T(\mathcal{M}_n) \oplus \mathcal{N}(\mathcal{M}_n) \tag{2.8.34}$$

leading to the following identity for the total Chern classes:

$$c_{total}\left(T(\mathbb{CP}_N)\right) = c_{total}\left(T(\mathcal{M}_n)\right) \wedge c_{total}\left(\mathcal{N}(\mathcal{M}_n)\right) \tag{2.8.35}$$

Clearly the normal bundle $\mathcal{N}(\mathcal{M}_n)$ has sections that are locally the direct sum of sections of the bundles $\mathcal{N}_\alpha$ whose total Chern class we have already calculated in (2.8.32). Indeed a vector field sticking out of the hypersurface is of the form

$$\mathbf{t}_{normal} = t_\alpha \frac{\partial}{\partial \mathcal{W}_\alpha} \tag{2.8.36}$$

where one has chosen a local coordinate system in $\mathbb{CP}_N$ of the form $y_1, \dots, y_n, \mathcal{W}_1, \dots, \mathcal{W}_r$. Hence we can conclude

$$\begin{aligned} c_{total}\left(\mathcal{N}(\mathcal{M}_n)\right) &= (1 + \nu_1 K) \wedge \cdots \wedge (1 + \nu_r K) \\ c_{total}\left(T(\mathcal{M}_n)\right) &= \frac{(1 + K)^{n+r+1}}{(1 + \nu_1 K) \wedge \cdots \wedge (1 + \nu_r K)} \end{aligned} \tag{2.8.37}$$

The last of Eqs. (2.8.37) is a formal identity providing us with a quick algorithm to compute the Chern classes of a complete intersection of homogeneous polynomial constraints as defined in Eq. (2.8.33). It suffices to formally expand the second of Eqs. (2.8.37) into a power series of the Kähler 2-form and identify the coefficients of each power. For instance the first Chern class is given by

$$c_1(\mathcal{M}_n) = \left( n + r + 1 - \sum_{\alpha=1}^{r} \nu_\alpha \right) K \tag{2.8.38}$$

and it vanishes if

$$\left( n + r + 1 - \sum_{\alpha=1}^{r} \nu_\alpha \right) = 0 \tag{2.8.39}$$

As we are going to see in Chapter 4, Calabi-Yau manifolds are defined as Kähler manifolds of vanishing first Chern class. Hence Eq. (2.8.39) selects the possible Calabi–Yau $n$-folds that can be constructed as polynomial intersections in a projective space (namely by Eq. (2.8.33)). For instance the simplest Calabi-Yau 3-fold is given by the quintic hypersurface $\nu = 5$ in $\mathbb{CP}_4$. Other constructions are based on the generalization of the above discussion to weighted projective spaces. Their notion will be naturally introduced in Chapter 5 after we have discussed N=2 superconformal theories and Landau–Ginzburg models. We come back to the discussion of the general properties of Calabi–Yau manifolds in Chapter 4. Prior to this, in the next Chapter we consider the Kähler geometry structure of the N=1 and N=2 matter-coupled supergravity lagrangians.

## 2.9 Bibliographical Note

*A sufficient set of references for the mathematical topics reviewed in the present Chapter is provided by the already quoted set of textbooks: [222], [255], [183], [253], [225], [224]. The reader interested in the original mathematical articles can retrieve them from the references quoted in the above textbooks.*

- *For the specific issue of the Atiyah–Singer index theorem, the original papers are [23, 22, 24]. A general account of the theory is given in [167], while [4, 157] contain a new proof of the theorem based on supersymmetry.*

# Chapter 3

# SUPERGRAVITY AND KÄHLER GEOMETRY

## 3.1 Introduction

In the previous chapters we have introduced the basic geometrical tools used in this book. In recent years the interest among physicists for these geometrical concepts has grown considerably, due to their role in string theory and two dimensional quantum field theories. Historically, however, most of these concepts were introduced in supergravity theories and in particular in the study of matter-coupled supergravity in four dimensions.

The main goal of the present chapter is to illustrate the crucial role played by Kähler geometry in supergravity theory. Basically we will review the coupling of the scalar and vector multiplets both to N=1 and N=2 supergravity, focusing on general aspects rather than entering into many technical details. The scalar fields of Wess–Zumino multiplets coupled to N=1 supergravity span an arbitrary Hodge–Kähler manifold [272, 93, 26] and the full structure of the lagrangian is determined by the Kähler metric and by the superpotential, which is just a section of the line bundle, whose first Chern class equals the Kähler class. On the other hand the complex scalar fields belonging to N=2 vector multiplets coupled to $N = 2$, $D = 4$ supergravity, span a Hodge–Kähler manifold, of a restricted type, namely a *special Kähler manifold.* The definition of *special Kähler geometry* and the illustration of its properties are among the main goals of the present chapter. Special Kähler geometry is a crucial concept in the development of our subject. It emerges in various contexts along our exposition. It is the geometry of the moduli space of Calabi–Yau manifolds, as we shall explain in Chapter 6, describing either the space of Kähler class deformations, or the space of complex structure deformations. Therefore special geometry appears while one is studying the flat directions of the scalar potential in those effective supergravity lagrangians that are obtained by compactifying heterotic strings on Calabi–Yau three-folds . It is also the crucial geometric structure pertaining to the marginal deformations in topological two-dimensional theories.

In this chapter, however, we want to show that special geometry is a pure outcome of the vector multiplet coupling in N=2, D=4 supergravity [111, 96, 75, 99]. A truth that is not always appreciated in the literature is that the set of special Kähler manifolds is larger than the set of moduli spaces of Calabi–Yau manifolds, (2,2)-superconformal field theories or topological theories. For instance there is the infinite family of homogeneous special Kähler manifolds whose members not all the time have the interpretation of moduli spaces of anything. Yet any special Kähler manifold can be used to write a consistent matter-coupled N=2 supergravity lagrangian. It is then clear that the supergravity viewpoint is the most natural and general in order to introduce this vital concept. Furthermore the relation between special geometry and N=2 supersymmetry in four dimension is the *a priori* argument that predicts this particular geometry for Calabi–Yau moduli spaces. Indeed since the same Calabi–Yau three-fold can be used to compactify the heterotic superstring (displaying N=1 target supersymmetry) and the type II superstring (displaying N=2 target supersymmetry), it follows that the moduli geometry must be consistent with the constraints of N=2 SUSY, namely special geometry. We shall elaborate on these concepts in later Chapters, after Calabi–Yau compactifications have been introduced, but we felt it proper to anticipate them here in order to give the right perspective to the role of special geometry: *it is a consequence of N=2 supersymmetry in four dimensions.* In the spirit of the present book that is the illustration of the geometrical implications of N=2 supersymmetry in two and four dimensions, this was quite obligatory.

Let us then begin with the study of the geometric structure of N=1 and N=2 four-dimensional supergravity.

## 3.2 The Geometric Structure of Standard N=1 Supergravity

The physical content of standard N=1 supergravity is given by the following entries (see [76] for a review and for notations):

i) Gravitational multiplet

$$V^a, \, \psi, \, \omega^{ab} \tag{3.2.1}$$

which corresponds to the vielbein, the spin 3/2 gravitino and the spin connection. These are 1-forms that gauge the super-Poincaré group. From the particle physics point of view the spin content of the multiplet (3.2.1) is described by $s = [2, 3/2]$.

ii) Scalar multiplet

$$\Lambda^i, A^i, B^i \quad i = 1, \dots, n \tag{3.2.2}$$

where $A^i, B^i$ are respectively a real scalar and a real pseudoscalar and the 0-forms $\Lambda^i$ are Majorana spinors, like the gravitino 1-form. This multiplet can be described by the spin assignments $s = [0^+, 0^-, 1/2]$.

iii) Gauge multiplet $s = [1, 1/2]$

$$A^\Sigma, \lambda^\Sigma \tag{3.2.3}$$

where $A^\Sigma$ is a 1-form (the gauge boson) and $\lambda^\Sigma$ is a Majorana spinor 0-form (the gaugino) both with values in the adjoint representation of the gauge group $G$ ($\Sigma = 1, \dots, \dim G$). This multiplet, in general, gauges a subgroup of the Kähler manifold isometry group.

We focus for the moment on the scalar and gravitational multiplets, without discussing the gauge one. We shall resume our discussion of such a multiplet when we study the isometries of the scalar manifold.

The $2n$ scalars can be combined into a set of $n$ complex scalars by setting

$$z^i = A^i + iB^i \tag{3.2.4}$$

The variables $z^i$ can be interpreted as coordinates of an $n$-dimensional complex manifold $\mathcal{M}$. On $\mathcal{M}$ we introduce a Kähler potential

$$\mathcal{K} = \mathcal{K}(z^i, \overline{z}^i) \tag{3.2.5}$$

which defines, via its derivatives, the metric, the connection and the curvature as in Eqs. (2.5.18), (2.5.19).

To match the change of basis performed on the bosons we introduce the chiral projections of the Majorana spinors $\Lambda^i$:

$$\Lambda^i = \chi^i + \chi^{i^*} \quad ; \quad \chi^i = \frac{1+\gamma_5}{2}\Lambda^i \quad ; \quad \chi^{i^*} = \frac{1-\gamma_5}{2}\Lambda^i \tag{3.2.6}$$

where $\gamma_5 = \prod_{a=1}^{4} \gamma_a$, and $\{\gamma_a\}$ is a basis of Dirac gamma matrices in four dimensions. The chiral components $\chi^i, \chi^{i^*}$ satisfy

$$\gamma_5 \chi^i = \chi^i \quad \gamma_5 \chi^{i^*} = -\chi^{i^*} \tag{3.2.7}$$

Similar projections are introduced for the gravitino 1-form:

$$\psi = \psi_\bullet + \psi^\bullet = \frac{1+\gamma_5}{2}\psi + \frac{1-\gamma_5}{2}\psi \tag{3.2.8}$$

As shown in (2.6.13), the Kähler metric is invariant under Kähler transformation

$$\mathcal{K}(z, \overline{z}) \to \mathcal{K}(z, \overline{z}) + \mathrm{Re}\, f(z) \tag{3.2.9}$$

where $f(z)$ is any holomorphic function of $z$. We can easily extend the action of the Kähler transformation to the fermion fields. On these fields this action can be defined as a chiral rotation. Explicitly we set

$$\begin{aligned}
\psi_\bullet &\to e^{\frac{1}{2}\mathrm{Im}\, f(z)} \psi_\bullet \\
\psi^\bullet &\to e^{-\frac{1}{2}\mathrm{Im}\, f(z)} \psi^\bullet \\
\chi^i &\to e^{-\frac{1}{2}\mathrm{Im}\, f(z)} \chi^i \\
\chi^{i^*} &\to e^{\frac{1}{2}\mathrm{Im}\, f(z)} \chi^{i^*}
\end{aligned} \tag{3.2.10}$$

In order to define derivatives that are covariant under Kähler transformations one needs a Kähler connection. It is provided by the following 1-form:

$$Q = \frac{1}{2i}(\partial_i \mathcal{K} dz^i - \partial_{i^*} \mathcal{K} d\bar{z}^i) \tag{3.2.11}$$

which, under a Kähler transformation, transforms as follows:

$$Q \to Q + d(Im\ f(z)) \tag{3.2.12}$$

The 1-form $Q$ is a connection on the line bundle $\mathcal{L}$ whose first Chern class equals the Kähler class, making the manifold of the scalar fields $\mathcal{M}$ a Hodge–Kähler manifold. Indeed by construction the curvature 2-form is

$$\Theta(Q) = dQ = 2\pi K \tag{3.2.13}$$

Utilizing $Q$ we can write the complete covariant derivatives of $\psi_\cdot$ and $\chi^i$ which take into account Lorentz, general covariant holomorphic transformations on the manifold $\mathcal{M}$ and Kähler transformations. They read

$$\nabla \psi_\cdot = d\psi_\cdot - \frac{1}{4}\omega^{ab} \wedge \gamma_{ab}\psi_\cdot + \frac{i}{2} Q \wedge \psi_\cdot \tag{3.2.14}$$

$$\nabla \chi^i = d\chi^i - \frac{1}{4}\omega^{ab}\gamma_{ab}\chi^i + \Gamma^i_{jk} dz^i \chi^k - \frac{i}{2} Q\chi^i \tag{3.2.15}$$

where $\gamma_{ab} \equiv \gamma_{[a}\gamma_{b]}$ and $\Gamma^i_{jk}$ is the Kähler Riemann connection. Equations (3.2.10), (3.2.14) and (3.2.15) simply state that the left-handed chiral gravitino $\psi_\cdot \in \Gamma(\mathcal{L}^{1/2}, \mathcal{M})$ is a section of the square root of the U(1)-bundle $\mathcal{L}$, while the fermions $\chi^i \in \Gamma(\mathcal{L}^{-1/2} \otimes T\mathcal{M})$ are sections of the specified tensor product of bundles.

The fields of the graviton multiplet and of the scalar multiplets are bound together in a single structure described by the following curvature definitions:

$$\begin{aligned}
R^a &= \mathcal{D}V^a - i\overline{\psi}^\cdot \gamma^a \psi_\cdot \quad \mathcal{D}V^a \equiv dV^a - \omega^{ab} \wedge V_b \\
R^{ab} &= d\omega^{ab} - \omega^{ac} \wedge \omega_c^{\ b} \\
\rho_\cdot &= \nabla \psi_\cdot \\
\nabla z^i &= dz^i
\end{aligned} \tag{3.2.16}$$

In addition one has the "curvatures" of the $\chi^i$ fields as defined in Eq. (3.2.15).

The Bianchi identities following from (3.2.16) are

$$\begin{aligned}
&\mathcal{D}R^a && + R^{ab} \wedge V_b + i\overline{\rho}^\cdot \gamma^a \psi_\cdot - i\overline{\psi}^\cdot \gamma^a \rho_\cdot = 0 \\
&\mathcal{D}R^{ab} && = 0 \\
&\nabla \rho_\cdot && + \frac{1}{4} R^{ab} \wedge \gamma_{ab}\psi_\cdot + \frac{i}{2} dQ \wedge \psi_\cdot = 0 \\
&d^2 z_i && = 0 \\
&\nabla^2 \chi^i && + \frac{1}{4} R^{ab} \wedge \gamma_{ab}\chi^i - \frac{i}{2} dQ\chi^i - R^i_{m^* jk} dz^{m^*} \wedge dz^j \chi^k = 0
\end{aligned} \tag{3.2.17}$$

In the rheonomic formalism, given the curvature definitions (3.2.15) and (3.2.16) one has to find the rheonomic parametrizations that are consistent with Bianchi identities (3.2.17). The rheonomic formalism was originally proposed in 1978–80 [226, 100], [98, 106, 107] as an alternative method with respect to the superfield formalism in studying supersymmetric theories. A complete account of its fundamentals along with its application to the formulation of all supergravity and superstring theories is given in the three-volume book [76]. The main advantage of this approach is the reduction of the computational effort for constructing supersymmetric theories to a simple series of geometrically meaningful steps. *Rheonomy* means "law of the flux" and refers to the fact that the supersymmetrization of a theory can be viewed as a Cauchy problem, in which space–time, described by the inner (bosonic) components $x$ of superspace, represents the boundary, while the outer (Grassmann) components $\theta$ represent the direction of "motion": the rheonomic principle represents the "equation of motion". At the end of the rheonomic procedure, the only free choice is the boundary condition, which is the space–time theory, namely the projection of the superspace theory onto the inner components. The fields (or more generally differential forms) are functions on superspace and the supersymmetry transformations are viewed as odd diffeormorphisms in superspace.

Another basic feature of the rheonomy formalism is that the construction of globally supersymmetric theories almost coincides with the construction of their coupling to supergravity. Indeed, in order to describe a supersymmetric theory, both in the global and in the local case, one needs the supervielbein multiplet, the difference being just the following one: in the global case the supercurvatures can be set to zero, while in the local case they have to be kept arbitrary.

In our case the rheonomic parametrizations for matter-coupled supergravity described by the curvatures (3.2.15) (without the gauge multiplet) are given by

$$\begin{aligned}
R^a &= 0 \\
R^{ab} &= R^{ab}_{mn} V^m \wedge V^n - (2i\overline{\psi}^{\cdot}\gamma^{[a}\rho^{b]c}_{\cdot} + 2i\overline{\psi}_{\cdot}\gamma^{[a}\rho^{b]c\cdot} \\
&- i\overline{\psi}^{\cdot}\gamma^c\rho^{ab}_{\cdot} - i\overline{\psi}_{\cdot}\gamma^c\rho^{ab\cdot}) \wedge V_c - iS^*\overline{\psi}_{\cdot} \wedge \gamma^{ab}\psi_{\cdot} \\
&+ iS\overline{\psi}^{\cdot} \wedge \gamma^{ab}\psi^{\cdot} - \frac{i}{4}A'_c\overline{\psi}^{\cdot} \wedge \gamma_d\psi_{\cdot}\epsilon^{abcd} \\
\rho_{\cdot} &= \rho^{ab}_{\cdot} V_a \wedge V_b + \frac{i}{8}A'_a\gamma^{ab}\psi_{\cdot} \wedge V_b + S\gamma_a\psi^{\cdot} \wedge V^a \\
dz^i &= \nabla_a z^i V^a + \overline{\chi}^i\psi_{\cdot} \\
\nabla\chi^i &= \nabla_a\chi^i V^a + i\nabla^i_a z^i\gamma^a\psi^{\cdot} + \mathcal{H}^i\psi_{\cdot}
\end{aligned} \tag{3.2.18}$$

where the intrisic derivatives $\nabla_a z^i$ are related to the space–time derivatives by

$$\nabla_a z^i = V^\mu_a(\partial_\mu z^i - \overline{\psi}_{\cdot\mu}\chi^i) \tag{3.2.19}$$

and similarly for $\nabla_a\chi^i$.

The general rule for obtaining the solution to the Bianchi identities is the rheonomic principle. We briefly describe it in three steps:

i) One expands the curvature 2-forms on a basis of superspace 2-forms: the "space-time" form $V^a \wedge V^b$ and the "superspace" forms, which can be fermionic (like $\psi_\cdot \wedge V^a$ and similar ones), or bosonic (like $\overline{\psi}_\cdot \wedge \psi_\cdot$ and similar ones).

ii) The coefficients of the space–time form $V^a \wedge V^b$ are independent ones: the rheonomic parametrizations can be viewed as a definition of them.

iii) The coefficients of the superspace forms, instead, can be arbitrary fields (auxiliary fields) or functions of the fields and of the coefficient of space–time forms, being determined by solving the Bianchi identities in superspace. Moreover the closure of Bianchi identities gives also the rheonomic parametrizations of the covariant derivatives of auxiliary fields. All the steps of this calculation are not reported here, we refer to [76] for the details. We shall give an explicit example of the rheonomic procedure in Chapter 4, for the case of two-dimensional N=2 supergravity.

In a formulation of N=1 supergravity without coupling to matter, the fields $S, \mathcal{H}, A'_a$ appearing in Eq. (3.2.18) play the role of generic auxiliary fields. When the Wess Zumino multiplet is added they acquire an explict dependence from matter fields. Explicitly we have

$$\begin{aligned} A'_a &= g_{ij^*}\overline{\chi}^i\gamma_a\chi^{j^*} \\ S &= ie^{[\frac{1}{2}G(z,\overline{z})]} \\ \mathcal{H}^i &= 2g^{ij^*}\partial_{j^*}Ge^{\frac{1}{2}G} \end{aligned} \tag{3.2.20}$$

where $G$ is a particular function of the scalar fields that will be specified in a moment.

Let us focus on the bosonic sector of the matter-coupled supergravity lagrangian. All the essential information is encoded in the bosonic part of such a lagrangian and we do not need the full expression. The bosonic lagrangian is

$$\mathcal{L}^{(N=1)}_{bosonic} = \sqrt{-g}\left[\mathcal{R} - g_{ij^*}\partial_\mu z^i\,\partial^\mu z^{j^*} - V(z,\overline{z})\right] \tag{3.2.21}$$

The scalar potential $V(z,\overline{z})$ has the following expression:

$$V(z,\overline{z}) = e^{G(z,\overline{z})}(\partial_i G\partial_{j^*}Gg^{ij^*} - 3) = -3SS^* + \frac{1}{4}\mathcal{H}^i\mathcal{H}^{j^*}g_{ij^*} \tag{3.2.22}$$

where

$$G(z,\overline{z}) = \mathcal{K}(z,\overline{z}) + log\,|W(z)| + log\,|\overline{W}(\overline{z})| \tag{3.2.23}$$

is the logarithm of the norm of a section $W(z) \in \Gamma(\mathcal{L},\mathcal{M})$ of the line bundle $\mathcal{L}$ whose first Chern class equals the Kähler class. This section $W(z)$ is named the superpotential, because it defines the nontrivial scalar interaction for the Wess-Zumino multiplets as it arises also in the flat case, where the coupling to four-dimensional supergravity is absent. Indeed if we restore the gravitational coupling constant $\kappa \sim \frac{1}{M_{Planck}}$ in (3.2.22):

$$V(z,\overline{z}) = e^{\kappa^2\mathcal{K}(z,\overline{z})}[g^{ij^*}(\partial_i W + \kappa^2 W\partial_i\mathcal{K})(\partial_{j^*}W + k^2\overline{W}\partial_{j^*}K) - 3\kappa^2|W|^2] \tag{3.2.24}$$

it is immediate to see that the only term surviving the limit $\kappa \to 0$ is the typical (flat) bosonic potential interaction $|\partial_i W|^2$.

The inclusion of the gauge multiplet $[1, 1/2]$ produces some nontrivial changes in the scalar lagrangian. As we already pointed out, the vector multiplet $\left\{A^\Lambda, \lambda^\Lambda\right\}$ gauges the isometries of the Kähler manifold $\mathcal{M}$. To study its coupling it is therefore necessary to analyse the isometries of a generic Kähler manifold.

### 3.2.1 Holomorphic Killing vectors on the scalar manifold and the momentum map

The Kähler metric $g_{ij^*}$ of the manifold $\mathcal{M}$ appears in the kinetic terms of the fields $(z^i, \overline{z}^{i^*})$:

$$\mathcal{L}_{kin} = \sqrt{-g} g_{ij^*} \partial_\mu z^i \partial_\nu \overline{z}^{i^*} g^{\mu\nu} \tag{3.2.25}$$

$g^{\mu\nu}$ being the metric of the space–time manifold. If the metric $g_{ij^*}$ has a non trivial group of isometries $G$ generated by Killing vectors $k^i_\Lambda$ ($\Lambda = 1, \ldots, \dim G$) then the lagrangian (3.2.25) admits $G$ as a group of global space–time symmetries. Indeed under an infinitesimal variation

$$z^i \to z^i + \epsilon^\Lambda k^i_\Lambda(z) \tag{3.2.26}$$

$\mathcal{L}_{kin}$ remains invariant. Furthermore if all the couplings of the scalar fields are performed in a diffeomorphic invariant way, then any isometry of $g_{ij^*}$ extends from a symmetry of $\mathcal{L}_{kin}$ to a symmetry of the whole lagrangian. Diffeomorphic invariance means that the scalar fields can appear only through the metric, the Christoffel symbol in the covariant derivative and through the curvature.

Let $k^i_\Lambda(z)$ be a basis of holomorphic Killing vectors for the metric $g_{ij^*}$. Holomorphicity means the following differential constraint:

$$\partial_{j^*} k^i_\Lambda(z) = 0 \leftrightarrow \partial_j k^{i^*}_\Lambda(\overline{z}) = 0 \tag{3.2.27}$$

while the generic Killing equation (suppressing the gauge index $\Lambda$):

$$\nabla_\mu k_\nu + \nabla_\mu k_\nu = 0 \tag{3.2.28}$$

in holomorphic indices reads as follows:

$$\begin{aligned} \nabla_i k_j + \nabla_j k_i &= 0 \\ \nabla_{i^*} k_j + \nabla_j k_{i^*} &= 0 \end{aligned} \tag{3.2.29}$$

where the covariant components are defined as $k_j = g_{ji^*} k^{i^*}$ (and similarly for $k_{i^*}$).

The vectors $k^i_\Lambda$ are generators of infinitesimal holomorphic coordinate transformations:

$$\delta z^i = \epsilon^\Lambda k^i_\Lambda(z) \tag{3.2.30}$$

which leave the metric invariant. In the same way as the metric is the derivative of a more fundamental object, the Killing vectors in a Kähler manifold are the derivatives of suitable prepotentials. Indeed the first equation in (3.2.29) is automatically satisfied by holomorphic vectors and the second equation reduces to the following one:

$$k^i_\Lambda = i g^{ij^*} \partial_{j^*} \mathcal{P}_\Lambda, \quad \mathcal{P}^*_\Lambda = \mathcal{P}_\Lambda \tag{3.2.31}$$

In other words if we can find a real function $\mathcal{P}^\Lambda$ such that the expression $i g^{ij^*} \partial_{j^*} \mathcal{P}_{(\Lambda)}$ is holomorphic, then (3.2.31) defines a Killing vector.

The construction of the Killing prepotential can be stated in a more precise geometrical formulation which involves the notion of *momentum map.* Let us give some details on this notion which will be of vital interest also in Chapter 5 and which reveals another deep connection between supersymmetry and geometry, namely the relationship between the supersymmetric gauging of a sigma model and the geometric construction of Kähler manifolds by means of a *Kähler quotient.*

*Momentum map*

Consider a Kählerian manifold $\mathcal{S}$ of real dimension $2n$ which, in the applications to supergravity theories, is to be identified with the scalar manifold $\mathcal{S} = \mathcal{M}_{scalar}$. Consider a compact Lie group $G$ acting on a $\mathcal{S}$ by means of Killing vector fields $\mathbf{X}$ which are holomorphic with respect to the complex structure $\mathcal{J}$ of $\mathcal{S}$; then these vector fields preserve also the Kähler 2-form

$$\left.\begin{array}{l} \mathcal{L}_\mathbf{X} g = 0 \leftrightarrow \nabla_{(\mu} X_{\nu)} = 0 \\ \mathcal{L}_\mathbf{X} \mathcal{J} = 0 \end{array}\right\} \Rightarrow 0 = \mathcal{L}_\mathbf{X} K = i_\mathbf{X} dK + d(i_\mathbf{X} K) = d(i_\mathbf{X} K) \tag{3.2.32}$$

Here $\mathcal{L}_\mathbf{X}$ and $i_\mathbf{X}$ denote respectively the Lie derivative along the vector field $\mathbf{X}$ and the contraction (of forms) with it.

If $\mathcal{S}$ is simply connected, $d(i_\mathbf{X} K) = 0$ implies the existence of a function $\mathcal{P}^\mathbf{X}$ such that

$$-\frac{1}{2\pi} d\mathcal{P}^\mathbf{X} = i_\mathbf{X} K \tag{3.2.33}$$

The function $\mathcal{P}^\mathbf{X}$ is defined up to a constant, which can be arranged so as to make it equivariant:

$$\mathbf{X}\mathcal{P}^\mathbf{Y} = \mathcal{P}^{[\mathbf{X},\mathbf{Y}]} \tag{3.2.34}$$

The $\mathcal{P}^\mathbf{X}$ constitutes then a *momentum map.* This can be regarded as a map

$$\mathcal{P} : \mathcal{S} \longrightarrow \mathbb{R} \otimes \mathcal{G}^* \tag{3.2.35}$$

where $\mathcal{G}^*$ denotes the dual of the Lie algebra $\mathcal{G}$ of the group $G$. Indeed let $x \in \mathcal{G}$ be the Lie algebra element corresponding to the Killing vector $\mathbf{X}$; then, for a given $m \in \mathcal{S}$

$$\mu(m) \, : \, x \longrightarrow \mathcal{P}^\mathbf{X}(m) \in \mathbb{C} \tag{3.2.36}$$

is a linear funtional on $\mathcal{G}$. If we expand $\mathbf{X} = x^\Lambda k_\Lambda$ in a basis of Killing vectors $k_\Lambda$ such that

$$[k_\Lambda, k_\Gamma] = f_{\Lambda\Gamma}{}^\Delta k_\Delta \tag{3.2.37}$$

we have also

$$\mathcal{P}^{\mathbf{X}} = x^{\Lambda}\mathcal{P}_{\Lambda} \tag{3.2.38}$$

In the following we will use the shorthand notation $\mathcal{L}_{\Lambda}, i_{\Lambda}$ for the Lie derivative (as defined in (3.2.32)) and contraction along the chosen basis of Killing vectors $k_{\Lambda}$.

From a geometrical point of view the prepotential, or momentum map, $\mathcal{P}_{\Lambda}$ is the Hamiltonian function providing the Poissonian realization of the Lie algebra on the Kähler manifold. This is just another way of stating the already mentioned *equivariance.* Indeed the very existence of the closed 2-form $K$ guarantees that every Kähler space is a sympletic manifold and that we can define a Poisson bracket.

Consider Eqs. (3.2.31). To every generator of the abstract Lie algebra $\mathcal{G}$ we have associated a function $\mathcal{P}_{\Lambda}$ on $\mathcal{S}$; the Poisson bracket of $\mathcal{P}_{\Lambda}$ with $\mathcal{P}_{\Sigma}$ is defined as follows:

$$\{\mathcal{P}_{\Lambda}, \mathcal{P}_{\Sigma}\} \equiv 4\pi K(\Lambda, \Sigma) \tag{3.2.39}$$

where $K(\Lambda, \Sigma) \equiv K(\vec{k}_{\Lambda}, \vec{k}_{\Sigma})$ is the value of $K$ along the pair of Killing vectors.

We now prove the following lemma.

**Lemma 3.2.1** *The following identity is true:*

$$\{\mathcal{P}_{\Lambda}, \mathcal{P}_{\Sigma}\} = f_{\Lambda\Sigma}{}^{\Gamma}\mathcal{P}_{\Gamma} + C_{\Lambda\Sigma} \tag{3.2.40}$$

*where $C_{\Lambda\Sigma}$ is a constant fulfilling the cocycle condition*

$$f_{\Lambda\Pi}{}^{\Gamma}C_{\Gamma\Sigma} + f_{\Pi\Sigma}{}^{\Gamma}C_{\Gamma\Lambda} + f_{\Sigma\Lambda}{}^{\Gamma}C_{\Gamma\Pi} = 0 \tag{3.2.41}$$

Proof: Let us set $f_{\Lambda\Sigma} \equiv K(\Lambda, \Sigma)$. Using Eq. (3.2.32) we get

$$\begin{aligned} 4\pi f_{\Lambda\Sigma} &= 4\pi f_{\Sigma\Lambda} = 2\pi i_{\Sigma} i_{\Lambda} K = -i_{\Sigma} d\mathcal{P}_{\Lambda} = i_{\Lambda} d\mathcal{P}_{\Sigma} = \\ &= \frac{1}{2}(\mathcal{L}_{\Lambda}\mathcal{P}_{\Sigma} - \mathcal{L}_{\Sigma}\mathcal{P}_{\Lambda}) \end{aligned} \tag{3.2.42}$$

Let us now calculate $df_{\Lambda\Sigma}$. Since the exterior derivative commutes with the Lie derivative we find

$$df_{\Lambda\Sigma} = \frac{1}{8\pi}(\mathcal{L}_{\Lambda} d\mathcal{P}_{\Sigma} - \mathcal{L}_{\Sigma} d\mathcal{P}_{\Lambda}) = \tag{3.2.43}$$

$$= \frac{1}{4}(-\mathcal{L}_{\Lambda} i_{\Sigma} K + \mathcal{L}_{\Sigma} i_{\Lambda} K) \tag{3.2.44}$$

Using now the identity

$$[i_{\Lambda}, \mathcal{L}_{\Sigma}] = i_{[\Lambda,\Sigma]} \tag{3.2.45}$$

and Eq. (3.2.37), from Eq. (3.2.42) we obtain

$$df_{\Lambda\Sigma} = \frac{1}{2} i_{[\Sigma,\Lambda]} K = -\frac{1}{2} f_{\Lambda\Sigma}{}^{\Gamma} i_{\Gamma} K = \tag{3.2.46}$$

$$= \frac{1}{4\pi} f_{\Lambda\Sigma}{}^{\Gamma} d\mathcal{P}_{\Gamma} \tag{3.2.47}$$

It follows that the difference

$$C_{\Lambda\Sigma} = \{\mathcal{P}_\Lambda, \mathcal{P}_\Sigma\} - f_{\Lambda\Sigma}{}^{\Gamma}\mathcal{P}_\Gamma \tag{3.2.48}$$

is a constant since we have shown that its exterior derivative vanishes: $dC_{\Lambda\Sigma} = 0$. The cocycle condition (3.2.41) follows from the Jacobi identities fulfilled by the Poisson bracket (3.2.40). This concludes the proof of the lemma.

If the Lie algebra $\mathcal{G}$ has a trivial second cohomology group $H^2(\mathcal{G}) = 0$, then the cocycle $C_{\Lambda\Sigma}$ is a coboundary; namely we have

$$C_{\Lambda\Sigma} = f_{\Lambda\Sigma}{}^{\Gamma} C_\Gamma \tag{3.2.49}$$

where $C_\Gamma$ are suitable constants. Hence, assuming $H^2(\Gamma) = 0$ we can reabsorb $C_\Gamma$ in the definition of $\mathcal{P}_\Lambda$:

$$\mathcal{P}_\Lambda \to \mathcal{P}_\Lambda + C_\Lambda \tag{3.2.50}$$

and we obtain the stronger equation

$$\{\mathcal{P}_\Lambda, \mathcal{P}_\Sigma\} = f_{\Lambda\Sigma}{}^{\Gamma}\mathcal{P}_\Gamma \tag{3.2.51}$$

Note that $H^2(\mathcal{G}) = 0$ is true for all semi-simple Lie algebras.

Using (3.2.40), Eq. (3.2.51) can be rewritten in components as follows:

$$\frac{i}{2} g_{ij^*}(k^i_\Lambda k^{j^*}_\Sigma - k^i_\Sigma k^{j^*}_\Lambda) = \frac{1}{2} f_{\Lambda\Sigma}{}^{\Gamma}\mathcal{P}_\Gamma \tag{3.2.52}$$

Equation (3.2.52) is identical with the equivariance condition (3.2.34).

### 3.2.2 The momentum map and the complete bosonic lagrangian of N=1 matter-coupled supergravity

We can now describe the complete form of the bosonic lagrangian of matter-coupled supergravity. Indeed the essential reason why the momentum map described in the previous section is relevant to supergravity theory is the following:

**Theorem 3.2.1** *The momentum map function $\mathcal{P}_\Lambda(z,\bar{z})$ is the on-shell value taken by the auxiliary field $P_\Lambda$ in the N=1 vector multiplet $\{A^\Lambda, \lambda^\Lambda, P_\Lambda\}$ that gauges the global symmetry generated by the Killing vector $k^i_\Lambda(z)$ of which $\mathcal{P}_\Lambda(z,\bar{z})$ is the prepotential.*

This theorem, whose proof is omitted (see [76] for more details), means essentially the following. Let $G$ be a group of isometries of the scalar Kähler manifold $\mathcal{S} = \mathcal{M}_{scalar}$ and let $\mathcal{G}$ be its Lie algebra realized on $\mathcal{S}$ by the holomorphic Killing vectors $k^i_\Lambda(z)$. Introduce an off-shell vector multiplet $\{A^\Lambda, \lambda^\Lambda, P_\Lambda\}$ for each of the considered isometries, where, in addition to the 1-form $A^\Lambda$, representing the gauge boson, and the spinor 0-form $\lambda^\Lambda$, representing the gaugino, the bosonic real 0-form $P_\Lambda$ is the auxiliary field. *By definition* it appears, as we are going to see, in the *rheonomic parametrization* of the *gaugino covariant differential.* Then the gauging of the global symmetry $G$ is performed by a change in the definition of the covariant derivatives and correspondingly in the curvature definitions of the matter fields:

$$\begin{aligned} \widehat{\nabla} z^i &= \nabla z^i + A^\Lambda k^i_\Lambda(z) \\ \widehat{\nabla} \chi^i &= \nabla \chi^i + A^\Lambda \partial_j k^i_\Lambda(z) \chi^j \end{aligned} \tag{3.2.53}$$

At the same time the curvature definitions for the gauge multiplet are

$$\begin{aligned} F^\Lambda &= dA^\Lambda + f_{\Sigma\Gamma}{}^\Lambda A^\Sigma \wedge A^\Gamma \\ \widehat{\nabla} \lambda^\Lambda_{\cdot} &= d\lambda^\Lambda_{\cdot} - \frac{1}{4}\omega^{ab}\gamma_{ab}\lambda^\Lambda_{\cdot} + f_{\Delta\Gamma}{}^\Lambda A^\Delta \lambda^\Gamma_{\cdot} + \frac{i}{2} Q \lambda^\Lambda_{\cdot} \end{aligned} \tag{3.2.54}$$

where the chiral projections $\lambda^\Lambda_{\cdot}, \lambda^{\cdot\Lambda}$ are defined as in (3.2.8) and the curvature definitions for the gravitational multiplet remain unmodified:

$$\begin{aligned} R^a &= \mathcal{D}V^a - i\overline{\psi}^{\cdot}\gamma^a\psi_{\cdot} \quad \mathcal{D}V^a \equiv dV^a - \omega^{ab} \wedge V_b \\ R^{ab} &= d\omega^{ab} - \omega^{ac} \wedge \omega_c^{\ b} \\ \rho_{\cdot} &= \widehat{\nabla}\psi_{\cdot} = \nabla\psi_{\cdot} \\ &= d\psi_{\cdot} - \frac{1}{4}\omega^{ab} \wedge \gamma_{ab}\psi_{\cdot} + \frac{i}{2} Q \wedge \psi_{\cdot} \end{aligned} \tag{3.2.55}$$

The procedure now is to replace $\nabla$ by $\widehat{\nabla}$, find the new rheonomic parametrizations, which solve the Bianchi identities associated with Eqs. (3.2.53), (3.2.54), (3.2.55) and then consider the new lagrangian with the inclusion of the gauge multiplet. In particular one has the rheonomic parametrizations of the fields belonging to the gauge multiplet:

$$\begin{aligned} F^\Sigma &= F^\Sigma_{ab} V^a V^b + \frac{i}{2}\overline{\lambda}^\Sigma_{\cdot}\gamma_m\psi^{\cdot} \wedge V^m + \frac{i}{2}\overline{\lambda}^{\Sigma\cdot}\gamma_m\psi_{\cdot} \wedge V^m \\ \widehat{\nabla}\lambda^\Lambda_{\cdot} &= \nabla_a\lambda^\Lambda_{\cdot} V^a + \overset{(+)\Lambda}{F}_{ab}\gamma^{ab}\psi_{\cdot} + i\left(W^\Lambda + P_\Lambda\right)\psi_{\cdot} \end{aligned} \tag{3.2.56}$$

where $\overset{(+)\Lambda}{F}_{ab}$ is the self-dual part of $F^\Lambda_{ab}$:

$$F^\Lambda_{ab} = \overset{(+)\Lambda}{F}_{ab} + \overset{(-)\Lambda}{F}_{ab} \qquad \varepsilon_{abcde}\overset{(\pm)\Lambda}{F}_{cd} = \pm 2i\overset{(\pm)\Lambda}{F}_{ab} \tag{3.2.57}$$

where $W^\Lambda$ denotes a bilinear in the fermions $\lambda^\Lambda$ and $\chi^i$, whose structure we do not discuss here, and where $P_\Lambda$ is the auxiliary field. The on-shell value of $P_\Lambda$ which is enforced by the coupling to the matter fields or, alternatively, by the construction of the action, is

$$P_\Lambda = \mathcal{P}_\Lambda(z,\overline{z}) \tag{3.2.58}$$

This is the content of theorem 3.2.1. Since, due to Ward identities of the N=2 supersymmetry, all the auxiliary fields contribute quadratically to the scalar potential (compare with Eq. (3.2.22), the momentum-map functions do appear in this way in the final bosonic lagrangian. Indeed the final result for this latter is:

$$\begin{aligned} \mathcal{L}^{(N=1)}_{bosonic} &= \sqrt{-g}\left[\mathcal{R} - g_{ij^*}\,\nabla_\mu z^i\,\nabla^\mu z^{j^*} - \frac{1}{4}\,Re f_{\Lambda\Sigma}(z) F^\Lambda_{\mu\nu}\,F^{\Sigma\mu\nu} - V(z,\overline{z})\right] \\ &- \frac{1}{8}\,Im f_{\Lambda\Sigma}(z)\,F^\Lambda_{\mu\nu}\,F^\Sigma_{\rho\sigma}\,\varepsilon^{\mu\nu\rho\sigma} \end{aligned} \quad (3.2.59)$$

where the bosonic potential, including the contributions from the gauge interactions, has the following expression:

$$V(z,\overline{z}) = \left(g^{ij^*}\,\partial_i G\,\partial_{j^*}G - 3\right)e^G + \frac{1}{4}\left([Ref]^{-1}\right)^{\Lambda\Sigma}\,\mathcal{P}_\Lambda\,\mathcal{P}_\Sigma \quad (3.2.60)$$

The requirement that the bosonic lagrangian be fully invariant under the gauge group $G$ implies that the function

$$G(z,\overline{z}) = \log||W(z)||^2 = \mathcal{K}(z,\overline{z}) + \log|W(z)|^2 \quad (3.2.61)$$

which is nothing else than the logarithm of the superpotential squared norm, should be exactly invariant under the Lie algebra $\mathcal{G}$, i.e.

$$\mathcal{L}_\Lambda G = k^i_\Lambda \partial_i G + k^{i^*}_\Lambda \partial_{i^*} G = 0 \quad (3.2.62)$$

Equation (3.2.62) is also implied by consistency with the Bianchi identities of matter-coupled supergravity. The "gauge coupling" function $f_{\Lambda\Sigma}$ appearing in Eqs. (3.2.60) and (3.2.59) is a holomorphic function of the scalar fields $z^i$, belonging to the symmetric $adj \otimes adj$ representation of the gauge group $G$. The kinetic term of the gauge multiplet, where the group index is contracted with $Re\, f_{\Lambda\Sigma}$, is the most general term compatible with N=1 supersymmetry [272, 25, 26, 93], [76]. The standard kinetic term is recovered by setting $Ref_{\Lambda\Sigma} = g^{-2}\delta_{\Lambda\Sigma}$, where $g$ denotes the gauge coupling constant.

If the gauge group has a linear action $\delta z^i = (T_\Lambda)^i_j\; z^j$ on the scalar fields, then the contribution to the scalar potential (3.2.60) that is proportional to the gauge coupling constant $g^2$ is given in terms of Killing vectors prepotentials of the form[1]

$$\mathcal{P}_\Lambda = -\,i\,\partial_i G (T_\Lambda)^i_j z^j \quad (3.2.63)$$

When the action of the gauge group is non-linear, the expression of $\mathcal{P}^\Lambda$ is more complicated, but we shall not be interested in this case.

### 3.2.3 Extrema of the potential and Kähler quotients

This is the general structure of N=1, D=4 standard supergravity. Shortly from now we shall consider the peculiarities of the N=1 supergravity action that emerges as low-energy theory when we compactify the heterotic superstring on a Calabi–Yau 3-fold.

[1]Notice that due to Eq. (3.2.62), the Killing prepotential (3.2.63) is automatically real.

Prior to that we want to elaborate further on the role of the momentum map functions. We consider the scalar potential of gauged supergravity in the infinite Planck mass limit $\kappa \longrightarrow 0$, where the gravitational multiplet decouples from the other multiplets. Recalling our previous discussion and comparing with Eq. (3.2.24) we see that in this flat limit the potential takes the form

$$V_{flat\ limit}(z,\overline{z}) = g^{ij^\star}\,\partial_i W(z)\,\partial_j \overline{W}(\overline{z}) + \frac{1}{4}\left([Ref]^{-1}\right)^{\Lambda\Sigma}\,\mathcal{P}_\Lambda\,\mathcal{P}_\Sigma \tag{3.2.64}$$

The classical vacua of the globally supersymmetric theory composed of the gauge multiplets and the Wess–Zumino multiplets correspond to the extrema of the potential (3.2.64). If we choose a positive definite Kähler metric $g_{ij^\star}$ and a minimal gauge coupling:

$$f^{\Lambda\Sigma} = g^{-2}\delta^{\Lambda\Sigma} \tag{3.2.65}$$

where $g$ is the gauge coupling constant, then the potential (3.2.64) reduces to a sum of squares:

$$V_{flat\ limit}(z,\overline{z}) = |\partial_i W(z)|^2 + \frac{1}{4}g^2\sum_{\Lambda=1}^{dimG}(\mathcal{P}_\Lambda)^2 \tag{3.2.66}$$

so that the extrema are given by the submanifold $\widehat{\mathcal{E}} \subset \mathcal{M}_{scalar}$ of the scalar manifold defined by the following conditions:

$$\partial_i W(z) = 0 \qquad \mathcal{P}_\Lambda(z,\overline{z}) = 0 \tag{3.2.67}$$

We also have to take into account the gauge invariance of the original model, so that the actual set of the classical vacua is not exactly the manifold $\widehat{\mathcal{E}}$; rather, it is the space of gauge orbits:

$$\mathcal{E} = \frac{\widehat{\mathcal{E}}}{G} \tag{3.2.68}$$

The submanifold $\mathcal{S} \subset \mathcal{M}_{scalar}$ defined by $\partial_i W(z) = 0$ is certainly a Kähler manifold since the Kähler 2-form defined on $\mathcal{M}_{scalar}$ can be pullbacked to the hypersurface $\mathcal{S}$. Then the set of operations (3.2.67,3.2.68) needed to derive the space of classical vacua corresponds to the geometric notion of *Kähler quotient*. It is a very interesting concept and it will surface again in Chapter 5 in connection with the vacuum structure of N=2 field theories on the world-sheet (another instance of the *N=2 world-sheet* $\leftrightarrow$ *N=1 target-space* correspondence). Let us then examine this geometrical notion in some detail.

*Kähler quotient*

It is a procedure that provides a way to construct from a Kähler manifold $\mathcal{S}$, supporting the holomorphic action of a Lie group $G$, a lower-dimensional Kähler manifold $\mathcal{M}$, obtained as follows. Let $\mathcal{Z}^\star \subset \mathcal{G}^\star$ be the dual of the centre of $\mathcal{G}$. For each $\zeta \in \mathbb{R}\otimes\mathcal{Z}^\star$, the level set of the momentum map

$$\mathcal{N} \equiv \mathcal{P}^{-1}(\zeta) \subset \mathcal{S} \tag{3.2.69}$$

which has dimension $\dim\mathcal{N} = \dim\mathcal{S} - \dim G$ is invariant under the action of $G$, due to the equivariance of $\mathcal{P}$. It is thus possible to take the quotient

$$\mathcal{M} = \mathcal{N}/G$$

$\mathcal{M}$ is a smooth manifold of dimension $\dim\mathcal{M} = \dim\mathcal{S} - 2\dim G$ as long as the action of $G$ on $\mathcal{N}$ has no fixed points. The 2-form $\widehat{K}$ on $\mathcal{M}$, defined via the restriction to $\mathcal{N} \subset \mathcal{S}$ of $K$ and the quotient projection from $\mathcal{N}$ to $\mathcal{M}$, is closed and thus provides $\mathcal{M}$ with a Kähler structure.

Let $\mathbf{Y}(s) = Y^\Lambda \mathbf{k}_\Lambda(s)$ be a Killing vector on $\mathcal{S}$, belonging to $\mathcal{G}$, the Lie algebra of the gauge group. Consider the vector field $I\mathbf{Y} \in \mathcal{G}^c$ (the complexified algebra), $I$ being the complex structure acting on $T\mathcal{S}$. This vector field is orthogonal to the hypersurface $\mathcal{P}^{-1}(\zeta)$, for any level $\zeta$; that is, it generates transformations that change the level of the surface. As just recalled, the Kähler quotient consists in starting from $\mathcal{S}$, restricting to $\mathcal{N} = \mathcal{P}^{-1}(\zeta)$ and taking the quotient $\mathcal{M} = \mathcal{N}/G$. The above remarks about the action of the complexified gauge group suggest that this is equivalent (at least if we skip the problems due to the non-compactness of $G^c$) to simply taking the quotient $\mathcal{S}/G^c$, the so-called "algebro-geometric" quotient.

The Kähler quotient allows one, in principle, to determine the expression of the Kähler form on $\mathcal{M}$ in terms of the original one on $\mathcal{S}$. Schematically, let $j$ be the inclusion map of $\mathcal{N}$ into $\mathcal{S}$, $p$ the projection from $\mathcal{N}$ to the quotient $\mathcal{M} = \mathcal{N}/G$, $K$ the Kähler form on $\mathcal{S}$ and $\widehat{K}$ the Kähler form on $\mathcal{M}$. It can be shown [191] that

$$\begin{array}{ccccc} \mathcal{S} & \stackrel{j}{\longleftarrow} & \mathcal{N} = \mathcal{P}^{-1}(\zeta) & \stackrel{p}{\longrightarrow} & \mathcal{M} = \mathcal{N}/G \\ K & \longrightarrow & j^*K = p^*\widehat{K} & \longleftarrow & \widehat{K} \end{array} \tag{3.2.70}$$

In the algebro-geometric setting, the holomorphic map that associates to a point $s \in \mathcal{S}$ its image $m \in \mathcal{M}$ is obtained as follows:
*i)* Bringing $s$ to $\mathcal{N}$ by means of the finite action infinitesimally generated by a vector field of the form $\mathbf{V} = I\mathbf{Y} = V^a \mathbf{k}_a$

$$\pi : \quad s \in \mathcal{S} \longrightarrow e^{-V} s \in \mathcal{P}^{-1}(\zeta) \tag{3.2.71}$$

*ii)* Projecting $e^{-V}$ to its image in the quotient $\mathcal{M} = \mathcal{N}/G$.

Thus we can consider the pull-back of the Kähler form $\widehat{K}$ through the map $p \cdot \pi$:

$$\begin{array}{ccccc} \mathcal{S} & \stackrel{\pi}{\longrightarrow} & \mathcal{N} = \mathcal{P}^{-1}(\zeta) & \stackrel{p}{\longrightarrow} & \mathcal{N}/G \\ \pi^*p^*\widehat{K} & \longleftarrow & p^*\widehat{K} & \longleftarrow & \widehat{K} \end{array} \tag{3.2.72}$$

Looking at (3.2.70) we see that $\pi^*p^*\widehat{K} = \pi^*j^*K$ so that, at the end of the day, in order to recover the pullback of $\widehat{K}$ to $\mathcal{S}$ it is sufficient:
*i)* to restrict $K$ to $\mathcal{N}$
*ii)* to pull back this restriction to $\mathcal{M}$ with respect to the map $\pi = e^{-V}$.

We see from (3.2.71) that the components of the vector field $\mathbf{V}$ must be determined by requiring

$$\mathcal{P}(e^{-V}s) = \zeta \tag{3.2.73}$$

The Kähler potential $\widehat{\mathcal{K}}$ for the manifold $\mathcal{M}$, such that

$$\widehat{K} = \frac{\mathrm{i}}{2\pi}\partial\bar{\partial}\widehat{\mathcal{K}} \tag{3.2.74}$$

is given by

$$\widehat{\mathcal{K}} = \mathcal{K}|_{\mathcal{N}} + V^a\zeta_a \tag{3.2.75}$$

Here $\mathcal{K}$ is the Kähler potential on $\mathcal{S}$; $\mathcal{K}|_{\mathcal{N}}$ is the restriction of $\mathcal{K}$ to $\mathcal{N}$, i.e. it is computed after acting on the point $s \in \mathcal{S}$ with the transformation $e^{-V}$ determined by Eq. (3.2.73); $V^a$ are the components of the vector field $\mathbf{V}$ along the $a$-th generator of the centre of the gauge algebra $\mathcal{Z}$, and $\zeta_a$ those of the level $\zeta$ of the momentum map.

### 3.2.4 Effective N=1 supergravities obtained from Calabi–Yau compactifications

As already mentioned, when we compactify the heterotic string on Calabi-Yau manifolds, we eventually obtain (3.2.60) as the low energy effective theory, since we preserve N=1 supersymmetry. In this case the gauge group is $E_6 \otimes E_8{}'$ and the scalar multiplets (all neutral under $E_8{}'$) are of six different types (see Chapter 4). In particular there is a number of (non-charged) moduli fields $M^A$ which span a moduli space $\mathcal{M}$ ($A = 1, \ldots, \dim\mathcal{M}$). In the context of low energy field theory, by moduli field one means the following. Consider the scalar potential (3.2.60) as a function of all the scalar fields $z, \overline{z}$. We decompose the set of scalar fileds in two subsets $\{z\} = \{M\} \cup \{P\}$, where the subset $\{M\}$ of the moduli fields is defined by the following property:

$$\frac{\partial V(M,\overline{M},P,\overline{P})}{\partial M}|_{P=0} = 0 \qquad \text{identically} \tag{3.2.76}$$

i.e. (3.2.76) is identically satisfied for all values of $M$ at $P = 0$, while in the subset $\{P\}$ of the non-moduli fields, the condition $\frac{\partial V}{\partial P} = 0$ is satisfied only for a specific value $P = P_0 = 0$. According to this definition the moduli fields correspond to the possible flat directions of the scalar potential. This means that each vacuum expectation value of the field theory is parametrized by points in the moduli space, i.e. a manifold with certain geometrical features to be studied case by case. The geometrical structure of the moduli space in the Calabi–Yau compactification is strongly constrained by the underlying superconformal theory. Indeed it can be shown that $\mathcal{M}$ is the direct product of two submanifolds, which correspond to the (1,1) and (2,1) moduli of the Calabi–Yau manifold. Moreover the geometry of each submanifold is "special" (see Chapter 6 for a proof of this statement). In correspondence with non-charged moduli there are charged scalar fields $C^\alpha, C^\mu$ ($\alpha =$

$1, \ldots, N_1$, $\mu = 1, \ldots, N_2$)[2], which belong respectively to the 27 and $\overline{27}$ representations of the gauge group $E_6$. Moreover there is a dilaton field $S$ and a set $\mathcal{Y}^u$ of non-moduli singlets ($u = 1, \ldots, N_3$). Let us label all these scalar fields with the capital index $I$ so that, in summary, we can write

$$z^I = \left\{ \begin{array}{lcl} S & = & \text{dilaton-axion field} \\ \mathcal{M}^A & = & (a = 1, \ldots, \dim \mathcal{M}) \\ \mathcal{C}^\alpha & = & 27\text{-charged fields } (a = 1, \ldots, N_1) \\ \mathcal{C}^i & = & \overline{27}\text{-charged fields } (i = 1, \ldots, N_2) \\ \mathcal{Y}^u & = & \text{non-moduli singlets } (u = 1, \ldots, N_3) \end{array} \right\} \tag{3.2.77}$$

For the case of the heterotic string compactification on a Calabi–Yau manifold we have $A = (a, i)$ with $a = 1, \ldots h^{1,1}$ and $i = 1, \ldots h^{2,1}$ and $N_1 = h^{1,1}, N_2 = h^{2,1}$, where $h^{1,1}, h^{2,1}$ are the number of Kähler and complex structure deformations. The number $N_3$ of non-moduli singlets is left undetermined, since we do not need its explicit value. The Kähler potential is also strongly constrained in its form by the compactification procedure: it has the form

$$\begin{aligned} \mathcal{K}(z^I, \overline{z}^I) &= -\frac{1}{\kappa^2} \ln(S + \overline{S}) + \widehat{\mathcal{K}}(M, \overline{M}) + G_{\alpha\overline{\beta}}(M, \overline{M}) \mathcal{C}^\alpha \overline{\mathcal{C}}^{\overline{\beta}} + G_{\mu\overline{\nu}}(M, \overline{M}) \mathcal{C}^\mu \overline{\mathcal{C}}^{\overline{\nu}} \\ &+ O(\mathcal{Y}\overline{\mathcal{Y}}) + O(\mathcal{C}^\alpha \overline{\mathcal{C}}^{\overline{\beta}}) + O(\mathcal{C}^\mu \overline{\mathcal{C}}^{\overline{\nu}}) + \cdots \end{aligned} \tag{3.2.78}$$

In (3.2.78) $\widehat{\mathcal{K}}$ is the Kähler potential of the moduli space and $\kappa$ is the gravitational coupling constant. $G_{\alpha\overline{\beta}}$ and $G_{\mu\overline{\nu}}$ are the moduli-dependent metrics for the 27 and $\overline{27}$ fields. Finally the dots in (3.2.78) mean "higher order powers in charged fields".

Moreover the superpotential can be written as follows:

$$W = \frac{1}{3} W_{\alpha\beta\gamma}(M) \mathcal{C}^\alpha \mathcal{C}^\beta \mathcal{C}^\gamma + \frac{1}{3} W_{\mu\nu\rho}(M) \mathcal{C}^\mu \mathcal{C}^\nu \mathcal{C}^\rho + O(\mathcal{C}^\alpha \mathcal{C}^\beta \mathcal{Y}) + O(\mathcal{Y}^3) + \cdots \tag{3.2.79}$$

where $W_{\alpha\beta\gamma} = \partial_\alpha \partial_\beta \partial_\gamma W$ and where $W_{\mu\nu\rho} = \partial_\mu \partial_\nu \partial_\rho W$ (with $\partial_\alpha = \frac{\partial}{\partial \mathcal{C}^\alpha}$ and $\partial_\mu = \frac{\partial}{\partial \mathcal{C}^\mu}$).

Indeed, for points in moduli space the superpotential vanishes with its first and second derivatives with respect to all massless fields. The coefficients $W_{\alpha\beta\gamma}, W_{\mu\nu\rho}$ are the 27 and $\overline{27}$ Yukawa couplings. This is so because in the complete N=1 supergravity lagrangian the fermions $\chi^I$ are coupled to charged scalar fields via terms like

$$\mathcal{L}_{Yukawa} = W_{IJK} \chi^I \chi^J \mathcal{C}^K \tag{3.2.80}$$

which are precisely proportional to the third derivative of the superpotential (in using the collective index $I$, it is understood that some of the Yukawa couplings introduced above can be zero). The coefficients $W_{\alpha\beta\gamma}, W_{\mu\nu\rho}$ are particular holomorphic functions of the moduli fields and, in the case of Calabi–Yau compactifications, they are the fundamental objects that enter the definition of special Kähler geometry (see Eq. (3.3.16)). Moreover, in Chapter 7 they will be related to the topological three-point functions of marginal topological deformations.

---

[2]**Our notation assigns the first Greek letters to the 27 and the last ones to the $\overline{27}$.**

## 3.3 Special Kähler Geometry

In previous sections we have studied the case of N=1 matter-coupled supergravity. The scalar Wess–Zumino multiplets coupled to the supergravity multiplet span a Kähler manifold. In the following subsections we give a definition of a very peculiar geometry, the special Kähler geometry, which arises in the coupling of vector multiplets to N=2, D=4 supergravity. In this case the complex scalar fields sitting in the vector multiplets span a manifold $\mathcal{M}$ which is not only Kählerian (as it is implied by N=1 supersymmetry) but *special Kählerian.* The new property resides in a particular constraint that has to be fulfilled by the Riemman tensor, as will be apparent in Eq. (3.3.16).

As in the previous sections we do not give a fully detailed treatement of the matter coupling procedure, rather we focus on the general formulae and on the basic concepts. However we insist on giving a geometrical description of the special Kähler manifolds, since this is the crucial structure that represents the link between most of the topics treated in this book.

We start our discussion with the definition of special Kähler geometry.

### 3.3.1 Special Kähler manifolds with special Killing vectors

To define a special Kähler manifold we recall some basic geometrical data that were introduced in Chapter 2.

A special Kähler manifold is a Kähler manifold of a restricted type (Hodge manifold). This means that it has a complex structure and a hermitean metric

$$ds^2 = g_{ij^*}(z,\overline{z})dz^i \otimes d\overline{z}^{j^*} \tag{3.3.1}$$

such that the (1,1)-form

$$K = \frac{i}{2\pi} g_{ij^*}(z,\overline{z})dz^i \wedge d\overline{z}^{j^*} \tag{3.3.2}$$

is closed ($dK = 0$). In every coordinate patch we can write the following formulae (recalling Eqs. (3.2.11) and (3.2.13)):

$$\begin{aligned} g_{ij^*} &= \partial_i \partial_{j^*} \mathcal{K} && (3.3.3)\\ K &= \frac{dQ}{2\pi} && (3.3.4)\\ Q &= -\frac{i}{2}(\partial_i \mathcal{K} dz^i - \partial_{i^*} \mathcal{K} d\overline{z}^{i^*}) && (3.3.5) \end{aligned}$$

where $\mathcal{K}$ is the Kähler potential. Under a Kähler transformation (2.6.13) the 1-form $Q$ transforms as

$$Q \to Q + d(\mathrm{Im} f) \tag{3.3.6}$$

$Q$ is a $U(1)$ connection.

The $U(1)$ covariant differential of a field $\Phi(z,\overline{z})$ of weight $p$ is defined by

$$\nabla\Phi = (d + ip\mathcal{Q})\Phi \tag{3.3.7}$$

or, in components,

$$\begin{aligned} \nabla_i\Phi &= (\partial_i + \frac{1}{2}p\partial_i\mathcal{K})\Phi & (3.3.8) \\ \nabla_{i^\star}\Phi &= (\partial_{i^\star} - \frac{1}{2}p\partial_{i^\star}\mathcal{K})\Phi & (3.3.9) \end{aligned}$$

A covariantly holomorphic field of weight $p$ is defined by

$$\nabla_{i^\star}\Phi = 0 \tag{3.3.10}$$

By a change of trivialization the $U(1)$-bundle can be reduced to a holomorphic line bundle $\mathcal{L}$. Indeed setting

$$\tilde{\Phi} = e^{-p\mathcal{K}/2}\Phi \tag{3.3.11}$$

we have

$$\begin{aligned} \nabla_i\tilde{\Phi} &= (\partial_i + p\partial_i\mathcal{K})\tilde{\Phi} & (3.3.12) \\ \nabla_{i^\star}\tilde{\Phi} &= \partial_{i^\star}\tilde{\Phi} & (3.3.13) \end{aligned}$$

In particular, if $\Phi$ is a covariantly holomorphic section with respect to the $\mathcal{Q}$-connection, $\tilde{\Phi}$ is a holomorphic section with respect to the holomorphic connection $\partial_i\mathcal{K}$.

If the $U(1)$ line bundle is such that the first Chern class $c_1(\mathcal{L})$ coincides with the Kähler class $[K]$ then the Kähler manifold is of a restricted type or a Hodge manifold. In addition to the $U(1)$-holomorphic connection $\partial_i\mathcal{K}$ one has the holomorphic Levi–Civita connection as decribed by Eqs. (2.5.18).

We are now able to give the following definition:

**Definition 3.3.1** *A restricted Kähler manifold is special Kählerian if there exists a completely symmetric holomorphic 3-index section $W_{ijk}$ of $(T^\star\mathcal{M})^3 \otimes \mathcal{L}^2$ (and its antiholomorphic conjugate $W_{i^\star j^\star k^\star}$) such that*

$$\begin{aligned} \partial_{m^\star}W_{ijk} &= 0 \quad \partial_m W_{i^\star j^\star k^\star} = 0 & (3.3.14) \\ \nabla_{[m}W_{i]jk} &= 0 \quad \nabla_{[m}W_{i^\star]j^\star k^\star} = 0 & (3.3.15) \\ \mathcal{R}_{i^\star j\ell^\star k} &= g_{\ell^\star j}g_{ki^\star} + g_{\ell^\star k}g_{ji^\star} - e^{2\mathcal{K}}W_{i^\star\ell^\star s^\star}W_{tkj}g^{s^\star t} & (3.3.16) \end{aligned}$$

In the above equations $\nabla$ denotes the covariant derivative with respect to both the Levi–Civita and the $U(1)$ holomorphic connection. In the case of $W_{ijk}$, the $U(1)$ weight is $p = 2$.

On an $n$-dimensional special Kähler manifold $\mathcal{M}$ one can always introduce an $(n+1)$-dimensional holomorphic vector bundle whose holomorphic sections we denote by $X^\Lambda$ ($\Lambda = 1, \ldots, n+1$), with

$$\partial_{i^*} X^\Lambda = 0 \tag{3.3.17}$$

and a function $F(L)$ which is holomorphic and homogeneous of degree two in the transformed section

$$L^\Lambda(z, \overline{z}) = e^{1/2\mathcal{K}(z,\overline{z})} X^\Lambda(z) \tag{3.3.18}$$

This means that $F(L) = e^{\mathcal{K}(z,\overline{z})} F(X(z))$, so that $F(X)$ is a holomorphic section of $\mathcal{L}^2$.

For the $L^\Lambda(z, \overline{z})$ we have

$$\begin{aligned} \nabla L^\Lambda &= dL^\Lambda + iQL^\Lambda = \nabla_i L^\Lambda dz + \nabla_{i^*} L^\Lambda dz^{i^*} & (3.3.19) \\ \nabla_{i^*} L^\Lambda &= \partial_{i^*} L^\Lambda - \frac{1}{2}\partial_{i^*}\mathcal{K} L^\Lambda = 0 & (3.3.20) \end{aligned}$$

Equation (3.3.20) follows from (3.3.17) and Eq. (3.3.18).

The geometry of the special manifold is completely determined by the sections $\{X^\Lambda\}$ and by the analytic function $F(L)$. If we define

$$F_{\Lambda_1 \ldots \Lambda_n} = \frac{\partial}{\partial L^{\Lambda_1}} \frac{\partial}{\partial L^{\Lambda_2}} \cdots \frac{\partial}{\partial L^{\Lambda_n}} F(L) \tag{3.3.21}$$

and set

$$\begin{aligned} N_{\Lambda\Sigma} &= F_{\Lambda\Sigma} - \overline{F}_{\Lambda\Sigma} & (3.3.22) \\ f_i^\Lambda &= \nabla_i L^\Lambda \equiv \partial_i L^\Lambda + \frac{1}{2}\mathcal{K}_i L^\Lambda \equiv e^{\mathcal{K}/2}(\delta_\Sigma^\Lambda - \frac{X^\Lambda (N\overline{X})_\Sigma}{XN\overline{X}}\partial_i X^\Sigma & (3.3.23) \\ \overline{f}_{i^*}^\Lambda &= \nabla_{i^*} \overline{L}^\Lambda = \partial_{i^*} \overline{L}^\Lambda + \frac{1}{2}\mathcal{K}_{i^*} \overline{L}^\Lambda \equiv e^{\mathcal{K}/2}(\delta_\Sigma^\Lambda - \frac{\overline{X}^\Lambda (NX)_\Sigma}{XN\overline{X}})\partial_{i^*} \overline{X}^\Sigma & (3.3.24) \\ S &= -\frac{1}{2} N_{\Lambda\Sigma} L^\Lambda L^\Sigma & (3.3.25) \end{aligned}$$

we get

$$\begin{aligned} g_{ij^*} &= i f_i^\Lambda \overline{f}_{j^*}^\Sigma N_{\Lambda\Sigma} = \partial_i X^\Lambda \partial_{j^*} \overline{X}^\Sigma \partial_\Lambda \partial_{\overline{\Sigma}} \mathcal{K}(X, \overline{X}) & (3.3.26) \\ e^{\mathcal{K}} W_{ijk} &= \nabla_i \nabla_j \nabla_k S = f_i^\Lambda f_j^\Sigma f_k^\Gamma F_{\Lambda\Sigma\Gamma} = & \\ &= e^{\mathcal{K}} \partial_i X^\Lambda \partial_j X^\Sigma \partial_k X^\Gamma F_{\Lambda\Sigma\Gamma}(X) & (3.3.27) \\ -i N_{\Lambda\Sigma} L^\Lambda \overline{L}^\Sigma &= -i e^{\mathcal{K}} N_{\Lambda\Sigma} X^\Lambda \overline{X}^\Sigma = 1 & (3.3.28) \\ f_i^\Lambda \overline{L}^\Sigma N_{\Lambda\Sigma} &= 0 & (3.3.29) \\ \overline{f}_{i^*}^\Lambda L^\Sigma N_{\Lambda\Sigma} &= 0 & (3.3.30) \\ \mathcal{U}^{\Lambda\Sigma} \equiv g^{ij^*} f_i^\Lambda \overline{f}_{j^*}^\Sigma &= -(N^{-1})^{\Lambda\Sigma} + L^\Lambda \overline{L}^\Sigma & (3.3.31) \\ N_{\Lambda\Sigma} U^{\Sigma\Pi} N_{\Pi\Gamma} &= e^{\mathcal{K}} \partial_\Lambda \partial_{\overline{\Sigma}} \mathcal{K} & (3.3.32) \end{aligned}$$

where $\partial_\Lambda \equiv \frac{\partial}{\partial X^\Lambda}, \partial_{\overline{\Lambda}} \equiv \frac{\partial}{\partial \overline{X}^\Lambda}$.

There are two other possible definitions [244] of a special Kähler manifold which are equivalent to (3.3.1). These definitions show that $\{X^\Lambda, \frac{\partial F}{\partial X^\Sigma}\}$ can be viewed as a section of a flat holomorphic, $Sp(2n+2)$ bundle, whose existence is a characteristic feature of a special Kähler manifold.

**Definition 3.3.2** *A special manifold is an n-dimensional Kähler manifold of a restricted type such that on each path $U_i$ of a good cover there exist complex projective coordinates $X_i^\Lambda$ and a homogeneous, degree two holomorphic function $F(X)_i$, related to the Kähler potential by*

$$\mathcal{K}_i = -log[-i(X_i^\Lambda \overline{\partial}_\Lambda \overline{F}_i - \overline{X}_i^\Lambda \partial_\Lambda F_i)] \tag{3.3.33}$$

*On the intersection of adiacent patches $U_i$ and $U_j$, $\partial_\Lambda F$ and $X^\Lambda$ are related by special coordinate transformations:*

$$\begin{pmatrix} X \\ \partial F \end{pmatrix}_i = e^{f_{ij}} M_{ij} \begin{pmatrix} X \\ \partial F \end{pmatrix}_j \tag{3.3.34}$$

*where the $f_{ij}$ are holomorphic and $M_{ij}$ is a constant element of $Sp(2n+2, \mathbb{R})$. The transition functions are subject to the usual cocycle condition on a triple overlap:*

$$\begin{aligned} e^{f_{ij}+f_{jk}+f_{ki}} &= 1 \\ M_{ij} M_{jk} M_{ki} &= 1 \end{aligned} \tag{3.3.35}$$

This definition refers to a particular coordinate system. A coordinate independent definition is given by

**Definition 3.3.3** *Let $\mathcal{L}$ denote the complex line bundle whose first Chern class equals the Kähler form $K$ of an n-dimensional Kähler manifold $\mathcal{M}$ of a restricted type. Let $\mathcal{H}$ denote a holomorphic $Sp(2n+2, \mathbb{R})$ vector bundle over $\mathcal{M}$ and $-i\langle \mid \rangle$ the compatible hermitean metric on $\mathcal{H}$. $\mathcal{M}$ is a special manifold if, for some choice of $\mathcal{H}$, there exists a holomorphic section $\Omega$ of $\mathcal{H} \otimes \mathcal{L}$ with the property*

$$K = \frac{i}{2\pi} \partial \overline{\partial} log \left( -i \langle \Omega \mid \overline{\Omega} \rangle \right). \tag{3.3.36}$$

Note that the transition functions of a holomorphic $Sp(2n+2, \mathbb{R})$ vector bundle are necessarily constant on each overlap. The compatible hermitean metric can be defined as

$$-i\langle \Omega \mid \overline{\Omega} \rangle = -i\Omega^\dagger \begin{pmatrix} 0 & \mathbb{1} \\ -\mathbb{1} & 0 \end{pmatrix} \Omega \tag{3.3.37}$$

The equivalence of all the definitions of special geometry we give is easily understood by thinking of the section $\Omega$ as being expressed by $\Omega = (X, \partial F)$ in each coordinate patch and by utilizing Eqs. (3.3.17)–(3.3.32).

Recalling Eqs. (3.3.22), (3.3.2) and (3.3.5) and the fact that $F(X)$ is homogeneous of degree two, Eqs. (3.3.36) and (3.3.33) are equivalent to the following formula relating the Kähler potential $\mathcal{K}$ to the norm of the holomorphic section $\Omega$ in the $Sp(2n+2,\mathbb{R})$ flat bundle:

$$\mathcal{K} = -\ln \parallel \Omega \parallel^2 \equiv -\ln(-i\langle\overline{\Omega}|\Omega\rangle) = -\ln(-iN_{\Lambda\Sigma}X^\Lambda\overline{X}^\Lambda) \tag{3.3.38}$$

There is a particular choice of coordinates in the manifold $\mathcal{M}$ for which the expression of $W_{ijk}$ becomes particularly simple. These are the so called *special coordinates*, which are defined by the choice

$$z^i = \frac{X^i}{X^0} \quad i = 1,\ldots,n \tag{3.3.39}$$

If we define

$$\mathcal{F}(z) = (X^0)^{-2}F(X) \tag{3.3.40}$$

then the Kähler potential is expressed by

$$\mathcal{K}(z,\overline{z}) = -\ln i[2(\mathcal{F}-\overline{\mathcal{F}}) - (\partial_i\mathcal{F} + \partial_{i^*}\overline{\mathcal{F}})(z^i - \overline{z}^{i^*})] \tag{3.3.41}$$

and

$$W_{ijk} = \partial_i\partial_j\partial_k\mathcal{F}(z) \tag{3.3.42}$$

Historically, special geometry was introduced in these special coordinates through Eq. (3.3.41), in studying the coupling of vector multiplets to N=2 supergravity. The function $\mathcal{F}(z)$ was called "prepotential".

Let us now consider the *special isometries* of $\mathcal{M}$. They are generated by holomorphic Killing vectors:

$$z^i \rightarrow z^i + \epsilon^\Lambda k^i_\Lambda(z) \tag{3.3.43}$$

$$\overline{z}^{i^*} \rightarrow z^{i^*} + \epsilon^\Lambda k^{i^*}_\Lambda(\overline{z}) \tag{3.3.44}$$

which must be compatible with both the Kähler and the special structure of our manifold.

Let $\mathcal{G}_{spec}$ be the Lie algebra spanned by the special Killing vectors under consideration:

$$\vec{k}_\Lambda = k^i_\Lambda\vec{\partial}_i + k^{i^*}_\Lambda\vec{\partial}_{i^*} \tag{3.3.45}$$

whose commutation relations (3.2.37) define the structure constants $f_{\Lambda\Sigma}^{\;\;\Gamma}$. In a general Kähler manifold the dimension $d = \dim\,\mathcal{G}$ of the isometry algebra has no particular relation with the dimension $n$ of the manifold. In a special Kähler manifold $\mathcal{M}$ one has

$$d_{spec} = n + 1 \tag{3.3.46}$$

where

$$d_{spec} \stackrel{\text{def}}{=} \dim\mathcal{G}_{spec} \tag{3.3.47}$$

and the Killing vectors are in one-to-one correspondence with the components of the holomorphic sections $\{X^\Lambda(z)\}$. It is for this reason that the special Killing vectors have been labelled with the same capital Greek index.

Invariance of the Kähler 2-form under the action of $\mathcal{G}_{spec}$ implies Eq. (3.2.32) (with $X = \vec{k}_\Lambda$). In every coordinate patch we can write Eq. (3.2.33) and

$$k^i_\Lambda(z) = ig^{ij^*}\partial_{j^*}\mathcal{P}_\Lambda(z,\overline{z}) \tag{3.3.48}$$
$$k^{i^*}_\Lambda(\overline{z}) = -ig^{i^*j}\partial_j\mathcal{P}_\Lambda(z,\overline{z}) \tag{3.3.49}$$

where $\mathcal{P}_\Lambda(z,\overline{z})$ is the prepotential (or momentum map) of the special Killing vector.

Now, all the Killing vector properties we have proved in the previous section are true on any Kähler manifold. In the case of a special Kähler manifold there is an additional condition to be imposed on a special $\vec{k}_\Lambda$. We must require that the sections $\{X^\Lambda\}$ of the holomorphic $(n+1)$ bundle should transform in the adjoint representation of $\mathcal{G}_{spec}$:

$$\mathcal{L}_\Lambda X^\Gamma = k^i_\Lambda\partial_i X^\Gamma = -f_{\Lambda\Sigma}{}^\Gamma X^\Sigma \tag{3.3.50}$$

In the sequel we shall require that the Kähler potential be exactly invariant under the Lie algebra $\mathcal{G}$:

$$\mathcal{L}_\Lambda\mathcal{K} \equiv k^i_\Lambda\partial_i\mathcal{K} + k^{i^*}_\Lambda\partial_{i^*}\mathcal{K} = 0 \tag{3.3.51}$$

Under this hypothesis Eq. (3.3.50) is equivalent to

$$\mathcal{L}_\Lambda L^\Gamma \equiv k^i_\Lambda\partial_i L^\Gamma + k^{i^*}_\Lambda\partial_{i^*}L^\Gamma = -f_{\Lambda\Sigma}{}^\Gamma L^\Sigma \tag{3.3.52}$$

Using the definition of $f^\Lambda_i$, from (3.3.52) we get

$$k^i_\Lambda f^\Lambda_i = \frac{1}{2}(k^i_\Lambda\partial_i\mathcal{K} - k^{i^*}_\Lambda\partial_{i^*}\mathcal{K})L^\Lambda - f^\Gamma_{\Lambda\Sigma}L^\Sigma \tag{3.3.53}$$

or, equivalently,

$$k^i_\Lambda f^\Gamma_i = i\mathcal{P}_\Lambda L^\Gamma - f^\Gamma_{\Lambda\Sigma}L^\Sigma \tag{3.3.54}$$

Indeed from Eqs. (3.3.49) and (3.3.51) it follows that the prepotential $\mathcal{P}$ has the following general form:

$$\mathcal{P}_\Lambda = -ik^i_\Lambda\partial\mathcal{K} + C_\Lambda = ik^{i^*}_\Lambda\partial_{i^*}\mathcal{K} + C_\Lambda \tag{3.3.55}$$

where $C_\Lambda$ is a real constant: Eq. (3.3.54) then follows. We can now prove the following relations:

$$k^i_\Lambda L^\Lambda = k^{i^*}_\Lambda\overline{L}^\Lambda = 0 \tag{3.3.56}$$
$$\mathcal{P}_\Lambda L^\Lambda = \overline{\mathcal{P}}_\Lambda\overline{L}^\Lambda = 0 \tag{3.3.57}$$
$$v^\Lambda \equiv f^\Lambda_i k^i_\Sigma\overline{L}^\Sigma = -(v^\Lambda)^* \tag{3.3.58}$$

To prove Eq. (3.3.56) let us contract Eq. (3.3.50) with $X^\Lambda$. We find

$$X^\Lambda\mathcal{L}_\Lambda X^\Gamma = X^\Lambda k^i_\Lambda\partial_i X^\Gamma = 0 \tag{3.3.59}$$

where we have used the anti-symmetry of the structure constants $f_{\Lambda\Sigma}{}^\Gamma$ in the lower indices. Equation (3.3.59) implies

$$X^\Lambda k^i_\Lambda\partial_i X^\Gamma\partial_{j^*}\overline{X}^\Sigma\partial_\Gamma\partial_\Sigma\mathcal{K} = 0 \tag{3.3.60}$$

and, using (3.3.26), we get

$$g_{ij^\star} X^\Lambda k^i_\Lambda = 0 \tag{3.3.61}$$

Since $g_{ij^\star}$ is nondegenerate, we find

$$X^\Lambda k^i_\Lambda = 0 = \overline{X}^\Lambda k^{i^\star}_\Lambda \tag{3.3.62}$$

where the second equality follows by complex conjugation. Equation (3.3.56) is then a consequence of the definition (3.3.18).

To prove Eqs. (3.3.57) we contract both sides of Eq. (3.3.54) with the matrix $N_{\Gamma\Sigma}\overline{L}^\Sigma L^\Lambda$. Then

$$L^\Lambda k^i_\Lambda N_{\Gamma\Sigma}\overline{L}^\Sigma f^\Gamma_i = iL^\Lambda \mathcal{P}_\Lambda N_{\Gamma\Sigma}\overline{L}^\Sigma L^\Gamma \tag{3.3.63}$$

Using the identities (3.3.28) and (3.3.29) we find the first of Eqs. (3.3.57) (the others follow by complex conjugation).

Let us note that by contracting Eqs. (3.3.56) with $\partial_i\mathcal{K}$ and using (3.3.55) we also find

$$(\mathcal{P}_\Lambda - C_\Lambda)L^\Lambda = (\mathcal{P}_\Lambda - C_\Lambda)\overline{L}^\Lambda = 0 \tag{3.3.64}$$

Therefore the constant vector $C_\Lambda$ must in any case be orthogonal to $L^\Lambda, \overline{L}^\Lambda$:

$$C_\Lambda L^\Lambda = C_\Lambda \overline{L}^\Lambda = 0 \tag{3.3.65}$$

in order for Eqs. (3.3.64) to be consistent with Eqs. (3.3.56). Finally, to prove Eq. (3.3.58) we observe that from Eq. (3.3.54) it follows that

$$k^\Lambda_i f^\Gamma_i \overline{L}^\Lambda = i\mathcal{P}_\Lambda L^\Gamma \overline{L}^\Lambda - f^\Gamma_{\Lambda\Sigma} L^\Sigma \overline{L}^\Lambda \tag{3.3.66}$$

Hence, using Eq. (3.3.57), one obtains

$$(v^\Lambda)^\star \equiv k^{i^\star}_\Sigma f^\Lambda_{i^\star} L^\Sigma = -f^\Lambda_{\Sigma\Pi} L^\Sigma \overline{L}^\Pi = -k^i_\Sigma f^\Lambda_i L^\Sigma = -v^\Lambda \tag{3.3.67}$$

This concludes our discussion about the special Killing vectors on special Kähler manifolds.

### 3.3.2 Special geometry and N=2, D=4 supergravity

So far we have given the geometrical definitions of special Kähler manifolds and we have studied their special isometries. In this section we show how this peculiar geometric structure appears in N=2, D=4 supergravity. As in the N=1 case, we are interested in the general ideas and we do not give all the technical details involved in the computations. Basically we describe the coupling of N=2 supegravity to $n$-vector multiplets.

The physical supergravity N=2 multiplet $s = [2, \frac{3}{2}, 1]$ is given by the 1-forms ($V^a$, $\Psi_A$, $A$, $\omega^{ab}$)), where $V^a$ is the vielbein, $\Psi_A$ are two gravitini, $A$ is the graviphoton and $\omega^{ab}$ is the usual spin connection 1-form. The N=2 vector multiplets $s = (1, \frac{1}{2}, 0^+, 0^-)$ contain the fields $(A^i_\mu, \lambda^{iA}, \lambda^{i^*}_A, z^i)$, where $A^i_\mu$ are the $n$-gauge vectors, $\lambda^{iA}, \lambda^{i^*}_A$ are the spin 1/2 components of the gauginos of positive and negative chirality, respectively, and $z^i$ are complex scalar fields that are obtained by the combination (3.2.4) of scalar and pseudoscalar fields. For simplicity we assume an abelian gauge group. This restriction does not influence the geometry of the Kähler manifold spanned by the scalars $z^i$. The Majorana spinors $\Psi_A$ are decomposed into their chiral and anti-chiral components according to

$$\psi_A \equiv \frac{1}{2}(1+\gamma_5)\Psi_A \quad \psi^A \equiv \frac{1}{2}(1-\gamma_5)\Psi_A \tag{3.3.68}$$

Therefore the upper and the lower position of the $SO(2)$ index $A = 1,2$ are related to the chirality of the spinor and this is true also for the gauginos. The manifold $\mathcal{M}$ of the scalar fields $z^i$ is assumed to be a Kähler manifold of complex dimension $n$ and Kähler potential $\mathcal{K}(z^i, z^{i^*})$. The curvature definitions are given by

$$\begin{aligned} R^a &= \mathcal{D}V^a - i\overline{\psi}_A \wedge \gamma^a \psi^A && (3.3.69)\\ \rho_A &= d\psi_A - \frac{1}{4}\gamma_{ab}\omega^{ab} \wedge \psi_A + \frac{i}{2}\mathcal{Q}\wedge\psi_A && (3.3.70)\\ \rho^A &= d\psi^A - \frac{1}{4}\gamma_{ab}\omega^{ab} \wedge \psi^A + -\frac{i}{2}\mathcal{Q}\wedge\psi^A && (3.3.71)\\ R^{ab} &= d\omega^{ab} - \omega^a_c \wedge \omega^{ab} && (3.3.72) \end{aligned}$$

For the vector multiplet we define together with the differentials $dz^i$, $d\overline{z}^{i^*}$ ("curvatures" of $z^i$, $\overline{z}^{i^*}$) the following superspace fieldstrengths:

$$\begin{aligned} \nabla\lambda^{iA} &\equiv d\lambda^{iA} - \frac{1}{4}\gamma_{ab}\omega^{ab}\lambda^{iA} - \frac{i}{2}\mathcal{Q}\lambda^{iA} - \Gamma^i{}_j\lambda^{jA} && (3.3.73)\\ \nabla\lambda^{i^*}_A &\equiv d\lambda^{i^*}_A - \frac{1}{4}\gamma_{ab}\omega^{ab}\lambda^{i^*}_A + \frac{i}{2}\mathcal{Q}\lambda^{i^*}_A - \Gamma^{i^*}{}_{j^*}\lambda^{j^*}_A && (3.3.74)\\ F^\Lambda &\equiv dA^\Lambda + \overline{L}^\Lambda\overline{\psi}_A \wedge \psi_B\epsilon^{AB} + L^\Lambda\overline{\psi}^A \wedge \psi^B\epsilon_{AB} && (3.3.75) \end{aligned}$$

where $\Gamma^i_j, \Gamma^{i^*}_{j^*}$ are defined as in (2.5.18), and where, in $A^\Lambda$, the index $\Lambda$ runs from 0 to $n$, the value 0 being associated with the graviphoton, and the values $1,\ldots,n$ being associated with the vectors of the $n$-vector multiplets. Finally the quantities $L^\Lambda, \overline{L}^\Lambda$ are a priori arbitrary functions of $z^i, z^{i^*}$ on the Kähler manifold, but the Bianchi identities constrain them in such a way thet they coincide with the object defined in Eq. (3.3.18), in terms of which the whole set up of special Kähler geometry can be derived. Note that the $U(1)$ weights of the fermion fields, associated with the connection $Q$, are $[\psi_A] = [\lambda^{i^*}_A] = -[\psi^A] = -[\lambda^{iA}] = 1/2$. It follows that $[L^\Lambda] = -[\overline{L}^\Lambda] = 1$.

From (3.3.69)–(3.3.75) one derives the Bianchi identities:

$$\mathcal{D}R^{ab} = 0 \tag{3.3.76}$$

$$\mathcal{D}R^a + R^{ab} \wedge V_b - i\overline{\psi}^A\gamma_a\rho_A + i\overline{\rho}^A \wedge \gamma_a\psi_A = 0 \tag{3.3.77}$$

$$\nabla \rho_A + \frac{1}{4}\gamma_{ab} R^{ab} \wedge \psi_A - \frac{i}{2}\widehat{K} \wedge \psi_A = 0 \quad (3.3.78)$$

$$\begin{aligned} dF^\Lambda &- \epsilon_{AB} \nabla L^\Lambda \overline{\psi}^A \wedge \psi^B + 2\epsilon_{AB} L^\Lambda \overline{\psi}^A \wedge \rho^B - \epsilon^{AB} \nabla \overline{L}^\Lambda \overline{\psi}_A \wedge \psi_B \\ &+ 2\epsilon^{AB} \overline{L}^\Lambda \overline{\psi}_A \wedge \rho_B = 0 \end{aligned} \quad (3.3.79)$$

$$\nabla^2 \lambda^{iA} + \frac{1}{4} R^{ab} \gamma_{ab} \lambda^{iA} + \frac{i}{2} \widehat{K} \lambda^{iA} + R^i{}_j(\Gamma) \lambda^{jA} = 0 \quad (3.3.80)$$

$$d^2 z^i = d^2 \overline{z}^{i*} = 0 \quad (3.3.81)$$

where $R^j_i = d\Gamma^i_j + \Gamma^i_k \wedge \Gamma^k_j$. Similar equations hold for $\rho^A$, $\nabla \lambda^{i*}_A$, with reversed chirality.

The complete parametrizations of the curvatures for an abelian gauge group are given by

$$\begin{aligned} R^a &= 0 && (3.3.82) \\ \rho_A &= \rho_{A|ab} V^a \wedge V^b + \{A_{A\,|b}{}^B \eta^{ab} + A'_{A\,|b}{}^B \gamma^{ab})\psi_B + \\ &+ [iS_{AB}\eta^{ab} + \epsilon_{AB}(T^+_{ab} - iF^+_{ab})]\gamma^a \psi^B\} \wedge V^b && (3.3.83) \\ \rho^A &= \rho^A_{|ab} V^a \wedge V^b + \{(\overline{A}^A{}_{B|b} \eta^{ab} + \overline{A}'^A{}_{B|b} \gamma^{ab})\psi_B \\ &+ [i\overline{S}^{AB} \eta^{ab} + \epsilon^{AB}(T^-_{ab} + iF^-_{ab})]\gamma^a \psi_B\} \wedge V^b && (3.3.84) \\ R^{ab} &= R^{ab}{}_{cd} V^c \wedge V^d - i(\overline{\psi}_A \theta^{A|ab}_c + \overline{\psi}^A \theta^{ab}_{A|c}) \wedge V^c + \epsilon^{abcf} \overline{\psi}^A \wedge \gamma_f \psi_B (A'^B_{A|c} - \overline{A}'^B{}_{A|c}) \\ &- \overline{\psi}_A \wedge \gamma^{ab} \psi_B \overline{S}^{AB} - \overline{\psi}^A \wedge \gamma^{ab} \psi^B S_{AB} + i\epsilon^{AB} \overline{\psi}_A \wedge \psi_B (T^{ab} + iF^{-ab}) \\ &-i\ \epsilon_{AB} \overline{\psi}^A \wedge \psi^B (T^{ab} - iF^{+ab}) && (3.3.85) \end{aligned}$$

$$\begin{aligned} F^\Lambda &= F^\Lambda_{ab} V^a \wedge V^b + (if^\Lambda_i \overline{\lambda}^{iA} \gamma^a \psi^B \epsilon_{AB} + if^\Lambda_{i*} \overline{\lambda}^{i*}_A \gamma^a \psi_B \epsilon^{AB}) \wedge V_a \\ \nabla \lambda^{iA} &= \nabla_a \lambda^{iA} V^a + iZ^i_a \gamma^a \psi^A + G^{+i}_{ab} \gamma^{ab} \psi_B \epsilon^{AB} + Y^{iAB} \psi_B \\ \nabla \lambda^{i*}_A &= \nabla_a \lambda^{i*}_A V^a + i\overline{Z}^{i*}_a \gamma^a \psi_A + G^{-i}_{ab} \gamma^{ab} \psi^B \epsilon_{AB} + Y^{i*}{}_{AB} \psi^B \\ dz^i &= Z^i_a V^a + \overline{\psi}_A \lambda^{iA} \\ d\overline{z}^{i*} &= Z^{i*}_a V^a + \overline{\psi}^A \lambda^{i*}_A \end{aligned} \quad (3.3.86)$$

$$\begin{aligned} \widehat{K} &\equiv ig_{ij^*} dz^i \wedge d\overline{z}^{j^*} = ig_{ij^*} Z^i_a \overline{Z}^{j^*}_b V^a \wedge V^b + \\ &+ ig_{ij^*}(Z^i_a \overline{\psi}^A \lambda^{j^*}_A + \overline{Z}^{j^*}_a \overline{\psi}_A \lambda^{iA}) \wedge V^a \end{aligned} \quad (3.3.87)$$

where

$$\begin{aligned} A_{A\,|a}{}^B &= -\frac{i}{4} g_{k^*\ell}(\overline{\lambda}^{k^*}_A \gamma_a \lambda^{\ell B} - \delta^B_A \overline{\lambda}^{k^*}_C \gamma_a \lambda^{\ell C}) && (3.3.88) \\ A'^B_{A\,|a} &= \frac{i}{4}(\overline{\lambda}^{k^*}_A \gamma_a \lambda^{\ell B} - \frac{1}{2}\delta^B_A \overline{\lambda}^{k^*}_C \gamma_a \lambda^{C\ell}) && (3.3.89) \end{aligned}$$

$$S_{AB} = \overline{S}^{AB} = 0 \tag{3.3.90}$$

$$\theta^{ab}_{cA} = 2\gamma^{[a}\rho^{b]c}_{A} + \gamma^{c}\rho^{ab}_{A}; \quad \theta^{ab\ A}_{c} = 2\gamma^{[a}\rho^{b]c|A} + \gamma^{c}\rho^{ab|A} \tag{3.3.91}$$

$$T^{+}_{ab} - iF^{+}_{ab} = \frac{i}{4}\overline{L}^{\Sigma}N_{\Lambda\Sigma}\overline{S}^{-1}\left\{F^{+\Lambda}_{ab} + \frac{1}{8}\nabla_{i}f^{\Lambda}_{j}\overline{\lambda}^{iA}\gamma_{ab}\lambda^{jB}\epsilon_{AB}\right\} \tag{3.3.92}$$

$$T^{-}_{ab} + iF^{-}_{ab} = -\frac{i}{4}L^{\Sigma}N_{\Lambda\Sigma}S^{-1}\left\{F^{-\Lambda}_{ab} + \frac{1}{8}\nabla_{i^*}f^{\Lambda}_{j^*}\overline{\lambda}^{i^*}_{A}\gamma_{ab}\lambda^{j^*}_{B}\epsilon^{AB}\right\} \tag{3.3.93}$$

$$G^{i+}_{ab} = \frac{1}{2}g^{ij^*}f^{\Gamma}_{j^*}N_{\Gamma\Sigma}\left(-\delta^{\Sigma}_{\Lambda} - \frac{1}{4}\overline{S}^{-1}N_{\Lambda\Delta}\overline{L}^{\Delta}\overline{L}^{\Sigma}\right) \times \times \left(F^{+\Lambda}_{ab} + \frac{1}{8}\nabla_{k}f^{\Lambda}_{l}\overline{\lambda}^{kA}\gamma_{ab}\lambda^{lB}\epsilon_{AB}\right) \tag{3.3.94}$$

$$G^{i-}_{ab} = \frac{1}{2}g^{i^*j}f^{\Gamma}_{j}N_{\Gamma\Sigma}\left(-\delta^{\Sigma}_{\Lambda} - \frac{1}{4}S^{-1}N_{\Lambda\Delta}L^{\Delta}L^{\Sigma}\right) \times \times \left(F^{-\Lambda}_{ab} + \frac{1}{8}\nabla_{k^*}f^{\Lambda}_{l^*}\overline{\lambda}^{k^*}_{A}\gamma_{ab}\lambda^{l^*}_{B}\epsilon^{AB}\right) \tag{3.3.95}$$

$$Y^{ABi} = g^{ij^*}C_{j^*k^*\ell^*}\overline{\lambda}^{k^*}_{C}\lambda^{\ell^*}_{D}\epsilon^{AC}\epsilon^{BD} \tag{3.3.96}$$

$$Y^{i^*}_{AB} = g^{i^*j}C_{ik\ell}\overline{\lambda}^{kC}\lambda^{\ell D}\epsilon_{AC}\epsilon_{BD} \tag{3.3.97}$$

where $S$ has been defined in Eq. (3.3.25) and the upper indices + and − on the antisymmetric tensor $T_{ab}$, $F^{\Lambda}_{ab}$ and $G^{i}_{ab}$ mean their self-dual and anti-selfdual projections respectively.

The special geometry gadgets $L^{\Lambda}, \overline{L}^{\Lambda}, f^{\Lambda}_{i}, f^{\Lambda}_{i^*}$ and the weight $p = \pm, 2$ tensors $C_{ijk}$, and $C_{i^*j^*k^*}$ turn out to be constrained by consistency of the Bianchi identities as it follows that

$$\nabla_{i^*}L^{\Lambda} = \nabla_{i}\overline{L}^{\Lambda} = 0 \tag{3.3.98}$$

$$f^{\Lambda}_{i} = \nabla_{i}L^{\Lambda}; \quad f^{\Lambda}_{i^*} = \nabla_{i^*}L^{\Lambda} \tag{3.3.99}$$

$$\nabla_{\ell^*}C_{ijk} = \nabla_{\ell}C_{i^*j^*k^*} = 0 \tag{3.3.100}$$

$$\nabla_{[\ell}C_{i]jk} = \nabla_{[\ell^*}C_{i^*]j^*k^*} = 0 \tag{3.3.101}$$

$$-ig^{i\ell^*}f^{\Lambda}_{\ell^*}C_{ijk} = \nabla_{j}f^{\Lambda}_{k} \tag{3.3.102}$$

We do not report the explicit calculation to prove the above equations, but we stress that they are fully determined by the Bianchi identities of N=2 supergravity. The solution for $C_{ijk}$ is best expressed in terms of the degree two homogeneous and holomorphic function $F(L^{\Lambda})$. Using (3.3.26), from (3.3.102) one finds

$$C_{ijk} = -N_{\Lambda\Sigma}f^{\Lambda}_{i}\nabla_{j}f^{\Sigma}_{k} \equiv \nabla_{i}\nabla_{j}\nabla_{k}S \tag{3.3.103}$$

where $N_{\Lambda\Sigma}$ and $S$ are given by (3.3.22) and (3.3.25). Inserting (3.3.103) into (3.3.100) one obtains

$$R_{i^\star jk^\star \ell} = g_{i^\star j} g_{\ell k^\star} + g_{i^\star \ell} g_{jk^\star} - C_{i^\star k^\star s^\star} C_{tj\ell} g^{s^\star t} \tag{3.3.104}$$

which implies that $C_{ijk}(C_{i^\star j^\star k^\star})$ are completely symmetric tensors.

The constraints (3.3.101) are then consequences of the Bianchi identities of the Riemann tensor $R_{i^\star jk^\star \ell}$. Since $C_{ijk}$ and $C_{i^\star j^\star k^\star}$ satisfy (3.3.100) with $U(1)$-Kähler weight $p = \pm 2$ respectively, we may define the holomorphic and anti-holomorphic sections

$$W_{ijk} = e^{\mathcal{K}}\, C_{ijk} \tag{3.3.105}$$
$$W_{i^\star j^\star k^\star} = e^{\mathcal{K}}\, C_{i^\star j^\star k^\star} \tag{3.3.106}$$

in terms of which Eqs. (3.3.100), (3.3.101) and (3.3.104) become identical to Eqs. (3.3.16) defining the set-up of special Kähler geometry.

Eqs. (3.3.103) and (3.3.104) show the full equivalence between the three definitions of special Käher geometry, and in particular the equivalence between the first two definitions.

## 3.4 Bibliographical Note

*The first discovery that the self-interaction of Wess–Zumino multiplets is governed by Kähler geometry is due to Zumino [272] (1979). Independently, the parametrization of the coupling of Wess–Zumino multiplets to supergravity in terms of a real function, later identified with the Kähler potential, was obtained in [95, 94] (1978), shortly after that supergravity had been discovered by Freedman, Ferrara and van Nieuwenhuizen [154] (1976) and recast in first order formalism by Deser and Zumino [113] (1976).*

*The complete form of standard N=1 supergravity, determined by means of the superconformal calculus, was obtained in [93] (1983), while the geometric interpretation of the coupling structure is due to Bagger and Witten [25, 26] (1983).*

*Special Kähler geometry in special coordinates was introduced in 1984–85 by B. de Wit et al. in [111, 109] and E. Cremmer et al. in [96], where the coupling of N=2 vector multiplets to N=2 supergravity was fully determined. The more intrinsic definition of special Kähler geometry in terms of symplectic bundles is due to Strominger [244] (1990), who obtained it in connection with the moduli spaces of Calabi–Yau compactifications. The coordinate-independent description and derivation of special Kähler geometry in the context of N=2 supergravity is due to Castellani, D'Auria, Ferrara [75] and to D'Auria, Ferrara, Fre' [99] (1991).*

- *For a mathematical background on Kähler geometry see: [255], [92] , [222].*

- *For the structure of matter-coupled N=1 supergravity see: [272, 25, 26, 93], [95, 94], [76].*

- *For the concept of momentum map and Kähler quotients see: [158, 191, 60], [132, 206], [46, 6], [45, 262].*
- *For special Kähler geometry in supergravity see: [99, 96, 111, 108, 75], [74, 109, 146, 153, 110], [78, 77, 112, 97], [82, 81, 147, 236].*

# Chapter 4

# COMPACTIFICATIONS ON CALABI–YAU MANIFOLDS

## 4.1 Introduction to Calabi–Yau Compactifications

Calabi–Yau manifolds [61, 268] have been introduced [70] in particle physics as candidate compactified vacua for heterotic string theory. Their proper mathematical definition is the following:

**Definition 4.1.1** *A Calabi–Yau n-fold $\mathcal{M}$ is a compact Kähler manifold of complex dimension $dim_{\mathbb{C}}\,\mathcal{M} = n$ whose first Chern class vanishes: $c_1(\mathcal{M}) = 0$.*

The case $n = 3$ is the one relevant to superstring theory, since a Calabi–Yau 3-fold can be used to represent the six-dimensional internal space when one compactifies from ten to four dimensions. More precisely one finds that:

**Theorem 4.1.1** *A Ricci-flat manifold of $SU(3)$ holonomy with the spin connection embedded in the gauge connection corresponds to an exact solution of the field theory limit of heterotic superstring theory.*

The compactification of heterotic superstring theory on such a class of solutions is characterized by the following desirable property: it preserves N=1 supersymmetry after dimensional reduction to D=4 and admits chiral fermion families [70]. This property is a yield of the $SU(3)$ holonomy, which, on the other hand, follows from the topological characterization $c_1 = 0$. We shall prove Theorem 4.1.1 by first deriving *minimal anomaly-free supergravity* [53, 102], namely the effective ten-dimensional supergravity that incorporates the Green–Schwarz anomaly cancellation mechanism [178], and then showing that its bosonic field equations admit the vacua described in 4.1.1 as exact classical solutions. Actually the complete structure of the *4D low energy effective supergravity lagrangian* one obtains by compactifying ten-dimensional supergravity on $\mathcal{M}_6$ is completely determined by the topology of the latter and in particular by the geometry of its moduli space. A major part of the subsequent chapters will be devoted to the study of these geometries using both algebraic geometric and topological field theoretical methods. In the present chapter we want to introduce Calabi–Yau manifolds and define their essential properties adopting a *Kaluza–Klein viewpoint.* Clearly one can doubt whether a classical solution of the effective field theory lagrangian is a true superstring vacuum, in other words whether the corresponding $\sigma$-model is truly conformal invariant. The answer to this problem is the subject of later chapters where it is shown how to associate a (2,2)-superconformal field theory with, at least, a large class of Calabi–Yau $n$-folds. In the present chapter we keep a purely Kaluza-Klein attitude.

### 4.1.1 D=10, N=1 matter-coupled supergravity

The heterotic string [186, 187] in ten-dimensions has a spectrum of light modes corresponding to the field content of 10D, N=1 supergravity coupled to the vector multiplet of a gauge group $\mathcal{G}$ equal to either $SO(32)/\mathbb{Z}_2$ or $E_8 \times E_8$ [103], [43], [91]. Namely, we have:

1) the vielbein $V^a_\mu$ (a bosonic vector-valued 1-form describing the spin 2 graviton)

2) the gravitino $\psi_\mu$ (a fermionic spinor-valued 1-form describing the spin $\frac{3}{2}$ gravitino)

3) the axion $B_{\mu\nu}$ (a bosonic 2-form with a gauge symmetry describing a particle of corresponding ten-dimensional spin)

4) the dilaton $\sigma$ (a bosonic 0-form describing a scalar particle)

5) the dilatino $\chi$ (a fermionic spinor-valued 0-form describing a spin $\frac{1}{2}$ particle)

6) the spin connection $\omega^{ab}_\mu$ (an auxiliary bosonic 1-form with values in the $SO(1,9)$ adjoint representation, that is eliminated through its own equation of motion)

7) the gauge bosons $A^\alpha_\mu$ (a bosonic 1-form with values in the adjoint representation of the gauge group $\mathcal{G}$ which describes a multiplet of spin 1 particles)

8) the gauginos $\lambda^\alpha$ (a fermionic 0-form with values in the adjoint representation of the gauge group $\mathcal{G}$ that describes a multiplet of spin $\frac{1}{2}$ particles)

where both the Latin and Greek indices are vectorial and run from 0 to 9, the former being flat, the latter curved. The spinorial indices, running from 1 to 32, are not explicitly written. The first six fields of the above list constitute the gravitational multiplet, while the last two correspond to the gauge multiplet. Following the notations of [76] (see Chapter VI.9 which treats D=10 supergravity in full detail), the supersymmetry transformation rules of these fields are as follows:

$$
\begin{aligned}
\delta_{SUSY} V^a_\mu &= \mathrm{i}\overline{\varepsilon}\,\Gamma^a \psi_\mu \\
\delta_{SUSY} \psi_\mu &= \mathcal{D}_\mu \varepsilon + \frac{1}{36}\Gamma_a \Gamma^{pqr} \varepsilon V^a_\mu T_{pqr} \\
\delta_{SUSY} \chi &= -2\mathrm{i}\,\Gamma_a \varepsilon \partial^a \sigma + \mathrm{i}\,\Gamma_{pqr} \varepsilon Z^{pqr} \\
\delta_{SUSY} \sigma &= \frac{1}{4}\overline{\chi}\varepsilon \\
\delta_{SUSY} B_{\mu\nu} &= \frac{1}{6} e^{\frac{4}{3}\sigma} \overline{\chi}\,\Gamma_a \varepsilon V^a_\mu V^b_\nu + \mathrm{i}e^{\frac{4}{3}\sigma} \overline{\varepsilon}\,\Gamma_a \psi_{[\mu} V^a_{\nu]} + \gamma_1 \times \text{corrections}
\end{aligned}
$$

$$\begin{aligned} \delta_{SUSY}\,\lambda &= -\frac{1}{4}\,\Gamma^{pq}\,\varepsilon\,F_{pq} \\ \delta_{SUSY}\,A_\mu &= 2\mathrm{i}\,\overline{\lambda}\,\Gamma_a\,\varepsilon\,V^a_\mu \end{aligned} \tag{4.1.1}$$

where, in the case of ordinary (anomalous) N=1 supergravity, the tensors $T_{pqr}$ and $Z_{pqr}$ are given by

$$\begin{aligned} T_{pqr} &= e^{-\frac{4}{3}\sigma}\left(-3\,H_{pqr} + 4\mathrm{i}\,\overline{\lambda}\,\Gamma_{pqr}\,\lambda\right) \\ Z_{pqr} &= e^{-\frac{4}{3}\sigma}\left(-\frac{1}{2}\,H_{pqr} + \frac{1}{3}\,\mathrm{i}\,\overline{\lambda}\,\Gamma_{pqr}\,\lambda\right) + \frac{\mathrm{i}}{288}\,\overline{\chi}\,\Gamma_{pqr}\,\chi \end{aligned} \tag{4.1.2}$$

in terms of the fermion bilinears and of the *supercovariantized axion field strength*:

$$\begin{aligned} H_{pqr} &= \left(V^{-1}\right)^{\mu}_{p}\left(V^{-1}\right)^{\nu}_{q}\left(V^{-1}\right)^{\rho}_{q}\left[H_{\mu\nu\rho} + e^{\frac{4}{3}\sigma}\left(-\mathrm{i}2\,\overline{\psi}_{[\mu}\Gamma_a\psi_\nu V^a_{\rho]} + \frac{1}{6}\overline{\chi}\Gamma_{ab}\psi_{[\mu}V^a_\nu V^b_{\rho]}\right)\right] \\ H_{\mu\nu\rho} &\stackrel{\text{def}}{=} \partial_{[\mu}\,B_{\nu\rho]} + \beta_1\,\Omega^{(Y.M.)}_{\mu\nu\rho}(A) \end{aligned} \tag{4.1.3}$$

the 3-form $\Omega^{(Y.M.)}(A) = Tr(dA\wedge A + \frac{2}{3}g\; A\wedge A\wedge A)$ being the Yang–Mills Chern–Simons form and $\beta_1$ a coupling parameter that can be fixed to $\beta_1 = -4$ by normalizing the kinetic terms of the gauge bosons in a canonical way. Also $\partial_a\sigma$ denotes the *supercovariantized derivative of the dilaton*,

$$\partial_a\sigma = \left(V^{-1}\right)^{\mu}_{a}\left(\partial_\mu\sigma + \frac{1}{4}\overline{\chi}\psi_\mu\right) \tag{4.1.4}$$

and $\mathcal{D}_\mu\varepsilon = \partial_\mu\varepsilon - \frac{1}{4}\,\omega^{ab}_\mu\,\Gamma^{ab}\,\varepsilon$ is the Lorentz covariant derivative of the SUSY parameter with a spin connection determined by the *torsion equation*:

$$dV^a - \omega^{ab}\,V_b + \frac{\mathrm{i}}{2}\overline{\psi}\Gamma^a\psi = T^{abc}\,V_b\wedge V_c \tag{4.1.5}$$

The parameter $\gamma_1$ appearing in Eq. (4.1.1) expresses the coupling of the Lorentz Chern–Simons term. Supersymmetry leaves its value free which is instead fixed by anomaly cancellation to

$$\gamma_1 = -\frac{1}{32} \tag{4.1.6}$$

Ordinary (anomalous) N=1 supergravity corresponds to the value $\gamma_1 = 0$, and for this specific value Eqs. (4.1.2) hold true. At $\gamma_1 \neq 0$ the values (4.1.2) of the tensors $T_{pqr}$ and $Z_{pqr}$ appearing in the gravitino and dilatino SUSY transformation rules are modified in a way that will be discussed in the next section. At $\gamma_1 = 0$, the lagrangian of ordinary matter-coupled N=1 supergravity is the Bergshoeff–de Roo–de Wit–van Nieuwenhuizen–Chapline–Manton lagrangian, which is reported, for instance, in Eqs. (VI.9.315)–(VI.9.316) of [76]. The fields appearing in that lagrangian are related to those used here by a simple field redefinition given in Eqs. (VI.9.290) and (VI.9.305) of [76]. We do not repeat such formulae here both for brevity and because the theory that admits Calabi–Yau vacua as exact solutions is not ordinary supergravity, but rather anomaly-free supergravity discussed in the next section.

### 4.1.2 Killing spinors and SU(3) holonomy

The notion of Killing spinors in supergravity theories is well established. One considers bosonic field configurations where all the spinorial fields are zero while the bosonic ones are non-trivial. Next one looks for supersymmetry parameters $\varepsilon(x)$ such that in performing a supersymmetry transformation in the given bosonic background with these parameters, the fermionic fields remain identically zero after the transformation. Parameters of this type are named *Killing spinors* since the commutator of two such supersymmetries is an infinitesimal isometry for the bosonic background, namely it is generated by a *Killing vector*. Usually the requirement that a bosonic background admits Killing spinors implies that it also satisfies the field equations, namely that it is a classical vacuum configuration. In the Kaluza–Klein interpretation of supergravity the number $n_{Ks}$ of Killing spinors admitted by a compactified background of the form

$$\mathcal{M}_D = M_4^{(Mink)} \otimes \mathcal{M}_{D-4} \qquad (4.1.7)$$

where $M_4^{(Mink)}$ is four-dimensional Minkowski space and $\mathcal{M}_{D-4}$ is a $(D-4)$-dimensional compact internal manifold, has a special significance. Indeed $n_{Ks} = N$ is the number of local supersymmetries mantained by the effective four-dimensional lagrangian $\mathcal{L}_{eff}$ one obtains by integrating the $D$-dimensional lagrangian $\mathcal{L}_D$ over the compact dimensions of $\mathcal{M}_{D-4}$. Indeed in the presence of a compactified background of type (4.1.7), naming $x$ the coordinates in $M_4^{(Mink)}$ and $y$ the coordinates in $\mathcal{M}_{D-4}$, any Killing spinor $\varepsilon^{\widehat{\alpha}}(x,y)$ (where $\widehat{\alpha}$ are the spinor indices in $D$-dimensions) can be written as follows:

$$\varepsilon^{\widehat{\alpha}}(x,y) = \sum_{I=1}^{N} \varepsilon_I^{\alpha}\, u_I^{t}(y) \qquad (4.1.8)$$

where $\varepsilon_I^{\alpha}$ are constant, anticommuting four-dimensional spinors and $u_I^t(y)$ are commuting $(D-4)$-dimensional spinors satisfying an appropriate equation deduced from the explicit form of the supersymmetry transformation rules. The index $I$ enumerates a basis of solutions of such an equation and in the Kaluza–Klein interpretation of the theory it enumerates the residual $N$ supersymmetries admitted by the compactified vacuum. In the present case, we consider bosonic backgrounds characterized by a vanishing axion field strength:

$$H_{abc} = 0 \qquad (4.1.9)$$

together with vanishing fermionic fields:

$$\psi_\mu = \lambda = \chi = 0 \qquad (4.1.10)$$

and we assume that Eqs. (4.1.9) and (4.1.10) imply

$$T_{abc} = Z_{abc} = 0 \qquad (4.1.11)$$

This is obviously true in ordinary supergravity as a consequence of Eqs. (4.1.2) but, as we are going to see, remains true also in anomaly-free supergravity. In a bosonic background of this sort the supersymmetry transformations of the fermionic fields reduce to

$$\begin{aligned} \delta\psi_\mu &= \mathcal{D}_\mu[\overset{\circ}{\omega}]\,\varepsilon \\ \delta\chi &= 0 \\ \delta\lambda^\alpha &= -\frac{1}{4}\Gamma^{ab}\,\varepsilon\,F^\alpha_{ab} \end{aligned} \tag{4.1.12}$$

Hence a *Killing spinor* $\varepsilon(x)$ is a covariantly constant spinor:

$$\mathcal{D}_\mu[\overset{\circ}{\omega}]\,\varepsilon = 0 \tag{4.1.13}$$

with respect to the Levi–Civita (zero torsion) spin connection $\overset{\circ}{\omega}$ which, in addition, satisfies the condition

$$-\frac{1}{4}\Gamma^{ab}\,\varepsilon\,F^\alpha_{ab} = 0 \tag{4.1.14}$$

Let us introduce the following convention. From now on the ten-dimensional indices ( curved, flat and spinorial) will receive a hat in order to distinguish them from the four-dimensional ones. In addition we set

$$\begin{aligned} \widehat{\mu} &= (\mu = 0,1,2,3) \oplus (\Lambda = 1,\dots,6) \\ \widehat{a} &= (a = 0,1,2,3) \oplus (A = 1,\dots 6) \\ \widehat{\alpha} &= (\alpha = 1,2,3,4) \otimes (t = 1,\dots,8) \end{aligned} \tag{4.1.15}$$

Then replacing the ansatz (4.1.8) into Eqs. (4.1.13) and (4.1.14) we obtain the following conditions on the six-dimensional spinor $u_I(y)$:

$$\begin{aligned} \overset{\circ}{\mathcal{D}}_\Lambda\, u_I(y) &= 0 \\ F_{\Lambda\Sigma}\left(e^{-1}\right)^\Lambda_{\ A}\left(e^{-1}\right)^\Sigma_{\ B}\Gamma^{AB}\,u_I &= 0 \end{aligned} \tag{4.1.16}$$

where $e^A_\Lambda(y)$ is the vielbein of the internal six-dimensional space, $\overset{\circ}{\mathcal{D}}_\Lambda$ is the covariant derivative associated with the corresponding Levi–Civita connection and $F_{\Lambda\Sigma}$ are the components of the gauge field strength on the internal manifold, finally, $\Gamma^A$ is a basis of gamma matrices for the $SO(6)$ Clifford algebra. In deriving (4.1.16) we have assumed that in the vacuum the only non-vanishing components of the gauge field strength are the internal ones. This is obligatory to preserve four-dimensional Lorentz invariance. The consequences of these equations are immediate. Iterating for a second time the first of Eqs. (4.1.16) one obtains the integrability condition:

$$\left[\overset{\circ}{\mathcal{D}}_\Lambda\,,\,\overset{\circ}{\mathcal{D}}_\Sigma\right]u_I(y) = -\frac{1}{4}\overset{\circ}{R}{}^{AB}_{\Lambda\Sigma}\,\Gamma_{AB}\,u_I = 0 \tag{4.1.17}$$

Equation (4.1.17) implies that the holonomy group of the Levi–Civita connection should be $Hol = SU(3) \subset SO(6)$ or smaller. The argument is simple. Equation (4.1.17) states that the spinor $u_I(y)$ is annihilated by all those generators of the $SO(6)$ Lie algebra that are of the form $-\frac{1}{4}\overset{o}{R}{}^{AB}_{\Lambda\Sigma}(y)\,\Gamma_{AB}$, the symbol $\overset{o}{R}{}^{AB}_{\Lambda\Sigma}(y)$ denoting the Riemann tensor in the same point where the spinor $u_I(y)$ is evaluated. The linear span of such generators closes onto a subalgebra of $SO(6)$ which is, by definition, the ***holonomy algebra*** of the connection. In order for non-vanishing solutions $u_I \neq 0$ to exist it is necessary that the irreducible **8**-dimensional spinor representation of SO(6) decomposes under *Hol* in such a way as to admit at least a singlet. Each singlet in this decomposition corresponds to an independent solution of Eq. (4.1.17) and might be integrated to an independent solution of Eqs. (4.1.16). If we want N=1 residual supersymmetry we need just one singlet. Now since $SO(6) \approx SU(4)$ and $\mathbf{8} \approx \mathbf{4} \oplus \overline{\mathbf{4}}$, if $Hol = SU(3)$ we have:

$$\mathbf{8} \overset{SU(3)}{\longrightarrow} \mathbf{1} \oplus \overline{\mathbf{1}} \oplus \mathbf{3} \oplus \overline{\mathbf{3}} \tag{4.1.18}$$

By $\mathbf{1} \oplus \overline{\mathbf{1}}$ we mean that we obtain a complex singlet. This singlet is that out of which we can construct a solution of Eqs. (4.1.16) and, as a consequence, the Killing spinor required by a residual N=1 supersymmetry in D=4. Thus $SU(3)$ holonomy is the basic condition to be imposed on the internal six-dimensional manifold. We are not finished, however; we still have to discuss the second of Eq.s (4.1.16). A simple solution is to choose a vanishing field strength

$$F_{\Lambda\Sigma} = 0 \tag{4.1.19}$$

This is what we are forced to do if we are looking for a solution of ordinary (anomalous) N=1 supergravity. Indeed, as we are going to show later, the following theorem is true.

**Theorem 4.1.2** *A manifold $\mathcal{M}_{2n}$ whose complex dimension is $n = \dim_{\mathbb{C}} \mathcal{M}_{2n}$ and which has holonomy $Hol(\mathcal{M}_{2n}) = SU(n)$ is Ricci-flat.*

Hence in the field equations of ordinary supergravity, a vanishing Ricci tensor implies that the stress–energy tensor of the gauge field should also vanish, which is just compatible with Eq. (4.1.19). Also the Bianchi identity of the axion,

$$dH = -\beta_1 \,\mathrm{tr}\,(F \wedge F) \tag{4.1.20}$$

with the choice (4.1.9), namely with $H = 0$, implies

$$\mathrm{tr}\,(F \wedge F) = 0 \quad \longrightarrow \quad F = 0 \tag{4.1.21}$$

which is the same as (4.1.19).

This situation has two disadvatanges:

i) The vacuum we have found is a solution to the wrong field theory limit of superstring theory (anomalous ordinary N=1 supergravity).

ii) If one looks at the spectrum of light modes one does not find chiral families.

In minimal anomaly-free supergravity both disadvantages are corrected. In this theory, which is anomaly-free as the parent string theory, the Bianchi identity (4.1.20) becomes

$$dH - \beta_1 \, Tr(F \wedge F) + \gamma_1 \, R^{ab} \wedge R_{ab} = 0 \tag{4.1.22}$$

so that with the choice (4.1.9), the condition (4.1.21) is replaced by

$$\beta_1 \, Tr(F \wedge F) = \gamma_1 \, R^{ab} \wedge R_{ab} \tag{4.1.23}$$

which is solved by embedding the spin connection into the gauge connection:

$$A^{\alpha}_{\Lambda}(y) = c^{\alpha}_{AB} \, \overset{o}{\omega}{}^{AB}_{\Lambda}(y) \tag{4.1.24}$$

where $c^{\alpha}_{AB}$ is a constant tensor which realizes the embedding of the holonomy subgroup $SU(3) \subset SO(6)$ into the gauge group $\mathcal{G} = E_8 \otimes E_8$. Miracolously this embedding, which solves the improved Bianchi identity (4.1.23), is also consistent with the improved Einstein equation. Indeed, as we are going to see in the next section, in minimal anomaly-free supergravity, the stress–energy tensor contains also quadratic terms in the curvature and vanishes upon use of Eq. (4.1.24). On the other hand, upon use of Eq. (4.1.24), the consistency condition (4.1.17) and the second of Eqs. (4.1.16) become identical. Indeed for a Levi–Civita connection the Riemann tensor has the property

$$R^{AB}_{CD} \stackrel{\text{def}}{=} \left(e^{-1}\right)^{\Lambda}_{C} \left(e^{-1}\right)^{\Sigma}_{D} R^{AB}_{\Lambda\Sigma} = R^{CD}_{AB} \tag{4.1.25}$$

and so

$$\left(e^{-1}\right)^{\Lambda}_{C} \left(e^{-1}\right)^{\Sigma}_{D} F^{\alpha}_{\Lambda\Sigma} \, c^{AB}_{\alpha} \, \Gamma^{CD} = R^{AB}_{CD} \, \Gamma^{CD} = R^{CD}_{AB} \, \Gamma^{CD} \tag{4.1.26}$$

As a bonus, the embedding of the spin connection into the gauge connection allows the emergence of chiral families. This will be seen later.

In conclusion, if we take anomaly-free supergravity as the correct field theory limit of heterotic superstring theory we find that Theorem 4.1.1 is true and can alternatively be viewed as the following proposition.

**Theorem 4.1.3** *Calabi–Yau 3-folds endowed with a Ricci-flat metric and the spin connection embedded in the gauge connection are the exact compactified solutions of the field equations that admit one Killing spinor and therefore lead to N=1 supersymmetry in the corresponding low energy four-dimensional theory.*

### 4.1.3 The plan of this chapter

In view of the previous discussion, the next section, 4.2, is devoted to a short presentation of minimal anomaly-free supergravity and to the proof of Theorems 4.1.1 and 4.1.3. Next in section 4.3 we consider the properties of Calabi–Yau $n$-folds. We show the relation between $SU(n)$-holonomy and the vanishing of the first Chern class. This justifies the topological definition 4.1.1 of Calabi–Yau manifolds. Furthermore we show that the vanishing of the first Chern class is equivalent to the existence of a unique holomorphic $n$-form $\Omega$ on the $n$-fold. Such a differential form will play a crucial role in all the subsequent chapters. Then we discuss the Dolbeault cohomology ring of Calabi–Yau 3-folds and we show that the Hodge diamond (see Chapter 1)

$$\begin{array}{ccccccc} & & & h^{(0,0)} & & & \\ & & h^{(1,0)} & & h^{(0,1)} & & \\ & h^{(2,0)} & & h^{(1,1)} & & h^{(0,2)} & \\ h^{(3,0)} & & h^{(2,1)} & & h^{(1,2)} & & h^{(0,3)} \\ & h^{(3,1)} & & h^{(2,2)} & & h^{(1,3)} & \\ & & h^{(3,2)} & & h^{(2,3)} & & \\ & & & h^{(3,3)} & & & \end{array} \tag{4.1.27}$$

has necessarily the following form:

$$\begin{array}{ccccccc} & & & 1 & & & \\ & & 0 & & 0 & & \\ & 0 & & h^{(1,1)} & & 0 & \\ 1 & & h^{(2,1)} & & h^{(2,1)} & & 1 \\ & 0 & & h^{(1,1)} & & 0 & \\ & & 0 & & 0 & & \\ & & & 1 & & & \end{array} \tag{4.1.28}$$

the free parameters being the Hodge number $h^{(1,1)}$ which enumerates the harmonic $(1,1)$-forms and provides the dimensionality of the space of *Kähler class deformations* and the Hodge number $h^{(2,1)}$ which enumerates the harmonic $(2,1)$-forms and provides, as we are going to see, the dimensionality of the space of *complex structure deformations*. The index of the de Rham complex, namely the Euler characteristic of Calabi–Yau 3-folds is therefore given by

$$\chi_{Euler} = 2\left(h^{(1,1)} - h^{(2,1)}\right) \tag{4.1.29}$$

The even integer $\chi_{Euler}$ has therefore a deep physical interpretation: its absolute value is twice the number of chiral fermion generations in the **27** (or $\overline{\mathbf{27}}$) of the observable $E_6$ gauge group. The reason is given in section 4.4, where we perform a Kaluza–Klein analysis of the light particle modes in a generic Calabi–Yau background. There we show that to each $(1,1)$ harmonic form we can associate both a neutral Wess–Zumino multiplet (the modulus field of the corresponding Kähler structure deformation) and a charged Wess–Zumino multiplet in the **27** representation of $E_6$, while to each $(2,1)$ harmonic form we associate both a neutral Wess–Zumino multiplet (the modulus field of the given complex-structure deformation) and a charged Wess–Zumino multiplet in the $\overline{\mathbf{27}}$ of $E_6$. This discussion justifies the field content and the structure of the N=1 low energy supergravity discussed in Chapter 3. As was pointed out there the key point in that lagrangian is the special Kähler geometry of both types of moduli fields which is determined by the moduli dependence of the Yukawa couplings. The derivation of moduli field special geometry will be discussed in Chapter 6. In section 4.4 we consider the Yukawa couplings from a Kaluza–Klein point of view. The exact calculations of these couplings in string theory are essentially the topics and the motivation of all the subsequent chapters. Finally in the last chapter we discuss how the numbers $h_{(1,1)}$, $h_{(2,1)}$ are explicitly calculated when the Calabi–Yau 3-fold is described as a projective algebraic surface.

## 4.2 D=10 Anomaly-Free Supergravity

As emphasized in the introduction 4.1, the heterotic string [186, 187] in ten dimensions has a spectrum of light modes corresponding to the field content of 10D N=1 supergravity coupled to the vector multiplet of a gauge group $\mathcal{G}$ equal to either $SO(32)/ZZ_2$ or $E_8 \times E_8$, yet the correct effective lagrangian describing the interactions of these fields is not the standard lagrangian of N=1 10D matter-coupled supergravity [103], [43], [91], whose structure was recalled above.

The reason for this fact is that standard N=1, D=10 supergravity is anomalous [5] while the heterotic string is not. Indeed, in order to cancel the anomaly one has to implement the Green–Schwarz mechanism [178] requiring the following modification of the axion field strength:

$$H_{\mu\nu\rho} = \partial_{[\mu} B_{\nu\rho]} + \beta_1 \, \Omega^{(Y.M.)}_{\mu\nu\rho}(A) \rightarrow \widehat{H}_{\mu\nu\rho} = H_{\mu\nu\rho} - \gamma_1 \, \Omega^{(L)}_{\mu\nu\rho}(\omega) \tag{4.2.1}$$

where $\Omega^{(L)}(\omega)$ is the Lorentz Chern–Simons 3 form. After this replacement the theory is no longer supersymmetric. Restoring supersymmetry, however, must be possible through additional convenient modifications since the parent theory (the heterotic string) is both supersymmetric and anomaly-free. The supersymmetrization of the coupling (4.2.1) was an outstanding problem, attacked by many research groups for a few years. It was finally solved in 1988 in independent work by Bonora–Pasti–Tonin [53], [50], [209] and by D'Auria–Fré–Raciti–Riva [102], [105]. This work was subsequentely completed by Pesando, who derived the complete form of the field equations [227, 228]. In this section we recall the structure of this solution, corresponding to the theory of *minimal anomaly free supergravity*, since, as we have already shown, it is intimately related to the existence of Calabi–Yau compactifications of the heterotic superstring.

### 4.2.1 The role of anomaly-free supergravity in the derivation of Calabi–Yau compactifications

In section 4.1 we defined Calabi–Yau $n$-folds as complex manifolds of the vanishing first Chern class. As already stressed, this is equivalent to the existence of a metric with $SU(n)$-holonomy. From the physical point of view this is the key property for the use of a Calabi–Yau 3-fold as a compactification of the six extra dimensions, since, as we have shown, such is the condition required to obtain a residual N=1 supersymmetry in D=4.

Relying on a celebrated theorem by Yau [268] one knows that every Kähler class on a Calabi–Yau 3-fold contains a Ricci-flat Kähler metric. Suppose that the matter-coupled supergravity theory implementing the Green–Schwarz mechanism contains an R, square term whose contribution to the stress–energy tensor can cancel that of the Yang–Mills field, then one has an exact solution on a Ricci-flat Calabi–Yau manifold with the spin connection embedded in the gauge connection. This is the solution originally proposed in the paper by Candelas et al. [70] that introduced Calabi–Yau manifolds into the physical literature. There the presence of a suitable R square term was conjectured and its coefficient was supposed to be related by supersymmetry to the coefficient of the Lorentz Chern–Simons term ($\gamma_1$). Indeed, calculations of string amplitudes revealed this conjecture to be correct [185]. Later work, both on the $\beta$-function of $\sigma$-models [184] and on string amplitudes [188] revealed that the effective lagrangian of the heterotic superstring contains higher curvature terms that do not vanish on Ricci-flat manifolds and whose coefficient is transcendental. This very fact shows that these terms correspond to independent higher curvature supersymmetry invariants unrelated to the supersymmetrization of the Lorentz Chern–Simons. There was a moment when the existence of these invariants jeopardized the Calabi–Yau vacua of superstring. The danger was immediately removed by Witten, who showed [258] that one can perturbatively readjust the metric so that the $\beta$-function vanishes at each order in $\alpha'$. The non-perturbative proof that Calabi–Yau compactifications do indeed exist, was finally given by Gepner [161], who directly constructed the (2,2) superconformal theories associated with given Calabi–Yau 3-folds. The subject of superstring compactifications on abstract (2,2) superconformal theories was later developed in many directions, using the Landau–Ginzburg approach and other techniques. In particular the identity was established between the chiral ring of the (2,2) theory and either the Hodge ring associated with the variations of the Hodge structure (see section 5.9) or with the quantum corrected Dolbeault cohomology ring of the Calabi–Yau 3-fold (see Chapters 6, 7 and 8). The moduli spaces of the complex structure and Kähler class deformations of the 3-fold were, in this way, identified with the moduli spaces of the (2,2) superconformal theory. We can refer to this as to the *microscopic or world-sheet* viewpoint on the *compactification*. It is the correct approach, which captures the typical stringy phenomena absent in field theory, and most of the subsequent chapters deal with it. On the contrary, in the present chapter we address Calabi–Yau compactifications from the *macroscopic, or Kaluza–Klein* viewpoint. In other words, in the present chapter we consider Calabi–Yau manifolds, with the spin connection embedded into the gauge connection, times 4D Minkowski space, as exact solutions of an effective field theory for the light modes of the heterotic superstring that should capture its essential features, namely:

a) *local supersymmetry*

b) *anomaly-freedom*

Such an effective field theory is *minimal anomaly-free supergravity* (shortened to MAFS), which we describe in this section. The effective 4D lagrangian, whose structure is the ultimate goal of all these studies, is then obtained by integrating the effective

10D lagrangian on the compact six-dimensional manifold. In this respect an important point to be stressed is the following. The structure of the effective D=4 supergravity lagrangian (the one relevant for phenomenology) depends only on the topology, and cohomology of the complex 3-fold, but not on its metric properties.

This shows that the supersymmetrization of the Lorentz Chern–Simons term leads to an effective lagrangian (minimal anomaly-free supergravity) whose solutions represent the universality classes of the correct string vacua. In other words, the metric satisfying the equations of motions of MAFS is not the correct metric fulfilling the complete string equations but it is in the same "universality" class as the good one; meaning, in the case of Calabi–Yau 3-folds, the same Kähler class and the same complex structure. Similar statements hold for the other fields. Furthermore, as we are going to see later, the result one obtains in the present Kaluza–Klein approach is actually correct for the case of complex structure moduli, while it corresponds only to a large radius approximation in the case of Kähler structure moduli. In the last chapter on mirror maps we show how mirror pairs of Calabi–Yau manifolds allow the calculation of the stringy effects on each manifold using the classical Kaluza–Klein results on the mirror partner.

We have gone through this discussion in order to emphasize four points:

*1) MAFS is to be regarded as an effective lagrangian and, as such, unitarity ,renormalizability or finiteness are properties not to be expected (compare this case with the relation between Fermi theory and Weinberg–Salam theory).*

*2) MAFS should admit Ricci-flat manifolds with the spin connection embedded into the gauge connection as exact solutions. As we report here, this is indeed the case.*

*3) The solution of Bianchi identities must involve a freedom apt to accommodate, besides MAFS, more complicated theories containing the higher curvature invariants revealed by string amplitude calculations.*

*4) The improved anomaly-free supergravity (IAFS), which includes all possible higher curvature SUSY invariants with the coefficients dictated by string theory, has to be unitary (ghost-free) Since these are in an infinite number (unlimited numbers of derivatives) IAFS is no longer a local field theory as MAFS is. Although we cannot give a formal proof of this statement, we will argue that a local, unitary, anomaly-free supergravity in D=10 does not exist. Indeed this could happen only if a finite number of higher curvature invariants was sufficient to remove the ghosts.*

### 4.2.2 Strategy to derive anomaly-free supergravity

Having clarified the meaning of the field theory which we want to consider, we go over to its derivation. The works on this subject can be divided into two categories, depending on the approach to the string corrections: those that undertake the perturbative expansion in $\gamma_1$ (the coupling constant of the Chern–Simons form) [233], [189], [38], [160], and those [141], [53], [102], [105], [50] that use a first order formalism in order to have, in an implicit form, the whole series in $\gamma_1$. It was by using this second approach that in [141], [53], [102], [105], [50], [231] the complete solution of the supersymmetrization problem was constructed. What was obtained is a consistent, closed form solution of the Bianchi identities yielding a complete set of transformation rules for all the fields of the theory. As stated, the catch of the method is first order formalism. Indeed the spin connection is solved in terms of the other fields once we know the torsion $T_{abc}$ and this latter is related to the axion field strength $H_{abc}$ by a differential equation [53], [102], [105], [50]. The original proposal of first order formalism as the key to the solution of this problem is due to Ferrara, Fre' and Porrati [141].

Since N=1, D=10 supergravity is only an on-shell theory, the determination of the SUSY transformation rules suffices to determine the complete classical theory. Indeed the field equations should emerge as consistency conditions from the closure of the algebra. This is precisely what happens while one is solving the superspace Bianchi identities. Once these are algebraically solved in the deeper sectors (those with the higher number of fermionic vielbeins), in the remaining upper sectors one finds differential constraints that are exactly identifiable as the expected field equations.

More specifically the logical steps for the derivation of anomaly free supergravity in D=10 are the following ones:

*1st step.* Identification of the underlying free differential algebra, definition of the curvatures and derivation of the corresponding Bianchi identities.

*2nd step.* Enucleation, in the system of Bianchi identities, of the subsystem corresponding to the pure super-Poincaré algebra (no axion field is included) and complete rheonomic solution of this subsystem in terms of the gravitino field strength, the Riemann tensor and of an auxiliary completely antisymmetric torsion $T_{abc}$ subject to the constraint $\mathcal{D}^a T_{abc} = 0$ . This solution was first found by the Atick, Dhar and Ratra in [19]. In [51], Lechner, Pasti and Tonin have shown that, up to field redefinitions this solution is also the most general one, once the canonical SUSY transformation rule of the Vielbein is assumed: $\delta V^a_\mu = i\bar{\epsilon}\gamma^a \psi_\mu$.

*3rd step.* Given the universal rheonomic parametrization of the super-Poincaré curvatures, discussed in *step 2*, one establishes a well-defined cohomology theory for superforms. The $p$-forms on superspace are decomposed into $(r,s)$ sectors according to the number of bosonic ($V^a$) and fermionic ($\psi^\alpha$) intrinsic differentials they contain. This is analogous to the decomposition of exterior forms one deals with on complex manifolds. Next one decomposes the exterior derivative $d$ into $\partial_{r,s}$ operators ($r+s=1$), increasing the degree of a $(p,q)$-form by $(r,s)$. Actually the $\partial_{r,s}$ are sums of two operators: a differential one acting on the form components and an algebraic one (determined by the curvature parametrization) acting on the supervielbein basis. Some of these operators are separately nilpotent and this allows an analysis of superspace cohomology into subcohomologies. This formalism was introduced by Bonora, Pasti and Tonin in [53].

*4th step.* Relying on this formalism the study of the 3-form Bianchi identity can now be addressed. Its most general solution can be exhibited as the sum of the general solution to the homogeneous equation plus a particular solution of the complete inhomogeneous one. In this case the role of source term is played by the 4-forms

$$Q(F) \stackrel{\text{def}}{=} \text{tr}\,(F \wedge F) \qquad Q(R) \stackrel{\text{def}}{=} \text{tr}\,(R \wedge R) \tag{4.2.2}$$

corresponding to the inclusion of the gauge and Lorentz Chern–Simons into the definition of the axion curvature. Studying the homogeneous equation (pure supergravity without Lorentz Chern–Simons) one finds that, up to field redefinitions, the freedom inherent to the general solution is codified in two spinor tensor superfields obeying a complicated spin differential constraint and appearing respectively in the (0,3) and (1,2) sectors of the axion field strength. If one sets $H_{(0,3)} = 0$ all the freedom is fixed and one retrieves standard, pure, N=1, D=10 supergravity with no higher derivative interaction. Hence the aforementioned tensor superfields codify all the possible higher curvature interactions one can introduce in pure supergravity. This result is due to Lechner, Pasti and Tonin [209].

*5th step.* Considering now the inhomogeneous equation, the crucial ingredient for obtaining a solution is provided by a cohomological theorem due to Bonora, Pasti and Tonin (BPT theorem) [53]. One notices that the very existence of standard N=1, D=10 supergravity is due to the particular structure of the 4-form $Q(F)$. One has $Q(F)_{(0,4)} = Q(F)_{(1,3)} = 0$ and the sector $Q(F)_{(2,2)}$ contains only a **120** representation, namely the antisymmetric tensor $tr(\overline{\lambda}\,\gamma_{abc}\,\lambda)$. This is important in order to reduce the (2,2) sector of the Bianchi identity to an equation determining the auxiliary torsion field $T_{abc}$. Such a structure of $Q(F)$ follows from the parametrization of the Yang–Mills curvature $F$, in particular from $F_{(0,2)} = 0$. In the case of the Lorentz curvature we have $R_{(0,2)} \neq 0$ so that $Q(R)$ has not the same structure as $Q(F)$. The BPT theorem, however, states that $Q(R)$ differs from a 4-form $K$ having the same structure as $Q(F)$ by an exact 4-form $dX$ where $X$ is a Lorentz invariant 3-form:

$$\begin{aligned} R^{ab} \wedge R_{ab} &= K + dX \\ K_{(0,4)} &= K_{(1,3)} = 0 \\ K_{(2,2)} &= i\, W^{(5)}_{abc}\, \overline{\psi} \wedge \gamma^{abcmn}\, \psi \wedge V_m \wedge V_n \end{aligned} \tag{4.2.3}$$

The relevance of this result is the following. The 3-form $X$ can be added to the 3-form $\widehat{H}$ which solves the homogeneous Bianchi identity and the sum is a rheonomic parametrization satisfying the inhomogeneous Bianchi identity. The 3-index tensor $W^{(5)}_{abc}$ together with that contained in $X_{(3,0)}$ determines the corrections to the torsion equation due to the Lorentz Chern–Simons.

The 3-form X is uniquely determined if one fixes $X_{(0,3)} = 0$. This choice picks up a particular solution of the inhomogeneous equation to which we should still add the general solution of the homogeneous one. If we specialize the general solution to the one corresponding to pure supergravity $\widehat{H}_{(0,3)} = 0$ we obtain a theory that in the limit $\gamma_1 = 0$, reduces to the Bergshoeff–de Roo–de Wit–van Nieuwenhuizen–Chapline–Manton theory. In other words the overall constraint $H_{(0,3)}$ amounts to superymmetrization of the single Lorentz Chern–Simons term with no other higher curvature SUSY invariant included. The resulting theory is named minimal anomaly-free supergravity (MAFS) and $H_{(0,3)} = 0$ is named the minimality constraint.

*6th step* Explicit construction of the 3-form $X$. The evaluation of the $X_{(1,2)}$ and $X_{(2,1)}$ sectors is straightforward but extremely long and tedious. The evaluation of $K_{(2,2)}$ and $X_{(3,0)}$ is also straightforward, but already poses a rather formidable task to man-powered computing. It was independently done by two groups of people including Bonora, Bregola, D' Auria, Fre', Lechner, Pasti, Raciti, Riva and Tonin [102], [105], [50]. In this way all the SUSY transformation rules for MAFS were determined and the torsion equation was worked out. In particular in $X_{(1,2)}$ one finds a term in the **210** representation of $SO(1,9)$ obtained by antisymmetrizing the derivative of the torsion $T$. The coefficient of this term is fixed by the (1,3) sector of the $H$-Bianchi and it is non-zero. It is this term that it is responsible for the appearance in MAFS of a ghost multiplet propagating at the Planck mass.

*7th step.* Explication of the field equations. A corollary to the BPT theorem states that the (3,1) sector of the 4-form $K$ contains only the **560**, **144** and **16** representations, the **1200** irrep, a priori possible, cancelling identically. Furthermore the **560** of $K_{(3,1)}$ cancels against the **560** contained in $d\widehat{H}_{(3,1)}$. The remaining irrepses, **144** and **16**, are the Lorentz Chern–Simons corrections to the gravitino and dilatino equations, respectively. Having the complete fermion field equations, by means of a further spinor derivative one obtains also the bosonic equations of motion. The programme is conceptually very simple but completely beyond the reach of man-powered computing because of its gigantic algebraic extension. As a consequence it was left unaccomplished in the 1987–88 papers.

This last step was completed by Pesando, who, for the purpose, utilized a dedicated symbolic manipulation programme [227], [228]. This symbolic programme is an improved version (LISP rewritten by Pesando) of the SUPERGRAV package due to Castellani [73]. As a by-product all the previous results have also been rechecked, finding total agreement.

In spite of the gigantic structure of the intermediate steps the final form of the field equations is particularly simple and elegant. Furthermore they have the non-trivial property that Ricci-flat, torsionless manifolds with the spin connection embedded in the gauge connection are exact solutions (Calabi–Yau vacua) [227], [228].

The structure of MAFS was described in full detail in the reference book [76] (see Chapter VI.9), but at the time of publication the final form of the field equations, namely the most relevant item in connection with Calabi–Yau compactifications, was not yet available. For this reason, in the present section we recall the main points of the construction and we report the final form of these equations that constitute our starting point for the discussion of compactifications.

### 4.2.3 The free differential algebra (*step 1*)

In what follows we utilize the rheonomic approach to supergravity and its standard conventions. The first step is easily done. First we can immediately write down the canonical curvatures of the purely gravitational sector:

$$\begin{aligned} T^a &= DV^a - \frac{i}{2}\overline{\psi} \wedge \gamma_a \psi \equiv dV^a - \omega^{ab} \wedge V_b - \frac{i}{2}\overline{\psi} \wedge \gamma_a \psi \\ \rho &= D\psi = d\psi - \frac{1}{4}\omega^{ab} \wedge \gamma_{ab}\psi \\ R^{ab} &= d\omega^{ab} - \omega^{ac} \wedge \omega_c{}^{.b} \end{aligned} \tag{4.2.4}$$

where $V^a$ is the vielbein 1-form, $\omega^{ab}$ is the 1-form spin connection and $\psi$ is the Majorana–Weyl 1-form describing the gravitino field:

$$C\overline{\psi}^T = \psi \qquad \gamma_{11}\psi = \psi \tag{4.2.5}$$

Secondly we can write the dilaton, axion and dilatino curvatures:

$$\begin{aligned} D\sigma &= d\sigma \\ H &= dB - \frac{i}{2}e^{\frac{4}{3}\sigma}\overline{\psi} \wedge \gamma_a \psi \wedge V^a + \beta_1 \Omega(A) - \gamma_1 \Omega(\omega) \\ D\chi &= D\chi^\alpha T_\alpha = d\chi - \frac{1}{4}\omega^{ab}\gamma_{ab}\chi - [A, \chi] \end{aligned} \tag{4.2.6}$$

where $B_{\mu\nu}$, $\sigma$ and $\chi$ are, indeed, the axion 2-form, the dilaton and dilatino 0-forms. The dilatino is the supersymmetric partner of $\sigma$ and of the two index photon $B$, satisfying the Majorana–anti-Weyl conditions:

$$C\overline{\chi}^T = \chi \qquad \gamma_{11}\chi = -\chi \tag{4.2.7}$$

Moreover $\Omega(A)$ and $\Omega(\omega)$ are the Chern–Simons 3-forms given by

$$\begin{aligned} \Omega(A) &= Tr(dA \wedge A + \frac{2}{3}g\ A \wedge A \wedge A) \\ \Omega(\omega) &= d\omega^{ab} \wedge \omega_{ab} - \frac{2}{3}\omega^{ab} \wedge \omega_{ac} \wedge \omega^c{}_{.b} \end{aligned} \tag{4.2.8}$$

Notice that the Chern–Simons can be coupled only to the $B$ field strength and that one has

$$\beta_1 = -4 \tag{4.2.9}$$

if one demands a canonical normalization of the gauge kinetic term in the action at $\gamma_1 = 0$ (see for instance [141], or the detailed account given in the book [76]). On the other hand the parameter $\gamma_1$ is dimensionful and it is free, as far as supersymmetry is concerned, but gets fixed by anomaly cancellation.

Thirdly we can write the canonical curvatures of the Yang–Mills multiplets as

$$\begin{aligned} F &= dA - A\wedge A = F^{\alpha}T_{\alpha} = (dA^{\alpha} - \frac{1}{2}f^{\alpha}_{\beta\gamma}\, A^{\beta}\wedge A^{\gamma})T_{\alpha} \\ D\lambda &= D\lambda^{\alpha}T_{\alpha} = d\lambda - \frac{1}{4}\omega^{ab}\gamma_{ab}\lambda - [A,\lambda] \end{aligned} \tag{4.2.10}$$

where $A = A^{\alpha}T_{\alpha}$ is the Lie-algebra-valued 1-form representing the gauge bosons ($T_{\alpha}$ are the anti-hermitean generators in the adjoint of a Lie algebra $\mathcal{G} = E_8 \times E_8$ or $SO(32)$ for heterotic string), and $\lambda$ is the gaugino, i.e. the 0-form Lie-algebra-valued field associated with $A$ and satisfying the Majorana and Weyl conditions:

$$C\overline{\lambda}^{T} = \lambda \qquad \gamma_{11}\lambda = \lambda \tag{4.2.11}$$

Associated with these curvatures are the Bianchi identities

$$\begin{gathered} DT^{a} + R^{ab}\wedge V_{b} - i\,\overline{\psi}\wedge\gamma^{a}\psi \equiv 0 \\ D\rho + \tfrac{1}{4}R^{ab}\wedge\gamma_{ab}\psi \equiv 0 \\ DR^{ab} \equiv 0 \\ dH + \tfrac{2i}{3}e^{\frac{4}{3}\sigma}d\sigma\wedge\overline{\psi}\wedge\gamma_{a}\psi\wedge V^{a} - i\,e^{\frac{4}{3}\sigma}\overline{\psi}\wedge\gamma_{a}\rho\wedge V^{a} \\ +\tfrac{i}{2}e^{\frac{4}{3}\sigma}\overline{\psi}\wedge\gamma_{a}\psi\wedge T^{a} - \beta_{1}\,Tr(F\wedge F) + \gamma_{1}\,R^{ab}\wedge R_{ab} \equiv 0 \\ DF = dF + A\wedge F - F\wedge A \equiv 0 \\ dd\sigma \equiv 0 \\ D\chi + \tfrac{1}{4}R^{ab}\wedge\gamma_{ab}\chi \equiv 0 \\ D\lambda + \tfrac{1}{4}R^{ab}\wedge\gamma_{ab}\lambda + [F,\lambda] \equiv 0 \end{gathered} \tag{4.2.12}$$

In the rheonomic approach to supersymmetric theories, the SUSY transformation rules are encoded in the rheonomic solution of the Bianchi identities. By this, one refers to an explicit parametrization of the curvatures where the outer components are expressed in terms of the inner ones (see [76] for details). When the theory is only on shell, the rheonomic parametrization implies through the Bianchi identities also the field equations.

### 4.2.4 Parametrization of the super-Poincaré curvatures (*step 2*)

One of the key points for constructing MAFS is to realize that one can solve the purely gravitational sector, independently of the other Bianchis; this was originally noticed by Atick, Dhar and Ratra in [19]. The task is accomplished by standard techniques and the result is

$$\begin{aligned} T^{a} &= T^{a}_{\ .bc}\, V^{b}\wedge V^{c} \\ R^{ab} &= R^{ab}_{cd}\, V^{c}\wedge V^{d} + \overline{\psi}\theta^{ab}_{c}\ \wedge V^{c} + \frac{5}{6}i\, T^{abc}\,\overline{\psi}\wedge\gamma_{c}\psi + \frac{1}{36}i\, T_{ijk}\,\overline{\psi}\wedge\gamma^{abijk}\psi \\ \rho &= \rho_{ab}V^{a}\wedge V^{b} + \frac{1}{36}\gamma_{a}\gamma_{ijk}\psi\wedge V^{a}T^{ijk} \end{aligned} \tag{4.2.13}$$

where $T_{abc}$ is a torsion auxiliary field (one imposes $T_{a|bc} = T^{(120)}_{abc}$), $\rho_{ab}$ is the gravitino field strength and $R^{ab}_{cd}$ is the Riemann tensor. Here and in what follows we adopt the convention to denote the irreducible components of a tensor by the number denoting the dimensionality of the corresponding irreducible Lorentz representation. Hence $T^{(120)}_{abc}$ denotes the completely antisymmetric part of the torsion tensor that, being a three-index antisymmetric tensor, corresponds to the irrededucible representation of $SO(1,9)$ of dimension **120**. Also, in solving Bianchi identities one finds the constraints

$$\begin{aligned}
\gamma_{ab}\,\rho^{ab} &= 0 \\
D_m T^{mab} &= 0 \\
\theta_{ab|c} &= 2i\;\gamma_c\rho_{ab} - 3i\;\gamma_{m[ab}\rho_{c]m} \\
R^{ab}_{cd} &= R^{(770)}_{\substack{ab\\cd}} + \frac{1}{2}\delta_{[a}R^{(54)}_{\substack{b]\\ [c\;\;d]}} + D_{[c}T_{d]ab} + 2\;T_{m[ab}T_{cd]m} + \frac{2}{135}\delta^{ab}_{cd}T^2 \qquad (4.2.14)
\end{aligned}$$

Hence the only free parts of the Riemann tensor are the Weyl tensor $R^{(770)}_{\substack{ab\\cd}}$ and the Ricci tensor $R^{(54)}_{\substack{a\\b}}$. One can wonder if by imposing $T_{a|bc} = T^{(120)}_{abc}$ it is restrictive; the answer is no, if we want the vielbein supersymmetric transformation rule in the form $\delta_{SUSY}V^a_\mu = i\bar{\epsilon}\gamma^a\psi_\mu$. This is demonstrated in [51] where it is also shown, by means of field redefinitions that the only ambiguity is the possibility of adding to the torsion parameterization a term of the kind $T^a_{(0,2)} = T^{(1050)}_{\substack{a_1a_2a_3a_4a_5\\a}}\;\overline{\psi}\wedge\gamma_{a_1a_2a_3a_4a_5}\psi$ which would change the SUSY transformation rule of the vielbein.

### 4.2.5 Cohomology of superforms (*step 3*)

Another key point towards the explicit construction of MAFS is the possibility of exploiting superspace subcohomologies of the cohomology generated by $d$ [53]. This is the same idea underlying the splitting $d = \partial + \bar{\partial}$ in complex spaces and the use of Dolbeault cohomology. In our case we have the splitting

$$d = \partial_{(1,0)} + \partial_{(0,1)} + \partial_{(-1,2)} + \partial_{(2,-1)} \qquad (4.2.15)$$

where the $\partial_{(r,s)}$ are defined by their action

$$\partial_{(r,s)}\;\omega_{(p,q)} = (d\omega)_{(p+r,q+s)} \qquad (4.2.16)$$

Moreover we can do the splitting

$$\partial_{(r,s)} = \nabla_{(r,s)} + \mu_{(r,s)} \qquad (4.2.17)$$

where both $\nabla_{(r,s)}$ and $\mu_{(r,s)}$ are derivations, but $\nabla_{(r,s)}$ acts on the anoholomic components of forms while $\mu_{(r,s)}$ acts on the supervielbein basis:

$$\begin{aligned}
&\partial_{(r,s)}(\omega_{a_1..a_p\alpha_1..\alpha_q}\;V^{a_1}..\wedge V^{a_p}\wedge\psi_{\alpha_1}..\wedge\psi_{\alpha_q}) \\
&= (\nabla_{(r,s)}\omega_{a_1..a_p\alpha_1..\alpha_q})\wedge V^{a_1}..\wedge V^{a_p}\wedge\psi_{\alpha_1}..\wedge\psi_{\alpha_q}) \\
&\quad + \omega_{a_1..a_p\alpha_1..\alpha_q}\;\mu_{(r,s)}(V^{a_1}..\wedge V^{a_p}\wedge\psi_{\alpha_1}..\wedge\psi_{\alpha_q}) \qquad (4.2.18)
\end{aligned}$$

In particular we have $\nabla_{(-1,2)} \equiv \nabla_{(2,-1)} = 0$ and $\mu_{(0,1)} = 0$. From the nilpotency of the exterior derivative ($d^2 = 0$), it follows that

$$\begin{aligned} \mu^2_{(-1,2)} = \mu^2_{(2,-1)} &= 0 \\ \left[ \mu_{(2,-1)}\, ,\, \partial_{(1,0)} \right]_+ = \left[ \mu_{(-1,2)}\, ,\, \partial_{(0,1)} \right]_+ &= 0 \\ \nabla^2_{(0,1)} + [\, \mu_{(-1,2)}\, ,\, \partial_{(1,0)}\, ]_+ &= 0 \end{aligned} \tag{4.2.19}$$

For the explicit action of the above operators, we refer the reader to [53], [102] or [76].

### 4.2.6 Discussion of the homogeneous $H$-Bianchi (*step 4*)

In addition to the general solution of the super-Poincaré Bianchi identities, we can also write the general parametrization of the dilaton and Yang–Mills field strengths, defining the dilatino $\chi$ and gaugino $\lambda$ spinor fields, respectively:

$$\begin{aligned} d\sigma &= \partial_a\sigma\, V^a + \frac{1}{4}\overline{\psi}\chi \\ F &= F_{ab}\, V^a \wedge V^b - 2i\, \overline{\lambda}\gamma_a\psi \wedge V^a \end{aligned} \tag{4.2.20}$$

After these parametrizations are introduced, the specification of any possible dynamics is determined by the solutions of the $H$-Bianchi, which we conveniently rewrite as follows:

$$d(\,H\, +\, U\,)\, -\, \beta_1\, Q(F)\, =\, -\,\gamma_1\, Q(R) \tag{4.2.21}$$

where $U\, =\, \frac{i}{2}\, e^{\frac{4}{3}\sigma}\, \overline{\psi} \wedge \gamma^a\, \psi \wedge V_a$ and $Q(R)$ was defined in Eq. (4.2.2). The general solution for the parametrization of $H$ is given by $H\, =\, H_g\, +\, H_s$, where $H_g$ is the general solution of the homogeneous equation obtained by setting $\gamma_1 = 0$ and $H_s$ is some particular solution of the inhomogeneous equation (4.2.21). The homogeneous equation admits the particular solution $H_g\, =\, \widehat{H}$, where

$$\begin{aligned} \widehat{H}_{(3,0)} &= (-\frac{1}{3}e^{\frac{4}{3}\sigma}T_{abc} - \frac{\beta_1}{3}i\, Tr(\overline{\lambda}\gamma_{abc}\lambda))V^a \wedge V^b \wedge V^c \\ \widehat{H}_{(2,1)} &= -\frac{1}{6}e^{\frac{4}{3}\sigma}\, \overline{\chi}\gamma_{ab}\psi \wedge V^a \wedge V^b \\ \widehat{H}_{(2,1)} &= 0 \qquad \widehat{H}_{(0,3)} = 0 \end{aligned} \tag{4.2.22}$$

One can verify that this solution is (up to field redefinitions) uniquely selected, once the constraint $\widehat{H}_{(0,3)} = 0$ is imposed. Choosing the conventional value $\beta_1\, =\, -\,4$ (corresponding to a canonical normalization of the gauge kinetic term), the solution (4.2.22) reproduces Bergshoeff–de Roo–de Wit–van Nieuwenhuizen–Chapline–Manton theory, while for $\beta_1\, =\, 0$ one retrieves pure N=1, D=10 supergravity. In [209] the problem of the general solution to the homogeneous equation (4.2.21) was addressed. Modding out the trivial freedom corresponding to field redefinitions of the axion 2-form $B$ and of the dilaton

0-form $\sigma$, the result of this analysis is the following. The $\widehat{H}_{(0,3)}$ sector is parametrized in terms of a **672** superfield:

$$\widehat{H}_{(0,3)} = \overline{\psi}\, \phi^{672}_{abcde}\, \overline{\psi}\, \gamma^{abcde}\, \psi \tag{4.2.23}$$

whose spinor derivative $\nabla_{(0,1)}\phi^{672}$ should not contain the **2772** representation. On the other hand $\phi^{672}$ should not be the spinor derivative $\nabla_{(0,1)}\pi^{126}$ of some **126** superfield $\pi^{126}_{abcde}$, a situation that would automatically enforce the previous constraint. Clearly these conditions are cohomologically flavoured. Furthermore the $H_{(1,2)}$ sector contains a **210** superfield, the **1440** part of whose spinor derivative must be identified with the **1440** part of the spatial derivative of $\phi^{672}$. The general solution of these superfield constraints codifies all the possible higher curvature interactions but nobody was so far able to construct it. The authors of [209] have shown how the $\zeta(3)\, R^4$ term of [231] can be included via the choice of a convenient $\phi^{672}$. More generally a convenient $\phi^{672}$ must exist that makes the $\beta$-functions vanish to all loops and restores exact conformal invariance of the $\sigma$-model. This $\phi^{672}$ will be an infinite series with an unlimited number of spatial derivatives and will lead to a non-local supergravity theory. We conjecture that this is the only $\phi^{672}$ that can remove the ghosts mentioned above.

### 4.2.7 The BPT-theorem (*step 5*)

Let us just sketch the proof of the Bonora-Pasti-Tonin theorem stated in Eqs. (4.2.3). It goes as follows. First one writes the explicit form of Eqs. (4.2.3) in the sectors (0,4) and (1,3):

$$\begin{aligned} Q_{(0,4)} &= \mu_{(-1,2)}\, X_{(1,2)} + \nabla_{(0,1)}\, X_{(0,3)} \\ Q_{(1,3)} &= \nabla_{(0,1)}\, X_{(1,2)} + \mu_{(-1,2)}\, X_{(2,1)} + (\, \nabla_{(1,0)} + \mu_{(1,0)}\,)\, X_{(0,3)} \end{aligned} \tag{4.2.24}$$

Next from the explicit form of $K_{(2,2)}$,

$$K_{(2,2)} = Q_{(2,2)} - (\, \nabla_{(1,0)} + \mu_{(1,0)}\,)\, X_{(1,2)} - \nabla_{(0,1)}\, X_{(2,1)} \tag{4.2.25}$$

utilizing Eqs. (4.2.24) and the commutation rules (4.2.19), one derives

$$\mu_{(-1,2)}\, K_{(2,2)} = \partial_{(-1,2)} Q_{(2,2)} + \partial_{(0,1)} Q_{(1,3)} + \partial_{(1,0)} Q_{(0,4)} = (dQ)_{(1,4)} = 0 \tag{4.2.26}$$

Hence $K_{(2,2)}$ is a cocycle of the coboundary operator $\mu_{(-1,2)}$. A priori a (2,2)-form contains the following $SO(1,9)$ irrepses: **3696 + 1728 + 126 +120 + 120' + 10**. The kernel of $\mu_{(-1,2)}$ is spanned by the irrepses **120** and **120'**, so that $K_{(2,2)}$ contains only these tensor structures (see [53], [102], [105], [50] for details, or the book [76]). In this way we have proved that the theorem is true in the (2,2) sector (last of Eqs. (4.2.3)) if it is true in the (0,4) and (1,3) sectors. To show it also in these sectors is a matter of explicit construction. Indeed, fixing the gauge condition $X_{(0,3)} = 0$, one can uniquely solve Eqs. (4.2.24) for $X_{(1,2)}$, $X_{(2,1)}$ and $X_{(3,0)}$.

### 4.2.8 Construction of the 3-form $X$ (*step 6*)

The explicit solution of Eqs. (4.2.3) determined in [102], [105], [50] has the form

$$
\begin{aligned}
X_{(0,3)} &= 0 \\
X_{(1,2)} &= \tfrac{14}{9} i\, T^{aij} T_{bij}\, \overline{\psi} \wedge \gamma_a \psi \wedge V^b + \tfrac{2}{27} i\, T_{[a_1 a_2 a_3} T_{a_4 a_5]b}\, \overline{\psi} \wedge \gamma^{a_1 a_2 a_3 a_4 a_5} \psi \wedge V^b \\
&\quad + i\left(\tfrac{1}{9} T_{i[a_1 a_2} T_{a_3 a_4]i} + \tfrac{1}{18} D_{[a_1} T_{a_2 a_3 a_4]}\right) \overline{\psi} \wedge \gamma_{a_1 a_2 a_3 a_4 b} \psi \wedge V^b \\
&\quad + i h_1\, T^2\, \overline{\psi} \wedge \gamma_a \psi \wedge V^a \\
X_{(2,1)} &= V^a \wedge V^b \wedge \overline{\psi}\big(-2\gamma_{i_0} D_{i_0} \rho_{ab} - 2\gamma_{[a} D^{i_2} \rho_{b]i_2} + 2\gamma_{i_2 i_3 [a} D^{i_3} \rho_{b]i_2} \\
&\quad - \tfrac{7}{18} \gamma_{i_2 i_3 i_4 i_5 [a} \rho_{b]i_5} T_{i_2 i_3 i_4} - \left(\tfrac{13}{18} + 12 h_1\right) \gamma_{i_2 i_3 [a} \rho_{b]i_4} T_{i_2 i_3 i_4} - \left(\tfrac{4}{9} + 24 h_1\right) \gamma_{i_2 i_3 [a} \rho^{i_3 i_4} T_{b]i_2 i_4} \\
&\quad + \left(\tfrac{28}{9} - 12 h_1\right) \gamma_{[a} \rho^{i_2 i_3} T_{b]i_2 i_3} + \tfrac{1}{2} \gamma_{i_2 i_3 i_4} \rho_{i_4 [a} T_{b]i_2 i_3} - \left(\tfrac{55}{9} + 24 h_1\right) \gamma_{i_2} \rho_{i_3 [a} T_{b]i_2 i_3} \\
&\quad - \left(\tfrac{1}{9} + 6 h_1\right) \gamma_{a b i_0 i_1 i_2} \rho_{i_2 i_3} T_{i_0 i_1 i_3} - \left(\tfrac{4}{9} + 6 h_1\right) \gamma_{a b i_0} \rho_{i_1 i_2} T_{i_0 i_1 i_2} - \tfrac{1}{6} \gamma_{i_0 i_1 i_2} \rho_{ab} T_{i_0 i_1 i_2} \\
&\quad + \left(\tfrac{2}{9} + 12 h_1\right) \gamma_{i_0} \rho_{i_0 i_1} T_{a b i_1}\big)
\end{aligned}
\tag{4.2.27}
$$

where $h_1$ is a free parameter that might be reabsorbed in the definition of the dilaton $\sigma$. In addition one finds:

$$
\begin{aligned}
W^{(5)}_{abc} &= \tfrac{1}{36} \Box T_{abc} - \left(\tfrac{1}{9} + \tfrac{3}{2} h_1\right) T_{ij[a} R^{ij}_{\;\;..\,bc]} + \left(\tfrac{2}{9} + 3 h_1\right) T_{i[ab} R^{i}_{\;.\,c]} \\
&\quad + \tfrac{1}{36} D_i T_{j[ab} T_{c]}^{\;\cdot\, ij} + \left(\tfrac{1}{4} + 3 h_1\right) T^{ij\cdot}_{\;\;\;[a} D_b T_{c]ij} \\
&\quad - \left(\tfrac{5}{54} + \tfrac{5}{3} h_1\right) \tfrac{1}{5!}\, \epsilon_{abc i_1 i_2 i_3 j_1 j_2 j_3 j_4} T^{i_1 i_2 i_3} D^{j_1} T^{j_2 j_3 j_4} \\
&\quad - \left(\tfrac{5}{18} + \tfrac{35}{6} h_1\right) \tfrac{1}{5!}\, \epsilon_{abc i_1 i_2 i_3 j_1 j_2 j_3 j_4} T^{i_1 i_2 i_3} T^{m j_1 j_2} T^{j_3 j_4 \cdot}_{\;\;\;\;..m} \\
&\quad + \left(\tfrac{5}{9} + 5 h_1\right) T_{lm[a} T^{m\;\cdot\, n}_{\;\;b} T^{\;\cdot\,.\, l}_{c]n} + \left(\tfrac{5}{18} + \tfrac{5}{2} h_1\right) T_{m[ab} T_{c]ij} T^{mij} - \left(\tfrac{1}{108} + \tfrac{7}{36} h_1\right) T_{abc} T^2 \\
&\quad + \left(\tfrac{7}{144} + \tfrac{3}{8} h_1\right) \overline{\rho}_{ij} \gamma_{jabck} \rho_{ki} + \left(\tfrac{7}{144} + \tfrac{3}{8} h_1\right) \overline{\rho}_{ij} \gamma_{abc} \rho_{ij} + \left(-\tfrac{1}{18} + 6 h_1\right) \overline{\rho}_{i[a} \gamma_{ibj} \rho_{c]i} \\
&\quad + \left(\tfrac{13}{72} + \tfrac{3}{4} h_1\right) \overline{\rho}_{ij} \gamma_{j[ab} \rho_{c]i} - \left(\tfrac{7}{72} + \tfrac{39}{4} h_1\right) \overline{\rho}_{i[a} \gamma_b \rho_{c]i} + \left(\tfrac{1}{3} + 18 h_1\right) \overline{\rho}_{ai} \gamma_i \rho_{bc}
\end{aligned}
\tag{4.2.28}
$$

where $W^{(5)}$ was defined in Eq. (4.2.3). Finally $X_{(3,0)}$ is also determined to be

$$
X_{(3,0)} = \frac{2}{3} W^{(1)}_{abc}\, V^a \wedge V^b \wedge V^c \tag{4.2.29}
$$

where $W^{(1)}_{abc}$ has a structure similar to that of $W^{(5)}$. We do not report it here just for brevity. The $H$-Bianchi identity is solved in the (2,2) sector by imposing the conditions

$$
\begin{aligned}
T_{abc} &= -3\, H_{abc}\, e^{-\frac{4}{3}\sigma} + 4i\, e^{-\frac{4}{3}\sigma} Tr(\overline{\lambda} \gamma_{abc} \lambda) - 2\gamma_1\, e^{-\frac{4}{3}\sigma} \left(W^{(1)}_{abc} + 6 W^{(5)}_{abc}\right) \\
Z_{abc} &= \frac{1}{6}\, T_{abc} - \frac{i}{3} e^{-\frac{4}{3}\sigma} Tr(\overline{\lambda} \gamma_{abc} \lambda) + \frac{i}{288}\, \overline{\chi} \gamma_{abc} \chi + 6\gamma_1\, e^{-\frac{4}{3}\sigma}\, W^{(5)}_{abc}
\end{aligned}
\tag{4.2.30}
$$

where $H_{abc}$ is the physical field strength of the axion ($H_{(3,0)} = H_{abc}\, V^a \wedge V^b \wedge V^c$ ), $T_{abc}$ is the torsion and $Z_{abc}$ is the dilatino auxiliary field appearing in the parametrization

$$
D\chi = D_a \chi\, V^a - 2i\, \gamma_a \psi\, \partial^a \sigma + i\, \gamma_{abc} \psi\, Z^{abc} \tag{4.2.31}
$$

The torsion equation (4.2.30) is made explicit by writing the form of the linear combination

$$
\begin{aligned}
W_{abc} &= W^{(1)}_{abc} + 6W^{(5)}_{abc} \\
&= \tfrac{1}{2}\Box T_{abc} + 3\, T_{ij[a}R^{ij}{}_{..bc]} + 3\, T_{i[ab}R^{i}{}_{.c]} + 4\, T_{lm[a}T^{m}{}_{b}{}^{.n}T_{c]n}{}^{.l} - (\tfrac{2}{27} + h_1)T_{abc}T^2 \\
&\quad -\tfrac{1}{2}\overline{\rho}_{ij}\gamma_{jabck}\rho_{ki} + 6\overline{\rho}_{i[a}\gamma_{ibj}\rho_{c]i} - 3\overline{\rho}_{ij}\gamma_{j[ab}\rho_{c]i} - 3\overline{\rho}_{i[a}\gamma_{b}\rho_{c]i} + 9\overline{\rho}_{ai}\gamma_{i}\rho_{bc} \qquad (4.2.32)
\end{aligned}
$$

Looking at Eq. (4.2.33) one sees the appearance of the term $\Box T_{abc}$ . In the later development of the equations of motion, it is just this term that is responsible for the appearance of ghosts propagating at the Planck mass. (This is for instance revealed by a $\Box R_{\mu\nu}$ term in the Einstein equation.) The origin of $\Box T_{abc}$ in Eqs. (4.2.32) is easily spotted at the preceding level. Indeed it originates from the term $\Omega = \frac{i}{18}\, D_{[a_1}T_{a_a a_3 a_4]}\, \overline{\psi}\wedge\gamma_{a_1 a_2 a_3 a_4 b}\psi\wedge V^b$ appearing in $X_{(1,2)}$. This latter is a cocycle of $\mu_{(-1,2)}$ and, as such, it has an arbitrary coefficient, as far as the solution of the first of Eqs. (4.2.24) is concerned (sector (0,4)). However, the coefficient of $\Omega$ is fixed by the next equation in (4.2.24) (sector (1,3)).

### 4.2.9 Field equations of MAFS (*step 7*)

One begins by finding the equations of motions of the gauge multiplet from the Bianchi identity of $F = F^{\alpha}T_{\alpha}$ in the sector (2,1). After extracting the 16 one gets

$$
i\, \gamma^a\nabla_a\lambda - \frac{1}{18}\gamma_{ab}\nabla_{01}F^{ab} - \frac{7}{54}i\, \gamma_{abc}\lambda T^{abc} = 0 \qquad (4.2.33)
$$

Considering the sector $(0,2)$ it is immediate to obtain the equation of motion for the gaugino:

$$
\gamma^a\nabla_a\lambda + \frac{1}{6}\gamma_{abc}\lambda T^{abc} = 0 \qquad (4.2.34)
$$

Taking the spinor derivative of this equation and looking at the coefficient of $\gamma_m\psi$ one obtains the equation of motion for the gauge field:

$$
\nabla_m F^{ma} = -4i\, \overline{\lambda}\gamma^a\lambda + 2i\, \overline{\rho}^{am}\gamma_m\lambda \qquad (4.2.35)
$$

Let us comment on this quite typical procedure: generally one finds the equations of motion for the fermionic fields since they appear as constraints on the outer directions (i.e. the directions of superspace orthogonal to those of space–time). After that, one obtains the bosonic equations taking the spinor derivative of the fermionic equations. This procedure can be followed to obtain the equations of the gravitino and graviton and of the dilatino and the dilaton. On the contrary the equation for the axion $B_{\mu\nu}$ is given by the constraint corresponding to the first of Eqs. (4.2.14), in which one inserts the first of Eqs. (4.2.30). The equations of motion of the gravitino and the dilatino can be extracted from the sector (3,1) of the $H$-Bianchi and correspond to its **144** and **16** part, respectively. The possibility of finding in this sector the equation of motion of the dilatino relies upon the fact that $\rho_{ab}$ has not the **16** part because of the constraint corresponding to the first of Eqs. (4.2.14). The result is

$$\begin{aligned} & i\,(\rho)^{(144)}_a = i\,\gamma^b \rho_{ab} = \\ & +\tfrac{1}{12}(\chi T)^{(144)}_a + \tfrac{1}{6}(D\chi)^{(144)}_a + \tfrac{2}{9}(\chi\partial\sigma)^{(144)}_a + 8i\, e^{-\frac{4}{3}\sigma}(\lambda F)^{(144)}_a - \tfrac{6}{7}\gamma_1\, e^{-\frac{4}{3}\sigma}\zeta_a \\ & \not{D}\chi = -\tfrac{2}{9}(\chi T)^{(16)} - \tfrac{4}{3}(\chi\partial\sigma)^{(16)} - 8i\, e^{-\frac{4}{3}\sigma}(\lambda F)^{(16)} - 2\gamma_1\, e^{-\frac{4}{3}\sigma}\zeta \end{aligned} \tag{4.2.36}$$

where the explicit corrections $\zeta_a$ and $\zeta$ were calculated and presented in [228]. Taking the spinor derivative of these equations one obtains the bosonic field equations for the graviton and the dilaton. The bosonic equations up to Fermi bilinears are given next. They were extracted from the complete result obtained in [228]. We have

$$\begin{aligned} & R^a_b - \tfrac{1}{2}\delta^a_b R = \\ & -\tfrac{4}{15}\delta^a_b T_{ijk}T^{ijk} - \tfrac{2}{3}(D_a\partial_b\sigma - \tfrac{1}{10}\delta^b_a D^i\partial_i\sigma) - \tfrac{8}{9}(\partial_a\sigma\partial_b\sigma - \tfrac{1}{10}\delta^b_a\partial^a\sigma\partial_i\sigma) \\ & +8\, e^{-\frac{4}{3}\sigma}Tr(F_{ai}F^{bi} - \tfrac{1}{10}\delta^b_a F_{ij}F^{ij}) \\ & +\gamma_1 e^{-\frac{4}{3}\sigma}(-\Box\, R^a_b - 4\, R^a_{i_0}\, R^b_{i_0} + 2\, R^{ai_0}_{i_1 i_2} R^{bi_0}_{i_1 i_2} \\ & +(\tfrac{4}{27} + 2\, h_1)R^a_b T^2 + 8\, T_{ai_0i_1}T_{bi_1i_2}T_{i_0i_3i_4}T_{i_2i_3i_4} + 16\, T_{ai_0i_1}T_{bi_2i_3}T_{i_0i_2i_4}T_{i_1i_3i_4} \\ & -4\, T_{ai_0i_1}T_{i_1i_2i_3}D_b T_{i_0i_2i_3} + 4\, T_{ai_0i_1}T_{i_1i_2i_3}D_{i_0}T_{bi_2i_3} + 8\, T_{ai_0i_1}T_{i_1i_2i_3}D_{i_3}T_{bi_0i_2} \\ & -(\tfrac{1}{54} - 2\, h_1)D_a T_{i_0i_1i_2}D_b T_{i_0i_1i_2} + D_a T_{i_0i_1i_2}D_{i_2}T_{bi_0i_1} - \tfrac{1}{2}D_{i_0}T_{ai_1i_2}D_{i_0}T_{bi_1i_2} \\ & -D_{i_0}T_{ai_1i_2}D_{i_2}T_{bi_0i_1} + (\tfrac{4}{27} + 2\, h_1)T_{i_0i_1i_2}D_a D_b T_{i_0i_1i_2})^{(54)} \end{aligned}$$

$$\begin{aligned} & D_a\partial^a\sigma = \\ & -\tfrac{1}{6}T^{abc}T_{abc} - \tfrac{4}{3}\partial_a\sigma\partial^a\sigma + 2\, e^{-\frac{4}{3}\sigma}Tr(F^{ab}F_{ab}) \\ & +\gamma_1 e^{-\frac{4}{3}\sigma}(-R^{i_1}_{i_2}\, R^{i_1}_{i_2} + \tfrac{1}{2}R^{i_0i_1}_{i_2i_3}\, R^{i_0i_1}_{i_2i_3} \\ & -2\, T_{i_0i_1i_2}T_{i_0i_1i_3}T_{i_2i_4i_5}T_{i_3i_4i_5} + 4\, T_{i_0i_1i_2}T_{i_0i_3i_4}T_{i_1i_3i_5}T_{i_2i_4i_5} \\ & +(\tfrac{2}{81} + \tfrac{1}{3}h_1)T^2_{i_0i_1i_2}T^2_{i_3i_4i_5} + 4\, T_{i_0i_1i_2}T_{i_2i_3i_4}D_{i_1}T_{i_0i_3i_4} + \tfrac{1}{2}D_{i_0}T_{i_1i_2i_3}D_{i_3}T_{i_0i_1i_2} \\ & +(-\tfrac{5}{18} + 3h_1)D_{i_0}T^2_{i_1i_2i_3} + (-\tfrac{1}{9} + 3h_1)T_{i_0i_1i_2}D_{i_3}D_{i_3}T_{i_0i_1i_2}. \end{aligned}$$

$$D_m F^{am} = 0$$

$$D_m H^{mab} - \tfrac{4}{3}\,\partial_m\sigma\, H^{mab} + \gamma_1\tfrac{2}{9}i\,(3\, D_m W^{mab} - 4\,\partial_m\sigma W^{mab}) = 0$$

$$T_{abc} =$$

$$
\begin{aligned}
&-3\, H_{abc}\, e^{-\frac{4}{3}\sigma} - 2\gamma_1\, e^{-\frac{4}{3}\sigma}\big(\tfrac{1}{2}\Box T_{abc} + 3\, T_{ij[a}R^{ij}{}_{..bc]} \\
&+3\, T_{i[ab}R^{i}{}_{.c]} + 4\, T_{lm[a}T^{m}{}_{b}{}^{.n}T_{c]n}{}^{.l} - (\tfrac{2}{27} + h_1)T_{abc}T^2\big)
\end{aligned}
$$

$$
dH - \beta_1\, Tr(F \wedge F) + \gamma_1\, R^{ab} \wedge R_{ab} = 0 \tag{4.2.37}
$$

where $W^{mab}$ was defined in Eq. (4.2.32).

### 4.2.10 Calabi–Yau compactifications as exact solutions of minimal anomaly-free supergravity

The bosonic equations (4.2.37) admit as an exact solution a field configuration where:

$$
\begin{aligned}
T_{abc} &= 0 \\
H_{abc} &= 0 \\
\sigma &= \text{const} \quad \Longrightarrow \quad \partial_a \sigma = 0 \\
R^a_b &= 0 \quad \text{(Ricci-flat space)} \\
R^{a_1 a_2}_{b_1 b_2} &= \text{const} \times c^{a_1 a_2}_{\alpha}\, F^{\alpha}_{b_1 b_2} \quad \text{(embedding of the spin connection} \\
&\qquad \text{into the gauge connection)}
\end{aligned} \tag{4.2.38}
$$

where $c^{a_1 a_2}_{\alpha}$ are suitable constants that express the embedding of an $SU(3) \subset SO(6) \subset SO(1,9)$ holonomy group into the gauge group $\mathcal{G} = E_8 \times E_8$. Group-theoretically, this is guaranteed by the existence of the subgroup:

$$
SU(3) \otimes E_6 \subset E_8 \tag{4.2.39}
$$

The configuration (4.2.38) is a solution just because, after putting $R^{(54)} = 0$, $T_{abc} = 0$, $H_{abc} = 0$ and $\partial_a \sigma = 0$ in all the non-identically vanishing equations, there does appear the same combination of $FF$ and $RR$. This is a quite non-trivial fact and follows from supersymmetry.

On the other hand a configuration such as which given in Eqs. (4.2.38) is a Calabi–Yau compactification. Indeed, by choosing the ten-dimensional manifold to be the product of four-dimensional Minkowski space with a Calabi–Yau 3-fold, we get an overall Ricci-flat space whose holonomy is $SU(3)$. Hence we can choose a gauge field equal to the spin connection that just lives in the compact six dimensions and we achieve the situation described by Eqs. (4.2.38). This is the proof of Theorems 4.1.1 and 4.1.3 anticipated in the introductory section.

## 4.3 Properties of Calabi–Yau Manifolds

By definition 4.1.1 a complex manifold $\mathcal{M}$ is a Calabi–Yau $n$-fold if its first Chern class vanishes:

$$
c_1(\mathcal{M}) = 0 \tag{4.3.1}
$$

As a differential form the first Chern class is the cohomology class of the Ricci form:

$$c_1 = \frac{i}{2\pi}[\mathcal{R}] \qquad \mathcal{R} = \frac{i}{2\pi} R_{ij^\star}\, dz^i \wedge dz^{j^\star} \tag{4.3.2}$$

where $R_{ij^\star} = \sum_{k=1}^{n} R^k_{kij^\star}$ is the Ricci tensor. Hence the Calabi–Yau condition $c_1 = 0$ implies that the Ricci form is exact:

$$R_{ij^\star}\, dz^i \wedge dz^{j^\star} = (\partial_i A_{j^\star} - \partial_{j^\star} A_i)\, dz^i \wedge dz^{j^\star} \tag{4.3.3}$$

where

$$\mathcal{A} = A_i dz^i + A_{i^\star} d\overline{z}^{i^\star} \tag{4.3.4}$$

is a globally defined 1-form.

The celebrated *Yau's theorem* [268], which we quote without proof, states the following.

**Theorem 4.3.1** *Let $\mathcal{M}$ be a Calabi–Yau complex n-fold, namely $c_1(\mathcal{M}) = 0$. For each cohomology class $\left[\omega^{(1,1)}\right] \in H^{(1,1)}$ there exists a Kähler metric $g_{ij^\star}$ such that:*

*i) the corresponding Kähler class is $\left[\omega^{(1,1)}\right]$, namely the Kähler 2-form $K = \frac{i}{2\pi} g_{ij^\star}\, dz^i \wedge dz^{j^\star}$ differs from $\omega^{(1,1)}$ by at most an exact form.*

*ii) The Ricci form vanishes $R_{ij^\star}\, dz^i \wedge dz^{j^\star} = 0$.*

In other words, for a Calabi–Yau $n$-fold in each Kähler class we can find a representative metric that is Ricci flat.

## 4.3.1 Ricci-flatness and $SU(n)$ holonomy

Given this fundamental theorem we can consider the relation between Ricci-flatness and holonomy that we have repeatedly invoked. On a complex manifold endowed with a hermitean metric, the Levi-Civita connection $\Gamma^i{}_j = \left\{ {i \atop jk} \right\} dz^k$ is $U(n)$ Lie-algebra-valued. What this really means is that we can introduce a vielbein form, well-adapted to the complex structure:

$$e^a = e^a_i\, dz^i \quad e^{\overline{a}}_{i^\star}\, d\overline{z}^{i^\star} = (e^a)^\star \quad (a = 1, \ldots, n) \tag{4.3.5}$$

such that it obeys the structural equation

$$T^a = de^a - \omega^a{}_b \wedge e^b = 0 \tag{4.3.6}$$

where $\omega^a{}_b = -\overline{\omega}^{\overline{b}}{}_{\overline{a}}$ is a $U(n)$ Lie-algebra-valued spin connection. The Levi-Civita connection is related to the spin connection in the usual way:

$$\begin{aligned} \partial_i\, e^a_j - \Gamma^k_{ij}\, e^a_k - \omega^a_{i\;b}\, e^b_j &= 0 \\ \partial_{i^\star} e^a_j - \omega^a_{i^\star\;b}\, e^b_j &= 0 \end{aligned} \tag{4.3.7}$$

while the metric is related to the vielbein by

$$g_{ij^\star} = e^a_i \, e^{\bar{b}}_{j^\star} \, \eta_{a\bar{b}} \tag{4.3.8}$$

$\eta_{a\bar{b}} = \mathrm{diag}\,(+,+,\ldots +,)$ being the flat hermitean metric left invariant by $U(n)$ transformations. The $U(n)$ Lie-algebra-valued curvature

$$R^a{}_b \stackrel{\mathrm{def}}{=} d\omega^a{}_b - \omega^a{}_c \wedge \omega^c{}_b \tag{4.3.9}$$

and the ordinary Riemann tensor are related in the following standard way:

$$R^a{}_b = R^i_{mn^\star j} \, dz^m \wedge d\bar{z}^{n^\star} \, e^a_i \, e^j_b \tag{4.3.10}$$

Since $e^a_i \, e^j_a = \delta^i_j$, the Ricci 2-form is equal to the $U(1)$ part of the $U(n)$ curvature. Indeed the $U(n)$ Lie algebra is not semi-simple and we can decompose the spin connection in a traceless part $\widehat{\omega}^a{}_b$ with values in the $SU(n)$ Lie algebra plus a trace $U(1)$ part $\mathcal{A}$:

$$\omega^a{}_b = \widehat{\omega}^a{}_b + \frac{1}{n} \, \delta^a_b \, \mathcal{A}_{U(1)} \tag{4.3.11}$$

and we obtain

$$\frac{i}{2\pi} \, R_{ij^\star} \, dzi \wedge dz^{j^\star} = \mathcal{F}_{U(1)} \stackrel{\mathrm{def}}{=} d \, \mathcal{A}_{U(1)} \tag{4.3.12}$$

Hence if the first Chern class vanishes, the curvature of the $U(1)$ connection is an exact (1,1)-form. Namely the gauge field $\mathcal{A}_{U(1)}$ is actually the globally defined 1-form quoted in Eq. (4.3.4). In particular, if we choose the Ricci-flat metric, whose existence is guaranteed by Yau's theorem 4.3.1, this curvature vanishes, namely the holonomy group is $SU(n)$ rather than $U(n)$. In this case the $U(1)$ potential $\mathcal{A}_{U(1)}$ is not only globally defined but also closed.

### 4.3.2 Harmonic forms and spinors

Given a complex $n$-fold, consider a well-adapted basis of gamma matrices:

$$\begin{aligned} \{\Gamma_a \,, \Gamma_b\} &= 0 \quad \{\Gamma_{\bar{a}} \,, \Gamma_{\bar{b}}\} \\ \{\Gamma_a \,, \Gamma_{\bar{b}}\} &= 2\,\eta_{a\bar{b}} \quad (a,b = 1,\ldots,n) \end{aligned} \tag{4.3.13}$$

and turn it into a curved basis by setting

$$\Gamma_i \stackrel{\mathrm{def}}{=} e^a_i \, \Gamma_a \qquad \Gamma_{i^\star} \stackrel{\mathrm{def}}{=} e^{\bar{a}}_{i^\star} \Gamma_{\bar{a}} \quad \Gamma^i = g^{ij^\star} \, \Gamma_{j^\star} \quad \Gamma^{i^\star} = g^{i^\star j} \, \Gamma_j \tag{4.3.14}$$

so that

$$\{\Gamma_i \,, \Gamma_j\} = \{\Gamma_{i^\star} \,, \Gamma_{j^\star}\} = 0 \quad \{\Gamma_i \,, \Gamma_{j^\star}\} = 2\,g_{ij^\star} \tag{4.3.15}$$

Spinor wave functions can now be constructed in the following way. One starts from an eight-component spinor $\zeta$ that obeys the subsidiary conditions

$$\Gamma_{i^\star} \, \zeta = 0 \quad i = 1, \ldots, n \tag{4.3.16}$$

and then constructs a generic spinor $\psi(z,\overline{z})$ by setting

$$\psi(z,\overline{z}) = \omega^{(0,0)}(z,\overline{z})\,\zeta + \omega^{(0,1)}_{i^\star}(z,\overline{z})\,\Gamma^{i^\star}\,\zeta + \ldots + \omega^{(0,n)}_{i_1^\star\ldots i_n^\star}(z,\overline{z})\,\Gamma^{i_1^\star\ldots i_n^\star}\,\zeta \tag{4.3.17}$$

where the coefficients $\omega^{(0,k)}_{i_1^\star\ldots i_k^\star}(z,\overline{z})$ transform under coordinate changes as $(0,k)$-differential forms. The reason for this construction is that the Clifford algebra (4.3.15) can be viewed as an algebra of fermionic harmonic oscillators, for which we can choose a Clifford vacuum that is annihilated by all the fermionic annihilation operators ($\Gamma_{i^\star} \approx \Gamma^i$). This Clifford vacuum is the spinor $\zeta$. On such a vacuum we can build the $2^n$ states spanning an irreducible representation of the algebra, by applying, successively, all the creation operators, namely $\Gamma_i \approx \Gamma^{i^\star}$. If we are interested in spinors of a fixed chirality we can apply on $\zeta$ $\Gamma$-matrix monomials only of a given parity. Suppose that $\zeta$ is left-handed, namely

$$\Gamma_{2n+1}\,\zeta = \zeta \tag{4.3.18}$$

then all the other left-handed spinors can be written as

$$\psi(z,\overline{z})_L = \omega^{(0,0)}(z,\overline{z})\,\zeta + \omega^{(0,2)}_{i^\star j^\star}(z,\overline{z})\,\Gamma^{i^\star j^\star}\,\zeta + \ldots\,\omega^{(0,2r_M)}_{i_1\ldots i_{2r_M}}(z,\overline{z})\,\Gamma^{i_1^\star\ldots i_{2r_M}^\star} \tag{4.3.19}$$

where, by definition, $2r_M$ is the maximal even number smaller than or equal to $n$. The right-handed spinors are instead of the form

$$\psi(z,\overline{z})_R = \omega^{(0,1)}_{i^\star}(z,\overline{z})\,\Gamma^{i^\star}\zeta + \omega^{(0,3)}_{i^\star j^\star k^\star}(z,\overline{z})\,\Gamma^{i^\star j^\star k^\star}\,\zeta + \ldots\,\omega^{(0,2s_M+1)}_{i_1^\star\ldots i_{2s_M+1}^\star}(z,\overline{z})\,\Gamma^{i_1^\star\ldots i_{2s_M+1}^\star} \tag{4.3.20}$$

where $2s_M+1$ is the maximal odd integer smaller than or equal to $n$. Given the representation (4.3.17) of a spinor let us consider the action of the Dirac operator, which we decompose as follows:

$$\nabla\!\!\!/ = \Gamma^i\,\partial_i + \Gamma^{i^\star}\,\partial_{i^\star} \stackrel{\text{def}}{=} \nabla\!\!\!/_+ + \nabla\!\!\!/_- \tag{4.3.21}$$

If the spinor $\zeta$ is also covariantly constant, then a simple calculation shows that the operators $\nabla\!\!\!/_\pm$ map the generic spinor (4.3.17) into a new one of the same form with

$$\begin{aligned}
\nabla\!\!\!/_- &: \omega^{(0,k)} \longrightarrow \overline{\partial}\omega^{(0,k)} \\
\nabla\!\!\!/_+ &: \omega^{(0,k)} \longrightarrow \overline{\partial}^\dagger\omega^{(0,k)}
\end{aligned} \tag{4.3.22}$$

where

$$\begin{aligned}
\overline{\partial}\omega^{(0,k)} &= \partial_{[i_1^\star}\omega^{(0,k)}_{i_2^\star\ldots i_{k+1}^\star]}\,d\overline{z}^{i_1^\star}\wedge\ldots\wedge d\overline{z}^{i_{k+1}^\star} \\
\overline{\partial}^\dagger\omega^{(0,k)} &= g^{ij^\star}\partial_i\omega^{(0,k)}_{j^\star i_1^\star\ldots i_{k_1}^\star}\,d\overline{z}^{i_2^\star}\wedge\ldots\wedge d\overline{z}^{i_{k-1}^\star}
\end{aligned} \tag{4.3.23}$$

Hence the zero-modes of the Dirac operator on a complex manifold are related to its Dolbeault cohomology groups $H^{(0,k)}$. Indeed an efficient way to obtain such zero-modes is provided by replacing the forms $\omega^{(0,k)}$ in Eq. (4.3.17) with holomorphic forms.

### 4.3.3 The covariantly constant spinor

We already saw that the existence of a Killing spinor, namely of a covariantly constant spinor $\zeta$ on the internal complex manifold $\mathcal{M}$, requires its holonomy to be $SU(n)$, which is the same thing as Ricci-flatness. We can reconsider this fact in a direct way by looking at the equation

$$\partial_\Lambda \zeta = 0 \tag{4.3.24}$$

and at its consistency condition:

$$R^{AB}_{CD}\, \Gamma_{AB}\, \zeta = 0 \tag{4.3.25}$$

Multiplying by $\Gamma^C$ we obtain

$$R^{CB}_{CD} \Gamma_B\, \zeta = 0 \tag{4.3.26}$$

which admits solutions if and only if the Ricci tensor $R^{CB}_{CD}$ vanishes.

### 4.3.4 The holomorphic n-form

The main property of Calabi–Yau $n$-folds is given by the following theorem:

**Theorem 4.3.2** *A compact Kähler manifold $\mathcal{M}$ has vanishing first Chern class if and only if it admits a unique non-vanishing holomorphic n-form:*

$$\Omega = \frac{1}{n!}\, \Omega_{i_1 \dots i_n}(z)\, dz^{i} \wedge \dots \wedge dz^{i_n} \tag{4.3.27}$$

*with the following properties:*
*i) $\Omega$ is harmonic*
*ii) The components of $\Omega$ are covariantly constant in the Ricci-flat metric*

From Theorem 4.3.2 immediately follows the following corollary.

**Corollary 4.3.1** *A Calabi–Yau n-fold has a one-dimensional Dolbeault cohomology group $H^{(n,0)}$, namely $h^{(n,0)} = 1$.*

Indeed the unique form $\Omega$, which is defined up to a multiplicative constant, is the generator of $H^{(n,0)}$.

*Proof of Theorem* 4.3.2

I) Let us assume that a holomorphic form $\Omega$ with the stated property does exist and let us show that the Ricci form is exact. For this purpose define the norm:

$$||\Omega||^2 = \frac{1}{n!}\, \Omega_{i_1 \dots i_n}\, \overline{\Omega}^{i_1 \dots i_n} \tag{4.3.28}$$

where the adjoint tensor $\overline{\Omega}^{i_1 \dots i_n}$ is given by

$$\overline{\Omega}^{i_1 \dots i_n} \stackrel{\text{def}}{=} \Omega^\star_{j_1^\star \dots j_n^\star}\, g^{i_1 j_1^\star} \dots g^{i_n j_n^\star} \tag{4.3.29}$$

and observe that in each coordinate patch

$$\Omega_{i_1 \dots i_n} = f(z)\, \varepsilon_{i_1 \dots i_n} \tag{4.3.30}$$

where $f(z)$ is a local holomorphic function that cannot vanish in the patch where it is defined. Hence

$$\overline{\Omega}^{i_1 \dots i_n} \stackrel{\text{def}}{=} \overline{f}(\overline{z})\, (\det g)^{-\frac{1}{2}}\, \varepsilon^{i_1 \dots i_n} \tag{4.3.31}$$

so that, using (4.3.31) in (4.3.28), we get

$$||\Omega||^2 = (\det g)^{-\frac{1}{2}} |f(z)|^2 \longrightarrow (\det g)^{\frac{1}{2}} = \frac{|f(z)|^2}{||\Omega||^2} \tag{4.3.32}$$

Hence inserting (4.3.32) in the standard expression for the Ricci 2-form:

$$\frac{i}{2\pi} \mathcal{R} = \frac{i}{2\pi} R_{ij^\star} dz^i \wedge d\overline{z}^{j^\star} = \frac{i}{2\pi} \partial \overline{\partial} \ln (\det g)^{\frac{1}{2}} \tag{4.3.33}$$

we get

$$\frac{i}{2\pi} \mathcal{R} = -\frac{i}{2\pi} \partial \overline{\partial} \ln ||\Omega||^2 \tag{4.3.34}$$

which shows that $\frac{i}{2\pi}\mathcal{R}$ is globally exact since, by definition, $||\Omega||^2$ is a true coordinate scalar. Indeed the reason why, in the general case, the Ricci 2-form is not globally exact, but only locally exact, is that the square root of the metric determinant $(\det g)^{\frac{1}{2}}$ transforms as a density and not as a scalar.

II) If $c_1(\mathcal{M}) = 0$, by definition the Ricci 2-form is exact. Hence there exists a 1-form:

$$\mathcal{A} = \mathcal{A}^{(1,0)} + \mathcal{A}^{(0,1)} = A_i\, dz^i + A_{i^\star}\, d\overline{z}^{i^\star} \tag{4.3.35}$$

such that

$$\begin{aligned} \partial \mathcal{A}^{(1,0)} = \overline{\partial} \mathcal{A}^{(0,1)} &= 0 \\ \overline{\partial} \mathcal{A}^{(1,0)} + \partial \mathcal{A}^{(0,1)} &= \mathrm{i}\mathcal{R} \end{aligned} \tag{4.3.36}$$

We shall shortly show that on a manifold of non-vanishing Euler characteristic the dimensions $h^{(1,0)} = h^{(0,1)}$ of the Dolbeault cohomology groups $H^{(1,0)} = H^{(0,1)}$ are zero. Hence, assuming $\chi_{Euler} \neq 0$, there cannot be harmonic $(1,0)$ or $(0,1)$ forms, so that we necessarily have

$$\mathcal{A}^{(1,0)} = \partial \alpha(z, \overline{z}) \qquad \mathcal{A}^{(0,1)} = \overline{\partial} \overline{\alpha}(z, \overline{z}) \tag{4.3.37}$$

where $\alpha(z, \overline{z})$ is a $(0,0)$-form, namely a globally defined function on the manifold. Then define the following $(n,0)$-form:

$$\begin{aligned} \Omega &= \Omega_{i_1 \dots i_n}\, dz^{i} \wedge \dots \wedge dz^{i^n} \\ \Omega_{i_1 \dots i_n} &= e^{-\mathrm{i}\alpha(z\overline{z})}\, \zeta^T \Gamma_{i_1 \dots i_n}\, \zeta \end{aligned} \tag{4.3.38}$$

where $\zeta$ is a complex spinor obeying the equation:

$$\nabla\zeta - \frac{i}{2}\mathcal{A}\zeta = 0 \quad\Longrightarrow\quad \begin{cases}\nabla_{k^\star}\zeta = \frac{i}{2}\partial_{k^\star}\overline{\alpha}\,\zeta \\ \nabla_k\zeta = \frac{i}{2}\partial_k\alpha\,\zeta\end{cases} \tag{4.3.39}$$

Equation (4.3.39) admits a solution precisely because its integrability condition is Eq. (4.3.36). Indeed, by iterating (4.3.39), we obtain

$$\left(-\frac{1}{4}R^{AB}_{CD}\,\Gamma_{AB} + \frac{i}{2}F_{CD}\right)\zeta = 0 \tag{4.3.40}$$

where $R^{AB}_{CD}$ are the intrinsic components of the Riemann tensor and $F_{CD}$ are the intrinsic components of $F = d\mathcal{A}$. If we multiply by $\Gamma^C$ Eq. (4.3.40) reduces to

$$\left(R^{CB}_{CD}\,\Gamma_B + iF_{BD}\Gamma^B\right)\zeta = 0 \tag{4.3.41}$$

Imposing the auxiliary condition $\Gamma_{\overline{a}}\zeta = 0$, already introduced in Eq. (4.3.16), Eq. (4.3.41) becomes

$$\left(R^b_a\,\Gamma_b + i\,\Gamma^{b^\star}F_{b^\star a}\right)\zeta = 0 \quad\Longrightarrow\quad iF_{b^\star a} = -\,R_{b^\star a} \tag{4.3.42}$$

the last line of (4.3.42) being nothing else than the second of Eqs. (4.3.36) rewritten in intrinsic components. The existence of the spinor $\zeta$ being established, a straightforward calculation shows that, due to Eq. (4.3.39), the components $\Omega_{i_1\ldots i_n}$ of the $(n,0)$-form (4.3.38) satisfy the equation

$$\nabla_{k^\star}\Omega_{i_1\ldots i_n} = 0 \tag{4.3.43}$$

Equation (4.3.43) implies that $\Omega$ is $\overline{\partial}$-closed:

$$\overline{\partial}\Omega = 0 \tag{4.3.44}$$

On the other hand, being a top form of the $(n,0)$-type, $\Omega$ is also $\partial$-closed:

$$\partial\Omega = 0 \tag{4.3.45}$$

and hence $d$-closed. In addition $\Omega$ is also co-closed, as a consequence of Eq. (4.3.43):

$$\nabla^j\,\Omega_{ji_2\ldots i_n} = g^{jk^\star}\nabla_{k^\star}\,\Omega_{ji_2\ldots i_n} = 0 \tag{4.3.46}$$

Therefore $\Omega$ is harmonic, as stated by the theorem we are proving. The components of $\Omega$ are covariantly constant, if in addition to Eq. (4.3.43) we also have

$$\nabla_k\,\Omega_{i_1\ldots i_n} = 0 \tag{4.3.47}$$

In general Eq. (4.3.47) is not true because the holomorphic derivative of the spinor $\zeta$ cannot be compensated by the holomorphic derivative of the prefactor $e^{-i\overline{\alpha}}$ as it happens in the antiholomorphic case. In the Ricci-flat metric, however, where $\alpha(z,\overline{z}) = 0$, Eq.

(4.3.47) becomes true and the holomorphic $(n,0)$-form (4.3.38) has covariantly constant components as stated by the theorem. Finally $\Omega$ never vanishes. Indeed, inserting (4.3.38) in the expression (4.3.28) for the norm we obtain, by Fierz rearrangement,

$$||\Omega||^2 = e^{i(\alpha-\overline{\alpha})}\,\zeta^\dagger\zeta > 0 \tag{4.3.48}$$

To complete the proof it remains for us only to show that the holomorphic $(n,0)$-form we have constructed is unique up to a multiplicative function. For any $(n,0)$-form $\tilde{\Omega}$ we can write

$$\tilde{\Omega} = h(z,\overline{z})\,dz^{i_1}\wedge\ldots.\wedge dz^{i_n}\,\varepsilon_{i_1\ldots i_n} \tag{4.3.49}$$

where $h(z)$ is a non-singular local function. Globally $h(z)$ is not a function; rather, it is a section of the line bundle:

$$\Omega^{(n,0)}(\mathcal{M}) \xrightarrow{\pi} \mathcal{M} \tag{4.3.50}$$

of $(n,0)$-forms. This is a quite important distinction. From one coordinate patch to another, $h(z)$ transforms with a transition function that is the determinant of $\det\frac{\partial z^{i'}}{\partial z^j}$. Correspondingly the curvature 2-form of this line bundle is just the trace of the curvature 2-form of the canonical bundle $T\mathcal{M}$, namely the Ricci 2-form. Hence the first Chern class of the line bundle (4.3.50) is just the first Chern class of the manifold $\mathcal{M}$. Hence for Calabi–Yau manifolds the line bundle (4.3.50) is trivial, having $c_1 = 0$ so that it admits a single global holomorphic section. This is just the uniqueness of the form $\Omega$ we have constructed. The theorem is completely proved.

**QED**

### 4.3.5 The Hodge diamond of Calabi–Yau 3-folds

The Calabi–Yau manifolds of interest to string theory are the 3-folds. In this subsection we want to show that their Hodge diamond has the particular structure mentioned in Eq. (4.1.28). For this purpose we have to introduce some mathematical results that we quote without proof.

On a hermitean complex manifold $\mathcal{M}$, besides the usual Laplacian,

$$\Delta \stackrel{\text{def}}{=} \left(dd^\dagger + d^\dagger d\right) \tag{4.3.51}$$

we can introduce other two laplacians:

$$\begin{aligned} \Delta_\partial &\stackrel{\text{def}}{=} \left(\partial\partial^\dagger + \partial^\dagger\partial\right) \\ \Delta_{\overline{\partial}} &\stackrel{\text{def}}{=} \left(\overline{\partial}\overline{\partial}^\dagger + \overline{\partial}^\dagger\overline{\partial}\right) \end{aligned} \tag{4.3.52}$$

and we can define the corresponding spaces of harmonic forms:

$$\begin{aligned} \mathrm{Harm}_\partial^{(r,s)} &\stackrel{\text{def}}{=} \left\{\omega \in \Omega^{(r,s)}(\mathcal{M})\,|\,\Delta_\partial\omega = 0\right\} \\ \mathrm{Harm}_{\overline{\partial}}^{(r,s)} &\stackrel{\text{def}}{=} \left\{\omega \in \Omega^{(r,s)}(\mathcal{M})\,|\,\Delta_{\overline{\partial}}\omega = 0\right\} \end{aligned} \tag{4.3.53}$$

Then one has the generalization of the Hodge theorem:

**Theorem 4.3.3** *The space $\Omega^{(r,s)}(\mathcal{M})$ of $(r,s)$-forms admits the two Hodge decompositions*

$$\begin{aligned} \Omega^{(r,s)}(\mathcal{M}) &= \bar{\partial}\Omega^{(r,s-1)}(\mathcal{M}) \oplus \bar{\partial}^{\dagger}\Omega^{(r,s+1)} \oplus \mathrm{Harm}_{\bar{\partial}}^{(r,s)} \\ \Omega^{(r,s)}(\mathcal{M}) &= \partial\Omega^{(r-1,s)}(\mathcal{M}) \oplus \bar{\partial}^{\dagger}\Omega^{(r+1,s)} \oplus \mathrm{Harm}_{\partial}^{(r,s)} \end{aligned} \tag{4.3.54}$$

*namely any $(r,s)$-form $\omega$ can be uniquely written as*

$$\omega = \bar{\partial}\alpha + \bar{\partial}^{\dagger}\beta + \gamma \tag{4.3.55}$$

*or as*

$$\omega = \partial\alpha' + \partial^{\dagger}\beta' + \gamma' \tag{4.3.56}$$

The proof is omitted. In a completly analogous way to the de Rham case one can also show that:

**Corollary 4.3.2** *The space of harmonic forms is isomorphic to the corresponding Dolbeault cohomology group, namely*

$$\begin{aligned} \mathrm{Harm}_{\bar{\partial}}^{(r,s)} &\approx H_{\bar{\partial}}^{(r,s)} \\ \mathrm{Harm}_{\partial}^{(r,s)} &\approx H_{\partial}^{(r,s)} \end{aligned} \tag{4.3.57}$$

When the manifold $\mathcal{M}$ is not only hermitean but also kählerian we have an additional property of far-reaching consequences. This is given by the following theorem:

**Theorem 4.3.4** *Let $\mathcal{M}$ be a Kähler manifold; then all the laplacians are proportional, namely*

$$\Delta = 2\,\Delta_{\partial} = 2\,\Delta_{\bar{\partial}} \tag{4.3.58}$$

The proof of this important theorem is once more omitted (see for instance [255]). It has immediate consequences on the Hodge numbers of a Kähler manifold. We have

**Theorem 4.3.5** *Let $\mathcal{M}$ be a Kähler manifold of complex dimension $\dim_{\mathbb{C}}\mathcal{M} = n$. Then the Hodge diamond has the following symmetries:*

*i)* $h^{(r,s)} = h^{(s,r)}$

*ii)* $h^{(r,s)} = h^{(n-r,n-s)}$

*proof:* i) If $\omega \in \Omega^{(r,s)}$ is harmonic it satisfies $\Delta_{\partial}\omega = \Delta_{\bar{\partial}}\omega = 0$. Since by complex conjugation we have

$$\overline{\Delta_{\partial}\,\omega} = \Delta_{\bar{\partial}}\overline{\omega} = \Delta_{\partial}\overline{\omega} \tag{4.3.59}$$

$\overline{\omega}$ is also harmonic. But $\overline{\omega}$ is a $(s,r)$-form. Hence there is a one-to-one correspondence between $(r,s)$ and $(s,r)$ harmonic forms that implies $h^{(r,s)} = h^{(s,r)}$.

ii) Let $\omega \in H_{\bar{\partial}}^{(r,s)}$ and $\tau \in H_{\bar{\partial}}^{(m-r,m-s)}$. We can define a non-singular pairing

$$H_{\bar{\partial}}^{(r,s)} \otimes H_{\bar{\partial}}^{(m-r,m-s)} \longrightarrow \mathbb{C} \tag{4.3.60}$$

via the integral $\int_{\mathcal{M}} \omega \wedge \tau$. Hence, being dual to each other, the vector spaces $H_{\overline{\partial}}^{(r,s)}$ and $H_{\overline{\partial}}^{(m-r,m-s)}$ are isomorphic and have the same dimensions.

QED

Using these results the most general form of the Hodge diamond (4.1.27) for a Kähler 3-fold reduces to

$$\begin{array}{ccccccc} & & & h^{(0,0)} & & & \\ & & h^{(1,0)} & & h^{(1,0)} & & \\ & h^{(2,0)} & & h^{(1,1)} & & h^{(2,0)} & \\ h^{(3,0)} & & h^{(2,1)} & & h^{(2,1)} & & h^{(3,0)} \\ & h^{(2,0)} & & h^{(1,1)} & & h^{(2,0)} & \\ & & h^{(1,0)} & & h^{(0,1)} & & \\ & & & h^{(0,0)} & & & \end{array} \tag{4.3.61}$$

and contains six independent entries.

If we assume the Calabi–Yau condition $c_1(\mathcal{M})$, then the Hodge diamond has a further symmetry that reduces the number of its independent entries. Indeed we have

**Theorem 4.3.6** *The Hodge numbers of a Calabi–Yau n-fold obey the relation*

$$h^{(p,0)} = h^{(0,n-p)} \tag{4.3.62}$$

*Proof:* Given a harmonic $(p,0)$-form $\omega$, whose components we denote by $\omega_{i_1 \dots i_p}$, we can construct a harmonic $(0, n-p)$-form $u$ of components $u_{j_1^\star \dots j_{n-3}^\star}$, via the relation

$$u_{j_1^\star \dots j_{n-3}^\star} = \frac{1}{p!} \prod_{r=1}^{n-p} g_{j_r^\star k_r} \, \overline{\Omega}^{k_1 \dots k_{n-p} i_1 \dots i_p} \, \omega_{i_1 \dots i_p} \tag{4.3.63}$$

where $\overline{\Omega}^{k_1 \dots k_{n-p} i_1 \dots i_p}$ are the components of the adjoint of $\Omega$ defined in Eq. (4.3.29). Equation (4.3.63) is invertible:

$$\omega_{i_1 \dots i_p} = \frac{1}{(n-p)!} \prod_{r=1}^{p} g_{i_r^\star k_r} \frac{\Omega^{k_1^\star \dots k^\star j_1^\star \dots j_{n-p}^\star}}{||\Omega||^2} u_{j_1^\star \dots j_{n-p}^\star} \tag{4.3.64}$$

and defines an isomorphism between the two vector spaces, which that have therefore the same dimension.

QED

Using this property the Hodge diamond of a Calabi–Yau 3-fold reduces to the form

$$\begin{array}{ccccccc} & & & 1 & & & \\ & & h^{(1,0)} & & h^{(1,0)} & & \\ & h^{(1,0)} & & h^{(1,1)} & & h^{(1,0)} & \\ 1 & & h^{(2,1)} & & h^{(2,1)} & & 1 \\ & h^{(1,0)} & & h^{(1,1)} & & h^{(1,0)} & \\ & & h^{(1,0)} & & h^{(1,0)} & & \\ & & & 1 & & & \end{array} \tag{4.3.65}$$

where we have used the information obtained in the previous subsection that $h^{(n,0)} = h^{(3,0)} = 1$, and also the following theorem:

**Theorem 4.3.7** *For compact connected complex manifolds the dimension of $H^{(0,0)}(\mathcal{M})$ is $h^{(0,0)} = 1$, namely the globally defined holomorphic maps*

$$\begin{aligned} h \;:\; \mathcal{M} &\longrightarrow \mathbb{C} \\ \bar{\partial} h &= 0 \end{aligned} \tag{4.3.66}$$

*are the constant maps $h = \text{const.}$*

This is just the generalization to higher dimensions of the Liouville theorem that holds true on the Riemann sphere. We omit the proof (see, for instance, [183]).

Hence the Hodge diamond of a general Calabi–Yau 3-fold has apparently three free entries: $h^{(2,1)}$, $h^{(1,1)}$ and $h^{(1,0)}$. There is, however, a further caveat. The Euler characteristic is given by

$$\begin{aligned} \chi_{Euler} &\stackrel{\text{def}}{=} \sum_{r=0}^{6} (-1)^r \, b^{(r)} = \sum_{r=0}^{6} (-1)^r \left( \sum_{k=0}^{r} h^{(r-k,k)} \right) \\ &= 2 \left( h^{(1,1)} - h^{(2,1)} \right) \end{aligned} \tag{4.3.67}$$

and a general theorem states that

**Theorem 4.3.8** *On a manifold of Euler character $\chi$ a non-singular 1-form has at least $|\chi|$ zeros.*

We omit the proof of Theorem 4.3.8 (see, for instance, [183]). To help the reader's intuition we just recall what happens on Riemann surfaces. There the canonical divisor $D_K$, namely the set of zeros, divided by the set of poles of a (1,0)-form:

$$D_K = z_1 \, z_2 \ldots z_N \, p_1^{-1} \ldots p_M^{-1} \tag{4.3.68}$$

has degree

$$\begin{aligned} \deg D_K &= \#\text{zeros} - \#\text{poles} \\ &= 2g - 2 = -\chi_{Euler} \end{aligned} \tag{4.3.69}$$

so that there happens precisely what Theorem 4.3.8 states. On the Riemann sphere $\chi_{Euler} = 2$ there are no non-singular 1-forms. On the torus $\chi_{Euler} = 0$ there is just one non-singular 1-form that never vanishes $(dz)$, while on genus $g \geq 2$ surfaces there are precisely $g$ holomorphic 1-forms (the abelian differentials) that have exactly $2g-2$ zeros. Indeed the Hodge diamond of a genus $g$ Riemann sphere is

$$\begin{array}{ccc} & 1 & \\ g & & g \\ & 1 & \end{array} \tag{4.3.70}$$

which, by the way, shows that the Calabi–Yau 1-fold are the tori $g = 1$.

Using now Theorem 4.3.8 we can show that

**Theorem 4.3.9** *For a Calabi–Yau 3-fold of non-vanishing Euler character*

$$\chi_{Euler} \neq 0 \quad \longrightarrow \quad h^{(1,1)} \neq h^{(2,1)} \tag{4.3.71}$$

*we necessarily have* $h^{(1,0)} = 0$.

*Proof:* By definition the Hodge number $h^{(1,0)}$ differs from zero if there exist harmonic 1-forms. Let $u_B$ be the intrinsic component of such a harmonic form. They obey the equation

$$\Delta_{Hodge-de-Rham}\, u_B \;=\; -\,\nabla^A \nabla_A u_B \,+\, R_B^{\ C}\, u_C \;=\; 0 \tag{4.3.72}$$

where $R_B^{\ C}$ is the Ricci tensor. On a Calabi–Yau $n$-fold, using the Ricci-flat metric we have

$$\nabla^A \nabla_A u_B \;=\; 0 \tag{4.3.73}$$

and hence, by partial integration,

$$0 \;=\; \int_{\mathcal{M}} g^{\frac{1}{2}} u^B \nabla^A \nabla_A u_B \, d^{2n}x \;=\; -\int_{\mathcal{M}} \nabla^A u^B \nabla_A u_B \, g^{\frac{1}{2}}\, d^{2n}x \;=\; -||\nabla^A u^B||^2 \tag{4.3.74}$$

so that the covariant vector $u_B$ is covariantly constant:

$$\nabla_A u_B \;=\; 0 \tag{4.3.75}$$

On the other hand if a covariantly constant vector has a zero, it vanishes identically; so if $\chi_{Euler} \neq 0$, $u_B = 0$ and no harmonic 1-form exists.

QED

In conclusion the Hodge diamond of a Calabi–Yau 3-folds has at most two free entries. If $\chi_{Euler} \neq 0$, then it has the form (4.1.28) and it is characterized by the two numbers $h^{(1,1)}$ and $h^{(2,1)}$, whose interpretation is that of complex dimensionality of the spaces of Kähler class and complex structure deformations, respectively. This will be discussed in the next section. If instead $\chi_{Euler} = 0$, then the Hodge diamond is characterized by $h^{(1,0)}$ and $h^{(1,1)}$. A typical example of a Calabi–Yau 3-fold with vanishing Euler character is provided by the complex three–torus $T_3$, whose Hodge diamond is the following:

$$\begin{array}{ccccccc} & & & 1 & & & \\ & & 3 & & 3 & & \\ & 3 & & 9 & & 3 & \\ 1 & & 9 & & 9 & & 1 \\ & 3 & & 9 & & 3 & \\ & & 3 & & 3 & & \\ & & & 1 & & & \end{array} \tag{4.3.76}$$

## 4.4 Kaluza–Klein zero-modes and Yukawa Couplings

We have seen that $h^{(1,1)}$ and $h^{(2,1)}$ are the numbers parametrizing the cohomology of a Calabi–Yau 3-fold. We shall shortly consider their geometric interpretation. We are now interested to see how they determine, in a Kaluza–Klein interpretation of 10D supergravity the number of Bose and Fermi zero-modes and hence the field content of the compactified lagrangian. Since we are guaranteed that our compactification is N=1 supersymmetric, we do not have to analyse the spectrum of the bosons and of the femions independently. Once the bosonic or the fermionic spectrum is worked out, the other follows by supersymmetry. We choose to work with the bosons. Hence we just have to look for the Kaluza–Klein zero-modes of the following fields:

$$
\begin{aligned}
\text{the metric } g_{\widehat{\mu\nu}} &= \begin{cases} g_{\mu\nu} \\ g_{\mu i} \;\; g_{\mu i^\star} \\ g_{ij} \;,\; g_{ij^\star} \;,\; g_{i^\star j^\star} \end{cases} \\
\text{the axion } B_{\widehat{\mu\nu}} &= \begin{cases} B_{\mu\nu} \\ B_{\mu i} \;\; B_{\mu i^\star} \\ B_{ij} \;,\; B_{ij^\star} \;,\; B_{i^\star j^\star} \end{cases} \\
\text{the dilaton } \sigma & \\
\text{the gauge field } A^{I_{E_8}}_{\hat{\mu}} &= \begin{cases} A^{(A_8,0)}_{\mu} \;,\; A^{(0,A_{78})}_{\mu} \; A^{(a,I_{27})}_{\mu} \;,\; A^{(\overline{a},\overline{I}_{\overline{27}})}_{\mu} \\ A^{(A_8,0)}_{i} \;,\; A^{(0,A_{78})}_{i} \; A^{(a,I_{27})}_{i} \;,\; A^{(\overline{a},\overline{I}_{\overline{27}})}_{i} \\ A^{(A_8,0)}_{i^\star} \;,\; A^{(0,A_{78})}_{i^\star} \; A^{(a,I_{27})}_{i^\star} \;,\; A^{(\overline{a},\overline{I}_{\overline{27}})}_{i^\star} \end{cases} \\
\text{the other gauge field } A^{I_{E_8'}}_{\hat{\mu}} &= \begin{cases} A^{I_{E_8'}}_{\mu} \\ A^{I_{E_8'}}_{i} \\ A^{I_{E_8'}}_{i^\star} \end{cases}
\end{aligned}
\tag{4.4.1}
$$

where we have decomposed the ten-dimensional indices in four-dimensional $(\mu, \nu, \dots)$ and six-dimensional ones $(i, i^\star, \dots)$. For the internal indices we have already taken advantage of the complex structure of the Calabi–Yau 3-fold and we have distinguished between holomorphic and antiholomorphic indices, both ranging from 1 to 3. In addition we have utilized the decomposition of the $E_8$ adjoint representation with respect to the subgroup $SU(3) \otimes E_6$:

$$
\mathbf{248} \;\stackrel{SU(3)\otimes E_6}{\longrightarrow}\; (\mathbf{8},\mathbf{1}) \oplus (\mathbf{1},\mathbf{78}) \oplus (\mathbf{3},\mathbf{27}) \oplus \left(\overline{\mathbf{3}},\overline{\mathbf{27}}\right)
\tag{4.4.2}
$$

The indices $A_8$, $A_{78}$ range in the adjoint representation of $SU(3)$ and $E_6$, respectively, while $I_{27}$ and $\overline{I}_{\overline{27}}$ range in the **27** and $\overline{\mathbf{27}}$ of $E_6$, respectively. On the other hand the indices $a = 1, 2, 3$ and $\overline{a} = \overline{1}, \overline{2}, \overline{3}$ range in the triplet and conjugate triplet of $SU(3)$. Since the Calabi–Yau solution of the field equations has identified the spin connection on the internal manifold with the gauge connection, we have used for these gauge indices the same notation as for the flat vector indices on the internal manifold. Indeed they have the same transformation properties with respect to the background fields. All the fields listed in (4.4.1) and here collectively denoted by $\Phi^{(I)}(x,z,\overline{z})$ are functions both of the space-time coordinates $x^\mu$ and of the internal coordinates $y = z^i, \overline{z}^{i^\star}$ which label points on the internal manifold. Hence they must be expanded

$$\Phi^{(I)}(x,z,\overline{z}) = \sum_{\alpha}^{\infty} \Phi^{(I)}_\alpha(x)\, Y^\alpha_{(I)}(y) \tag{4.4.3}$$

in harmonics $Y^\alpha_{(I)}(y)$, appropriate to their tensor character, denoted by $(I)$. The light fields appearing in the compactified four-dimensional lagrangian are the coefficents $\Phi^{(I)}_0(x)$ of those harmonics $Y^0_{(I)}(y)$ which are zero-modes, namely which belong to the kernel of the differential operator $\Delta_{(I)}$ which acts on them in the field equations, once these latter are linearized around the background solution. The operators $\Delta_{(I)}$ turn out to be Hodge–de Rham laplacian operators and therefore the zero-modes can be analysed in terms of harmonic forms and cohomologies.

### 4.4.1 Analysis of the gauge sector

To this effect we need some preparatory steps on $(2,1)$-forms and on vector-valued 1-forms. We anticipate some results which are discussed much more extensively in section 5.9. In particular we consider the isomorphism (5.9.38), which applied to the case of interest to us becomes

$$\begin{aligned} \rho &: H^{(2,1)}(\mathcal{M}_3) \longrightarrow H^{(1)}_{\overline{\partial}}\left(\mathcal{M}_3\,,\,T^{(1,0)}\mathcal{M}_3\right) \\ \rho^{-1} &: H^{(1)}_{\overline{\partial}}\left(\mathcal{M}_3\,,\,T^{(1,0)}\mathcal{M}_3\right) \longrightarrow H^{(2,1)}(\mathcal{M}_3) \end{aligned} \tag{4.4.4}$$

its explicit form (5.9.41) being, in the specific case,

$$\begin{aligned} \rho &: V^i_{j^\star} = \frac{1}{2}\,\overline{\Omega}^{il_1l_2}\,\omega_{l_1l_2,j^\star} \\ \rho^{-1} &: \omega_{l_1l_2j^\star} = \Omega_{il_1l_2}\,\frac{1}{\|\Omega\|^2}\,V^i_{j^\star} \end{aligned} \tag{4.4.5}$$

where $\omega_{l_1l_2j^\star}$ are the components of a harmonic $(2,1)$-form and $V^i_{j^\star}$ the components of a vector-valued harmonic (0,1)-form:

$$\mathbf{V} = V^i_{j^\star}\, d\overline{z}^{j^\star}\,\frac{\partial}{\partial z^i} \tag{4.4.6}$$

Similarly one can define the map

$$\overline{\rho} : H^{(1,2)}(\mathcal{M}_3) \sim H^{(2,1)}(\mathcal{M}_3) \to H^{(1)}_{\partial}\left(\mathcal{M}_3, T^{(0,1)}\mathcal{M}_3\right) \tag{4.4.7}$$

which relates $(2,1)$-forms with a basis $V_i^{j^*}\, dz^i \frac{\partial}{\partial z^{j^*}}$ of vector-valued harmonic (1,0)-forms. Using these results we obtain the following zero-mode expansions for the fluctuations of the gauge field of the $E_8$ gauge group into which the spin connection has been embedded:

$$
\delta A_{\widehat{\mu}}^{I_{E_8}} = \left\{ \begin{array}{lcl}
\delta A_\mu{}^{I_{E_8}} & = & \left\{ \begin{array}{lcl}
\delta A_\mu^{(A_8,0)} & = & \text{no zero-modes} \\
\delta A_\mu^{(0,A_{78})} & = & \mathcal{A}_\mu^{A_{78}}(x) \\
\delta A_\mu^{(a,I_{27})} & = & \text{no zero-modes} \\
\delta A_\mu^{(\overline{a},\overline{I}_{\overline{27}})} & = & \text{no zero-modes}
\end{array} \right. \\
\delta A_i^{I_{E_8}} & = & \left\{ \begin{array}{lcl}
\delta A_i^{(A_8,0)} & = & \sum_{u=1}^{\mathrm{N}_{EndT}} \mathcal{Y}^u(x) \left(\Gamma_{i|u}\right)^a{}_b \\
\delta A_i^{(0,A_{78})} & = & \text{no zero-modes} \\
\delta A_i^{(a,I_{27})} & = & \sum_{\alpha=1}^{h^{(1,1)}} \mathcal{C}^{(I_{27},\alpha)}(x)\, e_j^a\, g^{jk^*}\, \mathcal{U}_{\alpha|ik^*} \\
\delta A_i^{(\overline{a},\overline{I}_{\overline{27}})} & = & \sum_{\mu=1}^{h^{(2,1)}} \mathcal{C}^{(\overline{I}_{\overline{27}},\mu)}(x)\, e_{j^*}^{\overline{a}}\, V_{\mu|i}^{j^*}
\end{array} \right. \\
\delta A_{i^*}^{I_{E_8}} & = & \left\{ \begin{array}{lcl}
\delta A_{i^*}^{(A_8,0)} & = & \sum_{u=1}^{\mathrm{N}_{EndT}} \overline{\mathcal{Y}}^{\overline{u}}(x) \left(\overline{\Gamma}_{i^*|\overline{u}}\right)^{\overline{a}}{}_{\overline{b}} \\
\delta A_{i^*}^{(0,A_{78})} & = & \text{no zero-modes} \\
\delta A_{i^*}^{(a,I_{27})} & = & \sum_{\overline{\mu}=1}^{h^{(2,1)}} \overline{\mathcal{C}}^{(I_{27},\overline{\mu})}(x)\, e_j^a\, V_{\overline{\mu}|i^*}^j \\
\delta A_{i^*}^{(\overline{a},\overline{I}_{\overline{27}})} & = & \sum_{\overline{\alpha}=1}^{h^{(1,1)}} \overline{\mathcal{C}}^{(\overline{I}_{\overline{27}},\overline{\alpha})}(x)\, e_{j^*}^{\overline{a}}\, g^{j^*k}\, \mathcal{U}_{\overline{\alpha}|i^*k}
\end{array} \right.
\end{array} \right. \tag{4.4.8}
$$

where by

$$
\begin{aligned}
\omega_\alpha &= \mathcal{U}_{\alpha|i^*k}\, d\overline{z}^{i^*} \wedge dz^k \quad \{\alpha = 1, \ldots, h^{(1,1)}\} \\
\omega_{\overline{\alpha}} &= \mathcal{U}_{\overline{\alpha}|ik^*}\, dz^i \wedge d\overline{z}^{k^*} \quad \{\overline{\alpha} = 1, \ldots, h^{(1,1)}\}
\end{aligned} \tag{4.4.9}
$$

we have denoted a basis for the cohomology group $H^{(1,1)}(\mathcal{M}_3)$, while by

$$
\begin{aligned}
V_{\overline{\mu}} &= V_{\overline{\mu}|i^*}^j\, d\overline{z}^{i^*} \frac{\partial}{\partial z^j} \quad \{\overline{\mu} = 1, \ldots, h^{(2,1)}\} \\
V_\mu &= V_{\mu|i}^{j^*}\, dz^i \frac{\partial}{\partial \overline{z}^{j^*}} \quad \{\mu = 1, \ldots, h^{(2,1)}\}
\end{aligned} \tag{4.4.10}
$$

we have denoted the bases of $H_{\overline{\partial}}^1\left(T^{(1,0)}\mathcal{M}_3\right)$ and $H_{\partial}^1\left(T^{(0,1)}\mathcal{M}_3\right)$ respectively. Finally $e_j^a$ denotes the vielbein on the Calabi–Yau manifold and

$$
\Gamma_{u|}^a{}_b = \left(\Gamma_{i|u}\, dz^i\right)^a{}_b \quad \left\{u = 1, \ldots, N_{EndT} = \dim_{\mathbb{C}} H^1\left(End\ T^{(1,0)}\mathcal{M}_3\right)\right\} \tag{4.4.11}
$$

denotes a basis for the cohomology group $H_{\overline{\partial}}^1\left(End\ T^{(1,0)}(\mathcal{M}_3)\right)$. The elements of this group are $\overline{\partial}$-closed (1,0)-forms with values in the holomorphic bundle $End\ T^{(1,0)}\mathcal{M}_3$. The sections of this bundle are point-wise linear maps of the tangent spaces to $\mathcal{M}$ into themselves. The case of $A_{\widehat{\mu}}^{I_{E_8'}}$ will not play any significant role in the following discussion and it is omitted in formula (4.4.8). In any case we can easily summarize its zero-mode content: $A_i^{I_{E_8'}}, A_{i^*}^{I_{E_8'}}$ have no zero-modes, while $A_\mu^{I_{E_8'}}$ has a zero-mode for each

$E_8'$ index. In the literature on string-inspired phenomenology this gauge sector arising from $E_8'$ is referred to as the hidden gauge sector. All particles in this sector, having only gravitational interactions with the particles in the visible sector, are essentially transparent. However, the hidden sector can play a useful role if hidden gauginos do condensate through instantonic effects. In this case the gaugino condensate can provide a non-perturbative mechanism for supersymmetry breaking.

Let us give a qualitative explanation of the zero-mode expansion displayed in Eq. (4.4.8). The relevant field equation is the Yang–Mills linearized equation

$$\mathcal{D}_{\widehat{\mu}}(A_{bkg})\,\delta F^{\widehat{\mu\nu}} = 0 \tag{4.4.12}$$

$A_{bkg}$ is the background gauge field identified with the $SU(3)$ spin connection and $\delta F^{\widehat{\mu\nu}}$ is the fluctuation of the field strength. Decomposing Eq. (4.4.12) in the various sectors according to the values of the indices, as in Eq. (4.4.1), and separating the contributions of the space–time derivatives from those of the internal manifold derivatives, we find, as already pointed out, that the mass operator acting on each type of fields is a Hodge–de Rahm operator for a specific elliptic complex. The main question to answer is: Which elliptic complex corresponds to which field? The cases where there are no zero-modes are those where the relevant cohomology groups are trivial. In all the other cases one associates an $x$-space light field to each generator of the corresponding non-trivial cohomology group. Explicitly the correspondence between cohomology groups and fields that emerges from the above-outlined procedure and that explains Eqs. (4.4.8) is the following one:

$$\begin{array}{lcl}
\delta A_\mu^{(A_8,0)} & \leftrightarrow & H^0_{\bar\partial}(End(T\mathcal{M}_3)) \\
\delta A_\mu^{(0,A_{78})} & \leftrightarrow & H^{(0,0)}(\mathcal{M}_3) \\
\delta A_\mu^{(a,I_{27})} & \leftrightarrow & H^0_{\bar\partial}\left(T^{(1,0)}\mathcal{M}_3\right) \\
\delta A_\mu^{(\bar a,\bar I_{\overline{27}})} & \leftrightarrow & H^0_{\bar\partial}\left(T^{(0,1)}\mathcal{M}_3\right) \\
\delta A_{i^\star}^{(A_8,0)} & \leftrightarrow & H^1_{\bar\partial}(End(T\mathcal{M}_3)) \\
\delta A_{i^\star}^{(0,A_{78})} & \leftrightarrow & H^{(0,1)}(\mathcal{M}_3) \\
\delta A_{i^\star}^{(a,I_{27})} & \leftrightarrow & H^1_{\bar\partial}\left(T^{(1,0)}\mathcal{M}_3\right) \sim H^{(2,1)}(\mathcal{M}_3) \\
\delta A_{i^\star}^{(\bar a,\bar I_{\overline{27}})} & \leftrightarrow & H^1_{\bar\partial}\left(T^{(0,1)}\mathcal{M}_3\right) \sim H^{(1,1)}(\mathcal{M}_3)
\end{array} \tag{4.4.13}$$

and similar relations for the $E_8'$ cases and for the fields $\delta A_i$, which follow from complex conjugation. From the above list of correspondences we easily understand the absence of zero-modes for the fields $A_\mu^{(A_8,0)}$, $A_\mu^{(a,I_{27})}$, $A_\mu^{(\bar a,\bar I_{27})}$, $A_{i^\star}^{(0,A_{78})}$. Indeed the complex dimension of $H^0(End(T\mathcal{M}_3))$, $H^0(T^{(1,0)}\mathcal{M}_3)$ and $H^0(T^{(0,1)}\mathcal{M}_3)$ is zero since their bundles do not admit global holomorphic sections. On the other hand the fields $A_{i^\star}^{(0,A_{78})}$ have no zero-modes since $H^{(0,1)}(\mathcal{M}_3) = 0$ for all those Calabi–Yau 3-folds that have a non-vanishing Euler characteristic. This conclusion does not apply when $\chi_{Euler} = 0$, as it happens for instance in the case of the complex three torus (see Eq. (4.144)). The group $H^{(0,0)}(\mathcal{M}_3)$ instead has always complex dimension $dim_{\mathbb{C}}\, H^{(0,0)}(\mathcal{M}_3) = 1$, since the space of global holomorphic 0-forms is given by the constant functions. The remaining cohomology groups displayed in (4.4.13) have already been discussed and their dimensions are parametrized by the Hodge numbers $h^{(1,1)}, h^{(2,1)}$ and by $N_{End\,T}$.

While the Hodge numbers correspond to an intrinsic topological characterization of the manifold, the number $N_{End\,T}$ can in general depend on the complex structure chosen for $\mathcal{M}_3$ and therefore, as we are going to see, on the point in moduli spaces. The $N_{End\,T}$ massless $E_6$-singlets are therefore non-moduli fields, in the sense we have discussed in Chapter 2, namely they do not correspond to flat directions of the scalar potential in the effective four-dimensional lagrangian. These observations complete the analysis of the Kaluza-Klein light particle spectrum emerging from the gauge sector. We have just considered the bosons because $N = 1$ supersymmetry enables us to predict also the fermions, once the bosons are given. Indeed, summarizing, we have found that the D=10 gauge multiplet gives origin, in D=4, to the following $N = 1$ supermultiplets:

i) The gauge multiplet of the group $E_6$:

$$\{\mathcal{A}_\mu^{(0,A_{78})}, \lambda^{(0,A_{78})}\} \tag{4.4.14}$$

ii) $N_{End\,T}$ $E_6$-singlet Wess-Zumino multiplets:

$$\{\mathcal{Y}^u, \chi^u\} \tag{4.4.15}$$

iii) $h^{(1,1)}$ Wess-Zumino multiplets in the chiral 27 representation of $E_6$:

$$\{\mathcal{C}^{\alpha,I_{27}}, \chi^{\alpha,I_{27}}\} \tag{4.4.16}$$

iv) $h^{(2,1)}$ Wess-Zumino multiplets in the chiral $\overline{27}$ representation of $E_6$:

$$\{\mathcal{C}^{\mu,I_{\overline{27}}}, \chi^{\mu,I_{\overline{27}}}\} \tag{4.4.17}$$

Hence the chiral families of $E_6$ are in a number equal to the mismatch between 27 and $\overline{27}$ representations, namely $|h^{(2,1)} - h^{(1,1)}|$. Therefore the number of chiral generations is $\frac{1}{2}|\chi_{Euler}|$, where $\chi_{Euler}$ is the Euler character of $\mathcal{M}_3$ as calculated in Eq. (4.1.29).

### 4.4.2 Analysis of the gravitational sector

In the gravitational sector the light mode expansion of the fields is derived by analysing the linearization of the Einstein, dilaton and axion equations. The zero-modes one obtains are as follows:

$$\delta g_{\widehat{\mu\nu}} = \begin{cases} \delta g_{\mu\nu} = h_{\mu\nu}(x) = \text{the 4D graviton} \\ \delta g_{\mu i} = \text{no zero-modes} \\ \delta g_{\mu i^\star} = \text{no zero-modes} \\ \delta g_{ij} = \sum_{\mu=1}^{h^{(2,1)}} M^\mu(x)\mathcal{O}_{\mu ij} \\ \delta g_{i^\star j^\star} = \sum_{\overline{\mu}=1}^{h^{(2,1)}} \overline{M}^{\overline{\mu}}(x)\mathcal{O}_{\overline{\mu} i^\star j^\star} \\ \delta g_{ij^\star} = \sum_{\alpha=1}^{h^{(1,1)}} \mathrm{Im}\, M^\alpha \mathcal{U}_{\alpha ij^\star} \end{cases} \tag{4.4.18}$$

$$\delta B_{\widehat{\mu\nu}} = \begin{cases} \delta B_{\mu\nu} = b_{\mu\nu}(x) = \text{the 4D axion} \\ \delta B_{\mu i} = \text{no zero-modes} \\ \delta B_{\mu i^\star} = \text{no zero-modes} \\ \delta B_{ij} = \text{no zero-modes} \\ \delta B_{i^\star j^\star} = \text{no zero-modes} \\ \delta B_{ij^\star} = \sum_{\alpha=1}^{h_{(1,1)}} \mathrm{Re}\, M^\alpha \mathcal{U}_{\alpha ij^\star} \end{cases} \tag{4.4.19}$$

$$\sigma = \phi(x) = \text{the 4D dilaton} \tag{4.4.20}$$

where by

$$\mathcal{O}_{\mu ij} = -\frac{1}{||\Omega||^2}\Omega^{k_1^\star k_2^\star r^\star} g_{r^\star i}\omega_{\mu|k_1^\star k_2^\star j} \tag{4.4.21}$$

we have denoted the symmetric holomorphic 2-tensors (the antiholomorphic tensors are respectively denoted by $\mathcal{O}_{\overline{\mu} i^\star j^\star}$) belonging to the kernel of the Lichnerovitz operator

$$(\nabla^k\nabla_k + \nabla^{k^\star}\nabla_{k^\star})\mathcal{O} = 0 \tag{4.4.22}$$

which are in one-to-one correspondence with a basis of harmonic $(1,2)$ (respectively $(2,1)$ forms) $\omega_\mu, \omega_{\overline{\mu}}$ via the relations

$$\begin{aligned} \delta g_{lj} &= -\frac{1}{||\Omega||^2}\Omega^{k_1^\star k_2^\star}{}_l\, \omega_{k_1^\star k_2^\star j} \\ \delta g_{l^\star j^\star} &= -\frac{1}{||\Omega||^2}\overline{\Omega}^{k_1 k_2}{}_{l^\star}\omega_{k_1 k_2 j^\star} \end{aligned} \tag{4.4.23}$$

Moreover, on a Calabi–Yau 3-fold, the $(1,2)$-forms are in one-to-one correspondence with the $(2,1)$-forms so that the indices $\mu, \overline{\mu}$ run over the same set, $\mu, \overline{\mu} = 1, \ldots, h_{(2,1)}$. Equation (4.4.23), once generalized to an $n$-dimensional Calabi–Yau manifold, shows that the existence of the holomorphic $n$-form allows a one-to-one map between symmetric harmonic $(0,2)$-tensors (or $(2,0)$-tensors) and harmonic $(n-1,1)$-forms. The relation (4.4.23) is one-to-one because $H^{(0,2)}(\mathcal{M}_3) = H^{(2,0)}(\mathcal{M}_3) = 0$ so that a harmonic (0,2)-tensor in the image of the map (4.4.23) is necessarily symmetric. On the other hand, in view of the isomorphism (4.149) we can write

$$\begin{aligned} \mathcal{O}_{\mu lj} &= -\frac{2}{||\Omega||^2} g_{li^\star} V^{i^\star}_{\mu j} \\ \mathcal{O}_{\overline{\mu} l^\star j^\star} &= -\frac{2}{||\Omega||^2} g_{l^\star i} V^{i}_{\overline{\mu} j^\star} \end{aligned} \tag{4.4.24}$$

where $V^i_{\overline{\mu}|j^\star}$ are the components of the cohomology group $H^1_{\overline{\partial}}(T^{(1,0)}\mathcal{M}_3)$ generators.

Similarly $\mathcal{U}_{\alpha|ij^\star}$ appearing in the expansion of $\delta g_{ij^\star}$ and $\delta B_{ij^\star}$ are the components of the $(1,1)$-forms generating $H^{(1,1)}$. In perfect analogy with the gauge sector case the zero-mode expansion of Eqs. (4.4.20), (4.4.19), (4.4.18) is explained in terms of the following correspondence between fields and cohomology groups:

$$\begin{array}{rcl}
g_{\mu\nu} & \leftrightarrow & H^{(0,0)}(\mathcal{M}_3) \\
g_{\mu i} & \leftrightarrow & H^{(1,0)}(\mathcal{M}_3) \\
g_{ij} & \leftrightarrow & H^1_{\bar\partial}(T^{(0,1)}(\mathcal{M}_3)) \sim H^{1,2)}(\mathcal{M}_3) \\
g_{i^\star j^\star} & \leftrightarrow & H^1_{\bar\partial}(T^{(1,0)}(\mathcal{M}_3)) \sim H^{(2,1)}(\mathcal{M}_3) \\
g_{ij^\star} & \leftrightarrow & H^1_{\bar\partial}(T^{(0,1)}(\mathcal{M}_3)) \sim H^{(1,1)}(\mathcal{M}_3) \\
B_{\mu\nu} & \leftrightarrow & H^{(0,0)}(\mathcal{M}_3) \\
B_{\mu i} & \leftrightarrow & H^{(1,0)}(\mathcal{M}_3) \\
B_{ij} & \leftrightarrow & H^{(0,0)}(\mathcal{M}_3) \\
B_{ij^\star} & \leftrightarrow & H^{(1,1)}(\mathcal{M}_3) \\
\sigma & \leftrightarrow & H^{(0,0)}(\mathcal{M}_3)
\end{array} \tag{4.4.25}$$

As in the previous case the analysis of the bosonic sector suffices to predict (via supersymmetry) the spectrum of supermultiplets emerging in the low energy effective theory. We obtain:
i) Graviton multiplet
ii) $h^{(1,1)}$ Wess–Zumino $E_6$ singlet multiplets:

$$(M^\alpha, \chi^\alpha) \tag{4.4.26}$$

which emerge combining the contribution coming from $g_{ij^\star}$, and $B_{ij^\star}$.
iii) $h^{(2,1)}$ $E_6$–singlet Wess–Zumino multiplets:

$$(M^\mu, \chi^\mu) \tag{4.4.27}$$

which come from the contributions of $g_{ij}$ and $g_{ij^\star}$.
iv) A Wess–Zumino scalar multiplet:

$$(S = \phi + iP, \zeta) \tag{4.4.28}$$

which comes combining the dilaton and the axion via a duality transformation:

$$\mathcal{D}_\mu P = \varepsilon_{\mu\nu\rho\sigma} \mathcal{D}^\nu B^{\rho\sigma} \tag{4.4.29}$$

In this way we have just retrieved the spectrum of the $N = 1$ supergravity considered in section 3.2, and in particular displayed in Eq. (3.2.59). According to the definition of moduli fields given in section 3.2 we want to argue that the fields listed in (ii) and (iii) are indeed flat directions of the scalar potential in the effective N=1, D=4 field theory. This follows from their geometrical interpretation: their vacuum expectation values parametrize the spaces of Kähler and complex structure deformations. Let us now see what this means.

Using the Yau theorem we have chosen a particular Kähler class and within that class we have selected as background metric $g^0_{ij^*}$, the Ricci-flat one. The background metric $g^0_{ij^*}$ is associated with a Kähler 2-form

$$K = \frac{i}{2\pi} g^0_{ij^*} dz^i \wedge dz^{j^*} \tag{4.4.30}$$

whose cohomology class can be written as a linear combination of the cohomology classes $\omega^{(1,1)}_\alpha$ with certain coefficients $\mathrm{Im}\, t^\alpha$ which can be regarded as the imaginary part of a certain complex number $t^\alpha$:

$$K = \mathrm{Im}\, t^\alpha \omega^{(1,1)}_\alpha \tag{4.4.31}$$

At the same time the $B_{ij^*}$ field is not exactly zero but is a linear combination of the $(1,1)$-forms $\omega^{(1,1)}_\alpha$ with coefficients $\mathrm{Re}\, t^\alpha$ which we can regard as the real part of the same $t^\alpha$ parameters. The background metric components $g^0_{ij}$, $g^0_{i^*j^*}$ are zero in the complex structure we have chosen in order to write the solutions. Let us see what is the meaning of giving a vacuum expectation value $\langle M^\alpha \rangle$ to the fields $M^\alpha$ which appeared in Eqs. (4.4.18) and (4.4.19). It just corresponds to a shift of the Kähler class parameters

$$\begin{aligned} \mathrm{Re}\, t^\alpha &\to \mathrm{Re}\, t^\alpha + \mathrm{Re}\, \delta t^\alpha = \mathrm{Re}\, t^\alpha + \mathrm{Re}\, \langle M^\alpha \rangle \\ \mathrm{Im}\, t^\alpha &\to \mathrm{Im}\, t^\alpha + \mathrm{Im}\, \delta t^\alpha = \mathrm{Im}\, t^\alpha + \mathrm{Im}\, \langle M^\alpha \rangle \end{aligned} \tag{4.4.32}$$

This means that expectation values of $M^\alpha$ parametrize the possible Kähler class on a given Calabi–Yau manifold. However, any Ricci-flat metric is an exact solution of minimal anomaly-free supergravity. Hence any value of $M^\alpha$ must correspond to a possible vacuum of the effective low energy theory with zero potential energy. Therefore $M^\alpha$ should be a flat direction of the scalar potential. A similar argument applies to $M^\mu$, whose vacuum expectation values parametrize the manifold of the complex structure deformations of the given Calabi–Yau 3-fold. Since a Ricci-flat metric is a solution, irrespectively of the complex structure, once more $M^\mu$ represents flat directions of the superpotential.

One of the crucial points of this discussion is that, looking at the gauge and gravitational multiplets, we have established a one-to-one correspondence between moduli fields $M$ and charged scalar fields $\mathcal{C}$. In the Kaluza–Klein approach this is true because the zero-modes of the gravitational and gauge fluctuations are determined by the same cohomology groups. This remarkable correspondence will be used in Chapter 6 to prove, by the N=2 superconformal approach, that the geometry of the moduli space is special. From another point of view special geometry of the moduli space can be derived, from a purely geometrical approach, studying the Weyl–Petersson metric defined on the moduli space. In Chapter 6 we will indeed show how to construct for the moduli spaces the holomorphic homogeneous function $F(X)$, which is the defining structure of the special geometry. Moreover we will introduce, in this purely geometrical language, the three-index "tensors" (they are actually sections of certain bundles as explained in Chapter 2) $W_{\alpha\beta\gamma}$ and $W_{\mu\nu\rho}$, which, as we saw in Eq. (3.2.80), are related to the Yukawa couplings in the effective 4D lagrangian.

A natural question arises in the context of Kaluza–Klein compactification from ten to four dimensions. How can the Yukawa couplings of the effective 4d be deduced from the 10D parent theory? We answer this question in the following section.

### 4.4.3 Yukawa couplings

Recalling Eq. (3.2.79) we see that the superpotential, as a power series in the charged fields, begins with cubic terms. The coefficients of these terms are functions of the moduli fields and, as already pointed out, they are interpreted as Yukawa couplings of the effective 4D lagrangian. As we shall prove in Chapter 6, the coefficients $W_{\alpha\beta\gamma}$ and $W_{\mu\nu\rho}$ are precisely the holomorphic sections of $(T^*\mathcal{M})^3 \times L$ ($L$ is the line bundle whose first Chern class equals the Kähler class) for the moduli space of Kähler structures and for the moduli space of complex structures respectively. They appear in the special geometry formula (3.3.16).

At the present discussion level we are interested to see how $W_{\alpha\beta\gamma}$ and $W_{\mu\nu\rho}$ can be calculated as functions of the moduli in the Kaluza-Klein compactification context, by dimensional reduction from the effective D=10 lagrangian. The result that we shall obtain for (1,1)-forms in the Kaluza-Klein approximation is actually incorrect at the string level. It corresponds only to a limit which is valid for a large radius of the corresponding Calabi–Yau manifold used for the compactification. The exact result requires an infinite number of "quantum" corrections, due to world-sheet instantons [123, 124, 125]. In order to calculate these corrections one has to set up the formalism of topological field theories which reduces the explicit evaluation to a counting problem. The instanton corrections are computed by counting the number of rational curves embedded into the Calabi–Yau manifold one considers. Such a number is usually unknown in algebraic geometry so one is apparently stuck at this point. On the other hand the Kaluza–Klein result we obtain for the (2,1) moduli case is actually exact also at the string level: a conclusion that can once more be drawn from the formalism of topological field theories. The explicit evaluation of these (2,1) Yukawa couplings will be performed in Chapter 8, using Picard–Fuchs equations. Furthermore the notion of mirror symmetry introduced in that chapter allows the evaluation of the Yukawa couplings of (1,1)-forms of a given Calabi–Yau 3-fold as the Yukawa couplings of (2,1)-forms for the mirror manifold. So one is able, through mirror symmetry, to solve the problem of exactly calculating the "quantum-corrected" Yukawa couplings and, simultaneously, to solve the problem of counting the rational curves on a given Calabi-Yau manifold.

From this point of view all the subsequent chapters of this book are devoted to developing the machinery that allows the realization of this programme, namely the quantum string corrections of the Kaluza-Klein results we present in this section.

Let us now consider the gaugino kinetic term in the effective 10D lagrangian. Due to covariant derivatives with respect to the gauge fields $A^{I_{E_8}}_{\widehat{\mu}}$, such a term has a contribution of the type

$$\mathcal{L}_{\lambda A \lambda} = (\lambda^{I_{E_8}})^T C \Gamma^{\widehat{\mu}} A^{J_{E_8}}_{\widehat{\mu}} \lambda^{K_{E_8}} f^{I_{E_8} J_{E_8} K_{E_8}} \tag{4.4.33}$$

Yukawa couplings of the effective four-dimensional theory come from the reduction of (4.4.33). Let us see how this happens. We consider the case where the $E_8$ indices are chosen in the $(3,27)$ or $(\overline{3},\overline{27})$ representations of the $SU(3)\times E_6$ subgroup, i.e. $I_{E_8}=(a,I_{27})$ or $I_{E_8}=(\overline{a},\overline{I}_{27})$, so that

$$
\begin{aligned}
f^{I_{E_8}J_{E_8}K_{E_8}} &= \varepsilon^{abc}d^{I_{27}J_{27}K_{27}} \equiv \varepsilon^{abc}d^{IJK} \\
f^{I_{E_8}J_{E_8}K_{E_8}} &= \varepsilon^{\overline{abc}}d^{\overline{I}_{27}\overline{J}_{27}\overline{K}_{27}} \equiv \varepsilon^{\overline{abc}}d^{\overline{IJK}}
\end{aligned} \tag{4.4.34}
$$

where $d^{IJK}$ is the symmetric three-index tensor in the 27 representation. We consider the case where the index $\hat{\mu}$ runs over the internal manifold indices $(i,i^*)$. Moreover we set, for the gaugino field,

$$
\lambda^{aI}=\chi^I(x)\psi^a(y), \quad \lambda^{\overline{a}\overline{I}}=\chi^{\overline{I}}(x)\psi^{\overline{a}}(y) \tag{4.4.35}
$$

where $\chi$ is a four-dimensional Majorana spinor in the 27 and $\psi^a$ is an $SO(6)$ spinor, which is a zero-mode of the Dirac equation on the internal manifold. If we make an expansion along the zero modes, as indicated in Eq. (4.4.8) for the fields $A_i, A_{i^*}$, and at the same time we consider Eq. (4.3.17) for the internal spinor $\psi^a$, we arrive at the following expressions for the (1,1) and (2,1) Yukawa couplings:

$$
\begin{aligned}
W_{\overline{\alpha\beta\gamma}} &= \int_{\mathcal{M}} \omega^{(1,1)}_{\overline{\alpha}}\wedge\omega^{(1,1)}_{\overline{\beta}}\wedge\omega^{(1,1)}_{\overline{\gamma}} & W_{\alpha\beta\gamma} &= \int_{\mathcal{M}} \omega^{(1,1)}_{\alpha}\wedge\omega^{(1,1)}_{\beta}\wedge\omega^{(1,1)}_{\gamma}\equiv d_{\alpha\beta\gamma} \\
W_{\overline{\mu\nu\rho}} &= \int_{\mathcal{M}} \Omega^*\wedge V_{\overline{\mu}}\wedge V_{\overline{\nu}}\wedge V_{\overline{\rho}} & W_{\mu\nu\rho} &= \int_{\mathcal{M}} \Omega\wedge V_{\mu}\wedge V_{\nu}\wedge V_{\rho}
\end{aligned} \tag{4.4.36}
$$

where we have defined the three-index constant $d_{\alpha\beta\gamma}$ equal to the intersection form between the three (1,1)-forms (and its obvious complex conjugate). Equations (4.4.36) follow from the observation that in the decomposition (4.3.17) of the internal spinor, we have two separate contributions:

i) the contribution coming from the **27** representation $\omega^{a}_{i^*}\Gamma^{i^*}\zeta$, which can be decomposed along a basis of $H^1_{\overline{\partial}}(T^{(1,0)}\mathcal{M})\sim H^{(2,1)}$,

ii) the contribution coming from the $\overline{\mathbf{27}}$ representation $\omega^{\overline{a}}_{i^*}\Gamma^{i^*}\zeta$, which can be decomposed along a basis of $H^1_{\overline{\partial}}(T^{(0,1)}\mathcal{M})\sim H^{(1,1)}$. The same is true for the gauge fields $A_i, A_{i^*}$, which are expanded according to Eqs. (4.4.8). Moreover we use the fact that $\Omega$ can be represented as

$$
\Omega_{ijk}=\zeta^T\Gamma_{ijk}\zeta \tag{4.4.37}
$$

Finally we set the charge conjugation matrix $C$ to unity on the internal space.

## 4.5 Complete Intersection Calabi–Yau Manifolds

We conclude this chapter by considering some explicit examples of Calabi–Yau $n$-folds, in particular Calabi–Yau 3-folds. An interesting and large class of such manifolds can be constructed as a complete intersection of polynomial constraints in tensor products of complex projective spaces (or weighted projective spaces defined and considered in later chapters). For a brief analysis of these important examples, we resume the discussion of section 2.8. Consider an $n$-dimensional hypersurface $\mathcal{M}_n \subset \mathbb{CP}_N$ defined as the simultaneous vanishing locus of $r = N - n$ homogeneous polynomials $\mathcal{W}_\alpha(X)$ of degrees $\nu_\alpha$ (recall Eq. (2.8.33), of Chapter 2). This vanishing locus $\mathcal{M}_n$ is a true smooth, compact complex manifold if the intersection is complete. By this we mean that the differential form

$$\Theta = d\mathcal{W}_1 \wedge d\mathcal{W}_2 \wedge \cdots \wedge d\mathcal{W}_r \tag{4.5.1}$$

never vanishes on the locus $\mathcal{M}_n$. The meaning of this condition is intuitive. At every point of the surface the gradients of the polynomials should identify a complete non-degenerate normal space to the surface itself. Clearly the polynomials $\mathcal{W}_\alpha(X^\Lambda; \psi^{A_\alpha})$ depend both on the homogeneous coordinates $X^\Lambda$ $(\Lambda = 1, \ldots, N+1 = r+n+1)$ of $\mathbb{CP}_N$ and on their coefficients, which we collective name $\psi^{A_\alpha}$ $(A_\alpha = 1, \ldots, m_\alpha)$, the number $m_\alpha$ depending on the degree $\nu_\alpha$. Each choice of these coefficients $\psi^{A_\alpha} \in \mathbb{C}$ identifies a specific $n$-fold $\mathcal{M}_n(\psi)$ with a fixed complex structure. As topological spaces, however, the $n$-folds obtained by varying the coefficients $\psi^{A_\alpha}$ are all equivalent: indeed, as shown in Chapter 2, the Chern classes depend only on the degrees of the homogeneous polynomials. Hence we name

$$\mathbb{CP}_N[\nu_1, \nu_2, \ldots, \nu_r] \tag{4.5.2}$$

the class of manifolds $\mathcal{M}_n(\psi) \subset \mathbb{CP}_N$ obtained by considering all possible specific choices of the polynomials $\mathcal{W}_\alpha(X; \psi)$. The first Chern class of such loci was calculated in Eq. (2.8.38) and reads

$$c_1(\mathbb{CP}_N[\nu_1, \nu_2, \ldots, \nu_r]) = \left(n + r + 1 - \sum_{\alpha=1}^{r} \nu_\alpha\right) K \tag{4.5.3}$$

From the above result we conclude that examples of Calabi–Yau $n$-folds are obtained each time we find a solution of the following Diophantine equation:

$$\left(n + r + 1 - \sum_{\alpha=1}^{r} \nu_\alpha\right) = 0 \tag{4.5.4}$$

For any fixed value of $n$ there is a finite number of solutions to Eq. (4.5.4) if we require that $\nu_\alpha \geq 2$. Indeed we have

$$1 + n + r = \sum_{\alpha=1}^{r} \nu_\alpha \geq 2r \quad \longrightarrow \quad 1 + n \geq r \tag{4.5.5}$$

On the other hand, by imposing this constraint, we do not lose any generality since linear polynomials simply reduce the dimensionality of the ambient space. Given these preliminaries, the list of Calabi–Yau 3-folds that we can construct in this way is immediately found:

$$r = 1 \qquad \nu_1 = 5 \longrightarrow \mathbb{CP}_4[5]$$

$$\begin{array}{llll} r=2 & \nu_1+\nu_2=6 & \longrightarrow & \mathbb{CP}_5\,[2,4] \quad \mathbb{CP}_5\,[3,3] \\ r=3 & \nu_1+\nu_2+\nu_3=7 & \longrightarrow & \mathbb{CP}_6\,[2,2,3] \\ r=4 & \nu_1+\nu_2+\nu_3+\nu_4=8 & \longrightarrow & \mathbb{CP}_7\,[2,2,2,2] \end{array} \tag{4.5.6}$$

Many more examples can be constructed by considering polynomial constraints in a product of complex projective spaces. We confine our discussion to the previous examples since our aim is centred on general ideas rather than on model building. It is obvious that phenomenological reasons have motivated the search for and analysis of Calabi–Yau 3-folds with Euler number $\chi_{Euler} = 6$, since these predict three chiral fermion generations as observed in nature. Unfortunately none of the above-listed simple examples has such a property: three-generation manifolds were constructed by taking quotients of complete intersection manifolds with respect to discrete subgroups and resolving, when necessary the singularities this process gives rise to. To avoid any clumsyness, we omit these more complicated constructions. We rather focus on the calculation of the Euler character $\chi$ for the manifolds listed in (4.5.6) and on the derivation of the corresponding Hodge diamonds, as an illustration of the general concepts explained in previous sections. The Euler number $\chi$ for any complex 3-fold $\mathcal{M}_3$ is given by the integral of the third Chern class:

$$\chi_{Euler}\left(\mathcal{M}_3\right) \;=\; \int_{\mathcal{M}_3} c_3\left(\mathcal{M}_3\right) \tag{4.5.7}$$

Using Eq. (2.8.37) we obtain

$$\begin{aligned} c_3\left(\mathbb{CP}_{r+4}\left[\nu_1,\ldots,\nu_r\right]\right) &= f_{(r+4)}\left(\nu_1,\ldots,\nu_r\right) \; K\wedge K\wedge K \\ f_{(r+4)}\left(\nu_1,\ldots,\nu_r\right) &= \sum_{\substack{l_0+l_1+\ldots+l_r=3 \\ [l_0,l_1,\ldots l_r\in \mathbb{Z}_+]}} \binom{4+r}{l_0} \prod_{\alpha=1}^{r}\left(-\nu_\alpha\right)^{l_\alpha} \end{aligned} \tag{4.5.8}$$

where $K$ is the pull-back on $\mathbb{CP}_{r+4}\left[\nu_1,\ldots,\nu_r\right]$ of the Kähler 2-form of the ambient space $\mathbb{CP}_{r+4}$. Hence the integral

$$\int_{\mathbb{CP}_{r+4}[\nu_1,\ldots,\nu_r]} K\wedge K\wedge K \;=\; \prod_{\alpha=1}^{r}\nu_\alpha \tag{4.5.9}$$

yields the total degree of the embedded surface and the Euler number is given by

$$\chi_{Euler}\left(\mathbb{CP}_{r+4}\left[\nu_1,\ldots,\nu_r\right]\right) \;=\; f_{r+4}\left(\nu_1,\ldots,\nu_r\right)\prod_{\alpha=1}^{r}\nu_\alpha \tag{4.5.10}$$

From formula (4.5.10) we obtain

$$\begin{aligned} \chi\left(\mathbb{CP}_4\,[5]\right) &= -200 \\ \chi\left(\mathbb{CP}_5\,[2,4]\right) &= -176 \\ \chi\left(\mathbb{CP}_5\,[3,3]\right) &= -144 \\ \chi\left(\mathbb{CP}_6\,[2,2,3]\right) &= -144 \\ \chi\left(\mathbb{CP}_7\,[2,2,2,2]\right) &= -128 \end{aligned} \tag{4.5.11}$$

Since all these manifolds are projectively embedded in a single $\mathbb{CP}_N$ space which has a single harmonic (1,1)-form, namely the Kähler class $K$, their $H^{(1,1)}$ Dolbeault group is also one-dimensional so that $h^{(1,1)} = 1$ for all of them. The above calculation of the Euler numbers allows therefore the prediction also of $h^{(2,1)}$ and the complete determination of the corresponding Hodge diamonds. We have

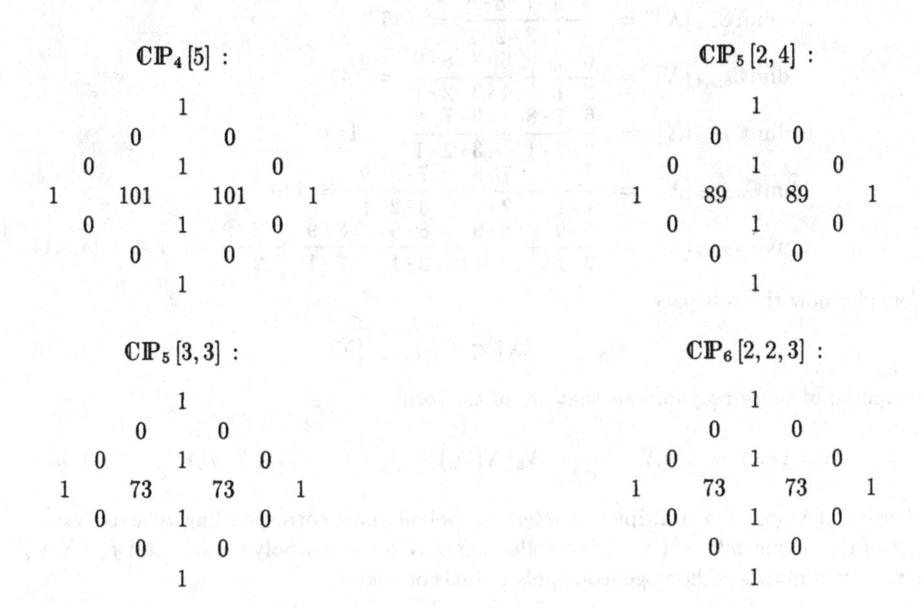


$\mathbb{CP}_7\,[2,2,2,2]$ :

$$\begin{array}{ccccccc} & & & 1 & & & \\ & & 0 & & 0 & & \\ & 0 & & 1 & & 0 & \\ 1 & & 65 & & 65 & & 1 \\ & 0 & & 1 & & 0 & \\ & & 0 & & 0 & & \\ & & & 1 & & & \end{array} \tag{4.5.12}$$

The first thing one notices is that $\mathbb{CP}_5[3,3]$ and $\mathbb{CP}_6[2,2,3]$, although they are different topological manifolds, have identical Hodge diamonds. This is a manifestation of the well-known fact that in dimension larger than two there is no finite number of topological invariants that can completely characterize the topology of a variety. Secondly one would like to get some understanding of the predicted number of complex structure deformations $h^{(2,1)} = 101, 73, 89, 65$. Let us observe that these numbers can be accounted for in the following way. Consider the the space of homogeneous polynomial constraints

$$\mathbb{C}_{N;\nu_1\cdots\nu_r}[X] = \{\mathcal{W}_\alpha(X;\psi)\,(\alpha=1,\dots,r)\,|\,\deg\mathcal{W}_\alpha(X;\psi)=\nu_\alpha\} \qquad (4.5.13)$$

These are vector spaces whose dimension is easily computed by counting the number of independent coefficients in each of the polynomials $\mathcal{W}_\alpha(X;\psi)$. We have

$$\begin{aligned}
\dim\mathbb{C}_{4;5}[X] &= \frac{5\cdot6\cdot7\cdot8\cdot9}{5\cdot4\cdot3\cdot2\cdot1} = 126\\
\dim\mathbb{C}_{5;2,4}[X] &= \frac{6\cdot7}{2\cdot1}+\frac{6\cdot7\cdot8\cdot9}{4\cdot3\cdot2\cdot1} = 147\\
\dim\mathbb{C}_{5;3,3}[X] &= \frac{6\cdot7\cdot8}{3\cdot2\cdot1}+\frac{6\cdot7\cdot8}{3\cdot2\cdot1} = 112\\
\dim\mathbb{C}_{6;2,2,3}[X] &= \frac{7\cdot8}{2\cdot1}+\frac{7\cdot8}{2\cdot1}+\frac{7\cdot8\cdot9}{3\cdot2\cdot1} = 140\\
\dim\mathbb{C}_{7;2,2,2,2}[X] &= \frac{8\cdot9}{2\cdot1}+\frac{8\cdot9}{2\cdot1}+\frac{8\cdot9}{2\cdot1}+\frac{8\cdot9}{2\cdot1}+\frac{8\cdot9}{2\cdot1} = 144
\end{aligned} \qquad (4.5.14)$$

Consider now the subspaces

$$\mathbb{Q}_{N;\nu_1\cdots\nu_r}[X] \subset \mathbb{C}_{N;\nu_1\cdots\nu_r}[X] \qquad (4.5.15)$$

composed of those polynomials that are of the form

$$\mathcal{W}_\alpha(X;\psi) = c^\Lambda(X,\psi)\frac{\partial}{\partial X_\Lambda}\mathcal{W}_\alpha(X;\psi_0) + f_\alpha{}^\beta(X,\psi)\,\mathcal{W}_\beta(X;\psi_0) \qquad (4.5.16)$$

where $\mathcal{W}_\alpha(X;\psi_0)$ is a multiplet of reference polynomials corresponding to fixed values $\psi_0^{A_\alpha}$ of the coefficients, $c^\Lambda(X,\psi)$ is a collection of $N+1$ linear polynomials and $f_\alpha{}^\beta(X,\psi)$ is an $r\times r$ matrix of homogeneous polynomials of degree

$$\deg f_\alpha{}^\beta(X,\psi) = \nu_\alpha - \nu_\beta \qquad (4.5.17)$$

When $\nu_\alpha - \nu_\beta < 0$ the polynomial $f_\alpha{}^\beta(X,\psi)$ is by definition zero. Furthermore the matrix of polynomials $f_\alpha{}^\beta(X,\psi)$ appearing in Eq. (4.5.16) should not be a constant times the identity matrix. Having defined the subspaces $\mathbb{Q}_{N;\nu_1\cdots\nu_\mathbf{R}}[X]$, their dimensionality is easily calculated by counting the number of coefficients in the polynomials $c^\Lambda(X,\psi)$ and $f_\alpha{}^\beta(X,\psi)$. For instance we have

$$\begin{aligned}
\dim\mathbb{Q}_{6;2,2,3} &= 7\times7\\
&\quad+\dim f_1{}^1+\dim f_1{}^2+\dim f_2{}^1+\dim f_2{}^2+\dim f_3{}^1+\dim f_3{}^2+\dim f_3{}^3\\
&\quad-1\\
&= 49\\
&\quad+1+1+1+1+7+7+1\\
&\quad-1\\
&= 67
\end{aligned} \qquad (4.5.18)$$

In a similar way we can complete the list which reads

$$\begin{aligned} \dim \mathbb{Q}_{4;5} &= 25 \quad \dim \mathbb{Q}_{5;2,4} = 58 \\ \dim \mathbb{Q}_{5;3,3} &= 39 \quad \dim \mathbb{Q}_{6;2,2,3} = 67 \\ \dim \mathbb{Q}_{7;2,2,2,2} &= 79 \end{aligned} \tag{4.5.19}$$

If now we define the spaces

$$\mathbb{P}_{N;\nu_1\cdots\nu_r}[X] \stackrel{\text{def}}{=} \frac{\mathbb{C}_{N;\nu_1\cdots\nu_r}[X]}{\mathbb{Q}_{N;\nu_1\cdots\nu_r}[X]} \tag{4.5.20}$$

of homogeneous polynomial constraints modulo the *trivial ones* contained in the spaces $\mathbb{Q}_{N;\nu_1\cdots\nu_r}$, we find that the dimensionality of these spaces:

$$\dim \mathbb{P}_{N;\nu_1\cdots\nu_r}[X] = \dim \mathbb{C}_{N;\nu_1\cdots\nu_r}[X] - \dim \mathbb{Q}_{N;\nu_1\cdots\nu_r}[X] \tag{4.5.21}$$

precisely coincides with the $h^{(2,1)}$ Hodge–numbers displayed in Eq. (4.5.12). What is the meaning of this result? It is easy to understand. As earlier remarked, the deformations of complex structures correspond to a variation of the defining polynomials and hence, given a reference set of polynomials $\mathcal{W}(X;\psi_0)$, their deformations $\mathcal{W}(X;\psi_0) + \delta\mathcal{W}(X;\psi)$ are given by elements $\delta\mathcal{W}(X;\psi) \in \mathbb{P}_{N;\nu_1\cdots\nu_r}[X]$ of the spaces (4.5.13). However, we have to be careful that linear transformations of the homogeneous coordinates in the ambient space are immaterial. Hence polynomial deformations generated by such transformations should not be counted as true deformations. These are the elements in $\mathbb{Q}_{N;\nu_1\cdots\nu_r}$ of the form $c^\Lambda(X,\psi)\frac{\partial}{\partial X_\Lambda}\mathcal{W}_\alpha(X;\psi_0)$. Similarly the hypersurfaces defined by polynomial constraints that are in the same ideal as that generated by $\mathcal{W}(X,\psi_0)$ are all the same hypersurface. This explains why the elements of $\mathbb{Q}_{N;\nu_1\cdots\nu_r}$ of the form $f_\alpha{}^\beta(X,\psi)\,\mathcal{W}_\beta(X;\psi_0)$ are also to be discarded as true deformations. In conclusion the space of genuine complex structure deformations is $\mathbb{P}_{N;\nu_1\cdots\nu_r}[X]$ as defined in Eq. (4.5.20). This explains the numerology of Hodge diamonds. There is, however, more than numerology in this correspondence between polynomial rings and Dolbeault cohomology groups: indeed, as we show in section 5.9, the harmonic $(n-k,k)$-forms on a projective algebraic Calabi–Yau $n$-fold can be constructed in terms of elements of the polynomial ring defined by the polynomial constraints $\mathcal{W}_\alpha(X;\psi)$ and such a construction is the mathematical basis for the development of the conformal field theory and the topological field theory approach to the study of Calabi–Yau compactifications. In the next chapter we turn to such a microscopic viewpoint.

## 4.6 Bibliographical Note

- *Calabi's conjecture and Yau's proof of the existence of a Ricci-flat metric in each Kähler class of a manifold with vanishing first Chern class are respectively contained in [61] and [268].*

- *Calabi–Yau compactifications of superstring theory were proposed by Candelas, Horowitz, Strominger and Witten in 1985 [70] shortly after the discovery of the Green–Schwarz anomaly cancellation mechanism [178]. On the properties of Calabi–Yau compactifications and on the explicit construction of Calabi–Yau 3-folds there exists a vast literature. Among others we cite the papers [64, 65], [258], [122], [71].*

- *Ordinary* $N = 1, D = 10$ *matter-coupled supergravity was formulated by Bergshoeff et al. [43] and by Chapline and Manton [91] in 1982. The coupling to the "gauge" Chern–Simons term had been first found by D'Auria, Fré and Da Silva in the same year [103]. The complete matter-coupled theory was shown to be anomalous by Alvarez-Gaumé and Witten [5].*

- *anomaly-free supergravity was constructed by Bonora, Pasti, Tonin and by D'Auria, Fre', Raciti, Riva. The vast literature on this topic has already been discussed analytically in the text. Summarizing, the basic references are [185, 188, 233, 189, 38, 160], [44, 141, 53, 102, 105, 50], [231, 19, 51, 209, 37, 49], [227, 228, 104, 83, 101].*

- *The issue of Yukawa couplings in Calabi–Yau compactifications has been discussed in many papers. Appropriate basic references are [63, 173, 243], [258].*

- *Form complete intersection Calabi–Yau manifolds (CICY) we cite in particular [64, 65] .*

# Chapter 5

# N=2 FIELD THEORIES IN TWO DIMENSIONS

## 5.1 Introduction

In this chapter we study the structure of two-dimensional field theories possessing an N=2 extended supersymmetry and their relation both with N=2 superconformal theories and with some classical problems of algebraic geometry. Conceptually Chapter 7 and the present one contain the most fundamental material and constitute the heart of our subject. Indeed, as we are going to see, there exists a deep relation between N=2 supersymmetric and topological field theories, which is realized by the so called topological twist. This is a projection algorithm that, from any N=2 quantum field theory, singles out a topological field theory model, whose correlators are just a subclass of the correlators in the N=2 mother theory. This amounts to saying that every N=2 field theory contains a topological sector where Green functions are point-independent and have a geometrical interpretation as intersection numbers of homology cycles in a suitable moduli space of suitable instantons. It also turns out that the topological sector is the most relevant one and almost characterizes the whole N=2 theory.

The string theory motivation to study N=2 field theories on a two-dimensional space-time is that N=1 target supersymmetry implies N=2 supersymmetry on the world-sheet. Hence a string compactification that preserves N=1 target supersymmetry is necessarily based on the tensor product of the $c = 6$ superconformal theory $SCT_{Mink}$ which describes the $\sigma$-model on Minkowski space with some $c = 9$, N=2 superconformal theory $SCT_{int}$ which describes the $\sigma$-model on the 6-dimensional internal space. Indeed, in order to define a true string vacuum, the $\sigma$-model must be conformal invariant and, in the presence of N=2 supersymmetry, this, inevitably, leads to a realization of the N=2 superconformal algebra. In the field theory limit we already saw that a compactification that preserves N=1 target supersymmetry is obtained by choosing the six-dimensional internal space to be a Calabi–Yau 3-fold. For this reason we expect that the N=2 two-dimensional $\sigma$-model on a Calabi–Yau 3-fold should lead to an N=2 superconformal theory at the quantum level. One can actually make a stronger statement: *the N=2 $\sigma$-model on every Calabi–Yau n-fold defines an N=2 superconformal theory with central charge* $c = 3n$ [162, 161]. The proof of this proposition can be obtained only via indirect arguments and at the price of introducing other N=2 field theories and a more general framework. However, by considering the general set of N=2 field theories and the way they relate to N=2 superconformal models one obtains a profound understanding of the interplay between N=2 supersymmetry and the geometrical features of complex $n$-folds of vanishing first Chern-class. Schematically the chain of arguments goes as follows. One considers two-dimensional N=2 scalar field theories, described by the the Landau–Ginzburg lagrangian presented in section 5.6. This theory, which is globally supersymmetric but not conformal invariant at the classical level, contains two geometrical data: the Kählerian metric which, as we will see, can be taken flat without losing any generality, and the superpotential $\mathcal{W}(X)$, the complex scalar fields being denoted by $X^i$. Under the action of the renormalization group the metric changes, but the superpotential remains unrenormalized. At the infrared critical point we obtain a superconformal field theory, whose properties are completely encoded in the quasi-homogeneous part of the superpotential $\mathcal{W}$. In a wide class of cases, this theory can be explicitly constructed by tensoring rational conformal field theories, namely N=2 minimal models [162, 161]. These are the unitary irreducible representations of the N=2 superalgebra belonging to the discrete series [136, 55, 115], characterized by a central charge of the form $c = \frac{3k}{k+2}$, $k = 0, 1, 2, \ldots$ and by a modular invariant partition function, whose possible structures have been classified [72] and set into relation with the ADE classification of modality zero singularities for quasi-homogeneous polynomials [14, 219, 248]. The level at which the identification of the abstract superconformal field theory with the corresponding Landau–Ginzburg model is most direct and significant is that of the *chiral ring*. By definition this is the ring $\mathcal{R}$ of primary chiral fields of the abstract N=2 theory, discussed in section 5.2, and it is isomorphic with the local ring $\frac{\mathbf{C}(X)}{\partial \mathcal{W}(X)}$ of polynomials in $X^i$ modulo the derivatives of the superpotential. By means of this token the structures of a class of abstract N=2 superconformal theories and those of the corresponding N=2 Landau–Ginzburg models can be related to each other. The relationship can be made completely explicit by recasting the Landau–Ginzburg lagrangian into a different formalism (see section 5.8) that uses a set of pseudo-ghost fields and exhibits superconformal invariance already at the classical level.

The relation between the N=2 Landau–Ginzburg model and the $\sigma$-model on the corresponding Calabi–Yau space can be exhibited in a rigorous way by showing that these two theories emerge as low energy effective lagrangians in two different phases of the same N=2 matter coupled gauge theory. This approach, due to Witten [262], is reviewed in section 5.5.

Alternatively, in a more heuristic approach, by considering the path integral definition of the partition function for the Landau–Ginzburg model, one realizes that, if certain conditions on the homogeneity weights of the scalar fields $X^i$ are verified, then, by means of a reparametrization of the integration variables, one can reduce the integration domain to the vanishing locus of the superpotential $\mathcal{W}(X)$. This locus identifies a complex $n$-fold which has vanishing first Chern class in force of the same conditions that guarantee unit-valuedness of the jacobian in the aforementioned reparametrization. Hence the superconformal field theory associated with a quasi-homogeneous Landau–Ginzburg superpotential $\mathcal{W}(X)$ is the same that should emerge from the N=2 $\sigma$-model on the corresponding Calabi–Yau $n$-fold, namely the vanishing locus of $\mathcal{W}(X)$. In certain cases this vanishing locus is not simply given by $\mathcal{W}(X) = 0$, rather it corresponds to the separate vanishing of the addends $\mathcal{W}_A(X)$ in a suitable decomposition $\mathcal{W}(X) = \sum_A \mathcal{W}_A(X)$ of the complete superpotential. Ignoring, for the time being, these additional complicacies we can anyhow state that the explicit identification of the N=2 superconformal field theory associated with a $c_1 = 0$ $n$-fold occurs via the Landau–Ginzburg picture and the study of the chiral ring.

According to the above viewpoint the plan of our exposition in the present chapter is the following one. We begin by defining the N=2 superconformal algebra and by discussing the general properties of an N=2 superconformal theory: these include the ring of chiral (antichiral) primary fields and the spectral flow from the Neveu–Schwarz to the Ramond sector. Then we turn to a discussion of the N=2 minimal models.

Subsequently we introduce the general set-up of N=2 field theories using the rheonomy framework. We discuss the general form of the N=2 superalgebra and we solve the Bianchi identities in terms of an off-shell graviton supermultiplet. This provides an off-shell parametrization for the superworld-sheet curvatures so that we can address the construction of any N=2 matter theory in the background of an arbitrary curved world-sheet. As an example we construct the action of $n$ chiral multiplets interacting via a superpotential $\mathcal{W}(X)$. Next we set the world-sheet curvatures to zero and in the resulting flat superspace (= global supersymmetry case) we construct the action and transformation rules of an N=2 gauge theory coupled to $n$ chiral multiplets. Choosing a special form for the superpotential this is the theory that admits the N=2 Landau–Ginzburg model and the N=2 $\sigma$-model as effective lagrangians in two different phases. These N=2 effective lagrangians are also independently constructed and their global R-symmetries are discussed. The relation between Landau–Ginzburg theories and Calabi–Yau manifolds is also presented in the heuristic way alluded to above. To see the emergence of superconformal symmetry in the Landau–Ginzburg phase, we then introduce another formulation of this model in terms of pseudoghosts that displays superconformal invariance already at the classical level.

Finally, in order to appreciate the mathematical structure underlying the physical identifications we discuss the *Griffiths residue mapping [182] and the Hodge filtration.* This is a very important topic because it provides the relation between the chiral ring of a quasi-homogeneous polynomial $\mathcal{W}(X)$ and the Dolbeault cohomology spaces $H^{(n-k,k)}$ of the corresponding Calabi–Yau $n$-fold $\mathcal{W}(X) = 0$. Basically one establishes a representation of the $(n-k,k)$-harmonic forms on $\mathcal{W}(X) = 0$ in terms of rational meromorphic $(n+1)$-forms in the ambient $\mathbb{CP}_{n+1}$ space. In such a representation the order of the pole at $\mathcal{W}(X) = 0$ equals the number $k+1$ in $H^{(n-k,k)}$ and the residue is a non-trivial polynomial $P(X) \in \frac{\mathbb{C}(X)}{\partial \mathcal{W}(X)}$ of order $deg\, P = k\, deg\, \mathcal{W}$. Since the local ring $\frac{\mathbb{C}(X)}{\partial \mathcal{W}(X)}$ has already been identified with the chiral ring of the corresponding N=2 conformal field theory, we can, by means of the Griffiths residue mapping, associate well-defined chiral–chiral primary fields to each $(n-k,k)$-harmonic form. In this way we can construct the emission vertices for all the relevant light fields of a Calabi–Yau compactification and calculate the associated Yukawa couplings, namely the intersection integrals discussed in section 4.4.3, as three-point correlators in the appropriate N=2 CFT. This will be discussed in Chapter 6, where we shall utilize $N = 2$ conformal field theories to study the moduli space geometry. The residue mapping, however, is important also in another respect. It provides the basis for the calculation of the periods of harmonic $(n-k,k)$-forms and for the derivation of the differential Picard–Fuchs equations satisfied by these periods (see Chapter 8).

The relation with topological field theories, to be discussed in Chapter 7, emerges precisely at the level of the Yukawa couplings. The correlation functions that describe these couplings belong to the topological sector of the N=2 field theory and their calculation can be addressed within the topological field theory framework.

## 5.2 Abstract N=2 Superconformal Theories

In this section we study, from an algebraic point of view, the N=2 superconformal theories [1, 2]. Our purpose is to introduce the basic instruments that are involved in this subject. We have no pretence of completeness and we skip many technical subtleties. The literature on this subject is immense. Here we follow essentially [214, 254].

The N=2 superalgebra is an extension of the Virasoro algebra [252, 177] by means of two anticommuting supercurrents $G^{\pm}(z)$ of conformal weight $h = 3/2$ and a bosonic $U(1)$ current $J(z)$, under whose action the two supercurrents have charges +1 and −1, respectively. The complete structure of the algebra is encoded in the following operator product expansions:

$$
\begin{aligned}
T(z)\,T(w) &= \frac{c}{2}\frac{1}{(z-w)^4} + \frac{2\,T(w)}{(z-w)^2} \\
&+ \frac{\partial T(w)}{(z-w)} + reg \\
T(z)\,G^{\pm}(w) &= \frac{3}{2}\frac{1}{(z-w)^2}G^{\pm}(w) + \frac{\partial G^{\pm}(w)}{(z-w)} + reg \\
T(z)\,J(w) &= \frac{1}{(z-w)^2}J(w) + \frac{1}{(z-w)}\partial J(w) + \\
G^{+}(z)\,G^{-}(w) &= \frac{2}{3}c\frac{1}{(z-w)^3} + 2\frac{J(w)}{(z-w)^2} + \\
& 2\,\frac{T(w)+\frac{1}{2}\partial J(w)}{(z-w)} + reg
\end{aligned}
$$

$$J(z)\,G^{\pm}(w) = \pm\frac{G^{\pm}(w)}{(z-w)} + reg$$

$$J(z)\,J(w) = \frac{c}{3}\frac{1}{(z-w)^2} + reg \qquad (5.2.1)$$

The Neveu–Schwarz and Ramond sectors of the algebra (5.2.1) are selected by requiring the following boundary conditions on the cylinder:

$$G^{\pm}(e^{2\pi i}z) = G^{\pm}(z) \qquad \text{Ramond sector}$$
$$G^{\pm}(e^{2\pi i}z) = -G^{\pm}(z) \qquad \text{Neveu–Schwarz sector} \qquad (5.2.2)$$

implying that $T(z)$ and $J(z)$ are periodic.

The commutation relations of the N=2 superconformal generators defined by the mode expansions of the currents appearing in (5.2.1) are given by

$$
\begin{aligned}
[L_m, L_n] &= (m-n)L_{m+n} + \frac{c}{12}m(m^2-1)\delta_{n+m,0} \\
\left[L_n, G_r^{\pm}\right] &= (\frac{n}{2}-r)G^{\pm}_{n+r} \\
[L_n, J_m] &= -mJ_{n+m} \\
[J_m, J_n] &= \frac{c}{3}m\delta_{n+m,0} \\
\{G_r^{-}, G_s^{+}\} &= 2L_{r+s} - (r-s)J_{r+s} + \frac{c}{3}(r^2-\frac{1}{4})\delta_{r+s,0} \\
\left[L_n, G_r^{\pm}\right] &= (\frac{n}{2}-r)G^{\pm}_{n+r} \\
\left[J_n, G_r^{\pm}\right] &= \pm G^{\pm}_{n+r}
\end{aligned} \qquad (5.2.3)
$$

where

$$L(z) = \sum \frac{L_n}{z^{n+2}}, \quad G^{\pm}(z) = \sum \frac{G_r^{\pm}}{z^{r+\frac{1}{2}}}, \quad J(z) = \sum \frac{J_n}{z^{n+1}} \qquad (5.2.4)$$

Formula (5.2.3) is valid both in the Neveu-Schwarz and in the Ramond sector, the only difference being that $r$ and $s$ run over half integers in the Neveu-Schwarz sector and over integers in the Ramond sector. The boundary conditions (5.2.2) are just particular cases of a one parameter twisting of the $N=2$ algebra (5.2.1) defined by the boundary conditions:

$$G_a^{\pm}(z) = e^{2\pi i a}G^{\pm}(e^{2\pi i}z) \qquad (5.2.5)$$

where $a$ is a real parameter. Setting $a = 0$ (mod 1) yields the Neveu-Schwarz sector, while $a = 1/2$ (mod 1) corresponds to the Ramond sector. In fact there is a continuous one-parameter spectrum of sectors, labelled by $a$, corresponding to different representations of the $N=2$ algebra, all of which, however, are isomorphic. This is a characteristic feature of the N=2 algebra, which allows one to connect continuously the different sectors of the theory. More precisely we can rewrite the last three equations appearing in (5.2.3) as depending on the continuous parameter $a$, i.e.

$$
\begin{aligned}
\{G^+_{n+a}, G^+_{m-a}\} &= 2L_{m+n} - (n-m+2a)J_{n+m} + \frac{c}{3}\left[(n+a)^2 - \frac{1}{4}\right]\delta_{n+m,0} \\
\left[L_n, G^\pm_{m\pm a}\right] &= [\frac{n}{2} - (m \pm a)]G^\pm_{m+n\pm a} \\
\left[J_n, G^\pm_{m\pm a}\right] &= \pm G^\pm_{m+n\pm a}
\end{aligned}
\tag{5.2.6}
$$

where $n, m$ are integer numbers and $a$ is a real number. If $a \in \mathbb{Z}$ or $a \in \mathbb{Z} + \frac{1}{2}$ then we recover the standard Ramond and Neveu–Schwarz sectors. Let $\mathcal{H}_a$ the Hilbert space associated with the N=2 algebra with parameter $a$. Let now $\mathcal{U}_\theta$ be the operator that maps $\mathcal{H}_a$ into $\mathcal{H}_{a+\theta}$. $U_\theta$ will act on a generic operator as

$$\mathcal{O}_{a+\theta} = U_\theta \mathcal{O}_a U_\theta^{-1} \tag{5.2.7}$$

Looking at Eqs. (5.2.6) we immediately recognize that $U_\theta$ acts on the fundamental operators in $\mathcal{H}_0$ as

$$\mathcal{U}_\theta L_n \mathcal{U}_\theta^{-1} = L_n + \theta J_n + \frac{c}{6}\theta^2 \delta_{n,0} \tag{5.2.8}$$

$$\mathcal{U}_\theta J_n \mathcal{U}_\theta^{-1} = J_n + \frac{c}{3}\theta \delta_{n,0} \tag{5.2.9}$$

$$\mathcal{U}_\theta G^+_r \mathcal{U}_\theta^{-1} = G^+_{r+\theta} \tag{5.2.10}$$

$$\mathcal{U}_\theta G^-_r \mathcal{U}_\theta^{-1} = G^-_{r-\theta} \tag{5.2.11}$$

For $\theta \in \mathbb{Z} + \frac{1}{2}$, $\mathcal{U}_\theta$ interpolates between the Neveu–Schwarz and Ramond sectors. For $\theta \in \mathbb{Z}$ it takes Neveu–Schwarz to Neveu–Schwarz and Ramond to Ramond. From Eq. (5.2.9) we realize that each state in $\mathcal{H}_\theta$ (or generically in $\mathcal{H}_{a+\theta}$) has the $U(1)$ charge shifted by $-\frac{c}{3}\theta$ with respect to the corresponding one in $\mathcal{H}_0$ ($\mathcal{H}_a$). Vice-versa, if we construct an operator that implements the transformation (5.2.9), then all the other transformations follow as a consequence. This means that the crucial property that allows one to contruct the spectral flow operator is completely encoded in the $U(1)$ symmetry of the theory. Equations (5.2.8)–(5.2.11) define the so called *spectral flow* of the $N = 2$ superconformal algebra. We will give some more detail on the spectral flow in the following.

We next turn our attention to the study of the irreducible representations of the N=2 algebra (5.2.3) (together with its obvious antiholomorphic counterpart defined by the mode expansions of $\overline{T}(\overline{z})$, $\overline{J}(\overline{z})$and $\overline{G}^\pm(\overline{z})$). In particular we are interested in the so-called highest weight representations, i.e. those that are generated by applying the negative mode expansions of the currents in (5.2.1) to a given state: the highest weight state or primary state. In the present discussion we consider only unitary $N = 2$ theories, and some of the considerations reported here should be carefully modified in non-unitary models.

Looking at the OPE (5.2.1), or equivalently at Eq. (5.2.3), we see that a generic state $|\psi\rangle$ of the theory can be diagonalized into simultaneous eigenvectors of $L_0$ ($\overline{L}_0$) and $J_0$ ($\overline{J}_0$), i.e.

$$\begin{aligned} L_0|\psi\rangle &= h|\psi\rangle, \quad \overline{L}_0|\psi\rangle = \overline{h}|\psi\rangle \\ J_0|\psi\rangle &= q|\psi\rangle, \quad \overline{J}_0|\psi\rangle = \overline{q}|\psi\rangle \end{aligned} \tag{5.2.12}$$

where $h, \overline{h}$ are conformal weights and $q, \overline{q}$ are the $U(1)$ charges. A primary state in the Neveu–Schwarz sector is defined by the following conditions:

$$G_r^{\pm}|\Phi\rangle = 0 \qquad r \geq \frac{1}{2} \tag{5.2.13}$$

$$L_n|\Phi\rangle = J_n|\Phi\rangle = 0 \qquad n \geq 1 \tag{5.2.14}$$

together with their antiholomorphic counterparts. Correspondingly we can define a primary field $\Phi(z,\overline{z})$, which is labelled, according to Eqs. (5.2.12) by the conformal weights and U(1) charges of the associated state, and which is denoted by

$$\Phi\begin{pmatrix} h & \overline{h} \\ q & \overline{q} \end{pmatrix}(z,\overline{z}) \tag{5.2.15}$$

Any primary field can be factorized into a left-moving part, depending on $z$, and into a right-moving component, depending on $\overline{z}$. Correspondingly any primary state can be written as $|\Phi\rangle = |\Phi\rangle_L \otimes |\Phi\rangle_R$. Let us focus on the left-moving sector (similar considerations are understood for the right-moving sector). It is not difficult to see that the condition (5.2.14) is a consequence of (5.2.13), due to the algebra (5.2.3), so that the first condition is sufficient to define a primary state.

The corresponding primary field $\Phi$ satisfies the following operator product expansions (OPEs) :

$$\begin{aligned} T(z)\,\Phi(w) &= \frac{h}{(z-w)^2}\Phi(w) + \frac{\partial_w\Phi(w)}{(z-w)} + reg \\ J(z)\,\Phi(w) &= \frac{q}{(z-w)}\Phi(w) \\ G^{\pm}(z)\,\Phi(w) &= \frac{1}{z-w}\Gamma^{\pm}(w) \end{aligned} \tag{5.2.16}$$

(5.2.17)

where the fields $\Gamma^{\pm}(w)$ are the superpartners of $\Phi(w)$.

By definition *chiral* and *antichiral* fields satisfy respectively the following conditions:

$$\begin{aligned} G^{+}(z)\Phi_c(w) &= \text{reg} \\ G^{-}(z)\Phi_{ac}(w) &= \text{reg} \end{aligned} \tag{5.2.18}$$

Equivalently a chiral state is a state in the Neveu–Schwarz Hilbert space which satisfies

$$G^{+}_{-1/2}|\Phi\rangle = 0 \tag{5.2.19}$$

and an antichiral state is defined by a similar condition with $G^+$ replaced by $G^-$.

The simultaneous requirement of the properties (5.2.13) and (5.2.19) define what is named a *chiral primary state*, i.e. $|\phi\rangle$ is chiral primary if it fulfills the following conditions:

$$\begin{aligned} G^+_{-1/2}|\phi\rangle &= 0 \\ G^\pm_r|\phi\rangle &= 0 \qquad r \geq \frac{1}{2} \end{aligned} \tag{5.2.20}$$

Unitary representations of the algebra (5.2.3) are those satisfying the hermiticity conditions $L_n^\dagger = L_{-n}$, $J_n^\dagger = J_{-n}$, $(G^\pm)_n^\dagger = (G^\mp)_{-n}^\dagger$ and satisfying the requirement that the internal product, in the Fock space, should be positive definite. As we will see in the following section this condition selects, for the values of central charge in the region $c \leq 3$, a discrete series of unitary models: the minimal models.

Given a primary $|\Phi\rangle$ and the descendent fields $G^\pm_{-1/2}|\psi\rangle$, we have for a unitary theory

$$0 \leq ||G^\pm_{-1/2}|\Phi\rangle||^2 \equiv \langle\Phi|\{G^\pm_{1/2}, G^\mp_{-1/2}\}|\Phi\rangle = \langle\Phi|2L_0 \pm J_0|\Phi\rangle \tag{5.2.21}$$

which implies that:

$$h \geq \frac{1}{2}|q| \tag{5.2.22}$$

It is also easy to specify (5.2.21) to the case of a chiral primary state, which satisfies the bound $h_\phi = \frac{1}{2}q$. The converse is also true, since, if a state has $h_\phi = \frac{1}{2}q$, then it is necessarily annihilated by $G^\pm_r$ for $r \geq \frac{1}{2}$ and by $G^+_{-1/2}$, as can be shown using (5.2.21) and the bound (5.2.22).

Given a generic state $|\widehat{\Phi}\rangle$ in the Hilbert space, with dimension and charge $(h, q)$, there exists a "Hodge" decomposition of $|\widehat{\Phi}\rangle$, which reads:

$$|\widehat{\Phi}\rangle = |\phi\rangle + G^+_{-1/2}|\phi_1\rangle + G^-_{1/2}|\phi_2\rangle \tag{5.2.23}$$

where $|\phi\rangle$ is a chiral primary state and $|\phi_1\rangle$, $|\phi_2\rangle$ are arbitrary states in the Hilbert space. We call Eq. (5.2.23) a "Hodge" decomposition, since we can interpret the operator $G^+_{-1/2}$ as an elliptic operator acting on the Fock space, with the kernel defined by the span of the chiral primary fields. Using Eq. (5.2.23), if we act on a chiral state with $G^+_{-1/2}$ we find

$$G^+_{-1/2}G^-_{1/2}|\phi_2\rangle = 0 \tag{5.2.24}$$

Formula (5.2.24) means that the norm of the state $G^-_{1/2}|\phi_2\rangle$ is zero.

Thus, for a chiral state the term proportional to $|\phi_2\rangle$ does not give any significant contribution and can be set to zero, so that

$$|\widehat{\Phi}_c\rangle = |\phi\rangle + G^+_{-1/2}|\phi_1\rangle \tag{5.2.25}$$

Formula (5.2.25) suggests an alternative definition for chiral primary fields. Chiral primary states can be defined as non trivial closed states under the action of the cohomology operator $G^+_{-\frac{1}{2}}$. This definition will become of primary importance in the topological version of the N=2 theories, where $G^+(z)$ is interpreted as the BRST charge. Moreover this is a preparatory definition w will be resumed in the Landau-Ginzburg approach to superconformal models, and in the study of the correspondence between chiral primary operators and cohomology classes of the Dolbeault operator $\overline{\partial}$ (defined on a suitable Calabi-Yau manifold).

Now consider a generic chiral primary state $|\phi\rangle$ and consider the norm of the state $G^+_{-3/2}|\phi\rangle$:

$$0 \leq \langle\phi|\{G^+_{-3/2}, G^-_{3/2}\}|\phi\rangle = \langle\phi|2L_0 - 3J_0 + \frac{2c}{3}|\phi\rangle \tag{5.2.26}$$

which implies, using $h = \frac{q}{2}$, that $h \leq \frac{c}{6}$. There exists one state (the top chiral primary state) that saturates the bound $h = \frac{c}{6}$.

It can be shown, using the spectral flow, that such a state is unique, up to an overall normalization (see below). The presence of this bound means that, in a nondegenerate theory, for which the spectrum of $L_0$ is discrete, there is only a finite number of primary chiral operators.

Another property which we discuss in this context is the peculiar operator algebra satisfied by chiral primary operators. Consider two chiral primary fields $\phi_1$ and $\phi_2$, it is immediate to check that the naive product:

$$\chi(z) = \lim_{z' \to z} \phi_1(z')\, \phi_2(z) \tag{5.2.27}$$

is nonsingular. This is because $h_\chi \geq \frac{1}{2}(q_{\phi_1} + q_{\phi_2}) = h_{\phi_1} + h_{\phi_2}$ so that any singularity of the kind $(z - z')^{h_\chi - h_{\phi_1} - h_{\phi_2}}$ is always absent due to the previous inequality. Moreover we observe that the product of two chiral fields is necessarily chiral (the check is also immediate) and can be primary only if $h_\chi = \frac{1}{2}q_\chi = \frac{1}{2}(q_{\phi_1} + q_{\phi_2})$, while for a non primary field $h_\chi \rangle \frac{1}{2}q_\chi = \frac{1}{2}(q_{\phi_1} + q_{\phi_2})$. In this last case, however, the product $\chi$ is automatically set to zero, since $\lim_{z \to z'}(z - z')^{h_\chi - h_{\phi_1} - h_{\phi_2}} \to 0$. In a nondegenerate N=2 conformal theory this discussion shows that the set of chiral primary fields defines a finite ring equipped with the product (5.2.27). The chiral ring of a generic unitary (2,2) conformal theory is a crucial structure characterizing this theory. In the Landau–Ginzburg formulation of (2,2) theories, as we will explain in section 5.6, the chiral ring of the N=2 theory is in one-to-one correspondence with the chiral ring associated with the defining potential $\mathcal{W}$. Moreover, through the Griffiths residue mapping, any element of the chiral ring is related to a $(n-k, k)$ form belonging to $H^{(n-k,k)}_{\overline{\partial}}(M)$, where $M$ is the Calabi–Yau manifold defined as the zero locus of the potential $\mathcal{W}$. These associations are of crucial importance in studying the correspondence between Landau-Ginzburg theories and superconformal theories and are the basic elements for interpreting the correlation functions of the topological version of these models.

To go on with our programme we now give some details on the spectral flow and consequently we study some basic properties of the states (or fields) in the Ramond sector.

The Ramond sector is characterized by the commutations relations (5.2.3), where $r, s \in \mathbb{Z}$. By considering the relation

$$\{G_0^-, G_0^+\} = 2L_0 - \frac{c}{12} \tag{5.2.28}$$

and repeating the analogous steps as in (5.2.21), we see that any state in the Ramond sector satisfies $h \geq \frac{c}{24}$. The equality is reached only if $G_0^+$ and $G_0^-$ annihilate the state of conformal weight $h$. States annihilated by $G_0^+$ and $G_0^-$ are the chiral primary states in the Ramond sector. They are in one to one correspondence with the chiral primary states $|\phi\rangle$ in the Neveu–Schwarz sector.

Indeed, if we consider a chiral primary state in the Neveu–Schwarz sector satisfying the conditions (5.2.13) and (5.2.19) and we act on it by the spectral flow operator $\mathcal{U}_{\frac{1}{2}}$, we find

$$\mathcal{U}_{\frac{1}{2}} G^+_{-\frac{1}{2}} \mathcal{U}^{-1}_{-\frac{1}{2}} \mathcal{U}_{\frac{1}{2}} |\phi\rangle = G_0^+ |\tilde{\phi}\rangle = 0 \tag{5.2.29}$$

where $|\tilde{\phi}\rangle$ is a state in the Ramond sector. Moreover, using (5.2.13 ) and the spectral flow we see that $|\tilde{\phi}\rangle$ satisfies

$$G_n^- |\tilde{\phi}\rangle = G_{n+1}^+ |\tilde{\phi}\rangle = 0 \quad \text{for } n \geq 0 \tag{5.2.30}$$

Equation (5.2.30) shows that $|\tilde{\phi}\rangle$ is annihilated also by $G_0^-$ and, as a consequence, is a chiral primary operator in the Ramond sector: any chiral primary state in the Ramond sector is the specral flow of a unique chiral primary state in the Neveu–Schwarz sector.

As is evident from the above discussion and from the explicit form of the spectral flow (5.2.8)–(5.2.11), all the chiral primary states in the Ramond sector have conformal weight $h = \frac{c}{24}$ and $U(1)$ charges which span the range $h = -\frac{c}{6}, \dots \frac{c}{6}$.

We now give a more explicit realization of the spectral flow operator $U_\theta$, by "bosonizing" the $U(1)$ current $J(z)$. Indeed let us write

$$\begin{aligned} J(z) &= i\sqrt{\frac{c}{3}}\partial_z \phi(z) \\ \overline{J}(\overline{z}) &= i\sqrt{\frac{c}{3}}\partial_{\overline{z}} \tilde{\phi}(\overline{z}) \end{aligned} \tag{5.2.31}$$

where $\phi$ and $\tilde{\phi}$ are free bosonic fields satisfying

$$\begin{aligned} \phi(z)\phi(w) &= -\ln(z-w) + \text{reg} \\ \tilde{\phi}(\overline{z})\phi(\overline{w}) &= -\ln(\overline{z}-\overline{w}) + \text{reg} \end{aligned} \tag{5.2.32}$$

and where the normalization $\sqrt{c/3}$ is fixed by the OPE of the $U(1)$ current in eq (5.2.1). Given a $U(1)$ neutral field $\chi(z)$ it is easy to verify that the new field $exp\left[i\sqrt{\frac{3}{c}}q\right]\phi$ has charge $q$. It is now immediate to construct, in term of the bosonic fields $\phi, \tilde{\phi}$ the operator that implements the $U(1)$ charge shift according to Eq. (5.2.9), i.e.

$$U_\theta = exp\left[-\frac{c}{3}\theta(\phi - \tilde{\phi})\right] \tag{5.2.33}$$

Since the spectral flow shifts the charges by $\frac{c}{3}\theta$, every state of charge $(q, \overline{q})$ flows to a new state of charges $(q - \frac{c}{3}\theta, \overline{q} - \frac{c}{3}\theta)$. This means that to every primary field of a (2,2) superconformal theory one can associate an entire family of primary fields obtained by shifting the $U(1)$ charges.

If we now choose a spectral flow operator with $\theta = 1$, we easily see that the chiral primary field with $h = 0, q = 0$ (the identity) flows to an antichiral field with $h = \frac{c}{6}, q = -\frac{c}{3}$, and the top chiral field flows to the identity field in the ring of antichiral fields. This means that any chiral primary field is in one-to-one correspondence with an antichiral field, through the spectral flow map $\mathcal{U}_\theta$, with $\theta = 1$.

Finally if we choose $\theta = -1$, we see that the (unique) identity operator flows to a (unique) state of charge $q = \frac{c}{3}$, which is a chiral primary operator that saturates the bound $h = \frac{c}{6}$. This shows that the top chiral primary operator is unique.

It is important to stress that the spectral flow $\mathcal{U}_\theta$ is performed simultaneously on the left and right-moving sectors. Thus the left top chiral field can be paired with the corresponding right moving state so that it satisfies, in this case, $h = \overline{h} = 2q = 2\overline{q} = \frac{c}{6}$. More precisely can define four different rings, by considering all the pairings between chiral and antichiral fields in both the left and right-moving sectors. What one obtains are the chiral–chiral ring $(c, c)$, the antichiral–chiral, chiral–antichiral rings $(a, c)$, $(c, a)$, and the antichiral–antichiral ring $(a, a)$.

## 5.3 N=2 Minimal Models

In this section we discuss minimal unitary N=2 models. By minimal N=2 series we mean the following. Given an N=2 superalgebra defined by (5.2.3), we can study the highest weight irreducible representations associated with this algebra. An irreducible highest weight representation of an N=2 algebra is specified by the numbers $(c, h, q)$. As we have pointed out in the previous section unitary representations are those satisfying the hermiticity conditions $L_n^\dagger = L_{-n}$, $J_n^\dagger = J_{-n}$, $(G^\pm)_n^\dagger = (G^\mp)_{-n}^\dagger$ and satisfying the requirement that the internal product, in the Fock space, be positive definite.

Minimal unitary N=2 series are specified by the following values of the central charge and by the following conformal and $U(1)$ weights for the primary fields:

$$\begin{aligned} c &= \frac{3k}{k+2} \\ q_{m,s} &= \frac{m}{k+2} + \frac{ks}{k+2} \\ h_{l,m,s} &= \frac{l(l+2)}{4(k+2)} - \frac{m^2}{4(k+2)} + \frac{s^2 + 2sm}{2(k+2)} \end{aligned} \tag{5.3.1}$$

where $k \in \mathbb{Z}$, $l = 0, \ldots, k$, $m = -l, -l+2\ldots, l-2, l$ and $s = 0$ or $s = \pm\frac{1}{2}$ in the Neveu–Schwarz or in the Ramond sector. Minimal unitary series (5.3.1) are not the only possible unitary $N = 2$ representations, but they are of primary importance to us, since, by taking tensor products of such representations we obtain $N = 2$ models with central charge $c = 9$ (or more generally $c = 3n$), which are used for the compactification of the heterotic string on Calabi–Yau manifolds.

In this section we give the so-called Goddard-Kent-Olive (GKO) construction [176] of the N=2 minimal unitary series. This is one of the simplest ways of constructing minimal unitary series in terms of unitary irreducible representation of suitable Kac–Moody algebras. We partially follow [115, 200].

To begin with we recall the definition of a Kac–Moody algebra of level $k$. Given a simple group $G$ with Lie algebra $\mathcal{G}$ and structure contants $f^{ABC}$ and a holomorphic current $J^A(z)$, a Kac–Moody algebra of level $k$ is defined by the following operator algebra:

$$J^A(z)J^B(w) = i\frac{f^{ABC}}{z-w}J^C(w) + \frac{k}{2}\frac{\delta^{AB}}{(z-w)^2} + \text{reg} \tag{5.3.2}$$

Given this algebra we can construct a stess–energy tensor $T(z)$ by the Sugawara formula:

$$T_G(z) = \frac{1}{k+c_A} : J^A J^A : (z) \tag{5.3.3}$$

where the ordered product is defined by subtracting singularities at coinciding points and $c_A$ is the Casimir in the adjoint representation [1].

Let us now consider the Kac–Moody algebra (5.3.2) together with spin 1/2 operators $\lambda^A(z)$ living in the adjoint representation of the group $G$, and having the OPE

$$\lambda^A(z)\lambda^B(w) = \frac{1}{k}\frac{\delta^{AB}}{(z-w)} \tag{5.3.4}$$

We construct the following operators:

$$\begin{aligned} T_G(z) &= \frac{1}{p+c_A} : J^A J^A : (z) - \frac{k}{2}\lambda^A\partial_z\lambda^A(z) \\ G(z) &= \sqrt{2}\sqrt{\frac{k}{k+c_a}}\left(\lambda^A(z)J^A(z) - i\frac{k}{6}f^{ABC}\lambda^A(z)\lambda^B(z)\lambda^C(z)\right) \end{aligned} \tag{5.3.5}$$

These two operators close an $N = 1$ algebra which is defined by the OPEs

$$\begin{aligned} T(z)T(w) &= \frac{c/2}{(z-w)^4} + \frac{2T(w)}{(z-w)^2} + \frac{\partial T(w)}{z-w} \\ T(z)G(w) &= \frac{3}{2}\frac{G(w)}{(z-w)^2} + \frac{\partial G(w)}{z-w} \\ G(z)G(w) &= \frac{2}{3}\frac{c}{(z-w)^3} + 2\frac{T(w)}{z-w} \end{aligned} \tag{5.3.6}$$

---

[1] $f^{ABC}f^{ABD} = c_A\delta^{CD}$

where

$$c = \frac{1}{2}d_G + \frac{kd_G}{k+c_A} \qquad (5.3.7)$$

Now let $H$ be a semi-simple subgroup of $G$ and consider a symmetric coset decomposition of the Lie algebra $\mathcal{G}$:

$$\mathcal{G} = \mathcal{H} \oplus \mathcal{K} \qquad (5.3.8)$$

with

$$\begin{aligned} [\mathcal{H},\mathcal{H}] &= \mathcal{H} \\ [\mathcal{K},\mathcal{K}] &= \mathcal{H} \\ [\mathcal{H},\mathcal{K}] &= \mathcal{K} \end{aligned} \qquad (5.3.9)$$

We use the following index notation: $a, b, c \in \mathcal{H}$, $\alpha, \beta, \gamma \in \mathcal{K}$, so that $f^{ABC} = \{f^{abc}, f^{a\beta\gamma}\}$ To define the (super) GKO currents we construct the following stress–energy tensor

$$T_{G/H} = \frac{1}{k+c_A(G)} J^\alpha J^\alpha + i\frac{k}{k+c_A(G)} f^{\alpha\beta a}\lambda^\alpha\lambda^\beta J^a - \frac{1}{2}\frac{k^2}{k+c_A(G)}\lambda^\alpha\partial\lambda^\alpha \qquad (5.3.10)$$

and the supercurrent:

$$G_{G/H} = \sqrt{2}\sqrt{\frac{k}{k+c_A}}\lambda^\alpha(z)J^\alpha(z) \qquad (5.3.11)$$

$T$ and $G$ close an N=1 superalgebra with central charge

$$c = \frac{3}{2}\frac{k}{k+c_A}\dim\frac{G}{H} \qquad (5.3.12)$$

To realize the N=2 superalgebra, we consider the special case where $G = SU(2)$ and $H$ is the maximal torus $U(1)$ whose Lie algebra is the Cartan subalgebra $\mathcal{H} \subset \mathcal{G}$. It is convenient to change the basis of the fields appearing in Eq. (5.3.10). We define

$$\begin{aligned} \chi^+(z) &= (J^1 + iJ^2) \\ \chi^-(z) &= (J^1 - iJ^2) \\ \sqrt{2}H(z) &= 2J^3 \\ \psi &= \sqrt{\frac{p}{2}}(\lambda^1 + i\lambda^2) \\ \psi^* &= \sqrt{\frac{p}{2}}(\lambda^1 - i\lambda^2) \end{aligned} \qquad (5.3.13)$$

It is immediate to verify that $\chi^\pm, H$ satisfy the OPEs

$$\begin{aligned} H(z)H(w) &= \frac{k}{(z-w)^2} \\ H(z)\chi^\pm(w) &= \frac{\pm\sqrt{2}}{(z-w)}\chi^\pm(w) \\ \chi^+(z)\chi^-(w) &= \frac{\sqrt{2}}{(z-w)}H(w) + \frac{k}{(z-w)^2} \end{aligned} \qquad (5.3.14)$$

and

$$\psi^*(z)\psi(w) = \frac{1}{(z-w)} \tag{5.3.15}$$

In terms of these new variables the stress–energy tensor can be written as

$$T(z) = \frac{1}{2(k+2)}\left[\chi^+\chi^- + \chi^-\chi^+\right] + \frac{1}{k+2}\sqrt{2}H(z)\psi^*\psi + \frac{k}{k+2}\left[\psi^*\partial\psi + \psi\partial\psi^*\right] \tag{5.3.16}$$

The N=2 algebra can be easily constructed by looking at the operator algebra of the field defined in (5.3.13) and by guessing the correct structure of the N=2 currents. We can indeed verify that the operators

$$\begin{aligned} J(z) &= \frac{1}{k+2}(\sqrt{2}H(z) + k\psi^*(z)\psi(z)) \\ G^+(z) &= \sqrt{\frac{2}{k+2}}\psi^*\chi^+(z) \\ G^-(z) &= \sqrt{\frac{2}{k+2}}\psi(z)\chi^-(z) \end{aligned} \tag{5.3.17}$$

together with the stress–energy tensor defined above close an N=2 algebra with central charge

$$c = \frac{3k}{k+2} \tag{5.3.18}$$

We have then finally realized the N=2 algebra in terms of an $SU(2)_k$ Kac–Moody algebra plus a complex free fermion. The choice of the possible boundary conditions on the cylinder, given by Eq. (5.2.2), is completely determined by the boundary conditions of the free fermion $\psi$. If $\psi$ is periodic we naturally select the Ramond sector; vice-versa, if $\psi$ is antiperiodic we select the Neveu–Schwarz sector.

The Hilbert space structure of the theory is given by the tensor product of the Fock spaces associated with the $SU(2)_k$ theory and with the free fermion theory.

This means that highest weight states are built up from the tensor product of primary fields of the $SU(2)_k$ theory with the Fock vacuum of the free fermion theory. This last, in the case of Ramond boundary conditions, carries a representation of the Clifford algebra. To make this statement more precise we define the following quantum numbers. Let $l = 2j$, where $j$ is the eigenvalue of the isospin for the $SU(2)$ representation ($l = 0, \ldots, k$). A primary field $O^l_m$ in the SU(2) theory is defined by the following conditions:

$$\begin{aligned} \sqrt{2}H(z)O^l_m(w) &= \frac{m}{z-w} + \text{reg} \\ T_{SU(2)}(z)O^l_m(w) &= \frac{\Delta_l}{(z-w)^2} + \frac{\partial O^l_m}{z-w} + \text{reg} \end{aligned} \tag{5.3.19}$$

where $T_{SU(2)}(z) = \frac{1}{k+2}J^iJ^i$ $(i = 1, 2, 3)$, $\Delta_l = \frac{l(l+2)}{4(k+2)}$ and $m = -l, -l+2, \ldots, l$.

A simple calculation now shows that the primary fields, corresponding to the N=2 minimal algebra (5.3.16)–(5.3.17), have the following values for the conformal weights $h$ and $U(1)$ charges $q$:

$$\begin{aligned} q_{m,s} &= \frac{m}{k+2} + \frac{ks}{k+2} \\ h_{l,m,s} &= \frac{l(l+2)}{4(k+2)} - \frac{m^2}{4(k+2)} + \frac{s^2+2sm}{2(k+2)} \end{aligned} \tag{5.3.20}$$

where $s$ is the eigenvalue of the zero-mode of the operator $\psi^*\psi$, which is zero in the Neveu–Schwarz sector, and takes values $\pm\frac{1}{2}$ in the Ramond sector. Equation (5.3.20) coincides with (5.3.1).

Chiral primary fields in the Neveu–Scwharz sector are identified by the equation

$$h_{l,m} = \frac{1}{2} q_{l,m} \tag{5.3.21}$$

which yields in (5.3.20) $l = m$, so that

$$h = \frac{l}{2(k+2)} \qquad l = 0 \dots k. \tag{5.3.22}$$

Similarly antichiral primary fields satisfy $l = -m$.

The fact that minimal conformal models can be studied in terms of the GKO construction with $G/H = SU(2)/U(1)$, is particularly relevant in the study of modular invariant partition functions associated with these algebras. We do not give any details on this point; we just remember that, for the $SU(2)$ algebra, there exists a classification of modular invariant partition functions on the torus (in terms of characters of the representations) that falls in the so called ADE classification [72]. This classification of modular invariants is related to the Dynkin classification of the semi-simple algebras $A_n$, $D_n$, $E_6$,$E_7$, $E_8$, and for this reason is nicknamed ADE classification. Correspondingly there are specific values of central charges (and of $k$) in the N=2 minimal models which are classified as ADE minimal series, i.e.

$$\begin{array}{lll} A_n & n \geq 1, & c = \frac{3n-3}{n+1} \\ D_n & n \geq 2, & c = \frac{3n-6}{n-1} \\ E_6 & & c = \frac{5}{2} \\ E_7 & & c = \frac{8}{3} \\ E_8 & & c = \frac{14}{5} \end{array} \tag{5.3.23}$$

Notice that all the values of $c$ in (5.3.23) correspond to $c = \frac{3k}{k+2}$ for given values of $k$. For example the $D_n$ series correspond to the values of $k = 2(n-2)$, which are the same labelling the $A_{2n-3}$ series. The difference between the two series of models resides in a $\mathbb{Z}_2$ orbifold on the $A_{2n-3}$ series which projects out some fields and changes the number of modular invariants on the torus.

## 5.4 The Rheonomy Framework for N=2 Field Theories

In this section, following mainly [12, 45], we introduce the rheonomy formalism for the construction of N=2 supersymmetric field theories in $1+1$ space–time dimensions. The starting point is provided by the definition of the curvatures of the Poincaré (or de Sitter) supergroup and by the solution of the corresponding Bianchi identities for the graviton supermultiplet. For the case of N=2 supersymmetry in two space–time dimensions this solution can be given off-shell, providing a generic background that can be utilized for the coupling to any kind of N=2 matter multiplets. There is, however, a preliminary crucial point to be discussed that concerns the very definition of the Poincaré supergroup.

### 5.4.1 N=2 2D supergravity and the super-Poincaré algebra

The formulation of supergravity is performed in the physical Minkowski signature $\eta_{ab} = \mathrm{diag}(+,-)$; yet, after Wick rotation we shall deal with supersymmetry defined on compact Riemann surfaces. These have a positive, null or negative curvature $\mathcal{R}$ depending on their genus $g$ ($\mathcal{R} > 0$ for $g = 0$, $\mathcal{R} = 0$ for $g = 1$ and $\mathcal{R} < 0$ for $g \geq 2$). The sign of the curvature in the Euclidean theory is inherited from the same signature in the Minkowskian formulation. Since eventually we would like to discuss the theory for all genera $g$, we need formulations of supergravity that can accommodate both signs of the curvature. From the group-theoretical point of view, curved superspace with $\mathcal{R} > 0$ is a continuous deformation of the supersymmetric version of the de Sitter space, whose isometry group is $SO(1, D)$, $D$ being the space–time dimensions. Hence, for $\mathcal{R} > 0$ the appropriate superalgebra to begin with is, if it exists, the $N$-extended supersymmetrization of $SO(1, D)$. Similarly, for $\mathcal{R} < 0$, curved superspace is a continuous deformation of the supersymmetric version of anti-de-Sitter space, whose isometry group is $SO(2, D-1)$. Hence, in this second case, the appropriate superalgebra to start from in the construction of supergravity is the $N$-extended supersymmetrization of $SO(2, D-1)$. Alternatively, one can start from the Poincaré superalgebra, which corresponds to an Inonü–Wigner contraction of either the de Sitter or the anti-de-Sitter algebra, and reobtain either one of the decontracted algebras as vacuum configurations alternative to the Minkowski one, by giving suitable expectation values to the auxiliary fields appearing in the rheonomic parametrizations of the Poincaré curvatures. Actually, in space–time dimensions different from D=2, the supersymmetric extensions of both the de Sitter and the anti-de-Sitter algebras are not guaranteed to exist. For instance, in the relevant D=4 case, the real orthosymplectic algebra $Osp(4/N)$ is the $N$-superextension of the anti-de-Sitter algebra

$SO(2,3)$ but a superextension of the de Sitter algebra $SO(1,4)$ does not exist. This is the group-theoretical rationale of some otherwise well known facts. In four dimensions all supergravity vacua with a positive sign of the cosmological constant (de Sitter vacua) break supersymmetry spontaneously, while the only possible supersymmetric vacua are either in Minkowski or in anti-de-Sitter space. Indeed, starting from a formulation of D=4 supergravity based on the Poincaré superalgebra, one obtains both de Sitter and anti-de-Sitter vacuum configurations through suitable expectation values of the auxiliary fields, but it is only in the anti-de-Sitter case that these constant expectation values are compatible with the Bianchi identities of a superalgebra, namely respect supersymmetry. In the de Sitter case the gravitino develops a mass and supersymmetry is broken. In other words, in D=4, supersymmetry chooses a definite sign for the curvature $\mathcal{R} < 0$ (see for example [76, 256]).

If this were the case also in $D = 2$, supersymmetric theories could not be constructed on all Riemann surfaces, but only either in genus $g \geq 2$ or in genus $g < 2$. Fortunately, for D=2 it happens that the de Sitter group $SO(1,2)$ and the anti-de-Sitter group $SO(2,1)$ are isomorphic. Hence, a supersymmetrization of one is also a supersymmetrization of the other, upon a suitable formal correspondence. Once we have fixed the conventions for what we call the physical zweibein, spin connection and gravitini, we obtain an off-shell formulation of supergravity where the sign of the curvature is fixed: it is either non-negative or non-positive. Through a field redefinition we can however make a transition from one case to the other, but a continuous deformation of the auxiliary field vacuum expectation value is not sufficient to make a shift from one case to the other. Furthermore, as we are going to see, the Inonü–Wigner contraction of the N=2 algebra displays also some interesting features with respect to the $U(1)$ generator associated with the graviphoton.

The most general $N = 2, D = 2$ superalgebra one can write down, through Maurer–Cartan equations, is obtained by setting to zero the following curvatures:

$$
\begin{aligned}
T^{+} &= de^{+} + \omega e^{+} - \frac{i}{2}\zeta^{+}\zeta^{-} \\
T^{-} &= d\epsilon^{-} - \omega\epsilon^{-} - \varepsilon\frac{i}{2}\tilde{\zeta}_{+}\tilde{\zeta}_{-} \\
\rho^{+} &= d\zeta^{+} + \frac{1}{2}\omega\zeta^{+} + \frac{ia_1}{4}A^{\bullet}\zeta^{+} + a_1 a_2 \tilde{\zeta}_{-}\epsilon^{+} \\
\rho^{-} &= d\zeta^{-} + \frac{1}{2}\omega\zeta^{-} - \frac{ia_1}{4}A^{\bullet}\zeta^{-} + a_1 a_2 \tilde{\zeta}_{+}\epsilon^{+} \\
\tilde{\rho}_{+} &= d\tilde{\zeta}_{+} - \frac{1}{2}\omega\tilde{\zeta}_{+} - \frac{ia_1}{4}A^{\bullet}\tilde{\zeta}_{+} - \varepsilon a_1 a_2 \zeta^{-}\epsilon^{-} \\
\tilde{\rho}_{-} &= d\tilde{\zeta}_{-} - \frac{1}{2}\omega\tilde{\zeta}_{-} + \frac{ia_1}{4}A^{\bullet}\tilde{\zeta}_{-} - \varepsilon a_1 a_2 \zeta^{+} e^{-} \\
R &= d\omega - 2\varepsilon a_1^2 a_2^2 e^{+}\epsilon^{-} - \frac{i}{2}a_1 a_2(\zeta^{+}\tilde{\zeta}_{+} + \zeta^{-}\tilde{\zeta}_{-}) \\
F^{\bullet} &= dA^{\bullet} - a_2(\zeta^{-}\tilde{\zeta}_{-} - \zeta^{+}\tilde{\zeta}_{+})
\end{aligned}
\tag{5.4.1}
$$

where $e^+$ and $e^-$ denote the two components (left and right-moving) of the world-sheet zweibein 1-form, while $\zeta^+$, $\tilde{\zeta}_+$ are the two components of the gravitino one form, $\zeta^-$, $\tilde{\zeta}_-$ are the two components of its complex conjugate. $\varepsilon$ can take the values $\pm 1$ and distinguishes the de Sitter ($\varepsilon = 1$) and anti-de-Sitter ($\varepsilon = -1$) cases. Formally, one can pass from positive to negative curvature by replacing $e^-$ and $T^-$ with $-e^-$ and $-T^-$. The algebra (5.4.1) is understood in the following way. Adopting the notations of [76], the explicit representation of the 2D gamma matrix algebra

$$\begin{aligned} \left\{\gamma^a, \gamma^b\right\} &= \eta^{ab} \quad ; \quad a, b = 0, 1 \\ \eta^{ab} &= \mathrm{diag}\,(+-) \end{aligned} \tag{5.4.2}$$

is as follows:

$$\begin{aligned} \gamma^0 &= \begin{pmatrix} 0 & -i \\ i & 0 \end{pmatrix} \\ \gamma^1 &= \begin{pmatrix} 0 & i \\ i & 0 \end{pmatrix} \end{aligned} \tag{5.4.3}$$

The charge conjugation matrix is defined by

$$C\,\gamma^a\,C^{-1} = -\,(\gamma^a)^T \tag{5.4.4}$$

with

$$C = \begin{pmatrix} 0 & 1 \\ -1 & 0 \end{pmatrix} \tag{5.4.5}$$

and we obtain

$$\gamma^{ab} = \frac{1}{2}\left[\gamma^a, \gamma^b\right] = \varepsilon^{ab}\,\gamma^3 \tag{5.4.6}$$

where

$$\begin{aligned} \gamma^3 &= \begin{pmatrix} 1 & 0 \\ 0 & -1 \end{pmatrix} \\ \varepsilon^{ab} &= \begin{pmatrix} 0 & 1 \\ -1 & 0 \end{pmatrix} \end{aligned} \tag{5.4.7}$$

Then, for the value $\varepsilon = 1$, the curvatures (5.4.1) are Lorentz-covariantly rewritten as follows:

$$\begin{aligned} T^a &= \mathcal{D}V^a - \frac{i}{2}\overline{\xi} \wedge \gamma^a \xi \\ \rho &= \mathcal{D}\xi + i\frac{a_1}{4}A^\bullet \wedge \xi + i\frac{a_1 a_2}{2}\gamma^a \xi \wedge V_a \\ R &= d\omega - \frac{1}{2}\,a_1^2 a_2^2\,\varepsilon_{ab}\,V^a \wedge V^b - \frac{1}{2}\,a_1 a_2 \overline{\xi} \wedge \gamma^3 \xi \\ F^\bullet &= dA^\bullet - i\,a_2\,\overline{\xi} \wedge \xi \end{aligned} \tag{5.4.8}$$

where $V^a$ is the zweibein, $\xi$ is the gravitino 1-form, $A^\bullet$ is the $U(1)$ connection, $\omega^{ab} = \varepsilon^{ab}\,\omega$ is the spin connection and $\mathcal{D}$ denotes the Lorentz-covariant derivative: $\mathcal{D}V^a = dV^a - \omega\,\varepsilon^{ab}\wedge V^b$, $\mathcal{D}\xi = d\xi - \frac{1}{2}\omega\,\gamma^3\xi$. The gravitino $\xi$ is a Dirac spinor. In general we can write

$$\xi = e^{-i\pi/4}\begin{pmatrix}\zeta^+\\ \tilde{\zeta}_-\end{pmatrix} \tag{5.4.9}$$

with $\zeta^+$ and $\tilde{\zeta}_-$ independent complex components. If we set

$$\begin{aligned}\zeta^- &= \left(\zeta^+\right)^*\\ \tilde{\zeta}_+ &= \left(\tilde{\zeta}_-\right)^*\end{aligned} \tag{5.4.10}$$

and if we define

$$e^\pm = \frac{1}{2}(V^0 \pm V^1) \tag{5.4.11}$$

then we obtain

$$T^a = dV^a - \omega^{ab}\wedge V^b - \frac{i}{2}\overline{\xi}\wedge\gamma^a\xi \tag{5.4.12}$$

or

$$T^\pm = de^\pm \pm \omega\wedge e^\pm - \frac{i}{2}\overline{\xi}\wedge\gamma^\pm\xi \tag{5.4.13}$$

where $\gamma^\pm \equiv \frac{1}{2}\gamma^0(1\pm\gamma_3)$. Using (5.4.9), we have

$$\begin{aligned}T^+ &= de^+ + \omega\wedge e^+ - \frac{i}{2}\zeta^+\wedge\zeta^-\\ T^- &= de^- + \omega\wedge e^- - \frac{i}{2}\tilde{\zeta}_+\wedge\tilde{\zeta}_-\end{aligned} \tag{5.4.14}$$

Similarly, with these notations, from Eq. (5.4.8) we reobtain all the other equations in (5.4.1) for the case $\varepsilon = 1$. Indeed it suffices to identify

$$\begin{aligned}\rho &= e^{-i\pi/4}\begin{pmatrix}\rho^+\\ \tilde{\rho}_-\end{pmatrix}\\ \overline{\rho} &= \rho^\dagger\gamma^0 = e^{-i\pi/4}\left(-\tilde{\rho}_+\,,\,-\rho^+\right)\end{aligned} \tag{5.4.15}$$

As already noted, the case of the other algebra $\varepsilon = -1$ is formally obtained by redefining $e^- \longrightarrow -e^-$. In the ordinary basis this corresponds to the exchange $V^0 \longrightarrow V^1$, $V^1 \longrightarrow V^0$ which is equivalent to a reversal of the Lorentzian signature from $\eta^{ab} = \mathrm{diag}\,(+-)$ to $\eta^{ab} = \mathrm{diag}\,(-+)$ and hence to a transition from $SO(2,1)$ to $SO(1,2)$. The algebra (5.4.8) contains two free (real) parameters $a_1, a_2$ in its structure constant. Choosing $a_1 = a_2 = \frac{a}{\sqrt{2}} \neq 0$ we have the usual curvature definitions for a de Sitter algebra with cosmological constant $\Lambda = \varepsilon a^2$, namely the superextension of the $SL(2,R)$ Lie algebra. In the limit $a_2 \to 1$ and $a_1 \to 0$ we get the usual $D = 2$ analogue of the $N = 2$ super

Poincaré Lie algebra, where, calling $L$ the $U(1)$ generator dual to the graviphoton $A^\bullet$, the supercharge $Q^\pm, \tilde{Q}^\pm$ are neutral under $L$:

$$[L, Q^\pm] = [L, \tilde{Q}^\pm] = 0 \tag{5.4.16}$$

From (5.4.16) we can easily interpret $L$ as a "central charge", since it appears in the supercharges anticommutators:

$$\{Q^-, \tilde{Q}^-\} \sim L \qquad \{Q^+, \tilde{Q}^+\} \sim L \tag{5.4.17}$$

Finally, in the limit $a_2 \to 0$, $a_1 \to 1$ we get a new kind of Poincaré superalgebra, which we name "charged Poincaré ", where the supercharges do rotate under the $U(1)$ action:

$$[L, Q^\pm] = \pm Q^\pm \qquad [L, \tilde{Q}^\pm] = \mp Q^\pm \tag{5.4.18}$$

In this case $L$ is not a central charge, since it does not appear in the supercharge anticommutators:

$$\begin{array}{ll} \{Q^+, Q^-\} = P & \{\tilde{Q}^+, \tilde{Q}^-\} = \tilde{P} \\ \{Q^+, \tilde{Q}^+\} = 0 & \{Q^-, \tilde{Q}^-\} = 0 \end{array} \tag{5.4.19}$$

$P$ and $\tilde{P}$ being the left and right translations dual to $e^+$ and $e^-$ respectively.

In the construction of global $N = 2$ supersymmetric theories we consider flat superspace described by the zero curvature condition

$$T^\pm = \rho = F^\bullet = R = 0 \tag{5.4.20}$$

In this case the use of the ordinary Poincaré superalgebra is indistinguishable from the use of the charged one. Indeed we can always choose the gauge $\omega = A^\bullet = 0$ and we can altogether forget about these 1-forms: in the solution of the Bianchi identities we simply have to respect the global Lorentz and $U(1)$ symmetries. However, at the level of curved superspace, there is a novelty that distinguishes $D = 2$ from higher dimensions. It turns out that the correct algebra is the charged one. For this reason, although in this book we confine ourselves to globally supersymmetric theories, in the present section we consider the solution of Bianchi identities for the N=2 supergravity multiplet. If we wanted to extend the analysis of topological sigma models to higher genus Riemann surfaces and couple it to topological gravity (as it is required in applications to full-fledged string theory), we had to use the supergravity theory we now present.

The field content of the off shell graviton multiplet is easily described. The zweibein describes one bosonic degree of freedom (four components restricted to one by the two-parameter diffeomorphisms and by the one-parameter Lorentz symmetry), while each gravitino describes two degrees of freedom (eight components restricted by the four-parameter supersymmetry). Finally, the graviphoton $A^\bullet$ yields one bosonic degree of freedom (two components restricted by the one-parameter $U(1)$ gauge symmetry). The mismatch of two bosonic degrees of freedom is filled by a complex scalar auxiliary field $S$

and by its conjugate $\overline{\mathcal{S}}$. The problem is therefore that of writing a rheonomic parametrization for the curvatures (5.4.1) using as free parameters their space–time components plus an auxiliary complex scalar $\mathcal{S}$.

As can be easily read from (5.4.1) the curvature two-forms satisfy

$$\begin{aligned}
\nabla T^{+} &= Re^{+} - \frac{i}{2}(\rho^{+}\zeta^{-} - \zeta^{+}\rho^{-}) \\
\nabla T^{-} &= -Re^{-} - \frac{i}{2}\varepsilon(\tilde{\rho}_{+}\tilde{\zeta}_{-} - \tilde{\zeta}_{+}\tilde{\rho}_{-}) \\
\nabla \rho^{\pm} &= \frac{1}{2}R\zeta^{\pm} \pm \frac{ia_1}{4}F^{\bullet}\zeta^{\pm} + a_1 a_2(\tilde{\rho}^{\mp}e^{+} - \tilde{\zeta}_{\mp}T^{+}) \\
\nabla \tilde{\rho}_{\pm} &= -\frac{1}{2}R\tilde{\zeta}_{\pm} \mp \frac{ia_1}{4}F^{\bullet}\tilde{\zeta}_{\pm} - a_1 a_2 \varepsilon(\rho^{\mp}e^{-} - \zeta^{\mp}T^{-}) \\
\nabla R &= -2a_1^2 a_2^2 \varepsilon(T^{+}e^{-} - e^{+}T^{-}) - \frac{i}{2}a_1 a_2(\rho^{+}\tilde{\zeta}_{+} - \zeta^{+}\tilde{\rho}_{+} + \rho^{-}\tilde{\zeta}_{-} - \zeta^{-}\tilde{\rho}^{-}) \\
\nabla F^{\bullet} &= -a_2(\rho^{-}\tilde{\zeta}_{-} - \zeta^{-}\tilde{\rho}_{-} - \rho^{+}\tilde{\zeta}_{+} + \zeta^{+}\tilde{\rho}_{+})
\end{aligned} \tag{5.4.21}$$

The general solution for the above Bianchi identities with vanishing torsion is

$$\begin{aligned}
&T^{+} = 0 \qquad\qquad T^{-} = 0 \\
&\rho^{+} = \tau^{+}e^{+}e^{-} - a_1(\mathcal{S} - a_2)\tilde{\zeta}_{-}e^{+} \qquad \tilde{\rho}_{+} = \tilde{\tau}_{+}e^{+}e^{-} + a_1\varepsilon(\mathcal{S} - a_2)\zeta^{-}e^{-} \\
&R = (\mathcal{R} - 2a_1^2 a_2^2 \varepsilon)e^{+}e^{-} + \tfrac{i}{2}\varepsilon e^{-}(\tau^{+}\zeta^{-} + \tau^{-}\zeta^{+}) + \tfrac{i}{2}e^{+}(\tilde{\tau}_{-}\tilde{\zeta}_{+} + \tilde{\tau}_{+}\tilde{\zeta}^{-}) \\
&\quad + \tfrac{ia_1}{2}\left[(\mathcal{S} - a_2)\zeta^{-}\tilde{\zeta}_{-} + (\overline{\mathcal{S}} - a_2)\zeta^{+}\tilde{\zeta}^{+}\right] \\
&F^{\bullet} = \mathcal{F}e^{+}e^{-} + (\mathcal{S} - a_2)\zeta^{-}\tilde{\zeta}_{-} - (\overline{\mathcal{S}} - a_2)\zeta^{+}\tilde{\zeta}_{+} - \tfrac{1}{a_1}\varepsilon(\tau^{+}\zeta^{-} - \tau^{-}\zeta^{+})e^{-} \\
&\quad + \tfrac{1}{a_1}(\tilde{\tau}_{-}\tilde{\zeta}_{+} - \tilde{\tau}_{+}\tilde{\zeta}_{-})e^{+}.
\end{aligned} \tag{5.4.22}$$

The formulae for $\rho^{-}$ and $\tilde{\rho}_{-}$ can be derived from those of $\rho^{+}$ and $\tilde{\rho}_{+}$ by complex conjugation. In doing this, one has to take into account that the complex conjugation reverses the order of the fields in a product of fermions.

It is immediate to see in Eqs. (5.4.22) that the limit $a_1 \to 0$ is singular, and this reflects the fact that we are not able to find the correct parametrizations for this case. On the contrary the limit $a_2 \to 0$ is perfectly consistent and we call it "charged Poincaré algebra".

From now on, to avoid any confusion in using the formulae for the curvature definition, *we will always refer to the symbols* $R, F^{\bullet}, \rho^{\pm}, \tilde{\rho}^{\pm}$ *as to those defined in (5.4.1) with* $a_1 = 1, a_2 = 0$.

The general rule for obtaining the solution to the Bianchi identities is the rheonomic principle. For this specific case we recall the three steps necessary to solve Bianchi identities (see also Chapter 3)

i) One expands the curvatures 2-forms $\rho^{\pm}$, $\tilde{\rho}^{\pm}$, $R$ and $F^{\bullet}$ in a basis of superspace 2-forms: the "space–time" form $e^{+}e^{-}$ and the "superspace" forms, which can be fermionic, like $\zeta^{\pm}e^{\pm}$ and $\tilde{\zeta}^{\pm}e^{\pm}$, or bosonic, like $\zeta^{\pm}\zeta^{\pm}$, $\zeta^{\pm}\tilde{\zeta}^{\pm}$ and $\tilde{\zeta}^{\pm}\tilde{\zeta}^{\pm}$.

ii) The coefficients $\tau^\pm$, $\tilde{\tau}^\pm$, $\mathcal{R}$ and $\mathcal{F}$ of the space–time form $e^+e^-$ are the independent ones: the rheonomic parametrizations (5.4.1) can be viewed as a definition of them. They are the supercovariantized derivatives of the fields. In particular, $\mathcal{R}$ is the supercurvature and $\mathcal{F}$ is the super-field-strength.

iii) The coefficients of the superspace forms, instead, are functions of the fields and of the supercovariantized derivatives $\tau^\pm$, $\tilde{\tau}^\pm$, $\mathcal{R}$ and $\mathcal{F}$. They are determined by solving the Bianchi identities (5.4.21) in superspace. Their form is strongly constrained by Lorentz invariance, global $U(1)$ invariance and scale invariance. These restrictions are such that the role of (5.4.21) is simply that of fixing some numerical coefficients, while providing also several self-consistency checks. Moreover, implementation of (5.4.21) also provides the rheonomic parametrizations of $\nabla\tau^\pm$, $\nabla\tilde{\tau}^\pm$, $\nabla\mathcal{R}$ and $\nabla\mathcal{F}$ and of the covariant derivatives $\nabla\mathcal{S}$ and $\nabla\overline{\mathcal{S}}$ of the auxiliary fields $\mathcal{S}$ and $\overline{\mathcal{S}}$, namely

$$\begin{aligned}
\nabla\tau^+ &= \nabla_+\tau^+e^+ + \nabla_-\tau^+e^- + \left(\frac{1}{2}\mathcal{R} + \frac{i}{4}\mathcal{F} - \varepsilon\mathcal{S}\overline{\mathcal{S}}\right)\zeta^+ + \nabla_-\mathcal{S}\tilde{\zeta}_- \\
\nabla\tilde{\tau}_+ &= \nabla_+\tilde{\tau}_+e^+ + \nabla_-\tilde{\tau}_+e^- - \left(\frac{1}{2}\mathcal{R} + \frac{i}{4}\mathcal{F} - \varepsilon\mathcal{S}\overline{\mathcal{S}}\right)\tilde{\zeta}_+ + \nabla_+\mathcal{S}\zeta_- \\
\nabla\mathcal{S} &= \nabla_+\mathcal{S}e^+ + \nabla_-\mathcal{S}e^- - \frac{i}{2}\varepsilon(\tilde{\tau}_+\zeta^+ + \tau^+\tilde{\zeta}_+) \\
\nabla\mathcal{R} &= \nabla_+\mathcal{R}e^+ + \nabla_-\mathcal{R}e^- + \frac{i}{2}(\varepsilon\nabla_-\tilde{\tau}_-\tilde{\zeta}_+ + \varepsilon\nabla_-\tilde{\tau}_+\tilde{\zeta}_- - \nabla_+\tau^-\zeta^+ - \nabla_+\tau^+\zeta^-) \\
&- i[\overline{\mathcal{S}}(\tau^+\tilde{\zeta}_+ + \tilde{\tau}_+\zeta^+) + \mathcal{S}(\tau^-\tilde{\zeta}_- + \tilde{\tau}_-\zeta^-)] \\
\nabla\mathcal{F} &= \nabla_+\mathcal{F}e^+ + \varepsilon\nabla_-\mathcal{F}e^- - \varepsilon\nabla_-\tilde{\tau}_-\tilde{\zeta}_+ + \nabla_-\tilde{\tau}_+\tilde{\zeta}_- + \nabla_+\tau^-\zeta^+ - \nabla_+\tau^+\zeta^- \quad (5.4.23)
\end{aligned}$$

These equations, in turn, are the definitions of the supercovariantized derivatives of $\tau^\pm$, $\tilde{\tau}^\pm$, $\mathcal{R}$ and $\mathcal{F}$. Finally, the $e^+$–$e^-$ sector of the Bianchi identities (5.4.21) gives the "space–time" counterparts of the Bianchi identities themselves, i.e. the formulæ for $[\nabla_+, \nabla_-]\Phi$ of the various fields $\Phi$.

The formal correspondence between de Sitter and anti-de-Sitter theories is summarized by

$$e^- \to -e^-, \quad \tau \to -\tau, \quad \mathcal{R} \to -\mathcal{R}, \quad \mathcal{F} \to -\mathcal{F}, \quad \nabla_- \to -\nabla_-. \qquad (5.4.24)$$

From (5.4.23) we can confirm that $\varepsilon = 1$ corresponds to positive curvature, while $\varepsilon = -1$ corresponds to negative curvature. Indeed, setting $\mathcal{R}$=const and $\mathcal{F} = 0$, the expressions of $\nabla\mathcal{R}$ and $\nabla\mathcal{F}$ imply either $\mathcal{S} = \overline{\mathcal{S}} = 0$ or $\tau^\pm = \tilde{\tau}^\pm = 0$. If $\mathcal{S} = \overline{\mathcal{S}} = 0$, then $\nabla\mathcal{S}$ and $\nabla\overline{\mathcal{S}}$ also imply $\tau^\pm = \tilde{\tau}^\pm = 0$. So, we can conclude that $\tau^\pm = \tilde{\tau}^\pm = 0$ is in any case true. Finally, $\nabla\tau$ implies $\mathcal{R} = 2\varepsilon\mathcal{S}\overline{\mathcal{S}}$ and $\mathcal{S}$ =const. This also shows that one cannot move from the de Sitter to the anti-de-Sitter case by a continuous deformation of the expectation value $\mathcal{S}, \overline{\mathcal{S}}$.

For simplicity, from now on we set $\varepsilon = +1$.

For convenience we also recall the rule for complex conjugation. Let $\psi_1, \psi_2$ be two forms of degree $p_1, p_2$ and statistics $F_1, F_2$ ($F = 0, 1$ for bosons or fermions) so that

$\psi_1\psi_2 = (-1)^{p_1p_2+F_1F_2}\,\psi_2\psi_1$; then we have:

$$(\psi_1\psi_2)^* = (-1)^{p_1p_2+F_1F_2}\,\psi_1^*\psi_2^* \tag{5.4.25}$$

Thus, for example, for the gravitinos we have

$$(\zeta^+\wedge\zeta^-)^* = (\zeta^+)^*\wedge(\zeta^-)^* = -\zeta^-\wedge\zeta^+ = -\zeta^+\wedge\zeta^- \tag{5.4.26}$$

### 5.4.2 Chiral multiplets in curved superspace

In the previous subsection we derived the off-shell structure of the graviton multiplet. The next step is to introduce matter multiplets. As we stressed several times, in this book we shall confine our analysis to globally supersymmetric theories. It is, however, useful to have an idea of how our formulae generalize if we include also the effects due to 2D-dimensional gravity, namely if we write our matter theories on a curved supersymmetric world-sheet. For this reason, in the present subsection we consider the rheonomic construction of chiral multiplets in a curved superspace environnment. Chiral multiplets are the basic matter multiplets that enter either the $\sigma$-model lagrangian or the Landau–Ginzburg lagrangian. The field content of an off-shell chiral multiplet is $X^I,\psi^I,\widetilde{\psi}^I,H^I$, where $X^I$ is a complex physical scalar, $(\psi^I,\widetilde{\psi}^I)$ are complex spin $\frac{1}{2}$ fields and $H^I$ is a complex auxiliary scalar. In the $\sigma$-model case the fields $X^I$ play the role of complex coordinates of the target Kähler manifold. In the Landau–Ginzburg case they are instead coordinates in $\mathbb{C}^n$ and describe scalar fields with a canonical kinetic term, but interacting via a holomorphic superpotential $\mathcal{W}(X^I)$. As already pointed out, the two theories can be seen as effective actions in two phases of the same theory. This happens, in flat superspace, if we couple chiral multiplets to gauge multiplets and if we introduce a suitable gauge-invariant superpotential $W(X^I)$ (distinct, as we are going to see, from the superpotential $\mathcal{W}(X^I)$ of the effective Landau–Ginzburg model). This construction will be considered in later subsections. In the present one we just consider chiral multiplets in a curved superspace self-interacting via a superpotential.

To start our programme we need the correct covariant derivatives for the matter fields (Landau–Ginzburg fields)[2], which are

$$\begin{aligned}
\nabla X^I &= dX^I \\
\nabla\psi^I &= d\psi^I - \frac{1}{2}\omega\psi^I + \frac{i}{4}A^\bullet\psi^I \\
\nabla\widetilde{\psi}^I &= d\widetilde{\psi}^I + \frac{1}{2}\omega\widetilde{\psi}^I - \frac{i}{4}A^\bullet\widetilde{\psi}^I \\
\nabla H^I &= dH^I
\end{aligned} \tag{5.4.27}$$

From the Bianchi identities, which are easily read off from (5.4.27), we find the following rheonomic parametrizations:

$$\nabla X^I = \nabla_+X^Ie^+ + \nabla_-X^Ie^- + \psi^I\zeta^- + \widetilde{\psi}^I\zeta_-$$

[2] Our notation for the covariant derivative is $\nabla\phi = d\phi - s\omega\phi - \frac{i}{2}qA^\bullet$, where $s,q$ are the spin and the $U(1)$ charge for the field $\phi$.

$$
\begin{aligned}
\nabla\psi^I &= \nabla_+\psi^I e^+ + \nabla_-\psi^I e^- - \frac{i}{2}\nabla_+ X^I \zeta^+ + H^I \tilde{\zeta}_- \\
\nabla\tilde{\psi}^I &= \nabla_+\tilde{\psi}^I e^+ + \nabla_-\tilde{\psi}^I e^- - \frac{i}{2}\nabla_- X^I \tilde{\zeta}_+ - H^I \zeta^- \\
\nabla H^I &= \nabla_+ H^I e^+ + \nabla_- H^I e^- - \frac{i}{2}\nabla_-\psi^I \tilde{\zeta}_+ + \frac{i}{2}\nabla_+\tilde{\psi}^I \zeta^+
\end{aligned} \tag{5.4.28}
$$

The usual choice for the auxiliary field in the Landau–Ginzburg matter system is

$$
H^I = \eta^{IJ^*}\partial_{J^*}\overline{W} \tag{5.4.29}
$$

$\eta_{IJ*}$ denoting a flat (constant) metric. If we explicitly make this choice, we also find the fermionic equation of motions from consistency of the parametrizations $\nabla H^I$:

$$
\begin{aligned}
&\tfrac{i}{2}\ \nabla_-\psi^I - \eta^{IJ^*}\partial_{M^*}\partial_{J^*}\overline{W}\tilde{\psi}^{M^*} = 0 \\
&\tfrac{i}{2}\ \nabla_+\tilde{\psi}^I + \eta^{IJ^*}\partial_{M^*}\partial_{J^*}\overline{W}\psi^{M^*} = 0
\end{aligned} \tag{5.4.30}
$$

Finally, from the supersymmetric variations of the fermionic field equation we find the bosonic field equation

$$
\begin{aligned}
[\nabla_-\nabla_+ &+ \nabla_+\nabla_-]X^I - 8\eta^{IJ^*}\partial_{M^*}\partial_{J^*}\partial_{L^*}\overline{W}\psi^{L^*}\tilde{\psi}^{M^*} + 8\eta^{IJ^*}\partial_{M^*}\partial_{J^*}\overline{W}\eta^{M^*L}\partial_L W \\
&- 4i\overline{\mathcal{S}}\eta^{IJ^*}\partial_{J^*}\overline{W} + \tau^-\psi^I - \tilde{\tau}_-\tilde{\psi}^I = 0
\end{aligned} \tag{5.4.31}
$$

The coupling of the Landau–Ginzburg matter with the N=2 supergravity is described by the following lagrangian, derived from the field equations (5.4.29), (5.4.30) and (5.4.31)

$$
\mathcal{L}_{chiral} = \mathcal{L}_{kin} + \mathcal{L}_W \tag{5.4.32}
$$

where $\mathcal{L}_{kin}$ and $\mathcal{L}_W$ are the kinetic and superpotential terms

$$
\begin{aligned}
\mathcal{L}_{kin} &= \eta_{IJ^*}(\nabla X^I - \psi^I\zeta^- - \tilde{\psi}^I\tilde{\zeta}_-)(\Pi_+^{J^*}e^+ - \Pi_-^{J^*}e^-) \\
&+ \eta_{IJ^*}(\nabla X^{J^*} + \psi^{J^*}\zeta^+ + \tilde{\psi}^{J^*}\tilde{\zeta}_+)(\Pi_+^I e^+ - \Pi_-^I e^-) \\
&+ \eta_{IJ^*}(\Pi_+^I\Pi_-^{J^*} + \Pi_-^I\Pi_+^{J^*})e^+e^- \\
&+ \eta_{IJ^*}(\nabla X^{J^*}\psi^I\zeta^- - \nabla X^I\psi^{J^*}\zeta^+ - \nabla X^{J^*}\tilde{\psi}^I\tilde{\zeta}_- + \nabla X^I\tilde{\psi}^{J^*}\tilde{\zeta}_+) \\
&+ 2i\eta_{IJ^*}(-\psi^I\nabla\psi^{J^*}\epsilon^+ - \psi^{J^*}\nabla\psi^I\epsilon^+ + \tilde{\psi}^I\nabla\tilde{\psi}^{J^*}\epsilon^- + \tilde{\psi}^{J^*}\nabla\tilde{\psi}^I\epsilon^-) \\
&+ \eta_{IJ^*}(\psi^I\tilde{\psi}^{J^*}\zeta^-\tilde{\zeta}_+ + \psi^{J^*}\tilde{\psi}^I\tilde{\zeta}_-\zeta^+) - 8\eta_{IJ^*}H^I H^{J^*}\epsilon^+\epsilon^- \\
\mathcal{L}_W &= 4i(\psi^I\partial_I W\tilde{\zeta}_+e^+ + \psi^{J^*}\partial_{J^*}\overline{W}\tilde{\zeta}_-\epsilon^+ + \tilde{\psi}^I\partial_I W\zeta^+\epsilon^- + \tilde{\psi}^{I^*}\partial_{I^*}\overline{W}\zeta^-\epsilon^-) \\
&+ 8[(\partial_I\partial_J W\psi^I\tilde{\psi}^J - \partial_{I^*}\partial_{J^*}\overline{W}\psi^{I^*}\tilde{\psi}^{J^*}]\epsilon^+\epsilon^- \\
&+ 4i(\mathcal{S}W - \overline{\mathcal{S}W})\epsilon^+\epsilon^- + 2\overline{W}\tilde{\zeta}_-\zeta^- - 2W\tilde{\zeta}_+\zeta^+ \\
&+ (8H^I\partial_I W + 8H^{I^*}\partial_{I^*}\overline{W})e^+e^-
\end{aligned} \tag{5.4.33}
$$

The fields $\Pi_\pm^I$ and $\Pi_\pm^{I*}$ are auxiliary fields for the first order formalism: their equation of motion equates them to the supercovariant derivatives of the $X$-fields,

$$
\Pi_\pm^I = \nabla_\pm X^I \qquad\qquad \Pi_\pm^{I*} = \nabla_\pm X^{I*} \tag{5.4.34}
$$

Substitution of these expressions in $\mathcal{L}_{kin}$ gives the usual second order lagrangian. The rheonomic parametrizations of $\nabla\Pi^I_\pm$ and $\nabla\Pi^{I*}_\pm$ are derived from the Bianchi identities and the rheonomic parametrizations (5.4.28) of the fields, in the same way as (5.4.23) are derived from the Bianchi identities (5.4.21) and the rheonomic parametrizations (5.4.22).

There are now two ways we can look at the above lagrangian, which correspond to the case of *dynamical* or of *external* supergravity. Let us discuss them separately.

If we are interested in the *dynamical* coupling to supergravity, we should consider the variations of (5.4.33) in the fields of the graviton multiplet along with its variations in the matter fields. These gravitational variations yield the stress–energy tensor $T_{ab}$, the supercurrent $G_a$, the $U(1)$ current $J$ and an auxiliary current $\mathcal{J}_{aux}$ associated with the variation in $\mathcal{S}$. All these currents are set to zero and their vanishing is the 2D analogue of the supergravity equations of motions. Obviously if there are other matter multiplets one has to take into account also their contributions to the supergravity currents and the supergravity field equations correspond to the vanishing of the total currents. There can also be additional couplings of the chiral fields to the curvature multiplet. This is the N=2 supersymmetrization of the dilaton coupling. Let us consider the chiral multiplet labelled with the index $I = 0$. We call it the "dilaton" supermultiplet, for a reason which will be immediately clear. For convenience, we relabel the dilaton multiplet as

$$(X^0, X^{0*}, \psi^0, \psi^{0*}, \widetilde{\psi}^0, \widetilde{\psi}^{0*}, H^0, H^{0*}) \rightarrow (X, \overline{X}, \lambda_-, \lambda_+, \widetilde{\lambda}^-, \widetilde{\lambda}^+, H, \overline{H}) \qquad (5.4.35)$$

The N=2 extension of the lagrangian $(X + \overline{X})R$ is given by

$$\begin{aligned}\frac{1}{\alpha}\mathcal{L}_{dil} &= (X+\overline{X})R - \frac{i}{2}(X-\overline{X})F^\bullet - 2\lambda_-\rho^- + 2\lambda_+\rho^+ + 2\widetilde{\lambda}^-\widetilde{\rho}_- - 2\widetilde{\lambda}^+\widetilde{\rho}_+ \\ &\quad -4i\overline{\mathcal{S}}He^+e^- + 4i\mathcal{S}\overline{H}e^+e^- \end{aligned} \qquad (5.4.36)$$

where $\alpha$ is the dilaton coupling constant. Let us recall how an off-shell supersymmetric lagrangian is constructed in the rheonomic framework. This is very simple, since it is sufficient to find an $\mathcal{L}$ that satisfies

$$\nabla\mathcal{L} = d\mathcal{L} = 0 \qquad (5.4.37)$$

In checking this equation one combines the use the rheonomic parametrizations for the covariant derivatives of the fields (5.4.22) with their definitions (5.4.1). $\mathcal{L}_{dilaton}$ was determined starting from the first term $(X + \overline{X})R$ and deriving the others in order to satisfy (5.4.37). A similar procedure is used for $\mathcal{L}_{cosm}$, which we discuss below.

One can pass from the second order formalism to the first order one by adding the term

$$\mathcal{L}_{torsion} = p_+T^+ + p_-T^- \qquad (5.4.38)$$

where $p_+, p_-$ are (bosonic) lagrangian multipliers implementing the torsion constraint $T^\pm = 0$. $\mathcal{L}_{torsion}$ is clearly supersymmetric (the supersymmetry variation of the spin connection is still determined from the variations of the vielbein and the gravitini: this is the so-called *1.5 order formalism*). Moreover we can add to Eq. (5.4.36) a "cosmological constant term" compatible with the N=2 local supersymmetry

$$\begin{aligned} \frac{1}{a}\mathcal{L}_{cosm} &= (\mathcal{S}X + \overline{\mathcal{S}X})e^+e^- + \lambda_-\overline{\zeta}_+e^+ - \lambda_+\overline{\zeta}_-e^+ + \overline{\lambda}^-\zeta^+e^- - \overline{\lambda}^+\zeta^-e^- \\ &+ \frac{i}{2}X\zeta^+\overline{\zeta}_+ + \frac{i}{2}\overline{X}\zeta^-\overline{\zeta}_- + 2i(\overline{H} - H)e^+e^- \end{aligned} \tag{5.4.39}$$

where $a$ is an additional dimensionful coupling constant that provides the scale of the curvature.

With these ingredients we can write the total lagrangian expressing the coupling of the chiral multiplets to supergravity as follows:

$$\begin{aligned} \mathcal{L}_{total} &= \mathcal{L}_{chiral} + \mathcal{L}_{Liouville} \\ \mathcal{L}_{chiral} &= \mathcal{L}_{kin} + \mathcal{L}_{W} \\ \mathcal{L}_{Liouville} &= \mathcal{L}_{dil} + \mathcal{L}_{torsion} + \mathcal{L}_{cosm} \end{aligned} \tag{5.4.40}$$

Let us now consider, in the case of dynamical supergravity, the field equations of the auxiliary fields, namely the variations in $H^I$ and $\mathcal{S}$. We get

$$\begin{aligned} H^{i^\star} &= \eta^{i^\star j}\,\partial_j\,W \\ H^{0^\star} &= \partial_0\,W - \frac{i}{2}\alpha\overline{\mathcal{S}} - 2i\,a \\ 0 &= 4i\,W + a\,X^0 + 4i\alpha\,H^{0^\star} \end{aligned} \tag{5.4.41}$$

where we distinguished the case of the index $I = 0$ from all the other cases $I = i$. To discuss these equations it is convenient to note that when in the matter lagrangian (5.4.33) the index $I$ takes the value 0 and $W = -\frac{i}{4}X^0$, $\mathcal{L}_W$ coincides with the cosmological term $\mathcal{L}_{cosm}$. Hence all the contributions in the parameter $a$ can be reabsorbed into the superpotential contributions by redefining $W(X)' = W(X) - \frac{i}{4}\,a\,X^0$. From this point of view there are just two regimes. If the dilaton coupling is zero $\alpha = 0$ we have, for all the fields

$$H^{I^\star} = \eta^{I^\star J}\,\partial_J\,W \tag{5.4.42}$$

but the variation in the gravitational field $\mathcal{S}$ enforces:

$$W(X) = 0 \tag{5.4.43}$$

while the value of $\mathcal{S}$ remains undetermined. If the dilaton coupling does not vanish, we have instead an equation for the supergravity auxiliary field:

$$\overline{\mathcal{S}} = -\frac{2i}{\alpha}\left(\partial_0\,W + \frac{W(X)}{\alpha}\right) \tag{5.4.44}$$

and for $H^{0^\star}$:

$$H^{0^\star} = -\frac{1}{\alpha}\,W(X) \tag{5.4.45}$$

Substituting back these field equations into the action we obtain an on-shell supersymmetric theory where the chiral multiplets exhibit a Liouville type of coupling to the two-dimensional curvature in the case $\alpha \neq 0$, while if $\alpha = 0$ one of them can be eliminitated in terms of the others through the constraint (5.4.43).

In the case of *external supergravity* the logic is different since we are not supposed to vary $\mathcal{L}_{chiral}$ in the fields of the gravitational multiplets. For any choice of the curvature $\mathcal{R}$ and gravitino field strength $\tau$ we have to determine the corresponding auxiliary field $\mathcal{S}$ consistent with the Bianchi identities and then substitute these data into the matter lagrangian $\mathcal{L}_{chiral}$. The simplest choice is that of flat superspace $\mathcal{R} = \tau = 0$. In this case we also have $\mathcal{S} = 0$, which is another way of saying that all the charged super-Poincaré curvatures are set to zero. If we do this in equations (5.4.33) we obtain the rheonomic action of the *rigid Landau–Ginzburg theory* discussed in later subsections.

Having clarified what are the implications of a non-flat supergravity background, in the sequel we choose to work in flat superspace, namely we concentrate on the case of global N=2 supersymmetry.

For notational ease in the case of flat superspace we pay less attention to the position of the gravitino indices and we set

$$\tilde{\zeta}^{\pm} = \tilde{\zeta}_{\pm} \tag{5.4.46}$$

In the subsequent sections all gravitino rigid forms have upper indices.

## 5.5 An N=2 Gauge Theory and Its Two Phases

As claimed in the introduction, the rigid N=2 Landau–Ginzburg model, associated with a convenient quasi-homogeneous superpotential $\mathcal{W}(X)$, and the N=2 $\sigma$-model on the Calabi–Yau manifold defined as the vanishing locus of $\mathcal{W}(X)$ in a suitable weighted projective space, can be seen as the effective low energy theories in two different phases of the same N=2 matter-coupled gauge theory. This fact provides the logical link between the two kinds of N=2 matter models and it is the reason why the superconformal field theory emerging from these two field theory models is the same.

In this section we construct the interpolating N=2 matter-coupled gauge theory mentioned above and the N=2 $\sigma$-model. Then we show how we can retrieve the latter as a low energy effective lagrangian in one phase of the gauge theory, while the N=2 rigid Landau–Ginzburg model is retrieved as the effective lagrangian in the other phase. Since the interpolating gauge theory is by itself a Landau–Ginzburg model with a local gauge symmetry, the rheonomic construction of the rigid N=2 Landau–Ginzburg model is obtained, as a by-product of our construction, by setting the gauge-coupling constant to zero. Its action is the same as that obtained in the previous section by switching off the gravitational fields in Eq. (5.4.33). Indeed Eq.(5.4.33) expresses the coupling of Landau–Ginzburg matter to N=2 supergravity. We begin now our construction of N=2 2D field theories starting with the vector multiplet. We follow in our approach [45, 262].

### 5.5.1 The N=2 abelian gauge multiplet

In this section we discuss the rheonomic construction of an N=2 abelian gauge theory in two dimensions. This study will provide a basis for our subsequent coupling of the N=2 gauge multiplet to an N=2 Landau–Ginzburg system invariant under the action of one or several $U(1)$ gauge groups or even of some non-abelian gauge group $G$.

In the N=2 case a vector multiplet is composed of a gauge boson $\mathcal{A}$, namely a world-sheet 1-form, two spin 1/2 gauginos, whose four components we denote by $\lambda^+,\lambda^-$, $\tilde{\lambda}^+,\tilde{\lambda}^-$, a complex physical scalar $M \neq M^*$ and a real auxiliary scalar $\mathcal{P}^* = \mathcal{P}$. Each of these fields is in the adjoint representation of the gauge group $G$ and carries an index of that representation that we have not written.

In the abelian case, defining the field strength

$$F = d\mathcal{A} \tag{5.5.1}$$

the rheonomic parametrizations that solve the Bianchi identities:

$$dF = d^2\lambda^- = d^2\tilde{\lambda}^+ = d^2\lambda^+ = d^2\lambda^- = d^2M = d^2\mathcal{P} = 0 \tag{5.5.2}$$

are given by

$$\begin{aligned}
F &= \mathcal{F}\, e^+e^- - \frac{i}{2}(\tilde{\lambda}^+\zeta^- + \lambda^-\zeta^+)\, e^- + \frac{i}{2}(\lambda^+\tilde{\zeta}^- + \lambda^-\tilde{\zeta}^+)\, e^+ + M\,\zeta^-\tilde{\zeta}^+ - M^*\,\zeta^+\tilde{\zeta}^- \\
dM &= \partial_+M\, e^+ + \partial_-M\, e^- - \frac{1}{4}(\lambda^-\zeta^+ - \tilde{\lambda}^+\tilde{\zeta}^-) \\
d\tilde{\lambda}^+ &= \partial_+\tilde{\lambda}^+\, e^+ + \partial_-\tilde{\lambda}^+\, e^- + (\frac{\mathcal{F}}{2} + i\mathcal{P})\,\zeta^+ - 2i\,\partial_-M\,\tilde{\zeta}^+ \\
d\lambda^- &= \partial_+\lambda^-\, e^+ + \partial_-\lambda^-\, e^- + (\frac{\mathcal{F}}{2} - i\mathcal{P})\,\zeta^- + 2i\partial_-M^*\tilde{\zeta}^- \\
d\lambda^+ &= \partial_+\lambda^+\, e^+ + \partial_-\lambda^+\, e^- + (\frac{\mathcal{F}}{2} - i\mathcal{P})\,\tilde{\zeta}^+ - 2i\,\partial_+M^*\,\zeta^+ \\
d\lambda^- &= \partial_+\lambda^-\, e^+ + \partial_-\lambda^-\, e^- + (\frac{\mathcal{F}}{2} + i\mathcal{P})\,\tilde{\zeta}^- + 2i\,\partial_+M\,\zeta^- \\
d\mathcal{P} &= \partial_+\mathcal{P}\, e^+ + \partial_-\mathcal{P}\, e^- - \frac{1}{4}(\partial_+\tilde{\lambda}^+\zeta^- - \partial_+\lambda^-\zeta^+ - \partial_-\lambda^+\tilde{\zeta}^- + \partial_-\lambda^-\tilde{\zeta}^+)
\end{aligned} \tag{5.5.3}$$

Given these parametrizations, we next write the rheonomic action whose variation yields the above parametrizations as field equations in superspace, together with the world-sheet equations of motion.

$$\begin{aligned}
\mathcal{L}^{(rheon)}_{gauge} &= \mathcal{F}\Big[F + \frac{i}{2}\left(\tilde{\lambda}^+\zeta^- + \lambda^-\zeta^+\right)\, e^- - \frac{i}{2}\left(\lambda^+\tilde{\zeta}^+ + \lambda^-\tilde{\zeta}^+\right)\, e^+ \\
&\quad - M\,\zeta^-\tilde{\zeta}^+ - M^*\,\zeta^+\tilde{\zeta}^-\Big] - \frac{1}{2}\mathcal{F}^2\, e^+e^-
\end{aligned}$$

$$
\begin{aligned}
&-\frac{1}{2}(\tilde{\lambda}^+ d\lambda^- + \lambda^- d\tilde{\lambda}^+)\, e^- + \frac{i}{2}(\lambda^+ d\lambda^- + \lambda^- d\lambda^+)\, e^+ \\
&-4\Big[dM^* - \frac{1}{4}(\lambda^+\zeta^- - \lambda^-\tilde{\zeta}^+)\Big](\mathcal{M}_+ e^+ - \mathcal{M}_- e^-) \\
&-4\Big[dM + \frac{1}{4}(\lambda^-\zeta^+ - \tilde{\lambda}^+\tilde{\zeta}^-)\Big](\mathcal{M}_+^* e^+ - \mathcal{M}_-^* e^-) \qquad (5.5.4) \\
&-4(\mathcal{M}_+^*\mathcal{M}_- + \mathcal{M}_-^*\mathcal{M}_+)\, e^+e^- - dM(\lambda^-\tilde{\zeta}^+ + \lambda^+\zeta^-) + dM^*(\tilde{\lambda}^+\tilde{\zeta}^- + \lambda^-\zeta^+) \\
&-\frac{1}{4}(\tilde{\lambda}^+\lambda^+\zeta^-\tilde{\zeta}^- + \lambda^-\lambda^-\zeta^+\tilde{\zeta}^+) + 2\mathcal{P}^2\, e^+e^- + 4i\,\frac{\partial\mathcal{U}}{\partial M}\Big(\frac{F}{2} + i\mathcal{P}\, e^+e^-\Big) \\
&-4i\,\frac{\partial\mathcal{U}^*}{\partial M^*}\Big(\frac{F}{2} - i\mathcal{P}\, e^+e^-\Big) - i\Big(\frac{\partial^2\mathcal{U}}{\partial M^2}\,\tilde{\lambda}^+\lambda^- + \frac{\partial^2\mathcal{U}^*}{\partial M^{*2}}\,\lambda^-\lambda^+\Big)\, e^+e^- \\
&+\Big(\frac{\partial\mathcal{U}}{\partial M} + \frac{\partial\mathcal{U}^*}{\partial M^*}\Big)\Big[(\tilde{\lambda}^+\zeta^- - \lambda^-\zeta^+)\, e^- + (\lambda^+\tilde{\zeta}^- - \lambda^-\tilde{\zeta}^+)\, e^+\Big] \\
&+2i\Big[2\mathcal{U} - M\Big(\frac{\partial\mathcal{U}}{\partial M} - \frac{\partial\mathcal{U}^*}{\partial M^*}\Big)\Big]\,\zeta^-\tilde{\zeta}^+ + 2i\Big[2\mathcal{U}^* - M^*\Big(\frac{\partial\mathcal{U}}{\partial M} - \frac{\partial\mathcal{U}^*}{\partial M^*}\Big)\Big]\,\zeta^+\tilde{\zeta}^-
\end{aligned}
\tag{5.5.5}
$$

The symbol $\mathcal{U}$ denotes a holomorphic function $\mathcal{U}(M)$ of the physical scalar $M$ which is named the gauge superpotential. It induces a self interaction of the scalar $M$ field and an interaction of this field with the gauge vector. The existence of arbitrariness in the choice of the vector multiplet dynamics is a consequence of the existence of the auxiliary field $\mathcal{P}$ in the solution of the Bianchi identities (5.5.2) and hence in the determination of the SUSY rules for this type of N=2 multiplet. In the superspace formalism the inclusion in the action of the terms containing the superpotential is effected by means of the use of the so called twisted chiral superfields. In the rheonomic framework there is no need for these distinctions: we just have an interaction codified by an arbitrary holomorphic superpotential.

Note that in Eqs. (5.5.3) and (5.5.5) we have suppressed the wedge product symbols for differential forms. This convention will be often adopted also in the sequel to avoid clumsiness. From the rheonomic action (5.5.3) we easily obtain the world-sheet action of the N=2 globally supersymmetric abelian vector multiplet, by deleting all the terms containing the gravitino 1-forms, replacing the first order fields $\mathcal{F}, \mathcal{M}_\pm$ with their values following from their own field equations, namely $\mathcal{F} = \frac{1}{2}(\partial_+\mathcal{A}_- - \partial_-\mathcal{A}_+)$, $\mathcal{M}_\pm = \partial_\pm M$, and by replacing $e^+ \wedge e^-$ with $d^2z$ which is factored out. In this way we get

$$
\begin{aligned}
\mathcal{L}^{(ws)}_{gauge} \;=\; & \frac{1}{2}\mathcal{F}^2 - i\,(\tilde{\lambda}^+\partial_+\lambda^- + \lambda^+\partial_-\lambda^-) - 4(\partial_+ M^*\partial_- M + \partial_- M^*\partial_+ M) + 2\mathcal{P}^2 \\
& + 4i\,\frac{\partial\mathcal{U}}{\partial M}\Big(\frac{\mathcal{F}}{2} + i\mathcal{P}\Big) - 4i\,\frac{\partial\mathcal{U}^*}{\partial M^*}\Big(\frac{\mathcal{F}}{2} - i\mathcal{P}\Big) - i\Big(\frac{\partial^2\mathcal{U}}{\partial M^2}\,\tilde{\lambda}^+\lambda^- + \frac{\partial^2\mathcal{U}^*}{\partial M^{2*}}\,\lambda^-\lambda^+\Big)
\end{aligned}
\tag{5.5.6}
$$

In the particular case of a linear superpotential

$$
\mathcal{U} = \frac{-it}{4}\,M\,,\ t \in \mathbb{C} \tag{5.5.7}
$$

setting

$$t = ir + \theta/2\pi \ , \ r \in \mathbb{R} \ , \ \theta \in [0, 2\pi] \tag{5.5.8}$$

the above expression reduces to

$$\begin{aligned} \mathcal{L}_{ws} &= \frac{1}{2}\mathcal{F}^2 - i\,(\tilde{\lambda}^+\partial_+\lambda^- + \lambda^+\partial_-\lambda^-) - 4(\partial_+M^*\partial_-M + \partial_-M^*\partial_+M) + \\ &+ 2\mathcal{P}^2 - 2r\mathcal{P} + \frac{\theta}{2\pi}\mathcal{F} \end{aligned} \tag{5.5.9}$$

The meaning of the parameters $r$ and $\theta$ introduced in the above lagrangian is clear. Indeed $r$, giving a vacuum expectation value $\mathcal{P} = \frac{r}{2}$ to the auxiliary field $\mathcal{P}$, induces a spontaneous breaking of supersymmetry and shows that the choice $\mathcal{U} = -\frac{r}{4}M$ corresponds to the insertion of a Fayet–Iliopoulos [140] term into the action. On the other hand the parameter $\theta$ is clearly a theta angle multiplying the first Chern class $\frac{1}{2\pi}\mathcal{F}$ of the gauge connection.

### 5.5.2 N=2 Landau–Ginzburg models with an abelian gauge symmetry

As stated above, our interest in the N=2 vector multiplet was instrumental to the study of an N=2 Landau-Ginzburg system possessing in addition to its own self interaction a minimal coupling to a gauge theory. This is the system studied by Witten in [45], using superspace techniques, rather than the rheonomy framework. By definition a Landau–Ginzburg system is a collection of N=2 chiral multiplets self-interacting via an analytic superpotential $W(X)$. As already discussed in section 5.4.2 each chiral multiplet is composed of a complex scalar field $(X^i)^* = X^{i^*}$ $(i = 1, ..., n)$, two spin 1/2 fermions, whose four components we denote by $\psi^i, \tilde{\psi}^i$ and $\psi^{i^*} = (\psi^i)^*, \tilde{\psi}^{i^*} = \left(\tilde{\psi}^i\right)^*$, together with a complex auxiliary field $H^i$ which is identified with the derivative of the holomorphic superpotential $\overline{W}(X)$, namely $H^i = \eta^{ij^*}\partial_{j^*}W^*$, $\eta^{ij^*}$ being the flat Kählerian metric on the complex manifold $\mathbb{C}^n$ of which the complex scalar fields $X^i$ are interpreted as the coordinates. Here the index enumerating the chiral multiplets is a lower case Latin letter, $i, j, k$, contrary to the notation of section 5.4.2, where it was a Capital latin letter, $I, J, K$. The reason is the following. The capital index runs on the values $I = 0, 1, 2, \dots, n$ where $X^0$ denotes the dilaton multiplet (that coupling to the 2D-curvature multiplet as in Eq. (5.4.36)), while $X^i$ $(i = 1, \dots, n)$ denotes all the other chiral multiplets. Using this system of fields, we can construct a rheonomic solution of the flat superspace Bianchi identities, a rheonomic action and a world-sheet action invariant under the supersymmetry transformations induced by the rheonomic parametrizations. In fact this solution is already at our disposal. It suffices to set $\mathcal{S} = \tau^{\pm} = \tilde{\tau}_{\pm} = \mathcal{R} = F^{\bullet} = 0$ in Eqs. (5.4.27), (5.4.28),(5.4.29), (5.4.30), (5.4.31),(5.4.33). In particular the value (5.4.29) of the auxiliary field $H^i$ is obtained as a field equation from the variation of the action (5.4.33). In this action that corresponds to the *rigid Landau–Ginzburg model* the kinetic terms are the canonical ones for a free field theory and the only interaction is that induced by the superpotential. In order to introduce our interpolating field theory rather than this action we need to study the same system in the presence of a minimal coupling to the gauge system studied in the previous section. In practice this amounts to solving the Bianchi identities for the gauge covariant derivatives rather than for the purely gravitational covariant derivatives (5.4.27), using as a background the rheonomic parametrizations of the gauge mulitiplet determined above. At the end of the construction, by setting the gauge coupling constant to zero, we also recover the formulation of the ordinary Landau–Ginzburg theory, above named the *rigid Landau–Ginzburg theory*. Indeed, switching off the gravitational or the gauge coupling, to nobody's surprise, we obtain the same result.

This being clarified, in flat superspace we choose the gauge $\omega = A^{\bullet} = 0$ and the coupling of the chiral multiplets to the gauge multiplet is defined through the gauge-covariant derivative

$$\nabla X^i \stackrel{\text{def}}{=} dX^i + i\mathcal{A}q^i{}_j X^j \tag{5.5.10}$$

where the hermitean matrix $q^i{}_j$ is the generator of the $U(1)$ action on the chiral matter. As a consequence, the Bianchi identities are of the form $\nabla^2 X^i = i\,Fq^i{}_j X^j$.

Let $W(X^i)$ be the holomorphic superpotential; then the rheonomic solution of the Bianchi identities is given by the following parametrizations:

$$\begin{aligned}
\nabla X^i &= \nabla_+ X^i\, e^+ + \nabla_- X^i\, e^- + \psi^i \zeta^- + \widetilde{\psi}^i \widetilde{\zeta}^- \\
\nabla X^{i^*} &= \nabla_+ X^{i^*}\, e^+ + \nabla_- X^{i^*}\, e^- - \psi^{i^*} \zeta^+ - \widetilde{\psi}^{i^*} \widetilde{\zeta}^+ \\
\nabla \psi^i &= \nabla_+ \psi^i\, e^+ + \nabla_- \psi^i\, e^- - \frac{i}{2}\nabla_+ X^i\, \zeta^+ + \eta^{ij^*}\partial_{j^*} W^*\, \widetilde{\zeta}^- + i\, M q^i{}_j X^j\, \widetilde{\zeta}^+ \\
\nabla \psi^{i^*} &= \nabla_+ \psi^{i^*}\, e^+ + \nabla_- \psi^{i^*}\, e^- + \frac{i}{2}\nabla_- X^{i^*}\, \zeta^- + \eta^{ji^*}\partial_j W\, \widetilde{\zeta}^+ - i\, M^* q^j{}_i X^{j^*}\, \widetilde{\zeta}^- \\
\nabla \widetilde{\psi}^i &= \nabla_+ \widetilde{\psi}^i\, e^+ + \nabla_- \widetilde{\psi}^i\, e^- - \frac{i}{2}\nabla_- X^i\, \widetilde{\zeta}^+ - \eta^{ij^*}\partial_{j^*} W^*\, \zeta^- - i\, M^* q^i{}_j X^j\, \zeta^+ \\
\nabla \widetilde{\psi}^{i^*} &= \nabla_+ \widetilde{\psi}^{i^*}\, e^+ + \nabla_- \widetilde{\psi}^{i^*}\, e^- + \frac{i}{2}\nabla_+ X^{i^*}\, \widetilde{\zeta}^- - \eta^{ji^*}\partial_j W\, \zeta^+ + i\, M q^j{}_i X^{j^*}\, \zeta^-
\end{aligned} \tag{5.5.11}$$

From the consistency of the above parametrizations with the Bianchi identities one also gets the following fermionic world-sheet equations of motion:

$$\begin{aligned}
\frac{i}{2}\nabla_- \psi^i - \eta^{ij^*}\partial_{l^*}\partial_{j^*} W^*\, \widetilde{\psi}^{l^*} + \frac{i}{4}\widetilde{\lambda}^+ q^i{}_j X^j + i\, M q^i{}_j \widetilde{\psi}^j &= 0 \\
\frac{i}{2}\nabla_+ \widetilde{\psi}^i + \eta^{ij^*}\partial_{l^*}\partial_{j^*} W^*\, \psi^{l^*} - \frac{i}{4}\lambda^+ q^i{}_j X^j - i\, M^* q^i{}_j \psi^j &= 0
\end{aligned} \tag{5.5.12}$$

and their complex conjugates for the other two fermions. Applying to Eqs. (5.5.12) a supersymmetry transformation, as it is determined by the parametrizations (5.5.11), we obtain the bosonic field equation:

$$\begin{aligned}
&\frac{1}{8}(\nabla_+\nabla_- X^i + \nabla_-\nabla_+ X^i) - \eta^{ik^*}\partial_{k^*}\partial_{j^*}\partial_{l^*} W^*\, \psi^{j^*} \widetilde{\psi}^{l^*} + \eta^{ik^*}\partial_{k^*}\partial_{j^*} W^*\, \eta^{lj^*}\partial_l W \\
&- \frac{i}{4}\lambda^- q^i{}_j \psi^j + \frac{i}{4}\lambda^- q^i{}_j \widetilde{\psi}^j + M^* M (q^2)^i_j X^j - \frac{1}{4}\mathcal{P} q^i{}_j X^j = 0
\end{aligned} \tag{5.5.13}$$

Equipped with this information, we can easily derive the rheonomic action from which the parametrizations (5.5.11) and the field equations (5.5.12), (5.5.13) follow as variational equations: it is

$$\begin{aligned}
\mathcal{L}^{(rheon)}_{chiral} &= \eta_{ij^*}\Big(\nabla X^i - \psi^i\zeta^- - \widetilde{\psi}^i\widetilde{\zeta}^-\Big)\Big(\Pi^{j^*}_+ e^+ - \Pi^{j^*}_- e^-\Big) \\
&+ \eta_{ij^*}(\nabla X^{j^*} + \psi^{j^*}\zeta^+ + \widetilde{\psi}^{j^*}\widetilde{\zeta}^+)(\Pi^i_+ e^+ - \Pi^i_- e^-\Big) \\
&+ \eta_{ij^*}(\Pi^i_+\Pi^{j^*}_- + \Pi^i_-\Pi^{j^*}_+)\, e^+e^- - 4i\,\eta_{ij^*}(\psi^i\nabla\psi^{j^*}\, e^+ - \widetilde{\psi}^i\nabla\widetilde{\psi}^{j^*}\, e^-) \\
&+ 4i(\psi^k\partial_k W\,\widetilde{\zeta}^+e^+ - \text{c.c.}\,) + 4i(\widetilde{\psi}^k\partial_k W\,\zeta^+e^- - \text{c.c.}\,) \\
&+ \eta_{ij^*}(\psi^i\widetilde{\psi}^{j^*}\,\zeta^-\widetilde{\zeta}^+ + \psi^{j^*}\widetilde{\psi}^i\,\widetilde{\zeta}^-\zeta^+) \\
&+ 8\Big((\partial_i\partial_j W\psi^i\widetilde{\psi}^j + \text{c.c.}\,) + \eta^{ij^*}\partial_i W\partial_{j^*}W^*\Big)\, e^+e^- \\
&- \eta_{ij^*}(\nabla X^i\psi^{j^*}\,\zeta^+ - \nabla X^i\widetilde{\psi}^{j^*}\,\widetilde{\zeta}^+ + \text{c.c.}\,) - (4M\psi^{j^*}\eta_{ij^*}q^i{}_kX^k\,\widetilde{\zeta}^+e^+ + \text{c.c.}\,) \\
&- (4M^*\widetilde{\psi}^{j^*}\eta_{ij^*}q^i{}_kX^k\,\zeta^+e^- + \text{c.c.}\,) - \Big(8i(M^*\widetilde{\psi}^{j^*}\eta_{ij^*}q^i{}_k\psi^k - \text{c.c.}\,) \\
&+ 2i(\widetilde{\lambda}^+\psi^{j^*}\eta_{ij^*}q^i{}_kX^k - \text{c.c.}\,) + 2i(\lambda^+\widetilde{\psi}^{j^*}\eta_{ij^*}q^i{}_kX^k - \text{c.c.}\,) \\
&- 2\mathcal{P}\eta_{ij^*}\,X^{j^*}q^i{}_kX^k + 8\,M^*M\,\eta_{ij^*}X^{j^*}(q^2)^i_kX^k\Big)\, e^+e^-
\end{aligned} \tag{5.5.14}$$

The world-sheet lagrangian for this system is now easily obtained through the same steps applied in the previous case. To write it, we introduce the following semplifications in our notation: a) we use a diagonal form for the flat $\mathbf{C}^n$ metric $\eta_{ij^*}X^iX^{j^*} \equiv X^iX^{i^*}$, b) we diagonalise the $U(1)$ generator, by setting $q^i{}_j \equiv q^i\delta^i_j$ ($q^i$ being the charge of the field $X^i$). Then we have

$$\begin{aligned}
\mathcal{L}^{(ws)}_{chiral} &= -(\nabla_+X^{i^*}\nabla_-X^i + \nabla_-X^{i^*}\nabla_+X^i) + 4i(\psi^i\nabla_-\psi^{i^*} + \widetilde{\psi}^i\nabla_+\widetilde{\psi}^{i^*}) \\
&+ 8\Big((\psi^i\widetilde{\psi}^j\partial_i\partial_j W + \text{c.c.}\,) + \partial_i W\partial_{i^*}W^*\Big) + 2i\sum_i q^i(\psi^i\lambda^-X^{i^*} - \widetilde{\psi}^i\lambda^-X^{i^*} - \text{c.c.}\,) \\
&+ 8i\left(M^*\sum_i q^i\psi^i\widetilde{\psi}^{i^*} - \text{c.c.}\right) + 8M^*M\sum_i (q^i)^2X^{i^*}X^i - 2\mathcal{P}\sum_i q^iX^{i^*}X^i
\end{aligned} \tag{5.5.15}$$

### 5.5.3 Structure of the scalar potential

We consider next the coupled system, whose lagrangian, with our conventions, is the difference of the two lagrangians we have just described:

$$\mathcal{L} = \mathcal{L}_{gauge} - \mathcal{L}_{chiral} \tag{5.5.16}$$

the relative sign being fixed by the requirement of positivity of the energy. The world-sheet form of the action (5.5.16) is the same, modulo trivial notational differences as the action (2.19)+(2.23)+(2.27) in Witten's paper [262]. We focus our attention on the potential energy of the bosonic fields: it is given by the expression

$$\begin{aligned} -U &= 2\mathcal{P}^2 - 4\mathcal{P}\left(\frac{\partial \mathcal{U}}{\partial M} + \frac{\partial \mathcal{U}^*}{\partial M^*}\right) + 2\mathcal{P}\sum_i q^i |X^i|^2 \\ & - 8\partial_i W \partial_{i^*} W^* - 8|M|^2 \sum_i (q^i)^2 |X^i|^2 \end{aligned} \tag{5.5.17}$$

The variation in the auxiliary field $\mathcal{P}$ yields the expression of $\mathcal{P}$ itself in terms of the physical scalars:

$$\mathcal{P} = \frac{\partial \mathcal{U}}{\partial M} + \frac{\partial \mathcal{U}^*}{\partial M^*} - \frac{1}{2}\sum_i q^i |X^i|^2 \tag{5.5.18}$$

In the above equation the expression $\mathcal{D}(X, X^*) = \sum_i q^i |X^i|^2$ is (proportional to) the momentum map function for the holomorphic action of the gauge group on the matter multiplets. Indeed if we denote by $\mathbf{X} = \mathrm{i}\sum_i q^i \left(X^i \partial_i - X^{i^*} \partial_{i^*}\right)$ the killing vector and by $\Omega = \sum_i dX^i \wedge dX^{i^*}$, then we have $\mathrm{i}d\mathcal{D} = i_{\mathbf{X}}\Omega$. As anticipated the auxiliary field $\mathcal{P}$ is identified with the momentum map function, plus the term $\frac{\partial \mathcal{U}}{\partial M} + \frac{\partial \mathcal{U}^*}{\partial M^*}$ due to the self-interaction of the vector multiplet. In the case of the linear superpotential of Eqs. (5.5.7) and (5.5.8), the auxiliary field is identified with

$$\mathcal{P} = -\frac{1}{2}(D(X, X^*) - r) \tag{5.5.19}$$

Eliminating $\mathcal{P}$ through Eq. (5.5.18), we obtain the final form for the scalar field potential in this kind of models, namely

$$U = 2\left[\left(\frac{\partial \mathcal{U}}{\partial M} + \frac{\partial \mathcal{U}^*}{\partial M^*}\right) - \frac{1}{2}\sum_i q^i |X^i|^2\right]^2 + |\partial_i W|^2 + 8|M|^2 \sum_i (q^i)^2 |X^i|^2 \tag{5.5.20}$$

In the case of the linear superpotential this reduces to

$$U = \frac{1}{2}\left[r - \sum_i q^i |X^i|^2\right]^2 + 8|\partial_i W|^2 + 8|M|^2 \sum_i (q^i)^2 |X^i|^2 \tag{5.5.21}$$

The theory characterized by the above scalar potential exhibits a two-phase structure as the parameter $r$ varies on the real line. This is the essential point in Witten's paper that allows an interpolation between an N=2 $\sigma$-model on a Calabi–Yau manifold and a rigid Landau–Ginzburg theory. The review of these two regimes is postponed to later sections. Here we note that the above results can be generalized to the case of a non-abelian vector multiplet or to the case of several abelian gauge multiplets.

### 5.5.4 Extension to non-abelian gauge symmetry

We fix our notations and conventions.

Consider a Lie algebra $\mathcal{G}$ with structure constants $f^{abc}$:

$$[t^a, t^b] = i f^{abc} t^c \tag{5.5.22}$$

In every representation the hermitean generators $t^a = (t^a)^\dagger$ are normalized in such a way that $\mathrm{Tr}\,(t^a t^b) = \delta^{ab}$. Let us name $T^a$ the generators of the adjoint representation, defined by $f^{abc} = i(T^a)^{bc}$.

Let us introduce the gauge vector field as a $\mathcal{G}$-valued 1-form:

$$\mathcal{A} = \mathcal{A}^a_\mu T^a dx^\mu \tag{5.5.23}$$

In the case we are interested in, the index $\mu$ takes two values and we can write $\mathcal{A} = \mathcal{A}^a_+ e^+ + \mathcal{A}^a_- e^-$. Note that $\mathcal{A}^\dagger = \mathcal{A}$. The field strength is defined as the 2-form

$$F = d\mathcal{A} + iA \wedge \mathcal{A} \tag{5.5.24}$$

The Bianchi identities read

$$\nabla F \stackrel{\text{def}}{=} dF + i(\mathcal{A} \wedge F - F \wedge \mathcal{A}) = 0 \tag{5.5.25}$$

The component expression of the field strength and of its associated Bianchi identity is

$$\begin{aligned} F^a_{\mu\nu} &= \partial_{[\mu}\mathcal{A}^a_{\nu]} - \frac{1}{2} f^{abc}\mathcal{A}^b_\mu \mathcal{A}^c_\nu \\ \partial_{[\rho}F^a_{\mu\nu]} - f^{abc}\mathcal{A}^b_{[\mu}F^c_{\rho\nu]} &= 0 \end{aligned} \tag{5.5.26}$$

Note that the Bianchi identity for a field $M = M^a T^a$ transforming in the adjoint representation is

$$\nabla^2 M = i[F, M] \tag{5.5.27}$$

The non-abelian analogue of the rheonomic parametrizations (5.5.3) is obtained in the following way. First we write the $\mathcal{G}$-valued parametrization of $F$:

$$F = \mathcal{F} e^+e^- - \frac{i}{2}(\tilde{\lambda}^+\zeta^- + \lambda^-\zeta^+)\,e^- + \frac{i}{2}(\lambda^+\tilde{\zeta}^+ + \lambda^-\tilde{\zeta}^+)\,e^+ + M\,\zeta^-\tilde{\zeta}^+ - M^*\,\zeta^+\tilde{\zeta}^- \tag{5.5.28}$$

In this way we have introduced the gauge scalars $M = M^a T^a$ and the gauginos $\lambda^\pm = \lambda^\pm_a T^a$, $\tilde{\lambda}^\pm = \tilde{\lambda}^\pm_a T^a$; their parametrizations are obtained by implementing the Bianchis for $F$, $\nabla F = 0$. One must also take into account the Bianchi identies for these fields: $\nabla^2 M = i[F, M]$ and $\nabla^2\lambda^\pm = i[F, \lambda^\pm]$ (analogously for the tilded gauginos). The rheonomic parametrizations fulfilling all these constraints turn out to be

$$\begin{aligned} F &= \mathcal{F} e^+e^- - \frac{i}{2}(\tilde{\lambda}^+\zeta^- + \lambda^-\zeta^+)\,e^- + \frac{i}{2}(\lambda^+\tilde{\zeta}^+ + \lambda^-\tilde{\zeta}^+)\,e^+ + M\,\zeta^-\tilde{\zeta}^+ - M^*\,\zeta^+\tilde{\zeta}^- \\ \nabla M &= \nabla_+ M\,e^+ + \nabla_- M\,e^- - \frac{1}{4}(\lambda^-\zeta^+ - \tilde{\lambda}^+\tilde{\zeta}^-) \\ \nabla\tilde{\lambda}^+ &= \nabla_+\tilde{\lambda}^+\,e^+ + \nabla_-\tilde{\lambda}^+\,e^- + \left(\frac{\mathcal{F}}{2} - 2i[M^*, M] + i\mathcal{P}\right)\zeta^+ - 2i\,\nabla_- M\,\tilde{\zeta}^+ \\ \nabla\lambda^+ &= \nabla_+\lambda^+\,e^+ + \nabla_-\lambda^+\,e^- + \left(\frac{\mathcal{F}}{2} - 2i[M^*, M] - i\mathcal{P}\right)\tilde{\zeta}^+ - 2i\,\nabla_+ M^*\,\zeta^+ \\ \nabla\mathcal{P} &= \nabla_+\mathcal{P}\,e^+ + \nabla_-\mathcal{P}\,e^- - \frac{1}{4}\Big[\Big(\nabla_+\tilde{\lambda}^+ - 2[\lambda^+, M]\Big)\zeta^- - \Big(\nabla_+\lambda^- + 2[\lambda^-, M^*]\Big)\zeta^+ \\ &\quad - \Big(\nabla_-\lambda^+ + 2[\lambda^-, M^*]\Big)\tilde{\zeta}^- + \Big(\nabla_-\lambda^- - 2[\tilde{\lambda}^+, M]\Big)\tilde{\zeta}^+\Big] \end{aligned} \tag{5.5.29}$$

We obtain the rheonomic action for the N=2 non-abelian gauge multiplet in two steps, setting

$$\mathcal{L}^{(rheon)}_{non\text{-}abelian} = \mathcal{L}_0 + \Delta\mathcal{L}_{int} \tag{5.5.30}$$

where $\mathcal{L}_0$ is the free part of the lagrangian whose associated equations of motion would set the auxiliary fields to zero: $\mathcal{P} = \mathcal{P}_a T^a = 0$. The insertion of the interaction term $\Delta\mathcal{L}_{int}$ corrects the equation of motion of the auxiliary fields, depending on a holomorphic function $\mathcal{U}(M)$ of the physical gauge scalars $M^b$, just as in the abelian case. The form of $\mathcal{L}_0$ is given below, where the trace is performed over the indices of the adjoint representation:

$$\begin{aligned}\mathcal{L} = \mathrm{Tr}\Big\{ & \mathcal{F}\Big[F + \frac{i}{2}(\tilde{\lambda}^+\zeta^- + \lambda^-\zeta^+)\,e^- - \frac{i}{2}(\lambda^+\tilde{\zeta}^+ + \lambda^-\tilde{\zeta}^+)\,e^+ - M\,\zeta^-\tilde{\zeta}^+ + M^\star\,\zeta^+\tilde{\zeta}^-\Big] \\ & - \frac{1}{2}\mathcal{F}^2\,e^+e^- - \frac{i}{2}(\tilde{\lambda}^+\,\nabla\lambda^- + \lambda^-\,\nabla\tilde{\lambda}^+)\,e^- + \frac{i}{2}(\lambda^+\,\nabla\lambda^- + \lambda^-\,\nabla\lambda^+)\,e^+ \\ & - 4\Big[\nabla M^\star - \frac{1}{4}(\lambda^+\zeta^- - \lambda^-\tilde{\zeta}^+)\Big](\mathcal{M}_+e^+ - \mathcal{M}_-e^-) \\ & - 4\Big[\nabla M + \frac{1}{4}(\lambda^-\zeta^+ - \tilde{\lambda}^+\tilde{\zeta}^-)\Big](\mathcal{M}^\star_+e^+ - \mathcal{M}^\star_-e^-) \\ & - 4(\mathcal{M}^\star_+\mathcal{M}_- + \mathcal{M}^\star_-\mathcal{M}_+)\,e^+e^- - \nabla M(\lambda^-\tilde{\zeta}^+ + \lambda^+\zeta^-) + \nabla M^\star(\tilde{\lambda}^+\tilde{\zeta}^- + \lambda^-\zeta^+) \\ & + 2[M^\star, M]\left((\lambda^+\,\tilde{\zeta}^- - \lambda^-\tilde{\zeta}^+)e^+ - (\tilde{\lambda}^+\zeta^- - \lambda^-\zeta^+)e^-\right) \\ & - \frac{1}{4}(\tilde{\lambda}^+\lambda^+\,\zeta^-\tilde{\zeta}^- + \lambda^-\lambda^-\,\zeta^+\tilde{\zeta}^+) + 2\mathcal{P}^2\,e^+e^-\Big\}\end{aligned} \tag{5.5.31}$$

As stated above, the variational equations associated with this action yield the rheonomic parametrizations (5.5.29) for the particular value $\mathcal{P}^a = 0$ of the auxiliary field. Furthermore they also imply $\mathcal{P}^a = 0$ as a field equation.

To determine the form of $\Delta\mathcal{L}_{int}$ we suppose that in the presence of this interaction the new field equation of $\mathcal{P}^a$ yields

$$\mathcal{P}^a = \frac{\partial\mathcal{U}(M)}{\partial M^a} + \left(\frac{\partial\mathcal{U}(M)}{\partial M^a}\right)^* = \frac{\partial\mathcal{U}(M)}{\partial M^a} + \frac{\partial\mathcal{U}^*(M^*)}{\partial M^{*a}}. \tag{5.5.32}$$

$\mathcal{U}$ is a holomorphic function of the scalars $M^a$ that characterizes their self-interaction. Then we can express $\nabla\mathcal{P}^a$ through the chain rule: $\nabla\mathcal{P}^a = \frac{\partial^2\mathcal{U}}{\partial M^a\partial M^b}\,\nabla M^b + \frac{\partial^2\mathcal{U}^*}{\partial M^{*a}\partial M^{*b}}\,\nabla M^{*b}$. Using the rheonomic parametrizations (5.5.29) for $\nabla M^b$ and comparing with the parametrization of $\nabla\mathcal{P}^a$ in the same equation (5.5.29), we get the fermionic equations of motion that the complete interacting lagrangian should imply as variational equations:

$$\begin{aligned}\nabla_+\tilde{\lambda}^+_a - 2if^{abc}\,\lambda^+_b\,M_c &= -\frac{\partial^2\mathcal{U}}{\partial M^{*a}\partial M^{*b}}\,\lambda^+_b \\ \nabla_-\lambda^+_a - 2if^{abc}\,\tilde{\lambda}^+_b\,M^*_c &= \frac{\partial^2\mathcal{U}}{\partial M^a\partial M^b}\,\tilde{\lambda}^+_b\end{aligned} \tag{5.5.33}$$

plus, of course, the complex conjugate equations. Furthermore the parametrization of $\nabla\mathcal{F}$ is also affected by having $\mathcal{P}^a$ a non-zero function of $M$. This can be seen from the parametrizations (5.5.29). Taking the covariant derivative of $\nabla\bar{\lambda}^+_a$ and focusing on the $\zeta^+\zeta^+$ sector, one can extract $\nabla_{\bullet+}\mathcal{F}^a$, the component of $\nabla\mathcal{F}^a$ along $\zeta^+$:

$$\nabla_{\bullet+}\mathcal{F}^a = f^{abc} M^\star_b \nabla_{\bullet+} M_c + \frac{i}{2}\frac{\partial^2 \mathcal{U}}{\partial M^a \partial M^b}\lambda^-_b \tag{5.5.34}$$

Analogously one gets the other fermionic components of $\nabla\mathcal{F}^a$.

Summarizing, in order to obtain $\mathcal{P}^a = \frac{\partial\mathcal{U}(M)}{\partial M^a} + \frac{\partial\mathcal{U}_*(M^\star)}{\partial M^{\star a}}$, to reproduce the fermionic field equations (5.5.33) and the last terms in the fermionic components of the parametrization (5.5.34) of $\nabla\mathcal{F}^a$, we have to set

$$\begin{aligned}
\Delta\mathcal{L}_0 &= 4i\frac{\partial\mathcal{U}}{\partial M^a}(\frac{F^a}{2} + i\mathcal{P}^a\, e^+e^-) - 4i\frac{\partial\mathcal{U}^*}{\partial M^{\star a}}(\frac{F^a}{2} - i\mathcal{P}^a\, e^+e^-) \\
&+ i\Big(\frac{\partial^2\mathcal{U}}{\partial M^a\partial M^b}\lambda^-_a\bar{\lambda}^+_b + \frac{\partial^2\mathcal{U}^*}{\partial M^{\star a}\partial M^{\star b}}\lambda^+_a\lambda^-_b\Big)e^+e^- \\
&+ \Big(\frac{\partial\mathcal{U}}{\partial M^a} = \frac{\partial\mathcal{U}^*}{\partial M^{\star a}}\Big)\Big[(\lambda^+_a\tilde{\zeta}^- - \lambda^-_a\tilde{\zeta}^+)e^+ + (\bar{\lambda}^+_a\zeta^- - \lambda^-_a\zeta^+)e^-\Big] \\
& 2i\Big[2\mathcal{U} - M^a\Big(\frac{\partial\mathcal{U}}{\partial M^a} - \frac{\partial\mathcal{U}^*}{\partial M^{\star a}}\Big)\Big]\zeta^-\tilde{\zeta}^+ + 2i\Big[2\mathcal{U}^* - M^{\star a}\Big(\frac{\partial\mathcal{U}^*}{\partial M^{\star a}} - \frac{\partial\mathcal{U}}{\partial M^a}\Big)\Big]\zeta^+\tilde{\zeta}^-
\end{aligned} \tag{5.5.35}$$

Note that $\mathcal{U}$ must be a gauge singlet. A linear potential of the type $\mathcal{U} = \sum_a c^a M^a$ with $c^a$ = const does not satisfy this requirement. Hence the "linear potential" of the abelian case, corresponding to the insertion of a Fayet–Iliopoulos term, has no non-abelian counterpart. Similarly a $\theta$-term is also ruled out in the non-abelian case. Indeed a term like $\frac{\theta^a}{2\pi}F^a$ would not be gauge-invariant, with a constant $\theta^a$. Also in this case, a term of this type would be implied by a linear superpotential $\mathcal{U}$, which is therefore excluded. The problem is that no linear function of the gauge scalars $M^a$ can be gauge-invariant.

In conclusion, if the Lie algebra $\mathcal{G}$ is not semi-simple, then for each of its $U(1)$ factors we can introduce a Fayet–Iliopoulos and a $\theta$-term. Fayet–Iliopoulos terms are associated only with abelian factors of the gauge group, namely with the center $\mathcal{Z} \subset \mathcal{G}$ of the gauge Lie-algebra. This yield of supersymmetry perfectly matches with the properties of Kähler quotients. Indeed we recall from Chapter 3 that the level set of the momentum map (see Eq.(3.2.69)) is well-defined for $\zeta \in \mathbb{R} \odot \mathcal{Z}^*$ in the Kähler case, $\mathcal{Z}^*$ being the center of the dual Lie algebra $\mathcal{G}^*$. Now the level parameters $\zeta$ are precisely identified with the parameters introduced into the lagrangian by the Fayet–Iliopoulos terms.

### 5.5.5 R-symmetries and the rigid Landau–Ginzburg model

In Chapter 7 we study topological field theories and we show how, by means of a procedure named the topological twist, we can always extract such a theory from any N=2 field theory. A crucial role in the topological twist is played by the so-called R-symmetries of the N=2 theory. These are global symmetries of the rheonomic parametrizations (namely automorphisms of the supersymmetry algebra) and of the action (both the rheonomic one and that concentrated on the bosonic world-sheet) that have a non trivial action also on the gravitino 1-forms (in the global theories this means on the supersymmetry parameters, but when extending the analysis to the locally supersymmetric case this means also on the world-sheet gravitinos). In the N=2 theories the R-symmetry group is $U(1)_L \otimes U(1)_R$, the first $U(1)_L$ acting as a phase rotation $\zeta^{\pm} \longrightarrow \zeta^{\pm} e^{\pm i\alpha_L}$ on the left-moving gravitinos, and leaving the right-moving gravitinos invariant, the second $U(1)_R$ factor rotating in the same way the right-moving gravitinos $\tilde{\zeta}^{\pm} \longrightarrow \tilde{\zeta}^{\pm} e^{\pm i\alpha_R}$ and leaving the left-moving ones invariant. In terms of the 2D spinor notation of section 5.4.1 we note that the subgroup $U_{local} \subset U(1)_L \otimes U(1)_R$ identified by the rotations of parameters $\alpha_L = -\alpha_R = \theta$ is the local $U(1)$ symmetry gauged by the graviphoton $A^{\bullet}$ ($\delta\xi = i\theta\,\xi$), while the other subgroup $U_{global} \subset U(1)_L \otimes U(1)_R$ identified by the rotations of parameters $\alpha_L = \alpha_R = \theta'$ is a global symmetry of flat superspace corresponding to the chiral transformation $\delta\xi = i\theta\,\gamma_3\,\xi$. Hence the R-symmetry group $U(1)_L \otimes U(1)_R$ corresponds to a possible symmetry of the globally supersymmetric theory, usually broken by the coupling to gravity. Its preservation at the quantum level is a necessary condition for the emergence of an N=2 superconformal theory.

In this section we study these R-symmetries starting from N=2 Landau–Ginzburg model with abelian gauge symmetries discussed in the previous sections. Next by taking a suitable limit we define the *rigid Landau–Ginzburg model* and consider its R-symmetries. They are intimately related with the $U(1)$ charges and conformal weights of the chiral primary fields in the corresponding superconformal model. Indeed R-symmetries are present only if the superpotential is a quasi-homogeneous function.

We begin by considering the the gauge-coupled case and we assume that the superpotential $W(X)$ is quasi-homogeneous of degree $d \in \mathbb{R}$ with scaling weights $\omega_i \in \mathbb{R}$ for the chiral scalar fields $X^i$. This means that if we rescale each $X^i$ according to the rule

$$X^i \longrightarrow \exp[\omega_i\lambda]\, X^i \qquad (5.5.36)$$

where $\lambda \in \mathbb{C}$ is some constant complex parameter, then the superpotential rescales as follows:

$$W\left(e^{\omega_i\lambda}X^i\right) = \exp[d\lambda]\, W\left(X^i\right) \qquad (5.5.37)$$

Under these assumptiosn, we can easily verify that the rheonomic parametrizations, the rheonomic and world-sheet action of the N=2 locally gauge-invariant Landau–Ginzburg model are also invariant under the following global $U(1)_L \otimes U(1)_R$ transformations:

$$
\begin{array}{lcllcl}
\zeta^{\pm} & \longrightarrow & \exp[\pm i\alpha_L]\,\zeta^{\pm} & \widetilde{\zeta}^{\pm} & \longrightarrow & \exp[\pm i\alpha_R]\widetilde{\zeta}^{\pm} \\
\lambda^{\pm} & \longrightarrow & \exp[\pm i\alpha_R]\,\lambda^{\pm} & \widetilde{\lambda}^{\pm} & \longrightarrow & \exp[\pm i\alpha_L]\widetilde{\lambda}^{\pm} \\
M & \longrightarrow & \exp[i(\alpha_L - \alpha_R)]\,M & M^{*} & \longrightarrow & \exp[-i(\alpha_L - \alpha_R)]M^{*} \\
 & & \mathcal{P} & \longrightarrow & \mathcal{P} & \\
 & & \mathcal{A} & \longrightarrow & \mathcal{A} & \\
X^{i} & \longrightarrow & \exp[-i\frac{\omega_i(\alpha_L+\alpha_R)}{d}]\,X^{i} & X^{i*} & \longrightarrow & \exp[i\frac{\omega_i(\alpha_L+\alpha_R)}{d}]\,X^{i*} \\
\psi^{i} & \longrightarrow & \exp[i\frac{(d-\omega_i)\alpha_L-\omega_i\alpha_R}{d}]\,\psi^{i} & \widetilde{\psi}^{i} & \longrightarrow & \exp[i\frac{(d-\omega_i)\alpha_R-\omega_i\alpha_L}{d}]\,\widetilde{\psi}^{i} \\
\psi^{i^*} & \longrightarrow & \exp[-i\frac{(d-\omega_i)\alpha_L-\omega_i\alpha_R}{d}]\,\psi^{i^*} & \widetilde{\psi}^{i^*} & \longrightarrow & \exp[-i\frac{(d-\omega_i)\alpha_R-\omega_i\alpha_L}{d}]\,\widetilde{\psi}^{i^*}
\end{array}
\tag{5.5.38}
$$

If we define the R-symmetry charges of a field $\varphi$ by means of the formula

$$
\varphi \longrightarrow \exp[\, i\ (q_L\,\alpha_L\ +\ q_R\,\alpha_R)\,]\,\varphi \tag{5.5.39}
$$

then the charge assignments of the locally gauge invariant N=2 Landau–Ginzburg model are easily readable from Eq. (5.5.38).

As many times anticipated we can also consider a *rigid N=2 Landau-Ginzburg* model. By this we mean a Landau–Ginzburg theory of the type described in the previous sections, where the coupling to the gauge fields has been suppressed. The structure of such a theory is easily retrieved from our general formulae (5.5.11), (5.5.14), (5.5.15) by setting the gauge coupling constant to zero: redefine $q^i_j \longrightarrow g\,\overline{q}^i_j$ and then let $g \longrightarrow 0$. In this limit the matter fields decouple from the gauge fields and we obtain the following world-sheet lagrangian:

$$
\begin{aligned}
\mathcal{L}^{(ws)}_{chiral} &= -(\partial_+X^{i^*}\partial_-X^{i} + \partial_-X^{i^*}\partial_+X^{i}) \\
&\quad + 2i(\psi^{i}\partial_-\psi^{i^*} + \widetilde{\psi}^{i}\partial_+\widetilde{\psi}^{i^*}) + 2i(\psi^{i^*}\partial_-\psi^{i} + \widetilde{\psi}^{i^*}\partial_+\widetilde{\psi}^{i}) \\
&\quad + 8\Big\{(\psi^{i}\widetilde{\psi}^{j}\partial_i\partial_j\mathcal{W} + \text{c.c.}) + \partial_i\mathcal{W}\partial_{i^*}\mathcal{W}^{*}\Big\}
\end{aligned}
\tag{5.5.40}
$$

where to emphasize that we are discussing a different theory we have used a curly letter $\mathcal{W}(X)$ to denote the superpotential. The action (5.5.40) defines a model extensively studied in the literature both for its own sake and in its topological version (see the Bibliographical Note of the present chapter and of Chapter 7). This action is invariant against the supersymmetry transformations that we derive from the rheonomic parametrizations (5.5.11) upon suppression of the gauge coupling ($g \longrightarrow 0$), namely from

$$
\begin{aligned}
\nabla X^{i} &= \partial_+X^{i}\,e^{+} + \partial_-X^{i}\,e^{-} + \psi^{i}\zeta^{-} + \widetilde{\psi}^{i}\widetilde{\zeta}^{-} \\
\nabla X^{i^*} &= \partial_+X^{i^*}\,e^{+} + \partial_-X^{i^*}\,e^{-} - \psi^{i^*}\zeta^{+} - \widetilde{\psi}^{i^*}\widetilde{\zeta}^{+} \\
\nabla\psi^{i} &= \partial_+\psi^{i}\,e^{+} + \partial_-\psi^{i}\,e^{-} - \frac{i}{2}\partial_+X^{i}\,\zeta^{+} + \eta^{ij^*}\partial_{j^*}\mathcal{W}^{*}\,\widetilde{\zeta}^{-} \\
\nabla\psi^{i^*} &= \partial_+\psi^{i^*}\,e^{+} + \partial_-\psi^{i^*}\,e^{-} + \frac{i}{2}\partial_-X^{i^*}\,\zeta^{-} + \eta^{ji^*}\partial_j\mathcal{W}\,\widetilde{\zeta}^{+}
\end{aligned}
$$

$$
\begin{aligned}
\nabla\widetilde{\psi}^{i} &= \partial_{+}\widetilde{\psi}^{i}\, e^{+} + \partial_{-}\widetilde{\psi}^{i}\, e^{-} - \frac{i}{2}\partial_{-}X^{i}\,\widetilde{\zeta}^{+} - \eta^{ij^{*}}\partial_{j^{*}}\mathcal{W}^{*}\,\zeta^{-} \\
\nabla\widetilde{\psi}^{i^{*}} &= \partial_{+}\widetilde{\psi}^{i^{*}}\, e^{+} + \partial_{-}\widetilde{\psi}^{i^{*}}\, e^{-} + \frac{i}{2}\partial_{+}X^{i^{*}}\,\widetilde{\zeta}^{-} - \eta^{ji^{*}}\partial_{j}\mathcal{W}\,\zeta^{+}
\end{aligned}
\tag{5.5.41}
$$

The explicit form of the supersymmetry transformations is therefore

$$
\begin{aligned}
\delta X^{i} &= -\varepsilon^{-}\psi^{i} - \widetilde{\varepsilon}^{-}\widetilde{\psi}^{i} \\
\delta X^{i^{*}} &= +\varepsilon^{+}\psi^{i^{*}} + \widetilde{\varepsilon}^{+}\widetilde{\psi}^{i^{*}} \\
\delta\psi^{i} &= -\frac{i}{2}\partial X^{i}\,\varepsilon^{+} + \eta^{ij^{*}}\partial_{j^{*}}\overline{\mathcal{W}}\,\widetilde{\varepsilon}^{-}\ , \\
\delta\widetilde{\psi}^{i} &= -\frac{i}{2}\overline{\partial}X^{i}\,\widetilde{\varepsilon}^{+} - \eta^{ij^{*}}\partial_{j^{*}}\overline{\mathcal{W}}\,\varepsilon^{-} \\
\delta\psi^{i^{*}} &= \frac{i}{2}\partial X^{i^{*}}\,\varepsilon^{-} + \eta^{ji^{*}}\partial_{j}\mathcal{W}\,\widetilde{\varepsilon}^{+} \\
\delta\widetilde{\psi}^{i^{*}} &= \frac{i}{2}\overline{\partial}X^{i^{*}}\,\widetilde{\varepsilon}^{-} - \eta^{ji^{*}}\partial_{j}\mathcal{W}\,\varepsilon^{+}
\end{aligned}
\tag{5.5.42}
$$

where we have identified $\overline{\partial} = \partial_{-}$ and $\partial = \partial_{+}$. Assuming that under the rescalings (5.5.36) the superpotential $\mathcal{W}(X)$ has the scaling property (5.5.37) with an appropriate $d = d_{\mathcal{W}}$ then the rigid Landau–Ginzburg model admits a $U(1)_L \otimes U(1)_R$ group of R-symmetries whose action on the fields is *formally* the restriction to the matter fields of the R-symmetries (5.5.38), namely

$$
\begin{array}{llll}
X^{i} \longrightarrow \exp[-i\frac{\omega_i(\alpha_L+\alpha_R)}{d_{\mathcal{W}}}]\, X^{i} & & X^{i*} \longrightarrow \exp[i\frac{\omega_i(\alpha_L+\alpha_R)}{d_{\mathcal{W}}}]\, X^{i^{*}} \\
\psi^{i} \longrightarrow \exp[i\frac{(d_{\mathcal{W}}-\omega_i)\alpha_L-\omega_i\alpha_R}{d_{\mathcal{W}}}]\, \psi^{i} & & \widetilde{\psi}^{i} \longrightarrow \exp[i\frac{(d_{\mathcal{W}}-\omega_i)\alpha_R-\omega_i\alpha_L}{d_{\mathcal{W}}}]\, \widetilde{\psi}^{i} \\
\psi^{i^{*}} \longrightarrow \exp[-i\frac{(d_{\mathcal{W}}-\omega_i)\alpha_L-\omega_i\alpha_R}{d_{\mathcal{W}}}]\, \psi^{i^{*}} & & \widetilde{\psi}^{i^{*}} \longrightarrow \exp[-i\frac{(d_{\mathcal{W}}-\omega_i)\alpha_R-\omega_i\alpha_L}{d_{\mathcal{W}}}]\, \widetilde{\psi}^{i^{*}}
\end{array}
\tag{5.5.43}
$$

One, however, has to be careful that the parameter $d_{\mathcal{W}}$ in Eqs. (5.5.43) is the scale dimension of the superpotential $\mathcal{W}(X^i)$ and not $d$, the scale dimension of the original $W(X)$ of the gauge coupled model. This discussion is relevant in view of the N=2 gauge model that allows an interpolation between a rigid N=2 Landau–Ginzburg theory and an N=2 $\sigma$-model, appearing as the low energy effective actions in two different phases of the same gauge theory. In this case the superpotential of the locally gauge invariant Landau–Ginzburg theory is chosen as follows:

$$
W\left(X^{I}\right) = X^{0}\,\mathcal{W}\left(X^{i}\right) \tag{5.5.44}
$$

where the index $i$ runs on $n$ values $i = 1, ..., n$, the index $I$ runs on $n+1$ values $I = 0, 1, ..., n$ and $\mathcal{W}(X^i)$ is a quasi-homogeneous holomorphic function of degree $d_{\mathcal{W}}$ under the rescalings (5.5.36) with appropriate choices of the $\omega_i$. Note also that the chiral multiplet corresponding to the field $X^0$ of Eq. (5.5.44) has nothing to do with the dilaton supermultiplet (also denoted by $X^0$) discussed in previous sections. Then choosing arbitrarily a scale weight $\omega_0$ for the field $X^0$, the complete superpotential $W(X^I)$ becomes a quasi-homogeneous function of degree $d = d_{\mathcal{W}} + \omega_0$. Now in this model, as we are going to see later in our discussion of the N=2 phases, there is a phase where the gauge multiplet becomes massive, together with the multiplet of $X^0$, while all the $X^i$-multiplets are massless and have vanishing vacuum expectation values. In this phase the low energy effective action is a rigid Landau–Ginzburg model with superpotential $\mathcal{W}(X^i)$. In this case, if we want to identify the R-symmetries of the effective action with those of the original theory, something which is important in the discussion of the topological twists, we have to be careful to choose $\omega_0 = 0$. Only in this case $d = d_{\mathcal{W}}$ and Eqs. (5.5.43) are truly the restriction of Eqs. (5.5.38).

### 5.5.6 N=2 sigma models

As a necessary term of comparison for our subsequent discussion of the effective low energy lagrangians of the N=2 matter-coupled gauge models and of their topological twists, in the present section we consider another important instance of a two-dimensional field theory possessing an N=2 supersymmetry, namely the N=2 $\sigma$-model. This is a theory of maps:

$$X \quad : \Sigma \quad \longrightarrow \quad \mathcal{M} \tag{5.5.45}$$

from a two-dimensional world sheet $\Sigma$ which, after Wick rotation, can be identified with a Riemann surface, to a Kähler manifold $\mathcal{M}$, whose first Chern number $c_1(\mathcal{M})$ is not necessarily vanishing. In the specific case where $\mathcal{M}$ is a Calabi–Yau $n$-fold ($c_1 = 0$) the $\sigma$-model leads to an N=2 superconformal field theory with central charge $c = 3n$ but, as far as ordinary N=2 supersymmetry is concerned, the Calabi–Yau condition is not required, the only restriction on the target manifold being that it is Kählerian. The relation of the N=2 $\sigma$-model with the already introduced $N = 2$ Landau–Ginzburg theory (see section 5.7) is discussed in a later section of the present chapter, while its relation, via the topological twists, with topological $\sigma$-models is addressed in Chapter 7: in the present section we restrict our attention to the construction of the $\sigma$-model lagrangian and supersymmetry transformation rules.

Our notation is as follows. The holomorphic coordinates of the Kählerian target manifold $\mathcal{M}$ are denoted by $X^i$ ($i = 1, ..., n$), and their complex conjugates by $\overline{X}^{i^*}$. The field content of the N=2 $\sigma$-model is identical with that of the $N = 2$ Landau-Ginzburg theory: in addition to the $X$-fields, which transform as world-sheet scalars, the spectrum contains four sets of of spin 1/2 fermions, $\psi^i$, $\widetilde{\psi}^i$, $\psi^{i^*}$, $\widetilde{\psi}^{i^*}$, that appear in the N=2 rheonomic parametrizations of $dX^i$ and $d\overline{X}^{i^*}$

$$\begin{aligned} dX^i &= \Pi^i_+ e^+ + \Pi^i_- e^- + \psi^i \zeta^- + \widetilde{\psi}^i \widetilde{\zeta}^- \\ d\overline{X}^{i^*} &= \Pi^{i^*}_+ e^+ + \Pi^{i^*}_- e^- - \psi^{i^*} \zeta^- - \widetilde{\psi}^{i^*} \widetilde{\zeta}^- \end{aligned} \tag{5.5.46}$$

The above equations are identical with the homologous rheonomic parametrizations of the Landau–Ginzburg theory (the first two of Eqs. (5.5.41)). The difference with the

Landau–Ginzburg case appears at the level of the rheonomic parametrizations of the fermion differentials. Rather than the last four of Eqs. (5.5.41) we write

$$\begin{aligned}
\nabla \psi^i &= \nabla_+ \psi^i e^+ + \nabla_- \psi^i e^- - \frac{i}{2} \Pi^i_+ \zeta^+ \\
\nabla \widetilde{\psi}^i &= \nabla_+ \widetilde{\psi}^i e^+ + \nabla_- \widetilde{\psi}^i e^- - \frac{i}{2} \Pi^i_- \widetilde{\zeta}^+ \\
\nabla \psi^{i^*} &= \nabla_+ \psi^{i^*} e^+ + \nabla_- \psi^{i^*} e^- + \frac{i}{2} \Pi^{i^*}_+ \zeta^- \\
\nabla \widetilde{\psi}^{i^*} &= \nabla_+ \widetilde{\psi}^{i^*} e^+ + \nabla_- \widetilde{\psi}^{i^*} e^- + \frac{i}{2} \Pi^{i^*}_- \widetilde{\zeta}^-
\end{aligned} \tag{5.5.47}$$

where the symbol $\nabla$ denotes the covariant derivative with respect to the target space Levi–Civita connection:

$$\begin{aligned}
\nabla \psi^i &= d\psi^i - \Gamma^i_{jk} \, dX^j \, \psi^k \\
\nabla \widetilde{\psi}^i &= d\widetilde{\psi}^i - \Gamma^i_{jk} \, dX^j \, \widetilde{\psi}^k \\
\nabla \psi^{i^*} &= d\psi^{i^*} - \Gamma^{i^*}_{j^*k^*} \, dX^{j^*} \, \psi^{k^*} \\
\nabla \widetilde{\psi}^{i^*} &= d\widetilde{\psi}^{i^*} - \Gamma^{i^*}_{j^*k^*} \, dX^{j^*} \, \widetilde{\psi}^{k^*}
\end{aligned} \tag{5.5.48}$$

In agreement with the conventions of Chapter 3, the metric, connection and curvature of the Kählerian target manifold are given by

$$\begin{aligned}
g_{ij^*} &= \frac{\partial}{\partial X^i} \frac{\partial}{\partial X^{j^*}} \mathcal{K} \\
\Gamma^i_{jk} &= -g^{il^*} \, \partial_j g_{kl^*} \\
\Gamma^{i^*}_{j^*k^*} &= -g^{i^*l} \, \partial_{j^*} g_{k^*l} \\
\Gamma^i{}_j &= \Gamma^i_{jk} \, dX^k \\
R_{i^*jk^*l} &= g_{ip^*} \, R^p{}_{jk^*l} \\
R^p{}_{jk^*l} &= \partial_{k^*} \Gamma^p_{jl} \\
R^i_j &= R^i{}_{jk^*l} \, dX^{k^*} \wedge dX^l
\end{aligned} \tag{5.5.49}$$

where $\mathcal{K}(X, \overline{X})$ denotes the Kähler potential. The parametrizations (5.5.46) and (5.5.47) are the unique solution to the Bianchi identities:

$$\begin{aligned}
d^2 X^i &= d^2 \overline{X}^{i^*} = 0 \\
\nabla^2 \psi^i &= -R^i{}_j \, \psi^j \\
\nabla^2 \widetilde{\psi}^i &= -R^i{}_j \, \widetilde{\psi}^j \\
\nabla^2 \psi^{i^*} &= -R^{i^*}{}_{j^*} \, \psi^{j^*} \\
\nabla^2 \widetilde{\psi}^{i^*} &= -R^{i^*}{}_{j^*} \, \widetilde{\psi}^{j^*}
\end{aligned} \tag{5.5.50}$$

The complete rheonomic action that yields these parametrizations as outer field equations is given by the following expression:

$$
\begin{aligned}
S_{rheonomic} &= \int \Big[ g_{ij^\star} \left( dX^i - \psi^i \zeta^- - \widetilde{\psi}^i \widetilde{\zeta}^- \right) \wedge \left( \Pi^{j^\star}_+ e^+ - \Pi^{j^\star}_- e^- \right) \\
&+ g_{ij^\star} \left( dX^{j^\star} + \psi^{j^\star} \zeta^+ + \widetilde{\psi}^{j^\star} \widetilde{\zeta}^+ \right) \wedge \left( \Pi^i_+ e^+ - \Pi^i_- e^- \right) \\
&+ g_{ij^\star} \left( \Pi^i_+ \Pi^{j^\star}_- + \Pi^i_- \Pi^{j^\star}_+ \right) e^+ \wedge e^- \\
&- 2i\, g_{ij^\star} \left( \psi^i \nabla \psi^{j^\star} \wedge e^+ - \widetilde{\psi}^i \nabla \widetilde{\psi}^{j^\star} \wedge e^- \right) \\
&- 2i\, g_{ij^\star} \left( \psi^{j^\star} \nabla \psi^i \wedge e^+ - \widetilde{\psi}^{j^\star} \nabla \widetilde{\psi}^i \wedge e^- \right) \\
&- g_{ij^\star} \left( dX^i \psi^{j^\star} \wedge \zeta^+ - dX^i \widetilde{\psi}^{j^\star} \wedge \widetilde{\zeta}^+ \right) \\
&+ g_{ij^\star} \left( dX^{j^\star} \psi^i \wedge \zeta^- - dX^{j^\star} \widetilde{\psi}^i \wedge \widetilde{\zeta}^- \right) \\
&+ g_{ij^\star} \left( \psi^i \psi^{j^\star} \zeta^+ \wedge \zeta^- + \widetilde{\psi}^i \widetilde{\psi}^{j^\star} \widetilde{\zeta}^+ \wedge \widetilde{\zeta}^- \right) \\
&+ 8\, R_{ij^\star kl^\star} \psi^i \psi^{j^\star} \widetilde{\psi}^k \widetilde{\psi}^{l^\star} e^+ \wedge e^- \Big]
\end{aligned}
\tag{5.5.51}
$$

From Eq. (5.5.51) we immediately obtain the world-sheet action in second order formalism, by deleting the terms containing the fermionic vielbeins $\zeta$ and by substituting back the value of the auxiliary fields $\Pi$ determined by their own field equations. The result is

$$
\begin{aligned}
S_{world\text{-}sheet} &= \int \Big[ - g_{ij^\star} \left( \partial_+ X^i \partial_- X^{j^\star} + \partial_- X^i \partial_+ X^{j^\star} \right) \\
&+ 2i\, g_{ij^\star} \left( \psi^i \nabla_- \psi^{j^\star} + \psi^{j^\star} \nabla_- \psi^i \right) \\
&+ 2i\, g_{ij^\star} \left( \widetilde{\psi}^i \nabla_+ \widetilde{\psi}^{j^\star} + \widetilde{\psi}^{j^\star} \nabla_+ \widetilde{\psi}^i \right) \\
&+ 8\, R_{ij^\star kl^\star} \psi^i \psi^{j^\star} \widetilde{\psi}^k \widetilde{\psi}^{l^\star} \Big] d^2 z
\end{aligned}
\tag{5.5.52}
$$

where we have denoted by

$$
\begin{aligned}
\nabla_\pm \psi^i &= \partial_\pm \psi^i - \Gamma^i_{jk} \partial_\pm X^j \psi^k \\
\nabla_\pm \psi^{i^\star} &= \partial_\pm \psi^{i^\star} - \Gamma^{i^\star}_{j^\star k^\star} \partial_\pm X^{j^\star} \psi^{k^\star}
\end{aligned}
\tag{5.5.53}
$$

the world-sheet components of the target space covariant derivatives: identical equations hold for the tilded fermions. The world-sheet action (5.5.52) is invariant against the supersymmetry transformation rules descending from the rheonomic parametrizations (5.5.46) and (5.5.47), namely

$$
\begin{aligned}
\delta \psi^i &= -\frac{i}{2} \partial_+ X^i \varepsilon^+ - \widetilde{\varepsilon}^- \Gamma^i_{jk} \widetilde{\psi}^j \psi^k \\
\delta \widetilde{\psi}^i &= -\frac{i}{2} \partial_- X^i \widetilde{\varepsilon}^+ - \varepsilon^- \Gamma^i_{jk} \psi^j \widetilde{\psi}^k \\
\delta \psi^{i^\star} &= +\frac{i}{2} \partial_+ X^{i^\star} \varepsilon^- + \widetilde{\varepsilon}^+ \Gamma^{i^\star}_{j^\star k^\star} \widetilde{\psi}^{j^\star} \psi^{k^\star} \\
\delta \widetilde{\psi}^{i^\star} &= +\frac{i}{2} \partial_- X^{i^\star} \widetilde{\varepsilon}^- + \varepsilon^+ \Gamma^{i^\star}_{j^\star k^\star} \psi^{j^\star} \widetilde{\psi}^{k^\star}
\end{aligned}
\tag{5.5.54}
$$

Comparing with the transformation rules defined by Eqs. (5.5.41) we see that in the variation of the fermionic fields, the term proportional to the derivative of the superpotential has been replaced with a fermion bilinear containing the Levi-Civita connection of the target manifold. Indeed one set of rules can be obtained from the other by means of the replacement

$$\begin{aligned} \eta^{ij^*}\,\partial_{j^*}\overline{W} &\longrightarrow -\Gamma^i_{jk}\,\widetilde{\psi}^j\,\psi^k \\ \eta^{i^*j}\,\partial_j W &\longrightarrow \Gamma^{i^*}_{j^*k^*}\,\widetilde{\psi}^{j^*}\,\psi^{k^*} \end{aligned} \tag{5.5.55}$$

This fact emphasizes that in the $\sigma$-model the form of the interaction and hence all the quantum properties of the theory are dictated by the Kähler structure, namely by the real, non holomorphic Kähler potential $\mathcal{K}(X,\overline{X})$, while in the Landau–Ginzburg case the structure of the interaction and the resulting quantum properties are governed by the holomorphic superpotential $\mathcal{W}(X)$. In spite of these differences, both types of models can yield at the infrared critical point an N=2 superconformal theory and can be related to the same Calabi–Yau manifold. In the case of the $\sigma$-model, the relation is most direct: it suffices to take, as target manifold $\mathcal{M}$, the very Calabi–Yau $n$-fold one is interested in and to choose for the Kähler metric $g_{ij^*}$ one representative in one of the available Kähler classes:

$$K = i\, g_{ij^*}\, dX^i \wedge d\overline{X}^{j^*} \in \left[K\right] \in H^{(1,1)}\,(\mathcal{M}) \tag{5.5.56}$$

If $c_1(\mathcal{M}) = 0$, within each Kähler class we can readjust the choice of the representative metric $g_{ij^*}$, so that at each perturbative order the beta function is made equal to zero. In this way we obtain conformal invariance and we associate an N=2 superconformal theory with any N=2 $\sigma$-model on a Calabi–Yau $n$-fold $\mathcal{M}$. The N=2 gauge model discussed in the previous sections interpolates between the $\sigma$-model and the Landau–Ginzburg theory with, as superpotential, the very function $\mathcal{W}(X)$ whose vanishing defines $\mathcal{M}$ as a hypersurface in a (weighted) projective space. More generally for the Landau–Ginzburg case the relation with Calabi–Yau $n$-folds is established by choosing as superpotential $\mathcal{W}(X) = \sum_A \mathcal{W}_A(X)$ the sum of the quasi-homogeneous polynomial constraints $\mathcal{W}_A(X) = 0$ which define $\mathcal{M}$ as an algebraic variety in a suitable weighted projective space. Indeed, in a later section 5.7 we show how the integral over the $X$-fields can be reduced to the vanishing locus $\mathcal{W}_A(X) = 0$ if the homogeneity weights are such that this locus is a variety with vanishing first Chern class. On the other hand, at the infrared critical point the quasi-homogeneous part of $\mathcal{W}(X)$ is the only surviving one and it determines an N=2 superconformal theory. The two superconformal theories, which derived from the $\sigma$-model and from the Landau–Ginzburg theory, are necessarily the same if the underlying Calabi–Yau manifold $\mathcal{M}$ is the same. What the two field theory formulations single out are two different types of deformations of the same superconformal field theory, geometrically of the same Calabi–Yau manifold. In fact the deformations of the $\sigma$-model correspond to the deformations of the Kähler structure, while the deformations of the Landau–Ginzburg theory correspond to the complex structure deformations of the same manifold. In the $n = 3$ case the $\sigma$-model accounts for the $(1,1)$-forms, while the

Landau–Ginzburg theory relates to the $(2,1)$-forms. This fact will be better appreciated after our discussion of mirror symmetry in Chapter 8 and the introduction of topological theories in Chapter 7.

As a matter of comparison a very important issue are the left-moving and right-moving R-symmetries of the $\sigma$-model. Indeed, also in this case, the rheonomic parametrizations, the rheonomic and world-sheet actions are invariant under a global $U(1)_L \otimes U(1)_R$ group. The action of this group on the $\sigma$-model fields, however, is different from that on the Landau–Ginzburg fields, namely we have

$$\begin{array}{ccccccc} \zeta^{\pm} & \longrightarrow & \exp[\pm i\alpha_L]\,\zeta^{\pm} & & \widetilde{\zeta}^{\pm} & \longrightarrow & \exp[\pm i\alpha_R]\widetilde{\zeta}^{\pm} \\ X^i & \longrightarrow & X^i & & X^{i*} & \longrightarrow & X^{i^*} \\ \psi^i & \longrightarrow & \exp[i\alpha_L]\,\psi^i & & \widetilde{\psi}^i & \longrightarrow & \exp[i\alpha_R]\,\widetilde{\psi}^i \\ \psi^{i^*} & \longrightarrow & \exp[-i\alpha_L]\,\psi^{i^*} & & \widetilde{\psi}^{i^*} & \longrightarrow & \exp[-i\alpha_R]\,\widetilde{\psi}^{i^*} \end{array} \tag{5.5.57}$$

where $\alpha_L$ and $\alpha_R$ are the two constant phase parameters. The crucial difference of Eqs. (5.5.57) with respect to Eqs. (5.5.43) resides in the R-invariance of the scalar fields $X^i$ which applies to the $\sigma$-model case, but not to the Landau–Ginzburg case. As a consequence, in the $\sigma$-model case the fermions have fixed integer R-symmetry charges, while in the Landau–Ginzburg case they acquire fractional R-charges depending on the homogeneity degree of the corresponding scalar field and of the superpotential.

### 5.5.7 Extrema of the N=2 scalar potential, phases of the gauge theory and reconstruction of the effective N=2 $\sigma$-model

Now we focus on the effective low energy theory emerging from the $N = 2$ gauge plus matter systems described in the above sections. Our considerations remain at a classical level. We are mostly interested in the case where the effective theory is an $N = 2$ $\sigma$-model. We show how the $N = 2$ $\sigma$-model lagrangian is technically retrieved, in a manner that is intimately related with the momentum map construction discussed in Chapter 2. Indeed this latter is just the geometrical counterpart of the physical concept of a low energy effective lagrangian. To be simple we perform our computations in the case where the target space of the low energy $\sigma$-model is the manifold $\mathbb{CP}_N$.

First of all we need to recall the structure of the classical vacua for a system decribed by the lagrangian (5.5.16), referring to the linear superpotential case $\mathcal{U} = (\frac{r}{4} - i\frac{\theta}{8\pi})M$; this structure was studied in Witten's paper [262]. We set the fermions to zero and we have to extremize the scalar potential (5.5.21). Since $U$ is given by a sum of moduli squared, this amounts to equating each term in (5.5.21) separately to zero. A articularly interesting situation arises when the Landau–Ginzburg potential has the form

$$W = X^0\,\mathcal{W}(X^i) \tag{5.5.58}$$

Here $\mathcal{W}(X^i)$ is a quasi-homogeneous function of degree $d$ of the fields $X^i$ which are assigned the weigths $q^i$, i.e. their charges with respect to the abelian gauge group. In the case where all the charges $q^i$ are equal (say all equal to 1, for simplicity) $\mathcal{W}(X^i)$ is homogeneous. $X^0$ is a scalar field of charge $-d$. $\mathcal{W}(X^i)$ must moreover be *transverse*: $\partial_i \mathcal{W} = 0$, $\forall i$ iff $X^i = 0$, $\forall i$.

In this case we have

$$U = \frac{1}{2}\Big(r + d|X^0|^2 - \sum_i q^i|X^i|^2\Big)^2 + 8|\mathcal{W}(X^i)|^2 + 8|X^0|^2|\partial_i \mathcal{W}|^2$$
$$+ 8|M|^2\Big(d^2|X^0|^2 + \sum_i (q^i)^2|X^i|^2\Big), \tag{5.5.59}$$

and two possibilities emerge.

- $r > 0$. In this case some of the $X^i$ must be different from zero. Due to the transversality of $\mathcal{W}$ it follows that $X^0 = 0$. The space of classical vacua is characterized not only by having $X^0 = 0$ and $M = 0$, but also by the condition $\sum_i q^i|X^i|^2 = r$. When $q^i = 1$ $\forall i$ this condition, together with the $U(1)$ gauge invariance, is equivalent to the statement that the $X^i$ represent homogeneous coordinates on $\mathbb{CP}_N$. In general, the $X^i$ are homogeneous coordinates on the weighted projective space $W\mathbb{CP}^N_{q^1 \dots q^{N+1}}$ (for the definition of $W\mathbb{CP}$ spaces see section 5.7). The last requirement, $\mathcal{W}(X^i) = 0$, defines the space of classical vacua as a transverse hypersurface embedded in $\mathbb{CP}_N$ or, in general, in $W\mathbb{CP}^N_{q^1 \dots q^{N+1}}$. The low energy theory around these vacua is expected to correspond to the $N = 2$ $\sigma$-model on such a hypersurface. Indeed, studying the quadratic fluctuations one sees that the gauge field $\mathcal{A}$ acquires a mass due to a Higgs phenomenon; the gauge scalar $M$ becomes massive together with those modes of the matter fields that are not tangent to the hypersurface. The only massless degrees of freedom, i.e. those described by the low energy theory, are the excitations tangent to the hypersurface. The fermionic partners behave consistently. We are in the *$\sigma$-model phase.*

- $r < 0$. In this case $X^0$ must be different from zero. Then it is necessary that $\partial_i \mathcal{W} = 0$ $\forall i$; this implies by transversality that all the $X^i$ vanish. The space of classical vacua is just a point. Indeed utilizing the gauge invariance we can reduce $X^0$ to be real, so that it is fixed to have the constant value $X^0 = \sqrt{\frac{-r}{d}}$. $M$ vanishes together with the $X^i$. The low energy theory can now be recognized to be a theory of massless fields, the $X^i$, governed by a Landau Ginzburg potential which is just $\mathcal{W}(x^i)$. We are in the *Landau-Ginzburg phase.*

Now we turn our attention to the $\mathbb{CP}_N$-model, which corresponds to the particular case in which all the charges are equal to 1 and $W = 0$. As is easy to see from the above discussion, in this case the only possible vacuum phase is the $\sigma$-model phase, i.e. one must have $r > 0$. We start by writing the complete rheonomic lagrangian of the system consisting of $N+1$ chiral multiplets with no self-interaction $(X^A, \psi^A, \tilde{\psi}^A)$, $A = 0, \dots, N$, coupled to an abelian gauge multiplet, each with charge 1. Differently to what we did in the previous sections, in this section we make the dependence on the gauge coupling constant $g$ explicit. To reinstall $g$ appropriately, after reinserting it into the covariant derivatives, $\nabla X^A = dx^A + ig\mathcal{A}X^A$, we redefine the fields of the gauge multiplet as follows:

$$\mathcal{A} \longrightarrow \frac{1}{g}\mathcal{A} \qquad M \longrightarrow \frac{1}{g}M \qquad \lambda \longrightarrow \frac{1}{g}\lambda \tag{5.5.60}$$

so that at the end no modification occurs in the matter lagrangian, while the gauge kinetic lagrangian is multiplied by $\frac{1}{g^2}$. Altogether we have

$$\begin{aligned}
\mathcal{L} = & \frac{\mathcal{F}}{g^2}\Big[F + \frac{i}{2}(\bar\lambda^+\zeta^- + \bar\lambda^-\zeta^+)\,e^- - \frac{i}{2}(\lambda^+\bar\zeta^+ + \lambda^-\bar\zeta^+)\,e^+ - M\,\zeta^-\bar\zeta^+ - M^*\,\zeta^+\bar\zeta^-\Big] \\
& - \frac{1}{2g^2}\mathcal{F}^2\,e^+e^- - \frac{i}{2g^2}(\bar\lambda^+\,d\bar\lambda^- + \bar\lambda^-\,d\bar\lambda^+)\,e^- + \frac{i}{2g^2}(\lambda^+\,d\lambda^- + \lambda^-\,d\lambda^+)\,e^+ \\
& - \frac{4}{g^2}\Big[dM^* - \frac{1}{4}(\lambda^+\zeta^- - \bar\lambda^-\bar\zeta^+)\Big](\mathcal{M}_+e^+ - \mathcal{M}_-e^-) \\
& - \frac{4}{g^2}\Big[dM + \frac{1}{4}(\lambda^-\zeta^+ - \bar\lambda^+\bar\zeta^-)\Big](\mathcal{M}^*_+e^+ - \mathcal{M}^*_-e^-) \\
& - \frac{4}{g^2}(\mathcal{M}^*_+\mathcal{M}_- + \mathcal{M}^*_-\mathcal{M}_+)\,e^+e^- - \frac{1}{g^2}dM(\bar\lambda^-\bar\zeta^+ + \lambda^+\zeta^-) + \frac{1}{g^2}dM^*(\bar\lambda^+\bar\zeta^- + \lambda^-\zeta^+) \\
& - \frac{1}{4g^2}(\bar\lambda^+\lambda^+\,\zeta^-\bar\zeta^- + \bar\lambda^-\lambda^-\,\zeta^+\bar\zeta^+) + \frac{2}{g^2}\mathcal{P}^2\,e^+e^- - 2r\mathcal{P}e^+e^- + \frac{\theta}{2\pi}F \\
& + \frac{r}{2g^2}\Big[(\bar\lambda^+\zeta^- - \bar\lambda^-\zeta^+)\,e^- + (\lambda^+\bar\zeta^- - \lambda^-\bar\zeta^+)\,e^+\Big] + i\frac{r}{g^2}\Big(M\zeta^-\bar\zeta^+ + M^*\zeta^+\bar\zeta^-\Big) \\
& - (\nabla X^A - \psi^A\zeta^- - \bar\psi^A\bar\zeta^-)(\Pi^{A^*}_+e^+ - \Pi^{A^*}_-e^-) \\
& - (\nabla X^{A^*} + \psi^{A^*}\zeta^+ + \bar\psi^{A^*}\bar\zeta^+)(\Pi^A_+e^+ - \Pi^A_-e^-) \\
& + (\Pi^{A^*}_+\Pi^A_- - \Pi^{A^*}_-\Pi^A_+)e^+e^- + 2i(\psi^A\nabla\psi^{A^*} + \psi^{a^*}\nabla\psi^A)e^+ \\
& - 2i(\bar\psi^A\nabla\bar\psi^{A^*} + \bar\psi^{A^*}\nabla\bar\psi^A)e^- - \psi^A\psi^{A^*}\zeta^-\zeta^+ + \bar\psi^A\bar\psi^{A^*}\bar\zeta^-\bar\zeta^+ \\
& - \psi^A\bar\psi^{A^*}\zeta^-\bar\zeta^+ - \psi^{A^*}\bar\psi^A\zeta^+\bar\zeta^- + \nabla X^A(\psi^{A^*}\zeta^+ - \bar\psi^{A^*}\bar\zeta^+) \\
& - \nabla X^{A^*}(\psi^A\zeta^- - \bar\psi^A\bar\zeta^-) + 4MX^A\psi^{A^*}\bar\zeta^+e^+ - 4M^*X^{A^*}\psi^A\bar\zeta^-e^+ \\
& + 4M^*X^A\bar\psi^{A^*}\zeta^+e^- - 4MX^{A^*}\bar\psi^A\zeta^-e^- \\
& + \Big\{8iM^*\bar\psi^{A^*}\psi^A + 8iM\bar\psi^A\psi^{A^*} + 2i\lambda^+\psi^{A^*}X^A + 2i\lambda^-\psi^AX^{A^*} \\
& - 2i\bar\lambda^+\bar\psi^{A^*}X^A - 2i\bar\lambda^-\bar\psi^AX^{A^*} + 2\mathcal{P}X^{A^*}X^A - 8M^*MX^{A^*}X^A\Big\}e^+e^-
\end{aligned} \tag{5.5.61}$$

The procedure that we follow to extract the effective lagrangian is the following. We let the gauge coupling constant go to infinity and we are left with a gauge-invariant lagrangian describing matter coupled to gauge fields that have no kinetic terms. Varying the action in these fields, the resulting equations of motion express the gauge fields in terms of the matter fields. Substituting back their expressions into the lagrangian we end up with a $\sigma$-model having as target manifold the quotient of the manifold spanned by the matter fields with respect to the action of the gauge group [191]. This procedure is nothing else, from the integral viewpoint, than the gaussian integration over the gauge multiplet in the limit $g \longrightarrow \infty$. To consider a gauge-coupled lagrangian without gauge kinetic terms is not a mere trick to implement the quotient procedure in a lagrangian formalism. Rather, it amounts to deriving the low energy effective action around the classical vacua of the complete, gauge plus matter system. Indeed we have seen that around these vacua the oscillations of the gauge fields are massive, and thus decouple from the low energy point of view. So we integrate over them: furthermore all masses are proportional to $\frac{1}{g}$ and the integration makes sense for energy scales $E << \frac{1}{g}$, namely in the limit $g \longrightarrow \infty$.

Here we show in detail how the above-sketched procedure works at the level of the rheonomic approach. In this way we retrieve the rheonomic lagrangian and the rheonomic parametrizations of the $N = 2$ $\sigma$-model, as described in section 5.5.6, the target space being $\mathbb{CP}_N$, equipped with the standard Fubini–Study metric. The whole procedure amounts geometrically to realizing $\mathbb{CP}_N$ as a Kähler quotient (see Chapter 3).

Let us consider the lagrangian (5.5.61), *in the limit* $g \longrightarrow \infty$, and let us perform the variations in the gauge fields.

The variations in $\tilde{\lambda}^-, \tilde{\lambda}^+, \lambda^-, \lambda^+$ give the following *fermionic constraints*:

$$X^A \psi^{A^*} = X^{A^*} \psi^A = X^A \tilde{\psi}^{A^*} = X^{A^*} \tilde{\psi}^{A^*} = 0 \tag{5.5.62}$$

Here the summation on the capital index $A$ is understood. In the following we use simplified notations, such as $X\psi^*$ for $X^A \psi^{A^*}$, and the like, everywhere it is possible without generating confusion.

The fermionic constraints (5.5.62) are explained by the bosonic constraint $X^*X = r$, for which the auxiliary field $\mathcal{P}$, in the limit $g \longrightarrow \infty$ becomes a Lagrange multiplier. Indeed taking the exterior derivative of this bosonic constraint we obtain $0 = d(X^*X) = X^*dX + XdX^*$ and substituting the rheonomic parametrizations (5.5.11) in the gravitino sectors this implies

$$X^*(\psi\zeta^- + \tilde{\psi}\tilde{\zeta}^-) - X(\psi^*\zeta^+ + \tilde{\psi}^*\tilde{\zeta}^+) = 0 \tag{5.5.63}$$

from which (5.5.62) follows.

The variation of the action with respect to $M^*$ in the gravitino sectors implies again the fermionic constraints (5.5.62). In the $e^+e^-$ sector we get the following equation of motion:

$$M = \frac{i\tilde{\psi}^*\psi}{X^*X} \tag{5.5.64}$$

The terms in the lagrangian (5.5.61) containing the connection $\mathcal{A}$ are hidden in the covariant derivatives. Explicitly they are

$$\begin{aligned}
&-i\mathcal{A}X^A(\Pi_+^{A^*}e^+ - \Pi_-^{A^*}e^-) + i\mathcal{A}X^{A^*}(\Pi_+^A e^+ - \Pi_-^A e^-) + 2i\psi^A(-i)\mathcal{A}\psi^{A^*}e^+ \\
&\quad + 2i\psi^{A^*}i\mathcal{A}\psi^A e^+ - 2i\tilde{\psi}^A(-i)\mathcal{A}\tilde{\psi}^{A^*}e^- - 2i\tilde{\psi}^{A^*}i\mathcal{A}\tilde{\psi}^A e^- \\
&\quad + i\mathcal{A}X^A(\psi^{A^*}\zeta^+ - \tilde{\psi}^{A^*}\tilde{\zeta}^+) + i\mathcal{A}X^{A^*}(\psi^A\zeta^- - \tilde{\psi}^A\tilde{\zeta}^-) + \frac{\theta}{2\pi}d\mathcal{A}
\end{aligned} \tag{5.5.65}$$

In the gravitino sector we again retrieve the constraints (5.5.62). In the $e^+$, $e^-$ sectors we respectively obtain:

$$\begin{aligned} iX^A\Pi_+^{A^*} - iX^{A^*}\Pi_+^A - 4\psi^A\psi^{A^*} &= 0 \\ -iX^A\Pi_-^{A^*} + iX^{A^*}\Pi_-^A + 4\tilde{\psi}^A\tilde{\psi}^{A^*} &= 0 \end{aligned} \tag{5.5.66}$$

At this point we take into account the variations with respect to the first order fields $\Pi$, which give $\Pi_+^A = \nabla_+ X^A = \nabla_+ X^A + i\mathcal{A}_+ X^A$, and so on. Substituting into Eqs. (5.5.66) and solving for $\mathcal{A}_+, \mathcal{A}_-$ we get

$$\begin{aligned} \mathcal{A}_+ &= \frac{-i(X\partial_+X^* - X^*\partial_+X) + 4\psi\psi^*}{2X^*X} \\ \mathcal{A}_- &= \frac{-i(X\partial_-X^* - X^*\partial_-X) + 4\tilde{\psi}\tilde{\psi}^*}{2X^*X} \end{aligned} \tag{5.5.67}$$

Substituting back the expression (5.5.64) for $M$ into the lagrangian (5.5.61) in the $g \longrightarrow \infty$ limit, we have

$$\begin{aligned} \mathcal{L} = &- \Big[dX^A + iX^A(\mathcal{A}_+e^+ + \mathcal{A}_-e^-) - \psi^A\zeta^- - \tilde{\psi}^a\tilde{\zeta}^-\Big](\Pi_+^{A^*}e^+ - \Pi_-^{A^*}e^-) \\ &- \Big[dX^{A^*} - iX^{A^*}(\mathcal{A}_+e^+ + \mathcal{A}_-e^-) + \psi^{A^*}\zeta^+ + \tilde{\psi}^{a^*}\tilde{\zeta}^+\Big](\Pi_+^Ae^+ - \Pi_-^Ae^-) \\ &- (\Pi_+^{A^*}\Pi_-^A + \Pi_-^{A^*}\Pi_+^A)e^+e^- + 2i(\psi^Ad\psi^{A^*} + \psi^{A^*}d\psi^A - 2i\mathcal{A}_-e^-\psi^A\psi^{A^*})e^+ \\ &- 2i(\tilde{\psi}^Ad\tilde{\psi}^{A^*} + \tilde{\psi}^{A^*}d\tilde{\psi}^A - 2i\mathcal{A}_+e^+\tilde{\psi}A\tilde{\psi}^{A^*})e^- - \psi^A\psi^{A^*}\zeta^-\zeta^+ \\ &- \tilde{\psi}^A\tilde{\psi}^{A^*}\tilde{\zeta}^-\tilde{\zeta}^+ - \psi^A\tilde{\psi}^{A^*}\zeta^-\tilde{\zeta}^+ - \psi^{A^*}\tilde{\psi}^A\zeta^+\tilde{\zeta}^- \\ &+ dX^A(\psi^{A^*}\zeta^+ - \tilde{\psi}^{A^*}\tilde{\zeta}^+) - dX^{A^*}(\psi^A\zeta^- - \tilde{\psi}^A\tilde{\zeta}^+) \\ &- 8\frac{\psi^A\psi^{B^*}\tilde{\psi}^B\tilde{\psi}^{A^*}}{X^*X}e^+e^- + 2\mathcal{P}(r - X^*X)e^+e^- \end{aligned} \tag{5.5.68}$$

where $\mathcal{A}_+$ and $\mathcal{A}_-$ are to be identified with their expressions (5.5.67). To obtain this expression we have also used the "fermionic constraints" (5.5.62). The $U(1)$ gauge invariance of the above lagrangian can be extended to a $\mathbb{C}^*$-invariance, where $\mathbb{C}^* \equiv \mathbb{C} - \{0\}$ is the complexification of the $U(1)$ gauge group, by introducing an extra scalar field $v$ transforming appropriately. Consider the $\mathbb{C}^*$ gauge transformation given by

$$\begin{array}{lllllll} X^A & \longrightarrow & e^{i\Phi}X^A & \psi^A & \longrightarrow & e^{i\Phi}\psi^A & \\ X^{A^*} & \longrightarrow & e^{-i\Phi^*}X^{A^*} & & \ldots & \ldots & (\Phi \in \mathbb{C}) \end{array} \tag{5.5.69}$$

which is just the complexification of the $U(1)$ transformation, the latter corresponding to the case $\Phi \in \mathbb{R}$, supplemented with

$$v \longrightarrow v + \frac{i}{2}(\Phi - \Phi^*) \tag{5.5.70}$$

One realizes that under the transformations (5.5.69), (5.5.70) the combinations $e^{-v}X^A$ (and similar ones) undergo just a $U(1)$ transformation:

$$e^{-v}X^A \longrightarrow e^{i\mathrm{Re}\Phi}e^{-v}X^A$$
$$e^{-v}X^{A^*} \longrightarrow e^{-i\mathrm{Re}\Phi}e^{-v}X^{A^*} \tag{5.5.71}$$

By substituting

$$X^A, \psi^A, \psi^{A^*}, \widetilde{\psi}^A, \widetilde{\psi}^{A^*}, \Pi^A_+, \ldots \longrightarrow e^{-v}X^A, e^{-v}\psi^A, e^{-v}\psi^{A^*}, \ldots \tag{5.5.72}$$

into the lagrangian (5.5.68) we obtain an expression which is invariant with respect to the $\mathbb{C}^*$-transformations (5.5.69),(5.5.70).

In particular the last term of (5.5.68) becomes

$$2\mathcal{P}(r - e^{-2v}X^*X) \tag{5.5.73}$$

If at this point we perform the so far delayed variation with respect to the auxiliary field $\mathcal{P}$, the resulting equation of motion identifies the extra scalar field $v$ in terms of the matter fields. If we introduce $\rho^2 \equiv r$ the result is that

$$e^{-v} = \frac{\rho}{\sqrt{X^*X}} \tag{5.5.74}$$

What is the geometrical meaning of the above "tricks" (introduction of the extra field $v$, consideration of the complexified gauge group)? The answer relies on the properties of the Kähler quotient construction; discussed in section 3.2.1 (see also [191] and [158, 60, 132]). Here we repeat the general discussion of the Kähler quotient given in that section, identifying at each step the explicit form pertaining to our example of the advocated geometrical structures.

Let $\mathbf{Y}(s) = Y^a\mathbf{k}_a(s)$ be a Killing vector on S (in our case $\mathbb{C}^{N+1}$), belonging to $\mathcal{G}$ (in our case $\mathbb{R}$), the algebra of the gauge group. In our case $\mathbf{Y}$ has a single component: $\mathbf{Y} = i\Phi(X^A\frac{\partial}{\partial X^A} - X^{A^*}\frac{\partial}{\partial X^{A^*}})$ ($\Phi \in \mathbb{R}$). The $X^A$'s are the coordinates on $\mathcal{S}$. Consider the vector field $I\mathbf{Y} \in \mathcal{G}^c$ (the complexified algebra), $I$ being the complex structure acting on $T\mathcal{S}$. In our case $I\mathbf{Y} = \Phi(X^A\frac{\partial}{\partial X^A} + X^{A^*}\frac{\partial}{\partial X^{A^*}})$. This vector field is orthogonal to the hypersurface $\mathcal{D}^{-1}(\zeta)$, for any level $\zeta$; that is, it generates transformations that change the level of the surface. In our case the surface $\mathcal{D}^{-1}(\rho^2) \in \mathbb{C}^{N+1}$ is defined by the equation $X^{A^*}X^A = \rho^2$. The infinitesimal transformation generated by $I\mathbf{Y}$ is $X^A \to (1+\Phi)X^A$, $X^{A^*} \to (1+\Phi)X^{A^*}$ so that the transormed $X^A$'s satisfy $X^{A^*}X^A = (1+2\Phi)\rho^2$. As recalled in section 3.2.1, the Kähler quotient consists in starting from $\mathcal{S}$, restricting to $\mathcal{N} = \mathcal{D}^{-1}(\zeta)$ and taking the quotient $\mathcal{M} = \mathcal{N}/G$. The above remarks about the action of the complexified gauge group suggest that this is equivalent (at least if we skip the problems due to the non-compactness of $G^c$) to simply taking the quotient $\mathcal{S}/G^c$, the so-called "algebro-geometric" quotient [158], [191, 206].

The Kähler quotient allows one, in principle, to determine the expression of the Kähler form on $\mathcal{M}$ in terms of the original one on $\mathcal{S}$. Schematically, let $j$ be the inclusion map of

$\mathcal{N}$ into $\mathcal{S}$, $p$ the projection from $\mathcal{N}$ to the quotient $\mathcal{M} = \mathcal{N}/G$, $\Omega$ the Kähler form on $\mathcal{S}$ and $\omega$ the Kähler form on $\mathcal{M}$. Then we have the situation illustrated by Eq.(3.2.70). In the algebro-geometric setting, the holomorphic map that associates to a point $s \in \mathcal{S}$ (for us, $\{X^A\} \in \mathbb{C}^{N+1}$) its image $m \in \mathcal{M}$ is obtained as explained in section 3.2.1 through the steps listed in Eqs. (3.2.71) and (3.2.72). Looking at (3.2.70) we see that $\pi^* p^* \omega = \pi^* j^* \Omega$ so that at the end of the day, in order to recover the pullback of $\omega$ to $\mathcal{S}$ it is sufficient:
*i)* to restrict $\Omega$ to $\mathcal{N}$
*ii)* to pull back this restriction to $\mathcal{M}$ with respect to the map $\pi = e^{-V}$.

We see from (3.2.71) that the components of the vector field $\mathbf{V}$ must be determined by requiring

$$\mathcal{D}(e^{-V} s) = \zeta \tag{5.5.75}$$

But this is precisely effected in the lagrangian context by the term having as Lagrange multiplier the auxiliary field $\mathcal{P}$ (see Eq. (5.5.74)), through the equation of motion of $\mathcal{P}$, once we have introduced the extra field $v$ (which is now interpreted as the unique component of the vector field $\mathbf{V}$) to make the lagrangian invariant under the complexified gauge group $\mathbb{C}^*$. The lagrangian formalism of $N = 2$ supersymmetry perfectly matches the key points of the momentum map construction. This allows us to determine the form of the map $\pi$: it corresponds to the transformations (5.5.72). The steps that we are going to discuss in treating the lagrangian just consist in implementing the Kähler quotient as in (3.2.72). Thus it is clear why at the end we obtain the $\sigma$-model on the target space $\mathcal{M}$ (in our case $\mathbb{CP}_N$) endowed with the Kähler metric corresponding to the Kähler form $\omega$. In our example such a metric is the Fubini–Study metric. Indeed as shown in section 3.2.1 the Kähler potential $\widehat{\mathcal{K}}$ for the manifold $\mathcal{M}$, such that $\widehat{K} = \frac{i}{2\pi}\partial\bar{\partial}\widehat{\mathcal{K}}$, is given by

$$\widehat{\mathcal{K}} = \mathcal{K}|_{\mathcal{N}} + V^a \zeta_a \tag{5.5.76}$$

Here $\mathcal{K}$ is the Kähler potential on $\mathcal{S}$; $\mathcal{K}|_{\mathcal{N}}$ is the restriction of $\mathcal{K}$ to $\mathcal{N}$, i.e. it is computed after acting on the point $s \in \mathcal{S}$ with the transformation $e^{-V}$ determined by Eq. (5.5.75); $V^a$ are the components of the vector field $\mathbf{V}$ along the $a^{\text{th}}$ generator of the gauge group, and $\zeta_a$ those of the level $\zeta$ of the momentum map. In our case we have the single component $v$ given by Eq. (5.5.74), and we named $\rho^2$ the single component of the level. The original Kähler potential on $\mathcal{S} = \mathbb{C}^{N+1}$ is $\mathcal{K} = \frac{1}{2} X^{A*} X^A$ so that when restricted to $\mathcal{D}^{-1}(\rho^2)$ it takes an irrelevant constant value $\frac{\rho^2}{2}$. Thus we deduce from (5.5.76) that the Kähler potential for $\mathcal{M} = \mathbb{CP}_N$ that we obtain is $\widehat{\mathcal{K}} = \frac{1}{2}\rho^2 \log(X^* X)$. Fixing a particular gauge to perform the quotient with respect to $\mathbb{C}^*$ (see later), we can rewrite this potential as $\widehat{\mathcal{K}} = \frac{1}{2}\rho^2 \log(1 + x^* x)$, namely the Fubini–Study potential.

Let us now proceed with our manipulations of the lagrangian. It is a trivial algebraic matter to rewrite the lagrangian (5.5.68) after the substitutions (5.5.72) with $e^{-v}$ given by Eq. (5.5.74). For convenience we divide the resulting expressions into three parts to be separately handled.

First we have what we can call the "bosonic kinetic terms":

$$
\begin{aligned}
\mathcal{L}_1 = & -\frac{\rho^2}{X^*X}\sum_A\Big\{\sum_B\Big[\Big(\delta_{AB}-\frac{X^AX^{B^*}}{2X^*X}\Big)dX^B-\frac{X^AX^B}{2X^*X}dX^{B^*}\Big] \\
& +iX^A\frac{-i(X\partial_+X^*-X^*\partial_+X)+4\psi\psi^*}{2X^*X}e^+ + iX^A\frac{-i(X\partial_-X^*-X^*\partial_-X)+4\widetilde{\psi}\widetilde{\psi}^*}{2X^*X}e^- \\
& -\psi^A\zeta^- -\widetilde{\psi}^A\widetilde{\zeta}^-\Big\}(\Pi_+^{A^*}e^+-\Pi_-^{A^*}e^-)+\text{c.c.}-\frac{\rho^2}{X^*X}\sum_A(\Pi_+^{A^*}\Pi_-^A+\Pi_-^{A^*}\Pi_+^A)e^+e^-
\end{aligned}
\tag{5.5.77}
$$

We would like to recognize in the above expressions the bosonic kinetic terms of an $N=2$ $\sigma$-model. By looking at the $\sigma$-model rheonomic lagrangian (5.5.51) we are inspired to perform a series of manipulations.

Collecting some suitable terms we can rewrite

$$
\begin{aligned}
X^A(X\partial_+X^*e^++X\partial_-X^*e^-) & \longrightarrow X^AXdX^* \\
X^A(X^*\partial_+Xe^++X^*\partial_-Xe^-) & \longrightarrow X^AX^*dX
\end{aligned}
\tag{5.5.78}
$$

due to the fact that the further terms in the rheonomic parametrizations of $dX, dX^*$, proportional to the gravitinos, give here a vanishing contribution in force of the constraints (5.5.62).

We introduce the following provisional notation:

$$
G_{AB^*}=\frac{\rho^2}{X^*X}\Big(\delta_{AB}-\frac{X^{A^*}X^B}{X^*X}\Big).
\tag{5.5.79}
$$

Noting that, because of the constraints (5.5.62),

$$
G_{AB^*}\psi^A=\frac{\rho^2}{X^*X}\psi^A
\tag{5.5.80}
$$

we can write

$$
\begin{aligned}
\mathcal{L}_1 = & -\Big[G_{AB^*}(dX^A-\psi^A\zeta^- -\widetilde{\psi}^A\widetilde{\zeta}^-)+2i\frac{\rho^2}{X^*X}X^B(\psi\psi^*e^+ \\
& +\widetilde{\psi}\widetilde{\psi}^*e^-)\Big](\Pi_+^{A^*}e^+\Pi_-^{A^*}e^-)-\text{c.c.}-\frac{\rho^2}{X^*X}(\Pi_+^{A^*}\Pi_-^A+\Pi_-^*\Pi_+^A)e^+e^-
\end{aligned}
\tag{5.5.81}
$$

In order to eliminate the terms containing the first order fields $\Pi$ multiplied by fermionic expressions we redefine the $\Pi$:

$$
\begin{aligned}
\Pi_-^A & \rightarrow \Pi_-^A+2iX^A\frac{\widetilde{\psi}\widetilde{\psi}^*}{X^*X} & \qquad \Pi_-^{A^*} & \rightarrow \Pi_-^{A^*}-2iX^{A^*}\frac{\widetilde{\psi}\widetilde{\psi}^*}{X^*X} \\
\Pi_+^A & \rightarrow \Pi_+^A+2iX^A\frac{\psi\psi^*}{X^*X} & \qquad \Pi_+^{A^*} & \rightarrow \Pi_+^{A^*}-2iX^{A^*}\frac{\psi\psi^*}{X^*X}
\end{aligned}
\tag{5.5.82}
$$

Then we perform a second redefinition of the $\Pi$:

$$\begin{aligned} \Pi^A_\pm &\rightarrow \left(\delta_{AB} \pm \frac{X^A X^{B^*}}{X^* X}\right)\Pi^B_\pm \\ \Pi^{A^*}_\pm &\rightarrow \left(\delta_{AB} \pm \frac{X^{A^*} X^B}{X^* X}\right)\Pi^{B^*}_\pm \end{aligned} \tag{5.5.83}$$

in such a way that the quadratic term in the first order fields takes the form

$$- G_{AB^*}(\Pi^A_+\Pi^{B^*}_- + \Pi^A_-\Pi^{B^*}_+)e^+e^- \tag{5.5.84}$$

After the redefinitions (5.5.82) and (5.5.83) we can rewrite the part $\mathcal{L}_1$ of the lagrangian in the following way; we take into account, besides the constraints (5.5.62), the fact that

$$G_{AB^*}a^A X^{B^*} \propto \left(\delta_{AB} - \frac{X^{A^*} X^B}{X^* X}\right)a^A X^{B^*} = 0 \tag{5.5.85}$$

and we obtain

$$\begin{aligned} \mathcal{L}_1 &= -G_{AB^*}(dX^A - \psi^A\zeta^- - \widetilde{\psi}^A\widetilde{\zeta}^-)(\Pi^{B^*}_+ e^+ - \Pi^{B^*}_- e^-) \\ &- G_{AB^*}(dX^{B^*} + \psi^{B^*}\zeta^+ + \widetilde{\psi}^{B^*}\widetilde{\zeta}^+)(\Pi^A_+ e^+ - \Pi^A_- e^-) \\ &- G_{AB^*}(\Pi^A_+\Pi^{B^*}_- + \Pi^A_-\Pi^{B^*}_+)e^+e^- + \frac{8\rho^2}{(X^*X)^2}\psi\psi^*\widetilde{\psi}\widetilde{\psi}^* e^+e^- \end{aligned} \tag{5.5.86}$$

Next we consider the fermionic kinetic terms in Eq. (5.5.68). Performing the substitutions (5.5.72) with $v$ given by Eq. (5.5.74) and using the fact that, for instance,

$$\frac{\rho^2}{X^*X}\psi^A d\psi^{A^*} = G_{AB^*}\psi^A d\psi^{B^*} \tag{5.5.87}$$

these terms are

$$\begin{aligned} \mathcal{L}_2 &= 2i\left\{G_{AB^*}(\psi^A d\psi^{B^*} + \psi^{B^*} d\psi^A) - \frac{\rho^2}{(X^*X)^2}\psi^A\psi^{A^*}(X\partial_- X^* - X^*\partial_- X)e^-\right\}e^+ \\ &- 2i\left\{G_{AB^*}(\widetilde{\psi}^A d\widetilde{\psi}^{B^*} + \widetilde{\psi}^{B^*} d\widetilde{\psi}^A) + \frac{\rho^2}{(X^*X)^2}\widetilde{\psi}^A\widetilde{\psi}^{A^*}(X\partial_+ X^* - X^*\partial_+ X)e^+\right\}e^- \\ &- 16\frac{\rho^2}{(X^*X)^2}\psi\psi^*\widetilde{\psi}\widetilde{\psi}^* e^+e^- \end{aligned} \tag{5.5.88}$$

Let us introduce another provisional notation:

$$\gamma^A_{BC} = \frac{1}{X^*X}(\delta^A_B X^{C^*} + \delta^A_C X^{B^*}) \tag{5.5.89}$$

It is not difficult to check that the expression (5.5.88) can be rewritten as follows:

$$\begin{aligned}\mathcal{L}_2 &= 2i\Big\{G_{AB^\star}\psi^A(d\psi^{B^\star} - \gamma^{B^\star}_{B^\star C^\star}\psi^{C^\star}dX^{D^\star}) + G_{AB^\star}\psi^{B^\star}(d\psi^A - \gamma^A_{CD}\psi^C dX^D)\Big\}e^+\\ &\quad - 2i\Big\{G_{AB^\star}\tilde{\psi}^A(d\tilde{\psi}^{B^\star} - \gamma^{B^\star}_{B^\star C^\star}\tilde{\psi}^{C^\star}dX^{D^\star}) + G_{AB^\star}\tilde{\psi}^{B^\star}(d\tilde{\psi}^A - \gamma^A_{CD}\tilde{\psi}^C dX^D)\Big\}e^-\\ &\quad - 16\frac{\rho^2}{(X^\star X)^2}(\psi\psi^\star)(\tilde{\psi}\tilde{\psi}^\star) \end{aligned} \tag{5.5.90}$$

The remaining terms in the lagrangian (5.5.68) become, after the substitutions (5.5.72)

$$\begin{aligned}\mathcal{L}_3 = &-\frac{8\rho^2}{(X^\star X)^2}\psi^A\psi^{B^\star}\tilde{\psi}^B\tilde{\psi}^{A^\star} - G_{AB^\star}(\psi^A\psi^{A^\star}\zeta^-\zeta^+ - \tilde{\psi}^A\tilde{\psi}^{B^\star}\tilde{\zeta}^-\tilde{\zeta}^+ + \psi^A\tilde{\psi}^{B^\star}\zeta^-\tilde{\zeta}^+\\ &+ \psi^{B^\star}\tilde{\psi}^A\zeta^+\tilde{\zeta}^-) - G_{AB^\star}dX^A(\psi^{B^\star}\zeta^+ - \tilde{\psi}^{B^\star}\tilde{\zeta}^+) + G_{AB^\star}dX^{B^\star}(\psi^A\zeta^- - \tilde{\psi}^A\tilde{\zeta}^-)\end{aligned} \tag{5.5.91}$$

We have succeeded so far in making the lagrangian (5.5.68) invariant under the $\mathbb{C}^\star$-transformations (5.5.69), and to write it in a nicer form consisting of the sum of the three parts $\mathcal{L}_1, \mathcal{L}_2, \mathcal{L}_3$ as given in Eqs. (5.5.86),(5.5.90),(5.5.91), respectively

$$\begin{aligned}\mathcal{L} &= -G_{AB^\star}(dX^A - \psi^A\zeta^- - \tilde{\psi}^A\tilde{\zeta}^-)(\Pi^{B^\star}_+ e^+ - \Pi^{B^\star}_- e^-)\\ &\quad - G_{AB^\star}(dX^{B^\star} + \psi^{B^\star}\zeta^+ + \tilde{\psi}^{B^\star}\tilde{\zeta}^+)(\Pi^A_+ e^+ - \Pi^A_- e^-)\\ &\quad - G_{AB^\star}(\Pi^A_+\Pi^{B^\star}_- + \Pi^A_-\Pi^{B^\star}_+)e^+e^-\\ &\quad + 2i\Big\{G_{AB^\star}\psi^A(d\psi^{B^\star} - \gamma^{B^\star}_{B^\star C^\star}\psi^{C^\star}dX^{D^\star}) + G_{AB^\star}\psi^{B^\star}(d\psi^A - \gamma^A_{CD}\psi^C dX^D)\Big\}e^+\\ &\quad - 2i\Big\{G_{AB^\star}\tilde{\psi}^A(d\tilde{\psi}^{B^\star} - \gamma^{B^\star}_{B^\star C^\star}\tilde{\psi}^{C^\star}dX^{D^\star}) + G_{AB^\star}\tilde{\psi}^{B^\star}(d\tilde{\psi}^A - \gamma^A_{CD}\tilde{\psi}^C dX^D)\Big\}e^-\\ &\quad - G_{AB^\star}(\psi^A\psi^{A^\star}\zeta^-\zeta^+ - \tilde{\psi}^A\tilde{\psi}^{B^\star}\tilde{\zeta}^-\tilde{\zeta}^+ + \psi^A\tilde{\psi}^{B^\star}\zeta^-\tilde{\zeta}^+ + \psi^{B^\star}\tilde{\psi}^A\zeta^+\tilde{\zeta}^-\\ &\quad - G_{AB^\star}dX^A(\psi^{B^\star}\zeta^+ - \tilde{\psi}^{B^\star}\tilde{\zeta}^+) + G_{AB^\star}dX^{B^\star}(\psi^A\zeta^- - \tilde{\psi}^A\tilde{\zeta}^-)\\ &\quad - 8\frac{\rho^2}{(X^\star X)^2}(\psi^A\psi^{A^\star}\tilde{\psi}^B\tilde{\psi}^{B^\star} + \psi^A\psi^{B^\star}\tilde{\psi}^B\tilde{\psi}^{A^\star})e^+e^-\end{aligned} \tag{5.5.92}$$

We can now utilize the gauge invariance to fix for instance (in the coordinate patch where $X^0 \neq 0$) $X^0 = 1$, fixing completely the gauge. In practice we perform the transformation

$$X^A \to e^{-\Phi}X^A = \frac{1}{X^0}X^A \tag{5.5.93}$$

i.e. we go from the homogeneous coordinates $(X^0, X^i)$ to the inhomogeneous coordinates $(1, x^i = X^i/X^0)$ on $\mathbb{CP}_N$.

Having chosen our gauge, we rewrite the lagrangian (5.5.92) in terms of the fields $x^i$ (and of their fermionic partners $\psi^i, \tilde{\psi}^i$). Note that now $dx^0 = 0$ implies (because of the rheonomic parametrizations) $\psi^0 = 0$ and $\widetilde{\psi}^0 = 0$. The expression $X^*X \equiv X^{A^*}X^A$ becomes $1 + x^{i^*}x^i \equiv 1 + x^*x$. Of the expressions $G_{AB^\star}$ and $\gamma^A_{BC}$ only the components not involving the index zero survive. We introduce the following notations:

$$\begin{array}{rclcrcl} G_{AB^\star} & \equiv & \frac{\rho^2}{X^*X}\left(\delta_{AB} - \frac{X^{A^*}X^B}{X^*X}\right) & \longrightarrow & g_{ij^\star} & \equiv & \frac{\rho^2}{1+x^*x}\left(\delta_{ij} - \frac{x^{i^*}x^j}{1+x^*x}\right) \\ \gamma^A_{BC} & \equiv & \frac{1}{X^*X}(\delta^A_B X^{C^*} + \delta^A_C X^{B^*}) & & \Gamma^i_{jk} & \equiv & \frac{1}{1+x^*x}(\delta^i_j x^{k^*} + \delta^i_k x^{j^*}) \end{array} \tag{5.5.94}$$

We see that $g_{ij^\star}$ is just the standard Fubini–Study metric on $\mathbb{CP}_N$, which is a Kähler metric of Kähler potential $\mathcal{K} = \rho^2 \log(1 + x^*x)$; $\Gamma^i_{jk}$ is just the purely holomorphic part of its associated Levi–Civita connection. Moreover the Riemann tensor for the Fubini–Study metric is given by

$$\begin{aligned} R_{ij^\star kl^\star} &= \frac{\rho^2}{(1+x^*x)}\Big\{\delta^i_j\delta^k_l + \delta^i_l\delta^k_j - \frac{1}{1+x^*x}\Big[(\delta^i_j x^l + \delta^i_l x^j)x^{k^*} \\ &\quad + \delta^k_j x^l + \delta^k_l x^j)x^{i^*})\Big] + 2\frac{x^{i^*}x^j x^{k^*}x^l}{(1+x^*x)^2}\Big\} \end{aligned} \tag{5.5.95}$$

and we see that using once more the fermionic constraints (5.5.62) the four-fermion terms in (5.5.92) can be rewritten as follows:

$$\frac{\rho^2}{(1+x^*x)^2}(\psi^i\psi^{i^*}\widetilde{\psi}^j\widetilde{\psi}^{j^*} + \psi^i\psi^{j^*}\widetilde{\psi}^j\widetilde{\psi}^{i^*}) = R_{ij^\star kl^\star}\psi^i\psi^{j^*}\widetilde{\psi}^k\widetilde{\psi}^{l^*} \tag{5.5.96}$$

Thus at the end of the above manipulations, corresponding to the procedure of obtaining $\mathbb{CP}_N$ as the Kähler quotient of $\mathbb{C}^{N+1}$, we have reduced our initial rheonomic lagrangian (5.5.61), in the limit $g \to \infty$, to a form which is that of the $N = 2$ $\sigma$-model as given in Eq. (5.5.51). The target space is $\mathbb{CP}_N$ equipped with the Kählerian Fubini–Study metric.

## 5.6 N=2 Landau–Ginzburg Models and N=2 Superconformal Theories

In section 5.5.5 we have defined the rigid Landau–Ginzburg action (5.5.40) and its symmetries. In the present section we show that the action (5.5.40), under suitable choice of the superpotential $\mathcal{W}$, corresponds, at large distance scales, to a well defined superconformal theory.

The properties of the conformal model are completely encoded in the superpotential $\mathcal{W}(X)$. The elements in the chiral ring of the N=2 algebra are in one-to-one correspondence with the elements belonging to the "local ring" of the superpotential, defined as the space of all monomials in $X^i$ modulo the "vanishing" relations $\partial_i\mathcal{W}(X) = 0$.

Our purpose is to give the general idea of these correspondences and we skip some conceptual and technical subtleties. In our approach we closely follow [214, 254] and also [248, 219].

The crucial idea beyond the Landau–Ginzburg formulation of N=2 superconformal models is the application of renormalization group techniques to well known lagrangians in field theory and statistical mechanics, in order to study their infrared fixed point. At the critical point the theory is scale invariant and can be viewed as a conformal model. In this context one can investigate whether it is possible to classify minimal conformal models by means of two-dimensional field theoretical models.

This idea has been applied for Landau–Ginzburg theories [198, 270] with no supersymmetries or with N=1 supersymmetry, but it gives very impressive results for N=2 models.

In general it is formidably difficult to follow the renormalization gruop flow, and to predict the characteristic features of the resulting conformal theory. The N=2 case possesses many beautiful properties which allow a precise correspondence between Landau–Ginzburg actions and minimal superconformal models. Indeed, in this case, the properties of the critical theory are immediately obtained. All the characteristic features of the superconformal algebras can be read from the starting Landau–Ginzburg lagrangian because they are completely governed by the superpotential of the Landau–Ginzburg model. Indeed the superpotential $\mathcal{W}$ does not suffer any change under the renormalization group flow, apart from a trivial wave function renormalization. This property is due to a non-renormalization theorem in N=2, D=2 field theories [194, 214, 254, 219, 248], which is the analogue of the non-renormalization theorem holding for N=1, D=4 theories (see for example [257, 195, 159]).

Let us now focus on the rigid N=2 Landau–Ginzburg action. The action (5.5.40) is not classically conformal invariant. An immediate way to see it is to consider the bosonic interaction term, which is given by

$$V(X) = 8|\partial_i \mathcal{W}(X)|^2 \tag{5.6.1}$$

We can see that when $\partial_i \mathcal{W}(X) = 0$, $V(X) = 0$ so that critical points of $\mathcal{W}(X)$ are energy minima for $V$: vacuum states of the theory are associated to the set of fields satisfying $\nabla \mathcal{W}(X) = 0$.

We therefore consider superpotential $\mathcal{W}$ which has at least one critical point, and we will always take one of these critical points at $X^i = 0$. We will also assume that the potential has no flat directions at $X^i = 0$, i.e. $\frac{\partial \mathcal{W}}{\partial X^i} = 0 \rightarrow X^i = 0$. In a more precise way we assume that a generic perturbation of $\mathcal{W}$ yields a potential that has in the neighbourhood of $X^i = 0$ a finite number $\mu$ of critical points at which $det\,(\frac{\partial^2 \mathcal{W}}{\partial X^i \partial X^j})$ does not vanish. The number $\mu$ is named the *Milnor number* of the potential $\mathcal{W}$ . From an intuitive physical point of view, $\mu$ can be interpreted as the number of different phases that meet together at the critical point $X^i = 0$ when the perturbation of $\mathcal{W}$ is switched off.

Let us now rescale the two-dimensional coordinates as $z \rightarrow \lambda^{-1} z$. Since the fields $X^i$ have zero canonical dimension, the bosonic potential does not rescale at all and the interaction term picks up a non-trivial scaling factor.

What we are interested in is the behaviour of such a model at its critical point.

In particular we shall consider the conformal theory that emerges from the Landau–Ginzburg model at large distance scales. The claim is that the superpotential $\mathcal{W}(X)$ is all we need to define the $N = 2$ SCFT. Two qualitative considerations support this statement. The first one can be discussed easily without going into difficult technical details.

In this context we find it convenient to re-express the Landau–Ginzburg lagrangian in superfield formalism. The rigid Landau–Ginzburg action in superfield formalism has the following expression:

$$\mathcal{A} = \int \mathcal{K}(X^i, \overline{X}^i) d^4\theta d^2 z + \int \mathcal{W}(X^i) d^2\theta^- d^2 z + \int \mathcal{W}(\overline{X}^i) d^2\overline{\theta}^+ d^2\overline{z} \tag{5.6.2}$$

where $X^i$, $i = 1, \ldots, n$, is a collection of $n$ complex N=2 chiral superfields, $\mathcal{K} = \overline{X}^i X^i$ is the "Kähler potential" defining the kinetic term of the action and $\mathcal{W}(X)$ is the superpotential, introduced previously as a function of the scalar fields and now regarded as a function of the chiral superfields $X^i$. The variables $\theta^\pm, \overline{\theta}^\pm$ are Grassmann variables and $d^2\theta^\pm = d\theta^\pm d\overline{\theta}^\pm$. In this context chirality means

$$D_+ X = \overline{D}_+ X = 0 \tag{5.6.3}$$

where

$$\begin{aligned} D_\pm &= \frac{\partial}{\partial\theta^\pm} + \theta^\mp \frac{\partial}{\partial z} \\ \overline{D}_\pm &= \frac{\partial}{\partial\overline{\theta}^\pm} + \overline{\theta}^\mp \frac{\partial}{\partial\overline{z}} \end{aligned} \tag{5.6.4}$$

Equations (5.6.3) imply that a generic chiral field X has the espansion[3]

$$\begin{aligned} X &= X(z,\overline{z}) + \psi(z,\overline{z})\theta^- + \overline{\psi}(z,\overline{z})\overline{\theta}^- + H(z,\overline{z})\theta^-\overline{\theta}^- - \partial_z X\theta^+\theta^- \\ &- \partial_{\overline{z}} X\overline{\theta}^+\overline{\theta}^- - (\tfrac{\partial\overline{\psi}}{\partial z})\theta^+\theta^-\overline{\theta}^- - (\tfrac{\partial\psi}{\partial\overline{z}})\theta^-\overline{\theta}^+\overline{\theta}^- + \theta^-\theta^+\overline{\theta}^-\overline{\theta}^+\partial\overline{\partial}X(z,\overline{z}) \end{aligned}$$

where $H$ is the auxiliary field introduced in Eqs. (5.4.28) and on-shell identified with the derivative of the superpotential (see Eq.(5.4.29)), through its own field equation. The action (5.6.2), re-expressed in components is exactly the same as Eq. (5.5.40) (modulo some irrelevant rescaling of the fields).

From a renormalization group point of view it is essential to establish whether an operator is relevant, irrelevant or marginal. At the critical point the relevance of an operator is completely determined by its conformal dimension. In the infrared regime one operator is relevant if it has conformal dimension $h+\overline{h} < 2$, it is irrelevant if $h+\overline{h} > 2$ and marginal if $h+\overline{h} = 2$, where $h, \overline{h}$ denote the left and right conformal dimensions.

In general, under the renormalization group flow, the fields $X^i, X^{i*}$ acquire anomalous dimensions $h_{tot} = h_X + \overline{h}_X$ and $h_{tot} = h_{X^*} + \overline{h}_{X^*}$ [4]. It is easy to see that the kinetic

[3] Here and in the following we use the same notation for the superfield and for the first bosonic component.

[4] The anomalous dimension of the superfield is the (anomalous) dimension of its first component.

term is an irrelevant operator at large scales, since its dimension is bigger than 2 (in the natural hypothesis $h_X, h_{X^*} \geq 0$, which we assume to be true for unitary theories). If the superpotential term has dimensions less than (or equal to) 2, then it becomes the only interesting object under the renormalization group flow. This simple reasoning shows that the fixed point field content is, in general, completely determined by the renormalization group behaviour of the superpotential $\mathcal{W}(X)$.

The second fundamental consideration is the following one: we assume that the superpotential $\mathcal{W}(X)$ does not renormalize, except for wave function renormalization, so that the superpotential is "invariant". This is a very strong assumption which is based on the non-renormalization theorem.

If $\mathcal{W}(X)$ is the only relevant operator then the fixed point will depend on $\mathcal{W}$. Moreover if $\mathcal{W}$ is invariant under renormalization group flow, then there is one conformal field theory for each superpotential $\mathcal{W}$.

This establishes a precise connection between holomorphic polynomial $\mathcal{W}(X)$ and N=2 superconformal field theories. A suitable classification of these polynomials should also give a characterization of N=2 SCFT, where by suitable we mean "up" to equivalent resulting superconformal theories. For example, any analytic change of variables will not affect the behaviour at the critical point, so we need to classify $\mathcal{W}(X)$ "up" to analytic field redefinitions.

Consider now, to fix ideas, a potential $\mathcal{W}(X) = X^{n+1}$, $n \geq 2$, depending on just one chiral field $X$. Consider a second chiral field $Y$ and the potential

$$\mathcal{W}(X,Y) = X^{n+1} + Y^2 \tag{5.6.5}$$

The quadratic term gives rise to a mass term in the bosonic interaction. In the large scale limit the mass scale of the theory becomes much smaller than the particle mass and the $Y$ particle does not change the infrared critical behaviour of the $X$-theory. This is true for a generic superpotential and N=2 theories can be classified by analytic functions $\mathcal{W}(X)$ up to field redefinitions and up to the addition of extra-dimensional quadratic interaction terms. This means that the two superpotentials:

$$\begin{aligned} &\mathcal{W}(X^i) \qquad i = 1, \ldots, n \\ \widehat{\mathcal{W}}(X^i, X^A) \;&=\; \mathcal{W}(X^i) + \sum_{A=n+1}^{m+n} (X^A)^2 \end{aligned} \tag{5.6.6}$$

are equivalent. The surprising aspect of this reasoning is that these physical arguments match perfectly with precise mathematical definitions in singularity theory (see, for instance, [14], [15, 56, 241]). These are the classifications of stable singularities, defined precisely as analytic functions up to field redefinitions and up to the addition of quadratic monomials in extra variables. As we are going to see, these are not the only points of contact with the singularity theory.

As we said, $\mathcal{W}(X)$ is invariant under the renormalization group flow, apart from wave function renormalization. In general, the fields $X_i$ acquire anomalous dimensions $\omega_i$ in such a way that they scale, at the critical point, as

$$X^i \to \lambda^{\omega_i} X^i \tag{5.6.7}$$

where $\omega_i = h_{X^i} + \bar{h}_{X^i}$. Since the integration measure $d^2z d^2\theta^-$ picks up a $\lambda^{-1}$ factor under the rescaling $z \to \lambda^{-1} z$, and $\theta \to \lambda^{1/2}\theta$, the superpotential $\mathcal{W}(X)$, at the critical point where the model is scale invariant, must rescale according to

$$\mathcal{W}(X) \to \lambda \mathcal{W}(X) \tag{5.6.8}$$

Notice that the scaling law (5.6.8) coincides with (5.5.37) provided that we normalize the degree of the potential to $d_{\mathcal{W}} = 1$.

For a generic potential $\mathcal{W}(X)$ conditions (5.6.8) and (5.6.7) mean that the surviving part at the infrared scale of $\mathcal{W}(X)$ should necessarily satisfy

$$\mathcal{W}(\lambda^{\omega_i} X^i) = \lambda \mathcal{W}(X_i) \tag{5.6.9}$$

By "surviving part" we mean the one which is truly relevant when $\lambda \to \infty$. An example should clarify this point. Consider the following potential:

$$\mathcal{W}(X) = g_1 X^n + g_2 X^m \qquad m > n \tag{5.6.10}$$

where $g_1, g_2$ are two coupling constants. Since at the critical point $\mathcal{W}(X) \to \lambda \mathcal{W}(X)$ and $X \to \lambda^\omega X$, then necessarily

$$\begin{aligned} g_1 &\to \lambda^{1-n\omega} g_1 \\ g_2 &\to \lambda^{1-m\omega} g_2 \end{aligned} \tag{5.6.11}$$
$$\tag{5.6.12}$$

Let us rewrite the superpotential as

$$W(X) = g_1 \left( X^n + \frac{g_2}{g_1} X^m \right) \tag{5.6.13}$$

Now, since $\frac{g_2}{g_1} \to \lambda^{(n-m)\omega} \to 0$ when $\lambda \to \infty$, the lowest order $X^n$ will dominate at the critical point, and necessarily the anomalous dimension acquired by $X$ will be $\omega = \frac{1}{n}$.

In general, at the fixed point the potential has to satisfy the condition (5.6.9). This is the mathematical definition of a quasi-homogeneous function with quasi-homogeneous weights $\omega_i$, as defined in section 5.5.5. The left–right conformal dimensions of the scalar fields[5] $X^i$, at the critical point, are $h_i = \bar{h}_i = \frac{\omega_i}{2}$ and for unitary theories they are all striclty positive. To completely characterize the SCFT arising from the Landau–Ginzburg description, we have to find the corresponding value of the central charge as well as the structure of the chiral ring.

[5]Since $X^i$ is a scalar field, then necessarily $h_i - \bar{h}_i = 0$.

Using the quasi-homogeneity property of the superpotential we can easily read the $U(1)$ charge content of the field of $X^i$. Indeed, at the critical point, we can find the $U(1)$ charge assignments from the R-symmetry charges appearing in Eq. (5.5.39), where we adopt the convention that the $U(1)$ charge is defined as the R charge in (5.5.39) with the opposite sign. From this equation we see that $X^i$ has $U(1)$ charges $q_i = \overline{q}_i = \omega_i$. In section 5.2 we have seen that a field which satisfies $h = \frac{1}{2}q$ is a chiral primary field. So the field $X^i$, in the critical limit, is in one-to-one correspondence with an element belonging to the $(c, c)$ ring of the N=2 theory.

Chiral primary fields in the N=2 conformal model form a finite ring. The natural question is: what is the corresponding chiral ring in the Landau–Ginzburg model?

Again singularity theory suggests what is the appropriate ring structure associated with an analytic polynomial $\mathcal{W}(X)$, which can be correctly associated with the N=2 ring. We define the *local ring* of $\mathcal{W}(X)$ as the space of all monomials in $X^i$ modulo the relation $\partial_i \mathcal{W}(X) = 0$, i.e.

$$\mathcal{R}_{\mathcal{W}} = \frac{\mathbb{C}[X^i]}{\partial_i \mathcal{W}(X^i)} \tag{5.6.14}$$

The local ring of $\mathcal{W}(X)$ is precisely isomorphic to the ring of primary chiral operators in the confomal field theory. To illustrate this correspondence, we consider the example

$$\mathcal{W}(X) = \frac{1}{5}X^5 \tag{5.6.15}$$

In this example we can easily verify the above-mentioned correspondence. We find that the local ring $\mathcal{R}_{\mathcal{W}}$ is given by

$$\mathcal{R}_{\mathcal{W}} = \{\mathbf{1}, X, X^2, X^3\} \tag{5.6.16}$$

At the critical point these fields corresponds to conformal operators with $\omega_i = h_i + \overline{h}_i = \frac{i}{5}$ $(i = 0, \ldots, 3)$. The top chiral field corresponds to the state with the highest conformal weight, i.e. $h_3 = \overline{h}_3 = \frac{3}{10}$. Since $h_3 = \frac{c}{6}$, we have $c = \frac{9}{5}$, which corresponds to a minimal conformal theory of level $k = 3$. The chiral ring of the $k = 3$ minimal theory is given by

$$\mathcal{R} = \{\mathbf{1}, \phi_1, \phi_2, \phi_3\} \tag{5.6.17}$$

with the following values of conformal and $U(1)$ weights: $q_i = \frac{h_i}{2} = \frac{i}{5}$. These fields are in one-to-one correspondence with the fields in the local ring $\mathcal{R}_{\mathcal{W}}$ defined by (5.6.15).

To make more precise our argument we have to identify, in the general case, the central charge of the N=2 theory defined by a generic $\mathcal{W}(X)$. For this purpose we define the Poincaré polynomial associated with a quasi-homogeneous function $\mathcal{W}(X)$. Let $\omega_i$ be the homogeneous weights of fields $X^i$. We write $\omega_i = p_i/N$, where $p_i$ and $N$ are relative prime integers. The Poincaré polynomial is defined as

$$P(t) = \sum_{k=1}^{\infty} n_k t^k = \Pi_i^n \left( \frac{1 - t^{N-p_j}}{1 - t^{p_j}} \right) \tag{5.6.18}$$

where $n_k$ is the number of monomials of scaling dimension $\frac{k}{N}$ in a basis for the local ring $\mathcal{R}_W$. In the last equality appearing in (5.6.18) the factor $1 - t^{N-p_j}$ performs the necessary subtraction for each $\partial_i W$ (which have precisely dimension $\frac{N-p_i}{N}$ ). We consider now in (5.6.18) the limit $t \to \infty$; the leading contribution is $t^{\sum(N-2p_j)} = t^N \sum_j (1-2\omega_j)$ and the coefficient is 1. Therefore there is a unique element $\rho$, of maximal dimension $2h = h + \overline{h} = \sum_j (1-2\omega_j)$ in the local ring $\mathcal{R}_W$. The field $\rho$ is defined by

$$\rho = \det \left( \frac{\partial W}{\partial X_i \partial X_j} \right) \tag{5.6.19}$$

Remembering that $h \leq c/6$ and that the bound is saturated by the highest scaling dimension element in the chiral ring of the N=2 theory, we immediately get

$$c = 3 \sum_j (1 - 2\omega_j) \tag{5.6.20}$$

From formula (5.6.20) we can construct a table of correspondence between the $N = 2$ minimal models and superpotential $\mathcal{W}(X)$ (up to stable equivalence) in singularity theory. The ADE classification which comes from the singularity theory agrees perfectly with the ADE classification of N=2 minimal models and their central charges computed using Eq. (5.6.20) are given by

$$\begin{array}{llll} A_n & X^{n+1} & n \geq 1 & c = \frac{3n-3}{n+1} \\ D_n & X^{n-1} + XY^2 & n \geq 2 & c = \frac{3n-6}{n-1} \\ E_6 & X^3 + Y^4 & & c = \frac{5}{2} \\ E_7 & X^3 + XY^3 & & c = \frac{8}{3} \\ E_8 & X^3 + Y^5 & & c = \frac{14}{5} \end{array} \tag{5.6.21}$$

For example, the correspondence between the level $k$ minimal model in the $A_n$ series is given by $k = n - 1$ (with $c = \frac{3k}{k+2}$). The ADE classification of superpotential is referred in the mathematical literature as the classification of modality zero quasi-homogeneous functions. A few words should be spent to clarify the concept of modality zero potential $\mathcal{W}(X)$. More precisely we give a physical intuitive definition of modal deformations for $\mathcal{W}(X)$ (we refer to mathematical books for a more precise definition [14]). A modal deformation for $\mathcal{W}(X)$ corresponds physically to either a marginal or irrelevant deformation of the N=2 SCFT. This means that adding such a deformation we preserve the multiplicity or multicritical structure of the theory. The important point is that such a deformation cannot be thought of as a change of variables in the $X_i$ fields. The modality of a singularity is easily counted by computing $\mathcal{R}_W$ and searching for operators with $h = \overline{h} \geq 1/2$, where $h = \overline{h} = 1/2$ is precisely a modulus (i.e. a truly marginal deformation) of the N=2 SCFT.

## 5.7 Landau–Ginzburg Models and Calabi–Yau Manifolds

In this section our aim is to connect the Landau–Ginzburg formulation of the N=2 minimal models to Calabi–Yau manifolds. This provides a very elegant interpretation of the Gepner construction [161, 162, 163], which puts into correspondence suitable orbifoldized tensor products of N=2 minimal models and Calabi–Yau manifolds.

The crucial idea is to use the Landau–Ginzburg potential to describe, through the algebraic construction given in Chapter 4, the Calabi–Yau variety and finally look at the corresponding SCFT defined by the Landau–Ginzburg superpotential. We will give some explicit computations for the simplest cases of tensor products of minimal models in the $A$ series, since the generalizations to other cases have no conceptual obstructions. To avoid any mathematical complicacy we will follow a path integral argument [181], which gives a simple heuristic proof of the correspondence. For more references on this point we refer to Bibliographical note 5.10.

We saw in Chapter 4 that the condition selecting the possible Calabi–Yau manifolds, which can be constructed as intersections in a projective space, $\mathbb{CP}_N$, is

$$(N+1-\sum_{\alpha=1}^{r}\nu_\alpha)=0 \tag{5.7.1}$$

where $N=n+r$, $r$ being the number of polynomial constraints, and $n$ the complex dimension of the Calabi–Yau manifold. As we mentioned in section 4.5, the case of the quintic potential in $\mathbb{CP}_4$

$$\mathcal{W}(X^\Lambda)=\sum_{\Lambda=0}^{5}(X^\Lambda)^5=0 \tag{5.7.2}$$

defines precisely a Calabi–Yau manifold in three complex dimensions, having $r=1$, $\nu=5$ and N=4. If we now look at the potential $\mathcal{W}(X^\Lambda)$ as a Landau–Ginzburg potential, we can wonder what is the corresponding N=2 SCFT which arises at the critical point of the Landau–Ginzburg model. If we are just cavalier in the interpretation of the $X$ coordinates[6] the answer is quite simple: at large distance scales $\mathcal{W}(X)$ describes the product of five copies of level $k=3$ minimal models. The total central charge is $c=\sum_{\Lambda=0}^{4}\frac{9}{5}=9$. From another point of view Gepner established, studying massless spectra and discrete symmetries, a well defined correspondence between the Calabi–Yau model defined by (5.7.2) and the tensor product of five copies of $k=3$ minimal models. Gepner construction is of course much more general, and involves many different examples of Calabi–Yau versus N=2 minimal models. But let us go back to our explicit example, and see whether it is possible to understand the correspondence N=2–Calabi–Yau models in the Landau–Ginzburg approach.

[6] We should indeed remember that $\mathcal{W}(X)$ in the Landau–Ginzburg model is a homogeneous potential in $\mathbb{C}^5$ while the polynomial constraint is defined in $\mathbb{CP}_4$.

Consider the path integral

$$\int dX_0 \dots dX_4 e^{i\int d^2z d^2\theta - W(X^\Lambda)} \tag{5.7.3}$$

where we have assumed that the kinetic term is very small. From the renormalization group point of view this is not a dangerous choice, since we have shown that the kinetic term is irrelevant at large distance scales. Now, if we perform the following change of variables:

$$\begin{aligned} \xi_0 &= X_0^5 \\ \xi_A &= \frac{X_A}{X_0} \qquad A = 1, \dots 4 \end{aligned} \tag{5.7.4}$$

we can write the path integral as

$$\int d\xi_0 \dots d\xi_4 e^{i\int d^2z d^2\theta \xi_0 (1+\xi_1^5+\dots \xi_4^5)} \tag{5.7.5}$$

where it is easily realized that the defined coordinate change has a constant jacobian. Now we just have to perform the $\xi_0$ integration to get a delta function which implements the constraint

$$1 + \sum_{A=1}^{4} \xi_A = 0 \tag{5.7.6}$$

which is the defining equation of the Calabi–Yau manifold (5.7.2) in inhomogeneous coordinates. This result is not just a magic of the particular model under consideration, but can be generalized for a wider class of Calabi–Yau manifolds.

Before giving some details on the more general construction we have to extend the construction given in section 4.5 to the so–called weighted projective spaces (WPSs). The defintion of a WPS, and precisely of $W\mathbb{CP}^N$, is almost immediate if we follow the guideline given by Eqs. (2.8.7),(2.8.8) in Chapter 2.

$W\mathbb{CP}^N$ is defined as the the set of equivalence relations

$$\{X^\Lambda\} \sim \{X^{\Lambda'}\} \quad \text{iff} \quad X^\Lambda = \lambda^{\omega_\Lambda} X^{\Lambda'} \quad \Lambda = 0, \dots N \tag{5.7.7}$$

where $\omega_\Lambda$ is a set of positive integers. Given the set $\{\omega_\Lambda\}$ and the equivalence (5.7.7) the standard notation for weighted projective spaces is $W\mathbb{CP}^N_{\omega_0 \dots \omega_N}$. In general the equivalence relation (5.7.7) has fixed points, so that $W\mathbb{CP}^N$ is a singular space. The total Chern class of $W\mathbb{CP}^N_{\omega_0 \dots \omega_N}$ is easily calculable in full analogy with the Eq. (2.8.37) in Chapter 2. Indeed we have

$$c_{\text{tot}}(W\mathbb{CP}^N_{\omega_0 \dots \omega_N}) = \Pi_0^N (1 + \omega_\Lambda K) \tag{5.7.8}$$

Let us consider now an algebraic $n$-dimensional variety $\mathcal{M}$ defined by the zero locus of $r = N - n$ polynomial costraints $\mathcal{W}_\alpha(X)$ of degrees $\nu_\alpha$, where $\mathcal{W}_\alpha$ are required to be quasi-homogeneous functions of degree

$$\mathcal{W}_\alpha(\lambda^{\omega_0} X^0 \dots \lambda^{\omega_N} X^N) = \lambda^{\nu_\alpha} \mathcal{W}_\alpha(X^0, \dots, X^N) \tag{5.7.9}$$

with $\nu_\alpha/\omega_\Lambda$ being positive integers.

Following Eq. (2.8.37) we can write

$$c(T(\mathcal{M})) = \frac{\Pi_0^N(1-\omega_\Lambda K)}{\Pi_1^r(1-\nu_\alpha K)} \tag{5.7.10}$$

so that

$$c_1(\mathcal{M}) = \sum_0^N \omega_\Lambda - \sum_1^r \nu_\alpha \tag{5.7.11}$$

The condition $c_1(M) = 0$ selects the possible CY manifolds that can be expressed as polynomial constraints in $W\mathbb{CP}^N$ space. For the case of one polynomial constraint we have

$$1 = \sum \frac{\omega_\Lambda}{\nu} = \sum_\Lambda \frac{1}{l_\Lambda} \tag{5.7.12}$$

$\{l_\Lambda\}$ being positive integers. For the simple example (5.7.2) we have $\nu = 5, l_\Lambda = 5$. Consider now the general case of a Landau–Ginzburg theory describing the tensor product of N+1 minimal models of level $k_\Lambda$:

$$W(X^\Lambda) = \sum_{\Lambda=0}^N (X^\Lambda)^{k_\Lambda+2} \tag{5.7.13}$$

and repeat the path integral argument given for the example (5.7.2). The change of variables is now expressed by

$$\begin{aligned} \xi_0 &= X_0^{k_0+2} \\ \xi_A &= \frac{X_A}{X_0^{\frac{k_0+2}{k_A+2}}} \end{aligned} \tag{5.7.14}$$

The jacobian in this case is explicitly given by

$$J(X,\xi) = \text{const } \xi_0^{\left(-\frac{k_0+1}{k_0+2}+\sum\frac{1}{k_A+2}\right)} \tag{5.7.15}$$

so that the condition for the jacobian to be a constant

$$\sum_{\Lambda=0}^N \frac{1}{k_\Lambda+2} = 1 \tag{5.7.16}$$

When the jacobian is constant the $\xi_0$ integration gives the delta function

$$\delta(1+\sum_A \xi_A^{k_A+2}) \tag{5.7.17}$$

i.e. the vanishing of the potential (5.7.13) in inhomogeneous coordinates. Since for any field $X_\Lambda$ in the potential (5.7.13), $l_\Lambda = k_\Lambda + 2$, we conclude that the condition requiring a constant jacobian coincides precisely with the conditions imposing the vanishing of the first Chern class, selecting the Calabi–Yau manifold defined by $\mathcal{W}(X) = 0$. There is a little subtlety that should be taken into account. The coordinate change defined in (5.7.14) is not single-valued. We have to require in addition the following identification:

$$(X_0, \dots X_N) \rightarrow (e^{\frac{2\pi i}{k_0+2}} X_0, \dots, e^{\frac{2\pi i}{k_N+2}} X_N) \qquad (5.7.18)$$

This means that the correspondence between Calabi–Yau and N=2 minimal models has to take into account the previous identification. We have to consider orbifolds of the N=2 tensor products, by dividing by the action (5.7.18). In the Gepner construction approach, relation (5.7.18) corresponds to the requirement of target space–time supersymmetry: to obtain a supersymmetric theory on the target space–time we have to impose a Gliozzi–Scherk–Olive [175] projection (selecting only supersymmetric states) that precisely coincides with the "orbifoldizations" defined in (5.7.18).

Let us now go back to the potential (5.7.13); the total central charge of the corresponding N=2 theory is given by

$$c = \sum_{\Lambda} \frac{3k_{\Lambda}}{k_{\Lambda} + 2} \qquad (5.7.19)$$

The vanishing first Chern class condition (5.7.16) coincides with the requirement $c = 3(N-1)$, and this shows a correspondence between $c = 3n$ superconformal theories and $n$-dimensional Calabi–Yau compactifications (with $n = N - 1$) for our one polynomial constraint defined by (5.7.13) ). Of course we can generalize the previous construction to a general multi-polynomial constraint to obtain the so called complete intersection Calabi–Yau (or CICY) manifolds (see the last section of Chapter 4), or tensoring different N=2 models through Landau–Ginzburg potential in the ADE classification, but the conceptual pattern of the theory is substantially the same, and we will not report it.

## 5.8 Landau–Ginzburg Potentials and Pseudo-Ghost First Order Systems

The correspondence between N=2 algebras and Landau–Ginzburg potentials can be studied in a different framework. Recently [265] there has been constructed an N=2 algebra with the expected central charge directly in terms of non-critical Landau–Ginzburg fields. This construction amounts to a new realization of the N=2 algebras in terms of "free fields". Indeed, in doing the OPEs between the generators of the algebra (depending on the Landau–Ginzburg fields), one can ignore the part of the lagrangian depending on the superpotential. This is because this contribution does not affect the singularities of the OPEs.

We can study the same "free field" algebra in terms of pseudo-ghost systems $b - c - \beta - \gamma$. In particular, in this case, we can naturally define a Landau–Ginzburg type interaction which does not spoil the superconformal invariance [152, 151]. The local ring of the Landau–Ginzburg potential is naturally associated with the chiral ring of the conformal theory. In this formulation the theory is conformal invariant also at the classical level.

In this section we then consider the realization of (2,2)-supersymmetric models in terms of quasi-free $(b, c, \beta, \gamma)$-systems. We show that it is possible to add a polynomial interaction $V$ of the Landau–Ginzburg type to a collection of free first order $(b, c, \beta, \gamma)$-systems (which we are going to define below) in such a way that, if $V$ is a quasi-homogeneous function, the theory possesses an N=2 superconformal symmetry already at the classical level. We also show that the interaction potential unambiguously fixes the weights of the pseudo-ghost fields. As in the standard Landau–Ginzburg case, here we can recover the ADE classification of the N=2 minimal models from ADE classification of the interaction potential; however, in this approach, the theory is always manifestly superconformal invariant. This formulation allows us to add all relevant perturbations and to study the renormalization group flows in a very simple way.

We start this programme by defining our model. We consider a collection of pseudo-ghost fields $\{b_\ell, c_\ell, \beta_\ell, \gamma_\ell; \tilde{b}_r, \tilde{c}_r, \tilde{\beta}_r, \tilde{\gamma}_r\}$, where $\ell = 1, \ldots, N_{\rm L}$ and $r = 1, \ldots, N_{\rm R}$. $\beta_\ell$ and $\gamma_\ell$ form a bosonic first order system with weights $\lambda_\ell$ and $1 - \lambda_\ell$ respectively whereas $b_\ell$ and $c_\ell$ form a fermionic first order system with weights $\lambda_\ell + \frac{1}{2}$ and $\frac{1}{2} - \lambda_\ell$ respectively. The same can be said for the tilded fields with $\lambda_\ell$ replaced by $\tilde{\lambda}_r$. The action is

$$S = \int d^2z\, \mathcal{L} = \int d^2z\, \left(\mathcal{L}_0 + \Delta\mathcal{L}\right) \tag{5.8.1}$$

where

$$\begin{aligned}\mathcal{L}_0 \;=\; & \sum_\ell \left[-\lambda_\ell\, \beta_\ell \bar{\partial}\gamma_\ell + (1-\lambda_\ell)\, \gamma_\ell \bar{\partial}\beta_\ell - \left(\lambda_\ell + \frac{1}{2}\right) b_\ell \bar{\partial} c_\ell + \left(\lambda_\ell - \frac{1}{2}\right) c_\ell \bar{\partial} b_\ell\right] \\ & + \sum_r \left[-\tilde{\lambda}_r\, \tilde{\beta}_r \partial\tilde{\gamma}_r + (1-\tilde{\lambda}_r)\, \tilde{\gamma}_r \partial\tilde{\beta}_r - (\tilde{\lambda}_r + \frac{1}{2}) \tilde{b}_r \partial\tilde{c}_r + (\tilde{\lambda}_r - \frac{1}{2}) \tilde{c}_r \partial\tilde{b}_r\right]\end{aligned} \tag{5.8.2}$$

and

$$\Delta\mathcal{L} = \sum_{\ell,r} b_\ell\, \tilde{b}_r\, \partial_\ell V(\beta)\, \tilde{\partial}_r \tilde{V}(\tilde{\beta}) \tag{5.8.3}$$

Here and in the following we use the shorthand notations $\partial_\ell \equiv \partial/\partial\beta_\ell$ and $\tilde{\partial}_r \equiv \partial/\partial\tilde{\beta}_r$. $\mathcal{L}_0$ in (5.8.2) represents the standard free lagrangian [152, 155] for first order systems of the given weights and $\Delta\mathcal{L}$ in (5.8.3) defines an interaction of the LG type when $V$ and $\tilde{V}$ are polynomial functions of $\beta_\ell$ and $\tilde{\beta}_r$ respectively. From (5.8.1) one can derive the following equations of motion

$$\begin{aligned}\bar{\partial}\beta_\ell &= 0 \qquad \bar{\partial} b_\ell = 0 \\ \partial\tilde{\beta}_\ell &= 0 \qquad \partial\tilde{b}_\ell = 0 \\ \bar{\partial} c_\ell &= \sum_r \tilde{b}_r \partial_\ell V(\beta) \tilde{\partial}_r \tilde{V}(\tilde{\beta})\end{aligned}$$

$$\begin{aligned}
\bar{\partial}\gamma_\ell &= \sum_{m,r} b_m \tilde{b}_r \partial_\ell \partial_m V(\beta) \tilde{\partial}_r \tilde{V}(\tilde{\beta}) \\
\partial \tilde{c}_r &= -\sum_\ell b_\ell \partial_\ell V(\beta) \tilde{\partial}_r \tilde{V}(\tilde{\beta}) \\
\partial \tilde{\gamma}_r &= \sum_{\ell,s} b_\ell \partial_\ell V(\beta) \tilde{b}_s \tilde{\partial}_r \tilde{\partial}_s \tilde{V}(\tilde{\beta})
\end{aligned} \tag{5.8.4}$$

The first two lines of (5.8.4) show that $\beta_\ell$, $b_\ell$, $\tilde{\beta}_r$ and $b_r$ satisfy the same equations as in the free case, whereas $c_\ell$, $\gamma_\ell$, $\tilde{c}_r$ and $\tilde{\gamma}_r$ have no longer a definite holomorphic or anti-holomorphic character in the presence of the interaction. We can write formal solutions to the equations for $c_\ell$ and $\gamma_\ell$ as follows:

$$c_\ell(z,\bar{z}) = c_\ell^0(z) + \int_\Delta \frac{d^2w}{2\pi \mathrm{i}} \frac{1}{w-z} \sum_r \tilde{b}_r(\bar{w}) \partial_\ell V(\beta(w)) \tilde{\partial}_r \tilde{V}(\tilde{\beta}(\bar{w})) \tag{5.8.5}$$

$$\gamma_\ell(z,\bar{z}) = \gamma_\ell^0(z) + \int_\Delta \frac{d^2w}{2\pi \mathrm{i}} \frac{1}{w-z} \sum_{m,r} b_m(w) \tilde{b}_r(\bar{w}) \partial_\ell \partial_m V(\beta(w)) \tilde{\partial}_r \tilde{V}(\tilde{\beta}(\bar{w})) \tag{5.8.6}$$

where $\Delta$ is a disc containing $w$, and $c_\ell^0$and $\gamma_\ell^0$ are arbitrary holomorphic fields. Similar formal expressions (with the obvious changes) hold also for $\tilde{c}_r$ and $\tilde{\gamma}_r$. It is fairly easy to realize that under canonical quantization of (5.8.1) the fundamental OPEs are the same as in the free case. Indeed, even in the presence of the interaction, we have

$$\begin{aligned}
\beta_\ell(z)\, \gamma_m(w,\bar{w}) &= -\frac{\delta_{\ell m}}{z-w} + \dots \\
b_\ell(z)\, c_m(w,\bar{w}) &= \frac{\delta_{\ell m}}{z-w} + \dots
\end{aligned} \tag{5.8.7}$$

and similarly for the tilded fields. Of course, the interaction is not immaterial and it has to be carefully analyzed in a complete quantum treatment. We shall come back to this point shortly.

The lagrangian $\mathcal{L}_0$ in (5.8.2) describes a (2,2)-superconformal field theory with central charges

$$c_\mathrm{L} = \sum_\ell (3 - 12\lambda_\ell) \qquad c_\mathrm{R} = \sum_r (3 - 12\tilde{\lambda}_r) \tag{5.8.8}$$

for the left and the right sectors respectively. The explicit form of the superconformal transformations, as well as the OPEs yielding the above values of the central charges, are not immediately reported, since they can be trivially obtained from the general case (see Eq. (5.8.11) and below ) setting to zero the interaction terms. We will indeed show that the addition of the interaction $\Delta\mathcal{L}$ does not destroy this (2,2)-superconformal invariance if $V$ and $\tilde{V}$ are quasi-homogeneous functions, i.e. if for any $a \in \mathbb{R}^+$,

$$V(a^{\omega_\ell}\beta_\ell) = a\, V(\beta_\ell) \qquad \tilde{V}(a^{\tilde{\omega}_r}\tilde{\beta}_r) = a\tilde{V}(\tilde{\beta}_r) \tag{5.8.9}$$

The parameters $\omega_\ell$ and $\tilde{\omega}_r$ are the homogeneous weights of $\beta_\ell$ and $\tilde{\beta}_r$ respectively. By enforcing the requirement that the interaction lagrangian $\Delta\mathcal{L}$ have the correct dimensions, one can see that

$$\omega_\ell = 2\lambda_\ell \qquad \tilde{\omega}_r = 2\tilde{\lambda}_r \tag{5.8.10}$$

The parameters $\lambda_\ell$ and $\tilde{\lambda}_r$ of the free lagrangian (5.8.2) are therefore *fixed* by the interaction terms. When (5.8.9) and (5.8.10) are satisfied, the action $S$ in (5.8.1) is invariant under the following N=2 holomorphic supersymmetry transformations:

$$\begin{aligned}
\delta\beta_\ell &= 2\sqrt{2}\epsilon^- b_\ell \\
\delta b_\ell &= \frac{1}{\sqrt{2}}\epsilon^+ \partial\beta_\ell + \sqrt{2}\lambda_\ell \partial\epsilon^+ \beta_\ell \\
\delta c_\ell &= 2\sqrt{2}\epsilon^- \gamma_\ell \\
\delta\gamma_\ell &= \frac{1}{\sqrt{2}}\epsilon^+ \partial c_\ell - \sqrt{2}\left(\lambda_\ell - \frac{1}{2}\right)\partial\epsilon^+ c_\ell \\
\delta\tilde{\beta}_r &= 0 \\
\delta\tilde{b}_r &= 0 \\
\delta\tilde{c}_r &= -\frac{1}{\sqrt{2}}\epsilon^+ V(\beta)\tilde{\partial}_r \tilde{V}(\tilde{\beta}) \\
\delta\tilde{\gamma}_r &= \frac{1}{\sqrt{2}}\epsilon^+ V(\beta)\sum_s \tilde{\partial}_r\tilde{\partial}_s \tilde{V}(\tilde{\beta})\tilde{b}_s
\end{aligned} \tag{5.8.11}$$

where $\epsilon^\pm$ are arbitrary holomorphic functions ($\bar{\partial}\epsilon^\pm = 0$). The action $S$ is also invariant under N=2 anti-holomorphic symmetries which are similar to those defined in (5.8.11), with the exchange of the tilded and untilded quantities, and the replacement of $\epsilon^\pm$ with arbitrary anti-holomorphic functions $\bar{\epsilon}^\pm$ ($\partial\bar{\epsilon}^\pm = 0$). Moreover, if we relax the hypothesis that $V$ and $\tilde{V}$ are quasi-homogeneous, the transformations (5.8.11) and their $\bar{\epsilon}$-analogues remain symmetries of (5.8.1) provided $\epsilon^\pm$ and $\bar{\epsilon}^\pm$ are constant parameters. This means that our model has a global N=2 supersymmetry for any choice of $V$ and $\tilde{V}$, and an N=2 superconformal invariance for quasi-homogeneous potentials. Setting $V, \tilde{V}$ to zero we recover the N=2 superconformal symmetries of the free $b - c - \beta - \gamma$ lagrangian.

Using Noether's theorem, we can calculate the conserved currents associated to (5.8.11).

Let $\mathcal{L}(\phi_i, \partial\phi_i, \bar{\partial}\phi_i)$ be a 2D-lagrangian for a collection of fields $\phi_i$ and let us assume that under a variation

$$\delta\phi_i = \epsilon_\Lambda T^\Lambda(\phi) \tag{5.8.12}$$

we have

$$\delta\mathcal{L} = \epsilon_\Lambda(\bar{\partial} f_z^\Lambda + \partial f_{\bar{z}}^\Lambda) \tag{5.8.13}$$

The corresponding currents are given by the formula:

$$\begin{aligned}
j_z^\Lambda &= T_i^\Lambda(\phi)\frac{\partial\mathcal{L}}{\partial(\bar{\partial}\phi)} - f_z^\Lambda \\
j_{\bar{z}}^\Lambda &= T_i^\Lambda(\phi)\frac{\partial\mathcal{L}}{\partial(\partial\phi)} - f_{\bar{z}}^\Lambda
\end{aligned} \tag{5.8.14}$$

and are conserved

$$\overline{\partial} j_z^A + \partial j_{\overline{z}}^A = 0 \tag{5.8.15}$$

If one of the two components of $j$ vanishes, the other is holomorphic (respectively anti-holomorphic).

For our lagrangian the above procedure leads to the conserved currents[7]

$$\begin{aligned} G_z^+ &= \sqrt{2}\sum_\ell \left[\left(\frac{1}{2} - \lambda_\ell\right) c_\ell \partial \beta_\ell - \lambda_\ell \beta_\ell \partial c_\ell\right] \\ G_{\overline{z}}^+ &= \sqrt{2}\sum_\ell \left[\lambda_\ell \beta_\ell \overline{\partial} c_\ell + \left(\lambda_\ell - \frac{1}{2}\right) \overline{\partial} \beta_\ell c_\ell\right] - \frac{1}{\sqrt{2}} \sum_r V(\beta) \tilde{b}_r \tilde{\partial}_r \tilde{V}(\tilde{\beta}) \\ G_z^- &= 2\sqrt{2} \sum_\ell \gamma_\ell b_\ell \\ G_{\overline{z}}^- &= 0 \end{aligned} \tag{5.8.16}$$

If we use the equations of motion (5.8.4) for quasi-homogeneous potentials, we see that $G_{\overline{z}}^+$ vanishes on-shell; thus from the conservation laws we deduce that $G_z^+$ and $G_z^-$ are holomorphic currents even if they contain the non-holomorphic fields $c_\ell$ and $\gamma_\ell$. We denote these currents by $G^{\pm}(z)$.

The general superconformal transformations (5.8.11) are retrieved from the structure of the supercurrents utilizing the general formula

$$\begin{aligned} \delta\phi(w,\overline{w}) &= \oint_w \frac{dz}{2\pi i}[\epsilon^+(z)G_z^+(z,\overline{z}) + \epsilon^-(z)G_z^-(z,\overline{z})]\phi(w,\overline{w}) \\ &+ \oint_{\overline{w}} \frac{d\overline{z}}{2\pi i}[\epsilon^+(z)G_{\overline{z}}^+(z,\overline{z}) + \epsilon^-(z)G_{\overline{z}}^-(z,\overline{z})]\phi(w,\overline{w}) \end{aligned} \tag{5.8.17}$$

which holds for any field $\phi(w,\overline{w})$.

The action (5.8.1) is also invariant under holomorphic conformal reparametrizations and $U(1)$ rescalings of the fields; the conserved Noether's currents associated to such symmetries are the stress–energy tensor $T_{\mu\nu}$ and the $U(1)$ current $J_\mu$. For homogeneous potentials it is not difficult to see that the trace of $T_{\mu\nu}$ and the $\overline{z}$-component of $J_\mu$ are zero on-shell. Therefore, from the conservation laws, we deduce that

$$T_{zz} = \sum_\ell \left[-\lambda_\ell \beta_\ell \partial \gamma_\ell + (1-\lambda_\ell)\gamma_\ell \partial \beta_\ell - \left(\lambda_\ell + \frac{1}{2}\right) b_\ell \partial c_\ell - \left(\frac{1}{2} - \lambda_\ell\right) c_\ell \partial b_\ell\right] \tag{5.8.18}$$

and

$$J_z = \sum_\ell [(2\lambda_\ell - 1)b_\ell c_\ell + 2\lambda_\ell \beta_\ell \gamma_\ell] \tag{5.8.19}$$

are holomorphic currents. We denote them by $T(z)$ and $J(z)$ respectively.

---

[7] From now on, to avoid repetition we will discuss only the left sector and understand that similar considerations can be made in the right sector, with some obvious change of signs.

Using the OPEs in (5.8.7), it is straightforward to check that $T(z)$, $G^{\pm}(z)$ and $J(z)$ close an N=2 superconformal algebra:

$$\begin{aligned}
T(z)\,T(w) &= \frac{c}{2}\frac{1}{(z-w)^4} + \frac{2\,T(w)}{(z-w)^2} + \frac{\partial T(w)}{(z-w)} + reg \\
T(z)\,G^{\pm}(w) &= \frac{3}{2}\frac{1}{(z-w)^2}G^{\pm}(w) + \frac{\partial G^{\pm}(w)}{(z-w)} + reg \\
T(z)\,J(w) &= \frac{1}{(z-w)^2}J(w) + \frac{1}{(z-w)}\partial J(w) + reg \\
G^{+}(z)\,G^{-}(w) &= \frac{2}{3}c\frac{1}{(z-w)^3} + 2\frac{J(w)}{(z-w)^2} + 2\frac{T(w)+\frac{1}{2}\partial J(w)}{(z-w)} + reg \\
J(z)\,G^{\pm}(w) &= \pm\frac{G^{\pm}(w)}{(z-w)} + reg \\
J(z)\,J(w) &= \frac{c}{3}\frac{1}{(z-w)^2} + reg
\end{aligned} \tag{5.8.20}$$

with central charge

$$c = c_L = \sum_{\ell}(3 - 12\lambda_\ell) \tag{5.8.21}$$

Thus, we have shown that the interaction $\Delta\mathcal{L}$ with homogeneous polynomials $V$ and $\tilde{V}$ does not spoil the superconformal properties of $\mathcal{L}_0$.

In our formulation the ADE classification of N=2 superconformal models is an immediate consequence of the ADE classification of homogeneous polynomials of zero modality. We just recall this classification, adapting it to our models:

$$\begin{aligned}
A_n &: \quad V = \frac{1}{n+1}\beta^{n+1} \quad \Rightarrow \lambda = \frac{1}{2n+2} \quad n \geq 1 \\
D_n &: \quad V = \frac{1}{n-1}\beta_1^{n-1} + \frac{1}{2}\beta_1\beta_2^2 \quad \Rightarrow \lambda_1 = \frac{1}{2n-2}\,,\ \lambda_2 = \frac{n-2}{4n-4} \quad n \geq 2 \\
E_6 &: \quad V = \frac{1}{3}\beta_1^3 + \frac{1}{4}\beta_2^4 \quad \Rightarrow \lambda_1 = \frac{1}{6}\,.\ \lambda_2 = \frac{1}{8} \\
E_7 &: \quad V = \frac{1}{3}\beta_1^3 + \frac{1}{3}\beta_1\beta_2^3 \quad \Rightarrow \lambda_1 = \frac{1}{6}\,.\ \lambda_2 = \frac{1}{9} \\
E_8 &: \quad V = \frac{1}{3}\beta_1^3 + \frac{1}{5}\beta_2^5 \quad \Rightarrow \lambda_1 = \frac{1}{6}\,.\ \lambda_2 = \frac{1}{10}
\end{aligned} \tag{5.8.22}$$

We remark that the values of $\lambda_\ell$'s listed in (5.8.22) are fixed by the homogeneous weights of $\beta_\ell$'s according to (5.8.10). If we now insert such values into (5.8.21) we obtain the correct central charges for the N=2 minimal models in the ADE classification, namely

$$c(A_n) = \frac{3n-3}{n+1}\,, c(D_n) = \frac{3n-6}{n-1}\,,\ c(E_6) = \frac{5}{2}\,,\ c(E_7) = \frac{8}{3}\,,\ c(E_8) = \frac{14}{5} \tag{5.8.23}$$

It is also interesting to observe that the ring determined by the potential $V$, which contains all polynomials in $\beta_\ell$'s modulo the vanishing relations $\partial_\ell V = 0$, coincides with the ring of chiral primary operators of the $n=2$ minimal model associated to $V$. Indeed, using (5.8.18) and (5.8.19), one can easily check that the $U(1)$ charge of $(\beta_\ell)^n$ is twice its conformal dimension.

Let us now analyze the quantum properties of our action. The presence of interactions can in principle spoil the conformal invariance at the quantum level and one has eventually to restore it after a suitable renormalization.

For the sake of clarity, we begin by considering a single left–right–symmetric $(b,c,\beta,\gamma)$-system of weight $\lambda=\tilde{\lambda}$ with potential

$$W=\frac{1}{(n+1)^2}(\beta\tilde{\beta})^{n+1} \tag{5.8.24}$$

This corresponds to the $A_n$ minimal model of the $n=2$ discrete series if $\lambda=1/(2n+2)$ (see (5.8.22)). We are going to give now two different "proofs" that the interaction (5.8.24) preserves all the classical properties of the $(2,2)$ action, and in particular the conformal invariance. The first one is supported by a so-called "criterion for integrability" of a marginal operator that is used in the literature (see for example [168]). It gives the recipe for stating when a perturbing operator is "truly marginal" (i.e. it does not modify its conformal dimension at quantum level). Let us explain it.

From a purely conformal field theory point of view, once requiring that the interaction term should have the right (1,1) dimension, we are perturbing the theory with a "candidate" marginal operator. The mere existence of a (1,1) operator is not sufficient to guarantee the existence of a fixed line. We have to require additional "integrability" conditions, so that the perturbation generated by the marginal operator does not act to change its own conformal weight from (1,1). In our case, where we have collected the potential term as a single marginal operator $O$ with coupling $g$, this reduces, at one-loop level, to the requirement that in the OPE of $O$ with itself there are no terms of the form

$$O(z,\bar{z})O(w,\bar{w})=c_{OOO}(z-w)^{-1}(\bar{z}-\bar{w})^{-1}O \tag{5.8.25}$$

If this condition is not satisfied the conformal weight of $O$ is shifted by a quantity proportional to the three-point function $c_{OOO}$: $O$ would not remain marginal away from the point of departure, and could not be used to generate a family of conformal field theories. To higher orders we need to require also the vanishing of integrals of the $(n+2)$-point functions of the $O$'s. If this is the case $O$ is called a "truly marginal operator". It is easily verified that our interaction satisfies this requirement (even in the general case), due to the OPE of the $b,\beta$ fields.

The second "proof" is more direct; we explicitly compute the stress–energy tensor and show that the interaction considered does not give any loop correction to it. For the time being we leave the weight $\lambda$ unfixed. The lagrangian for this system is $\mathcal{L}=\mathcal{L}_0+\Delta\mathcal{L}$, where $\mathcal{L}_0$ is as in (5.8.2) and the interaction term is

$$\Delta\mathcal{L}=g\; b\tilde{b}\,\beta^n\tilde{\beta}^n \tag{5.8.26}$$

where $g$ is a coupling constant. Since the weight of $\beta$ and $\widetilde{\beta}$ is arbitrary, $g$ is a quantity with dimension

$$[g] = (1 - 2\lambda(n+1)) \tag{5.8.27}$$

To study the scaling properties of this system, we compute the trace of the stress–energy tensor, which turns out to be [8]

$$T_{z\bar{z}} = \left[-\lambda\beta\bar{\partial}\gamma + (1-\lambda)\gamma\bar{\partial}\beta - (\lambda+\frac{1}{2})b\bar{\partial}c - (\frac{1}{2}-\lambda)c\bar{\partial}b + g\; b\tilde{b}\beta^n\widetilde{\beta}^n\right] + \text{c.c.} \tag{5.8.28}$$

After using the equations of motion (5.8.4), we have

$$\Theta \equiv -T_{z\bar{z}} = g(2(n+1)\lambda - 1)\, b\tilde{b}\, \beta^n\widetilde{\beta}^n \tag{5.8.29}$$

so that our system is classically invariant under scale transformations (*i.e.* $\Theta = 0$) either if

$$g = 0 \quad \text{for any} \quad \lambda \tag{5.8.30}$$

or if

$$\lambda = \frac{1}{2(n+1)} \quad \text{for} \;\; g \neq 0 \tag{5.8.31}$$

Discarding the case (5.8.30), which corresponds to a free theory, we see from (5.8.31) that $\lambda$ must be fixed by the homogeneous weight of the potential (cf. (5.8.10)); when (5.8.31) is satisfied of course $g$ becomes dimensionless and the operator $b\tilde{b}\,\beta\widetilde{\beta}$ becomes marginal, so that no dimensionful parameters are left in the model.

Let us now quantize this system by using perturbation theory in $g$. From the explicit expression of the lagrangian $\mathcal{L}$, we see that the propagators are

$$\begin{aligned} \langle\gamma(z,\bar{z})\beta(w,\bar{w})\rangle &= \langle b(z,\bar{z})c(w,\bar{w})\rangle = \frac{1}{z-w} \\ \langle\widetilde{\gamma}(z,\bar{z})\widetilde{\beta}(w,\bar{w})\rangle &= \langle\tilde{b}(z,\bar{z})\tilde{c}(w,\bar{w})\rangle = \frac{1}{\bar{z}-\bar{w}} \end{aligned} \tag{5.8.32}$$

so that it is obvious that even when the interaction (5.8.26) is present, it is impossible to form loops. Therefore we conclude that there are no (perturbative) quantum corrections to the classical results simply because there are no loops! These considerations imply in particular that $\Theta$ in (5.8.29) is also the *quantum* trace of the stress–energy tensor and hence the coefficient of the spinless operator $b\tilde{b}\beta^n\widetilde{\beta}^n$ appearing in (5.8.29) can be interpreted as a renormalization group $\beta$-function [269], namely

$$\beta(g) = g\,(2(n+1)\lambda - 1) \tag{5.8.33}$$

The zeroes of $\beta(g)$ identify the conformal fixed points and these are given precisely by (5.8.30) and (5.8.31).

[8] Here and in the following "c.c." means exchanging the untilded fields with the tilded ones and $\partial$ with $\bar{\partial}$.

The extension of these results to the generic case of quasi-homogeneous potentials is an easy task. To this end let us first recall that if $f$ is a quasi-homogeneous polynomial in $N$ variables with weights $(\omega_1, \ldots \omega_N)$ ($\omega_i \in Q$, $\omega_i > 0$) and

$$f = \sum_\rho a_\rho x^\rho \tag{5.8.34}$$

where $\rho \equiv (\rho_1, \ldots \rho_N)$, $X^\rho \equiv X_1^{\rho_1} \ldots X_N^{\rho_N}$, $\rho_i \in Z^+$ and $a_\rho \neq 0$, then

$$\rho_1\omega_1 + \ldots \rho_N\omega_N = 1 \tag{5.8.35}$$

Let us now consider the interaction term

$$\Delta\mathcal{L} = g\sum_{i,j} b_i\tilde{b}_j\partial_i V\tilde{\partial}_j\tilde{V} \tag{5.8.36}$$

where $V(\beta)$ and $\tilde{V}(\tilde{\beta})$ are quasi-homogeneous potentials satisfying (5.8.34) and (5.8.35). For simplicity, we take $V(\beta) = \tilde{V}(\tilde{\beta})$ and assume that the weights $\lambda_i = \tilde{\lambda}_i$ are unconstrained. Then, the trace of the stress–energy tensor, upon using the equations of motion (5.8.4), turns out to be

$$\Theta = -T_{z\bar{z}} = - g\sum_{i,j}\left(b_i\tilde{b}_j\partial_i V\tilde{\partial}_j\tilde{V} - 2\lambda_i\beta_i\tilde{b}_j\partial_i V\tilde{\partial}_j\tilde{V}\right) \tag{5.8.37}$$

$$- \sum_{i,j,l}\left(\lambda_i\beta_i b_l\tilde{b}_j\partial_i\partial_l V\tilde{\partial}_j\tilde{V} - \lambda_i\tilde{\beta}_i b_j\tilde{b}_l\tilde{\partial}_i\tilde{\partial}_l\tilde{V}\partial_j V\right) \tag{5.8.38}$$

Using (5.8.34), after some algebra, the trace (5.8.38) can be rewritten as

$$\Theta = g\left(2\sum_k\lambda_k\rho_k - 1\right)\left(\sum_{i,\rho} a_\rho b_i\rho_i\beta_1^{\rho_1}\ldots\beta_N^{\rho_N}\right)\left(\sum_{j,\rho}\tilde{a}_\rho\tilde{b}_j\rho_j\tilde{\beta}_1^{\rho_1}\ldots\tilde{\beta}_N^{\rho_N}\right) \tag{5.8.39}$$

If $g \neq 0$, the system is invariant under scale transformations only if

$$\sum_i \lambda_i\rho_i = \frac{1}{2} \tag{5.8.40}$$

Comparing (5.8.40) with (5.8.34) and (5.8.35) we see that the weights $\lambda_i$ must be one half of the homogeneous weights of the potential $\omega_i$. Since there are no loop corrections, this result extends automatically to the quantum theory. Furthermore, we point out that the same conclusion is obtained in a similar way when $V(\beta)$ and $\tilde{V}(\tilde{\beta})$ are different, or when the interaction depends on a single quasi-homogeneous function $W$ in the variables $X_i = \beta_i\tilde{\beta}_i$ with weight $\omega_i$.

Finally, in our formulation it is easy to realize that the potential

$$\hat{V}(\beta^{(i,A)}) = V(\beta^i) + \sum_{A=n+1}^{m+n}(\beta^A)^2 \tag{5.8.41}$$

defines the same conformal theory as the potential $V$. Indeed, a $(b, c, \beta, \gamma)$-system with $\lambda_A = \frac{1}{4}$ gives a $c = 0$ conformal field theory.

## 5.9 The Griffiths Residue Mapping and the Chiral Ring

In view of the motivations discussed in section 5.1 we address now the description of the residue mapping, namely of the representation of the harmonic $(n-k,k)$-forms on a Calabi–Yau $n$-fold $\mathcal{M}_n$, by means of rational meromorphic $(n+1)$-forms in the ambient space where $\mathcal{M}_n$ is defined as a hypersurface. We could work in a general weighted projective space and consider the case where our $n$-fold is the intersection of several polynomial constraints; yet, for the sake of clarity, we confine ourselves to the simplest case, where $\mathcal{M}_n$ is given by the zero locus, in $\mathbb{CP}_{n+1}$, of a single homogeneous polynomial $\mathcal{W}(X)$ of the $n+2$ homogeneous coordinates $X^\Lambda$, namely

$$X^\Lambda \in \mathcal{M}_n \quad \Longrightarrow \quad \mathcal{W}(X) = 0 \tag{5.9.1}$$

Griffiths residue mapping, namely the construction we are going to describe, can be applied to a general hypersurface with an arbitrary first Chern number $c_1$; however, we shall mostly restrict our attention to the case of a Calabi–Yau $n$-fold, having $c_1 = 0$. As we know from Chapter 4, naming $\nu = deg\, \mathcal{W}$ the degree of the polynomial constraint $\mathcal{W}(X)$, one finds

$$c_1 = \nu - (n+2) \tag{5.9.2}$$

so that in order for the first Chern class $c_1(\mathcal{M}_n)$ to vanish, the degree $\nu$ must be equal to the number of homogeneous coordinates:

$$\nu = n + 2 \tag{5.9.3}$$

The resulting Calabi–Yau $n$-fold is named $\mathbb{CP}_{n+1}[n+2]$ and admits a number of complex structure deformations (see section 4.5) equal to the number of coefficients in a homogeneous polynomial of degree $n+2$, minus the number of parameters in a linear transformation of the homogeneous coordinates. Explicitly we have

$$\#\,\text{complex structure parameters in } \mathbb{CP}_{n+1}[n+2] \;=\; \frac{(2n+3)!}{(n+2)!} - (n+2)^2 \tag{5.9.4}$$

The number of Kähler class deformations is instead equal to one for all the $\mathbb{CP}_{n+1}[n+2]$ manifolds since the only harmonic $(1,1)$-form is that inherited from the ambient $\mathbb{CP}_{n+1}$-space. Later we shall consider the case of the mirror $M\mathbb{CP}_{n+1}[n+2]$ manifolds which have many Kähler class deformations and just one complex structure deformation: these spaces are nothing else than the $\mathbb{CP}_{n+1}[n+2]$ surfaces modded out by a discrete group $H = (\mathbb{Z}_p)^{p-2}$ whose action on the homogeneous coordinates is defined in Eq. (8.1.50) (see also the discussion of mirror symmetry in Chapter 8). In the case of $M\mathbb{CP}_{n+1}[n+2]$ hypersurfaces the explicit form of the polynomial constraint $\mathcal{W}(X)$ can be easily written as a function of both the homogeneous coordinates $X^\Lambda$ and the single complex structure parameter $\psi$. We have

$$\mathcal{W}(\psi\,;X) = \frac{1}{p}\left(\sum_{\Lambda=1}^{p} X_\Lambda^p\right) - \psi \prod_{\Lambda=1}^{p} X_\Lambda \tag{5.9.5}$$

In what follows, although we do not mention it explicitly, we always assume that $\mathcal{W}(X)$ is a function of its own deformation parameters and we consider the locus $\mathcal{M}_n = 0$ to be a generic one. Furthermore we always denote the degree of $\mathcal{W}(X)$ by $\nu$ to remind the reader that what we are doing is general and applies also to hypersurfaces that are not Calabi–Yau $n$-folds.

### 5.9.1 Rational meromorphic (n + 1)-forms and the Hodge filtration

In the ambient $\mathbb{CP}_{n+1}$ manifold we consider the linear spaces $\mathcal{A}_k^q$ of meromorphic q-forms that admit $\mathcal{M}_n$ as a polar locus of order $k+1$. These differential forms have the following structure:

$$\mathcal{A}_k^q \ni \Omega_{(k)}^{(q)} = \frac{X^\Omega \, \Phi^{\Lambda_1 \dots \Lambda_{n-q}}(X)}{(\mathcal{W}(X))^{k+1}} dX^{\Sigma_1} \wedge \dots \wedge dX^{\Sigma_q} \, \varepsilon_{\Omega\Lambda_1\dots\Lambda_{n-k}\Sigma_1\dots\Sigma_q} \tag{5.9.6}$$

where $\Phi^{\Lambda_1 \dots \Lambda_{n-q}}(X)$ is an antisymmetric $n-q$ tensor, which depends polynomially on the $X^\Lambda$ and is homogeneous of degree

$$deg\, \Phi = (k+1)\nu - q - 1 \tag{5.9.7}$$

Condition (5.9.7) guarantees that the differential form (5.9.6) has total scaling weight zero under the rescaling $X^\Lambda \longrightarrow \lambda X^\Lambda$ and is therefore well-defined on $\mathbb{CP}_{n+1}$. Clearly the $\mathcal{A}_k^q$ spaces form a nested sequence of inclusions:

$$\mathcal{A}_0^q \subset \mathcal{A}_1^q \subset \dots \subset \mathcal{A}_k^q \subset \mathcal{A}_{k+1}^q \dots \tag{5.9.8}$$

since among the homogeneous $n-k$ tensors $\Phi^{\Lambda_1 \dots \Lambda_{n-q}}(X)$ of degree $(k+1)\nu - q - 1$ there are also those of the form $\mathcal{W}(X)\, \widetilde{\Phi}^{\Lambda_1 \dots \Lambda_{n-q}}(X)$ where $deg\, \widetilde{\Phi} = k\nu - q - 1$ . We are specifically interested in the cases $q = n+1$ and $q = n$. According to (5.9.6) a meromorphic $(n+1)$-form, namely an element of $\mathcal{A}_k^{(n+1)}$, can be written as follows:

$$\Omega_k^{(n+1)} = \frac{P_{k|\nu}(X)}{(\mathcal{W}(X))^{k+1}}\, \omega \tag{5.9.9}$$

where

$$\omega = X^\Omega dX^{\Lambda_1} \wedge \dots \wedge dX^{\Lambda_{n+1}} \varepsilon_{\Omega\Lambda_1\dots\Lambda_{n+1}} \tag{5.9.10}$$

and where $P_{k|\nu}(X)$ is a homogeneous polynomial of degree

$$deg\, P_{k|\nu}(X) = k|\nu \stackrel{\text{def}}{=} (k+1)\,\nu - (n+2) \tag{5.9.11}$$

In the case of a Calabi–Yau $n$-fold, using (5.9.3) we obtain

$$k|\nu = k(n+2) = k\,\nu \tag{5.9.12}$$

On the other hand a meromorphic $n$-form has the following expression

$$\mathcal{A}^n_{k-1} \ni \Theta^{(n)}_{k-1} = \frac{X^\Omega Y^\Lambda(X)}{(\mathcal{W}(X))^k} dX^{\Sigma_1} \wedge \ldots \wedge dX^{\Sigma_n} \, \varepsilon_{\Omega\Lambda\Sigma_1\ldots\Sigma_n} \tag{5.9.13}$$

where $Y^\Lambda(X)$ is a vector of homogeneous polynomials of degree

$$deg\, Y^\Lambda = k\nu - (n+1) \tag{5.9.14}$$

Computing the exterior differential of $\Theta^{(n)}_{k-1}$ we obtain

$$d\Theta^{(n)}_{k-1} = \frac{1}{\mathcal{W}^{k+1}} \left[ \frac{1}{n+1} \mathcal{W} \partial_\Lambda Y^\Lambda - \frac{k}{n+1} Y^\Lambda \partial_\Lambda \mathcal{W} \right] \omega \tag{5.9.15}$$

This elementary result has several important consequences. To see its implications we introduce the definitions

$$\begin{aligned}
Q(X)_{(k-1)|\nu} &= \tfrac{1}{n+1} \partial_\Lambda Y^\Lambda \quad [deg\, Q = k\nu - (n+2)] \\
\widetilde{P}(X)_{k|\nu} &= \tfrac{k}{n+1} Y^\Lambda, \partial_\Lambda \mathcal{W} \quad \left[deg\, \widetilde{P}_{k|\nu} = (k+1)\nu - (n+2)\right] \\
\widetilde{\Omega}^{n+1}_k &= \tfrac{\widetilde{P}_{k|\nu}(X)}{(\mathcal{W}(X))^{k+1}}\, \omega \\
\widetilde{\Omega}^{n+1}_{k-1} &= \tfrac{Q_{(k-1)|\nu}(X)}{(\mathcal{W}(X)^k)}\, \omega
\end{aligned} \tag{5.9.16}$$

and we rewrite Eq. (5.9.15) as follows

$$\widetilde{\Omega}^{n+1}_k = \widetilde{\Omega}^{n+1}_{k-1} - d\Theta^{(n)}_{k-1} \tag{5.9.17}$$

Then we name $\mathcal{Z}^q_k$ the kernels of the exterior derivative operator in the elliptic complex:

$$\ldots \mathcal{A}^{(q)}_k \xrightarrow{d} \mathcal{A}^{(q+1)}_{k+1} \xrightarrow{d} \mathcal{A}^{(q+1)}_{k+1} \ldots \tag{5.9.18}$$

and we introduce the cohomology groups:

$$\mathcal{H}_k \stackrel{\text{def}}{=} \frac{\mathcal{Z}^{(n+1)}_k}{d\mathcal{A}^n_{k-1}} = \frac{\mathcal{A}^{(n+1)}_k}{d\mathcal{A}^n_{k-1}} \tag{5.9.19}$$

In view of Eq. (5.9.8) it is clear that also the groups $\mathcal{H}_k$ form a nested sequence of inclusions:

$$\mathcal{H}_0 \subset \mathcal{H}_1 \subset \ldots \subset \mathcal{H}_k \subset \mathcal{H}_{k+1} \ldots \subset \mathcal{H}_n \tag{5.9.20}$$

and from Eq. (5.9.17) we can deduce the following isomorphism:

$$\frac{\mathcal{H}_k}{\mathcal{H}_{k-1}} \approx \mathcal{R}_{k|\nu}(\mathcal{W}) \tag{5.9.21}$$

with the degree $k|\nu$ subspaces of the chiral ring. By definition, $\mathcal{R}_j$ is the linear space of the homogeneous polynomials of degree $j$, modulo the vanishing relations

$$\partial_\Lambda \mathcal{W}(X) \approx 0 \tag{5.9.22}$$

and we have the following graded decomposition of the chiral ring:

$$\mathcal{R}(\mathcal{W}) = \frac{\mathbb{C}[X]}{\partial\mathcal{W}} = \sum_{j=0}^{n\nu} \mathcal{R}_j \tag{5.9.23}$$

Indeed in degree $n\nu$ there is just one nontrivial element of the chiral ring given by the determinant $det\,(\partial_\Lambda\partial_\Sigma\mathcal{W}(X))$ of the hessian and any polynomial with degree larger than $n\nu$ is necessarily proportional to the vanishing relations. The isomorphism (5.9.21) is seen as follows. First one notes that $\mathcal{Z}_k^{(n+1)} = \mathcal{A}_k^{(n+1)}$ since all the rational meromorphic $(n+1)$-forms are automatically closed, having maximum degree. Then using Eq. (5.9.17) we conclude that a rational $(n+1)$-form with a pole of order $k+1$ is cohomologous to an $(n+1)$-form with a pole of order $k$ (modulo an exact form) if and only if its residue at $\mathcal{W} = 0$ is a trivial element of the chiral ring, namely a polynomial $\tilde{P}(X)_{k\nu} = \frac{k}{n+1} Y^\Lambda \partial_\Lambda \mathcal{W}$ proportional to the vanishing relations (5.9.22). Hence we have a one-to-one correspondence between the polynomials of $\mathcal{R}_{k|\nu}$ and the rational $n+1$-forms whose pole is irreducible of order $k+1$ (i.e. cannot be reduced to $k$ by the addition of an exact form).

The isomorphism (5.9.21) being established, we want now to show the isomorphism between the filtration (5.9.20) of the space $\mathcal{H}_n$ of rational meromorphic $(n+1)$-forms on $\mathbb{CP}_{n+1}$ and the Hodge filtration

$$\mathcal{F}_{(n,0)} \subset \mathcal{F}_{(n-1,1)} \subset \ldots \subset \mathcal{F}_{(n-k,k)} \subset \mathcal{F}_{(n-k-1,k+1)} \cdots \subset \mathcal{F}_{(0,n)} = H^{(n)}(\mathcal{W}=0) \tag{5.9.24}$$

of the $n$-th order de-Rahm cohomology $H^{(n)}_{DR}(\mathcal{M}_n)$. By definition the flag manifolds $\mathcal{F}_{(n-k,k)}$ appearing in Eq. (5.9.24) are given by the sum of the first $k$ Dolbeault cohomology groups:

$$\mathcal{F}_{(n-k,k)} = H^{(n,0)} \oplus H^{(n-1,1)} \oplus \ldots \oplus H^{(n-k,k)} \tag{5.9.25}$$

appearing in the decomposition:

$$H^{(n)}_{DR}(\mathcal{M}_n) = H^{(n,0)} \oplus H^{(n-1,1)} \oplus \ldots \oplus H^{(n-k,k)} \oplus \ldots \oplus H^{(0,n)} \tag{5.9.26}$$

and containing the closed $(n-k,k)$ forms on the $n$-fold, modulo the exact ones with the same structure. In view of (5.9.21) the isomorphism of the Hodge filtration (5.9.25) with the filtration (5.9.20) will imply a one-to-one correspondence between the polynomials of $\mathcal{R}_{k|\nu}$ and the harmonic $(n-k,k)$-forms. This one-to-one correspondence can also be shown in a direct way, as we are going to see at the end of the present section. The advantage of retrieving this relation through the identification of the two filtrations (5.9.20) and (5.9.25) is that, in this way, we obtain, at the same time, an algorithm to calculate the periods of the harmonic $(n-k,k)$-forms on the homology $n$-cycles as holomorphic integrals on the complete $\mathbb{CP}_{n+1}$ space. Such a representation is the basic tool for deriving the Picard–Fuchs equations studied in later chapters.

Let us then derive the relation between the two filtrations we have discussed. To this end we introduce a new sequence of linear spaces $\mathcal{B}^{(q,k)}(l)$. By definition an element $\varphi \in \mathcal{B}^{(q,k)}(l)$ is a differential $q$-form on $\mathbb{CP}_{n+1}$ which has holomorphic type $(q,0)+(q-1,1)+(q-2,2)+\ldots+(q-k,k)$, is regular on $\mathbb{CP}_{n+1} - \mathcal{M}_n$ and has a pole on $\mathcal{M}_n$ not stronger than $l \in \mathbb{Z}_+$. More precisely we require that $\mathcal{W}^l\varphi$ and $\mathcal{W}^{l-1}\,d\mathcal{W}\wedge\varphi$ be regular forms on the whole $\mathbb{CP}_{n+1}$. A very useful lemma can be immediately proved.

**Lemma 5.9.1** *If $\varphi \in \mathcal{B}^{(q,k)}(l)$, then we can write*

$$\begin{aligned} \varphi &= \eta + d\psi \\ \eta &\in \mathcal{B}^{(q,k+1)}(l-1) \\ \psi &\in \mathcal{B}^{(q-1,k)}(l-1) \end{aligned} \tag{5.9.27}$$

Proof: Choosing in $\mathbb{CP}_{n+1}$ a local holomorphic coordinate frame $\{z_1,\ldots,z_n,\mathcal{W}\}$ such that $\mathcal{W}=0$ is the polar locus of our rational forms, for any $\varphi \in \mathcal{B}^{(q,k)}(l)$ we can always write

$$\varphi = \frac{1}{\mathcal{W}^l}\,\alpha\wedge d\mathcal{W} + \frac{1}{\mathcal{W}^{l-1}}\,\beta \tag{5.9.28}$$

where both $\alpha$ and $\beta$ are differential forms involving only the $dz^i$ differentials and not $d\mathcal{W}$. Setting

$$\begin{aligned} \eta &= \frac{1}{\mathcal{W}^{l-1}}\left(\beta - \frac{1}{l-1}\,d\alpha\right) \\ \psi &= \frac{1}{(l-1)\,\mathcal{W}^{l-1}}\,\alpha \end{aligned} \tag{5.9.29}$$

we obtain the proof of the lemma. Indeed by construction the forms (5.9.29) fulfill Eqs. (5.9.27). In particular note that $\eta \in \mathcal{B}^{(q,k+1)}(l-1)$ because $d\alpha$ can contain one more anti-holomorphic differential than $\alpha\wedge d\mathcal{W}$ did.

Let now $\mathcal{Z}^{(q,k)}(l) \subset \mathcal{B}^{(q,k)}(l)$ be the subset of closed forms in the linear space we have defined. Using the above lemma, to each $\varphi \in \mathcal{Z}^{(q,k)}(l)$ we can associate the corresponding $\eta \in \mathcal{Z}^{(q,k+1)}(l-1)$ (note that $d\varphi = 0$ implies $d\eta = 0$). In this way we obtain a natural linear mapping:

$$T : \frac{\mathcal{Z}^{(q,k)}(l)}{d\mathcal{B}^{(q-1,k-1)}(l)} \longrightarrow \frac{\mathcal{Z}^{(q,k+1)}(l-1)}{d\mathcal{B}^{(q-1,k)}(l-1)} \tag{5.9.30}$$

which, modulo exact forms, associates to every closed $\varphi$ its corresponding $\eta$ given, for instance, by Eq. (5.9.29). Repeating this procedure many times we obtain a sequence of linear isomorphisms:

$$\mathcal{H}_k \approx \frac{\mathcal{A}_k^{(n+1)}}{d\mathcal{A}_{k-1}^n} \xrightarrow{T} \frac{\mathcal{Z}^{(n+1,1)}(k)}{d\mathcal{B}^{(n,0)}(k)} \xrightarrow{T} \ldots \xrightarrow{T} \frac{\mathcal{Z}^{(n+1,k)}(1)}{d\mathcal{B}^{(q-1,k-1)}(1)} \tag{5.9.31}$$

In other words, Eq. (5.9.31) means that, for each meromorphic $(n+1)$-form $\Omega^{(n+1)}_{(k)}$, having an irreducible pole of order $k+1$ on $\mathcal{M}_n$, there exist ,$k$ $(n+1)$-forms $\psi_i$ $(i=1,\ldots,k)$, of type $(n+1,0)+(n,1)+\ldots..+(n-i,i)$ such that the form defined below:

$$T(\Omega^{(n+1)}_{(k)}) \stackrel{\text{def}}{=} \Omega^{(n+1)}_{(k)} - d(\psi_1 + \ldots + \psi_k) \tag{5.9.32}$$

is of type $(n+1,0)+(n,1)+\ldots+(n-k+1,k)$ and has a pole of order 1 on $\mathcal{M}_n$. As we did above, in appropriate coordinates we can always write

$$T(\Omega^{(n+1)}_{(k)}) = \frac{\gamma}{\mathcal{W}} \wedge d\mathcal{W} + \delta \tag{5.9.33}$$

where $\delta$ and $\gamma$ are regular on $\mathcal{M}_n$, $\gamma$ being of type $(n,0)+(n-1,1)+\ldots+(n-k,k)$. The restriction of $\gamma$ to the hypersurface has the the same holomorphic type, $(n,0)+(n-1,1)+\ldots+(n-k,k)$, and it is called the residue of $\Omega^{(n+1)}_{(k)}$:

$$Res\{\Omega^{(n+1)}_{(k)}\} = \gamma|_{\mathcal{M}_n} \tag{5.9.34}$$

Indeed, using Eqs. (5.9.32), (5.9.33), we can formally write

$$\int_c \Omega^{(n+1)}_{(k)} = \int_c T(\Omega^{(n+1)}_{(k)}) = \int_c \frac{\gamma}{\mathcal{W}} \wedge d\mathcal{W} = \gamma|_{\mathcal{M}_n} \tag{5.9.35}$$

where $c$ is a contour that encircles the hypersurface $\mathcal{M}_n$.

### 5.9.2 Interpretation of the residue map in N=2 conformal field theory

Summarizing, we have found an algorithm that associates an $(n-k,k)$-harmonic form on $\mathcal{M}_n = [\mathcal{W}=0]$ to each element of degree $k|\nu$ in the polynomial ring $\frac{\mathbb{C}(X)}{\partial\mathcal{W}(X)}$. This provides us with an easy way to count these harmonic forms and establishes their representation as operators in the associated N=2 conformal field theory. Indeed, recalling the results of section 5.8, we already know how to write a chiral–chiral primary field corresponding to each non-trivial polynomial $P_j(X)$ of degree $j$. This operator is simply given by the same polynomial $P_j(X)$ regarded as a function of the variable $X^\Lambda = \beta^\Lambda\,\tilde{\beta}^\Lambda$. From the N=2 CFT viewpoint $P_j(X)$ has conformal weights $h = \tilde{h} = \frac{j}{2\nu}$ and $U(1)$ charges $q = \tilde{q} = \frac{j}{\nu}$. In the case $j = k|\nu$ the same operator $P_{k|\nu}(X)$ describes also an $(n-k,k)$-harmonic form on $\mathcal{M}_n = [\mathcal{W}=0]$.

As the residue mapping holds true for a surface with an arbitrary first Chern number $c_1$, the identification

$$(n-k,k)\text{-harmonic forms} \Longrightarrow \begin{cases} \text{chiral primary fields} \\ \text{with } U(1) \text{ charge} \\ q = \frac{k|\nu}{\nu} \end{cases} \tag{5.9.36}$$

is also true for an arbitrary value of the homogeneity degree $\nu$ of the superpotential $\mathcal{W}(X)$. Recalling the results of section 5.8, $\nu$ is related to the conformal weight of the $\beta$-fields by $\lambda = \frac{1}{2\nu}$ and to the central charge by

$$c = (n+2)\left(3 - \frac{6}{\nu}\right) \tag{5.9.37}$$

Indeed the CFT model contains as many identical ($b$-$c$-$\beta$-$\gamma$-systems as there are homogeneous coordinates $X^\Lambda$. As one notices, for arbitrary values of $\nu$ the $U(1)$ charges in Eq. (5.9.36) are not integer-valued, and neither is the central charge in Eq. (5.9.37). For the Calabi–Yau case, however, by means of Eq. (5.9.3) everything simplifies dramatically: the $U(1)$ charges are integer-valued and just equal to $k$, while the central charge becomes $c = 3n$. In this case, if we take a GSO projection of the N=2 model to $U(1)$ integer-charged fields, then the surviving elements of the chiral ring are in one-to-one correspondence with the Hodge filtration (5.9.24). Furthermore the operators representing the $(n-1,1)$-forms are chiral–chiral primary fields with $U(1)$-charge $q = 1$, so that, according to the results of section 5.2, the last components in their supermultiplet are neutral fields with conformal weight $h = 1$. As already stressed, such operators are moduli of the N=2 CFT, namely they can be used to deform the world-sheet lagrangian and all the correlation functions. This is simply the CFT counterpart of the geometrical statement that for a Calabi–Yau $n$-fold the deformations of the complex structure are in one-to-one correspondence with the harmonic $(n-1,1)$-forms. This result is shown, in purely geometrical terms, in the next chapter.

### 5.9.3 Explicit construction of the harmonic $(\mathbf{n-k,k})$-forms and the realization of the chiral ring on the Hodge filtration

By means of the residue mapping we have associated each element $P^{(\alpha)}_{k|\nu}(X) \in \mathcal{R}(\mathcal{W})$ of the chiral ring to the cohomology class of some harmonic $(n-k,k)$-form $\omega^{(n-k,k)}_{(\alpha)}$. This has two implications: the first is that we should be able to write an explicit representation of the harmonic $(n-k,k)$-forms in terms of the polynomials $P^{(\alpha)}_{k|\nu}(X)$, the second is that the Hodge-filtration of the $n$-th de Rahm cohomology group (see Eq.(5.9.26)) should carry a representation of the chiral ring. The first thing to note, in relation with the second implication, is that the product operation that realizes the chiral ring on the $(n-k,k)$-forms is not the ordinary wedge product. This is quite obvious since, in the Dolbeault cohomology ring, the product of an $(n-k_1,k_1)$-form with an $(n-k_2,k_2)$-form is a form of type $(2n-k_1-k_2, k_1+k_2)$, while here we expect it to be an $(n-k_1-k_2, k_1+k_2)$-form. We shall presently show how to write the $(n-k,k)$-form associated with any $P^{(\alpha)}_{k|\nu}(X)$-polynomial, but, prior to that, without any reference to its explicit representation in terms of polynomials, we prove the existence of a ring structure on the Hodge filtration of the middle cohomology group $H^{(n)}_{DR}(\mathcal{M}_n)$ for a Calabi–Yau $n$-fold $\mathcal{M}_n$. The key point in the proof relies on the existence, for $c_1(\mathcal{M}_n) = 0$ manifolds of a linear isomorphism:

$$\begin{aligned} \rho \,:\, H^{(n-k,k)}(\mathcal{M}_n) &\longrightarrow H^{(k)}_{\bar\partial}\left(\mathcal{M}_n , \wedge^k T^{(1,0)}\mathcal{M}_n\right) \\ \rho^{-1} \,:\, H^{(k)}_{\bar\partial}\left(\mathcal{M}_n , \wedge^k T^{(1,0)}\mathcal{M}_n\right) &\longrightarrow H^{(n-k,k)} \end{aligned} \tag{5.9.38}$$

where $H^{(n-k,k)}(\mathcal{M}_n)$ is the $(n-k,k)$-th Dolbeault cohomology group of the $n$-fold, while $H^{(p)}_{\bar\partial}\left(\mathcal{M}_n , \wedge^q T^{(1,0)}\mathcal{M}_n\right)$ is the $p$-th cohomology group in the $\wedge^q T^{(1,0)}\mathcal{M}_n$-twisted Dolbeault complex:

$$\ldots \xrightarrow{\bar\partial} \Omega^{(0,p)} \otimes \wedge^q T^{(1,0)}\mathcal{M}_n \xrightarrow{\bar\partial} \Omega^{(0,p+1)} \otimes \wedge^q T^{(1,0)}\mathcal{M}_n \xrightarrow{\bar\partial} \ldots \tag{5.9.39}$$

By $\wedge^q T^{(1,0)}\mathcal{M}_n$ we denote the antisymmetric $q$-th tensor power of the holomorphic tangent bundle $T^{(1,0)}\mathcal{M}_n$. A section $V^{(0,p)}_{(q,0)} \in \Omega^{(0,p)} \otimes \wedge^q T^{(1,0)}\mathcal{M}_n$ is therefore a $(0,p)$-form with values in $\wedge^q T^{(1,0)}\mathcal{M}_n$, namely

$$V = d\bar z^{i_1^\star} \wedge \ldots \wedge d\bar z^{i_p^\star} V^{j_1\ldots j_q}_{i_1^\star\ldots i_p^\star} \frac{\partial}{\partial z^{j_1}} \wedge \ldots \wedge \frac{\partial}{\partial z^{j_q}} \tag{5.9.40}$$

the mixed tensor $V^{j_1\ldots j_q}_{i_1^\star\ldots i_p^\star}$ being completely antisymmetric both in the holomorphic and anti-holomorphic set of indices. An element of $H^{(p)}_{\bar\partial}\left(\mathcal{M}_n , \wedge^q T^{(1,0)}\mathcal{M}_n\right)$ is the cohomology class of a $V^{(0,p)}_{(q,0)}$-form that is $\bar\partial$-closed. The one-to-one map of Eq. (5.9.38) is explicitly realized by writing the invertible relations

$$\begin{aligned} \rho \,:\, V^{i_1\ldots i_k}_{j_1^\star\ldots j_k^\star} &= \frac{1}{(n-k)!} \overline{\Omega}^{i_1\ldots i_k l_1,\ldots,l_{n-k}} \omega_{l_1,\ldots,l_{n-k},j_1^\star\ldots j_k^\star} \\ \rho^{-1} \,:\, \omega_{l_1,\ldots,l_{n-k},j_1^\star\ldots j_k^\star} &= \Omega_{i_1\ldots i_k l_1,\ldots,l_{n-k}} \frac{1}{||\,\Omega\,||^2} V^{i_1\ldots i_k}_{j_1^\star\ldots j_k^\star} \end{aligned} \tag{5.9.41}$$

where $\omega_{l_1,\ldots,l_{n-k},j_1^\star\ldots j_k^\star} dz^{k_1} \wedge \ldots \wedge dz^{k_{n-k}} \wedge d\bar z^{j_1^\star} \wedge \ldots \wedge d\bar z^{j_k^\star}$ is a harmonic $(n-k,k)$-form and where $\Omega_{i_1\ldots i_k l_1,\ldots,l_{n-k}}$ denote the components of the unique $(n,0)$-form $\Omega^{(n,0)}$ whose uniqueness does indeed characterize Calabi–Yau $n$-folds. The bundles $\Omega^{(0,p)} \otimes \wedge^q T^{(1,0)}\mathcal{M}_n$ admit a natural wedge product:

$$\begin{aligned} \left[\Omega^{(0,p_1)} \otimes \wedge^{q_1} T^{(1,0)}\mathcal{M}_n\right] \wedge \left[\Omega^{(0,p_2)} \otimes \wedge^{q_2} T^{(1,0)}\mathcal{M}_n\right] &\subset \Omega^{(0,p_1+p_2)} \otimes \wedge^{q_1+q_2} T^{(1,0)}\mathcal{M}_n \\ V^{(0,p_1)}_{(0,q_1)} \wedge S^{(0,p_2)}_{(0,q_2)} &= d\bar z^{i_1^\star} \wedge \ldots \wedge d\bar z^{i^\star_{p_1+p_2}} V^{j_1\ldots j_{q_1}}_{i_1^\star\ldots i_{p_1}^\star} S^{j_1\ldots j_{q_2}}_{i_1^\star\ldots i_{p_2}^\star} \frac{\partial}{\partial z^{j_1}} \wedge \ldots \wedge \frac{\partial}{\partial z^{j_{q_1+q_2}}} \end{aligned} \tag{5.9.42}$$

and combining this operation with the $\rho$ isomorphism of Eq. (5.9.38) we can write the product operation:

$$\begin{aligned} \star \,:\, H^{(n-k_1,k_1)} \otimes H^{(n-k_2,k_2)} &\longrightarrow H^{(n-k_1-k_2,k_1+k_2)} \\ \omega^{(n-k_1,k_1)} \star \omega^{(n-k_2,k_2)} &\stackrel{\text{def}}{=} \rho^{-1}\left(\rho\left(\omega^{(n-k_1,k_1)}\right) \wedge \rho\left(\omega^{(n-k_2,k_2)}\right)\right) \end{aligned} \tag{5.9.43}$$

which is bilinear and symmetric. Equipped with the $\star$-product defined by Eq.(5.9.43), the Hodge filtration of the $n$-th De-Rham cohomology group (5.9.26) becomes a ring, named by us the *Hodge ring*:

$$\begin{aligned} Hod^\star(\mathcal{M}_n) &\stackrel{\text{def}}{=} \oplus_{k=0}^{n} H^{(n-k,k)} \\ H^{(n-k_1,k_1)} \star H^{(n-k_2,k_2)} &\subset H^{(n-k_1-k_2,k_1+k_2)} \end{aligned} \tag{5.9.44}$$

which is just isomorphic to the polynomial ring of the defining polynomial $\mathcal{W}(X)$, namely

$$Hod^\star(\mathcal{M}_n) \approx \mathcal{R}(\mathcal{W}) = \frac{\mathbb{C}[X]}{\partial\mathcal{W}(X)} \tag{5.9.45}$$

The isomorphism is verified by writing, as promised, the explicit representation of the $(n-k,k)$ forms in terms of polynomials $P_{k|\nu}(X) \in \mathcal{R}(\mathcal{W})$. To this end let us introduce the following definitions. Let

$$\begin{aligned} \mathcal{W}_\Lambda &= \frac{\partial}{\partial X^\Lambda}\mathcal{W}(X) \\ \overline{\mathcal{W}}_{\Lambda^\star} &= \frac{\partial}{\partial X^{\Lambda^\star}}\overline{\mathcal{W}}(\overline{X}) \\ ||\mathcal{W}||^2 &= \sum_{\Lambda=0}^{n+1} \mathcal{W}_\Lambda \overline{\mathcal{W}}_{\Lambda^\star} \\ \pi^\Gamma &= \eta^{\Gamma\Lambda^\star}\frac{\overline{\mathcal{W}}_{\Lambda^\star}}{||\mathcal{W}||^2} \end{aligned} \tag{5.9.46}$$

and note that the degree of homegeneity of the functions $\pi_{\Lambda^\star}$ is

$$\deg \pi^\Gamma = \nu - 1 - 2(\nu - 1) = -(\nu\ 1) \tag{5.9.47}$$

Consider a holomorphic vector field $\mathbf{t} \in T\mathbb{CP}_{n+1}$ in the ambient projective space where the Calabi–Yau $n$-fold is defined as the vanishing locus of the polynomial constraint $\mathcal{W}(X)$. The general expression of $\mathbf{t}$ is

$$\mathbf{t} = t^\Lambda(X)\frac{\partial}{\partial X^\Lambda} \tag{5.9.48}$$

where the components $t^\Lambda(X)$ are homogeneous functions of degree $\deg t^\Lambda(X) = 1$, in order for the operator $\mathbf{t}$ to have degree zero and be well-defined in the projective space. Using the notations (5.9.46) we can easily decompose the vector field $\mathbf{t}$ in a tangent plus a normal part to the $n$-fold. We introduce the projector:

$$\left(p^{||}\right)^\Lambda_\Sigma = \left(\delta^\Lambda_\Sigma - W_\Sigma\,\pi^\Lambda\right) \tag{5.9.49}$$

and we set

$$\begin{aligned}
\partial_\Lambda &= \partial^{||}_\Gamma + \partial^{\perp}_\Gamma \\
\partial^{||}_\Gamma &= \left(p^{||}\right)_\Gamma^{\ \Lambda} \partial_\Lambda \\
\partial^{\perp}_\Gamma &= \mathcal{W}_\Gamma \, \pi^\Delta \, \partial_\Delta \\
\mathbf{t}^{||} &\stackrel{\text{def}}{=} t^\Lambda(X)\, \partial^{||}_\Lambda \\
\mathbf{t}^{\perp} &\stackrel{\text{def}}{=} t^\Lambda(X)\, \partial^{\perp}_\Lambda
\end{aligned} \tag{5.9.50}$$

The operator $\partial^{||}_\Sigma$ is tangent to the hypersurface $\mathcal{W}(X) = 0$, because, by definition, it is orthogonal to the gradient $d\mathcal{W}$. Indeed $d\mathcal{W}\left(\partial^{||}_\Sigma\right) = \mathcal{W}_\Lambda \left(p^{||}\right)_\Sigma^{\ \Lambda} = 0$. The orthogonal complement $\partial^{\perp}_\Sigma$ is normal to the hypersurface. In a similar way we can decompose the differential $dX^\Lambda$ of the homogeneous coordinates into tangent and normal parts:

$$\begin{aligned}
dX^\Lambda &= dX^\Lambda_{||} + dX^\Lambda_\perp \\
dX^\Lambda_{||} &= \left(\delta^\Lambda_\Sigma - \pi^\Lambda \mathcal{W}_\Sigma\right) dX^\Sigma \\
dX^\Lambda_\perp &= \pi^\Lambda \, d\mathcal{W}
\end{aligned} \tag{5.9.51}$$

With these conventions we can introduce the Cauchy–Riemann operators in the ambient space and on the the Calabi–Yau $n$-fold by setting

$$\begin{aligned}
\partial &\stackrel{\text{def}}{=} dX^\Lambda \, \partial_\Lambda \\
\overline{\partial} &\stackrel{\text{def}}{=} d\overline{X}^{\Lambda^*} \, \partial_{\Lambda^*} \\
\partial\,\partial &= 0 \\
\overline{\partial}\,\overline{\partial} &= 0 \\
\partial\overline{\partial} + \overline{\partial}\partial &= 0 \\
\partial^{||} &\stackrel{\text{def}}{=} dX^\Lambda_{||} \, \partial_\Lambda = dX^\Lambda \, \partial^{||}_\Lambda \\
\overline{\partial}^{||} &\stackrel{\text{def}}{=} d\overline{X}^{\Lambda^*}_{||} \, \partial_{\Lambda^*} = d\overline{X}^{\Lambda^*} \, \partial^{||}_{\Lambda^*} \\
\partial^{||}\partial^{||} &= 0 \\
\overline{\partial}^{||}\overline{\partial}^{||} &= 0 \\
\partial^{||}\overline{\partial}^{||} + \overline{\partial}^{||}\partial^{||} &= 0
\end{aligned} \tag{5.9.52}$$

and we can also formally identify

$$dX^\Lambda_{||} = \partial^{||} X^\Lambda \tag{5.9.53}$$

Consider next the following $(n-k, k)$ differential expression:

$$\begin{aligned}
\tau^{(k)}_\Omega &= \varepsilon_{\Omega\Sigma\Gamma_1\ldots\Gamma_{n-k}\Delta_1\ldots\Delta_k} X^\Sigma \, dX^{\Gamma_1} \wedge \ldots \wedge dX^{\Gamma_{n-q}} \wedge \overline{\partial}\pi^{\Delta_1} \wedge \ldots \wedge \overline{\partial}\pi^{\Delta_k} \\
&\stackrel{\text{def}}{=} \varepsilon_{\Omega\Sigma\Gamma_1\ldots\Gamma_{n-k}\Delta_1\ldots\Delta_k} U^{\Sigma\Gamma_1\ldots\Gamma_{n-k}\Delta_1\ldots\Delta_k}
\end{aligned} \tag{5.9.54}$$

By construction, on the surface $\mathcal{W} = 0$, the differential expression $U^{\Theta_1 \dots \Theta_{n+1}}$ is orthogonal to $\mathcal{W}_\Lambda$:

$$W_\Lambda U^{\Lambda \Theta_2 \dots \Theta_{n+1}} = 0 \tag{5.9.55}$$

hence its dual $\tau^{(k)}_\Omega$ is proportional to $\mathcal{W}_\Lambda$; fixing by consistency the normalization factor we obtain the following identity which is true only on the hypersurface $\mathcal{W} = 0$

$$\varepsilon_{\Omega\Sigma\Gamma_1 \dots \Gamma_{n-k}\Delta_1 \dots \Delta_k} X^\Sigma dX^{\Gamma_1} \wedge \dots \wedge dX^{\Gamma_{n-q}} \wedge \overline{\partial}\pi^{\Delta_1} \wedge \dots \wedge \overline{\partial}\pi^{\Delta_k} = \mathcal{W}_\Omega \Theta^k \tag{5.9.56}$$

where

$$\Theta^k \stackrel{\text{def}}{=} \varepsilon_{\Lambda\Sigma\Gamma_1 \dots \Gamma_{n-k}\Delta_1 \dots \Delta_k} \pi^\Lambda X^\Sigma dX^{\Gamma_1} \wedge \dots \wedge dX^{\Gamma_{n-q}} \wedge \overline{\partial}\pi^{\Delta_1} \wedge \dots \wedge \overline{\partial}\pi^{\Delta_k} \tag{5.9.57}$$

From Eq. (5.9.56) we also derive

$$\overline{\partial}\Theta^k = 0 \tag{5.9.58}$$

Consider next the differential expression:

$$\omega^{(n-k,k)}[P] \stackrel{\text{def}}{=} P(X)\Theta^k \tag{5.9.59}$$

where $P(X)$ is a polynomial in the homogeneous coordinates. This differential expression is a globally defined $(n-k,k)$ differential form on $\mathbb{CP}_{n+1}$ if and only if its total homogeneity degree is zero. This implies

$$0 = \deg P(X) + 1 + (n-k) - (k+1)(\nu - 1) \tag{5.9.60}$$

which is the same condition as Eq. (5.9.11). Together with the Calabi–Yau condition $\nu = n+2$, from Eq.(5.9.60) we reobtain Eq.(5.9.12). When Eq. (5.9.12) is satisfied, the differential form (5.9.59) can be restricted to the hypersurface $\mathcal{W}(X) = 0$ by replacing each $dX^\Lambda$ and each $\overline{\partial}\pi^\Sigma$ differential with their tangent parts $dX^\Lambda_{||}$, $\overline{\partial}^{||}\pi^\Sigma$, at the same time restricting the $X^\Delta$-coordinates to the vanishing locus of $\mathcal{W}$. In this way we have an explicit construction of an $(n-k,k)$-form on the Calabi–Yau $n$-fold that is $\overline{\partial}$-closed. Indeed, due to Eq. (5.9.58), the form (5.9.59) is $\overline{\partial}$-closed on the whole ambient space $\mathbb{CP}_{n+1}$. The question is: When does it happen that, on the vanishing locus of $\mathcal{W}(X)$, the form (5.9.59) is exact? The answer is simple. This happens when the polynomial $P^\alpha_{k|\nu}(X)$ is a trivial element of the chiral ring. Indeed suppose that

$$P^\alpha_{k|\nu}(X) = \frac{k}{n+1} Y^\Lambda \mathcal{W}_\Lambda \tag{5.9.61}$$

where $Y^\Lambda(X)$ is a vector of homogeneous polynomials of degree as given in Eq. (5.9.14). Then, using Eq.(5.9.56), we conclude that on the vanishing locus of $\mathcal{W}(X)$ we have

$$\omega^{(n-k,k)}\left[P_{k|\nu}\right] = \overline{\partial}u^{(n-k,k-1)} \tag{5.9.62}$$

the differential form

$$u^{(n-k,k-1)} \stackrel{\text{def}}{=} \frac{k}{n+1} \varepsilon_{\Lambda\Sigma\Gamma_1\ldots\Gamma_{n-k}\Delta_1\ldots\Delta_k} X^{\Sigma} Y^{\Lambda} dX^{\Gamma_1} \wedge \ldots \wedge dX^{\Gamma_{n-k}} \wedge \overline{\partial}\pi^{\Delta_2} \wedge \ldots \wedge \overline{\partial}\pi^{\Delta_k} \pi^{\Delta_1} \tag{5.9.63}$$

being well-defined on $\mathbb{CP}_{n+1}$ since it has total homogeneity degree equal to zero. In this way we have verified that the same conditions that provide a correspondence between elements of the chiral ring of degree $k|\nu$ and rational meromorphic $(n+1)$-forms with a pole of order $k+1$ provide also a correspondence of the same chiral ring elements with cohomology classes of $(n-k, k)$-forms on the polar locus. Actually the restriction to $\mathcal{W}(X) = 0$ of the forms (5.9.59) is an explicit construction of the Griffith residue $Res\left\{\Omega^{(n+1)}_{(k)}\right\}$. Finally we can write

$$\left[\omega^{(n-k_1,k_1)}\left[P^{\alpha}_{k_1|\nu}\right] \star \omega^{(n-k_1,k_1)}\left[P^{\beta}_{k_2|\nu}\right]\right] = \left[\omega^{(n-k_1-k_2,k_1+k_2)}\left[\left(P^{\alpha}_{k_1|\nu} \cdot P^{\beta}_{k_2|\nu} \bmod \partial\mathcal{W}\right)\right]\right] \tag{5.9.64}$$

where $[\ldots]$ denotes a cohomology class and $\cdot$ the ordinary commutative pointwise product of polynomials. Equation (5.9.64) is the statement of the isomorphism of the chiral ring with the *Hodge ring* defined at the beginning of the present subsection.

Through the mathematical analysis of the present section the chiral ring of N=2 superconformal theories with $c = 3n$ has been identified with the *Hodge ring* that deals with the variation of the complex structure in the target manifold of a $\sigma$-model and with the *polynomial ring* of a Landau–Ginzburg model. This is the basic geometrical reason that allows interplay between $\sigma$–models, Landau–Ginzburg models and N=2 superconformal field theories. Such interplay will be even more manifest while one is dealing with the topological twist of these theories.

## 5.10 Bibliographical Note

- *The N=2 superconformal algebra was discovered by Ademollo et al in the context of superstring theory in [1, 2, 3].*

- *The relation between N=1 target supersymmetry and N=2 world-sheet supersymmetry was clarified in general terms by Sen in [239, 240] and by Banks, Dixon, Friedan in [30, 29].*

- *On N=2 superconformal field theories there exists a vast literature. Among others we cite the papers [134, 135, 136, 133], [55, 223, 115, 200, 156]. In particular the abstract structure of N=2 unitary representations with c < 3 was independently studied by Boucher, Friedan and Kent [55], Nam [223], and by Di Vecchia, Petersen and Zheng [115].*

- *Virasoro algebra was discovered by Virasoro [252] in the early work on dual resonance models, but the systematic description of string vacua and amplitudes in terms of conformal and superconformal field theories is mostly due to the work by Friedan, Martinec and Schenker [156]. General references on the modern view-point on conformal field theories that takes into account their application to the theory of critical phenomena are, for instance, [197] and [168]. In our specific discussion of N=2 superconformal field theories provided by section 5.2 we have mostly followed [214, 254].*

- *General references on Kac–Moody algebras and on the Sugawara construction of the corresponding Virasoro algebras are, for instance, [176, 177, 165].*

- *The notion of chiral ring in N=2 superconformal theories is due to Lerche, Vafa and Warner [214]. Other references are [80] and [79].*

- *The construction of Calabi–Yau compactifications in terms of tensor products of minimal models was introduced by Gepner in [161] and there exists a vast literature on the subject. Among others we cite the papers [162, 217, 216, 271, 163].*

- *The rheonomic construction of N=2, D=2 field theories is contained in ref.s [12], [45].*

- *The discussion of the $\sigma$-model and Landau-Ginzburg phases of an N=2 gauge model contained in section 5.5 follows ideas contained in [45, 262]. The phenomenon was pointed out by Witten in [262].*

- *The connection between (rigid) Landau-Ginzburg theories, Calabi–Yau manifolds, N=2 SCFT and singularity theory is analysed, for example, in [214, 84, 85, 254, 181, 219, 248], [194, 196, 245, 246, 265], [80, 79]. In particular the first rigorous proof, at non perturbative level, that an N=2 Landau-Ginzburg field theory flows to a superconformal model at the critical point was provided by Cecotti, Girardello and Pasquinucci in [84, 85]. Sections 5.6, 5.7 follow [214, 254, 219, 248, 181].*

- *The b–c–$\beta$–$\gamma$ description of N=2 Landau–Ginzburg models is due to Fré, Gliozzi, Monteiro and Piras [152] and to Fré, Girardello, Lerda and Soriani [151], starting from an initial observation by Friedan, Martinec and Schenker [155] . In section 5.8 we follow closely [151].*

- *The Griffiths residue mapping was described by Griffiths in [182], which we follow closely in section 5.9. Finally the notion of Hodge ring is introduced by us in section 5.9. It was contained in disguise in [80] although there it was not given any special name by the author (Cecotti).*

# Chapter 6

# MODULI SPACES AND SPECIAL GEOMETRY

## 6.1 Introduction

The concept of "modulus" can be introduced in five different contexts, namely
i) deformations of background fields in string theories;
ii) marginal deformations of abstract (2,2) conformal models;
iii) deformation parameters of the Kähler class and of the complex structure in a Calabi–Yau manifold;
iv) flat directions of the scalar potential in supergravity effective lagrangians;
v) coupling constants of ghost number zero in topological field theories.
All these contexts are deeply connected and the progress in understanding the structure of moduli space in any of them is a source of development for the others. The item v) is the main subject of Chapters 7 and 8 and it is not touched on the present chapter. Here we are concerned with the first four view-points for defining a modulus. Let us begin by considering them separately.

A natural environment for introducing moduli spaces is string theory. If we consider an N=1 superstring moving in a non-trivial metric and antisymmetric tensor background we can write its action, keeping only the bosonic degrees of freedom, as follows:

$$S = -\frac{1}{4\pi\alpha'}\int d^2\sigma\left[\sqrt{-h}h^{\alpha\beta}\partial_\alpha X^{\widehat{\mu}}\partial_\beta X^{\widehat{\nu}} g_{\widehat{\mu}\widehat{\nu}}(X) + \epsilon^{\alpha\beta}\partial_\alpha X^{\widehat{\mu}}\partial_\beta X^{\widehat{\nu}} B_{\widehat{\mu}\widehat{\nu}}(X)\right] \tag{6.1.1}$$

where $\alpha'$ is the string tension, $h_{\alpha\beta}$ is the two-dimensional metric and

$$X^{\widehat{\mu}}(\sigma) : \Sigma_g \rightarrow \mathcal{M}_{target} \tag{6.1.2}$$

is a map from a genus $g$ Riemann surface $\Sigma_g$ to a target space $\mathcal{M}_{target}$ (some Riemann manifold of dimension $d = 10$), whose background geometry is described by the metric $g_{\widehat{\mu}\widehat{\nu}}$ and by the axion $B_{\widehat{\mu}\widehat{\nu}}$. For the sake of the present discussion the index $\widehat{\mu}$ runs over a range $\widehat{\mu} = (\mu, i)$, where $\mu$ is a four-dimensional space-time index and $i$ is an internal six-dimensional index. In other words $\mathcal{M}_{target} = \mathcal{M}_4 \times \mathcal{M}_{int}$, where typically $\mathcal{M}_{int}$ is a Calabi–Yau manifold [70].

In order for the action (6.1.1) to describe a consistent string propagation, one has to require conformal invariance of the two-dimensional quantum field theory. This requirement imposes, via the vanishing condition for the beta functions, geometrical constraints on the internal metric $g_{ij}$ and the axion $B_{ij}$. Given a background $g^0_{ij}, B^0_{ij}$, which fulfills this requirement one usually finds that there are deformations of this background $g^0_{ij} + \delta g_{ij}$, $B^0_{ij} + \delta B_{ij}$ that leave the two-dimensional physics unchanged [150, 62]. These deformations provide the concept of modulus in the string theory context, i.e. the moduli space is naturally defined as the space parametrized by the allowed deformations of a given background on the internal manifold.

From a purely geometrical point of view we have seen in Chapter 4, studying Calabi–Yau manifolds, that moduli arise as possible deformations of complex and Kähler structure of a given Calabi–Yau manifold. Once we have chosen a Ricci flat metric, this solution is not uniquely determined: there are families of solutions parametrized by the moduli spaces.

We turn next to the concept of moduli in the context of the low energy theory associated with the string compactification. The compactification can be viewed either geometrically [70], as a compactification on a Calabi–Yau manifold in Eq. (6.1.1), or more abstractly as a compactification on a (2,2) superconformal field theory with $c = 9$ [126]. The relation between these two points of view has been analysed in previous chapters. Roughly speaking, the moduli space is just the manifold of classical vacua for the low energy theory. Moduli appear as massless neutral scalar fields $M_i$ whose vacuum expectation values $\langle M_i \rangle$ are left undetermined by the equations of motion and they represent the free parameters for the internal metric and anti-symmetric tensor field. This is precisely the point of view already discussed in Chapter 4, in the Kaluza–Klein approach to the compactification. As we have shown in that chapter, the moduli, seen as flat directions of the scalar potential, are in correspondence with moduli seen as complex and Kähler structure deformations.

If we consider the compactification from the abstract point of view, we can give a definition of moduli space using the underlying (2,2) superconformal theory. In the general context of conformal field theories, moduli are truly marginal operators, i.e. operators with conformal dimension $(h, \bar{h}) = (1, 1)$, with the property that perturbations generated by these operators do not act to change their own dimensions. A very interesting aspect of two-dimensional superconformal field theories is the possibility of describing the abstract space of such theories in standard geometrical terms. In particular the space of conformal field theories is equipped with a natural Riemannian structure, where the metric is identified with the two-point function between operators describing the moduli fields. We shall give further remarks on this point shortly.

To study the moduli space of (2,2) superconformal systems it is more convenient [147] to use them as internal spaces for type II rather than heterotic superstring. The reason is that in this case we get $N = 2$ space-time supersymmetry, and thus stronger

SUSY constraints. In the context of type II string theories this means that the internal conformal field theory with central charge $(c, \bar{c}) = (9,9)$ has left and right-moving $N = 2$ superconformal symmetries.

Such a theory can be mapped to a heterotic theory with the same internal superconformal system [138, 210, 142], [164]. The missing 13 units to the left-moving conformal anomaly are provided by 13 free bosons moving on the maximal torus of $E_8 \times SO(10)$. The $U(1)$ current of the left-moving N=2 algebra combines with the SO(10) current algebra to produce $E_6$. The gauge group of heterotic Calabi-Yau compactifications is thus in general $E_8 \times E_6$, as predicted by the naïve Kaluza-Klein approach presented in Chapter 4. The left-moving N=2 algebra establishes a one-to-one correpondence between those massless multiplets whose scalar components are moduli, and the matter multiplets charged under the gauge group [236, 126]. This relation is the counterpart, in the superconformal approach to the compactification, of the relation we have established studying the cohomology groups in the gauge and gravitational sectors of the light modes in the Kaluza-Klein approach.

The link between the definition of modulus as a flat direction of the scalar low energy potential and as an allowed deformation of the given background is provided by the requirement of vanishing for the beta functions associated with the background field couplings. In general, when we compactify the string on a Calabi-Yau manifold, the sigma model contains also four-dimensional scalar fields, which, as it happens for the background metric and antisymmetric tensor field, have their own beta functions to be considered while imposing conformal invariance.

For a generic scalar field $\phi$ in the $\sigma$-model approach, the requirement of conformal invariance implies

$$\beta_\phi = 0 \quad \rightarrow \quad \frac{\partial V}{\partial \phi} = 0 \text{ for } \phi = \langle \phi \rangle \,. \tag{6.1.3}$$

As we have pointed out in Chapter 3, Eq. (6.1.3) is identically satisfied by moduli fields, i.e. their expectation values parametrize a space of flat directions of the scalar potential. If we call $\mathcal{O}_{M_A}$ the marginal operators which correspond to moduli massless excitations, then $\mathcal{O}_{M_A}$ are conformally invariant perturbations for all $M_A$. The possible conformal field theories with the same field content fill a manifold that can be identified with the manifold of the coupling constants associated with the moduli operators $\mathcal{O}$. Let us assume that this space has some differentiable structure. Zamolodchikov has shown [269] that it can be regarded as a Riemannian space with metric given by

$$\langle \mathcal{O}_{M_A}(1) \mathcal{O}_{M_B}(0) \rangle = g_{AB}(M) \,. \tag{6.1.4}$$

It can be shown that in the effective lagrangian the moduli kinetic term is given by

$$g_{AB} \partial_\mu M_A \partial_\mu M_B \tag{6.1.5}$$

Moreover the scalar potential in the effective lagrangian satisfies

$$V(M) = 0 \tag{6.1.6}$$

when all the other non-moduli fields are set to their expectation values.

From a different point of view the fact that the kinetic term in the lagrangian has the expression (6.1.5) can also be inferred by dimensional reduction of the 10D low energy effective lagrangian. In particular, if we consider the Einstein term

$$\mathcal{L}_{Eins} = \sqrt{g}\widehat{R} \tag{6.1.7}$$

and we expand the curvature scalar $\widehat{R}$ for small metric fluctuations, in the moduli space sector we get the expression

$$\mathcal{L}_{kin} = g^{(WP)}_{AB^\star}\partial_\mu M^A \partial_\mu M^{B^\star} \tag{6.1.8}$$

where

$$\begin{aligned} g^{(WP)}_{AB^\star} &= \text{const}\frac{1}{\text{Vol}} \times \int_{\mathcal{M}} g^{pl^\star} g^{kj^\star}\, \delta g^{(A)}_{l^\star j^\star}\, \delta g^{(B^\star)}_{pk}\, (det\, g)^{\frac{1}{2}} \\ \text{Vol} &= \int_{\mathcal{M}} d^n z\, d^n \overline{z}\, (det\, g)^{\frac{1}{2}} \end{aligned} \tag{6.1.9}$$

is the Weyl–Petersson metric on the moduli manifold. The Weyl–Petersson metric is a Kaluza-Klein (classical) result, while the Zamolodchikov metric, obtained by compactifying the string on an abstract (2,2) superconformal model, is a "stringy" result. This means that we have to expect that the classical metric gets "quantum" corrections and therefore it differs from the Zamolodchikov one. This is indeed true for the metric defined in the moduli space of Kähler class deformations, while for the metric of the complex structure moduli, the classical result is true also at the string level.

In this chapter we first study the Weyl–Petersson metric (6.1.9) for the cases of (1,1) and (2,1) moduli $A = \alpha, \mu$ and we show that they are both special Kähler metrics [244, 66]. Then we study, using the superconformal approach, the Zamolodchikov metric for (1,1) and (2,1) moduli and we prove that also at the quantum level, these metrics are special. We perform this proof by comparing the four scattering amplitudes computed from the effective lagrangian and the string amplitudes obtained from the (2,2) conformal theory [126]. However, it has to be stressed that, in this context, we are not able to find the explicit form of the quantum-corrected metric; rather, we only prove that the geometrical relation obeyed by a special Kähler metric is satisfied. The explicit evaluation of the Zamolodchikov metric is provided in Chapter 8, for some specific examples.

## 6.2 The Special Geometry of $(2,1)$-Forms

In this section we consider the natural Weyl–Peterson metric defined on the space of complex structure deformations of a complex Calabi–Yau $n$-fold $\mathcal{M}_n$ and we show that it is Kählerian, the Kähler potential being the logarithm of the norm of $\Omega$, the holomorphic $n$-form. In the particular case $n = 3$, this Kähler metric is *special*, according to the definition of special Kähler geometry given in Chapter 3 (see section 3.3). For Calabi–Yau

3-folds the same special Kähler geometry characterizes also the space of Kähler structure deformations, as we show in the next section. Hence special Kähler geometry emerges as the characterizing feature of the moduli spaces in superstring compactifications on Calabi–Yau manifolds, namely on $c = 9$, $(2,2)$-theories. As we have stressed in the previous section, we actually should distinguish between the Weyl–Petersson metric, which is a geometrical concept, arising as the $\sigma$-model metric in Kaluza–Klein compactification and the Zamolodchikov metric, seen as the moduli two-point function in the string compactification on a Calabi–Yau manifold. The two metrics always coincide for the (2,1)-forms, and for the (1,1)-forms they coincide only in the large radius limit of the Calabi–Yau manifold.

Let us begin with the case of (2,1)-forms. We have seen in Chapter 5 that, given an $n$-dimensional Calabi–Yau manifold defined as the polar locus of a defining degree $\nu = n + 2$ polynomial $\mathcal{W}(X)$ in $\mathbb{CP}_{n+1}$, we can establish a one-to-one correspondence between the chiral ring $\mathcal{R}_{\mathcal{W}}$ of the polynomial and the cohomology groups $H^{n-k,k}$ of the Calabi–Yau manifold. This construction can be easily generalized to the case of complex $n$-folds defined by a set of polynomial constraints in a product of $\mathbb{CP}_m$ spaces. For the case of a Calabi–Yau $n$-fold we have in particular

$$\omega^{(n-1,1)}[P] = P(X)\Theta^1 \tag{6.2.1}$$

where $P(X)$ is a polynomial in the chiral ring $\mathcal{R}_{\mathcal{W}}$ with degree equal to the degree of the defining polynomial $\mathcal{W}$ and $\Theta^1$ is defined as in (5.9.57).

If we consider a set of polynomial constraints $\mathcal{W}_\alpha\,(X_1, \ldots, X_N; \psi^1 \ldots, \psi^r) = 0$ depending on $r$ complex parameters $\psi^\mu$, we can write a generic variation of the constraints as

$$Q_\mu \stackrel{\text{def}}{=} \frac{\partial \mathcal{W}_\alpha}{\partial \psi^\mu}\,\delta\psi^\mu \tag{6.2.2}$$

Equation (6.2.2) corresponds to a non-trivial element of degree $d = \nu$ in the chiral ring $\mathcal{R}_{\mathcal{W}}$, and via Eq. (6.2.1) we can associate it with an $(n-1,1)$-form on the Calabi–Yau $n$-fold. On the other hand we can associate a $(n-1,1)$-form with a unique element in $H^1_{\bar\partial}(\mathcal{M}, T^{(1,0)}\mathcal{M})$, via the map $\rho$ defined in Eq.s (5.9.38) and (5.9.41).

As we have seen in Chapter 4, for the particular case $n = 3$ (but the relation is immediately generalizable to $n$-dimensional Calabi–Yau manifolds), any element in $H_{\bar\partial}(\mathcal{M}^1, T^{(1,0)}\mathcal{M})$ is uniquely related to a variation $\delta g_{ij}$ of the complex structure of the Calabi–Yau manifold. This means that a variation on the parameters $\psi^\mu$ induces a variation on the complex structure of the corresponding Calabi–Yau manifold.

In order to establish the geometry of the moduli spaces of $(n-1,1)$ forms, the key observation is the following. Let

$$\Omega^{(n,0)} \;=\; \Omega\,(z,\psi)_{k_1\ldots k_n}\, dz^{k_1} \wedge \ldots \wedge dz^{k_n} \tag{6.2.3}$$

be the holomorphic $n$-form regarded as a function of the $z^i$ ($i = 1, \ldots n$), namely the complex coordinates of the $n$-fold and of $\psi^\mu$ ($\mu = 1, \ldots, h^{(n-1,1)}$), the coordinates of the complex structure space; then we have

$$\frac{\partial}{\partial \psi^\mu} \Omega^{(n,0)} = c_\mu \Omega^{(n,0)} + \omega_\mu^{(n-1,1)} \tag{6.2.4}$$

where $c_\mu$ is some constant parameter and $\omega_\mu^{(n-1,1)} = \omega_{\mu i_1 \dots i_{n-1} j^\star} dz^{i_1} \wedge \dots \wedge dz^{i_{n-1}} \wedge d\overline{z}^{j^\star}$ is an element of the Dolbeault cohomology group $H^{(n-1,1)}$. Equation (6.2.4) is easily understood by considering that, in taking the derivative with respect to the parameters $\psi^\mu$, we are supposed to differentiate not only the coefficient $\Omega(z,\psi)_{k_1 \dots k_n}$ but also the differentials $dz^i$: for each of them we have $\frac{\partial}{\partial \psi^\mu} dz^i = f^i_j \, dz^j + f^i_{j^\star} d\overline{z}^{j^\star}$, so that we increase the number of anti-holomorphic differentials by at most one unit. Equation (6.2.4) is just a particular instance of a more general equation. Let us recall from Chapter 5, section 5.9 the isomorphism between the filtration (5.9.8) of the space $\mathcal{H}_n$ of rational meromorphic $(n+1)$-forms on $\mathbb{CP}_{n+1}$ and the Hodge filtration

$$\mathcal{F}_{(n,0)} \subset \mathcal{F}_{(n-1,1)} \subset \dots \subset \mathcal{F}_{(n-k,k)} \subset \mathcal{F}_{(n-k-1,k+1)} \dots \subset \mathcal{F}_{(0,n)} = H^{(n)}_{DR}(\mathcal{M}_n) \tag{6.2.5}$$

of the $n$-th order de Rahm cohomology group of the $n$-fold. We also recall that by definition the flag manifolds $\mathcal{F}_{(n-k,k)}$ appearing in Eq. (6.2.5) are given by the sum of the first $k$ Dolbeault cohomology groups:

$$\mathcal{F}_{(n-k,k)} = H^{(n,0)} \oplus H^{(n-1,1)} \oplus \dots \oplus H^{(n-k,k)} \tag{6.2.6}$$

appearing in the decomposition:

$$H^{(n)}_{DR}(\mathcal{M}_n) = H^{(n,0)} \oplus H^{(n-1,1)} \oplus \dots \oplus H^{(n-k,k)} \oplus \dots \oplus H^{(0,n)} \tag{6.2.7}$$

If we consider a parametrized family of Calabi–Yau $n$-folds, for instance by considering the polynomial constraint $\mathcal{W}(X,\psi)$ as a function not only of the $n+2$ homogeneous $\mathbb{CP}_{n+1}$ coordinates but also of a convenient set of parameters $\psi^\mu$ entering the expression of the coefficients, then we always have

$$\frac{\partial}{\partial \psi^\mu} \mathcal{F}_{(n-k,k)} \subset \mathcal{F}_{(n-k-1,k+1)} \tag{6.2.8}$$

which for $k=0$ is nothing else than equation (6.2.4). The argument to prove (6.2.8) is the obvious generalization of that utilized to justify (6.2.4).

Consider now the following metric in the parameter space of complex structures, identified with the parameter space of $(n-1,1)$-forms via Eq. (5.9.38):

$$g^{(WP)}_{\mu\nu^\star}(\psi) = \frac{\int_{\mathcal{M}} \omega_\mu \wedge \overline{\omega}_{\nu^\star}}{\int_{\mathcal{M}} \Omega \wedge \overline{\Omega}} \tag{6.2.9}$$

where $\int_{\mathcal{M}}$ is the integral on the Calabi–Yau $n$-fold. Using the identifications between the metric deformations $\delta g^{(\mu)}_{l^\star j^\star}$, $\delta g^{(\nu^\star)}_{pk}$ and the corresponding $(n-1,1)$ or $(1,n-1)$-forms, respectively provided by Eq.s (4.4.23) and by their complex conjugates, we can prove that the metric $g^{(WP)}_{\mu\nu^\star}$ is just the Weyl–Petersson metric:

$$g^{(WP)}_{\mu\nu^\star} = \text{const}\frac{1}{\text{Vol}} \times \int_{\mathcal{M}} g^{pl^\star} g^{kj^\star} \delta g^{(\mu)}_{l^\star j^\star} \delta g^{(\nu^\star)}_{pk} (det\, g)^{\frac{1}{2}}$$
$$\text{Vol} = \int_{\mathcal{M}} d^n z\, d^n \overline{z}\, (det\, g)^{\frac{1}{2}} \tag{6.2.10}$$

To see this it suffices to recall from Chapter 4 (see Theorem 4.3.2) and Eq. (4.3.32) that:
a) in a Ricci-flat metric $\Omega^{(n,0)}$ has covariantly constant components, so that $||\Omega||^2 = \text{const}$, and $\int_{\mathcal{M}} \Omega \wedge \overline{\Omega} = n!\, \text{Vol}\, ||\Omega||^2$.
b) $(det\, g)^{\frac{1}{2}} = \frac{|f|^2}{||\Omega||^2}$, where the holomorphic function $f(z)$ is defined by

$$\frac{\Omega^{i_1^\star \dots i_n^\star}}{||\Omega||^2} = \frac{1}{f(z)} \varepsilon^{i_1^\star \dots i_n^\star} \tag{6.2.11}$$

We want to show that $g^{(WP)}$ is Kählerian, with Kähler potential given by

$$\mathcal{K}\left(\psi, \overline{\psi}\right) = -\log\left[-i \int_{\mathcal{M}} \Omega \wedge \overline{\Omega}\right] \tag{6.2.12}$$

Indeed from Eq. (6.2.12), by taking two derivatives we obtain

$$\partial_\mu \partial_{\nu^\star} \mathcal{K} = -\frac{1}{\left[\int_{\mathcal{M}} \Omega \wedge \overline{\Omega}\right]^2} \Big\{ \int_{\mathcal{M}} \partial_\mu \Omega \wedge \partial_{\nu^\star} \overline{\Omega} \int_{\mathcal{M}} \Omega \wedge \overline{\Omega}$$
$$- \int_{\mathcal{M}} \partial_\mu \Omega \wedge \overline{\Omega} \int_{\mathcal{M}} \Omega \wedge \partial_{\nu^\star} \overline{\Omega} \Big\} \tag{6.2.13}$$

and, expanding according to Eq. (6.2.4) we get

$$\partial_\mu \partial_{\nu^\star} \mathcal{K} = \frac{\int_{\mathcal{M}} \omega_\mu \wedge \overline{\omega}_{\nu^\star}}{\int_{\mathcal{M}} \Omega \wedge \overline{\Omega}} \tag{6.2.14}$$

since

$$\int_{\mathcal{M}} \omega_\mu^{(n-1,1)} \wedge \overline{\Omega}^{(0,n)} = \int_{\mathcal{M}} \Omega^{(n,0)} \wedge \overline{\omega}_{\nu^\star}^{(1,n-1)} = 0 \tag{6.2.15}$$

Hence the Weyl–Petersson metric (6.2.9) is Kählerian for any value of $n$. For $n = 3$ something very peculiar occurs and the Weyl–Petersson metric becomes a *special Kählerian metric* according to the definition of Special Kähler geometry given in Chapter 3. To show this let us introduce a basis for the $n$-homology cycles. Looking at Eq. (6.2.7) we see that the $n$-th Betti number $b_n = \dim_{\mathbb{R}} H^{(n)}_{DR}$, which is also the number of these cycles, is given by the expression

$$b_n = 2 + h^{(n-1,1)} + x$$
$$x = \sum_{k=2}^{n-2} h^{(n-k,k)} \tag{6.2.16}$$

In every dimension $h^{(n-1,1)} = \dim_{\mathbb{C}} \mathcal{M}_{cs}$, where $\mathcal{M}_{cs}$ is the moduli space of complex structure deformations, but this differs from the quantity $\frac{1}{2}(b_n - 2)$ by the amount $\frac{x}{2}$. It is only for $n = 3$ that we have $x = 0$ and that we can write the crucial equation

$$b_n = b_3 = 2 + 2\dim_{\mathbb{C}} \mathcal{M}_{cs} \tag{6.2.17}$$

In view of the definition of special Kähler geometry (see Chapter 3 and Eq.s (3.3.36)), Eq. (6.2.17) is very suggestive. Indeed we know that the key feature of Special Kählerian manifolds $\mathcal{SK}$ is the existence of a flat $Sp\left(2 + 2\dim_{\mathbb{C}}\mathcal{SK}, \mathbb{R}\right)$ bundle, whose sections have the structure $\left\{ X^\Lambda, \frac{\partial F(X)}{\partial X^\Sigma} \right\}$, where the $1 + \dim_{\mathbb{C}}\mathcal{SK}$ complex numbers constitute a set of homogeneous complex coordinates for $\mathcal{SK}$ (*the special coordinates*) and $F(X)$ is a holomorphic generating function that is homogeneous of degree two. Now, when $b_n =$ even, homology bases with a given intersection matrix are related to each other by symplectic transformations of $Sp(b_n)$, so that it is tempting to assume that, for $n = 3$, $\mathcal{M}_{cs}$ is indeed a special manifold, the typical section $\left\{ X^\Lambda, \frac{\partial F(X)}{\partial X^\Sigma} \right\}$ of the flat $Sp\left(2 + 2\dim_{\mathbb{C}}\mathcal{M}_{cs}, \mathbb{R}\right)$ bundle being provided by the periods $\Pi_A = \int_{C_A} \Omega$ $(A = 1, .., b_3)$ of the holomorphic 3-form along a basis of 3-cycles. We show that this is indeed the case.

Choose a set of $2 + 2h^{(2,1)}$ 3-cycles $\left(A^\Lambda, B_\Sigma\right)$ with index convention:

$$\Lambda = (0, \mu) \qquad \mu = 1, \ldots, h^{(2,1)} \tag{6.2.18}$$

and such that

$$\begin{aligned} A^\Lambda \cap A^\Sigma = B^\Lambda \cap B^\Sigma &= 0 \\ A^\Lambda \cap B_\Sigma &= \delta^\Lambda_\Sigma \end{aligned} \tag{6.2.19}$$

If we name $\left(\alpha_\Lambda, \beta^\Sigma\right)$ the dual harmonic 3-forms with intersection matrix

$$\begin{aligned} \int_{A^\Lambda} \alpha_\Sigma = \int_{\mathcal{M}} \alpha_\Sigma \wedge \beta^\Lambda &= \delta^\Lambda_\Sigma \\ \int_{B_\Sigma} \beta^\Lambda = \int_{\mathcal{M}} \beta^\Lambda \wedge \alpha_\Sigma &= -\delta^\Lambda_\Sigma \\ \int_{A^\Lambda} \beta^\Sigma = \int_{B_\Sigma} \alpha_\Lambda &= 0 \end{aligned} \tag{6.2.20}$$

then we can name the periods of the holomorphic 3-form as follows

$$\begin{aligned} X^\Lambda &= \int_{A^\Lambda} \Omega = \int_{\mathcal{M}} \Omega \wedge \beta^\Lambda \\ F_\Sigma &= \int_{B_\Sigma} \Omega = \int_{\mathcal{M}} \Omega \wedge \alpha_\Sigma \end{aligned} \tag{6.2.21}$$

Both $X^\Lambda(\psi)$ and $F_\Sigma(\psi)$ are functions of the complex structure parameters, but they are an over-complete set to be taken as a set of coordinates for the moduli space $\mathcal{M}_{cs}$.

However, if we select the subset $X^\Lambda$ we can identify it with the complex structure parameters $\psi$ up to an overall constant. Indeed, if we rescale $\Omega \longrightarrow \lambda\Omega$ by a constant $\lambda$, the periods $X^\Lambda$ undergo the same rescaling, $X^\Lambda \longrightarrow \lambda X^\Lambda$. Then we can think that the holomorphic 3-form $\Omega(\psi)$ is rather a function of the periods $X^\Lambda$, homogeneous of degree one: $\Omega(\lambda X) = \lambda\Omega(X)$. In this way we conclude that the second set of periods $F_\Sigma(X)$ are functions of the first set. If we are able to show that

$$F_\Sigma = \frac{\partial}{\partial X^\Lambda} F(X) \tag{6.2.22}$$

where $F(X)$ is a homogeneous degree two function, then the proof of special geometry is achieved. To this effect we note that, in order to reproduce Eq.s (6.2.21), the holomorphic 3-form $\Omega^{(3,0)}$ must be written, in the harmonic basis $\{\alpha_\Lambda, \beta^\Sigma\}$, as follows:

$$\Omega = X^\Lambda \alpha_\Lambda - F_\Sigma \beta^\Sigma \tag{6.2.23}$$

Now, because of Eq.(6.2.4) we have

$$\int_{\mathcal{M}} \Omega \wedge \frac{\partial}{\partial\psi^\mu}\Omega = \frac{\partial X^\Lambda}{\partial\psi^\mu}\int_{\mathcal{M}} \Omega \wedge \frac{\partial\Omega}{\partial X^\Lambda} = 0 \tag{6.2.24}$$

since $\int_{\mathcal{M}} \Omega^{(3,0)} \wedge \Omega^{(3,0)} = \int_{\mathcal{M}} \Omega^{(3,0)} \wedge \omega^{(2,1)} = 0$. Combining Eq.s (6.2.24) and (6.2.23), we obtain

$$\begin{aligned} 0 &= \int_{\mathcal{M}} \left(X^\Lambda \alpha_\Lambda - F_\Sigma \beta^\Sigma\right) \wedge \left(\alpha_\Gamma - \partial_\Gamma F_\Sigma \beta^\Sigma\right) \\ &= -X^\Lambda \partial_\Gamma F_\Lambda + F_\Gamma \end{aligned} \tag{6.2.25}$$

i.e.

$$2F_\Gamma = \partial_\Gamma \left(X^\Lambda F_\Lambda\right) \tag{6.2.26}$$

From Eq. (6.2.26) we learn that $F_\Gamma = \partial_\Gamma F(X)$ is the derivative of a function $F(X) \stackrel{\text{def}}{=} \frac{1}{2} X^\Lambda F_\Lambda = \frac{1}{2} X^\Lambda \partial_\Lambda F$ that is homogeneous of degree two:

$$F(\lambda X) = \lambda^2 F(X) \tag{6.2.27}$$

Hence special Kähler geometry is proved for the Weyl–Petersson metric of complex structure moduli in the case of a Calabi–Yau 3-fold. In particular if we insert Eq. (6.2.23) into the definition (6.2.12) of the Kähler potential, we retrieve the special geometry formula (3.3.33):

$$\mathcal{K} = -\log i\left(X^\Lambda \overline{F}_\Lambda - \overline{X}^\Lambda F_\Lambda\right) \tag{6.2.28}$$

## 6.3 The Special Geometry of $(1,1)$-Forms

In this section we study, in analogy with the prèvious one, the geometrical properties of the moduli space of (1,1)-forms. We consider the case of a Calabi–Yau manifold of complex dimension $n = 3$ and we write the following integral:

$$g_{\alpha\beta} = const \frac{\int \mathcal{U}_{\alpha|ij^*}\mathcal{U}_{\beta|lm^*}g^{im^*}g^{j^*l}(det\, g)^{1/2}d^6 z}{(det\, g)^{1/2}d^6 z} \tag{6.3.1}$$

where $g_{ij^*} = \mathcal{K}_{ij^*}$ is the Kähler metric on the Calabi–Yau manifold, while $\omega_\alpha = \mathcal{U}_{\alpha|ij^*}dz^i \wedge dz^{j^*}$ is the basis of (1,1)-forms as introduced in Eq. (4.4.9). Equation (6.3.1) is the Weyl–Petersson metric defined in the space of (1,1) forms, which, in the Kaluza–Klein approach, is naturally associated with the $\sigma$-model metric for the corresponding light fields. For later convenience we choose the constant in Eq. (6.3.1) equal to $\pi^2$. In this section we show that the metric (6.3.1) is special, and the associated holomorphic homogeneous function has the form

$$F(X) = \frac{1}{3!}d_{\alpha\beta\gamma}\frac{X^\alpha X^\beta X^\gamma}{X^0} \tag{6.3.2}$$

where $d_{\alpha\beta\gamma}$ are defined in Eq. (4.4.36) of Chapter 4. The Yukawa couplings associated with (6.3.2) are constant and precisely equal to $d_{\alpha\beta\gamma}$. As we have many times stressed, this result will be corrected, at the string level, by instanton contributions.

We can now rewrite (6.3.1) as

$$g_{\alpha\beta} = \pi^2 \frac{\int \omega_\alpha \wedge {}^\star\omega_\beta}{\int (det\, g)^{1/2}d^6 z} = \frac{1}{\int (det\, g)^{1/2}d^6 z}\int \mathcal{U}_{\alpha|ij^*}\, {}^\star\mathcal{U}_{\beta|pqr^*s^*}\epsilon^{ipq}\epsilon^{j^*r^*s^*}d^6 z \tag{6.3.3}$$

where ${}^\star\omega_\beta$ is a $(2,2)$-form associated with the $(1,1)$-form $\omega_\beta$ via

$${}^\star\omega_\beta = {}^\star\mathcal{U}_{\beta|pqr^*s^*}dz^p \wedge dz^q \wedge dz^{r^*} \wedge dz^{s^*} \tag{6.3.4}$$

with

$${}^\star\mathcal{U}_{\beta|pqr^*s^*} = \frac{1}{4}(det\, g)^{\frac{1}{2}}\epsilon_{t^*r^*s^*}\epsilon_{upq}g^{m^*u}g^{lt^*}\mathcal{U}_{\beta|lm^*} \tag{6.3.5}$$

Moreover, since

$$\epsilon^{p^*q^*r^*}g_{ip^*}g_{jq^*}g_{kr^*} = (det g)^{1/2}\epsilon_{ijk} \tag{6.3.6}$$

we can write

$$\begin{aligned}
{}^\star\mathcal{U}_{\beta|pqr^*s^*} &= \frac{1}{4}\epsilon_{t^*r^*s^*}\epsilon^{f^*g^*h^*}g_{uf^*}g_{pg^*}g_{qh^*}g^{m^*u}g^{lt^*}\mathcal{U}_{\beta|lm^*} \\
&= \frac{3!}{4}\delta^{[f^*}_{t^*}\delta^{g^*}_{r^*}\delta^{h^*]}_{s^*}g_{uf^*}g_{pg^*}g_{qh^*}g^{m^*u}g^{lt^*}\mathcal{U}_{\beta|lm^*} \\
&= \frac{1}{2}(g^{m^*l}\mathcal{U}_{\beta|lm^*})g_{pr^*}g_{qs^*} - g_{qs^*}\mathcal{U}_{\beta|pr^*}
\end{aligned} \tag{6.3.7}$$

Hence from (6.3.7) we find

$$^*\omega_\beta = -\frac{1}{2}(\frac{2\pi}{i})^3(K.\omega^\beta)K\wedge K + \frac{2\pi}{i}K\wedge\omega_\beta \tag{6.3.8}$$

where $K$ is the Kähler 2-form of the Calabi–Yau manifold and we have defined

$$K.\omega_\beta = \frac{i}{2\pi}g^{m^*l}\mathcal{U}_{\beta|lm^*} \tag{6.3.9}$$

If $\omega_\beta$ is a harmonic form then $d(g^{m^*l}\mathcal{U}_{\beta|lm^*}) = 0$. This means that $\frac{2\pi}{i}K.\omega_\beta = const = c_\beta$. Consider now the following identities:

$$\int g^{m^*l}\mathcal{U}_{\beta|lm^*}(detg)^{1/2}d^6z = c_\beta\int(detg)^{1/2}d^6z = -\frac{ic_\beta}{3!}(2\pi)^3\int K\wedge K\wedge K = -2\pi i\int K\wedge {}^*\omega_\beta \tag{6.3.10}$$

where we have used

$$\frac{(2\pi)^3 i}{3!}\int K\wedge K\wedge K = \int(detg)^{1/2}d^6z \tag{6.3.11}$$

From Eq.s (6.3.8) and (6.3.10) we obtain

$$\frac{i}{2\pi}c_\beta = K.\omega_\beta = -\frac{3}{(2\pi)^2}\frac{\int K\wedge K\wedge\omega_\beta}{K\wedge K\wedge K} \tag{6.3.12}$$

so that

$$g_{\alpha\beta} = \frac{3}{2}\left(\frac{\int\omega_\alpha\wedge\omega_\beta\wedge K}{\int K\wedge K\wedge K} - \frac{3}{2}\frac{(\int\omega_\alpha\wedge K\wedge K)(\int\omega_\beta\wedge K\wedge K)}{(\int K\wedge K\wedge K)^2}\right) \tag{6.3.13}$$

If we now recall that

$$[K] = \mathrm{Im}t^\alpha\omega_\alpha \tag{6.3.14}$$

$$\int\omega_\alpha\wedge\omega_\beta\wedge\omega_\gamma = d_{\alpha\beta\gamma} \tag{6.3.15}$$

we immediately get

$$\int K\wedge K\wedge K = d_{\alpha\beta\gamma}\mathrm{Im}\,t^\alpha\mathrm{Im}\,t^\beta\mathrm{Im}\,t^\gamma \tag{6.3.16}$$

$$\int\omega_\alpha\wedge K\wedge K = d_{\alpha\beta\gamma}\mathrm{Im}\,t^\beta\mathrm{Im}\,t^\gamma \tag{6.3.17}$$

$$\int\omega_\alpha\wedge\omega_\beta\wedge K = d_{\alpha\beta\gamma}\mathrm{Im}\,t^\gamma \tag{6.3.18}$$

where $t^\alpha$ is the complex coordinate introduced in chapter 4, parametrizing the Kähler structure space. Consider the following function:

$$\mathcal{K}_1 = -\ln iY \tag{6.3.19}$$

with

$$Y \equiv \frac{(2i)^3}{3!} d_{\alpha\beta\gamma} \mathrm{Im}\, t^\alpha \mathrm{Im}\, t^\beta \mathrm{Im}\, t^\gamma \tag{6.3.20}$$

From (6.3.19) we get

$$g_{\alpha\beta} = \frac{\partial}{\partial t^\alpha}\frac{\partial}{\partial \overline{t}^\beta}\mathcal{K}_1 = -\frac{\partial}{\partial t^\alpha}\frac{\partial}{\partial \overline{t}^\beta} \ln \frac{4}{3}\int K \wedge K \wedge K \tag{6.3.21}$$

Now we observe that $Y$ can be written as

$$Y = X^\Lambda \overline{\partial}_\Lambda \overline{F} - \overline{X}^\Lambda \partial_\Lambda F \tag{6.3.22}$$

with

$$\begin{aligned} X^0 &= 1 \quad X^\alpha = t^\alpha \\ F(X) &= \frac{1}{3!} d_{\alpha\beta\gamma} \frac{X^\alpha X^\beta X^\gamma}{X^0} \end{aligned} \tag{6.3.23}$$

Hence the geometry of the Kähler structure is special and we have the explicit form of the prepotential $F(X)$. The intersection forms $d_{\alpha\beta\gamma}$ coincide with the Yukawa couplings, as shown in Chapter 4, and if this result is correct then a simple topological information on the moduli space manifold will be sufficient to determine the Yukawa couplings associated with (1,1) moduli.

## 6.4 Special Geometry from N=2 World Sheet Supersymmetry

As we have pointed out in the introduction one of the main goals of this chapter is to find an explicit relation between the field-theoretical amplitudes calculated from the effective lagrangian (3.2.59) and the string amplitudes calculated using an abstract N=2 superconformal theory to describe the internal degrees of freedom. Indeed, as many times emphasised, the string compactification on a Calabi–Yau manifold can be described by representing the internal degrees of freedom with an abstract N=2, c=9 superconformal field theory.

The effective lagrangian (3.2.59) describes the interaction on a four dimensional space-time of the massless modes of the string. In Eq. (3.2.77) we have listed the spectrum of the scalar fields. Each of them can be put into correspondence with an emission vertex constructed using the internal superconformal theory. So we have two ways to compute a scattering amplitude between massless modes. The first one is field theoretical; from the lagrangian (3.2.59) we read the Feynman diagrams and we compute the scattering amplitudes we are interested in. The second one is stringy: we first write the emission vertices in the conformal theory corresponding to the massless fields under consideration and compute the amplitude as in an ordinary 2-dimensional quantum field theory. What can be done is to utilize the constraints imposed on the field theoretical amplitudes by the Ward identities of the N=2 superconformal theory in order to derive the main geometrical properties of the manifold spanned by the massless moduli scalars. With this procedure we prove that the geometry spanned by the massless moduli scalars is a special Kähler geometry. This is a stringy derivation, for the case of (2,2) compactification, of the special geometry relations (3.3.16). The derivation presented in this section closely follows the main lines of a paper by Dixon, Klapunowsky and Louis [126], where this procedure was originally introduced.

We start with the action (3.2.59). We rewrite it for convenience as

$$\mathcal{L}_{bosonic} = \sqrt{-g}\left[\mathcal{R} - g_{IJ^\star}\nabla_\mu z^I \nabla^\mu z^{J^\star} - \frac{1}{4} Ref_{\Lambda\Sigma}(z) F^\Lambda_{\mu\nu} F^{\Sigma\mu\nu} - V(z,\bar{z})\right] + \dots \tag{6.4.1}$$

where in the following we omit the axion coupling $F\tilde{F}$ and we choose, for the Killing vector prepotential, the expression (3.2.63), so that the explicit form of the potential $V(z,\bar{z})$ is

$$\begin{aligned} V(z,\bar{z}) &= e^{\kappa^2\mathcal{K}}\left[g^{IJ^*}(\partial_I W + \kappa^2 W\partial_I\mathcal{K})(\partial_{J^*}\overline{W} + \kappa^2\overline{W}\partial_{J^*}\mathcal{K}) - 3\kappa^2|W|^2\right] \\ &+ [Re\, F_{\Lambda\Sigma}]^{-1}/8(\partial_I\mathcal{K}T^\Lambda z^I + \partial_{I^*}\mathcal{K}T^\Lambda\bar{z}^{I^*})^2 \end{aligned} \tag{6.4.2}$$

Notice that the last term in the scalar potential is expressed in terms of the Kähler potential $\mathcal{K}$ instead of the the full $G$ function as in Eq. (3.2.63). This is because the superpotential $W$ is explicitly gauge invariant when the gauge group action is linear and satisfies

$$\mathcal{L}_\Lambda W = \partial_i W k^i_\Lambda = 0 \tag{6.4.3}$$

The "gauge coupling" function $\mathrm{Re}\, F_{\Lambda\Sigma}$, the superpotential $W$ and the Kähler potential $\mathcal{K}$ are in general completely independent of each other. In the case of the heterotic string compactification, however, it turns out that the correct choice of Re $F_{\Lambda\Sigma}$ is

$$\mathrm{Re}\, F_{\Lambda\Sigma} = \mathrm{Re}\, S\delta_{\Lambda\Sigma} \equiv \delta_{\Lambda\Sigma} g^{-2} \tag{6.4.4}$$

where $S$ is the dilaton field and where by $g^2 = \frac{2}{S+\bar{S}}$ we denote the gauge coupling [122]. The relation between the gauge coupling and the dilaton field is a fundamental string result, which actually relates the gauge coupling to the gravitational constant $\kappa$ (see for instance [179]). Indeed, in string theory, one sees that the vacuum expectation value of the dilaton field satisfies $\langle \mathrm{Re}\, S\rangle = \frac{\alpha'}{\kappa^2}$. Moreover we are going to prove that $\mathcal{K}$ and $W$ are not independent of each other but they are constrained by (2,2) superconformal symmetry in such a way that the special geometry relations (3.3.16) are satisfied.

Given the formula for $\mathcal{K}(z,\bar{z})$ and $W(z,\bar{z})$ as in (3.2.78) and (3.2.79) and using (6.4.4) we can write down the expression of the scalar potential $V$ for the charged fields $\mathcal{C}^\alpha, \mathcal{C}^\mu$:

$$\begin{aligned} V(\mathcal{C},\overline{\mathcal{C}}) &= \frac{g^2}{2}(G_{\alpha\bar{\beta}}\overline{\mathcal{C}}^{\bar{\beta}}T^\Lambda\mathcal{C}^\alpha + G_{\mu\bar{\nu}}\overline{\mathcal{C}}^{\bar{\nu}}T^\Lambda\mathcal{C}^\mu)^2 + \frac{g^2}{2}e^{\kappa^2\hat{\mathcal{K}}}[W_{\alpha\beta\gamma}G^{\gamma\bar{\eta}}\overline{W}_{\bar{\eta}\bar{\gamma}\bar{\delta}}(\mathcal{C}^\alpha\mathcal{C}^\beta)_{\overline{27}}(\overline{\mathcal{C}}^{\bar{\gamma}}\overline{\mathcal{C}}^{\bar{\delta}})_{27} \\ &+ W_{\kappa\lambda\rho}G^{\rho\bar{\sigma}}\overline{W}_{\bar{\sigma}\bar{\mu}\bar{\nu}}(\mathcal{C}^\kappa\mathcal{C}^\lambda)_{27}(\overline{\mathcal{C}}^{\bar{\mu}}\overline{\mathcal{C}}^{\bar{\nu}})_{\overline{27}} + X_{\alpha\nu\bar{\beta}\bar{\nu}}(\mathcal{C}^\alpha\mathcal{C}^\nu)_1(\overline{\mathcal{C}}^{\bar{\beta}}\overline{\mathcal{C}}^{\bar{\nu}})_1 + \dots] \end{aligned} \tag{6.4.5}$$

expanded up to fourth order in the charged fields. We immediately explain the notation: $(\mathcal{C}^\alpha\mathcal{C}^\beta)_{\overline{27}}$ stands for the part of the product between the two 27 fields that trasforms as a $\overline{27}$ according to the decomposition

$$27 \times 27 = \overline{27} + 351 + 351' \tag{6.4.6}$$

Analogously $(\mathcal{C}^\alpha\mathcal{C}^\mu)$ stands for the singlet part of the product between the $\mathcal{C}^\alpha$ and $\mathcal{C}^\mu$. The tensor $X_{\alpha\mu\overline{\beta}\nu}$ appears when non-moduli massless singlets $\mathcal{Y}$ are present. Since the contribution due to $\mathcal{Y}$ fields will be inessential for determining the constraints of special geometry, we do not need the explicit form of $X_{\alpha\mu\overline{\beta}\overline{\nu}}$; neither do we include any term proportional to these fields in the scalar potential. Finally the dots in Eq. (6.4.5) stand for terms of order higher than quartic in the charged fields.

In this way we have given all the ingredients that parametrize the four-dimensional field theory: we have the lagrangian and the explicit form of the scalar potential. We are interested in calculating at tree level a particular set of four-point scattering amplitudes that are suited for our purposes, namely the four-modulus scattering, the two-modulus–two-charged-field scattering and the four-charged-field scattering. In particular we first consider the four-modulus amplitude, which we denote by

$$A(M^A, M^B, \overline{M}^C, \overline{M}^D) \tag{6.4.7}$$

where $A, \overline{B}$ denote collective indices which run on the whole moduli space. Since moduli have no gauge and potential interaction, they interact only via $\sigma$-model couplings described by the usual kinetic term:

$$\mathcal{L}_{moduli} = -g_{A\overline{B}}(M, \overline{M})\partial_\mu M^A \partial_\nu \overline{M}^{\overline{B}} g^{\mu\nu}\sqrt{-g} \tag{6.4.8}$$

where $g_{A\overline{B}} \doteq \partial_A\partial_{\overline{B}}\mathcal{K}$. To get from (6.4.8) the Feynman rules up to quartic terms, we expand (6.4.8) around a reference point $(M_0, \overline{M}_0)$ in the moduli space.

$$M^A = M_0^A + m^A \tag{6.4.9}$$

where the $m^A$ represent the fluctuations. Moreover we set

$$g_{\mu\nu} = \eta_{\mu\nu} + \kappa h_{\mu\nu} \tag{6.4.10}$$

where $h_{\mu\nu}$ is the quantum fluctuation of the space-time metric. By expanding (6.4.8) we easily get the following fundamental Feynman vertices (up to four external moduli legs).

- Three-modulus vertices (Fig. (6.1a,b)):

$$i\partial_A\partial_{\overline{D}}\partial_{\overline{C}}\hat{\mathcal{K}}(k_1.k_4 + k_1.k_3) \tag{6.4.11}$$

  and

$$i\partial_B\partial_A\partial_{\overline{C}}\hat{\mathcal{K}}(k_1.k_3 + k_3.k_2) \tag{6.4.12}$$

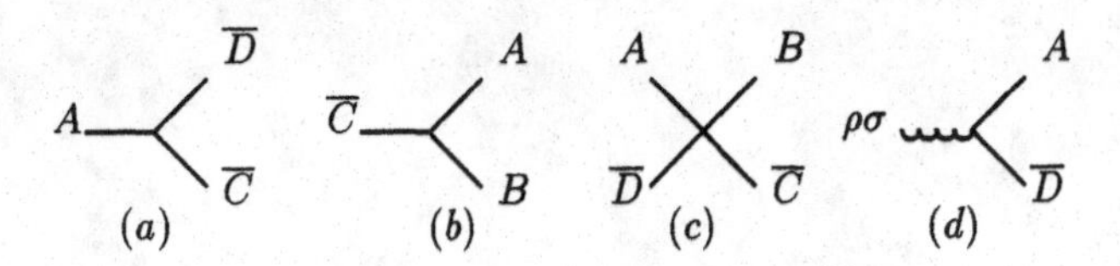


Figure 6.1: Moduli vertices

- Four-modulus vertex (Fig. (6.1c)):

$$\begin{aligned} &i\partial_B\partial_{\overline{C}}g_{A\overline{D}}k_1.k_2 + i\partial_A\partial_{\overline{D}}g_{B\overline{C}}k_2.k_3 + i\partial_B\partial_{\overline{D}}g_{A\overline{C}}k_1.k_3 + i\partial_A\partial_{\overline{C}}g_{B\overline{D}}k_2.k_4 \\ &= i\partial_A\partial_B\partial_{\overline{C}}\partial_{\overline{D}}\widehat{\mathcal{K}}(k_1.k_2 + k_1.k_3 + k_3.k_2 + k_4.k_2) \end{aligned} \tag{6.4.13}$$

- Two-modulus-graviton vertex (Fig. (6.1d))

$$-i\kappa g_{A\overline{D}}(k_1^\rho k_4^\sigma + k_4^\sigma k_1^\rho) \tag{6.4.14}$$

We assign momentum $k_1$ to $M^A$, $k_2$ to $M^B$, $k_3$ to $M^{\overline{C}}$ and $k_4$ to $M^{\overline{D}}$. The kinematics is easily given; we have

$$k_1 + k_2 + k_3 + k_4 = 0 \qquad k_i^2 = 0 \;\; i = 1, \dots 4 \tag{6.4.15}$$

and we introduce the usual Mandelstam variables:

$$\begin{aligned} s &= -(k_1 + k_2)^2 = -2k_1.k_2 = -2k_3k_4 \\ t &= -(k_1 + k_4)^2 = -2k_1.k_4 = -2k_2k_3 \\ u &= -(k_1 + k_3)^2 = -2k_1.k_3 = -2k_2k_4 \end{aligned} \tag{6.4.16}$$

with

$$u + s + t = 0 \tag{6.4.17}$$

Having fixed the notation let us calculate the amplitude (6.4.5). We are interested only in tree level amplitudes, since this approximation will be sufficient for our purposes. At tree level (6.4.7) gets contribution from the $\sigma$-model graphs depicted in Fig.(6.2). It is easy to show that the last two diagrams in figure (6.2) are zero due to (6.4.17), so that the total $\sigma$ model contribution to the scattering amplitude is given by

$$A(M^A, M^B, \overline{M}^{\overline{C}}, \overline{M}^{\overline{D}}) = is\partial_A\partial_B\partial_{\overline{C}}\partial_{\overline{D}}\widehat{\mathcal{K}} - is\partial_A\partial_B\partial_{\overline{E}}\widehat{\mathcal{K}}g^{\overline{E}F}\partial_F\partial_{\overline{C}}\partial_{\overline{D}}\widehat{\mathcal{K}} = isR_{A\overline{C}B\overline{D}} \tag{6.4.18}$$

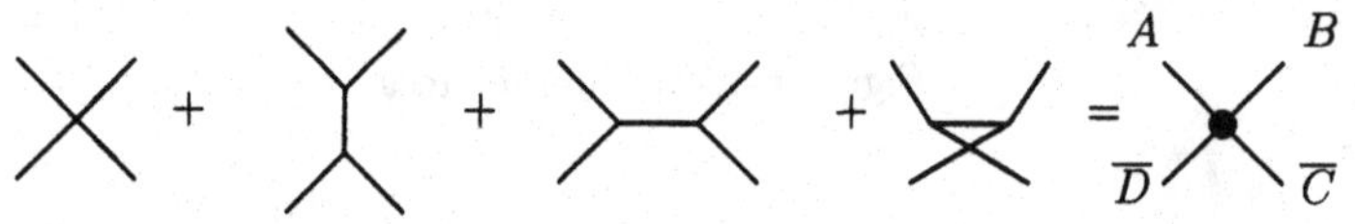

Figure 6.2: Sigma model contributions to four-modulus scattering

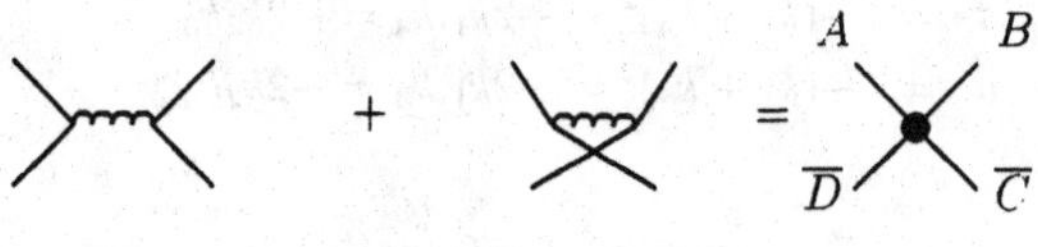

Figure 6.3: Gravitational contribution to four-modulus scattering

where $R_{A\overline{C}B\overline{D}}$ is the Riemann tensor for the Kähler metric $G_{A\overline{B}}$. The gravitational scattering contributes to the four-modulus scattering with the diagrams depicted in Fig (6.3). [1] the first one is easily computed and gives:

$$i\kappa^2 \frac{us}{t} g_{A\overline{D}} g_{B\overline{C}} \tag{6.4.19}$$

while the other one yields

$$i\kappa^2 \frac{ts}{u} g_{A\overline{C}} g_{B\overline{D}} \tag{6.4.20}$$

so that

$$A(M^A, M^B, \overline{M}^{\overline{C}}, \overline{M}^{\overline{D}}) = isR_{A\overline{C}B\overline{D}} + i\kappa^2 \frac{us}{t} g_{A\overline{D}} g_{B\overline{C}} + i\kappa^2 \frac{ts}{u} g_{A\overline{C}} g_{B\overline{D}} \tag{6.4.21}$$

We now focus on the scattering amplitude involving two moduli (one holomorphic and one anti-holomorphic) and two charged fields (one holomorphic and one anti-holomorphic) in the 27 representation of $E_6$. Since the 27 and the $\overline{27}$ fields are exactly massless for any value of the moduli fields, the scalar potential does not contribute at tree level to a scattering amplitude which involves only two charged fields. The same is true for the gauge interactions. This means that Feynman vertices are similar to those we have introduced for moduli fields: we have a two-modulus-two-charged-field vertex (with the derivatives of the full Kähler function $\mathcal{K}$ which involves the charged field indices), and one-modulus(anti-modulus)-two-charged-field vertices. Then we find

$$A(M^a, \mathcal{C}^\beta, \overline{\mathcal{C}}^{\overline{\gamma}}, \overline{M}^{\overline{D}}) = isR_{\beta\overline{\gamma}\overline{D}A} + i\kappa^2 \frac{us}{t} G_{\beta\overline{\gamma}} g_{A\overline{D}} \tag{6.4.22}$$

and similarly

$$A(M^a, \mathcal{C}^\lambda, \overline{\mathcal{C}}^{\overline{\mu}}, \overline{M}^{\overline{D}}) = +isR_{\lambda\overline{\mu}\overline{D}A} + i\kappa^2 \frac{us}{t} G_{\lambda\overline{\mu}} g_{A\overline{D}} \tag{6.4.23}$$

where

$$\begin{aligned} R_{\beta\overline{\gamma}A\overline{D}} &= \partial_A \partial_\beta \partial_{\overline{\gamma}} \partial_{\overline{D}} \mathcal{K} - \partial_A \partial_\beta \partial_{\overline{\epsilon}} \mathcal{K} G^{\overline{\epsilon}\eta} \partial_\eta \partial_{\overline{\gamma}} \partial_{\overline{D}} \mathcal{K} \\ R_{\lambda\overline{\mu}A\overline{D}} &= \partial_A \partial_\lambda \partial_{\overline{\mu}} \partial_{\overline{D}} \mathcal{K} - \partial_A \partial_\lambda \partial_{\overline{\rho}} \mathcal{K} G^{\overline{\rho}\kappa} \partial_\kappa \partial_{\overline{\mu}} \partial_{\overline{D}} \mathcal{K} \end{aligned} \tag{6.4.24}$$

Finally we are interested in the scattering amplitude involving four charged fields: two fields transforming in the 27 representation and two in the $\overline{27}$. For the sake of simplicity we restrict our attention to the case where all the charged fields belong to the vector representation of the $SO(10) \times U(1)$ subgroup of $E_6$. This is sufficient for our purposes and we do not lose generality since all the $E_6$ amplitudes can be reconstructed from the

[1] We use the following expression for the graviton propagator: $\frac{i}{2q^2}(\delta_{\mu\rho}\delta_{\nu\sigma} + \delta_{\mu\sigma}\delta_{\nu\rho} - \delta_{\mu\nu}\delta_{\rho\sigma})$, where $q$ is the graviton momentum.

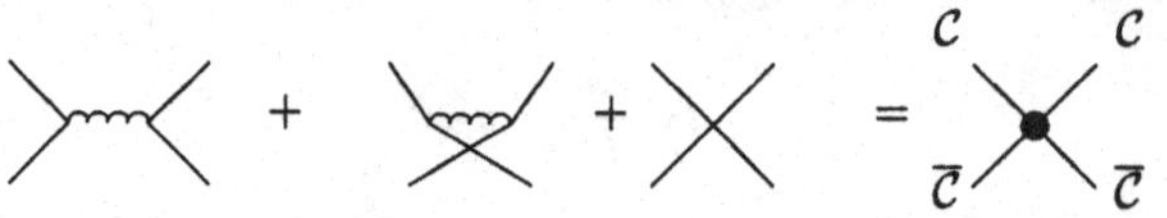

Figure 6.4: Gauge scattering and contact interaction contributing to four-charged-field scattering (at low energies)

$SO(10)$ ones by requiring $E_6$ invariance. Indeed we have the following decomposition of the 27 and $\overline{27}$ representation with respect to $SO(10) \times U(1)$:

$$\begin{aligned} 27 &= (10, q=1) \oplus (16, q=-\frac{1}{2}) \oplus (1, q=-2) \\ \overline{27} &= (\overline{10}, q=-1) \oplus (\overline{16}, q=\frac{1}{2}) \oplus (1, q=-2) \end{aligned} \tag{6.4.25}$$

The $U(1)$ subgroup of $E_6$ which we are considering is generated by the $U(1)$ current belonging to the left-moving N=2 algebra. So we have two gauge couplings involved in the scattering amplitude: the $SO(10)$ and the $U(1)$ gauge coupling. They are related to each other and we normalize these couplings in such a way that $g^2_{SO(10)} \equiv g^2 = 3g^2_{U(1)}$.

The scattering between charged fields is much more complicated with respect to the cases previously considered. However, at low energy scales we can considerably simplify the computation by considering only the leading contributions at such scales. In our case the four-charged-field scattering amplitude is dominated by the gauge scattering and the four-contact term interaction due to the scalar potential (6.4.5). The Feynman graphs that contribute to the amplitude are represented in Fig. (6.4) (we use thin straight lines to denote charged fields and thin wiggly lines for the gauge fields, to distinguish them from the thick lines denoting the moduli and the graviton fields). The fundamental Feynman rule that involves two charged fields and the gauge field has the following standard expression:

$$-iG_{\alpha\bar{\delta}}(k_1^\mu - k_4^\mu)\delta_p^{[m}\delta_s^{n]}g - iG_{\alpha\bar{\delta}}(k_1^\mu - k_4^\mu)\delta_{ps}g_{U(1)} \tag{6.4.26}$$

where $\delta_p^{[m}\delta_s^{n]}$ are the $SO(10)$ generators. Moreover we have to consider contact term interactions due to the potential (6.4.5). Assuming that the gauge group is exactly $E_6$ (no accidental enhancements), the total amplitudes due to gauge scattering and contact term interaction are given by

$$\frac{i}{2g^2}A(C_p^\alpha, C_q^\beta, \overline{C}_r^{\bar{\gamma}}, \overline{C}_s^{\bar{\delta}}) = \frac{s}{t}G_{\alpha\bar{\delta}}G_{\beta\bar{\gamma}}(\delta^{pq}\delta^{rs} - \delta^{pr}\delta^{qs} - \frac{1}{3}\delta^{qr}\delta^{ps})$$

$$\begin{aligned}
& & + \ \frac{s}{u}G_{\alpha\overline{\gamma}}G_{\beta\overline{\delta}}(\delta^{pq}\delta^{rs} - \delta^{ps}\delta^{qr} - \frac{1}{3}\delta^{pr}\delta^{qs}) \\
& & + \ e^{\kappa^2\widehat{\mathcal{K}}}W_{\alpha\beta\epsilon}G^{\epsilon\overline{\eta}}\overline{W}_{\overline{\eta}\overline{\gamma}\overline{\delta}}\delta^{pq}\delta^{rs} + O(k^2) \\
\frac{i}{2g^2}A(\mathcal{C}^{\alpha}_p,\mathcal{C}^{\lambda}_q,\overline{\mathcal{C}}^{\overline{\mu}}_r,\overline{\mathcal{C}}^{\overline{\delta}}_s) &=& \frac{s}{t}G_{\alpha\overline{\delta}}G_{\lambda\overline{\mu}}(\delta^{pq}\delta^{rs} - \delta^{pr}\delta^{qs} + \frac{1}{3}\delta^{ps}\delta^{qr}) \\
& & + \ e^{\kappa^2\widehat{\mathcal{K}}}X_{\alpha\lambda\overline{\mu}\overline{\delta}}\delta^{pq}\delta^{rs} + O(k^2) \\
\frac{i}{2g^2}A(\mathcal{C}^{\rho}_p,\mathcal{C}^{\lambda}_q,\overline{\mathcal{C}}^{\overline{\mu}}_r,\overline{\mathcal{C}}^{\overline{\nu}}_s) &=& \frac{s}{t}G_{\rho\overline{\nu}}G_{\lambda\overline{\mu}}(\delta^{pq}\delta^{rs} - \delta^{pr}\delta^{qs} - \frac{1}{3}\delta^{qr}\delta^{ps}) \\
& & + \ \frac{s}{u}G_{\rho\overline{\mu}}G_{\lambda\overline{\nu}}(\delta^{pq}\delta^{rs} - \delta^{ps}\delta^{qr} - \frac{1}{3}\delta^{pr}\delta^{qs}) \\
& & + \ e^{\kappa^2\widehat{\mathcal{K}}}W_{\rho\lambda\sigma}G^{\sigma\overline{\kappa}}\overline{W}_{\overline{\kappa}\overline{\mu}\overline{\nu}}\delta^{pq}\delta^{rs} + O(k^2) \qquad (6.4.27)
\end{aligned}$$

Where we denote by $O(k^2)$ all the contributions (due for example to moduli exchanges, gravitational scattering etc.) that are subleading at low energy scales. In (6.4.27) we have used the relations $2/(S+\overline{S}) = g^2$ and $u+s+t=0$. The factors $\frac{1}{3}$ in (6.4.27) come from the normalization of the $U(1)$ gauge coupling.

Now we come to the main point of this section: we are interested in the comparison between the amplitudes computed in the low energy effective theory and the amplitudes computed from the N=2 superconformal field theory. Hence we need a correspondence between fields in the effective theory and operators (or vertex operators) in the $2D$ quantum field theory. In other words we need "emission vertex" operators describing moduli and charged fields in the two-dimensional world. Building blocks of this construction are specific chiral (antichiral) primary operators of the N=2 theory. Indeed consider two primary operators with weights $h = \overline{h} = \frac{1}{2}$ and $U(1)$ charges $q = \overline{q} = 1$ and $q = -\overline{q} = 1$ respectively. We denote them by

$$\widehat{\psi}^{+}\begin{pmatrix} \frac{1}{2} & \frac{1}{2} \\ 1 & 1 \end{pmatrix} \quad ; \quad \widehat{\psi}^{-}\begin{pmatrix} \frac{1}{2} & \frac{1}{2} \\ -1 & 1 \end{pmatrix} \qquad (6.4.28)$$

The first one, in the normalization of the N=2 superalgebra (5.2.1), is a chiral operator, the second one a (left) anti-chiral operator. By acting on $\widehat{\psi}^{\pm}$ with the left-moving supercurrent $G^{\mp}(z)$ we get

$$\begin{aligned}
G^{\mp}(z)\widehat{\psi}^{\pm}(w,\overline{w}) &=& \frac{1}{z-w}\widehat{\phi}^{\pm}(w,\overline{w}) \qquad (6.4.29) \\
G^{\pm}(z)\widehat{\psi}^{\pm}(w,\overline{w}) &=& reg \qquad (6.4.30) \\
G^{\pm}(z)\widehat{\phi}^{\pm}(w,\overline{w}) &=& \frac{\partial}{\partial w}\left(\frac{\widehat{\psi}^{\pm}(w,\overline{w})}{z-w}\right) \qquad (6.4.31) \\
G^{\mp}(z)\widehat{\phi}^{\pm}(w.\overline{w}) &=& reg. \qquad (6.4.32)
\end{aligned}$$

The fields $\widehat{\phi}^{\pm} \equiv \widehat{\phi}^{\pm}\begin{pmatrix} 1 & \frac{1}{2} \\ 0 & 1 \end{pmatrix}$ can be seen as "upper components" (i.e. obtained by acting with the supercurrents) of $N = 2$ (super)multiplets whose lower components are $\widehat{\psi}^{\pm}$.

The operators $\widehat{\psi}^{\pm}$ are associated with the vertices of the 27 and $\overline{27}$ charged fields. In particular, charged fields belonging to decouplets of $SO(10) \subset E_6$ are associated with emission vertices of the form[2]

$$\begin{aligned} \mathcal{C}_p^{\alpha} &\stackrel{27}{\rightarrow} \lambda^p(z) e^{i\sqrt{2\alpha'}k\cdot X(z,\bar{z})} \psi_{\alpha}^{+}(z,\bar{z}) \equiv e^{i\phi^{sg}(\bar{z})} e^{i\sqrt{2\alpha'}k\cdot X(z,\bar{z})} \lambda^p(z) \widehat{\psi}_{\alpha}^{+} \begin{pmatrix} \frac{1}{2} & \frac{1}{2} \\ 1 & 1 \end{pmatrix} (z,\bar{z}) \\ \mathcal{C}_p^{\mu} &\stackrel{\overline{27}}{\rightarrow} \lambda^p(z) e^{i\sqrt{2\alpha'}k\cdot X(z,\bar{z})} \psi_{\mu}^{-}(z\bar{z}) \equiv e^{i\phi^{sg}(\bar{z})} e^{i\sqrt{2\alpha'}k\cdot X(z,\bar{z})} \lambda^p(z) \widehat{\psi}_{\mu}^{-} \begin{pmatrix} \frac{1}{2} & \frac{1}{2} \\ -1 & 1 \end{pmatrix} (z,\bar{z}) \end{aligned} \tag{6.4.33}$$

($\alpha = 1, \ldots h^{(1,1)}$, $\mu = 1, \ldots h^{(2,1)}$), where $\phi^{sg}(\bar{z})$ is a free field contribution emerging from the superghost bosonization in the right moving sector. The operator $e^{i\sqrt{2\alpha'}k\cdot X(z,\bar{z})}$, where $k \cdot X(z,\bar{z})$ denotes the scalar product in four dimensions, gives a nonzero momentum $k$ (with $k^2 = 0$) to the emission vertex located in $z$. In Eq. (6.4.33) we have introduced the shorthand notation

$$\psi_{\alpha(\mu)}(z,\bar{z}) = e^{i\phi^{sg}(\bar{z})} \widehat{\psi}_{\alpha(\mu)}^{+(-)} \begin{pmatrix} \frac{1}{2} & \frac{1}{2} \\ 1 & 1(-1) \end{pmatrix} (z,\bar{z}) \tag{6.4.34}$$

The operators $\psi_{\alpha(\mu)}^{\pm}(z,\bar{z})$ have conformal weights $(h,\bar{h}) = (1/2, 1)$, the additional 1/2 contribution to the right-moving conformal weight coming from $e^{i\phi^{sg}(\bar{z})}$. Finally in eq (6.4.33) $p$ is the $SO(10)$ vector index and $\lambda^p(z)$ are the left-moving fermions that generate the $SO(10)$ Kac–Moody algebra. The currents generating the $SO(10)$ Kac–Moody algebra are given by bilinears $\lambda^p\lambda^q$, $p < q$. This algebra can be enlarged to $E_6$ by adding to it the left-moving $U(1)$ current $J$ of the N=2 algebra (which generates the $U(1)$ subgroup of $E_6$ which commutes with $SO(10)$) and particular Ramond operators that are products of $SO(10)$ spin fields with operators of conformal weight $h = \frac{3}{8}$ and $U(1)$ charges $q = \pm\frac{3}{2}$ (obtained by acting with the spectral flow on the unit operator in the N=2 superalgebra [238]). The $SO(10)$ spin fields $S^I \equiv (S^A, S^{\dot{A}})$ ($I = 1, \ldots 32$) enter explicitly the emission vertices belonging to the $(16, q = -\frac{1}{2})$ and $(\overline{16}, q = \frac{1}{2})$ representations in the decomposition (6.4.25) (for additional details see Chapter VI.10 of [76]), i.e.

$$\begin{aligned} (16, q = -\frac{1}{2}) &\rightarrow e^{i\phi^{sg}(\bar{z})} \widehat{\psi}_{\alpha}^{+} \begin{pmatrix} \frac{3}{8} & \frac{1}{2} \\ -\frac{1}{2} & 1 \end{pmatrix} S_A(z) e^{i\sqrt{2\alpha'}k.X(z,\bar{z})} \\ (\overline{16}, q = \frac{1}{2}) &\rightarrow e^{i\phi^{sg}(\bar{z})} \widehat{\psi}_{\mu}^{-} \begin{pmatrix} \frac{3}{8} & \frac{1}{2} \\ \frac{1}{2} & 1 \end{pmatrix} S^{\dot{A}}(z) e^{i\sqrt{2\alpha'}k.X(z,\bar{z})} \end{aligned} \tag{6.4.35}$$

where the $\widehat{\psi}$ fields appearing in Eq. (6.4.35) are obtained from $\widehat{\psi}^{\pm}\begin{pmatrix} \frac{1}{2} & \frac{1}{2} \\ \pm 1 & 1 \end{pmatrix}$ by acting on them with the spectral flow operator in the left-moving N=2 algebra. The fields $\lambda^p$ satisfy the following fundamental OPE:

$$\lambda^p(z)\lambda^q(w) = \frac{\delta^{pq}}{z - z'} + \text{reg} \tag{6.4.36}$$

[2] **In this section we are interested in showing that the geometry of the moduli space is *special*. The association between chiral-chiral operators (anti-chiral-chiral) operators and vertices in the 27 ($\overline{27}$) is purely conventional. Here, for convenience, we follow the conventions of [126] and also of [76].**

In a similar way we can define the anti-27 and anti-$\overline{27}$ fields:

$$\begin{aligned} \mathcal{C}_p^{\overline{\alpha}} &\overset{anti\,27}{\longrightarrow} \lambda^p(z) e^{i\sqrt{2\alpha'}k.X(z,\overline{z})} \psi_{\overline{\alpha}}^-(z,\overline{z}) \\ \mathcal{C}_p^{\overline{\mu}} &\overset{anti\,\overline{27}}{\longrightarrow} \lambda^p(z) e^{i\sqrt{2\alpha'}k.X(z,\overline{z})} \psi_{\overline{\mu}}^+(z,\overline{z}) \end{aligned} \tag{6.4.37}$$

Notice that both the vertex operators of $\mathcal{C}^\alpha$ and $\overline{\mathcal{C}}^\mu$ involve $\psi^+$. This is because both fields should transform in the 27. However, from the target space point of view, $\psi_\alpha^+$ is associated with a holomorphic scalar field while $\psi_{\overline{\mu}}^+$ is associated with an anti-holomorphic one carrying a barred index.

Consider now the operators $\widehat{\phi}^\pm \begin{pmatrix} 1 & 1/2 \\ 0 & 1 \end{pmatrix}$ and the following vertex operators:

$$\begin{aligned} e^{i\phi^{sg}(\overline{z})} \widehat{\phi}_\alpha^+ \begin{pmatrix} 1 & 1/2 \\ 0 & 1 \end{pmatrix} e^{i\sqrt{2\alpha'}k\cdot X(z,\overline{z})} &\equiv \phi_\alpha^+(z,\overline{z}) e^{i\sqrt{2\alpha'}k\cdot X(z,\overline{z})} \\ e^{i\phi^{sg}(\overline{z})} \widehat{\phi}_\mu^- \begin{pmatrix} 1 & 1/2 \\ 0 & 1 \end{pmatrix} e^{i\sqrt{2\alpha'}k\cdot X(z,\overline{z})} &\equiv \phi_\mu^-(z,\overline{z}) e^{i\sqrt{2\alpha'}k\cdot X(z,\overline{z})} \end{aligned} \tag{6.4.38}$$

where again we have reabsorbed the superghost factor in the definition of the operators $\phi^\pm(z,\overline{z})$. Since these operators are neutral and marginal ($h = 1, q = 0$) they can be used as vertex operators of massless scalars that are $E_6$ singlets. Moreover, being upper components of N=2 supermultiplets, they can be added to the world-sheet lagrangian without spoiling the left moving N=2 superconformal symmetry [121]. This means that the space-time fields associated with $\phi^\pm$ can have arbitrary vacuum expectation value without generating a potential . The moduli fields are associated with emission vertices that involve the operators $\phi^\pm$, i.e. for a general coordinate system in the moduli space we can write the following association:

$$\begin{aligned} M^A &\rightarrow U_A^\alpha \phi_\alpha^+ + U_A^\mu \phi_\mu^- \\ \overline{M}^{\overline{A}} &\rightarrow \overline{U}_{\overline{A}}^{\overline{\alpha}} \phi_{\overline{\alpha}}^- + \overline{U}_{\overline{A}}^{\overline{\mu}} \phi_{\overline{\mu}}^+ \end{aligned} \tag{6.4.39}$$

In the expressions (6.4.39) $U_A^\alpha$ are moduli-dependent matrices, but from the world sheet point of view they are $\mathbb{C}$-numbers. For particular points in the moduli space we can find local coordinates $M^\alpha, M^\mu$ which trivialize the $U$-matrices, but this cannot be done on the whole moduli space. However, we know from the Kaluza–Klein approach that the moduli space has a rather simple structure: it is the direct product of two separate manifolds, corresponding to (1,1) and (2,1)-moduli. According to this it should be possible to define separate moduli fields $M^a, M^m$ in such a way that the matrix elements $U_m^\alpha, U_a^\mu, \overline{U}_{\overline{m}}^{\overline{\alpha}}$ and $\overline{U}_{\overline{a}}^{\overline{\mu}}$ all vanish in a finite patch. Notice that the above field separation associates in a unique way a modulus to a charged field. This is exactly what we already learned, by the analysis performed in Chapter 4, in the Kaluza–Klein approach to effective 4D lagrangians. However, in this section, we study the problem from a completely new point of view and we start by not assuming such a correspondence. If we use a local basis that distinguishes between the two types of moduli we replace (6.4.39) with

$$\begin{array}{llll} M^a & \rightarrow & U^\alpha_a \phi^+_\alpha & a=1,\dots h_{1,1} \quad (1,1)\,\textit{moduli} \\ M^m & \rightarrow & U^\mu_m \phi^-_\mu & m=1,\dots h_{2,1} \quad (2,1)\,\textit{moduli} \\ \overline{M}^{\overline{a}} & \rightarrow & \overline{U}^{\overline{\alpha}}_{\overline{a}} \phi^{+}_{\overline{\alpha}} & \overline{a}=1,\dots h_{1,1} \quad (1,1)\,\textit{anti-moduli} \\ \overline{M}^{\overline{m}} & \rightarrow & \overline{U}^{\overline{\mu}}_{\overline{m}} \phi^{-}_{\overline{\mu}} & \overline{m}=1,\dots h_{2,1} \quad (2,1)\,\textit{anti-moduli} \end{array} \tag{6.4.40}$$

Now we have the building blocks to compute scattering amplitudes in the two-dimensional quantum field theory. Before doing that there are some preliminary considerations that have to be done. Let us consider the following two-point functions

$$\langle \psi^+_\alpha(z,\overline{z})\psi^-_{\overline{\beta}}(w,\overline{w})\rangle = \frac{G_{\alpha\overline{\beta}}}{(z-w)(\overline{z}-\overline{w})^2} \tag{6.4.41}$$

$$U^\alpha_a \overline{U}^{\overline{\beta}}_{\overline{b}} \langle \phi^+_\alpha(z,\overline{z})\phi^-_{\overline{\beta}}(w,\overline{w})\rangle = \frac{g_{a\overline{b}}}{|z-w|^4} \tag{6.4.42}$$

The first one is related to the two point function between charged fields and the second one (using Eq. (6.4.36) to the two-point function between two moduli fields. $g_{a\overline{b}}$ and $G_{\alpha\overline{\beta}}$ are named, in the two-dimensional theory, the Zamolodchikov metrics of the coupling constant space spanned by the moduli and by the charged scalar fields, respectively. They precisely coincide with the metrics appearing in the effective lagrangian (3.2.59), as follows from the very definition of the effective theory. If we now consider

$$\begin{aligned} \langle \phi^+_\alpha(z,\overline{z})\phi^-_{\overline{\beta}}(w.\overline{w})\rangle &= \oint \frac{d\xi}{2\pi i}\langle G^-(\xi)\psi^+_\alpha \phi^-_{\overline{\beta}}\rangle \\ &= \oint \frac{d\xi}{2\pi i}\langle \psi^+_\alpha G^- \phi^-_{\overline{\beta}}\rangle = \frac{\partial}{\partial w}\langle \psi^+_\alpha(z,\overline{z})\psi^-_{\overline{\beta}}(w,\overline{w})\rangle \end{aligned} \tag{6.4.43}$$

where in the first equality the contour encircles $z$, but not $w$, and in the second one we consider the same contour in the reverse order, encircling $w$ and $\infty$. Now using (6.4.43), (6.4.42) and (6.4.41) we get

$$g_{a\overline{b}} = U^\alpha_a G_{\alpha\overline{\beta}} \overline{U}^{\overline{\beta}}_{\overline{b}} \tag{6.4.44}$$

and similarly

$$g_{m\overline{n}} = U^\mu_m G_{\mu\overline{\nu}} \overline{U}^{\overline{\nu}}_{\overline{n}} \tag{6.4.45}$$

If we consider next the correlation function between a (1,1) and a (2,1)-modulus we obtain

$$\begin{aligned} \langle M^a \overline{M}^{\overline{n}}\rangle &\sim \frac{g_{a\overline{n}}}{|z-w|^4} = U^\alpha_a \overline{U}^{\overline{\mu}}_{\overline{n}} \langle \phi^+_\alpha(z,\overline{z})\phi^+_{\overline{\mu}}(w,\overline{w})\rangle \\ &= U^\alpha_a \overline{U}^{\overline{\mu}}_{\overline{n}} \oint \frac{d\xi}{2\pi i}\langle G^-(\xi)\psi^+_\alpha \phi^+_{\overline{\nu}}\rangle = 0 \end{aligned} \tag{6.4.46}$$

where the last equality comes form the fact that the OPE between $G^-$ and $\phi^+_{\bar{\mu}}$ is regular. Since $g_{a\bar{n}} = \hat{\mathcal{K}}_{a\bar{n}} \equiv \partial_a\partial_{\bar{n}}\mathcal{K} = 0$ (and also $g_{\bar{m}b} = 0$), Eq. (6.4.46) implies

$$\hat{\mathcal{K}}(M^A, \overline{M}^{\overline{A}}) = \mathcal{K}_1(M^a, \overline{M}^{\overline{a}}) + \mathcal{K}_2(M^m, \overline{M}^{\overline{m}}) \tag{6.4.47}$$

so that, if the basis which separates the two types of moduli as in (6.4.40) can be defined in a finite patch, then the moduli manifold is a direct product of two separate spaces respectively associated with (1,1) and (2,1)-moduli. Notice that, in this context, the explicit proof that the moduli space is a direct product is given by Eq. (6.4.65).

Let us now go on with our programme. We have to compute the scattering amplitudes from the point of view of two-dimensional string theory. We start with $A(M^a, C^\beta, \overline{C}^{\overline{\gamma}}, \overline{M}^{\overline{d}})$. For a four-point function in string theory we can use the $SL(2,\mathbb{C})$ invariance to fix three locations for the coordinates $z_1, z_2, z_3, z_4$ of the vertex operators. Let $J(z_1, z_2, z_3)$ denote the jacobian for this fixing (we do not need its explicit expression) and let $E(z_i, z_j)$ be the correlation function of the exponentials $e^{i\sqrt{2\alpha'}k_j\cdot X}$ which has to be attached to a vertex operator at non zero momentum $k_j$:

$$E(z_i, z_j) = \langle \prod_{j=1}^{4} e^{i\sqrt{2\alpha'}k_j\cdot X}\rangle = \prod_{i<j} |z_{ij}|^{\alpha' k_i\cdot k_j} \equiv |z_{12}z_{34}|^{-\alpha' s/2}|z_{14}z_{23}|^{-\alpha' t/2}|z_{13}z_{24}|^{-\alpha' u/2} \tag{6.4.48}$$

where $z_{ij} = z_i - z_j$. The first amplitude we write down is

$$\begin{aligned} A(M^a, C^\beta_p, \overline{C}^{\overline{\gamma}}_q, \overline{M}^{\overline{d}}) &= U^\alpha_a \overline{U}^{\overline{\delta}}_{\overline{d}} |J(z_1, z_2, z_3)|^2 \int d^2 z_4 E(z_i, z_j) \langle \lambda^p(z_2)\lambda^q(z_3)\rangle \\ &\times \langle \phi^+_\alpha(z_1,\bar{z}_1)\psi^+_\beta(z_2,\bar{z}_2)\psi^-_{\overline{\gamma}}(z_3,\bar{z}_3)\phi^-_{\overline{\delta}}(z_4,\bar{z}_4)\rangle \end{aligned} \tag{6.4.49}$$

We now apply the current algebra to the last block in the correlation (6.4.49). We can write

$$\langle \phi^+_1\psi^+_2\psi^-_3\phi^-_4\rangle = \oint_{z_1} dw \frac{w-\xi}{z_1-\xi}\langle G^-(w)\psi^+_1\psi^+_2\psi^-_3\phi^-_4\rangle \tag{6.4.50}$$

where $\xi$ is an arbitrary complex number, the contour encircles only $z_1$ and we have used obvious shorthand notations for the operators $\phi^\pm, \psi^\pm$. As usual we can interpret the same circuit encircling $z_1$ anti-clockwise as the one "encircling" $z_2, z_3, z_4, \infty$ clockwise, leaving out $z_1$. Now we can choose $\xi = z_2$ in such a way that the integrand has no singularities at $z_2$. Moreover no singularities are present at $z_3$ due to the OPE (6.4.32), and also at infinity the integrand is regular. Hence the only contribution to the integral comes from

$$\oint_{z_4} \frac{dw}{2\pi i}\frac{w-z_1}{z_1-z_2}G^-(w)\phi_4 = \frac{\partial}{\partial z_4}(\frac{z_4-z_2}{z_1-z_2}\psi^-_4) \tag{6.4.51}$$

so that the correlation (6.4.49) becomes

$$\begin{aligned} & U^\alpha_a \overline{U}^{\overline{\delta}}_{\overline{d}} |J(z_1, z_2, z_3)|^2 \int d^2 z_4 E \frac{\delta^{pq}}{z_{23}} \partial_4 (\frac{z_{42}}{z_{12}} \langle \psi^+_1\psi^+_2\psi^-_3\psi^-_4\rangle) \\ &= U^\alpha_a \overline{U}^{\overline{\delta}}_{\overline{d}} |J(z_1, z_2, z_3)|^2 \delta^{pq} \int_{z_4} E \langle \psi^+_1\psi^+_2\psi^-_3\psi^-_4\rangle \left(\frac{\alpha' s/4}{z_{12}z_{34}} - \frac{\alpha' t/4}{z_{14}z_{23}}\right) \end{aligned} \tag{6.4.52}$$

where we have integrated by parts and used $s+t+u=0$.

The amplitude between four charged fields is given by

$$A(\mathcal{C}^{\alpha}_{p},\mathcal{C}^{\beta}_{q},\overline{\mathcal{C}}^{\overline{\gamma}}_{r}\overline{\mathcal{C}}^{\overline{\delta}}_{s}) = |J|^2 \int d^2 z_4 E\langle \psi^+_1 \psi^+_2 \psi^-_3 \psi^+_4 \rangle \langle \lambda^p_1 \lambda^q_2 \lambda^r_3 \lambda^s_4 \rangle \tag{6.4.53}$$

where

$$\langle \lambda^p_1 \lambda^q_2 \lambda^r_3 \lambda^s_4 \rangle = \left( \frac{\delta^{pq}\delta^{rs}}{z_{12}z_{34}} + \frac{\delta^{ps}\delta^{qr}}{z_{14}z_{23}} - \frac{\delta^{pr}\delta^{qs}}{z_{13}z_{24}} \right) \tag{6.4.54}$$

Now we are able to compare the amplitudes (6.4.53) and (6.4.49):

$$A(M^a,\mathcal{C}^{\beta}_{p},\overline{\mathcal{C}}^{\overline{\gamma}}_{p},\overline{M}^{\overline{d}}) = \frac{1}{4}\alpha' U^{\alpha}_{a}\overline{U}^{\overline{\delta}}_{\overline{d}} \left[ sA(\mathcal{C}^{\alpha}_{p},\mathcal{C}^{\beta}_{p},\overline{\mathcal{C}}^{\overline{\gamma}}_{q},\overline{\mathcal{C}}^{\overline{\delta}}_{q}) - tA(\mathcal{C}^{\alpha}_{p},\mathcal{C}^{\beta}_{q},\overline{\mathcal{C}}^{\overline{\gamma}}_{q},\overline{\mathcal{C}}^{\overline{\delta}}_{p}) \right] \tag{6.4.55}$$

where the repeated indices $p,q$ are not summed ($p \neq q$).

There are other possible string amplitudes between moduli and charged fields. In full analogy with (6.4.55) we write

$$\begin{aligned}
A(M^k,\mathcal{C}^{\beta}_{p},\overline{\mathcal{C}}^{\overline{\gamma}}_{p},\overline{M}^{\overline{n}}) &= \frac{1}{4}\alpha' U^{\mu}_{k}\overline{U}^{\overline{\nu}}_{\overline{n}} \left[ uA(\mathcal{C}^{\mu}_{p},\mathcal{C}^{\beta}_{q},\overline{\mathcal{C}}^{\overline{\gamma}}_{p},\overline{\mathcal{C}}^{\overline{\nu}}_{q}) - tA(\mathcal{C}^{\mu}_{p},\mathcal{C}^{\beta}_{q},\overline{\mathcal{C}}^{\overline{\gamma}}_{q},\overline{\mathcal{C}}^{\overline{\nu}}_{p}) \right] \\
A(M^a,\mathcal{C}^{\lambda}_{p},\overline{\mathcal{C}}^{\overline{\mu}}_{p},\overline{M}^{\overline{d}}) &= \frac{1}{4}\alpha' U^{\alpha}_{a}\overline{U}^{\overline{\delta}}_{\overline{d}} \left[ uA(\mathcal{C}^{\alpha}_{p},\mathcal{C}^{\lambda}_{q},\overline{\mathcal{C}}^{\overline{\mu}}_{p},\overline{\mathcal{C}}^{\overline{\delta}}_{q}) - tA(\mathcal{C}^{\alpha}_{p},\mathcal{C}^{\lambda}_{q},\overline{\mathcal{C}}^{\overline{\mu}}_{q},\overline{\mathcal{C}}^{\overline{\delta}}_{p}) \right] \\
A(M^k,\mathcal{C}^{\lambda}_{p},\overline{\mathcal{C}}^{\overline{\mu}}_{p},\overline{M}^{\overline{n}}) &= \frac{1}{4}\alpha' U^{\mu}_{k}\overline{U}^{\overline{\kappa}}_{\overline{n}} \left[ sA(\mathcal{C}^{\mu}_{p},\mathcal{C}^{\lambda}_{q},\overline{\mathcal{C}}^{\overline{\kappa}}_{p},\overline{\mathcal{C}}^{\overline{\nu}}_{q}) - tA(\mathcal{C}^{\mu}_{p},\mathcal{C}^{\lambda}_{q},\overline{\mathcal{C}}^{\overline{\kappa}}_{p},\overline{\mathcal{C}}^{\overline{\nu}}_{q}) \right]
\end{aligned} \tag{6.4.56}$$

Amplitudes that involve two moduli of different types vanish identically, since they are proportional to correlators of the form $\langle \phi^+ \psi^+ \psi^- \phi^+ \rangle$ (or complex conjugate ones) which can be written as

$$\langle \phi^+_1 \psi^+_2 \psi^-_3 \phi^+_4 \rangle = \oint \frac{dw}{2\pi i} \frac{w - z_2}{z_1 - z_2} \langle G^-(w) \psi^+_1 \psi^+_2 \psi^-_3 \phi^+_4 \rangle = 0 \tag{6.4.57}$$

which, following the same argument as for Eq. (6.4.50), are zero due to the fact that the OPE between $G^-$ and $\phi^+_4$ is regular.

Finally we consider the four-modulus scattering amplitudes. It is clear that such amplitudes involve the correlation between four $\phi^{\pm}$ fields. Following an argument similar to that leading to Eq. (6.4.56) it is easy to show that correlations involving terms like $\phi^+\phi^+\phi^+\phi^+$ or $\phi^+\phi^+\phi^+\phi^-$ do vanish. The nonvanishing four-modulus amplitudes are those involving $\phi^+\phi^+\phi^-\phi^-$. Consider first

$$A(M^a, M^b, \overline{M}^{\overline{c}}, \overline{M}^{\overline{d}}) = U^{\alpha}_{a} U^{\beta}_{b} \overline{U}^{\overline{\gamma}}_{\overline{c}} \overline{U}^{\overline{\delta}}_{\overline{d}} |J|^2 \int d^2 z_4 E \langle \phi^+_1 \phi^+_2 \phi^-_3 \phi^-_4 \rangle \tag{6.4.58}$$

Again we write $\phi^+_1 = \oint G^- \psi^+_1$ and we find

$$\langle \phi^+_1 \phi^+_2 \phi^-_3 \phi^-_4 \rangle = \frac{\partial}{\partial z_3} \langle \psi^+_1 \phi^+_2 \psi^-_3 \phi^-_4 \rangle + \frac{\partial}{\partial z_4} \langle \psi^+_1 \phi^+_2 \phi^-_3 \psi^-_4 \rangle \tag{6.4.59}$$

and by similar steps as in (6.4.50)–(6.4.51):

$$\langle\phi_1^+\phi_2^+\phi_3^-\phi_4^-\rangle = -\frac{\partial}{\partial z_3}\frac{\partial}{\partial z_4}\left(\frac{z_{34}}{z_{12}}\langle\psi_1^+\psi_2^+\psi_3^-\psi_4^-\rangle\right) \tag{6.4.60}$$

We can now substitute Eq. (6.4.60) into (6.4.58) and the final result is

$$\begin{aligned} &\int d^2z_4 E\langle\phi_1^+\phi_2^+\phi_3^-\phi_4^-\rangle = \\ &\int d^2z_4 E\langle\psi_1^+\psi_2^+\psi_3^-\psi_4^-\rangle\left(\frac{\alpha' s/4+(\alpha' s/4)^2}{z_{12}z_{34}} + \frac{(\alpha' t/4)^2}{z_{14}z_{23}} - \frac{(\alpha' u/4)^2}{z_{13}z_{24}}\right) \end{aligned} \tag{6.4.61}$$

To perform the computation we have integrated by parts both on the $z_4$ and $z_3$ coordinates. This is possible since the $SL(2,\mathbb{C})$ gauge-fixing procedure allows us to fix any three of the variables $z_i$. The remaining coordinate is the integration variable. If the integrand is of the form $X\partial_j Y$ it is possible to keep as integration variable $z_j$, fixing the other three coordinates. This permits to integrate by parts any of the $z_i$ coordinates. Equation (6.4.61) allows us to compare the four-modulus amplitude with the four-charged-field amplitude. If we neglect the subleading contribution (higher order terms in $\alpha'^2 k^4$), we obtain

$$A(M^a, M^b, \overline{M}^{\bar c}, \overline{M}^{\bar d}) = \frac{\alpha' s}{4} U_a^\alpha U_b^\beta \overline{U}_{\bar c}^{\bar\gamma} \overline{U}_{\bar d}^{\bar\delta} A(\mathcal{C}_p^\alpha, \mathcal{C}_p^\beta, \overline{\mathcal{C}}_q^{\bar\gamma}, \overline{\mathcal{C}}_q^{\bar\delta}) \tag{6.4.62}$$

The other non-vanishing four-modulus amplitudes are

$$\begin{aligned} A(M^a, M^l, \overline{M}^{\bar m}, \overline{M}^{\bar d}) &= \frac{\alpha' u}{4} U_a^\alpha U_l^\lambda \overline{U}_{\bar m}^{\bar\mu} \overline{U}_{\bar d}^{\bar\delta} A(\mathcal{C}_p^\alpha, \mathcal{C}_q^\lambda, \overline{\mathcal{C}}_p^{\bar\mu}, \overline{\mathcal{C}}_q^{\bar\delta}) \\ A(M^k, M^l, \overline{M}^{\bar m}, \overline{M}^{\bar n}) &= \frac{\alpha' s}{4} U_k^\kappa U_l^\lambda \overline{U}_{\bar m}^{\bar\mu} \overline{U}_{\bar n}^{\bar\nu} A(\mathcal{C}_p^\kappa, \mathcal{C}_p^\lambda, \overline{\mathcal{C}}_q^{\bar\mu}, \overline{\mathcal{C}}_q^{\bar\nu}) \end{aligned} \tag{6.4.63}$$

The relations we have just derived between the four-particle scattering impose a very powerful constraint on the geometry of the field space. As a first step we consider the amplitude

$$A(M^a, M^l, \overline{M}^{\bar m}, \overline{M}^{\bar d}) = i\frac{g^2\alpha'}{2}\frac{su}{t} U_a^\alpha G_{\alpha\bar\delta} \overline{U}_{\bar d}^{\bar\delta} U_l^\lambda G_{\lambda\bar\mu} \overline{U}_{\bar m}^{\bar\mu} = i\kappa^2 \frac{su}{t} g_{a\bar d} g_{l\bar m} \tag{6.4.64}$$

where we have used (6.4.63), (6.4.21) and the relation between the gauge and gravitational coupling which is given by $\alpha' g^2 = 2\kappa^2$. Comparing (6.4.64) with (6.4.17) we conclude that

$$R_{a\bar m l\bar d} = 0 \tag{6.4.65}$$

Hence the only non-vanishing components of the Riemann tensor $R_{A\bar C B\bar D}$ are $R_{a\bar c b\bar d}$ and $R_{k\bar m l\bar n}$. This fact shows that the manifold is the product of two moduli spaces corresponding to (1,1) and (2,1)-moduli respectively. Indeed if we define the restricted holonomy group of a Riemannian manifold as the group generated by parallel trasport along contractible loops in the manifold, then the associated algebra is generated by the components of the Riemann tensor viewed as matrices in the first two indices of $R$. The relations that we have just found imply that the restricted holonomy group is contained in $U(h_{1,1}) \times U(h_{2,1})$. But if the holonomy group of an $(n+m)$-dimensional manifold is a subgroup of $U(m) \times U(n)$, then the manifold is also a direct product of two submanifolds of complex dimensions $m$ and $n$.

Now, from (6.4.63) and (6.4.17) we easily get

$$\begin{aligned} R_{a\bar{c}b\bar{d}}\kappa^2 + \frac{u}{t} g_{a\bar{d}} g_{b\bar{c}} \ &+ \ \kappa^2 \frac{t}{u} g_{a\bar{c}} g_{b\bar{d}} = \\ -\frac{1}{2}\alpha' g^2 [\frac{s}{t} g_{a\bar{d}} g_{b\bar{c}} \ &+ \ \frac{s}{u} g_{a\bar{c}} g_{b\bar{d}} + e^{\kappa^2 \hat{\mathcal{K}}} (WUUU)_{abe} g^{e\bar{f}} (\overline{WUUU})_{\bar{f}\bar{c}\bar{d}}] \end{aligned} \quad (6.4.66)$$

and the analogous one for (2,1)-moduli, so that

$$\frac{1}{\kappa^2} R_{a\bar{c}b\bar{d}} = g_{a\bar{c}} g_{b\bar{d}} + g_{b\bar{c}} g_{a\bar{d}} - e^{\kappa^2 \hat{\mathcal{K}}} (WUUU)_{abe} g^{e\bar{f}} (\overline{WUUU})_{\bar{f}\bar{c}\bar{d}} \quad (6.4.67)$$

$$\frac{1}{\kappa^2} R_{k\bar{m}l\bar{n}} = g_{k\bar{m}} g_{l\bar{n}} + g_{k\bar{n}} g_{l\bar{m}} - e^{\kappa^2 \hat{\mathcal{K}}} (WUUU)_{kli} g^{i\bar{j}} (\overline{WUUU})_{\bar{j}\bar{m}\bar{n}} \quad (6.4.68)$$

where $(WUUU)_{abe} = W_{\alpha\beta\epsilon} U^\alpha_a U^\beta_b U^\epsilon_e$. To conclude the calculation we also need the explicit form of the $U$ matrices. To find it we have to use the relations (6.4.55), (6.4.56) and the other vanishing relations involving two moduli and two charged fields. This yields

$$\frac{1}{\kappa^2} R_{\beta\bar{\gamma}a\bar{d}} = U^\alpha_a G_{\alpha\bar{\gamma}} G_{\beta\bar{\delta}} \overline{U}^{\bar{\delta}}_{\bar{d}} + \frac{2}{3} G_{\beta\bar{\gamma}} g_{a\bar{d}} - e^{\kappa^2 \hat{\mathcal{K}}} (WUU)_{\beta a e} g^{e\bar{f}} (\overline{WUU})_{\bar{f}\bar{d}\bar{\gamma}} \quad (6.4.69)$$

and

$$\frac{1}{\kappa^2} R_{\beta\bar{\gamma}k\bar{n}} = \frac{1}{3} G_{\beta\bar{\gamma}} g_{k\bar{n}}, \quad R_{\beta\bar{\gamma}a\bar{n}} = 0, \quad R_{\beta\bar{\gamma}k\bar{d}} = 0 \quad (6.4.70)$$

and similar equations for $R_{\lambda\bar{\mu}A\bar{D}}$. Let us now denote by $U$ the matrix $U^\alpha_a$ and by $g, G$ the matrices representing the metrics $g_{a\bar{b}}, G_{\alpha\bar{\beta}}$ respectively (or $g_{m\bar{n}}, G_{\mu\bar{\nu}}$). Equation (6.4.69) can be rephrased in the following way:

$$\begin{aligned} \frac{1}{\kappa^2} R^{\bar{\beta}}_{\ \bar{\gamma}a\bar{d}} &= \frac{1}{\kappa^2} G^{\bar{\beta}\beta} R_{\beta\bar{\gamma}a\bar{d}} = \frac{1}{\kappa^2} \partial_a \left[ U^\dagger g^{-1} U \partial_{\bar{d}} (U^{-1} g U^{\dagger -1}) \right]^{\bar{\beta}}_{\bar{\gamma}} \\ &= U^\alpha_a G_{\alpha\bar{\gamma}} \overline{U}^{\bar{\beta}}_{\bar{d}} + \frac{2}{3} \delta^{\bar{\beta}}_{\bar{\gamma}} g_{a\bar{d}} - e^{\kappa^2 \hat{\mathcal{K}}} G^{\bar{\beta}\beta} (WUU)_{\beta a \epsilon} g^{\epsilon\bar{f}} (\overline{WUU})_{\bar{f}\bar{d}\bar{\gamma}} \\ &= \overline{U}^{\bar{\beta}}_{\bar{b}} g^{b\bar{b}} [g_{a\bar{c}} g_{b\bar{d}} + g_{b\bar{c}} g_{a\bar{d}} - e^{\kappa^2 \hat{\mathcal{K}}} (WUUU)_{abe} g^{e\bar{f}} (\overline{WUUU})_{\bar{f}\bar{c}\bar{d}}] (\overline{U}^{-1})^{\bar{c}}_{\bar{\gamma}} \\ &- \frac{1}{3} g_{a\bar{d}} \delta^{\bar{\beta}}_{\bar{\gamma}} \\ &= \frac{1}{\kappa^2} \overline{U}^{\bar{\beta}}_{\bar{b}} (U^{-1})^{\bar{c}}_{\bar{\gamma}} R^{\bar{b}}_{\ \bar{c}a\bar{d}} - \frac{1}{3} g_{a\bar{d}} \delta^{\bar{\beta}}_{\bar{\gamma}} \\ &\equiv \frac{1}{\kappa^2} \left[ U^\dagger \partial_a (g^{-1} \partial_{\bar{d}} g) U^{\dagger -1} \right]^{\bar{\beta}}_{\bar{\gamma}} - \frac{1}{3} g_{a\bar{d}} \delta^{\bar{\beta}}_{\bar{\gamma}} \end{aligned} \quad (6.4.71)$$

where we have used (6.4.67) and

$$\begin{aligned} R^{\bar{b}}_{\bar{c}a\bar{d}} &= [\partial_a(g^{-1}\partial_{\bar{d}}g)]^{\bar{b}}_{\bar{c}} \\ R^{\bar{\beta}}_{\bar{\gamma}a\bar{d}} &= \partial_a[G^{-1}\partial_{\bar{d}}G]^{\bar{\beta}}_{\bar{\gamma}} \\ g_{a\bar{b}} &= [UGU^\dagger]_{a\bar{b}} \end{aligned} \tag{6.4.72}$$

More generally we can write (6.4.69) and (6.4.70) as

$$\partial_C[U^\dagger g^{-1} U \partial_{\bar{d}}(U^{-1} g U^{\dagger -1})] = U^\dagger \partial_C(g^{-1}\partial_{\bar{d}} g) U^{\dagger -1} + \frac{1}{3}\kappa^2 \partial_C \partial_{\bar{D}}(\mathcal{K}_2 - \mathcal{K}_1) \times \mathbb{1} \tag{6.4.73}$$

where $\mathbb{1}$ is the unit matrix.

The general solution for the matrix $U$ is

$$U^\alpha_a(M,\overline{M}) = V^\alpha_a(M) e^{\frac{1}{6}\kappa^2(\mathcal{K}_1 - \mathcal{K}_2)} \tag{6.4.74}$$

where $V(M)$ is an arbitrary matrix valued holomorphic function of the moduli fields $M^a, M^m$. Similarly

$$U^\mu_m(M,\overline{M}) = V^\mu_m(M) e^{\frac{1}{6}\kappa^2(\mathcal{K}_2 - \mathcal{K}_1)} \tag{6.4.75}$$

The holomorphic matrices $V(M)$ can be reabsorbed by changing the basis of the charged fields, i.e by defining $A^a \equiv (V^{-1})^a_\alpha A^\alpha$, so that the $U$ matrices become proportional to the unit matrix. With this convention we can write the final fundamental relations:

$$\begin{aligned} G_{a\bar{b}} &= g_{a\bar{b}} e^{\frac{1}{3}\kappa^2(\mathcal{K}_2 - \mathcal{K}_1)} \\ G_{m\bar{n}} &= g_{m\bar{n}} e^{\frac{1}{3}\kappa^2(\mathcal{K}_1 - \mathcal{K}_2)} \end{aligned} \tag{6.4.76}$$

and

$$\frac{1}{\kappa^2} R_{a\bar{c}b\bar{d}} = g_{a\bar{c}} g_{b\bar{d}} + g_{b\bar{c}} g_{a\bar{d}} - e^{2\kappa^2 \mathcal{K}_1} W_{abe} g^{e\bar{f}} \overline{W}_{\bar{f}\bar{c}\bar{d}} \tag{6.4.77}$$

$$\frac{1}{\kappa^2} R_{k\bar{m}l\bar{n}} = g_{k\bar{m}} g_{l\bar{n}} + g_{k\bar{n}} g_{l\bar{m}} - e^{2\kappa^2 \mathcal{K}_2} W_{kli} g^{e\bar{j}} \overline{W}_{\bar{j}\bar{m}\bar{n}} \tag{6.4.78}$$

Notice that the form (6.4.76) of the $U$-matrices is essential for giving the correct moduli dependence in (6.4.77) and (6.4.78). Indeed no (2,1)-moduli dependence should be present in (6.4.77) (and similarly for the other). Equations (6.4.77), (6.4.78) are the "special geometry" constraints. They precisely coincide with Eqs. (3.3.16) if we set $\kappa^2 = 1$.

The proof we have just presented shows that special geometry is a crucial property of c=9, (2,2) superconformal theories. Moreover, as in the Kaluza–Klein approach, there is a one-to-one correspondence between charged fields in the 27 and $\overline{27}$ representation, which are related to each other by the $V$-matrices. As we have previously pointed out, we can choose a different basis for the charged fields in such a way that we can label with the same indices $a$ ($a = 1, \ldots h^{1,1}$) the (1,1) charged and moduli fields and with $m$ ($m = 1, \ldots h^{2,1}$) the (2,1) charged and moduli fields. From now on, to avoid any notational clumsiness, we will pay no more attention to the distinction between $\alpha, a = 1, \ldots h_{1,1}$ and $\mu, m = 1, \ldots h_{2,1}$.

## 6.5 Concluding Remarks

In this chapter we have shown that the geometry associated with the moduli space of Kähler and complex structure deformation is a special Kähler geometry. This is proved either at the geometrical level, by studying the Weyl–Petersson metrics associated with these deformations on a given Calabi–Yau 3-fold, or at the abstract level, by studying the Zamolodchikov metrics associated with the marginal deformations of the underlying superconformal $c = 9$ system.

However, the superconformal approach does not give any recipe for computeing the metric explicitly: it gives only a relation between the metric and the Yukawa couplings. On the other hand, the geometrical approach allows in principle an explicit calculation of the Weyl–Petersson metric through the $F$ function, once the Yukawa couplings are given. However there is an additional subtlety in the calculation of the metric. What one actually need is the expression of the Yukawa couplings in special coordinates. This problem is addressed and solved in some explicit examples in Chapter 8.

## 6.6 Bibliographical Note

- *The concept of Zamolodchikov metric dates back to 1986 and was introduced by A. Zamolodchikov in [269].*

- *The first general discussion of the moduli spaces for* $(2,2)$*-compactifications was given in 1989 by Cecotti, Ferrara and Girardello in [82].*

- *The moduli space Kähler geometry of* $(1,1)$*-forms was first derived by Strominger in 1985 in [243]. The analogous derivation for the case of* $(2,1)$*-forms was instead given by Candelas and de la Ossa in 1991 in [66].*

- *The first proof that the moduli space geometry of Calabi–Yau 3-folds should be of the special Kählerian type, proof based on the h-map from type II to heterotic superstring, was given by Ferrara and Strominger in [146] in 1989. The formulae for the Kähler potential were introduced by Cecotti, Ferrara and Girardello in [81] (1988).*

- *The basic abstract definition of special geometry, as applied to the case of Calabi–Yau moduli spaces, was given in 1990 by Strominger [244].*

- *The derivation of the special geometry identities on the Riemann tensor from the Ward indentities of the N=2 superconformal algebra is due to Dixon, Kaplunovski and Louis [126], whose paper is closely followed by us in section 6.4.*

- *Basic references on background fields and moduli in string theory are [147, 146, 236], [210, 142].*

- *Additional useful references on the above topics are [57, 63, 173, 243].*

# Chapter 7

# TOPOLOGICAL FIELD THEORIES

## 7.1 Introduction

In this chapter we consider a very special class of quantum field theories that go under the name of *topological or cohomological field theories.* Their distinctive property is that the *correlation functions* of local operators:

$$G(x_1, \ldots, x_n) = \langle \mathcal{O}(x_1)\, \mathcal{O}(x_2) \ldots \mathcal{O}(x_n) \rangle \tag{7.1.1}$$

are actually *correlation constants*, in the sense that they do not depend on the locations $x_i$ of the operators $\mathcal{O}(x_i)$:

$$\frac{\partial}{\partial x_i} G(x_1, \ldots, x_n) = 0 \quad \forall x_i \tag{7.1.2}$$

This rather miracolous situation occurs because of the *topological nature* of the theory: roughly speaking this is the independence of the involved functional integration from the *metric* $g_{\mu\nu}$ of the *world manifold* $\mathcal{M}$ on which the theory is defined and to which the points $x_i \in \mathcal{M}$ belong. Essentially these field theories are in correspondence with ordinary quantum field theories, but, through mechanisms that we are going to describe, select the *instanton configurations* of these latter as the only field configurations. In this way, the ordinarily infinite dimensional functional integral is reduced to a finite dimensional integral over the *instanton moduli space* by means of the *semi-classical approximation* which, in these theories, becomes exact. The emergence of *moduli space* as the proper *substratum* of topological field theories is the first indication that they must be intimately related with the topics we have been so far discussing. The second reason why they constitute a fundamental conceptual instrument for the understanding of our subject is their *symbiotic relation* with N=2 supersymmetric field theories. Through a *dichotomic algorithm* that is named *the topological twist* each N=2 field theory in D=2 can be associated with two topological field theories, the A-model and the B-model which shed light on complementary properties of the parent N=2 field theory. In the cases related to Calabi–Yau manifolds, the A-model and B-model deal with the Kähler class and complex structure deformations respectively. Topological field theories do exist also in dimensions higher than two and at least in four dimensions they are also related to N=2 field theories through a topological twist. In this book we just focus on two-dimensional cases.

Having listed without explanation the reasons why *topological field theories* should enter the stage, let us try to illustrate and explain in more detail the *basic idea* and *nature* of these remarkable field theories .

The realm of topological field theories can be divided, according to Witten [261], into two broad classes: the *cohomological, or semi-classical theories*, whose prototypes are either the topological Yang–Mills theory [33] or the topological $\sigma$-model [260] (both discussed in later sections of this chapter), and the *quantum theories*, whose prototypes are the BF theory or the abelian Chern–Simons theory [234, 47].

In this book, by *topological field theories* we just mean the the *cohomological* ones, which were introduced by Witten in 1988 [259], to which we confine our attention. In Witten's words, cohomological field theories are concerned with *sophisticated counting problems*. The fundamental idea is that a generic correlation function of $n$ physical observables $\{\mathcal{O}_1, \ldots, \mathcal{O}_n\}$ has an interpretation as the *intersection number*

$$\langle \mathcal{O}_1 \mathcal{O}_2 \ldots \mathcal{O}_n \rangle = \#(H_1 \cap H_2 \cap \ldots \cap H_n) \tag{7.1.3}$$

of $n$ *homology cycles* $H_i \subset \mathcal{M}_{mod}$ in the *moduli space* $\mathcal{M}_{mod}$ of suitable *instanton* configurations $\Im[\phi(x)]$ of the basic fields $\phi$ of the theory. For example in the topological $\sigma$-model the basic fields are the maps

$$X : \Sigma_g \longrightarrow \mathcal{M}_K \tag{7.1.4}$$

from a genus $g$ Riemann surface $\Sigma_g$ into a Kählerian manifold $\mathcal{M}_K$. In this case the instantons $\Im[X(z,\bar{z})]$ are the holomorphic maps $\partial_+ X = \partial_- \overline{X} = 0$ and the moduli space is, for each homotopy class of holomorphic embeddings of degree $k$, the parameter space $\mathcal{M}_{\{k\}}$ of such a class of maps. The degree $k$ is defined by $\int_{\Sigma_g} X^* K = k$, where $X^* K$ is the pull-back of the Kähler 2-form $K$ of $\mathcal{M}_K$. The observables $\mathcal{O}_{A_i}(z_i)$ are in one-to-one correspondence with the de Rahm cohomology classes $A_i \in H^p(\mathcal{M}_K)$ of the target manifold $\mathcal{M}_K$ and the homology cycles $H_i$ are defined as the subvarieties of $\mathcal{M}_{\{k\}}$ that contain all those instantons such that $X(z_i) \in [A_i]^*$. In this definition we have denoted by $[A_i]^* \subset \mathcal{M}_K$ the Poincaré dual of the cohomology class $A_i \in H^p(\mathcal{M}_K)$.

It is clear that topological field theories can been defined in completely geometrical terms. However, in every topological model, the right hand side of Eq. (7.1.3) should admit an independent definition as a functional integral in a suitable lagrangian quantum field theory, in order to be of physical interest. The basic feature of the classical lagrangian is that of possessing a very large group of gauge symmetries, the *topological symmetry*, which is the most general continuous deformation of the classical fields. The topological symmetry is treated through the standard techniques of BRST quantization and the instanton equations are imposed as a gauge-fixing. In this way, Eq. (7.1.3), rather than a definition, becomes a map between a *physical* and a *mathematical* problem, and this viewpoint is the main source of interest for topological field theories.

From the physical point of view, the basic properties of a topological field theory are encoded in the BRST algebra $\mathcal{B}$ and the anomaly of the ghost number.

The moduli space cohomology gives rise to the cohomology of the BRST operator $s$: the left hand side of Eq. (7.1.3) is the vacuum expectation value of the product of $n$ representatives $\mathcal{O}_i$ of non-trivial BRST cohomology classes

$$s\mathcal{O}_i = 0, \qquad \mathcal{O}_i \neq s\{\text{anything}\}. \tag{7.1.5}$$

Correspondingly, the right hand side of Eq. (7.1.3) can be expressed as an integral of a product of cocycles over the moduli space.

In full generality, the BRST algebra $\mathcal{B}$ can be decomposed as

$$\mathcal{B} = \mathcal{B}_{gauge\text{-}free} \oplus \mathcal{B}_{gauge\text{-}fixing}, \tag{7.1.6}$$

where $\mathcal{B}_{gauge\text{-}free} \subset \mathcal{B}$ is the subalgebra that contains only the physical fields and the ghosts (fields of non-negative ghost number), while $\mathcal{B}_{gauge\text{-}fixing}$ is the extension of $\mathcal{B}_{gauge\text{-}free}$ by means of anti-ghosts and Lagrange multipliers (or the corresponding gauge-fixing conditions), of non-positive ghost number. Usually, $\mathcal{B}_{gauge\text{-}fixing}$ is trivial, but this is not always the case.

Next we consider the mathematical meaning of the other basic aspect of the field theoretical approach, i.e. the anomaly of the ghost number. The left hand side of Eq. (7.1.3) can be non-zero only if

$$\sum_i d_i = \Delta U = \int \partial^\mu J_\mu^{(ghost)} d^D x, \tag{7.1.7}$$

where $J_\mu^{(ghost)}$ is the ghost number current, $\Delta U$ is its integrated anomaly and $d_i = gh[\mathcal{O}_i]$ is the ghost number of $\mathcal{O}_i$. The divergence of the ghost current has an interpretation as index density for some elliptic operator $\nabla$ which appears in the quantum action through the kinetic term of the ghost($C$)–anti-ghost($\overline{C}$) system:

$$S_{quantum} = \int (\ldots + \overline{C}\nabla C + \ldots). \tag{7.1.8}$$

We have

$$\text{index}_\nabla = \#\text{ zero modes of ghosts } - \#\text{ zero modes of anti-ghosts}. \tag{7.1.9}$$

On the other hand, the right hand side of Eq. (7.1.3) can be non-zero only if the sum of the codimensions of the homology cycles $H_i$ adds up to the total dimension of the moduli space:

$$\sum_i \text{codim}\, H_i \;=\; \dim \mathcal{M}_{moduli}. \tag{7.1.10}$$

In other words, the physical observables must reduce, after functional integration on the irrelevant degrees of freedom, to closed forms $\Omega_i$ of degree $d_i$ on the moduli space $\mathcal{M}_{mod}$ (the Poincaré duals of the cycles $H_i$) and their wedge product must be a top form. This means

$$\Delta U = \dim \mathcal{M}_{moduli}. \tag{7.1.11}$$

Such an equation is understood in the following way. In the background of an instanton, namely of a gauge-fixed configuration, the zero-modes of the topological ghosts correspond to the residual infinitesimal deformations that preserve the gauge condition. Their number is therefore the dimension of the tangent space to the parameter space of the instanton. The zero-modes of the anti-ghosts correspond, instead, to potential global obstructions to the integration of these infinitesimal deformations. The index $\Delta U$ is therefore named the formal dimension of the moduli space $\mathcal{M}_{mod}$. The true dimension of the moduli space is larger than or equal to its formal dimension,

$$\dim^{true} \mathcal{M}_{moduli} \geq \dim^{formal} \mathcal{M}_{moduli} = \Delta U, \tag{7.1.12}$$

depending on whether the potential obstructions become real obstructions or not.

There are therefore two different viewpoints on topological field theories that depend on whether one reads Eq. (7.1.3) from the left (= physics) to the right (= mathematics), or from the right to the left.

i) From the right to the left, equation (7.1.3) means:

*Given a well-defined mathematical problem (intersection theory on the moduli space of the instantons of some class of maps), find a quantum field theory that represents the intersection forms as physical amplitudes (averages of products of physical observables).*

This problem is solved by BRST-quantizing the most general continuous deformations of the classical fields (which sort of fields depending on the class of maps that is under consideration) and imposing the instantonic equations as a gauge-fixing.

ii) From the left to the right, (7.1.3) means:

*given a physically well-defined topological quantum field theory, find the mathematical problem (maps, instantons, intersection theory) that it represents.*

Of course, there is no general recipe for solving this second problem. In particular, which topological field theories one should consider is not evident a priori. However, the second viewpoint becomes particularly powerful and significant if one discovers the existence of the *topological twist.* This is an algorithm that, both in two and in four space–time dimensions, associates a well-defined gauge-fixed topological field theory to every N=2 supersymmetric field theory. The algorithm consists, after Wick rotation, of a redefinition of the Euclidean Lorentz group, of the BRST charge and of the ghost number. The new Lorentz group is a diagonal subgroup of the old Lorentz group and of the supersymmetry automorphism group, the new BRST charge is a linear combination of the old BRST charge with one of the supersymmetry charges and the new ghost number operator is the sum of the old one with the generator of an R symmetry of the N=2 theory. The whole procedure turns out to define a sophisticated projection of the old N=2 theory. Indeed, in the topologically twisted theory, the only non-vanishing correlators are those of operators that correspond to non-trivial cohomology classes of the new BRST charge. These are nothing else than a subset of all the correlators of the original N=2 theory. This means that all N=2 field theories admit a topological sector of correlation functions that are constant and represent intersection numbers in the moduli space of suitable instantons. In addition, as already anticipated, the twisting procedure, in two dimensions, is dichotomic, in the sense that each N=2 theory admits two *topological projections*, the A-model and the B-model, and, correspondingly, two classes of correlation functions with an interpretation as *intersection numbers on moduli spaces*.

In the case of most interest to us, namely that of an N=2 sigma model on a Calabi–Yau 3-fold, the two classes of topological correlators contain two old acquaintances of ours:

I) In the A-model we have, as a three-point topological correlator, the Yukawa-coupling:

$$W_{\alpha\beta\gamma} \tag{7.1.13}$$

of the **27** families, associated with Kähler class (1,1)-moduli.

II) In the B-model we have, as a three-point topological correlator the Yukawa-coupling,

$$W_{\mu\nu\rho} \tag{7.1.14}$$

of the $\overline{\mathbf{27}}$ anti-families associated with (2,1)-moduli.

Hence a very strong motivation for the development of topological field theories is, in the present context, the calculation of the Yukawa couplings and, eventually, through their knowledge, the derivation of the *special Kähler geometry* of moduli spaces relevant to string compactifications. The completion of such a programme will be actually done in Chapter 8 by means of mirror symmetry, yet it is only through a proper understanding of topological field theories that the very idea of mirror symmetry can be formulated. Indeed this is the interchange of the A-model and B-model for a mirror pair of manifolds.

Yukawa couplings, however, are just two instances of topological correlators and, in the present chapter, we have the more general aim of clarifying the role of the topological twist in our understanding of supersymmetric N=2 theories.

The chapter is organized as follows.

In section 7.2 we discuss the general geometric formalism for the treatment of the BRST symmetry. In particular we show how the rheonomic formulation of supersymmetric field theories is especially suited for their BRST-quantization. Furthermore we lay down the rules for writing the $\mathcal{B}_{gauge\text{-}free}$ BRST algebra of any topological field theory of interest to us.

In section 7.3 we consider the direct BRST construction of topological Yang–Mills theory in four dimensions. This model, which selects the Yang–Mills instantons as the relevant configurations, is not only the first theory of this type to have been introduced, but it is also *our paradigm* of a topological field theory.

In section 7.4, making a return to two dimensions, which is the focus of our interest, we perform the direct BRST-construction of topological sigma models, taking as classical action the pull-back of the Kähler 2-form

$$S_{class} = -\pi i \int_{\Sigma_g} X^\star K \qquad (7.1.15)$$

and as gauge-fixing of the topological symmetry the holomorphic instanton conditions

$$\partial_- X^i = \partial_+ X^{i^\star} = 0 \qquad (7.1.16)$$

At the end of the construction we realize that the quantum action is identical with the classical action of the N=2 sigma model if we reinterpret the ghosts and anti-ghosts as ordinary spin $\frac{1}{2}$ fermions. This is the suggestion that leads to the idea of topological twist.

In section 7.5 we derive the general algorithm for the topological twist of any N=2 supersymmetric theory. Roughly speaking this is based on

a) a redefinition of the spin of the particles by mixing the old spin with the R symmetry charges,

b) a reinterpretation of the fermions as ghosts and anti-ghosts by a redefinition of the ghost number that mixes the old ghost number with the R symmetry charges,

c) a reinterpretation of some combination of the N=2 supersymmetry charges as a BRST charge.

Since the above three steps can be performed in two different consistent combinations one has the emergence of two twisted theories: either the A-model or the B-model.

In section 7.6 the formalism of topological A and B twist is applied to the two-phase N=2 gauge theory discussed in Chapter 5 that interpolates between the N=2 sigmA model on Calabi–Yau $n$-folds and the N=2 Landau–Ginzburg model. This study leads to an identification of the instantons relevant in the two cases that are either holomorphic maps (for the A-twisted model) or constant maps (in the B-twisted model) from the world-sheet to the extrema of the potential. Following this logical development we arrive at the reconsideration of topological sigmA models in the A-case and at the definition of topological Landau–Ginzburg models in the B-case.

In section 7.7 we address the explicit calculation of correlators in topological field theories, starting with the topological sigmA model. In the A-twisted model we show that the topological coupling constants are the *moduli* of Kähler class deformations and that the ghost-anomaly expresses, for instantons of instanton number $k$ the dimensionality of the moduli space of holomorphic maps $\Sigma_g \longrightarrow \mathcal{M}_K$. In the case of Calabi–Yau $n$-folds this anomaly is independent from the instanton number $k$ and this has a far reaching consequence. The non-vanishing topological correlators pick up contributions from all instanton numbers and are trascendental functions of the moduli. This, in particular, occurs for the Yukawa couplings of (1,1)-forms on a Calabi–Yau 3-fold. There is a ring structure underlying the topological A-twisted sigma model and this is a *quantum deformation* of the classical Dolbeault cohomology ring. In the B-twisted model things are simpler: the topological coupling constants are the *moduli* of the complex structure deformations and the underlying ring structure is that of the *Hodge ring*, defined in

Chapter 5, which, for complex hypersurfaces is isomorphic to the *chiral ring* $\mathcal{R}_{\mathcal{W}}$ of the defining polynomial $\mathcal{W}(X)$.

In section 7.8 we consider the special case of topological two-dimensional field theories that are also conformal. These are related by twisting to N=2 superconformal models. Their analysis is particularly important because it introduces the notion of *flat coordinates* $t_i$ in the space of topological coupling constants. These are coordinates in which the topological metric $\eta_{ij}$, defined as the two-point correlation function, is constant and the many-point correlators are all obtained as multiple derivatives with respect to $t_i$ of a generating function $F(t)$. Such a generating function bears a deep relation with the homogeneous degree two function of special geometry $F(X)$. As several times remarked, an N=2 sigma model on a Calabi–Yau $n$-fold, or its associated N=2 Landau–Ginzburg model, will eventually define, at the critical point, an N=2 superconformal model with central charge

$$c = 3n \tag{7.1.17}$$

After twisting such a model becomes a topological conformal model with $U(1)$ anomaly:

$$\frac{1}{2}\Delta U = n(1-g) \tag{7.1.18}$$

where $g$ is the genus of the world-sheet. Hence the existence of *flat coordinates* is a general property of topological models and they are associated with the conformal point. The search for flat coordinates will be one of the main issues of Chapter 8, dealing with Picard–Fuchs equations.

In section 7.9 we consider the explicit calculation of correlators in the topological Landau–Ginzburg model. We discuss how they reflect the chiral ring structure $\mathcal{R}_{\mathcal{W}}$ of the superpotential $\mathcal{W}(X)$. Furthermore we show that a subclass of these correlators (those of the operators with integer $U(1)$ charges) is essentially equivalent to the correlators of the B-twisted sigma model on the Calabi–Yau $n$-fold defined as the vanishing locus of $\mathcal{W}(X)$. Some explicit examples of calculations are given.

In the last section 7.10 we reconsider the topological twists of the two-phase N=2 theory interpolating between Landau-Ginzburg and sigma models. We discuss the interplay between the phase structure and the definition of instantons in the two phases, concluding with a more careful analysis of some subtleties.

## 7.2 The Geometric Formulation of BRST Symmetry

In this section, as a preliminary step toward topological field theories, we discuss the general geometric formulation of BRST symmetry, independently introduced by Bonora, Pasti, Tonin [52] and by Baulieu [31]. As already stressed in the introductory section 7.1 this approach, which reproduces the standard results in the case of ordinary gauge theories, is particularly suited for the discussion of topological field theories and reveals their conceptual foundations.

We begin with the example of ordinary Yang–Mills theory and then we extend the formalism to the case of a general field theory, possessing an arbitrary closed algebra of local symmetries.

The catch of the method relies on the extension of ordinary space–time, whose coordinates we denote by $x^\mu$, by means of two additional Lorentz scalar anti-commuting coordinates $\theta$ and $\overline{\theta}$:

$$\theta^2 = \overline{\theta}^2 = \theta\overline{\theta} + \overline{\theta}\theta = 0 \tag{7.2.1}$$

which we associate to the unphysical ghost and anti-ghost directions. The resulting superspace carries a natural representation of BRST symmetry, in the same way as ordinary superspace (where the thetas are Lorentz spinors) carries a natural representation of supersymmetry. In both cases a coordinate free framework in terms of differential forms is both possible and most clarifying [76]. Hence we introduce the following exterior derivatives or coboundary operators:

$$\begin{aligned} d &= dx^\mu \frac{\partial}{\partial x^\mu} \\ s &= d\theta \frac{\partial}{\partial \theta} \\ \overline{s} &= d\overline{\theta} \frac{\partial}{\partial \overline{\theta}} \end{aligned} \tag{7.2.2}$$

satisfying the algebra

$$\begin{aligned} d^2 = s^2 = \overline{s}^2 &= 0 \\ s\overline{s} + \overline{s}s &= 0 \end{aligned} \tag{7.2.3}$$

and then we define the extended 1-form:

$$\widetilde{A}^\alpha = A^\alpha_\mu\, dx^\mu + C^\alpha\, d\theta + \overline{C}^\alpha\, d\overline{\theta} \stackrel{\text{def}}{=} A^\alpha + c^\alpha + \overline{c}^\alpha \tag{7.2.4}$$

where $A^\alpha_\mu(x)$ is the ordinary Yang–Mills gauge field, $C^\alpha(x)$, $\overline{C}^\alpha(x)$ are respectively the ghost and anti-ghost fields and $\alpha$ is the adjoint index of the gauge group $G$, whose generators we denote $T_\alpha$. In more condensed notation we can write

$$\widetilde{A} \stackrel{\text{def}}{=} \widetilde{A}^\alpha T_\alpha = A + c + \overline{c} \tag{7.2.5}$$

and we can introduce the extended curvature 2-form, via the following definition:

$$\widetilde{F} = \widetilde{d}\widetilde{A} + \frac{1}{2}\left[\widetilde{A}, \widetilde{A}\right] \tag{7.2.6}$$

The extended derivative $\widetilde{d}$ appearing above is given by

$$\widetilde{d} = d + s + \overline{s} \tag{7.2.7}$$

and, upon use of Eqs. (7.2.2), it immediately follows that it is nilpotent:

$$\tilde{d}^2 = 0 \tag{7.2.8}$$

As a consequence of Eq. (7.2.8) the extended curvature (7.2.6) satisfies a Bianchi identity completely analogous to the usual Bianchi identity satisfied by the ordinary Yang–Mills field strength, namely

$$\tilde{d}\tilde{F} + \left[\tilde{A}\,,\tilde{F}\right] = 0. \tag{7.2.9}$$

What we have done by means of the above definitions is to extend the standard differential forms of Yang–Mills theory to ghost–anti-ghost forms $\Omega^{(f,g,\overline{g})}$ which are characterized by a ghost–anti-ghost type label $(f,g,\overline{g})$. The number $f$ is a positive integer specifying the degree of $\Omega$ as a differential form in $x$-space; the second number $g$, which is also a positive integer, specifies its degree as a differential form in ghost space, namely it expresses the number of $d\theta$ differentials contained by $\Omega$; finally the third positive integer number $\overline{g}$, equals the degree of $\Omega$ as a differential form in anti-ghost space. In other words $\overline{g}$ is the number of $d\overline{\theta}$ differentials. The total ghost number is defined as follows:

$$N_g\left[\Omega^{(f,g,\overline{g})}\right] = g - \overline{g} \tag{7.2.10}$$

According to this scheme the extended curvature 2-form can be decomposed into six sectors:

$$\tilde{F} = F_{(2,0,0)} + F_{(1,1,0)} + F_{(1,0,1)} + F_{(0,2,0)} + F_{(0,0,2)} + F_{(0,1,1)} \tag{7.2.11}$$

and, from the definition (7.2.6), we obtain

$$\begin{aligned}
F_{(2,0,0)} &= dA + \frac{1}{2}\,[A\,,A] \\
F_{(1,1,0)} &= sA + dc + [A\,,c] \\
F_{(1,0,1)} &= \overline{s}A + d\overline{c} + [A\,,\overline{c}] \\
F_{(0,2,0)} &= sc + \frac{1}{2}\,[c\,,c]\,, \\
F_{(0,0,2)} &= \overline{s}\overline{c} + \frac{1}{2}\,[\overline{c}\,,\overline{c}] \\
F_{(0,1,1)} &= s\overline{c} + \overline{s}c + [\overline{c}\,,c]
\end{aligned} \tag{7.2.12}$$

The BRST (anti-BRST) transformations one deals with in the quantization of ordinary Yang–Mills theory can now be reinterpreted as ghost-like or anti-ghost like differentiations in ghost–anti-ghost superspace, provided one makes the assumption that the ghost–anti-ghost curvature 2-form $\tilde{F}$ is horizontal, namely has non-vanishing components only in the $(2,0,0)$-sector. Indeed if we identify the action of the BRST (anti-BRST) charge with the action of the Slavnov derivative $s$ (respectively anti-Slavnov derivative $\overline{s}$) and if we set

$$F_{(1,1,0)} = F_{(1,0,1)} = F_{(0,2,0)} = F_{(0,0,2)} = F_{(0,1,1)} = 0 \tag{7.2.13}$$

then, using Eqs. (7.2.12), we obtain:

$$\begin{aligned} s\,A^\alpha_\mu &= D_\mu\, c^\alpha \\ s\,c^\alpha &= -\frac{1}{2} f^\alpha_{\beta\gamma}\, c^\beta\, c^\gamma \\ \overline{s}\,A^\alpha_\mu &= D_\mu\, \overline{c}^\alpha \\ \overline{s}\,\overline{c}^\alpha &= -\frac{1}{2} f^\alpha_{\beta\gamma}\, \overline{c}^\beta\, \overline{c}^\gamma \\ s\,\overline{c}^\alpha &= b^\alpha \\ \overline{s}\,c^\alpha &= -\,b^\alpha - \frac{1}{2} f^\alpha_{\beta\gamma}\, \overline{c}^\beta\, c^\gamma \\ s\,b^\alpha &= 0 \end{aligned} \tag{7.2.14}$$

where $f^\alpha_{\beta\gamma}$ denote the structure constants of the gauge group, $D_\mu \;=\; \partial_\mu + [\,A_\mu\,,\;\,]$ denotes the ordinary covariant derivative and the auxiliary field $b^\alpha$ is simply the name given to the Slavnov derivative of the anti-ghost, introduced to resolve the ambiguity inherent to the condition $F_{(0,1,1)} = 0$, which fixes only the value of the sum $s\,\overline{c}^\alpha + \overline{s}\,c^\alpha$.

Equation (7.2.14) encode the complete BRST–anti-BRST algebra of standard Yang–Mills theory. The statement that this is a closed algebra is nothing but the statement that the horizontality conditions (7.2.13) are consistent with the Bianchi identity (7.2.9), as can be checked by direct substitution of (7.2.13) into (7.2.9). The specific form of the Slavnov variation of the physical field ( namely $s\,A_\mu$) reveals that the classical symmetry one is quantizing is the standard gauge transformation:

$$\delta_{class}\, A_\mu \;=\; D_\mu\, \lambda \tag{7.2.15}$$

Indeed the BRST variation of any physical field is given by its classical symmetry transformation upon replacement of the local parameter with ghost fields.

We now extend this formalism to the case of a generic local symmetry.

We consider a classical theory containing a set $\{\,\phi^i(x)\,\}$ of fields and described by a lagrangian $\mathcal{L}(\,\phi\,)$. The lagrangian is invariant (up to total divergences) against certain gauge transformations:

$$\delta_{class}\, \phi^i(x) \;=\; R^i_\alpha\,(\,\phi\,,\,\partial\,)\;\varepsilon^\alpha(x) \tag{7.2.16}$$

where $R^i_\alpha\,(\,\phi\,,\,\partial\,)$ is an operator containing derivatives and fields. For instance, for a Yang–Mills gauge field we have

$$\delta_{class}\, A^\alpha_\mu(x) \;=\; R^\alpha_{\mu\,\beta}\,(\,A\,,\,\partial\,)\;\varepsilon^\beta(x) \;=\; \left(\,\delta^\alpha_\beta\,\partial_\mu \,+\, f^\alpha_{\gamma\beta}\,A^\gamma_\mu(x)\,\right)\,\varepsilon^\beta(x) \tag{7.2.17}$$

We define also the first functional derivative of the symmetry variation:

$$R^i_{\alpha\,,\,j} \;\stackrel{\text{def}}{=}\; \frac{\delta}{\delta\,\phi^j(x)}\, R^i_\alpha \tag{7.2.18}$$

which in the Yang–Mills case is proportional to the structure constants of the gauge group:

$$R^{\alpha\,\nu}_{\mu\beta\,,\,\gamma} \;=\; \frac{\delta}{\delta\,A^\gamma_\nu}\, R^\alpha_{\mu\,\beta} \;=\; \delta^\nu_\mu\, f^\alpha_{\gamma\beta} \;=\; \text{constant in this case} \tag{7.2.19}$$

so that all further functional derivatives vanish. However, in the general case, $R^i_{\alpha,j}$ is not necessarily a constant and higher functional derivatives of the variation do indeed exist. Keeping this fact in mind we write the closure condition of the classical algebra of gauge transformations:

$$[\,\delta_{\varepsilon_1}\,,\,\delta_{\varepsilon_2}\,]\,\phi^i = \left(R^i_{\alpha\,,j}\,R^j_\beta - R^i_{\beta\,,j}\,R^j_\alpha\right)\,\varepsilon_1^\beta\,\varepsilon_2^\alpha \;=\; R^i_\gamma\,f^\gamma_{\alpha\beta}\,\varepsilon_1^\beta\,\varepsilon_2^\alpha \tag{7.2.20}$$

which yields

$$\left(R^i_{\alpha\,,j}\,R^j_\beta - R^i_{\beta\,,j}\,R^j_\alpha\right) \;=\; -\,R^i_\gamma\,f^\gamma_{\alpha\beta} \tag{7.2.21}$$

the symbol $f^\gamma_{\alpha\beta}$ denoting now structure functions, rather than structure constants. The Jacobi identity

$$[\,\delta_{\varepsilon_1}\,,\,[\,\delta_{\varepsilon_2}\,,\,\delta_{\varepsilon_3}\,]\,]\;\phi + \text{cyclic perm.} \;=\; 0 \tag{7.2.22}$$

yields the relation

$$f^\mu_{\delta\gamma}\,f^\gamma_{\alpha\beta} + f^\mu_{\alpha\gamma}\,f^\gamma_{\beta\delta} + f^\mu_{\beta\gamma}\,f^\gamma_{\delta\alpha} - f^\gamma_{\alpha\beta\,,j}\,R^j_\alpha - f^\mu_{\delta\alpha\,,j}\,R^j_\beta \tag{7.2.23}$$

The above classical gauge algebra can be quantized by replacing each of the classical parameters $\varepsilon^\alpha(x)$ with a pair of a ghost $c^\alpha$ and an anti-ghost $\overline{c}^\alpha$, and then, by defining the action of the Slavnov (anti-Slavnov) derivatives as follows:

$$\begin{aligned}
s\,\phi^i &= R^i_\alpha\,c^\alpha \\
s\,c^\alpha &= -\frac{1}{2}f^\alpha_{\beta\gamma}\,c^\beta\,c^\gamma \\
s\,\overline{c}^\alpha &= b^\alpha \\
s\,b^\alpha &= 0 \\
\overline{s}\phi^i &= R^i_\alpha\,\overline{c}^\alpha \\
\overline{s}\,\overline{c}^\alpha &= -\frac{1}{2}f^\alpha_{\beta\gamma}\,\overline{c}^\beta\,\overline{c}^\gamma \\
\overline{s}\,c^\alpha &= -\,b^\alpha - \frac{1}{2}f^\alpha_{\beta\gamma}\,\overline{c}^\beta\,c^\gamma \\
\overline{s}b^\alpha &= -\,f^\alpha_{\beta\gamma}\,\overline{c}^\beta\,b^\gamma - \frac{1}{2}\,R^i_\delta\,f^\alpha_{\beta\gamma\,,i}\,c^\delta\,\overline{c}^\beta\,c^\gamma
\end{aligned} \tag{7.2.24}$$

Equations (7.2.21) and (7.2.23) guarantee that the algebraic relations (7.2.3) of nilpotency and anti-commutativity are satisfied. Indeed Eqs. (7.2.24) encode the correct generalization of the BRST algebra (7.2.14) to the case where the classical gauge algebra closes with structure functions rather than with structure constants.

It is instructive for our subsequent developments to discuss the appearance of structure functions in the BRST algebra from the point of view of the horizontality conditions (7.2.13). What one needs to realize is that the concept underlying the construction of the BRST algebra associated with any classical symmetry is that of the quantum-extension of the rheonomic parametrization of classical curvatures. In the case of Yang–Mills theory, classical rheonomy simply corresponds to horizontality, so that Eqs. (7.2) lead to the correct BRST algebra: in more complicated theories classical rheonomy is more involved and also its quantum version differs from the pure horizontality conditions (7.2.13). To explain these statements we recall that in a geometric field theory one always starts from a set of differential $p$-forms $\omega^p$ gauging some (super)-Lie algebra or free differential algebra. In the case of a (super)-Lie algebra all the gauge fields are just 1-forms. Given the (super)-algebra structure constants we easily construct the corresponding curvatures. For Yang–Mills theory we have $F = dA + \frac{1}{2}[A, A]$, while for Einstein gravity we write

$$\begin{aligned} R^a &= \mathcal{D}V^a = dV^a - \omega^a{}_b \wedge V^b \\ R^{ab} &= d\omega^{ab} - \omega^a{}_c \wedge \omega^{cb} \end{aligned} \tag{7.2.25}$$

$V^a$ $(a = 0, 1, 2, 3)$ being the vielbein 1-form and $\omega^{ab} = -\omega^{ba}$ the spin connection. If we consider a supersymmetric example, like the case of N=2 simple supergravity [8], we have to further extend the set of 1-forms by introducing also the doublet of fermionic (commuting) spinor-valued gravitino 1-forms $\psi_A = \psi_{A\mu}\, dx^\mu$ and the graviphoton $A = A_\mu\, dx^\mu$. $N = 2$ matter-coupled supergravity was considered in Chapter 3. For the sake of the present discussion we just focus on simple $N = 2$ supergravity where there is just the graviphoton and no additional vector multiplets. Furthermore for present convenience we use Majorana 4 component gravitino 1-form $C\overline{\psi}_A = \psi_A$ so that we do not distinguish between upper and lower $A, B$ indices as we did in Chapter 3. In these notations and for simple supergravity the curvature definitions are given by (see [76])

$$\begin{aligned} R^a &= \mathcal{D}V^a - \frac{i}{2}\overline{\psi}_A \wedge \gamma^a \psi_A = dV^a - \omega^a{}_b \wedge V^b - \frac{i}{2}\overline{\psi}_A \wedge \gamma^a \psi_A \\ R^{ab} &= d\omega^{ab} - \omega^a{}_c \wedge \omega^{cb} \\ \rho_A &= \mathcal{D}\psi_A = d\psi_A - \frac{1}{2}\omega^{ab} \wedge \sigma_{ab}\psi_A \\ R^{\otimes} &= F + \epsilon_{AB}\overline{\psi}_A \wedge \psi_B \end{aligned} \tag{7.2.26}$$

where $F \equiv dA$ is the curl of the graviphoton, $\mathcal{D}$ denotes the Lorentz-covariant exterior derivative and $\sigma^{ab} \equiv \frac{1}{4}[\gamma^a, \gamma^b]$; finally $\epsilon_{AB}$ is the completely anti-symmetric tensor with two indices. In order to obtain the classical symmetry transformation rules, which are uniformly interpreted as Lie derivatives in all geometrical theories, one must supply a *rheonomic parametrization of the classical curvatures*. By this one means an explicit expression for the 2-form curvature components in the anholonomic basis provided by the (super)-Vielbein (Vielbein $V^a$ plus gravitino $\psi_A$, in the supersymmetric case) that satisfies the following three requirements:

*i) It is consistent with the Bianchi identity.*

*ii) It is such that the outer components (those along all other super-Vielbein components but $V^c \wedge V^d$) are expressed as functions of the inner ones (those along $V^c \wedge V^d$).*

*iii) The outer components are generic apart from certain $x$-space differential constraints that may follow from Bianchi identities (when these constraints occur they are nothing but the field equations of the classical field theory).*

In the above examples the *classical rheonomic parametrizations* are the following ones:
*Yang–Mills theory*

$$F = F_{ab}\, V^a \wedge V^b \tag{7.2.27}$$

*Einstein gravity*

$$\begin{aligned} R^a &= 0 \\ R^{ab} &= R^{ab}_{cd}\, V^c \wedge V^d \end{aligned} \tag{7.2.28}$$

*N=2 simple supergravity*

$$\begin{aligned} R^a &= 0 \\ R^{ab} &= R^{ab}{}_{cd} V^c \wedge V^d + \overline{\theta}^{ab}_{A|c}\psi_A \wedge V^c - \frac{1}{2}\overline{\psi}_A \wedge \mathcal{F}^{ab}\psi_B \epsilon_{AB} \\ \rho_A &= \rho^{ab}_A V_a \wedge V_b + \frac{1}{2} i\gamma^a \mathcal{F}_{ab}\psi_B \wedge V^b \epsilon_{AB} \\ R^{\otimes} &= F_{ab} V^a \wedge V^b \end{aligned} \tag{7.2.29}$$

where $\mathcal{F}^{ab} \equiv F^{ab} + \frac{i}{2}\gamma_5 F_{cd}\varepsilon^{abcd}$ and $\overline{\theta}^{ab|c}_A = 2i\overline{\rho}^{c[a}_A\gamma^{b]} - i\overline{\rho}^{ab}_A\gamma^c$, the square brakets denoting anti-symmetrization. To quantize the theory and to obtain the BRST algebra one first extends each gauge field 1-form to a *ghost–anti-ghost form*. So, in the above three examples one performs the following extensions:
*Yang–Mills theory*

$$\begin{aligned} \tilde{A}^\alpha &= A^\alpha + c^\alpha + \overline{c}^\alpha \\ \tilde{V}^a &= V^a \end{aligned} \tag{7.2.30}$$

*Einstein gravity*

$$\begin{aligned} \tilde{V}^a &= V^a + \varepsilon^a + \overline{\varepsilon}^a \\ \tilde{\omega}^{ab} &= \omega^{ab} + \varepsilon^{ab} + \overline{\varepsilon}^{ab} \end{aligned} \tag{7.2.31}$$

*N=2 simple supergravity*

$$\begin{aligned} \tilde{V}^a &= V^a + \varepsilon^a + \overline{\varepsilon}^a \\ \tilde{\omega}^{ab} &= \omega^{ab} + \varepsilon^{ab} + \overline{\varepsilon}^{ab} \\ \tilde{\psi}_A &= \psi_A + c_A + \overline{c}_A \\ \tilde{A} &= A + c + \overline{c} \end{aligned} \tag{7.2.32}$$

where $c^\alpha$, $\varepsilon^a$, $\varepsilon^{ab}$, $c_A$ and $c$ (respectively $\overline{c}^\alpha$, $\overline{\varepsilon}^a$, $\overline{\varepsilon}^{ab}$, $\overline{c}_A$ and $\overline{c}$ ) are the ghosts (respectively anti-ghosts) of Yang–Mills transformations, diffeomorphisms, Lorentz rotations, supersymmetries and Maxwell transformations, respectively. It is important to note that in the case of Yang–Mills theory the vielbein $V^a$ does not contain any ghost part and remains equal to the classical vielbein. This is so because, in this case, diffeomorphisms are not a local symmetry and $V^a$ is not treated as a gauge field. However, in all cases, after extension to ghost–anti-ghost forms, the uniform prescription to obtain the correct BRST–anti-BRST algebra is that of promoting the classical rheonomic parametrization to a quantum one. By definition the *quantum rheonomic parametrization* is obtained from the *classical rheonomic* parametrization by replacing the classical cotangent basis of differential forms with the corresponding extended one. In this way the components of the extended curvatures in the extended basis are the same as the components of the classical curvatures in the classical basis. For instance in the above three examples, we have

*Yang–Mills theory*

$$\widetilde{F} = F_{ab}\, \widetilde{V}^a \wedge \widetilde{V}^b = F_{ab}\, V^a \wedge V^b \tag{7.2.33}$$

*Einstein gravity*

$$\begin{aligned} \widetilde{R}^a &= 0 \\ \widetilde{R}^{ab} &= R^{ab}_{cd}\, \widetilde{V}^c \wedge \widetilde{V}^d \end{aligned} \tag{7.2.34}$$

this time the vielbein being quantum-extended

*N=2 supergravity*

$$\begin{aligned} \widetilde{R}^a &= 0 \\ \widetilde{R}^{ab} &= R^{ab}{}_{cd}\widetilde{V}^c \wedge \widetilde{V}^d + \overline{\theta}^{ab}_{A|c}\overline{\psi}_A \wedge \widetilde{V}^c - \frac{1}{2}\overline{\widetilde{\psi}}_A \wedge \mathcal{F}^{ab}\widetilde{\psi}_B\epsilon_{AB} \\ \widetilde{\rho}_A &= \rho^{ab}_A\widetilde{V}_a \wedge \widetilde{V}_b + \frac{1}{2}i\gamma^a\mathcal{F}_{ab}\widetilde{\psi}_B \wedge \widetilde{V}^b\epsilon_{AB} \\ \widetilde{R}^{\otimes} &= F_{ab}\widetilde{V}^a \wedge \widetilde{V}^b \end{aligned} \tag{7.2.35}$$

this time the gravitino also being quantum-extended. As one sees from Eq. (7.2.33), one retrieves the horizontality conditions (7.2.13) of the quantum Yang–Mills curvature. However, in the other two examples, the ghost parts of the curvatures are not zero; rather, they can be expressed in terms of the $x$-space ones via the quantum rheonomic conditions. Using these conditions in the same way as we used the horizontality conditions in the Yang–Mills case, we obtain the action of the Slavnov (respectively anti-Slavnov) derivative $s$ (respectively $\overline{s}$) on all the fields. The result yields the transformation rules (7.2.24), where the structure functions $f^{\alpha}_{\beta\gamma}$ are the structure constants of the corresponding (super)-Lie algebra ($C^A_{BC}$) modified by the addition of terms proportional to the curvature components ($R^A_{BC}$), rheonomically expressed in terms of the inner subset of components.

These remarks are very important in order to appreciate the relation between the BRST algebra associated with the topological symmetry of a given physical system and the BRST algebra associated with the ordinary symmetries of the same physical system. This will be seen in full detail in the next section (7.3): before coming to topological symmetries we want to complete our general survey of BRST symmetry by discussing the *gauge-fixing* terms and the quantum lagrangian.

A quantum lagrangian which is both BRST and anti-BRST invariant is of the form

$$\mathcal{L}_{quantum} = \mathcal{L}_{class}(\phi) + \frac{1}{2} s\,\bar{s}\,P \tag{7.2.36}$$

where $P$ is a local polynomial in $(\phi, c, \bar{c}, b)$, namely in the physical, ghost, anti-ghost and auxiliary fields. This expression can be traced back to a gauge fixing in the following way. Suppose we want to *gauge-fix* the classical lagrangian by means of a term

$$\mathcal{L}_{gauge\text{-}fixing} = \mathcal{G}^\alpha\, \mathcal{G}^\alpha \tag{7.2.37}$$

where the gauge function is given by

$$\mathcal{G}^\alpha = \mathcal{O}_{ij}\, R^i_\alpha\, \phi^j \tag{7.2.38}$$

Then we can write

$$P = \phi^i\, \mathcal{O}_{ij}\, \phi^j + \bar{c}^\alpha\, c^\alpha \tag{7.2.39}$$

and we obtain

$$s\,\bar{s}\,P = 2 R^i_\beta\, c^\beta\, \mathcal{O}_{ij}\, R^j_\alpha\, c^\alpha + \ldots + 2\phi^i\, \mathcal{O}_{ij}\, R^j_\alpha\, b^\alpha + \ldots - b^\alpha\, b^\alpha \tag{7.2.40}$$

As one sees, from the elimination of the auxiliary fields, which is possible because their field equation is algebraic, one obtains

$$b^\alpha = \mathcal{G}^\alpha \tag{7.2.41}$$

and, upon substitution into the quantum lagrangian, one retrieves the quadratic gauge-fixing term (7.2.37).

Usually, insisting on BRST–anti-BRST invariance of the quantum action is an excessive requirement, since BRST invariance alone suffices to yield all of the Ward identity on the correlation functions. Hence, relaxing anti-BRST invariance, one can write the quantum lagrangian in the simpler form

$$\mathcal{L}_{quantum} = \mathcal{L}_{class}(\phi) + s\left(\bar{c}^\alpha\, \mathcal{G}^\alpha + \frac{1}{2}\bar{c}^\alpha\, b^\alpha\right) \tag{7.2.42}$$

where the quantity in brackets, which appears under Slavnov differentiation, is called the gauge fermion $\Psi_{gauge} = \left(\bar{c}^\alpha\, \mathcal{G}^\alpha + \frac{1}{2}\bar{c}^\alpha\, b^\alpha\right)$. Both in Eq. (7.2.36) and in Eq. (7.2.42) the quantum lagrangian is BRST-closed, $s\,\mathcal{L}_{quantum} = 0$ (respectively anti-BRST closed), because of the assumed classical symmetry of the classical lagrangian and of the BRST-exactness of the added terms. From Eq. (7.2.42), eliminating the auxiliary fields after $s$-differentiation, one obtains the second order form of the quantum lagrangian:

$$\mathcal{L}_{quantum} = \mathcal{L}_{class}(\phi) - \frac{1}{2}\mathcal{G}^\alpha\, \mathcal{G}^\alpha - \bar{c}^\alpha \frac{\delta\, \mathcal{G}^\alpha}{\delta\, \phi^i} R^i_\alpha\, c^\alpha \tag{7.2.43}$$

which puts in evidence the original interpretation of the Faddeev–Popov ghosts as a way of realizing the functional Faddeev–Popov jacobian due to the change of functional integration variables. In the ordinary Yang–Mills case one retrieves the standard well-known formulae setting $\mathcal{G}^\alpha = \partial^\mu A_\mu$. With this choice we find the usual action:

$$\mathcal{L}^{YM}_{quantum} = -\frac{1}{4} F^\alpha_{\mu\nu} F^\alpha_{\mu\nu} + \partial^\mu \overline{c}^\alpha \mathcal{D}_\mu c^\alpha - \frac{1}{2} (\partial^\mu A_\mu)^2 \qquad (7.2.44)$$

It is now time to specialize the above general BRST framework to the case of topolgical symmetries.

## 7.3 Topological Yang–Mills Theories

Topological Yang–Mills theory in four dimensions, sometimes also named Donaldson theory because of its relation with Donaldson polynomials [127], is the first example of a topological field theory that was discovered. Witten introduced it into the literature in 1988 [259] and, by means of this example, he gave the paradigms of a topological field theory. However his way of deriving the model was based on a procedure, later named *topological twist* [137, 259, 8, 7], that provides an already *gauged-fixed* topological lagrangian, starting from an N=2 supersymmetric lagrangian, in this case that of N=2 super-Yang–Mills theory. It was only through the work of Baulieu and Singer [33] that the general conceptual basis of topological field theories was clarified in the general context of BRST quantization.

We follow their point of view, leaving the discussion of the topological twist to a later stage of our exposition, where the relation between topological field theories and N=2 models will be addressed.

In the case of topological Yang–Mills theory the classical action is the integral over the four-dimensional base manifold $M_4$ of the first Chern class of a principal bundle:

$$P \longrightarrow M_4 \qquad (7.3.1)$$

having the gauge group $G_{gauge}$ as standard fibre. Explicitly one writes

$$S_{class} = \int_{M_4} tr\,(F \wedge F) \qquad (7.3.2)$$

where $F = F^i t_i = F^i_{\mu\nu} t_i dx^\mu \wedge dx^\nu$ is the field strength of the connection $A = A^i_\mu t_i dx^\mu$:

$$F = dA + \frac{1}{2}[A, A] \qquad (7.3.3)$$

and $t_i$ are the generators of the gauge group, satisfying the Lie algebra:

$$[t_i, t_j] = f^k_{ij} t_k \qquad (7.3.4)$$

In Eq. (7.3.3) $[A, A]$ denotes the graded commutator, which, for 1-forms, corresponds to the anti-commutator. In general, if $A$, $B$ are two matrix-valued forms of gradings $a$ and $b$, respectively, then their graded commutator is

$$[A, B] = A \wedge B - (-)^{ab} B \wedge A \tag{7.3.5}$$

The action (7.3.2) is invariant against a very large group of symmetries. Indeed, since $S_{class}$ is nothing but the integral of a cohomology class, its value expresses a winding number and it does not depend on the choice of a representative within each class, but only on the class itself. Arbitrary, continuous deformations of the gauge connection leave $S_{class}$ invariant. Taking into account also the ordinary gauge symmetry, the most general infinitesimal variation of $A$ that leaves the classical action (7.3.2) invariant can be written as follows:

$$\delta A = D\varepsilon + u \tag{7.3.6}$$

where $\varepsilon = \varepsilon^i(x)\, t_i$ is a Lie algebra-valued 0-form and $u = u^i_\mu\, t_i\, dx^\mu$ is a Lie algebra valued 1-form. By $D$ we have denoted the covariant derivative $D\varepsilon = d\varepsilon + [A, \varepsilon]$. Both $\varepsilon$ and $u$ are regarded as infinitesimal. Under these transformations we obtain

$$\delta F = -[\varepsilon, F] + Du \tag{7.3.7}$$

and indeed, using Stokes' lemma, we verify that

$$\delta\, S_{class} = 2 \int_{\partial M_4} tr\,(u \wedge F) = 0 \tag{7.3.8}$$

if $u$ vanishes on the boundary $\partial M_4$.

Topological Yang–Mills theory emerges from the BRST quantization of the above symmetries. Following the framework discussed in the previous section we introduce a BRST–anti-BRST complex formed by the fields listed in Table 7.1. This complex is derived in the following way. By looking at Eq. (7.3.6) we see that the action of the Slavnov operator on the physical gauge connection is necessarily of the following form

$$s\, A = -(Dc + \psi) \tag{7.3.9}$$

or, equivalently,

$$s\, A^i_\mu = D_\mu\, c^i + \psi^i_\mu \tag{7.3.10}$$

where the 0-form $c = c^i\, t_i$ and the 1-form $\psi = \psi^i_\mu\, dx^\mu\, tz_i$, both of ghost number $gh = 1$, are the ghosts of the ordinary gauge transformations ($\varepsilon$ parameter) and of the topological symmetry ($u$-parameter), respectively. From Eq. (7.3.6) it is also clear that the parameter of the topological transformation is defined only up to a gauge transformation $u \longrightarrow u + D\lambda$: hence the topological ghost $\psi$ is, by itself, a gauge field and requires its own ghost. This will be a 0-form $\phi = \phi^i\, t_i$ of ghost number $gh = 2$ and, taking into

Table 7.1: *Field content of topological YM theory*

| *Formdegree* / *ghostnumber* | 0 | 1 | 2 |
|---|---|---|---|
| −2 | $\overline{\phi}$ | | |
| −1 | $\overline{c}\ ,\ \overline{\eta}$ | $\overline{\psi}_\mu$ | $\overline{\chi}_{\mu\nu}$ |
| 0 | $b,\ L$ | $A_\mu\ ,\ T_\mu$ | $F_{\mu\nu}\ ,\ B_{\mu\nu}$ |
| 1 | $c\ ,\ \eta$ | $\psi_\mu$ | $\chi_{\mu\nu}$ |
| 2 | $\phi$ | | |

account these ghost number assignments, the most general ansatz for the action of the Slavnov operator on the ghost fields is the following:

$$\begin{aligned} s\,c &= \phi + \alpha_1\,[c,c] \\ s\,\psi &= \alpha_2\,D\phi + \alpha_3\,[c,\psi] \\ s\,\phi &= \alpha_4\,[c,\phi] \end{aligned} \tag{7.3.11}$$

where $\alpha_1$, $\alpha_2$, $\alpha_3$, $\alpha_4$ are C-number coefficients. Enforcing the nilpotency of the Slavnov operator $s^2 = 0$, we obtain :

$$\alpha_1 = -\frac{1}{2} \qquad \alpha_2 = 1 \qquad \alpha_3 = -1 \qquad \alpha_4 = -1 \tag{7.3.12}$$

so that we can write the *gauge-free* BRST topological Yang–Mills algebra in the following way:

$$\begin{aligned} s\,A &= -\,(Dc + \psi) \\ s\,F &= D\psi - [c,F] \\ s\,c &= \phi - \frac{1}{2}\,[c,c] \\ s\,\psi &= D\psi - [c,\psi] \\ s\,\phi &= -[c,\phi] \end{aligned} \tag{7.3.13}$$

In a completely similar way we can also write the *gauge-free* anti-BRST topological algebra:

$$\begin{aligned} \bar{s}\,A &= -\left(D\bar{c} + \overline{\psi}\right) \\ \bar{s}\,F &= D\overline{\psi} - [\bar{c},F] \\ \bar{s}\,\bar{c} &= \overline{\phi} - \frac{1}{2}\,[\bar{c},\bar{c}] \\ \bar{s}\,\psi &= D\overline{\psi} - \left[\bar{c},\overline{\psi}\right] \\ \bar{s}\,\overline{\phi} &= -\left[\bar{c},\overline{\phi}\right] \end{aligned} \tag{7.3.14}$$

These equations explain the role of most of the fields appearing in Table 7.1, but we still have to illustrate the interpretation of the fields $\eta$, $\overline{\eta}$, $b$, $L$, $T_\mu$, $\Upsilon_{\mu\nu}$, $\lambda_{\mu\nu}$ and $B_{\mu\nu}$. These latter are auxiliary fields and appear in the BRST variation of the anti-ghosts or in the anti-BRST variation of the ghosts. Indeed we can complete the closure of the BRST–anti-BRST topological algebra by adjoining to Eqs. (7.3.13) and (7.3.14) the following

$$\begin{aligned} s\,\bar{c} &= b \\ s\,\overline{\psi} &= T \\ s\,\overline{\phi} &= \eta \end{aligned}$$

$$\begin{aligned}
\overline{s}\,c &= L - b - [\overline{c}\,,\,c] \\
\overline{s}\,L &= -\overline{\eta} - [\overline{c}\,,\,L] - \left[c\,,\,\overline{\phi}\right] \\
\overline{s}\,\psi &= -T - \left[c\,,\,\overline{\psi}\right] - [\overline{c}\,,\,\psi] - DL \\
\overline{s}\,\phi &= \eta \\
s\,L &= -\eta - [\overline{c}\,,\,\phi] - [c\,,\,L]
\end{aligned} \tag{7.3.15}$$

and defining

$$\begin{aligned}
\overline{\chi} &= dx^\mu \wedge dx^\nu\,\overline{\chi}_{\mu\nu} = D\overline{\psi} \\
\chi &= dx^\mu \wedge dx^\nu\,\chi_{\mu\nu} = D\,\psi \\
B &= dx^\mu \wedge dx^\nu\,B_{\mu\nu} = -DT - \left[Dc\,,\,\overline{\psi}\right] - \left[\psi\,,\,\overline{\psi}\right]
\end{aligned} \tag{7.3.16}$$

so that

$$s\overline{\chi} = B \tag{7.3.17}$$

is an identity.

In the quantum action, $b$ will be the lagrangian multiplier for the gauge-fixing of the ordinary gauge transformations, while $T_\mu$ (or rather its functional $B_{\mu\nu}$) will be the lagrangian multiplier associated with the gauge-fixing of the topological symmetry. Finally $\overline{\eta}$ will be utilized to gauge fix the gauge invariance of the topological ghost $\psi_\mu$. Before addressing the construction of the quantum lagrangian, which involves the breaking of anti-BRST symmetry in favour of BRST invariance, exactly as it happens for ordinary gauge theories, it is worthwhile to appreciate the geometrical meaning of the full BRST–anti-BRST topological algebra encoded in Eqs. (7.3.13) , (7.3.14) , (7.3.15) and (7.3.16). We do this by comparing the above algebra with Eqs. (7.2.12) and (7.2.9). From the first comparison we obtain the following identification of the topological ghosts–anti-ghosts with the components of the extended curvature 2-form $\widetilde{F}$:

$$\begin{aligned}
\psi &= -F_{(1,1,0)} \\
\overline{\psi} &= -F_{(1,0,1)} \\
\phi &= -F_{(0,2,0)} \\
\overline{\phi} &= -F_{(0,0,2)} \\
L &= F_{(0,1,1)}
\end{aligned} \tag{7.3.18}$$

while from the second comparison we conclude that the transformation rules (7.3.15) are nothing else than the yield of the extended Bianchi identity (7.2.9), the auxiliary fields being the names given to the indeterminacy in the resolution of these identities. In other words, to obtain the topological BRST–anti-BRST algebra we have simply *lifted* the quantum rheonomic constraints of the extended curvature (see Eqs. (7.2.13)). In the ordinary gauge theory these constraints yield the BRST–anti-BRST algebra associated with the usual gauge symmetry; in the topological theory the extended curvature is left unconstrained and its exterior sectors provide the ghosts–anti-ghosts for the larger topological symmetries. This is not an accident of topological Yang–Mills theory, rather it is the rule in all topological field theories. We are now ready to discuss the quantum action. This has the following structure:

$$S_{quantum} = S_{class} + \int_{M_4} s\left(\Psi_{topol} + \Psi_{gauge} + \Psi_{ghost}\right) \tag{7.3.19}$$

where the gauge fermion is the sum of a gauge fermion fixing the topological symmetry ($\Psi_{topol}$), plus one fixing the ordinary gauge symmetry ($\Psi_{gauge}$), plus a last one fixing the gauge of the ghosts. The topological gauge-fixing must break the invariance under continuous deformations of the connection still preserving ordinary gauge invariance. A convenient gauge condition that satisfies this requirement is provided by enforcing self-duality of the field strength (the instanton condition). Define

$$\mathcal{G}^{\pm}_{\mu\nu} = F_{\mu\nu} \pm \frac{1}{2}\varepsilon_{\mu\nu\rho\sigma}F^{\rho\sigma} \tag{7.3.20}$$

The topological gauge-fixing can be chosen to be

$$\mathcal{G}^{+}_{\mu\nu} = 0 \tag{7.3.21}$$

This makes sense under the assumption that, within the set of all gauge connections characterized by the same first Chern class (value of the classical action), there is always at least one that has a self-dual field strength. Hence we set

$$\begin{aligned}
\Psi_{topol} &= tr\left\{\overline{\chi}_{\rho\sigma}\left(\mathcal{G}^{+}_{\mu\nu} + B_{\mu\nu}\right)g^{\rho\mu}g^{\sigma\nu}\right\} \\
\Psi_{gauge} &= tr\left\{\overline{c}\left(\partial_\rho A_\mu g^{\rho\mu} + b\right)\right\} \\
\Psi_{ghost} &= tr\left\{\overline{\phi}\partial_\rho \psi_\mu g^{\rho\mu}\right\}
\end{aligned} \tag{7.3.22}$$

fixing the ordinary gauge symmetry of the physical gauge-boson $A_\mu$ and of the topological ghost $\psi_\mu$ by means of the Lorentz gauge. The partition-function is given by the path integral

$$Z = \int [d\varphi]\exp\left[-S_{quantum}\right] \tag{7.3.23}$$

where the BRST-invariant integration measure is the product

$$[d\varphi] = [dA_\mu]\,[dc]\,[d\overline{c}]\,[d\psi_\mu]\,[d\overline{\chi}_{\mu\nu}]\,[d\phi]\,[d\overline{\phi}]\,[dB_{\mu\nu}]\,[db]\,[d\overline{\eta}] \tag{7.3.24}$$

all the fields being in the adjoint representation of the gauge group. As one sees, the BRST-invariant partition function is constructed utilizing, as anti-ghost fields, $\overline{c}$, $\overline{\phi}$, $\overline{\chi}_{\mu\nu}$ rather than $\overline{c}$, $\overline{\phi}$, $\overline{\psi}_\mu$. This choice is motivated by our previous choice of the gauge-invariant topological gauge-fixing (7.3.21). Furthermore, having deleted anti-BRST symmetry we can also restrict our attention to the intermediate situation where the Yang–Mills connection and curvature are extended only in the ghost direction, but not in the anti-ghost one. We do this by replacing Eqs. (7.2.5) , (7.2.6), (7.2.7) with

$$\begin{aligned} \widehat{A} &= A + c \\ \widehat{F} &= \widehat{d}\widehat{A} + \frac{1}{2}\left[\widehat{A}, \widehat{A}\right] \\ \widehat{d} &= d + s \end{aligned} \qquad (7.3.25)$$

In this case, Eq. (7.2.11) is also replaced by the decomposition

$$\widehat{F} = F_{(2,0)} + F_{(1,1)} + F_{(0,2)} \qquad (7.3.26)$$

yielding the identifications

$$\begin{aligned} \psi &= -F_{(1,1)} \\ \phi &= -F_{(0,2)} \end{aligned} \qquad (7.3.27)$$

which replace those of Eqs. (7.3.18). The extended Bianchi identity (7.3.9) is substituted with the analogous semi-extended one:

$$\widehat{d}\widehat{F} + \left[\widehat{A}, \widehat{F}\right] = 0 \qquad (7.3.28)$$

which follows from $\widehat{d}^2 = 0$. Using Eq. (7.3.28) we can derive the so-called *descent equations* which are the key instrument for deriving the structure of the *physical observables* of our *topological field theory*. As is the case in any BRST quantized theory, here the only operators $\mathcal{O}_i$ that we are allowed to insert in any correlation function are the BRST-invariant ones, namely those which are $s$-closed: $s\,\mathcal{O}_i = 0$. Furthermore, due to the BRST invariance of the path integral (7.3.23), the correlation function can be non-zero only if the $\mathcal{O}_i$ are not $s$-exact: $\mathcal{O}_i \neq s$ something. Hence the *physical observables* are in correspondence with the BRST-cohomology classes. The descent equations allow us to construct representatives of these classes given by well-defined local and integrated composite operators. We start from the *first Chern class* of the semi-extended connection:

$$\widehat{c}_1 \stackrel{\text{def}}{=} \widehat{\Delta} = tr\left\{\widehat{F} \wedge \widehat{F}\right\} \qquad (7.3.29)$$

and, as we already did for the extended curvature, we expand it into sectors of type $(p,q)$ (see Eq. (7.3.26)):

$$\begin{aligned} \widehat{\Delta} &= \Delta^{(4,0)} + \Delta^{(3,1)} + \Delta^{(2,2)} + \Delta^{(1,3)} + \Delta^{(0,4)} \\ \Delta^{(3,1)} &= -2tr\left\{\psi \wedge F\right\} \\ \Delta^{(2,2)} &= tr\left\{2\phi F + \psi \wedge \psi\right\} \\ \Delta^{(1,3)} &= -2tr\left\{\phi\psi\right\} \\ \Delta^{(4,0)} &= tr\left\{\phi\phi\right\} \end{aligned} \qquad (7.3.30)$$

From the identity

$$\hat{d}\hat{\Delta} = 0 \tag{7.3.31}$$

which follows from (7.3.28), we immediately obtain:

$$s\,\Delta^{(4-g,g)} = -\,d\Delta^{(3-g,g+1)} \tag{7.3.32}$$

which are the descent equations. Let now $c_i^{(4-n)}$ $(i = 1, \ldots, b_{(4-n)})$ be a set of $(4-n)$-cycles forming a basis for $H_{(4-n)}(M_4)$, the $(4-n)$-th homology group of the base manifold $M_4$ (by definition $b_{(4-n)} = dim_{\mathbb{R}}\, H_{(4-n)}(M_4)$ is the $(4-n)$-th Betti number). Define the operators

$$\mathcal{I}^{(n)}\left(c_i^{(4-n)}\right) \stackrel{\text{def}}{=} \int_{c_i^{(4-n)}} \Delta^{(4-n,n)} \tag{7.3.33}$$

Using Stoke's lemma, the descent equations and $\partial\, c_i^{(4-n)} = 0$, we conclude that the $\mathcal{I}_i^{(n)}$ are BRST-closed. Indeed

$$\begin{aligned} s\,\mathcal{I}^{(n)}\left(c_i^{(4-n)}\right) &= \int_{c_i^{(4-n)}} s\,\Delta^{(4-n,n)} = -\int_{c_i^{(4-n)}} d\,\Delta^{(3-n,n+1)} \\ &= -\int_{\partial\, c_i^{(4-n)}} \Delta^{(3-n,n+1)} = 0 \end{aligned} \tag{7.3.34}$$

These operators are the *physical observables* of the topological Yang–Mills theory. Correspondingly, a generic N-point function of the this quantum field theory is of the form

$$\begin{aligned} &\langle\, \mathcal{I}^{n_1}\left(c_{i_1}^{(4-n_1)}\right) \ldots \mathcal{I}^{n_N}\left(c_{i_N}^{(4-n_N)}\right) \rangle \stackrel{\text{def}}{=} \\ &\int [d\varphi] \exp\left[\, S_{quantum}\,\right] \mathcal{I}^{n_1}\left(c_{i_1}^{(4-n_1)}\right) \ldots \mathcal{I}^{n_N}\left(c_{i_N}^{(4-n_N)}\right) \end{aligned} \tag{7.3.35}$$

These Green functions have the distinguished properties that characterize a topological field theory:

*i) The correlators (7.3.35) depend only on the homology class of the cycles $c_{i_j}^{(4-n_j)}$ and not on the individual representatives.* Indeed we have

$$\begin{aligned} \mathcal{I}^{(n)}\left(c_i^{(4-n)} + \partial\gamma^{4-n+1}\right) &= \mathcal{I}^{(n)}\left(c_i^{(4-n)}\right) - s\,\mathcal{K}^{(n)}\left(\gamma^{(4-n)}\right) \\ \mathcal{K}^{(n)}\left(\gamma^{(4-n)}\right) &= \int_{\gamma_i^{(4-n+1)}} \Delta^{(4-n+1,n-1)} \end{aligned} \tag{7.3.36}$$

and the BRST invariance of the path integral (7.3.35) guarantees that

$$\langle\, \mathcal{I}^{n_1}\left(c_{i_1}^{(4-n_1)}\right) \ldots s\,\mathcal{K}^{n}\left(\gamma^{(4-n)}\right) \ldots \mathcal{I}^{n_N}\left(c_{i_N}^{(4-n_N)}\right) \rangle = 0 \tag{7.3.37}$$

In particular, if we consider the case of the local observable correlators:

$$\begin{aligned} \langle\, \mathcal{I}^4\left(c_{i_1}^{(0)}\right) \ldots \mathcal{I}^4\left(c_{i_N}^{(0)}\right) \rangle &= \langle\, tr\,\{\phi(x_1)\phi(x_1)\} \ldots tr\,\{\phi(x_N)\phi(x_N)\} \rangle \\ &= c\,(\,x_1 \ldots x_N) \end{aligned} \tag{7.3.38}$$

we see that they do not depend on the locations $x_1, \ldots, x_N$ in $M_4$, since the difference of any two points $x_i - y_i$ can always be seen as the boundary of a 1-chain. Hence the correlators $c\,(x_1 \ldots x_N)$ are constants.

*ii) The N-point correlators (7.3.35) do not depend on the choice of a metric $g_{\mu\nu}$ for the base manifold $M_4$.* To see this it suffices to note that the quantum action (7.3.19) depends on the base-manifold metric $g$ only through the gauge-fixing term which, by definition, is BRST-exact. Hence if we calculate the stress-energy tensor:

$$T_{\mu\nu}(x) \stackrel{\text{def}}{=} \frac{\delta S_{quantum}}{\delta g_{\mu\nu}(x)} = s\, K_{\mu\nu}(x) \tag{7.3.39}$$

we find that it is BRST-exact. Denoting by $\langle \ldots \rangle_g$ any correlator calculated using the metric $g$, under a small variation of this latter we get:

$$\langle \ldots \rangle_{g+\delta g} = \langle \ldots \rangle_g + \delta g^{\mu\nu}(x) \langle \ldots T_{\mu\nu}(x) \rangle_g = \langle \ldots \rangle_g \tag{7.3.40}$$

since

$$\langle \ldots T_{\mu\nu}(x) \rangle_g = \langle \ldots s\, K_{\mu\nu}(x) \rangle_g = 0 \tag{7.3.41}$$

As stressed in section 7.1 the BRST-exactness of the stress–energy tensor is the hallmark of topological field theories and can be used as their very definition. The topological Yang–Mills theory we have outlined in the present section fits into the scheme and provides the paradigma for this peculiar type of quantum field theories. The actual calculation of the correlators (7.3.35) is the goal of the formalism we have described. Because of the extremely large topological symmetry and in view of the chosen gauge fixing, the functional integration over the gauge orbits reduces to a finite dimensional integration over the instanton moduli space and the explicit evaluation of the correlators (7.3.35) reduces to the calculation of an *intersection integral* of *cohomology classes* on instanton moduli space. This is also a general property of topological field theories. Since the main interest in these book is in two-dimensional topological field theories, we postpone the illustration of this aspect to the discussion of topological $\sigma$-models. Another interesting feature is that the quantum action (7.3.19), once worked out explicitly, contains the ordinary Yang–Mills kinetic term $tr\,\{F_{\mu\nu}\,F^{\mu\nu}\}$ plus a long sequence of ghost–anti-ghost terms: actually the entire action is nothing but the action of the BRST-quantized N=2 super Yang–Mills theory after the spins of the fields have been redefined according to the topological twist [259, 8, 7], the topological ghost $\psi_\mu$ being the redefinition of the gaugino. Indeed the topological twist is the redefinition of the Euclidean Lorentz-group $SO(4) \approx SU_L(2) \otimes SU_R(2)$ as $SO(4)' \approx SU'_L(2) \otimes SU_R(2)$, where $SU'_L(2)$ is the diagonal of the old $SU_L(2)$ with the $SU_I(2)$ automorphism group of N=2 supersymmetry. A similar phenomenon happens also in two-dimensional topological theories: here the topological twist corresponds to a redefinition of the $SO(2)$ Lorentz group as the diagonal of the old $SO(2)$ with the $U(1)$ automorphism group of N=2 world-sheet supersymmetry. We also postpone the discussion of the topological twist to the case of two-dimensional theories. Indeed we have just used four-dimensional topological Yang–Mills theories as a particularly clean example where we could introduce the main ideas underlying the construction of this new class of quantum field theories. In particular they have been quite handy in clarifying the role of the geometric approach to BRST symmetry and the nature of the topological BRST algebra (lifting of the quantum rheonomic constraints): now shall revert to two dimensions, i.e. the appointed arena for our further developments.

## 7.4 Topological Sigma Models

In this section we present the construction of two-dimensional topological $\sigma$-models [260], using the Baulieu-Singer BRST approach.

Let $\Sigma_g$ be a Riemann surface of genus $g$ and $\mathcal{M}_K$ a Kählerian target manifold. We consider mappings

$$X \; : \; \Sigma_g \longrightarrow \mathcal{M}_K \tag{7.4.1}$$

Let $K$ be the Kähler 2-form on $\mathcal{M}_K$:

$$\begin{aligned} K &= \frac{i}{2\pi} g_{ij^*}\left(X,\overline{X}\right) dX^i \wedge d\overline{X}^{j^*} \\ dK &= 0 \end{aligned} \tag{7.4.2}$$

The classical action of the topological $\sigma$-model is the integral on the Riemann surface $\Sigma_g$ of the pull-back through the map $X$ of the Kähler 2-form

$$S_{class} = -\pi i \int_{\Sigma_g} X^* K = \int d^2\xi \, g_{ij^*}\left(X,\overline{X}\right) \partial_\alpha X^i \partial_\beta X^{j^*} \varepsilon^{\alpha\beta} \tag{7.4.3}$$

where we have denoted by $\xi^\alpha = (z, \overline{z})$ the two coordinates on the Riemann surface. Equation (7.4.3) can also be conveniently rewritten as follows:

$$S_{class}[X] = \frac{1}{2} \int g_{ij^*}\left(\partial_+ X^i \partial_- X^{j^*} - \partial_- X^i \partial_+ X^{j^*}\right) e^+ \wedge e^- \tag{7.4.4}$$

and it is clear that this classical action is invariant both with respect to reparametrizations of the Riemann surface:

$$\xi^\alpha \longrightarrow \xi^{\alpha'} \tag{7.4.5}$$

and to arbitrary variations of the embedding map:

$$X^i(\xi) \longrightarrow X^i(\xi) + \delta X^i(\xi) \tag{7.4.6}$$

within the same homotopy class.

Having identified the classical symmetries, in particular the deformation symmetry (7.4.6), our next step is the construction of the *BRST algebra*. Following our previous treatment of the topological Yang–Mills theories, we determine the spectrum of ghosts, anti-ghosts, physical fields and auxiliary fields that constitute the BRST double elliptic complex of the topological $\sigma$-model. They are summarized below:

| *Formdegree* / *ghostnumber* | 0 |
|---|---|
| $-1$ | $\bar{c}^i$ , $\bar{c}^{i^*}$ |
| $0$ | $X^i$ , $X^{i^*}$ , $b^i$ , $b^{i^*}$ |
| $1$ | $c^i$ , $c^{i^*}$ |

and satisfy the BRST algebra:

$$\begin{aligned} s\,X^i &= c^i \\ s\,X^{i^*} &= c^{i^*} \\ s\,c^i = s\,c^{i^*} &= 0 \\ s\,\bar{c}^i &= b^i \\ s\,\bar{c}^{i^*} &= b^{i^*} \\ s\,b^i = s\,b^{i^*} &= 0 \end{aligned} \tag{7.4.7}$$

Using the algebra (7.4.7) it is quite easy to solve the analogue of the descent equations (7.3.32) and realize that the observables of the topological model are in one-to-one correspondence with the elements of the Dolbeault cohomology groups of the target manifold. Indeed, let

$$\omega^{(p,q)} = \omega^{(p,q)}_{i_1,\ldots,i_p,j_1^*,\ldots,j_q^*}\left(X,\overline{X}\right) dX^{i_1} \wedge \cdots \wedge dX^{i_p} \wedge dX^{j_1^*} \wedge \cdots \wedge dX^{j_p^*} \tag{7.4.8}$$

be a closed $(p,q)$-form

$$d\omega^{(p,q)} = \partial\omega^{(p,q)} = \overline{\partial}\omega^{(p,q)} = 0 \tag{7.4.9}$$

representing some non-trivial element of the $(p,q)$-th Dolbeault cohomology group of the target manifold,

$$\omega^{(p,q)} \in H^{(p,q)}(\mathcal{M}_K) \tag{7.4.10}$$

Then we easily promote it to a ghost form $\widehat{\omega}^{(p,q)}$ by performing the following replacements:

$$\begin{aligned} dX^i &\longrightarrow \widehat{d}X^i \stackrel{\text{def}}{=} (d+s)\,X^i = dX^i + c^i \\ dX^{i^*} &\longrightarrow \widehat{d}X^{i^*} \stackrel{\text{def}}{=} (d+s)\,X^{i^*} = dX^{i^*} + c^{i^*} \end{aligned} \tag{7.4.11}$$

in its definition (7.4.8). Using Eqs. (7.4.7) and (7.4.9), we verify that:

$$\hat{d}\,\widehat{\omega}^{(p,q)} = 0 \tag{7.4.12}$$

Considering the pull-back of $\widehat{\omega}^{(p,q)}$ on the Riemann surface via the embedding map (7.4.1) and expanding it into addends $\Theta^{(d)}_{(g)}$ of definite form degree $d$ (on the Riemann surface) and definite ghost number $g$, we obtain

$$\begin{aligned}
\widehat{\omega}^{(p,q)} &= \Theta^{(2)}_{(p+q-2)}\left[\omega^{(p,q)}\right] - \Theta^{(1)}_{(p+q-1)}\left[\omega^{(p,q)}\right] + \Theta^{(0)}_{(p+q)}\left[\omega^{(p,q)}\right] \\
s\Theta^{(2)}_{(p+q-2)}\left[\omega^{(p,q)}\right] &= d\Theta^{(1)}_{(p+q-1)}\left[\omega^{(p,q)}\right] \\
s\Theta^{(1)}_{(p+q-1)}\left[\omega^{(p,q)}\right] &= d\Theta^{(0)}_{(p+q)}\left[\omega^{(p,q)}\right] \\
s\,\Theta^{(0)}_{(p+q)}\left[\omega^{(p,q)}\right] &= 0
\end{aligned} \tag{7.4.13}$$

where

$$\begin{aligned}
\Theta^{(0)}_{(p+q)}\left[\omega^{(p,q)}\right] &= \omega^{(p,q)}_{i_1,\dots,i_p,j^\star_1,\dots,j^\star_q}\, c^{i_1}\dots c^{i_p}\, c^{j^\star_1}\dots c^{j^\star_p} \\
\Theta^{(1)}_{(p+q-1)}\left[\omega^{(p,q)}\right] &= -\,p\,\omega^{(p,q)}_{i_1,\dots,i_p,j^\star_1,\dots,j^\star_q}\,\partial_\alpha X^{i_1} d\xi^\alpha c^{i_2}\dots c^{i_p}\, c^{j^\star_1}\dots c^{j^\star_p} \\
&- \,q\,\omega^{(p,q)}_{i_1,\dots,i_p,j^\star_1,\dots,j^\star_q}\, c^{i_1}\dots c^{i_p}\,\partial_\alpha X^{j^\star_1} d\xi^\alpha c^{j^\star_2}\dots c^{j^\star_p} \\
\Theta^{(2)}_{(p+q-2)}\left[\omega^{(p,q)}\right] &= p(p-1)\,\omega^{(p,q)}_{i_1,\dots,i_p,j^\star_1,\dots,j^\star_q}\,\partial_\alpha X^{i_1} d\xi^\alpha \wedge \partial_\beta X^{i_2} d\xi^\beta c^{i_3}\dots c^{i_p}\, c^{j^\star_1}\dots c^{j^\star_p} \\
&+ \,q(q-1)\,\omega^{(p,q)}_{i_1,\dots,i_p,j^\star_1,\dots,j^\star_q}\, c^{i_1}\dots c^{i_p}\,\partial_\alpha X^{j^\star_1} d\xi^\alpha \wedge \partial_\beta X^{j^\star_2} d\xi^\beta c^{j^\star_3}\dots c^{j^\star_p} \\
&+ \,p\,q\,\omega^{(p,q)}_{i_1,\dots,i_p,j^\star_1,\dots,j^\star_q}\,\partial_\alpha X^{i_1} d\xi^\alpha c^{i_2}\dots c^{i_p} \wedge \partial_\beta X^{j^\star_1} d\xi^\beta c^{j^\star_2}\dots c^{j^\star_p}
\end{aligned} \tag{7.4.14}$$

The last three of Eqs. (7.4.13) are a specialization of Eq. (7.4.12).

Once we have determined a quantum action $S_{quantum}$, by appropriate choice of the gauge-fixings, we can topologically deform it according to

$$S_{quantum} \longrightarrow S_{quantum} + \sum_{p+q\le \dim\mathcal{M}_K}\ \sum_{\left[\omega^{(p,q)}\in H^{(p,q)}\right]} t_{[\omega^{(p,q)}]}\int \Theta^{(2)}_{(p+q-2)}\left[\omega^{(p,q)}\right] \tag{7.4.15}$$

where $t_{[\omega^{(p,q)}]}$ are continuous parameters (*the topological coupling constants*), and then we can try to calculate the topological $k$-point correlators:

$$\begin{aligned}
&\langle \Theta^{(0)}_{(p_1+q_1)}\left[\omega_1{}^{(p_1,q_1)}\right]\dots\Theta^{(0)}_{(p_k+q_k)}\left[\omega_k{}^{(p_k,q_k)}\right] \\
&\exp\left[-\textstyle\sum_{p+q\le\dim\mathcal{M}_K}\ \sum_{\left[\omega^{(p,q)}\in H^{(p,q)}\right]} t_{[\omega^{(p,q)}]}\int \Theta^{(2)}_{(p+q-2)}\left[\omega^{(p,q)}\right]\right]\rangle
\end{aligned} \tag{7.4.16}$$

where the average $\langle\dots\rangle$ denotes the functional integral with respect to the measure $\exp\left[-S_{quantum}\right]$ determined by the undeformed quantum action. The problem is that of computing $S_{quantum}$ by choosing an appropriate gauge fermion. A convenient way of gauge fixing the topological symmetry (7.4.6) is that of restricting one's attention to the holomorphic embeddings satisfying the condition:

$$\begin{aligned} \partial_- X^i &\stackrel{\text{def}}{=} \frac{\partial}{\partial \bar{z}} X^i = 0 \\ \partial_+ X^{i^*} &\stackrel{\text{def}}{=} \frac{\partial}{\partial z} X^{i^*} = 0 \end{aligned} \tag{7.4.17}$$

In particular in genus $g = 0$ these are the rational curves one can embed in the target manifold $\mathcal{M}_K$. In any case the holomorphic embeddings (7.4.17) are the appropriate *instantons* for the topological $\sigma$-model and the instanton number is the value of the classical action

$$S_{classical} = -\pi i \int_{\Sigma_g} X^\star K = \text{const}\, k \;\; (k \in \mathbf{Z})\,. \tag{7.4.18}$$

At any value of $k$ there will be a discrete or continuous family (which can also have zero elements) of instantons (7.4.16). In the case of a continuous family, which is the most interesting, the parameters labelling these instantons are the *moduli* filling a certain finite dimensional *moduli space* $\mathcal{M}^{(k)}_{moduli}$. The topological correlators (7.4.16) will turn out to be intersection integrals of elements of the cohomology ring $H^\star\left(\mathcal{M}^{(k)}_{moduli}\right)$. In view of the choice (7.4.17) of the gauge-fixing, as gauge functions we choose

$$\begin{aligned} L^i_\alpha &= \partial_- X^i + \alpha\, \Gamma^i_{jk}\, \bar{c}^j\, c^k \\ L^{i^*}_\alpha &= \partial_+ X^{i^*} + \alpha\, \Gamma^{i^*}_{j^*k^*}\, \bar{c}^{j^*}\, c^{k^*} \end{aligned} \tag{7.4.19}$$

where $\Gamma^i_{jk}$, $\Gamma^{i^*}_{j^*k^*}$ denote the Levi-Civita connection coefficients of the Kähler target manifold, according to the conventions of Eqs. (5.5.49) and $\alpha$ is some numerical real parameter. For any value of $\alpha$ the gauge-fixing (7.4.17) is modified only by terms proportional to ghosts and anti-ghosts that do not affect classical configurations (all fermions equal to zero) and do not change the geometrical interpretation of the gauge fixing. Hence from this point of view the value of $\alpha$ is immaterial and we could also set $\alpha = 0$. However, as we shall realize at the end of our calculation, we can choose a special value of $\alpha$ such that the quantum action is entirely expressed in terms of target space tensors and covariant derivatives: this has formal advantages and will eventually allow an identification of the topological $\sigma$-model with the N=2 $\sigma$-model of section 5.5.6. Having fixed the gauge functions we write the quantum action as

$$S_{quantum} = S_{classical} + \int s\, \Psi_{(\alpha,\beta)} \tag{7.4.20}$$

where the gauge fermion is

$$\Psi_{(\alpha,\beta)} = -2\left[\bar{c}^{l^*} g_{l^*i}\left(L^i_\alpha + \beta\, b^i\right) + \bar{c}^l g_{li^*}\left(L^{i^*}_\alpha + \beta\, b^{i^*}\right)\right] \tag{7.4.21}$$

$\beta$ being an additional numerical parameter to be fixed later and amounting simply to a normalization of the auxiliary fields $b^i$, $b^{i^*}$ already conventionally fixed by setting $s\,\bar{c}^i = b^i$

in Eq. (7.4.7). Using the algebra (7.4.7) and the definitions (5.5.49) of the connection coefficients and of the curvature tensor, by means of a lengthy but straightforward calculation, we obtain

$$\begin{aligned} s\Psi_{(\alpha,\beta)} &= -4\beta\, g_{l^*i}\, b^{l^*}\, b^i + 2\, g_{l^*i}\, b^{l^*}\, X^i + 2\, g_{l^*i}\, X^{l^*}\, 2\, b^i \\ &+ 2\,\overline{c}^{l^*}\, g_{l^*i}\left(\partial_- c^i - \Gamma^i_{jm}\partial_- X^j c^m - g^{ik^*}\, g_{jp^*}\,\Gamma^{p^*}_{k^*m^*}\,\partial_- X^j c^{m^*}\right) \\ &+ 2\,\overline{c}^l\, g_{li^*}\left(\partial_+ c^{i^*} - \Gamma^{i^*}_{j^*m^*}\partial_+ X^{j^*} c^{m^*} - g^{i^*k}\, g_{j^*p}\,\Gamma^p_{km}\,\partial_+ X^{j^*} c^m\right) \\ &+ 4\,\alpha\,\overline{c}^{l^*}\, c^{m^*}\,\overline{c}^j\, c^k\left(R_{l^*jm^*k} - g_{p^*i}\Gamma^{p^*}_{m^*l^*}\,\Gamma^i_{jk}\right) \end{aligned} \tag{7.4.22}$$

where we have denoted

$$\begin{aligned} X^i_{(\alpha,\beta)} &= -\partial_- X^i - (\alpha+\beta)\,\Gamma^i_{jk}\,\overline{c}^j\, c^k + (\alpha-\beta)\, g^{iq^*}\, g_{jp^*}\,\Gamma^{p^*}_{q^*k^*}\,\overline{c}^j\, c^{k^*} \\ X^{i^*}_{(\alpha,\beta)} &= -\partial_+ X^{i^*} + (\alpha+\beta)\,\Gamma^{i^*}_{j^*k^*}\,\overline{c}^*\, c^{k^*} - (\alpha-\beta)\, g^{i^*q}\, g_{j^*p}\,\Gamma^p_{qk}\,\overline{c}^{j^*}\, c^k \end{aligned} \tag{7.4.23}$$

Eliminating the auxiliary fields $b^i$ and $b^{i^*}$ through their own equation of motion we obtain $b^i = \frac{1}{2}X^i_{(\alpha,\beta)}$, $b^{i^*} = \frac{1}{2}X^*_{(\alpha,\beta)}$ and we see that the expression for these fields simplifies if we choose the normalization

$$\beta = \alpha \tag{7.4.24}$$

Hence we set Eq. (7.4.24) and we obtain

$$\begin{aligned} b^i &= -\frac{1}{2}\partial_- X^i - \alpha\,\Gamma^i_{jk}\,\overline{c}^j\, c^k \\ b^{i^*} &= -\frac{1}{2}\partial_+ X^{i^*} + \alpha\,\Gamma^{i^*}_{j^*k^*}\,\overline{c}^*\, c^{k^*} \end{aligned} \tag{7.4.25}$$

If we substitute Eqs. (7.4.25), (7.4.24) and (7.4.23) back into Eq. (7.4.22), the BRST variation of the gauge fermion becomes

$$\begin{aligned} s\Psi_{(\alpha,\beta)} &= \frac{1}{\alpha}\, A\cdot\overline{A} \\ &+ \left(\frac{1}{\alpha}-\alpha\right)\left(4\,B\cdot\overline{B} + \left(A\cdot\overline{B} + B\cdot\overline{A}\right)\right) \\ &+ 2\,\overline{c}^{l^*}\, g_{l^*i}\left(\partial_- c^i - \Gamma^i_{jm}\partial_- X^j\, c^m\right) \\ &+ 2\,\overline{c}^l g_{li^*}\left(\partial_+ c^{i^*} - \Gamma^{i^*}_{j^*m^*}\partial_+ X^{j^*}\, c^{m^*}\right) \\ &+ 4\,\alpha\,\overline{c}^{l^*}\, c^{m^*}\,\overline{c}^j\, c^k\, R_{l^*jm^*k} \end{aligned} \tag{7.4.26}$$

where we have introduced the shorthand notation $V\cdot\overline{W} \stackrel{\text{def}}{=} V^i\, g_{ij^*}\, W^{j^*}$ and where

$$\begin{aligned} A^i &= \partial_- X^i \\ A^{i^*} &= \partial_+ X^{i^*} \\ B^i &= \Gamma^i_{jk}\,\overline{c}^j\, c^k \\ B^{i^*} &= \Gamma^{i^*}_{j^*k^*}\,\overline{c}^{j^*}\, c^{k^*} \end{aligned} \tag{7.4.27}$$

As is evident from Eq. (7.4.26), if we choose

$$\alpha = 1 \tag{7.4.28}$$

the terms in $B^i, B^{i^\star}$ disappear and the final form of the quantum action (7.4.20) is

$$\begin{aligned} S_{quantum} &= \int_{\Sigma_g} \left[ \frac{1}{2} g_{ij^\star} \left( \partial_+ X^i \, \partial_- X^{j^\star} + \partial_- X^i \, \partial_+ X^{j^\star} \right) \right. \\ &+ 2 g_{ij^\star} \left( \bar{c}^i \, \nabla_+ c^{j^\star} + \bar{c}^{j^\star} \, \nabla_- c^i \right) \\ &+ \left. 4 \bar{c}^{i^\star} \, c^{j^\star} \, \bar{c}^k \, c^l \, R_{i^\star k j^\star l} \right] \end{aligned} \tag{7.4.29}$$

where we have introduced the covariant derivatives:

$$\begin{aligned} \nabla_- c^i &= \partial_- c^i - \Gamma^i_{jk} \, \partial_- X^j \, c^k \\ \nabla_+ c^i &= \partial_+ c^{i^\star} - \Gamma^{i^\star}_{j^\star k^\star} \, \partial_+ X^{j^\star} \, c^{k^\star} \end{aligned} \tag{7.4.30}$$

At this point we realize that the action of the topological $\sigma$-model and the action of the N=2 $\sigma$-model (see Eq. (5.5.52)) are just proportional

$$S^{(topological)}_{quantum} = \frac{1}{2} S^{N=2} \tag{7.4.31}$$

provided one makes the following identifications of the fermions with the ghost and anti-ghost fields:

$$\begin{aligned} \tilde{\psi}^i &= i \, \bar{c}^i \\ \psi^i &= c^i \\ \tilde{\psi}^{i^\star} &= c^{i^\star} \\ \psi^{i^\star} &= i \, \bar{c}^{i^\star} \end{aligned} \tag{7.4.32}$$

Comparing Eqs. (5.5.54) with Eqs. (7.4.7), upon use of the identifications (7.4.25) and (7.4.32), we see that the BRST transformations (7.4.7) are a subset of the N=2 supersymmetry transformations, provided we set:

$$\begin{aligned} \varepsilon^+ &= \tilde{\varepsilon}^- = 0 \\ -\varepsilon^- &= \tilde{\varepsilon}^+ = \Lambda \end{aligned} \tag{7.4.33}$$

where by $\Lambda$ we have denoted the anti-commuting, nilpotent ($\Lambda^2 = 0$) BRST parameter. Equations (7.4.32) and (7.4.33) provide the first instance of the *topological twist* of an N=2 theory: ***a subset of the N=2 supersymmetry transformations can be reinterpreted as BRST transformations associated with a topological symmetry.*** It must be noted that while the supersymmetry charge $Q_{SUSY}$ has spin 1/2, the BRST charge $Q_{BRST}$ has spin zero. Hence the topological reinterpretation of the N=2 supersymmetry algebra requires a procedure of spin redefinition so that, after this redefinition the SUSY charges $Q^-$ and $\tilde{Q}^+$ acquire spin zero. On the other hand, by looking at the BRST algebra (7.4.7), (7.4.25) we see that the ghosts $c^i$, $c^{i*}$ have spin zero while the anti-ghosts $\bar{c}^i$, $\bar{c}^{i*}$ have spin 1 and $-1$, respectively. Hence the spin redefinition must assign spin $s = 0$ to $\psi^i$, $\tilde{\psi}^{i*}$ and spin $s = \pm 1$ to $\tilde{\psi}^i$, $\psi^{i*}$. If we recall our discussion of the R symmetries of the N=2 $\sigma$-model (see Eqs. (5.5.57)) then we see that a convenient redefinition of the 2D spin satisfying the desired requirement is provided by the formula

$$s' = s + \frac{1}{2}\left[q_R + q_L\right] \tag{7.4.34}$$

where $s$ is the eigenvalue of $J_S$, the generator of the $O(1,1)$ Lorentz group ($U(1)$ after Wick rotation) and $q_L$, $q_R$ are the eigenvalues of $J_L$, $J_R$, the two generators of the $U(1)_L \otimes U(1)_R$ R symmetry group. Equation (7.4.34) means that in the topological twist one redefines the Lorentz group as the diagonal of the old Lorentz group with the internal R symmetry group (i.e. the automorphism group of the N=2 supersymmetry algebra) according to the following equation:

$$J_S' = J_S + \frac{1}{2}\left[J_R + J_L\right] \tag{7.4.35}$$

This procedure is very general and any N=2 field theory actually admits two distinct topological twists, named by Witten the A and B models. This issue is addressed in the next section.

## 7.5 The A and B Topological Twists of an N=2 Field Theory

In the previous section we have seen that the topological $\sigma$-model, constructed according to the general philosophy of BRST quantization, turns out to have a quantum action identical with that of the N=2 $\sigma$-model discussed in section 5.5.6. Just the spin of the anti-commuting fermions is changed. This suggests the possibility that any N=2 field theory can be suitably reinterpreted as a topological field theory. Indeed one can codify a general procedure that provides such a reinterpretation for any N=2 model and which is based on a redefinition of the two-dimensional Euclidean Lorentz group: this procedure is named the *topological twist*. The goal is that of changing the spin of the fermions in order to reinterpret them as ghosts of some topological symmetry, at the same time reinterpreting some of the N=2 supersymmetry transformations as BRST transformations.

There are actually two different topological twists of the same N=2 theory: in Witten's nomenclature the A and B twists. The idea is that the topological twist extracts, from any N=2 supersymmetric theory, a topological field theory that is already *gauge-fixed*, namely where the BRST algebra already contains the anti-ghosts, whose Slavnov variation is proportional to the gauge-fixings. The appropriate *instanton conditions* that play the role of gauge-fixings for the topological symmetry are thus automatically selected when the topological field theory is obtained via the topological twist. This latter consists of the following steps:

*i) First one BRST quantizes the ordinary N=2 theory.* (This step is relevant when the ordinary N=2 theory is locally supersymmetric (supergravity) and/or it contains gauge fields. For rigid N=2 theories containing only matter multiplets as the N=2 $\sigma$-model or the rigid N=2 Landau–Ginzburg models, this step is empty. It is a relevant step for the N=2 gauge theory described in section 5.5 or for N=2 supergravity.)

*ii) Then one redefines the spins of all the fields taking as new Lorentz group the diagonal of the old Lorentz group with the internal automorphism group of N=2 supersymmetry.*

*iii) After this redefinition, one recognizes that at least one component of the N=2 multiplet of supercharges, say $Q_0$, has spin zero, is nilpotent and anti-commutes with the old BRST charge $Q^{(old)}_{BRST}\, Q_0 + Q_0\, Q^{(old)}_{BRST} = 0$. Then one defines the new BRST charge as $Q^{(new)} = Q_0 + Q^{(old)}_{BRST}$.*

*iv) Next one redefines the ghost number $gh^{(new)} = gh^{(old)} + F$, where $F$ is some appropriate fermion number, in such a way that the operator $(-1)^{gh^{(new)}}$ anti-commutes with the new BRST charge, in the same way as the operator $(-1)^{gh^{(old)}}$ did anti-commute with the old BRST charge. In this way all the fields of the BRST-quantized N=2 theory acquire a new well-defined ghost number.*

*v) Reading the ghost numbers one separates the physical fields from the ghosts and the anti-ghosts, the BRST variation of these latter yielding, after elimination of the lagrangian multipliers, the gauge fixing instanton equations. The gauge-free BRST algebra (that involving no anti-ghosts) should, at this point, be recognizable as that associated with a well-defined topological symmetry: for instance the continuous deformations of the vielbein (topological gravity), the continuous deformations of the gauge connection (topological Yang–Mills theory), the continuous deformations of the embedding functions (topological $\sigma$-model) and so on.*

*Step 1*

One BRST quantizes the local symmetries of the original N=2 model according to the standard procedures described in section 7.2.

*Step 2*

The second step is the delicate one. In two dimensions the Lorentz group is $O(1,1)$ which becomes $O(2)$ after Wick rotation. Let us name $J_S$ the Lorentz generator: the eigenvalues $s^i$ of this operator are the spins of the various fields $\varphi^i$. The number $s^i$ appears in the Lorentz covariant derivative of the field $\varphi^i$:

$$\nabla \varphi^i = d\varphi^i - s^i \omega \varphi^i \tag{7.5.1}$$

The automorphism group of the supersymmetry algebra that can be used to redefine the Lorentz group is the R symmetry group $U(1)_L \otimes U(1)_R$. Hence a crucial requirement imposed on the original N=2 model in order to perform a successful topological twist is that it should be R symmetric. This is true of the N=2 $\sigma$-model and of the Landau-Ginzburg model (with or without local gauge interactions) if the superpotential is quasi-homogeneous.

Denoting by $J_L$, $J_R$ the two R symmetry generators, we redefine the Lorentz generator according to the formula

$$J_S^{'} = J_S + \frac{1}{2}[J_R \pm J_L] \qquad (7.5.2)$$

Correspondingly the new spin quantum number is given by

$$s^{'} = s + \frac{1}{2}[q_R \pm q_L] \qquad (7.5.3)$$

The choice of sign in Eqs. (7.5.2), (7.5.3) corresponds to the existence of two different topological twists for the same N=2 theory. Following Witten they will be named the A-twist, leading to the A-model (upper choice of the sign) and the B-twist, leading to the B-model (lower choice of the sign). It might seem arbitrary to restrict the possible linear combinations of the operators $J_S$, $J_L$ and $J_R$ to those in Eqs. (7.5.2), (7.5.3), but, actually, these are the only possible ones if we take into account the following requirements. In the gravitational sector the spin redefinition must transform N=2 supergravity into topological gravity, hence the spins of the vielbein $e^{\pm}$ must remain the same before and after the twists: this fixes the coefficient of $J_S$ to be equal to one as in Eq. (7.5.2). Furthermore, of the four gravitino 1-forms $\zeta^+, \zeta^-, \tilde{\zeta}^+, \tilde{\zeta}^-$, two must acquire spin $s = 1$ and $s = -1$, respectively, and the other two must have spin zero. This is so because two of the gravitinos have to become the topological ghosts corresponding to continuous deformations of the vielbein (so they must have the same spins as the vielbein) while the other two must be the gauge fields of those supersymmetry charges that, acquiring spin zero, can be used to redefine the BRST charge. These constraints have two solutions: indeed they fix the coefficients of $J_L$ and $J_R$ to the values displayed in Eqs. (7.5.2), (7.5.3), the choice of sign distinguishing the two solutions.

*Step 3*

Naming $Q_{BRS}$ the BRST charge of the original gauge theory and $Q^{\pm}$, $\tilde{Q}^{\pm}$ the supersymmetry charges generating the transformations of parameters $\varepsilon^{\pm}, \tilde{\varepsilon}^{\pm}$, whose corresponding gauge fields are the gravitinos $\zeta^{\pm}, \tilde{\zeta}^{\pm}$, we realize that in the A-twist the spinless supercharges are $Q^-$ and $\tilde{Q}^+$ while in the B-twist they are $Q^+$ and $\tilde{Q}^+$. In both cases the two spinless supercharges anti-commute among themselves and with the BRST charge so that we can define the new BRST charge of the topological theory according to the formula

$$Q^{'}_{BRST} = Q_{BRST} \mp Q^{\mp} + \tilde{Q}^{+} \qquad (7.5.4)$$

The choice of signs in Eq. (7.5.4) is just a matter of convention: once more the upper choice of sign corresponds to the A-twist while the lower corresponds to the B-twist. The physical states of the topological theory are the cohomology classes of the operators (7.5.4).

*Step 4*

What matters in the definition of the ghost number are the differences of ghost numbers for the fields related by a BRST transformation. Indeed ghost number is one of the two gradings in a double elliptic complex. Hence to all the fields we must assign an integer grading which has to be increased of one unit by the application of the BRST charge (or Slavnov operator). In other words $Q'_{BRS}$ must have ghost number $gh = 1$. These requirements are satisfied if, for the redefinition of the ghost number $gh' = gh + F$, we use the generator $F$ of some $U(1)$ symmetry of the original N=2 theory with respect to which the new BRST generator (7.6.3) has charge $q_{BRST} = 1$ and such that the two gravitinos that acquire the same spin as the vielbein and become the ghosts of topological gravity have $gh' = 1$. In this case, the action, being invariant under the chosen symmetry, has ghost number $gh' = 0$.

In the A-twist case, naming $\#gh$ the ghost number of the original gauge theory, we fulfill all the desired properties if we define the ghost number of the topological theory according to the formula

$$\#gh' = \#gh + q_L - q_R \tag{7.5.5}$$

In this case the $U(1)$ symmetry utilized to redefine the ghost number is generated by $F = J_L - J_R$ and it is a subgroup of the R symmetry group $U(1)_L \otimes U(1)_R$, which becomes a local symmetry group after coupling to supergravity (see section 5.4.2 and in particular Eq. (5.4.27) where one sees that the graviphoton gauges precisely the subgroup generated by $F$. In the A-twist case, however, it is the new Lorentz group that it is not a linear combination of two local symmetry groups of the original N=2 theory).

In the B-twist case, the new ghost numbers are defined as follows:

$$\#gh' = \#gh - q_L - q_R \tag{7.5.6}$$

In this case the $U(1)$ symmetry utilized to redefine the ghost number is generated by $F' = J_L + J_R$ and it is a subgroup of the R symmetry group $U(1)_L \otimes U(1)_R$, which remains a global symmetry group also after coupling to supergravity. In the B-case, however, the new Lorentz symmetry is a linear combination of the old Lorentz symmetry with the other local $U(1)$ symmetry of the theory gauged by the graviphoton. From this point of view, in the local supersymmetry case, the B-twist is more natural than the A-twist. Indeed the ghost number symmetry is not required to be a local symmetry, while the Lorentz symmetry is. It should be noted, however, that the breaking of local Lorentz symmetry implicit in the A-twist is only an artifact of the gauge-fixing and does not affect the Lorentz invariance of the physical correlators.

*Step 5*

This depends on the explicit case considered and there are no general rules. The strategy relies on first identifying the $\mathcal{B}_{gauge\text{-}free}$ part of the BRST algebra so that one knows what are the topological symmetries one deals with, and secondly on inspecting $\mathcal{B}_{gauge\text{-}fixing}$ in order to extract the definition of the involved *instanton* conditions.

As an illustration of the general procedure outlined above, in the next section we consider the twists of the N=2 matter-coupled gauge model discussed in section 5.5. Since it interpolates between the N=2 $\sigma$-model and the N=2 Landau–Ginzburg theory

**the discussion of the twists of this gauge theory brings in the twists also of these latter. In particular through the discussion of the A-twist we retrieve the relation between the N=2 and the topological $\sigma$-model, already pointed out, while through the discussion of the B-twist we introduce the concept of topological Landau–Ginzburg model. This latter will be the subject of further investigation in later sections of this chapter. Furthermore the choice of this model as a case study for the topological twists has an additional motivation: it is a gauge theory and as such it provides also an illustration of step 1 that would be unclear if we started from a purely matter theory having no BRST charge to begin with.**

## 7.6 Twists of the Two-Phase N=2 Gauge Theory

**As mentioned we discuss the two possible topolological twists (A- and B-models) of the N=2 Landau–Ginzburg theories with local gauge symmetries presented in section 5.5. We focus on the formal aspects of the topological twist procedure, aiming at a clarification of the involved steps.**

### 7.6.1 The topological BRST algebra

*Step 1*

The first step is straightforward. The case of interest to us just involves an ordinary gauge symmetry. Hence we just make the shift

$$\mathcal{A} \longrightarrow \widehat{\mathcal{A}} = \mathcal{A} + c^{(gauge)} \tag{7.6.1}$$

where $c^{(gauge)}$ are ordinary Yang–Mills ghosts. Imposing the BRST-rheonomic conditions (see the first of Eq. (5.5.3):

$$\begin{aligned}
\widehat{F} &\stackrel{\text{def}}{=} \widehat{d}\widehat{\mathcal{A}} + \widehat{\mathcal{A}} \wedge \widehat{\mathcal{A}} \\
&= (d+s)\left(\mathcal{A} + c^{(gauge)}\right) + \left(\mathcal{A} + c^{(gauge)}\right) \wedge \left(\mathcal{A} + c^{(gauge)}\right) \\
&= \mathcal{F}\, e^+ e^- - \frac{i}{2}(\tilde{\lambda}^+ \zeta^- + \lambda^- \zeta^+)\, e^- + \frac{i}{2}(\lambda^+ \tilde{\zeta}^+ + \lambda^- \tilde{\zeta}^+)\, e^+ + M\, \zeta^- \tilde{\zeta}^+ - M^*\, \zeta^+ \tilde{\zeta}^-
\end{aligned} \tag{7.6.2}$$

we obtain the ordinary BRST algebra of an N=2 supersymmetric gauge theory. We do not dwell on this trivial point.

*Steps 2, 3, 4*

They are implemented in a straightforward way, utilizing in Eqs. (7.5.3), (7.5.4), (7.5.5), (7.5.6) the values of the R symmetry charges as defined in Eqs. (5.5.38).

*Step 5*

As a preparation for this step, namely for the identification of the topological BRST algebras and theories generated by the twists, we consider the explicit form of the BRST transformations of all the fields. In view of a very simple and powerful fixed point theorem due to Witten [263, 262], we also recall that the topological theory, besides being BRST-invariant with respect to the supercharge (7.5.4), has also a supergroup (0|2) of fermionic symmetries commuting with the BRST transformations and generated by the two spinless supercharges utilized to redefine the BRST charge. Hence while writing the topological BRST transformations we write also the (0|2)-transformations. As Witten pointed out, the topological functional integral is concentrated on those configurations that are a fixed point of the (0|2)-transformations: these are the true *instantons* of our theory and can be read from the formulae we are going to list. The result that the path integral is concentrated on the instanton configurations can also be obtained by showing that in the topological theories the semi-classical approximation becomes exact. This is a physically more easily understandable argument and will be illustrated in later sections dealing both with the topological $\sigma$-model and with the topological Landau–Ginzburg models. However, the fixed point argument is very convenient for extracting the definition of the instantons directly from the $\mathcal{B}_{gauge\text{-}fixing}$ part of the BRST algebra obtained through the twist procedure.

In the A-twisted case the BRST charge is given by

$$Q^{(A)}_{BRS} = Q^{(gauge)}_{BRS} - Q^{-} + \tilde{Q}^{+} \tag{7.6.3}$$

Correspondingly we rename the supersymmetry parameters as follows:

$$\begin{aligned} -\varepsilon^{-} &= \alpha \\ \tilde{\varepsilon}^{+} &= \alpha' \\ \alpha^{(A)} &= \tilde{\varepsilon}^{+} = \varepsilon^{-} = \alpha_g \end{aligned} \tag{7.6.4}$$

where $\alpha_g$ is the nilpotent BRST parameter associated with the original gauge symmetry and $\alpha^{(A)}$ is the BRST parameter of the A-twisted model. The parameters $\alpha$ and $\alpha'$ correspond to the two fermionic nilpotent transformations, commuting with the BRST transformations and generating the (0|2) supergroup of exact symmetries of the topological action. Using the above conventions the form of the BRST transformations and of the (0|2)-symmetries in the A-twisted version of the N=2 gauge-coupled Landau–Ginzburg model is given by the following formulae:

$$\begin{aligned} \delta\mathcal{A}_{+} &= \alpha^{(A)}\left(-\frac{i}{2}\lambda^{-} + \partial_{+}c^{gauge}\right) = -\frac{i}{2}\alpha\,\lambda^{-} + \alpha_g\,\partial_{+}c^{gauge} \\ \delta\mathcal{A}_{-} &= \alpha^{(A)}\left(-\frac{i}{2}\tilde{\lambda}^{+} + \partial_{-}c^{gauge}\right) = -\frac{i}{2}\alpha'\,\tilde{\lambda}^{+} + \alpha_g\,\partial_{-}c^{gauge} \\ \delta M &= 0 \\ \delta M^{*} &= \frac{1}{4}\alpha^{(A)}\left(\lambda^{+} + \tilde{\lambda}^{-}\right) = \frac{1}{4}\left(\alpha\,\lambda^{+} + \alpha'\,\tilde{\lambda}^{-}\right) \end{aligned}$$

$$
\begin{aligned}
\delta\lambda^{+} &= \alpha^{(A)}\left(\frac{\mathcal{F}}{2} - i\mathcal{P}\right) = \alpha^{'}\left(\frac{\mathcal{F}}{2} - i\mathcal{P}\right) \\
\delta\lambda^{-} &= -2i\,\alpha^{(A)}\,\partial_{+} M = -2i\,\alpha\,\partial_{+} M \\
\delta\widetilde{\lambda}^{+} &= -2i\,\alpha^{(A)}\,\partial_{-} M = -2i\,\alpha^{'}\,\partial_{-} M \\
\delta\widetilde{\lambda}^{-} &= \alpha^{(A)}\left(\frac{\mathcal{F}}{2} - i\mathcal{P}\right) = \alpha\left(\frac{\mathcal{F}}{2} - i\mathcal{P}\right) \\
\delta\mathcal{P} &= \alpha^{(A)}\frac{1}{4}\Big[-\partial_{+}\widetilde{\lambda}^{+} + \partial_{-}\lambda^{-}\Big] = \frac{1}{4}\Big[-\alpha\,\partial_{+}\widetilde{\lambda}^{+} + \alpha^{'}\,\partial_{-}\lambda^{-}\Big] \\
\delta X^{i} &= \alpha^{(A)}\left(\psi^{i} + i\,c^{gauge}\,q^{i}_{j}X^{j}\right) = \left(\alpha\psi^{i} + i\,\alpha_{g}c^{gauge}\,q^{i}_{j}X^{j}\right) \\
\delta X^{i^{*}} &= \alpha^{(A)}\left(\widetilde{\psi}^{i^{*}} - i\,c^{gauge}\,q^{i}_{j}X^{j^{*}}\right) = \left(\alpha^{'}\widetilde{\psi}^{i^{*}} - i\,\alpha_{g}c^{gauge}\,q^{i}_{j}X^{j^{*}}\right) \\
\delta\psi^{i} &= \alpha^{(A)}\left(i\,M\,q^{i}_{j}X^{j} + i\,c^{gauge}\,q^{i}_{j}\psi^{j}\right) \\
&= \left(i\,\alpha^{'}M\,q^{i}_{j}X^{j} + i\,\alpha_{g}\,c^{gauge}\,q^{i}_{j}\psi^{j}\right) \\
\delta\widetilde{\psi}^{i} &= \alpha^{(A)}\left(\frac{i}{2}\nabla_{-}X^{i} + \eta^{ij^{*}}\,\partial_{j^{*}}W^{*} + i\,c^{gauge}\,q^{i}_{j}\widetilde{\psi}^{j}\right) \\
&= \left(\alpha^{'}\,\frac{i}{2}\nabla_{-}X^{i} + \alpha\,\eta^{ij^{*}}\,\partial_{j^{*}}W^{*} + i\,\alpha_{g}\,c^{gauge}\,q^{i}_{j}\widetilde{\psi}^{j}\right) \\
\delta\psi^{i^{*}} &= \alpha^{(A)}\left(-\frac{i}{2}\nabla_{+}X^{i} + \eta^{i^{*}j}\,\partial_{j}W - i\,c^{gauge}\,q^{i^{*}}_{j^{*}}\psi^{j^{*}}\right) \\
&= \left(-\alpha\,\frac{i}{2}\nabla_{+}X^{i} + \alpha^{'}\,\eta^{i^{*}j}\,\partial_{j}W - i\,\alpha_{g}\,c^{gauge}\,q^{i^{*}}_{j^{*}}\psi^{j^{*}}\right) \\
\delta\widetilde{\psi}^{i^{*}} &= \alpha^{(A)}\left(i\,M\,q^{i^{*}}_{j^{*}}X^{j^{*}} + i\,c^{gauge}\,q^{i^{*}}_{j^{*}}\widetilde{\psi}^{j^{*}}\right) \\
&= \left(-i\,\alpha\,M\,q^{i^{*}}_{j^{*}}X^{j^{*}} + i\,\alpha_{g}\,c^{gauge}\,q^{i^{*}}_{j^{*}}\widetilde{\psi}^{j^{*}}\right)
\end{aligned}
\tag{7.6.5}
$$

On the other hand in the B-twisted version of the same N=2 theory, the BRST charge is given by

$$
Q^{(A)}_{BRS} = Q^{(gauge)}_{BRS} + Q^{+} + \widetilde{Q}^{+} \tag{7.6.6}
$$

In view of Eq. (7.6.6) and of our previous discussion of the ghost number, in the B-twist case, we rename the supersymmetry parameters as follows:

$$
\begin{aligned}
\frac{1}{2}\left(\varepsilon^{+} + \widetilde{\varepsilon}^{+}\right) &= \alpha \\
\frac{1}{2}\left(\varepsilon^{+} - \widetilde{\varepsilon}^{+}\right) &= \alpha^{'} \\
\alpha^{(B)} &= \widetilde{\varepsilon}^{+} = \varepsilon^{+} = \alpha_{g}
\end{aligned}
\tag{7.6.7}
$$

$\alpha^{(B)}$ being the new BRST parameter and $\alpha$, $\alpha^{'}$ the parameters of the $(0|2)$ fermionic supergroup relevant to this case. With these notations the BRST transformations and

(0|2)-symmetries of the B-model are the following:

$$\begin{aligned}
\delta\mathcal{A}_+ &= \alpha^{(B)}\left(-\frac{i}{2}\lambda^- + \partial_+ c^{gauge}\right) = -\frac{i}{2}(\alpha-\alpha')\,\lambda^- + \alpha_g\,\partial_+ c^{gauge} \\
\delta\mathcal{A}_- &= \alpha^{(B)}\left(\frac{i}{2}\widetilde{\lambda}^- + \partial_- c^{gauge}\right) = -\frac{i}{2}(\alpha-\alpha')\,\widetilde{\lambda}^- + \alpha_g\,\partial_- c^{gauge} \\
\delta M &= \alpha^{(B)}\frac{1}{4}\lambda^- = (\alpha+\alpha')\frac{1}{4}\lambda^- \\
\delta M^* &= \alpha^{(B)}\frac{1}{4}\widetilde{\lambda}^- = (\alpha-\alpha')\frac{1}{4}\widetilde{\lambda}^- \\
\delta\lambda^+ &= \alpha^{(B)}\Big[(\mathcal{F} - i\mathcal{P}) - 2i\partial_+ M^*\Big] \\
&= \Big[(\alpha-\alpha')\,(\mathcal{F} - i\mathcal{P}) - 2i\,(\alpha+\alpha')\,\partial_+ M^*\Big] \\
\delta\lambda^- &= 0 \\
\delta\widetilde{\lambda}^+ &= \alpha^{(B)}\Big[(\mathcal{F} + i\mathcal{P}) - 2i\partial_- M\Big] = \Big[(\alpha+\alpha')\,(\mathcal{F} - i\mathcal{P}) - 2i\,(\alpha-\alpha')\,\partial_- M\Big] \\
\delta\widetilde{\lambda}^- &= 0 \\
\delta\mathcal{P} &= \alpha^{(B)}\frac{1}{4}\Big[-\partial_+\widetilde{\lambda}^- + \partial_-\lambda^-\Big] = \frac{1}{4}\Big[-(\alpha+\alpha')\,\partial_+\widetilde{\lambda}^- + (\alpha-\alpha')\,\partial_-\lambda^-\Big] \\
\delta X^i &= \alpha^{(B)}\Big[i\,c^{gauge}\,q^i_j X^j\Big] = \Big[i\,\alpha_g c^{gauge}\,q^i_j X^j\Big] \\
\delta X^{i^*} &= \alpha^{(B)}\Big[\psi^{i^*} + \widetilde{\psi}^{i^*} + i\,c^{gauge}\,q^i_j X^{j^*}\Big] \\
&= \Big[(\alpha+\alpha')\psi^{i^*} + (\alpha-\alpha')\widetilde{\psi}^{i^*} + i\,\alpha_g c^{gauge}\,q^i_j X^{j^*}\Big] \\
\delta\psi^i &= \alpha^{(B)}\Big[-i\frac{1}{2}\nabla_+ X^i + i\,M\,q^i_j X^j + i\,c^{gauge}\,q^i_j\psi^j\Big] \\
&= \Big[-i\,(\alpha+\alpha')\frac{1}{2}\nabla_+ X^i + i\,(\alpha-\alpha')M\,q^i_j X^j + i\,\alpha_g\,c^{gauge}\,q^i_j\psi^j\Big] \\
\delta\widetilde{\psi}^i &= \alpha^{(B)}\Big[-i\frac{1}{2}\nabla_- X^i - i\,M^*\,q^i_j X^j + i\,c^{gauge}\,q^i_j\widetilde{\psi}^j\Big] \\
&= \Big[-i\,(\alpha-\alpha')\frac{1}{2}\nabla_- X^i + i\,(\alpha+\alpha')M\,q^i_j X^j + i\,\alpha_g\,c^{gauge}\,q^i_j\psi^j\Big] \\
\delta\psi^{i^*} &= \alpha^{(B)}\Big[\eta^{i^*j}\,\partial_j W - i\,c^{gauge}\,q^{i^*}_{j^*}\psi^{j^*}\Big] \\
&= \Big[(\alpha-\alpha')\,\eta^{i^*j}\,\partial_j W - i\,\alpha_g\,c^{gauge}\,q^{i^*}_{j^*}\widetilde{\psi}^{j^*}\Big]
\end{aligned}$$

$$\begin{aligned}\delta\widetilde{\psi}^{i^*} &= \alpha^{(B)}\left[-\eta^{i^*j}\,\partial_j W + i\,c^{gauge}\,q^{i^*}_{j^*}\widetilde{\psi}^{j^*}\right] \\ &= \left[(\alpha+\alpha')\,\eta^{i^*j}\,\partial_j W + i\,\alpha_g\,c^{gauge}\,q^{i^*}_{j^*}\widetilde{\psi}^{j^*}\right]\end{aligned} \tag{7.6.8}$$

The next point is the analysis of *Step 5*: we devote the following subsections to this task.

### 7.6.2 Interpretation of the A-model and topological $\sigma$-models

In this and the following subsection we consider the interpretation of the topological field theories described by the A- and B-models of the N=2 gauge-coupled Landau–Ginzburg theory. We begin with the A-model. To this end we recall the structure of a pure topological Yang–Mills theory described in section 7.3. In any space–time dimensions the field content of this theory is given by Table 7.1, where $A = A_\mu\, dx^\mu$ is the gauge field, $\psi = \psi_\mu\, dx^\mu$ the ghost of the topological symmetry, $c = c^{gauge}$ the ghost of the ordinary gauge symmetry and $\phi$ the ghost for the ghosts (indeed the ghost 1-form $\psi$ is, by itself a gauge field). These fields enter the *gauge-free* topological BRST algebra (7.3.13), which follows from the ghost form Bianchi identities (7.3.28), by removing the BRST-rheonomic conditions:

$$\widehat{F} = F_{ab}\, V^a \wedge V^b \tag{7.6.9}$$

and introducing the identifications (7.3.27) in the ghost number expansion (7.3.26) of the semi-extended field strength. The other fields appearing in Table 7.1 are either anti-ghosts or auxiliary fields. Indeed the complete, *gauge-fixed* topological BRST algebra is obtained by adjoining to Eqs. (7.3.1) the following:

$$\begin{aligned} s\,\overline{c} &= b \\ s\,\overline{\psi} &= T \\ s\,\overline{\phi} &= \overline{\eta} \end{aligned} \tag{7.6.10}$$

which are a subset of Eqs. (7.3.15). In (7.6.10) $\overline{c}$, $\overline{\psi}$, $\overline{\phi}$ are the anti-ghosts and $T$ and $\overline{\eta}$ are the auxiliary fields, as displayed in Table 7.1. Actually, rather than $\overline{\psi}$ and $T$ it is more convenient to use, as anti-ghost and auxiliary field, the functionals $\overline{\Upsilon}$ and $B$ defined by Eqs. (7.3.16), in such a way that

$$s\,\overline{\Upsilon} = B \tag{7.6.11}$$

is an identity.

In the quantum action, $b$ is the Lagrange multiplier for the gauge-fixing of the ordinary gauge transformations, while $T_\mu$ (or rather its functional $B_{\mu\nu}$) is the Lagrange multiplier associated with the gauge-fixing of the topological symmetry. Finally $\overline{\eta}$ is utilized to gauge-fix the gauge invariance of the topological ghost $\psi_\mu$. Indeed the quantum action has the form (7.3.19), where the gauge fermion is the sum of a gauge fermion fixing the topological symmetry ($\Psi_{topol}$), plus one fixing the ordinary gauge symmetry ($\Psi_{gauge}$), plus a last one fixing the gauge of the ghosts. In D=4 the classical action is the integral of the first Chern class $S_{class} = \int tr\ (F \wedge F)$ (see Eq. (7.3.2)), while in D=2 as classical action one takes the integral of the field strength in the direction of the centre of the gauge Lie algebra $S_{class} = \frac{\theta}{2\pi} \int F_{centre}$. The topological gauge-fixing must break the invariance under continuous deformations of the connection still preserving ordinary gauge-invariance. In four dimensions a convenient gauge condition that satisfies this requirement is provided by enforcing self-duality of the field strength, namely the instanton conditions (see Eqs. (7.3.20) and (7.3.21). In two dimensions, these equations have no meaning and the topological gauge-fixing (7.3.20), (7.3.21) can just be replaced by the condition of constant curvature ($\mathcal{F}_{centre} = const$) where, as already stated, $\mathcal{F}_{centre}$ denotes the field strength of the gauge group restricted to the centre of the Lie algebra, namely the only components of the field strength that, being gauge-invariant, can be given a constant value. This makes sense under the assumption that, within the set of all gauge connections characterized by the same first Chern class $\int F_{centre}$ (value of the classical action), there is always at least one that has a constant field strength in the centre-direction and a vanishing field strength in the other directions (almost flat connections). Hence we set

$$\begin{aligned} \Psi_{topol} &= tr \left\{ \overline{\chi}_{\rho\sigma} \left( \mathcal{F}_{\mu\nu} + const\, \varepsilon_{\mu\nu} \right) g^{\rho\mu} g^{\sigma\nu} \right\} \\ \Psi_{gauge} &= tr \left\{ \overline{c} \left( \partial_\rho A_\mu g^{\rho\mu} + b \right) \right\} \\ \Psi_{ghost} &= tr \left\{ \overline{\phi}\, \partial_\rho \psi_\mu g^{\rho\mu} \right\} \end{aligned} \tag{7.6.12}$$

fixing the ordinary gauge symmetry of the physical gauge boson $A_\mu$ and of the topological ghost $\psi_\mu$ by means of the Lorentz gauge. As we see, $\overline{c}$ is the anti-ghost of ordinary gauge symmetry, while $\overline{\phi}$ is the anti-ghost for the gauge symmetry of the topological ghost $\psi$.

In the case where the topological Yang–Mills theory is coupled to some topological matter system, the gauge-fixing of the topological gauge symmetry can be achieved by requiring that the field strength $\mathcal{F} = F_{+-}^{(centre)}$ be equal to some appropriate function of the matter fields:

$$\mathcal{F} = 2i\, \mathcal{P}(X) \tag{7.6.13}$$

In this case we can also suppress the auxiliary field $B$ and replace the anti-ghost part of the BRST algebra with the equations

$$\begin{aligned} s\overline{c} &= b \\ s\overline{\chi}_{+-} &= \left( \frac{1}{2}\mathcal{F} - i\mathcal{P} \right) \\ s\overline{\phi} &= \overline{\eta} \end{aligned} \tag{7.6.14}$$

Correspondingly, the gauge fermion $\Psi_{topol}$ of Eq. (7.6.12) can be replaced with

$$\Psi_{topol} = 2\overline{\chi}_{+-} \left( \frac{1}{2}\mathcal{F} + i\mathcal{P} \right) \tag{7.6.15}$$

It is now worth noting that, for consistency with the BRST algebra (7.3.13), if we define the 2-form $\Theta^{(2)} = 2i\mathcal{P}(X,\overline{X})\, e^+ \wedge e^-$, we must have $s\,\Theta^{(2)} = d\psi_{(centre)}$. Indeed, by restriction to the centre of the Lie algebra we obtain an abelian topological gauge theory, for which $s\,F = d\psi$. Reconsidering the A-twisted BRST transformation rules of the gauge multiplet, included in Eq. (7.6.5), we realize that the property required for the function $\mathcal{P}(X,\overline{X})$ is satisfied by the auxiliary field $\mathcal{P}$ of the gauge multiplet, provided we identify $\psi = \frac{i}{2}\left(\tilde{\lambda}^- e^- + \lambda^+ e^+\right)$. This is correct since, by looking at Eqs. (7.6.5), we recognize that a subset of the fields does indeed describe a topological Yang–Mills theory upon the identifications

$$\begin{aligned} \psi &= \frac{i}{2}\left(\tilde{\lambda}^- e^- + \lambda^+ e^+\right) \\ \phi &= M \\ \overline{\phi} &= M^* \\ \overline{\eta} &= \frac{1}{2}\left(\tilde{\lambda}^+ + \lambda^+\right) \\ \overline{\chi}_{+-} &= \frac{1}{2}\left(\tilde{\lambda}^+ - \lambda^-\right) \end{aligned} \tag{7.6.16}$$

We also see that the descent equations

$$\begin{aligned} s\Theta^{(2)} &= d\Theta^{(1)} \\ s\Theta^{(1)} &= d\Theta^{(0)} \\ s\Theta^{(0)} &= 0 \end{aligned} \tag{7.6.17}$$

are solved by the position

$$\begin{aligned} \Theta^{(2)} &= 2i\mathcal{P}\, e^+ \wedge e^- = 2i\mathcal{P}(X,\overline{X})\, e^+ \wedge e^- \\ \Theta^{(1)} &= \psi = \frac{i}{2}\left(\tilde{\lambda}^- e^- + \lambda^+ e^+\right) \\ \Theta^{(0)} &= \phi = M \end{aligned} \tag{7.6.18}$$

so that the quantum action of the topological gauge theory can be topologically deformed by

$$S_{quantum} \longrightarrow S_{quantum} - i\,r \int \Theta^{(2)} \tag{7.6.19}$$

Altogether we see that the classical action $S_{class} = \frac{\theta}{2\pi}\int F_{centre}$, plus the topological deformation $-i\,r\int \Theta^{(2)}$, constitute the Fayet–Iliopoulos term, while the remaining terms in the action (5.5.9) are BRST-exact and come from the gauge-fixings:

$$s\int\left[\overline{\chi}_{+-}\left(\frac{\mathcal{F}}{2} - i\mathcal{P}\right) + \overline{\phi}\left(\partial_+\psi_- + \partial_-\psi_+\right)\right] \tag{7.6.20}$$

On the other hand the matter multiplets with their fermions span a *flat topological $\sigma$-model* (see section 7.4) coupled to the topological gauge system. As discussed in section

7.4, the topological symmetry in the case of a $\sigma$-model is the possibility of deforming the embedding functions $X^i(z,\bar{z})$ in an arbitrary way. Correspondingly, in the absence of gauge couplings, the *gauge-free* topological BRST algebra is very simple and it is given by Eqs. (7.4.7). In the presence of a coupling to a topological gauge theory, defined by the covariant derivative:

$$\nabla X^i = dX^i + i\mathcal{A}\, q^i_j\, X^j \tag{7.6.21}$$

the *gauge-free* BRST algebra of the matter system becomes

$$\begin{aligned} s\,X^i &= c^i + i\,c^g\,q^i_j\,X^j \\ s\,X^{i^*} &= c^{i^*} - i\,c^g\,q^i_j\,X^{j^*} \\ s\,c^i &= -i\,q^i_j\left(c^j\,c^g + X^j\,\phi\right) \\ s\,c^{i^*} &= -i\,q^i_j\left(c^{j^*}\,c^g + X^{j^*}\,\phi\right) \end{aligned} \tag{7.6.22}$$

the last two of Eqs. (7.6.22) being uniquely fixed by the nilpotency $s^2 = 0$ of the Slavnov operator.

Comparing with Eqs. (7.6.5) we see that, indeed, Eqs. (7.6.22) are reproduced if we make the following idenfications:

$$\begin{aligned} c^i &= \psi^i \\ c^{i^*} &= \tilde{\psi}^{i^*} \end{aligned} \tag{7.6.23}$$

The remaining two fermions are to be identified with the anti-ghosts:

$$\begin{aligned} \bar{c}^i &= \tilde{\psi}^i \\ \bar{c}^{i^*} &= \psi^{i^*} \end{aligned} \tag{7.6.24}$$

and their BRST variation, following from Eqs. (7.6.5), yields the topological gauge-fixing of the matter sector:

$$\begin{aligned} s\,\bar{c}^i &= -i\,q^i_j\,\bar{c}^j\,c^g + \eta^{ij^*}\,\partial_{j^*}\,W^* + \frac{i}{2}\nabla_- X^i \\ s\,\bar{c}^{i^*} &= i\,q^i_j\,\bar{c}^{j^*}\,c^g + \eta^{i^*j}\,\partial_{j^*}\,W^* - \frac{i}{2}\nabla_+ X^i \end{aligned} \tag{7.6.25}$$

Following Witten [263, 262] we easily recover the interpretation of the "*instantons*" encoded in the topological gauge-fixings dictated by Eqs. (7.6.25) and (7.6.14). Indeed we just recall that the functional integral is concentrated on those configurations that are a fixed point of the (0|2) supergroup transformations. Looking at Eqs. (7.6.5) we see that such configurations have all the ghosts and anti-ghosts equal to zero while the bosonic fields satisfy the following conditions:

$$\begin{aligned} \eta^{ij^*}\,\partial_{j^*}\,W^*(X^*) &= 0 \\ \eta^{i^*j}\,\partial_j\,W(X) &= 0 \\ \nabla_- X^i &= 0 \\ \nabla_+ X^{i^*} &= 0 \\ \mathcal{F} &= 2i\mathcal{P} = -i\left(\mathcal{D}(X,X^*) - r\right) \end{aligned} \tag{7.6.26}$$

where $\mathcal{D}(X, X^*) = \sum_i q^i |X^i|^2$ is the momentum map function defined in section 5.5. Hence the instantons are holomorphic maps from the world-sheet to a locus in $\mathbb{C}^n$ characterized by the equations $\eta^{i^* j} \partial_j W(X) = 0$. In the case of the theory that interpolates between $\sigma$-models and Landau–Ginzburg models (see section 5.5), we have $W(X) = X^0 \mathcal{W}(X^i)$ and the aforementioned locus is the hypersurface $\mathcal{W}(X^i) = 0$, $(X^0 = 0)$ in a weighted projective space $W\mathbb{CP}_{n-2}$, the weights of the homogeneous coordinates $X^i$ being their charges. In other words the instantons are holomorphic solutions of the corresponding $N = 2$ $\sigma$-model. The value of the action on these *instantons* has been calculated by Witten and the result is easily retrieved in our notations. Indeed the lagrangian (5.5.16) restricted to the bosonic fields of zero ghost number is given by

$$\begin{aligned} \mathcal{L}_{no\text{-}ghost} &= \frac{1}{2}\mathcal{F}^2 + 2\mathcal{P}^2 - 2r\mathcal{P} + \frac{\theta}{2\pi}\mathcal{F} \\ &+ (\nabla_+ X^{i^*} \nabla_- X^i + \nabla_- X^{i^*} \nabla_+ X^i) + 2\mathcal{P}\,\mathcal{D}(X, X^*) \end{aligned} \tag{7.6.27}$$

Using Eqs. (7.6.26) and $[\nabla_-, \nabla_+] X^i = i\mathcal{F} q^i_j X^j$, we obtain

$$\int \mathcal{L}_{no\text{-}ghost} = \left(\frac{\theta}{2\pi} + ri\right) \int \mathcal{F} = -2\pi i\, t\, N \tag{7.6.28}$$

where $N = \frac{i}{2\pi} \int \mathcal{F}$ is the winding number and the parameter $t = \frac{\theta}{2\pi} + ri$ was defined in Eq. (5.5.8).

In section 7.10 we analyse the meaning of the instanton conditions (7.6.26) and the way the various instanton sectors, labelled by the winding number $N$, contribute to the calculation of topological correlation functions. We shall do that after having considered the general problem of evaluating topological correlators in the topological $\sigma$-model. There we shall also appreciate the subtle difference between the A-twist of an N=2 $\sigma$-model and the A-twist of the N=2 gauge theory which admits the $\sigma$-model as a low energy effective lagrangian. Prior to this discussion, we consider the interpretation of the B-model emerging from the same theory.

### 7.6.3 Interpretation of the B-model and topological Landau–Ginzburg theories

In order to identify the system described by the B-model we discuss the structure of a topological Landau–Ginzburg theory coupled to an ordinary abelian gauge theory. To this end we begin with the structure of a *topological rigid Landau-Ginzburg theory.* The rigid Landau–Ginzburg model was defined in section 5.5.5 and it is described by the action (5.5.40). It has the R symmetries (5.5.43) and it is N=2 supersymmetric under the transformations following from the rheonomic parametrizations (5.5.41). The rigid topological Landau–Ginzburg model has the same action (5.5.40), but the spin of the fields is changed, namely it is that obtained by B-twisting: the scalar fields $X^i$ and $X^{i^*}$ mantain spin 0, as in the ordinary model, while the spin 1/2 fermions acquire either spin

0 or spin $\pm 1$. Specifically $\psi^{i^*}$, $\widetilde{\psi}^{i^*}$ both have spin 0, while $\psi^i$ and $\widetilde{\psi}^i$ have spin $s = 1$ and $s = -1$, respectively. In view of this fact it is convenient to introduce the new variables

$$\begin{aligned} C^{i^*} &= \psi^{i^*} + \widetilde{\psi}^{i^*} \\ \overline{C}^i &= \left( \overline{C}^i_+ \, e^+ + \overline{C}^i_- \, e^- \right) \\ &= \left( \psi^i \, e^+ + \widetilde{\psi}^i \, e^- \right) \\ \theta^{i^*} &= \psi^{i^*} - \widetilde{\psi}^{i^*} \end{aligned} \tag{7.6.29}$$

and rewrite the action (5.5.40) in the form

$$\begin{aligned} \mathcal{L}_{(topol\ LG)} &= - \left( \partial_+ X^{i^*} \, \partial_- X^i + \partial_- X^{i^*} \, \partial_+ X^i \right) \\ &+ 2i \left( \overline{C}^i_+ \, \partial_- C^{i^*} + \overline{C}^i_- \, \partial_+ C^{i^*} \right) \\ &+ 2i \left( \overline{C}^i_+ \, \partial_- \theta^{i^*} - \overline{C}^i_- \, \partial_+ \theta^{i^*} \right) \\ &+ 8 \overline{C}^i_+ \overline{C}^j_- \, \partial_i \partial_j \mathcal{W} + 4 \, C^{i^*} \, \theta^{j^*} \, \partial_{i^*} \partial_{j^*} \overline{\mathcal{W}} \\ &+ 8 \, \partial_i \mathcal{W} \, \partial_{i^*} \overline{\mathcal{W}} \end{aligned} \tag{7.6.30}$$

If we denote by $[\Omega]_* = \Omega_+ e^+ - \Omega_- e^-$ the Hodge dual of the 1-form $\Omega = \Omega_+ e^+ + \Omega_- e^-$, then the action (7.6.30) can be rewritten in the following more condensed form:

$$\begin{aligned} S_{topol\ LG} &= \int \mathcal{L}_{(topol\ LG)} \, e^+ \wedge e^- \\ &= \int \Big\{ dX^i \wedge \left[ dX^{i^*} \right]_* - 2i \, \overline{C}^i \wedge \left[ dC^{i^*} \right]_* \\ &- 4 \overline{C}^i \wedge \overline{C}^j \, \partial_i \partial_j \mathcal{W} - 2i \, d\overline{C}^i \, \theta^{i^*} \\ &+ 4 \left\{ C^{i^*} \, \theta^{j^*} \, \partial_{i^*} \partial_{j^*} \overline{\mathcal{W}} + 2 \, \partial_i \mathcal{W} \, \partial_{i^*} \overline{\mathcal{W}} \right\} e^+ \wedge e^- \Big\} \end{aligned} \tag{7.6.31}$$

and it is closed under the following BRST transformations:

$$\begin{aligned} s \, X^i &= 0 \\ s \, X^{i^*} &= C^{i^*} \\ s \, C^{i^*} &= 0 \\ s \, \theta^{i^*} &= 2 \eta^{i^* j} \, \partial_j \mathcal{W} \\ s \, \overline{C}^i &= -\frac{i}{2} \, dX^i \end{aligned} \tag{7.6.32}$$

The ghost numbers assigned to the above fields are those emerging from the B-twist. The *scalar fields* $X^i$ and $X^{i^*}$ have ghost number $\#gh = \frac{2\omega^i}{d}$ and $\#gh = -\frac{2\omega^i}{d}$, respectively. They behave as physical fields. On the other hand $C^{i^*}$ has ghost number $\#gh = 1 - \frac{2\omega^i}{d}$ and behaves as a ghost, while $\theta^{i^*}$ and the 1-form $\overline{C}^i$, which have ghost numbers respectively given by $\#gh = 1 - \frac{2\omega^i}{d}$ and $\#gh = -1 + \frac{2\omega^i}{d}$, behave as anti-ghosts. As one sees in this case the ghost numbers are fractional and the subdivision of the fields into physical, ghosts and anti-ghosts cannot be done by looking simply at the values of the corresponding ghost numbers. However, if we decide that the scalar fields $X$ are physical, then the interpretation of the remaining fields as ghosts and anti-ghosts is fixed by the structure of the BRST transformations. In particular the *gauge-free BRST algebra* is given by the first three of Eqs. (7.6.32): it quantizes a symmetry which corresponds to a deformation of the complex structure of the target coordinates $X^i, X^{i*}$. The variation of the anti-ghosts defines the gauge-fixings:

$$\begin{aligned} \partial_i \mathcal{W}(X) &= 0 \\ dX^i &= 0 \end{aligned} \qquad (7.6.33)$$

which select, as "*instantons*", the constant maps ($dX^i = 0$) from the world-sheet to the critical points ($\partial_i \mathcal{W}(X_0) = 0$) of the superpotential $\mathcal{W}$. The action (7.6.31) is the sum of a BRST non-trivial part:

$$\Omega[W] = -\int \left[ 4\overline{C}^i \wedge \overline{C}^j \partial_i \partial_j \mathcal{W} + 2i\overline{C}^i \wedge d\theta^{i*} \right] \qquad (7.6.34)$$

which is closed ($s\Omega = 0$), but not exact ($\Omega \neq s(something)$) plus two BRST exact terms:

$$\begin{aligned} K^{(Kin)} &= \int \left\{ dX^i \wedge \left[ dX^{i*} \right]_\star - 2i\overline{C}^i \wedge \left[ dC^{i*} \right]_\star \right\} \\ &= s\int \Psi^{(Kin)} = s\int 2i\overline{C}^i \wedge \left[ dX^{i*} \right]_\star \\ K^{(W)} &= \int 4\left\{ C^{i*}\theta^{j*} \partial_{i*}\partial_{j*}\overline{\mathcal{W}} + 2\partial_i \mathcal{W}\, \partial_{i*}\overline{\mathcal{W}} \right\} e^+ \wedge e^- \\ &= s\int \Psi^{(W)} = s\int 4\partial_{j*}\overline{\mathcal{W}}\,\theta^{j*} e^+ \wedge e^- \end{aligned} \qquad (7.6.35)$$

which correspond to the BRST variation of the gauge fermions associated with the two gauge-fixings (7.6.33).The *rigid topological Landau–Ginzburg model* has been extensively studied in the literature. In the following sections we are going to make a detailed study of its correlation functions also as a preparation for the study of Picard–Fuchs equations. Shortly from now we shall also address the questions of the topological observables and of the ghost current anomaly of this model. Prior to this, however, we want to conclude the programme we have begun and find out the interpretation of the B-model emerging from the two-phase N=2 gauge theory we have been considering. For this purpose we are interested in the case where the *topological Landau–Ginzburg model* is *coupled to an ordinary abelian gauge theory.* Under this circonstance the BRST algebra (7.6.32) is replaced by

$$\begin{aligned} s\,A' &= dc^g \\ s\,F' &= 0 \end{aligned}$$

$$
\begin{aligned}
s\,c^g &= 0 \\
s\,X^i &= -i\,c^g\,q^i_j\,X^j \\
s\,X^{i^\star} &= C^{i^\star} + i\,c^g\,q^i_j\,X^{j^\star} \\
s\,C^{i^\star} &= -i\,c^g\,q^i_j\,C^{j^\star} \\
s\,\theta^{i^\star} &= 2\eta^{i^\star j}\,\partial_j W + i\,c^g\,q^i_j\,\theta^{j^\star} \\
s\,\overline{C}^i &= -\frac{i}{2}\nabla X^i - i\,c^g\,q^i_j\,\overline{C}^j
\end{aligned}
\qquad (7.6.36)
$$

where $\nabla(\ldots)^i = d(\ldots)^i + i\,A'\,q^i_j\,(\ldots)^j$ denotes the gauge-covariant derivative and the superpotential $\mathcal{W}(X)$ of the rigid theory has been replaced by $W(X)$, namely the superpotential of the gauged-coupled model. The action (7.6.31) is also replaced by a similar expression where the ordinary derivatives are converted into covariant derivatives.

The topological system emerging from the B-twist of the N=2 model discussed in section 5.5 is precisely a Landau–Ginzburg model of this type: in particular, differently from the case of the A-twist, there is no *topological gauge theory*, rather an ordinary gauge theory plus a topological massive vector. The identification is better discussed at the level of the BRST algebra comparing Eqs. (7.6.36) with Eqs. (7.6.8) after setting

$$
\begin{aligned}
\mathcal{A}' &= [\mathcal{A}_+ + 2i\,M]\,e^+ + [\mathcal{A}_- - 2i\,M^\star]\,e^- \\
\mathcal{B} &= M\,e^+ + M^\star\,e^- \\
\psi^{(mass)} &= \frac{1}{4}\left(\lambda^-\,e^+ + \tilde{\lambda}^-\,e^-\right) \\
\overline{\chi} &= \lambda^+ + \tilde{\lambda}^+ \\
\overline{\chi}^{(mass)} &= \frac{i}{2}\left[\lambda^+ - \tilde{\lambda}^+\right] \\
C^{i^\star} &= \psi^{i^\star} + \tilde{\psi}^{i^\star} \\
\theta^i &= \psi^{i^\star} - \tilde{\psi}^{i^\star} \\
\overline{C}^i &= \psi^i\,e^+ + \tilde{\psi}^i\,e^-
\end{aligned}
\qquad (7.6.37)
$$

With these definitions the BRST transformations of Eqs. (7.6.8) become indeed identical with those of Eqs. (7.6.36) plus the following

$$
\begin{aligned}
s\,\mathcal{B} &= \psi^{(mass)} \\
s\,\psi^{(mass)} &= 0 \\
s\,\overline{\chi}^{(mass)} &= \mathcal{P}(X,\overline{X}) + (\partial_+\mathcal{B}_- - \partial_-\mathcal{B}_+) \\
s\,\overline{\chi} &= \mathcal{F}'
\end{aligned}
\qquad (7.6.38)
$$

The first two of Eqs. (7.6.38) correspond to the *gauge-free* BRST algebra of the topological massive vector, $\psi^{(mass)}$ being the 1-form ghost associated with the continuous deformation symmetry of the vector $\mathcal{B}$. The second two of Eqs. (7.6.38) are BRST transformations of anti-ghosts and the left hand side defines the gauge-fixings of the massive vector and gauge vector, respectively, namely

$$\begin{aligned} \mathcal{P}(X,\overline{X}) + (\partial_+\mathcal{B}_- - \partial_-\mathcal{B}_+) &= 0 \\ \mathcal{F}' = \partial_+\mathcal{A}_- - \partial_+\mathcal{A}_+ &= 0 \end{aligned} \tag{7.6.39}$$

Actually, looking at Eqs. (7.6.8) we realize that the configurations corresponding to a *fixed point* of the (0|2) supergroup are characterized by all the fermions ( = ghosts + anti-ghosts) equal to zero and by

$$\begin{aligned} M = M^\star &= 0 \implies \mathcal{B} = 0 \\ \mathcal{F} &= 0 \implies \mathcal{F}' = 0 \\ \mathcal{P}(X,\overline{X}) &= 0 \\ \eta^{i^\star j}\,\partial_j W(X) &= 0 \\ dX^i &= 0 \end{aligned} \tag{7.6.40}$$

Hence in the B-twist the functional integral is concentrated on the constant maps from the world-sheet to the extrema of the classical scalar potential (5.5.20). As we have seen, in the A-twist the functional integral was concentrated on the holomorphic maps to such extrema: furthermore, in the A-twist the classical extrema were somewhat modified by the winding number effect since the equation $\mathcal{P} = 0$ was replaced by $\mathcal{P} = -\frac{i}{2}\mathcal{F}$. In the B-twist no *instantonic* effects modify the definition of classical extremum. The extrema of the scalar potential can be a point (Landau–Ginzburg phase) or a manifold ($\sigma$-model phase). The B-twist selects the constant maps in either case, and the A-twist selects the holomorphic maps in either case. However, in the Landau–Ginzburg phase the holomorphic maps to a point are the same thing as the constant maps, so that in this phase the instantons of the A-model coincide with those of the B-model.

## 7.7 Correlators of the Topological Sigma Model

We come now to the explicit calculation of correlators in topological field theories, beginning with the topological sigma model, which, as shown, can be viewed as the A-twist of the N=2 sigma model. Later we shall also consider the B-twist of the same N=2 field theory, where the underlying topological symmetry is the deformation of the target space complex structure, rather than the deformation of the embedding map.

To address the problem of correlator calculation, let us first summarize what we have so far learned about the basic structure of topological theories.

Given a collection of physical operators in a quantum field theory, their correlation functions depend, in general, on the positions of the operators and on the metric defined on the world manifold $\mathcal{M}$, on which we consider the correlators. The basic feature of a topological field theory is that the correlation functions of the observables are independent of the metric on $\mathcal{M}$, and therefore also independent of the position of the operators. The fact that the physical correlation functions are metric-independent is the consequence of a symmetry of the topological theory which reduces the Hilbert space $\mathcal{H}$ to the space $\mathcal{H}_{phys}$ of physical states, and causes the stress tensor $T_{\alpha\beta}$ to decouple from physical correlation functions. The symmetry which is responsible for all this is generated by a nilpotent BRST charge $Q_{BRST}$:

$$Q^2_{BRST} = 0 \tag{7.7.1}$$

which commuted with any field $\phi$, produces its *Slavnov* variation:

$$\{ Q_{BRST} \, , \, \phi \} \; = \; s\,\phi \tag{7.7.2}$$

The physical states are characterised by the cohomology of the operator $Q_{BRST}$. The space $\mathcal{H}_{phys}$ of the physical states is equal to

$$\mathcal{H} = \frac{\text{ker Q}}{\text{im Q}} \tag{7.7.3}$$

so that physical observables are defined up to $Q_{BRST}$ commutators:

$$\phi_i \equiv \phi_i + \{Q, \phi_i\} \tag{7.7.4}$$

The $Q_{BRST}$ invariance of the theory implies that physical correlators are independent of the representative of each $\phi_i$. Moreover the stress–energy tensor of topological theories is a $Q_{BRST}$ commutator:

$$T_{\alpha\beta} = \{Q_{BRST}, G_{\alpha\beta}\} \tag{7.7.5}$$

and thus vanishes inside the correlation functions. This ensures that the physical correlators are indeed independent of the two-dimensional metric $g_{\alpha\beta}$. The second important property of topological field theory is that correlation functions can be factorized by inserting a complete set of states in the intermediate channels. This amounts to the equation

$$\mathbb{1}_{phys} = \sum_{i,j} |\phi_i \langle \eta^{ij} \rangle \phi_j| \tag{7.7.6}$$

where $\eta^{ij}$ is the metric in $\mathcal{H}_{phys}$ defined by the inverse of the two point function on the sphere (see for example [116])

$$\langle \phi_i \phi_j \rangle = \eta_{ij} \tag{7.7.7}$$

In view of these facts, in any 2D topological field theory the key objects are the solutions of the descent equations (7.6.17). Indeed they provide the means to deform the topological action according to the generalization of Eq. (7.6.19):

$$S_{quantum} \longrightarrow S_{quantum} + \sum_A t_A \int \Theta^{(2)}_A \tag{7.7.8}$$

$\Theta^{(2)}_A$ being a complete base of solutions to Eq. (7.6.17), and to study the deformed correlation functions:

$$c_{A_1, A_2, \ldots, A_N}(t) = \langle\, \Theta^{(0)}_{A_1}, \ldots, \Theta^{(0)}_{A_N} \exp\left[ \sum_A t_A \int \Theta^{(2)}_A \right] \rangle \tag{7.7.9}$$

where the 0-forms $\Theta^{(0)}_{A_i}$ play the role of a complete set of physical observables. Hence in each 2D topological theory the final goal is that of computing the correlators (7.7.9) as functions of the coupling parameters $t_A$.

### 7.7.1 The topological $\sigma$-model or A-twist case

In the case of the topological sigma model the solution of the descent equations (7.6.17) was given in Eqs. (7.4.14) and was traced back to the Dolbeault cohomology of the target space. Indeed, by means of Eq. (7.4.14) one realizes an injection:

$$H^{\star}_{Dolbeault}\left(\mathcal{M}\right) \longrightarrow \mathcal{H}^{\star}_{BRST} \tag{7.7.10}$$

of the Dolbeault cohomology ring into the BRST cohomology ring. This injection is not an isomorphism since the product operations in the two rings turn out to be different. Let $[\omega_i], [\omega_j] \in H^{\star}_{Dolbeault}$ be two Dolbeault cohomology classes and $\Theta^{(0)}_{\omega_i}, \Theta^{(0)}_{\omega_j} \in \mathcal{H}^{\star}_{BRST}$ be the corresponding topological observables; then we have

$$\Theta^{(0)}_{\omega_i} \cdot \Theta^{(0)}_{\omega_j} \neq \Theta^{(0)}_{\omega_i \wedge \omega_j} \tag{7.7.11}$$

where the product in the BRST ring can be defined via the three-point function

$$c_{ijk}(t) = \langle\, \Theta^{(0)}_{\omega_i}\, \Theta^{(0)}_{\omega_j}\, \Theta^{(0)}_{\omega_k} \,\rangle \tag{7.7.12}$$

Indeed, using the inverse of the metric $\eta_{ij}$ defined by the two-point function (7.7.7):

$$\eta_{ij}(t) = \langle\, \Theta^{(0)}_{\omega_i}\, \Theta^{(0)}_{\omega_j} \,\rangle \tag{7.7.13}$$

and setting

$$\Theta^{(0)}_{\omega_i} \cdot \Theta^{(0)}_{\omega_j} = \overset{(BRST)}{f^{k}_{ij}}(t)\, \Theta^{(0)}_{\omega_k} \tag{7.7.14}$$

where the structure functions of the BRST cohomology ring are

$$\overset{(BRST)}{f^{l}_{ij}}(t) = c_{ijr}(t) \left(\eta^{-1}\right)^{rl}(t) \tag{7.7.15}$$

we obtain the self-consistency check

$$\begin{aligned} c_{ijk}(t) &= \langle\, \Theta^{(0)}_{\omega_i}\, \Theta^{(0)}_{\omega_j}\, \Theta^{(0)}_{\omega_k} \,\rangle \\ &= \overset{(BRST)}{f^{l}_{ij}}(t)\, \langle\, \Theta^{(0)}_{\omega_l}\, \Theta^{(0)}_{\omega_k} \,\rangle \\ &= \overset{(BRST)}{f^{l}_{ij}}(t)\, \eta_{lk}(t) \end{aligned} \tag{7.7.16}$$

Naming $\overset{(Dbt)}{f^{l}_{ij}}$ the structure constants of the Dolbeault cohomology ring:

$$\omega_i \wedge \omega_j = \overset{(Dbt)}{f^{k}_{ij}}\, \omega_k \tag{7.7.17}$$

we find that the topological structure functions $\overset{(BRST)}{f_{ij}^{l}}(t)$ differ from the classical structure constants $\overset{(Dbt)}{f_{ij}^{l}}$. Actually it will turn out that $\overset{(BRST)}{f_{ij}^{l}}(t)$ can be expanded in power series of the parameters

$$q_A = \exp\left[2\pi i t_A\right] \tag{7.7.18}$$

where $t_A$ are topological coupling constants and

$$\operatorname{Im} t_A = r_A \tag{7.7.19}$$

have the interpretation of radii of the target manifold $\mathcal{M}$, the 0-th order term in this series expansion being the structure constants of the Dolbeault cohomology ring. Hence the A-twisted topological sigma model defines a sort of *quantum deformation* of the Dolbeault cohomology ring that approaches the classical ring in the large radius limit:

$$r_A \longrightarrow \infty \tag{7.7.20}$$

To verify this structure we turn to the explicit calculation of the correlators.

Recalling Eqs. (7.4.8), (7.4.9), (7.4.10), (7.4.11) and introducing a complex coupling parameter $-2t$ to rescale the complete quantum action $S_{quantum} \longrightarrow -2t\, S_{quantum}$, the correlation function of any set of local observables $\Theta^{(0)}_{A_1}(x_1), \ldots, \Theta^{(0)}_{A_N}(x_N)$ is given by the following path integral:

$$c_{A_1,A_2,\ldots,A_N}(t) =$$

$$\int \mathcal{D}X\, \mathcal{D}c\, \mathcal{D}\bar{c}\, \exp\left[-2t\left(S_{class} + \left\{Q_{BRST}, \int \Psi_{(1,1)}\right\}\right)\right] \prod_{i=1}^{N} \Theta^{(0)}_{A_i}(x_i) \tag{7.7.21}$$

where $S_{class} = -i \int_{\Sigma_g} X^\star(K)$ is proportional to the pull-back on the Riemann surface $\Sigma_g$ of the Kähler 2-form $K$ and $\Psi_{(1,1)}$ is given by Eqs. (7.4.3), (7.4.21) with $\alpha = \beta = 1$. As already remarked, the value of $S_{class}$ depends only on the cohomology class of the Kähler 2-form $K$ and on the homotopy class of the map $X : \Sigma_g \longrightarrow \mathcal{M}_K$. In particular, let the second cohomology group of the target space be $H^2(\mathcal{M}_K, \mathbb{Z}) \approx \mathbb{Z}$ and let us normalize the Kähler metric in such a way that the periods of the Kähler 2-form $K = \frac{i}{2\pi} g_{ij^\star} dX^i \wedge dX^{j^\star}$, namely its integrals along a basis of non-contractible 2-cycles $C_A$ ( $A = 1, \ldots, b^2$), are integer numbers:

$$\int_{C_A} K = n_A \quad (n_A \in \mathbb{Z}) \tag{7.7.22}$$

Then we have $\int_{\Sigma_g} X^\star(K) = k$ where the integer $k$ is the *degree* of the map $X$ or, in the physicist's jargon, its *instanton number*. The field space decomposes into disconnected sectors of fixed instanton number for each of which we have $\exp\left[-2tS_{class}\right] = e^{2\pi i t k}$. More generally, if $\omega^{(1,1)}_a$ $(a = 1, \ldots, h^{(1,1)})$ is a basis of harmonic $(1,1)$-forms providing a set of integral generators for the Dolbeault group $H^{(1,1)}(\mathcal{M}_K)$, then the cohomology class of the Kähler 2-form is necessarily of the form

$$[K] = \sum_{a=1}^{h^{(1,1)}} c^a \left[\omega^{(1,1)}_a\right] \tag{7.7.23}$$

where $c_a$ are a set of parameters that are real in order for the Kähler metric to be hermitean. However for the sake of the topological $\sigma$-model these parameters can be extended to the complex domain without any harm since the rescaled parameters

$$t^a = t\,c^a \tag{7.7.24}$$

provide an instance of the complex topological coupling constants $t_{A_i}$ introduced in Eq. (7.7.9) and such that the correlators are analytic functions of these variables. Indeed let us consider, as a starting point, the degenerate classical action $S^0_{class} = 0$, which corresponds to a Kähler form of vanishing cohomology class $[K] = 0 \implies c^a = 0$. Define the vacuum expectation value $\langle \ldots \rangle_0$ as the path integral with respect to the quantum action $S^0_{quantum} = 0 + s\Psi_{(1,1)}$ associated with such a degenerate classical action:

$$\begin{aligned} &\langle \Theta^{(0)}_{A_1}(x_1) \ldots \Theta^{(0)}_{A_N}(x_N) \rangle_{(0)} = \\ &\int \mathcal{D}X\,\mathcal{D}c\,\mathcal{D}\bar{c}\, \exp\left[-2t\left(\left\{Q_{BRST}, \int \Psi_{(1,1)}\right\}\right)\right] \prod_{i=1}^{N} \Theta^{(0)}_{A_i}(x_i) \end{aligned} \tag{7.7.25}$$

Then the correlators (7.7.21) have the form (7.7.9) of deformed correlators, where the deformation is precisely provided by the classical action for a generic Kähler class:

$$-2t\,S_{class} = 2\pi i t \int_{\Sigma_g} X^\star(K) = 2\pi i \sum_{a=1}^{h^{(1,1)}} t^a \int_{\Sigma_g} X^\star(\omega^{(1,1)}_a) = 2\pi i \sum_{a=1}^{h^{(1,1)}} t^a \int_{\Sigma_g} \Theta^{(2)}_a \tag{7.7.26}$$

the pull-backs $X^\star(\omega^{(1,1)}_a)$ of the harmonic $(1,1)$-forms being the 2-forms observables $\Theta^{(2)}_a$. Consider now the image $X(\Sigma_g)$ of the world-sheet immersed in the target space $\mathcal{M}_K$. The two-dimensional submanifold $X(\Sigma_g) \subset \mathcal{M}_K$ is a 2-cycle and, as such, it belongs to a homology class:

$$[X(\Sigma_g)] = \sum_{A=1}^{2h^{(2,0)}+h^{(1,1)}} k_A C^A \tag{7.7.27}$$

where $k_A \in \mathbb{Z}$ are integer coefficients and where we have taken into account that the second Betti number is given by

$$b^2 = 2h^{(2,0)} + h^{(1,1)} \tag{7.7.28}$$

Indeed each of the harmonic (2,0)-, (1,1)- and (0,2)-forms is dual to a homology 2-cycle. As a consequence we obtain

$$\begin{aligned} \int_{\Sigma_g} X^\star(K) &= \sum_{a=1}^{h^{(1,1)}} c^a\,k_a \\ -2t\,S_{class} &= 2\pi i \sum_{a=1}^{h^{(1,1)}} t^a\,k_a \end{aligned} \tag{7.7.29}$$

having normalized the homology basis so that

$$\int_{C_a} \omega_b^{(1,1)} = \delta_b^a \qquad (a, b = 1, \ldots, h^{(1,1)}) \tag{7.7.30}$$

Each multiplet of integers $\{k_A\}$ characterizes a disconnected sector $\mathcal{B}_{\{k_A\}}$ of the field configuration space and the functional integral (7.7.21) can be decomposed into a sum of integrals over such instanton sectors:

$$\begin{aligned} &c_{A_1,A_2,\ldots,A_N}(t) = \\ &\int \mathcal{D}X\,\mathcal{D}c\mathcal{D}\overline{c}\,\exp\left[-2t\left(S_{class} + \left\{Q_{BRST}, \int \Psi_{(1,1)}\right\}\right)\right] \prod_{i=1}^{N} \Theta_{A_i}^{(0)}(x_i) = \\ &\sum_{\{k_A\}} \prod_{a=1}^{h^{(1,1)}} q_a^{k_a} \langle \prod_{i=1}^{N} \Theta_{A_i}^{(0)}(x_i) \rangle_{(0,\{k_A\})} \end{aligned} \tag{7.7.31}$$

where, specializing Eq. (7.7.25) to the various instanton sectors, we have defined the reduced correlators:

$$\begin{aligned} &\langle \prod_{i=1}^{N} \Theta_{A_i}^{(0)}(x_i) \rangle_{(0,\{k_A\})} = \\ &\int_{\mathcal{B}_{\{k_A\}}} \mathcal{D}X\,\mathcal{D}c\mathcal{D}\overline{c}\,\exp\left[-2t\left\{Q_{BRST}, \int \Psi_{(1,1)}\right\}\right] \prod_{i=1}^{N} \Theta_{A_i}^{(0)}(x_i) \end{aligned} \tag{7.7.32}$$

and where, in analogy with the position that is customary in the theory of modular forms, the parameters $q_a$ have been defined as follows:

$$q_a = \exp\left[2\pi i t_a\right] \tag{7.7.33}$$

After this preliminary analysis, the actual evaluation of the path integral relies on the observation that the reduced correlators (7.7.32) are actually independent of the coupling parameter $t$. Indeed, if we take the $\frac{\partial}{\partial t}$ derivative of a reduced correlator, what we obtain is the insertion of a BRST-exact factor $\left\{Q_{BRST}, \int \Psi_{(1,1)}\right\}$, which makes zero inside any correlation function. Hence we have the priviledge to evaluate the reduced path integral (7.7.32) at any convenient value of $t$ and in particular in the limit $\mathrm{Re}\,t \longrightarrow \infty$. In this limit the only configurations that contribute to the path integral are those that have a vanishing reduced action $\left\{Q_{BRST}, \int \Psi_{(1,1)}\right\}$: these are the sigma model instantons. Indeed recalling Eq. (7.4.26) we have

$$\begin{aligned} \left\{Q_{BRST}, \int \Psi_{(1,1)}\right\} &= ||\partial_- X^i||^2 \\ &+ 2\overline{c}^{l^*} g_{l^* i} \nabla_- c^i \\ &+ 2\overline{c}^{l} g_{l i^*} \nabla_+ c^{i^*} \\ &+ 4\overline{c}^{l^*} c^{m^*} \overline{c}^j c^k R_{l^* j m^* k} \end{aligned} \tag{7.7.34}$$

so that the exponential factor weighing the path integral has the structure

$$\exp\left[-2t\,||\partial_- X^i||^2 + \text{fermions}\right] \tag{7.7.35}$$

and, in the limit $\mathrm{Re}\,t \longrightarrow \infty$, non-vanishing contributions are provided only by the field configurations satisfying the instanton condition:

$$\partial_- X^i = 0 \tag{7.7.36}$$

Hence as far as the bosonic degrees of freedom are concerned the functional integral $\int_{\mathcal{B}_{\{k_A\}}} \mathcal{D}X \ldots$ reduces to an integral over the space of holomorphic instantons of the given homotopy class. In full generality we can describe the holomorphic instanton by writing the complex coordinates $X^i$ of the target space as analytic functions of the complex coordinate $z$ of the world-sheet and of a certain number $n$ of additional complex parameters $m^\alpha$, named the *moduli*:

$$X^i = U^i(z, m^\alpha) \tag{7.7.37}$$

For each fixed value $m_0^\alpha$ of the moduli parameters the functions $U^i(z, m_0^\alpha)$ describe a holomorphic embedding of the Riemann surface $\Sigma_g$ into the target manifold $\mathcal{M}_K$ with the same homotopy class. Furthermore the maps corresponding to different values of the moduli parameters are, by definition, inequivalent under holomorphic diffeomorphisms of either the target space or the world-sheet. In this way we have

$$\int_{\mathcal{B}_{\{k_A\}}} \mathcal{D}X \ldots \approx \int_{\mathcal{M}^{\{k_A\}}_{moduli}} \mu(m) \wedge_{\alpha=1}^n dm^\alpha \tag{7.7.38}$$

where $\mu(m)$ is a convenient measure on the moduli space $\mathcal{M}^{\{k_A\}}_{moduli}$. The dimension $n_{\{k_A\}}$ of this moduli space is governed by an index theorem. Let us consider a specific holomorphic instanton $X^i = U^i(z)$ corresponding to a given value of the moduli parameters and let us define $\mathcal{E}_U = U^*(T\mathcal{M}_K)$ the pull-back through $U$ of the tangent bundle to the target Kählerian manifold. Recalling the construction of subsection 2.2.6. $\mathcal{E} \xrightarrow{\pi} \Sigma_g$ is a holomorphic vector bundle over the Riemann surface $\Sigma_g$ of rank equal to the complex dimension $N = dim_{\mathbb{C}}\mathcal{M}_K$ of the target space. The field $X^i$ takes values in $\mathcal{M}_K$ and it is not a section of $\mathcal{E}_U$; however, any small variation $\delta X^i$ can be considered as a deformation of $\mathcal{M}_K$ and therefore such a field takes values in the pull-back of the tangent bundle, namely in $\mathcal{E}_U$. Hence if we take the instanton configuration $X^i = U^i(z)$ and we deform it by setting $X^i = U^i(z) + \delta U^i(z)$, the variation $\delta U^i(z) \in \Gamma(\mathcal{E}_U, \Sigma_g)$ is a section of the $\mathcal{E}_U$-bundle. In a similar way the ghost $c^i$ is an anti-commuting section of $\mathcal{E}_U$, while the anti-ghost $\bar{c}^i$, which has world-sheet spin $s = 1$, is a $\Gamma(\mathcal{E}_U, \Sigma_g)$-valued $(0,1)$-form. The infinitesimally deformed field $X^i = U^i(z) + \delta U^i(z)$ is still a holomorphic instanton if the holomorphic covariant derivative of the bundle $\mathcal{E}_U$ has a vanishing action on $\delta U^i(z)$:

$$\partial_- X^i = 0 \implies \nabla_- \delta U^i = 0\,; \quad \nabla \delta U^i = d\delta U^i - \Gamma^i_{jk}\, dU^j \delta U^k \tag{7.7.39}$$

This result can be understood also by recalling that, in real coordinates, the holomorphicity condition of the map $X : \Sigma_g \longrightarrow \mathcal{M}_K$ can be described as the commutativity of the following diagram:

$$\begin{array}{ccc} T\Sigma_g & \stackrel{X_\star}{\longrightarrow} & T\mathcal{M}_K \\ j\downarrow & & \uparrow J \\ T\Sigma_g & \stackrel{X_\star}{\longrightarrow} & T\mathcal{M}_K \end{array} \tag{7.7.40}$$

namely

$$X_\star - j \cdot X_\star \cdot J = 0$$

or

$$\partial_\alpha X^\mu - j_\alpha^\beta \partial_\beta X^\nu J_\nu^\mu = 0 \tag{7.7.41}$$

where the covariantly constant tensor $j_\alpha^\beta$ with square equal to -1 is the complex structure of the world-sheet while the tensor $J_\nu^\mu$, which has the same properties, is the complex structure of the target manifold and $X_\star : T\Sigma_g \longrightarrow T\mathcal{M}_K$ denotes the push-forward of the map $X$. If we perform an infinitesimal variation $\delta X^\mu$ of a solution to Eq. (7.7.41), we must be careful to differentiate also the complex structure tensor $\left(J_\nu^\mu(X+\delta X) = J_\nu^\mu(X) + \partial_\lambda J_\nu^\mu(X)\,\delta X^\lambda\right)$, and utilizing the covariant constancy of $J_\nu^\mu$ we retrieve the condition (7.7.39). In this way we come to the conclusion that the space of possible infinitesimal deformations of a holomorphic instanton $U$ is given by the space $H^0(\mathcal{E}_U)$ of covariantly holomorphic sections of the bundle $\mathcal{E}_U$. If each of these infinitesimal deformations can be integrated to a finite deformation, then the complex dimension $n_{\{k_A\}}$ of the moduli space coincides with the dimension of $H^0(\mathcal{E}_U)$. Indeed recalling Eq. (7.7.37), we have

$$\delta m^\alpha \frac{\partial U^i(z,m)}{\partial m^\alpha} \stackrel{\text{def}}{=} \delta m^\alpha u^i_\alpha(z) \in H^0(\mathcal{E}_U) \tag{7.7.42}$$

At this point it is also worth noting that the field equation of the ghost field $c^i$ is identical in form to Eq. (7.7.39) satisfied by the holomorphic instanton deformations. It follows that we have

$$\#\,\text{zero} - \text{modes of } c^i = \dim H^0(\mathcal{E}_U) \tag{7.7.43}$$

Looking at the field-equation for the anti-ghost that follows from the action (7.7.34) and recalling that this field is a $\Gamma(\mathcal{E}_U, \Sigma_g)$-valued $(0,1)$-form we conclude that

$$\#\,\text{zero} - \text{modes of } \bar{c}^i = \dim H^1(\mathcal{E}_U) \tag{7.7.44}$$

Twice the difference of these two numbers of zero-modes is, by definition, the integrated anomaly of the ghost current:

$$\begin{aligned} \Delta U \stackrel{\text{def}}{=} \int \partial^\alpha J_\beta^{(ghost)} d^2x &= \#\,\text{zero} - \text{modes of } \bar{c}^i \; - \#\,\text{zero} - \text{modes of } \bar{c}^i \\ &+ \#\,\text{zero} - \text{modes of } \bar{c}^{i\star} \; - \; \#\,\text{zero} - \text{modes of } \bar{c}^{i\star} \\ &= 2\left(\dim H^0(\mathcal{E}_U) - \dim H^1(\mathcal{E}_U)\right) \end{aligned} \tag{7.7.45}$$

Due to the last line of the above equation, the ghost anomaly $\Delta U$ can be calculated through the Hirzebruch–Riemann–Roch theorem. Indeed the last line of (7.7.45) shows that $\Delta U$ is the index of the $\mathcal{E}_U$-twisted Dolbeault complex. Recalling Eq. (2.7.76) we have

$$\begin{aligned} \frac{1}{2}\Delta U &= \int_{\Sigma_g} \mathrm{ch}\,(\mathcal{E}_U) \wedge \mathrm{td}\,(T\Sigma_g) \\ &= (1-g)\dim_{\mathbb{C}} \mathcal{M}_K + \int_{\Sigma_g} X^\star \left[c_1\left(\mathcal{M}_K\right)\right] \end{aligned} \tag{7.7.46}$$

As remarked in the introduction to the present chapter the integrated ghost anomaly $\Delta U$ is named the formal dimension of the moduli space:

$$\begin{aligned} \frac{1}{2}\Delta U &= \dim_{\mathbb{C}}^{formal} \mathcal{M}^{\{k_A\}}_{moduli} \\ &= \frac{1}{2}\dim_{\mathbb{R}}^{formal} \mathcal{M}^{\{k_A\}}_{moduli} \end{aligned} \tag{7.7.47}$$

The reason is simple. In a sufficiently generic situation one expects that if $\Delta U \rangle 0$ then $\dim H^1(\mathcal{E}_U) = 0$ and $\frac{1}{2}\Delta U = \dim H^0(\mathcal{E}_U) = \dim_{\mathbb{C}} \mathcal{M}^{\{k_A\}}_{moduli}$. If $\dim H^1(\mathcal{E}_U) \rangle 0$ the existence of these zero-modes is usually a signal that an equal number of the infinitesimal deformations contained in $H^0(\mathcal{E}_U)$ cannot be integrated to finite deformations, so that the actual dimension of the moduli space tends to be equal to the formal dimension although this is not guaranteed. An instructive illustration of the use of the index theorem (7.7.46) is provided by the simplest of all topological $\sigma$-models, namely that where the target manifold is a 2-sphere, i.e. $\mathbb{CP}_1$. Consider the case of a genus zero world-sheet $\Sigma_0 \approx \mathbb{CP}_1$. Under these circumstances the holomorphic instanton is a holomorphic map of the Riemann sphere into itself:

$$X : \mathbb{CP}_1 \xrightarrow{hol} \mathbb{CP}_1 \tag{7.7.48}$$

Such a map is represented by a meromorphic function of *degree* $k$ of the form

$$X = U_k(z) = \frac{P^n(z)}{P^m(z)} \tag{7.7.49}$$

where $X$ is a local complex coordinate on the target space $\mathbb{CP}_1$, $z$ is a local coordinate on the world-sheet $\mathbb{CP}_1$, and $P^n(z)$, $P^m(z)$ denote two polynomials respectively of degree $n$ and $m$. The degree $k$ of the map is given by the largest among the two integers $n$ and $m$:

$$k = \max\{n, m\} \tag{7.7.50}$$

Indeed, due to the fundamental theorem of algebra, each point $X \in \mathbb{CP}_1$ on the target space has exactly $k$ pre-images on the world-sheet. Hence the map $U_k$ is a $k$-folded holomorphic cover of the 2-sphere. By means of an $SL(2,\mathbb{C})$ transformation

$$X' = \frac{aX+b}{cX+d} \tag{7.7.51}$$

on the target manifold we can always reduce the holomorphic map of Eq. (7.7.49) to the form

$$X = \alpha \frac{\prod_{i=1}^{k} (z - \beta_i)}{\prod_{i=1}^{k} (z - \gamma_i)} \tag{7.7.52}$$

where $\beta_i$ are the $k$ roots of the polynomial in the numerator, $\gamma_i$ are the $k$ roots of the polynomial in the denominator and $\alpha$ is the value of $X$ when $z \longrightarrow \infty$. Altogether these are just $2k + 1$ complex parameters and they constitute the moduli of the degree $k$ holomorphic instanton. This counting matches with the counting provided by the index theorem. Recalling Eq. (2.8.6) the total Chern class of $\mathbb{CP}_1$ is $c_{tot}(\mathbb{CP}_1) = (1+K)^2$ so that

$$c_1(\mathbb{CP}_1) = 2K \tag{7.7.53}$$

and hence we get

$$\frac{1}{2}\Delta U = 1 + 2\int_{\Sigma_0} U_k^\star(K) = 1 + 2k \tag{7.7.54}$$

Indeed as already remarked $\int_{\Sigma_0} U_k^\star(K)$ is precisely the degree of the map $U_k$. This can be seen by explicit calculation, substituting Eq.(7.7.52) into

$$K = \frac{i}{2\pi} \frac{dX \wedge dX^\star}{(1 + |X|^2)^2} \tag{7.7.55}$$

As one sees from this example, the ghost anomaly or formal dimension of the moduli space, in general, depends on the instanton number. This has a profound implication for the structure of the correlation functions. In order for the reduced correlator (7.7.32) to be non-vanishing, it is necessary that the total ghost number of the product of observables inside the correlator be equal to the ghost anomaly:

$$\# gh \left[ \prod_{i=1}^{N} \Theta_{A_i}^{(0)}(x_i) \right] = \Delta U_{\{k\}} \tag{7.7.56}$$

This is forced by ghost-number conservation and will be further clarified, when we show that, in the evaluation of the path integral, the observables $\Theta_{A_i}^{(0)}(x_i)$ become differential forms on moduli space of degree equal to their ghost-number. The selection rule (7.7.56) implies that in the calculation of the complete correlator (7.7.31) of a given product of observables only a finite number of instanton sectors contribute, namely those for which Eq. (7.7.56) is satisfied. Indeed Eq. (7.7.56) can be regarded as a Diophanti-ne equation for the integers $k_A$ labelling the homology class of $U^\star(K)$. As a consequence, in a general situation, the topological sigma model correlators are polynomial functions of the $q_a$-parameters that correspond to a coordinate on the moduli space of the Kähler class deformations of $\mathcal{M}_K$. There is just one exception to this general situation. It occurs when the target manifold $\mathcal{M}_K$ is a *Calabi–Yau n-fold.* In this case $c_1(\mathcal{M}_K) = 0$ and the integrated anomaly is independent of the instanton number:

$$\frac{1}{2}\Delta U = (1 - g)n \tag{7.7.57}$$

In this case Eq. (7.7.56) becomes a selection rule for the product of observables that can have non-vanishing topological correlators:

$$\#gh \left[ \prod_{i=1}^{N} \Theta^{(0)}_{A_i}(x_i) \right] = 2(1-g)n \tag{7.7.58}$$

If (7.7.58) is not satisfied then $\langle \prod_{i=1}^{N} \Theta^{(0)}_{A_i}(x_i) \rangle = 0$. If (7.7.58) is satisfied then the correlator $\langle \prod_{i=1}^{N} \Theta^{(0)}_{A_i}(x_i) \rangle$ does not vanish and picks contributions from all instanton sectors. Hence, in the Calabi–Yau case, the non-vanishing topological correlators are *transcendental functions* of moduli parameters $q_a$ of Kähler class deformations. An even more specific situation occurs when the target space is a Calabi–Yau 3-fold, the case relevant to string theory compactifications. For n=3 and in the case of a genus zero world-sheet, the formal dimension of moduli space is just $\frac{1}{2}\Delta U = 3$. These are just the degrees of freedom of the automorphism group of the 2-sphere (or rational curve) that we are immersing into the Calabi–Yau 3-fold $\mathcal{M}_K$. Modding this out by fixing three points of $\mathcal{M}_K$ that must belong to the rational curve, no deformation is left. We conclude that *the rational curves on a Calabi–Yau 3-fold that contain three marked points of the 3-fold are isolated: the only thing which matters is their number* $a_{\{k_A\}}$. The calculation of the integers $a_{\{k_A\}}$ is a difficult mathematical problem that can be solved by the topological field theories, using *mirror maps.* This will become clear in the sequel. In particular, Chapter 8 is entirely devoted to this issue. For the time being let us go on with our general discussion of the topological sigma model correlators.

We have already provided arguments to prove that the path integral defining the reduced correlators (7.7.32) is dominated by the instanton configurations satisfying (7.7.36) and that the integral on $\mathcal{D}X$ reduces essentially to an integral over the moduli space of instantons (see Eq. (7.7.38)). This is not completely true. In the semi-classical approximation, which becomes exact in the limit $\mathrm{Re}t \longrightarrow \infty$ the integration on $\mathcal{D}X$ splits into an integral on the *zero-modes,* namely the instantons (so on the instanton moduli space) and, for each instanton, on the *non-zero-modes.* These are the fluctuations $\delta U^i(z)$, which, as already remarked, are sections of the bundle $\mathcal{E}_U$ defined above.

Hence we have

$$\begin{aligned} &\int_{\mathcal{B}_{\{k_A\}}} \mathcal{D}X \exp\left[-2t|\partial_- X|^2\right] \approx \\ &\int_{\mathcal{M}^{\{k_A\}}_{moduli}} \mu(m) \wedge_{\alpha=1}^{n} dm^{\alpha} i \times \int_{\Gamma(\mathcal{E}_{U(m)},\Sigma_g)} \mathcal{D}\delta U \exp\left[-2t\langle \delta U \,|\, \nabla_-^{\dagger}\nabla_- \,|\, \delta U\rangle\right] \\ &= \int_{\mathcal{M}^{\{k_A\}}_{moduli}} \mu(m) \wedge_{\alpha=1}^{n} dm^{\alpha} i \times \left[ Det\left(2t\,\nabla_-^{\dagger}\nabla_-\right)\right]^{-1} \end{aligned} \tag{7.7.59}$$

where we have used the standard path integral representation of the functional determinants.

We consider next the path integral on the fermionic fields. We focus on the case where $\frac{1}{2}\Delta U = \dim H^0(\mathcal{E}_U)$ and $\dim H^1(\mathcal{E}_U) = 0$. Under these conditions there are no anti-ghost zero-modes while the ghost zero-modes are as many as the instanton moduli parameters. Correspondingly we can expand these fields as follows:

$$\begin{aligned} c^i &= \sum_{\alpha=1}^{\Delta U} \theta^\alpha \, u^i_\alpha(z,m) + \delta c^i \\ \overline{c}^i &= \delta \overline{c}^i \end{aligned} \qquad (7.7.60)$$

where $\theta^\alpha$ are $\Delta U$ anti-commuting parameters (the *supermoduli*) and $\delta c^i \in \Gamma\left(\mathcal{E}_{U(m)}, \Sigma_g\right)$ is a section of the $\mathcal{E}_{U(m)}$-bundle. On the other hand $\delta \overline{c}^i$ is a $(0,1)$-form with values in some other section $s \in \Gamma\left(\mathcal{E}_{U(m)}, \Sigma_g\right)$ of the same bundle. Hence the path integral on these fields becomes

$$\begin{aligned} &\int \mathcal{D}c^i \, \mathcal{D}\overline{c}^i \, \exp\left[-2t\left(2\overline{c}^{l^\star} g_{l^\star i} \, \nabla_- c^i + 2\overline{c}^l g_{li^\star} \, \nabla_+ c^{i^\star}\right)\right] \approx \\ &\qquad \int \textstyle\prod_{\alpha=1}^{\Delta U} d\theta^\alpha \times \\ &\int \mathcal{D}\delta c^i \, \mathcal{D}\delta\overline{c}^i \, \exp\left[-2t\left(2\left\langle\delta\overline{c}\right| \nabla_- \left|\delta c^i\right\rangle + 2\left\langle\delta c\right| \nabla_-^\dagger \left|\delta\overline{c}\right\rangle\right)\right] \\ &\qquad = \int \textstyle\prod_{\alpha=1}^{\Delta U} d\theta^\alpha \times \left[Det\left(2t\nabla_-\right)\right] \left[Det\left(2t\nabla_-^\dagger\right)\right] \end{aligned} \qquad (7.7.61)$$

Putting together Eq.(7.7.61) and Eq.(7.7.59) we realize that the bosonic and fermionic determinants due to the integration on the non-zero-modes cancel against each other and we are left only with the integration over the moduli and supermoduli. Let us now consider the observables inserted into the path integral. Recalling Eq.(7.4.14) we know that each of the $\Theta^{(0)}_{A_r}$ is associated with a harmonic $(p_r, q_r)$-form $\omega^{(p_r,q_r)}_{A_r} \in H^{(p_r,q_r)}\left(\mathcal{M}_K\right)$ and in terms of that differential form has the following structure:

$$\Theta^{(0)}_{A_r} = \omega^{(p_r,q_r)}_{(A_r)\, i_1,\dots,i_{p_r}, j^\star_1,\dots,j^\star_{q_r}} \left(X, \overline{X}\right) c^{i_1} \dots c^{i_{p_r}} \, c^{j^\star_1} \dots c^{j^\star_{q_r}} \qquad (7.7.62)$$

In the semiclassical evaluation of the path integral, the expression (7.7.62) must be localized on the bosonic instanton and on the fermionic zero-modes, so that we obtain

$$\begin{aligned} \Theta^{(0)}_{A_r} &\approx \omega^{(p_r,q_r)}_{(A_r)\, i_1,\dots,i_{p_r}, j^\star_1,\dots,j^\star_{q_r}} \left(U(z,m), \overline{U}(\overline{z},\overline{m})\right) \times \\ &\theta^{\alpha_1} u^{i_1}_{\alpha_1}(z,m) \dots \theta^{\alpha_{p_r}} u^{i_{p_r}}_{\alpha_{p_r}}(z,m) \, \overline{\theta}^{\beta_1} \overline{u}^{j^\star_1}_{\beta_1}(\overline{z},\overline{m}) \dots \overline{\theta}^{\beta_{q_r}} \overline{u}^{j^\star_{q_r}}_{\beta_{q_r}}(\overline{z},\overline{m}) \end{aligned} \qquad (7.7.63)$$

At this point we can perform the fermionic integration on the supermoduli, utilizing the standard rule $\int \theta^\alpha \, d\theta^\beta = \delta^{\alpha\beta}$. The net result of such an integration is that of replacing each supermodulus $\theta^\alpha$ appearing in each observable (7.7.63) with the corresponding modulus differential $dm^\alpha$. Indeed we just have to recall that the total integration measure was $\cdots \wedge_{\alpha=1}^{\Delta U} dm^\alpha \, d\theta^\alpha$. After all these manipulations the final expression for the reduced correlators (7.7.32) is given by an integral over the $\{k_A\}$-instanton moduli space of an exterior product of cohomology classes $\mathcal{O}_{A_r}(x_r)$ of that space, namely

$$\left\langle \prod_{r=1}^{N} \Theta^{(0)}_{A_r}(x_r) \right\rangle_{(0,\{k_A\})} = \int_{\mathcal{M}^{\{k_A\}}_{moduli}} \mathcal{O}_{A_1}(x_1) \wedge \dots \wedge \mathcal{O}_{A_N}(x_N) \qquad (7.7.64)$$

where the $(p_r, q_r)$-differential form $\mathcal{O}_{A_r}(x_r)$ is defined by

$$\begin{aligned}\mathcal{O}^{(0)}_{A_r}(x_r) &= \omega^{(p_r,q_r)}_{(A_r)\, i_1,\dots,i_{p_r},j^\star_1,\dots,j^\star_{q_r}}\left(U(z,m),\overline{U}(\overline{z},m)\right) \times \\ & dm^{\alpha_1}\frac{\partial U^{i_1}}{\partial m^{\alpha_1}}(z,m)\wedge\cdots\wedge dm^{\alpha_{p_r}}\frac{\partial U^{i_{p_r}}}{\partial m^{\alpha_{p_r}}}(z,m)\times \\ & d\overline{m}^{\beta_1}\frac{\partial \overline{U}^{j^\star_1}}{\partial \overline{m}^{\beta_1}}(\overline{z},\overline{m})\wedge\cdots\wedge d\overline{m}^{\beta_{q_r}}\frac{\partial \overline{U}^{j^\star_{q_r}}}{\partial \overline{m}^{\beta_{q_r}}}(\overline{z},\overline{m})\end{aligned} \tag{7.7.65}$$

$x_i \in \Sigma_g$ being a point on the Riemann surface, whose local holomorphic coordinate we have called $z$. What is the meaning of the forms $\mathcal{O}^{(0)}_{A_r}(x_r)$? This is understood by regarding the family $X^i = U^i_{\{k\}}(z,m)$ of holomorphic instantons of degree $\{k_A\}$ as a holomorphic map:

$$U_{\{k\}} : \Sigma_g \otimes \mathcal{M}^{\{k_A\}}_{moduli} \longrightarrow \mathcal{M}_K \tag{7.7.66}$$

from the tensor product of the world-sheet Riemann surface with the instanton moduli space to the target manifold. Adopting this point of view, for any point $x_i \in \Sigma_g$, the function $U_{\{k\}}(x_i, m)$ is a map:

$$U_{\{k\};x_i} : \mathcal{M}^{\{k_A\}}_{moduli} \longrightarrow \mathcal{M}_K \tag{7.7.67}$$

from the moduli space to the target manifold. The differential form $\mathcal{O}^{(0)}_{A_r}(x_r)$ is nothing but the pull-back of the original Dolbeault cohomology class $\omega^{(p_r,q_r)}_{(A_r)} \in H^{(p_r,q_r)}(\mathcal{M}_K)$ through such a map:

$$\mathcal{O}^{(0)}_{A_r}(x_r) = U^\star_{\{k\};x_i}\left[\omega^{(p_r,q_r)}_{(A_r)}\right] \tag{7.7.68}$$

This interpretation leads to the final evaluation of the reduced correlator as an intersection number of homology cycles in moduli space. The strategy is the following. Let $M_{n-p} \subset \mathcal{M}_K$ be a homology $(n-p)$-cycle of the target manifold $(\partial M_{n-p} = 0)$, the number $n$ denoting now the real dimension of the latter. $M_{n-p}$ is a subvariety of codimension $p$. The Poincaré dual of $M_{n-p}$ is the cohomology class of a closed $p$-form $A^{(p)}$ such that, for any closed $(n-p)$-form $\Gamma^{(n-p)}$, we have

$$\int_{\mathcal{M}_K} A^{(p)} \wedge \Gamma^{(n-p)} = \int_{M_{n-p}} \Gamma^{(n-p)} \tag{7.7.69}$$

Every closed form, in particular the representative $\omega^{(p_r,q_r)}_{(A_r)} \in H^{(p_r,q_r)}$ of a Dolbeault cohomology class, is cohomologous to a linear combination of Poincaré duals of suitable submanifolds $M^{A_r}_{n-p_r-q_r}$. Let then

$$M^{A_r}_{n-p_r-q_r} \Longleftrightarrow \omega^{(p_r,q_r)}_{(A_r)} \tag{7.7.70}$$

be Poincaré duals and let

$$H_r = U^{-1}_{\{k\};x_r}\left[M^{A_r}_{n-p_r-q_r}\right] \subset \mathcal{M}^{\{k_A\}}_{moduli} \tag{7.7.71}$$

be the submanifold of moduli space containing all those instantons such that $U(x_r, m) \in M^{A_r}_{n-p_r-q_r}$. If $U(x_r, m)$ is not in the submanifold $M^{A_r}_{n-p_r-q_r}$ then $U^\star_{\{k\};x_i}\left[\omega^{(p_r,q_r)}_{(A_r)}\right] = 0$ since $\omega^{(p_r,q_r)}_{(A_r)}$ has delta function support on its Poincaré dual $M^{A_r}_{n-p_r-q_r}$. Therefore the reduced correlator (7.7.64) receives contributions only from the submanifold of moduli space given by the complete intersection of the submanifolds $H_r$. Each of the $H_r$ is a cycle of $\mathcal{M}^{\{k_A\}}_{moduli}$ since it is the image through a diffeomorphism of a cycle of the target space. Actually $H_r$ is the Poincaré dual of the closed form $U^\star_{\{k\};x_i}\left[\omega^{(p_r,q_r)}_{(A_r)}\right]$ and as such it is a $\left(2\Delta U_{\{k\}} - p_r - q_r\right)$-cycle. Due to the selection rule (7.7.56), the complete intersection $H_1 \cap H_2 \cap \ldots \cap H_N$ has dimension zero and the value of the reduced correlator (7.7.64) is equal to the intersection number. Hence we finally get the formula

$$\begin{aligned}\langle \prod_{i=1}^{N} \Theta^{(0)}_{A_i}(x_i) \rangle_{(0,\{k_A\})} &= \#(H_1 \cap H_2 \cap \cdots \cap H_N) \\ &\stackrel{\text{def}}{=} \#\left(U^{-1}_{\{k\};x_1}\left[M^{A_1}_{n-p_1-q_1}\right] \cap \cdots \cap U^{-1}_{\{k\};x_N}\left[M^{A_N}_{n-p_N-q_N}\right]\right)\end{aligned} \tag{7.7.72}$$

We can illustrate the meaning of Eq. (7.7.72) with the example of the $\mathbb{CP}_1$-model of Eq.(7.7.48). In this case the only non-trivial Dobeault cohomology class is given by the Kähler class $K$ defined in Eq. (7.7.55). The top form $K$ is the Poincaré dual of an arbitrary point $X_0 \in \mathbb{CP}_1^{(target)}$. Hence for each $z_r \in \mathbb{CP}_1^{(world\text{-}sheet)}$ pick a point $X_r \in \mathbb{CP}_1^{(target)}$ such that

$$X_r = \frac{\prod_{i=1}^{k}(z_r - \beta_i)}{\prod_{i=1}^{k}(z_r - \gamma_i)} \tag{7.7.73}$$

From Eqs. (7.7.54) and (7.7.56) we know that the only non vanishing reduced correlator is

$$\langle \prod_{r=1}^{2k+1} \Theta^{(0)}_{K}(x_r) \rangle_{\mathcal{B}_k} \tag{7.7.74}$$

so that there are exactly $2k+1$ conditions of the type (7.7.73) to be imposed. This is the number of parameters in the instanton and these conditions fix them completely. Correspondingly there is just only one instanton in the complete intersection and the value of the reduced correlator (7.7.74) is 1. To be more explicit, let us, for instance, choose $X_r = 0$ for $r = 1, \ldots k$, $X_r = \infty$ for $r = k+1, \ldots 2k$ and $X_r = 1$ for $r = 2k+1$. So doing the complete intersection in moduli space we are looking for corresponds to the instantons that have zeros at $z_1, z_2, \ldots, z_k$, simple poles at $z_{k+1}, z_{k+2}, \ldots, z_{2k}$ and take the value 1 at $z_{2k+1}$. There is just one such an instanton, namely

$$X = \frac{\prod_{i=1}^{k}(z - z_i)}{\prod_{i=1}^{k}(z - z_{i+k})} \frac{\prod_{i=1}^{k}(z_{2k+1} - z_{i+k})}{\prod_{i=1}^{k}(z_{2k+1} - z_i)} \tag{7.7.75}$$

Introducing the parameter $q = \exp[2\pi t]$, where $t$ is the modulus of the Kähler class, we have succeeded in solving the $\mathbb{CP}_1$ topological sigma model completely. All the even correlation functions are zero, while the odd ones are equal to $q^{2k+1}$:

$$\langle \prod_{r=1}^{N} \Theta_K^{(0)}(x_r) \rangle = \begin{cases} 0 & \text{if } N = 2k \\ q^{2k+1} & \text{if } N = 2k+1 \end{cases} \tag{7.7.76}$$

Having clarified the general strategy, we devote the next subsection to the case of Calabi–Yau 3-folds.

### 7.7.2 Topological $\sigma$-models on Calabi–Yau 3-folds

As pointed out in the previous subsection, Calabi–Yau n-folds correspond to the unique case where the moduli space of holomorphic instantons has a dimension which is independent of the instanton number, namely of the homotopy class of the map $U : \Sigma_g \longrightarrow \mathcal{M}_K$. As a consequence those topological correlators that do not vanish pick contributions from all instanton numbers and are transcendental functions of the exponentials (7.7.33) of the Kähler class deformation parameters $t_a$. Because of their relevance in superstring compactifications (see Chapter 4), in the present subsection we consider the case of Calabi–Yau 3-folds. Here the complex dimension of the holomorphic map moduli space, predicted by the index theorem (7.7.46), is

$$\frac{1}{2}\Delta U = \dim_{\mathbb{C}} \mathcal{M}_{moduli} = 3(1-g) \tag{7.7.77}$$

and the selection rule (7.7.58) becomes

$$\#gh \left[ \prod_{i=1}^{N} \Theta_{A_i}^{(0)}(x_i) \right] = 6(1-g) \tag{7.7.78}$$

In the following we restrict our attention to the case of genus zero world-sheets, namely we consider the holomorphic embeddings:

$$U : \Sigma_0 \longrightarrow \mathcal{M}_3 \tag{7.7.79}$$

of a Riemann sphere $\Sigma_0 \sim \mathbb{CP}_1$ into a Calabi–Yau 3-fold $\mathcal{M}_3$. Such maps define a set of analytic 1-folds $C^U$ immersed in the 3-fold

$$U(\mathbb{CP}_1) \stackrel{\text{def}}{=} C^U \subset \mathcal{M}_3 \tag{7.7.80}$$

which are named *rational curves.* The complex dimensionality three of the moduli space of rational curves predicted by Eq.(7.7.77) is easily understood in elementary terms. Let $z$ denote a complex coordinate of a point $p \in \mathbb{CP}_1$ and let $Z^i_\alpha \{i = 1,2,3\}$ be a set of complex coordinates labelling the points $P$ of the 3-fold in some open chart

$$\phi_\alpha : \mathcal{U}_\alpha \longrightarrow \mathbb{C}^3 \quad \mathcal{U}_\alpha \subset \mathcal{M}_3 \tag{7.7.81}$$

In such a coordinate patch a rational curve is described by

$$Z^i_\alpha(z) = U^i_\alpha(z) \tag{7.7.82}$$

where $U^i_\alpha(z)$ are three local holomorphic functions on $\mathbb{CP}_1$. Given any such rational curve, we can immediately construct another one by setting

$$Z^i_\alpha(z) = U^i_\alpha\left(\frac{\alpha z + \beta}{\gamma z + \delta}\right) \tag{7.7.83}$$

where

$$\begin{pmatrix} \alpha & \beta \\ \beta & \gamma \end{pmatrix} = \mathbf{g} \in SL(2,\mathbb{C}) \tag{7.7.84}$$

is an arbitrary element of the automorphism group of the Riemann sphere $\mathbb{CP}_1$. The essential idea behind Eq.(7.7.83) is that, given a holomorphic map as in Eq.(7.7.80), we can construct a three-parameter family of such maps by composing it with the $SL(2,\mathbb{C})$ transformations that map $\mathbb{CP}_1$ into itself:

$$\begin{aligned} \mathbb{CP}_1 &\stackrel{\mathbf{g}\in SL(2,C)}{\Longrightarrow} \mathbb{CP}_1 \stackrel{U}{\Longrightarrow} \mathcal{M}_3 \\ U_{\mathbf{g}} &\stackrel{\text{def}}{=} U \circ \mathbf{g} : \mathbb{CP}_1 \Longrightarrow \mathcal{M}_3 \end{aligned} \tag{7.7.85}$$

The index theorem (7.7.77) states that any such family is the most general continuous family of holomorphic instantons, the local coordinates on moduli space being the three complex parameters that appear in an $SL(2,\mathbb{C})$ group element. They are described by the entries $\alpha, \beta, \gamma, \delta$ of the $2 \times 2$ matrix (7.7.82) with the additional constraint:

$$\alpha\delta - \beta\gamma = 1 \tag{7.7.86}$$

A natural compactification of this local moduli space is obtained by regarding $\alpha, \beta, \gamma, \delta$ as homogeneous coordinates in $\mathbb{CP}_3$. In this way the compactification divisor, namely the extra boundary locus that we have added, is the algebraic variety

$$\alpha\delta - \beta\gamma = 0 \tag{7.7.87}$$

which is isomorphic to $\mathbb{CP}_1 \odot \mathbb{CP}_1$. We have insisted on the qualification of local moduli space, because the total space of holomorphic instantons is composed by a discrete sequence of such three-parameter continuous families:

$$\text{set of rational curves} = \cup_{\{k\}=\{0\}}^{\{\infty\}} \cup_{r=1}^{r=n_{\{k\}}} U_{\{k\};r} \circ \mathbf{g}(\alpha, \beta, \gamma, \delta) \tag{7.7.88}$$

where by $\{k\}$ we have denoted the degree of the holomorphic map, a concept that we analyse just below, and where $r$ is an index which labels the disjoint families of holomorphic maps of a given degree. The number of such families:

$$n_{\{k\}} \in \mathbb{Z}_+ \tag{7.7.89}$$

is an integer number, *the number of rational curves of degree* $\{k\}$, which plays an important role in the sequel.

Let us then discuss the degree of the rational curves. Recalling the results of Chapter 4 and in particular the structure of the Hodge–diamond (4.1.28), we conclude that for a Calabi–Yau 3-fold the Betti numbers $b^2$ and $b^4$ are both equal to $h^{(1,1)}$, since $h^{(2,0)} = h^{(0,2)} = 0$. Therefore we can choose a basis of homology 2-cycles $C_2^a$ $(a = 1, \ldots, h^{(1,1)})$ that is dual to the generators of the Dolbeault cohomology group $H^{(1,1)}$:

$$\int_{C_2^a} \omega_b^{(1,1)} = \delta_b^a \tag{7.7.90}$$

What Eq. (7.7.90) means is that we have selected a basis $\omega_a^{(2,2)}$ of harmonic (2,2)-forms dual to the chosen basis of the harmonic (1,1)-forms:

$$\int_{\mathcal{M}_3} \omega_a^{(1,1)} \wedge \omega_b^{(2,2)} = \delta_{ab} \tag{7.7.91}$$

and that the 2-cycles $C_2^a$ are the Poincaré duals of the (2,2)-forms $\omega_a^{(2,2)}$. At the same time there exists a basis of homology 4-cycles $C_4^a$, the Poincaré duals of the (1,1)-forms $\omega_a^{(1,1)}$, such that

$$\int_{C_4^a} \omega_b^{(2,2)} = \delta_b^a \tag{7.7.92}$$

Using these dual bases we can write

$$\#\left(C_4^a \cap C_2^b\right) = \int_{\mathcal{M}_3} \omega_a^{(1,1)} \wedge \omega_b^{(2,2)} = \delta_{ab} \tag{7.7.93}$$

Every rational curve $\mathcal{C}^U$, being the diffeomorphic image of a 2-sphere, is a closed surface, namely a 2-cycle and, as such, it has an integer homology class:

$$\left[\mathcal{C}^U\right] = \sum_{a=1}^{h^{(1,1)}} k_a \left[C_2^a\right] \qquad k_a \in \mathbb{Z} \tag{7.7.94}$$

The collection of integer numbers $\{k_a\}$ is what defines the *degree* of the *rational* curve $\mathcal{C}^U$. According to this definition we have

$$C_4^a \cap \mathcal{C}^U = k_a \tag{7.7.95}$$

Let us now consider the calculation of topological correlators. In genus zero, a natural set of correlation functions that saturate the selection rule (7.7.78) is given by the three-point correlators

$$W_{abc}(t) = \langle \Theta_a^{(0)}(x_1)\, \Theta_b^{(0)}(x_2)\, \Theta_c^{(0)}(x_3) \rangle \tag{7.7.96}$$

where the observables $\Theta_a^{(0)}(x)$, which have ghost number two, are the BRST cohomology classes associated with the (1,1)-forms:

$$\Theta_a^{(0)} = \omega_{a|ij^\star}^{(1,1)} c^i\, c^{j^\star} \tag{7.7.97}$$

The reason why this specific topological correlator has been named $W_{abc}(t)$ is that it is nothing but the Yukawa coupling of the 27 chiral families of $E_6$ associated with the $(1,1)$-forms. In Chapter 4 we saw that the Kaluza–Klein approximation yields a constant result for such a coupling, namely

$$W_{abc}(t) = d_{abc} = \int_{\mathcal{M}_3} \omega_a^{(1,1)} \wedge \omega_b^{(1,1)} \wedge \omega_c^{(1,1)} \tag{7.7.98}$$

In the present section we show that Eq.(7.7.98) corresponds to the contribution of the instantons of zero instanton number (the constant maps). In addition one has to take into account the infinite number of corrections due to the instantons of non-trivial homotopy. Using the general result (7.7.72), the reduced correlator has the following expression:

$$\begin{aligned} &\langle \Theta_a^{(0)}(x_1)\, \Theta_b^{(0)}(x_2)\, \Theta_c^{(0)}(x_3) \rangle_{(0,\{k\})} = \\ &\Sigma_{r=1}^{n\{k\}} \# \left( (U \circ \mathbf{g})^{-1}_{\{k\},r;x_1} [C_4^a] \cap (U \circ \mathbf{g})^{-1}_{\{k\},r;x_2} \left[C_4^b\right] \cap (U \circ \mathbf{g})^{-1}_{\{k\},r;x_3} [C_4^c] \right) \end{aligned} \tag{7.7.99}$$

where, in line with Eq.(7.7.66), the instanton $(U \circ \mathbf{g})_{\{k\},r}$ is regarded as a map

$$(U \circ \mathbf{g})_{\{k\},r} : \mathbb{CP}_1 \otimes \overline{SL(2,\mathbb{C})} \longrightarrow \mathcal{M}_3 \tag{7.7.100}$$

of the compactified moduli space $\overline{SL(2,\mathbb{C})} = \mathbb{CP}_3$ times the world-sheet $\mathbb{CP}_1$ into the Calabi–Yau 3-fold. The index $\{k\}, r$, means that for each value of $r$ and for each $\mathbf{g} \in \overline{SL(2,\mathbb{C})}$, the image

$$C^{U_{\{k\},r} \circ \mathbf{g}} \stackrel{\text{def}}{=} (U \circ \mathbf{g})_{\{k\},r} (\mathbb{CP}_1) \tag{7.7.101}$$

is a rational curve with homology class $\{k\}$ as defined in Eq. (7.7.94). In other words, for each continuous family of instantons of a given degree, we get a contribution to the cycle

$$H_a \stackrel{\text{def}}{=} (U \circ \mathbf{g})^{-1}_{\{k\},r;x} [C_4^a] \tag{7.7.102}$$

of $\mathbb{CP}_3$ moduli space from those instantons that take a particular point $x \in \mathbb{CP}_1$ to a point within the 4-cycle $C_4^a$. Which particular point in $\mathbb{CP}_1$ is chosen for each operator $\Theta_a^{(0)}$ does not matter since topological correlators are independent of the location of operator insertions. Hence we get a contribution to $H_a$ for each point of intersection of the 2-cycle $C^{U \circ \mathbf{g}}$ with the 4-cycle $C_4^a$. Such a condition is a linear constraint on the homogeneous coordinates $\alpha, \beta, \gamma, \delta$ and defines a hyperplane. Hence $H_a$, $H_b$, $H_c$ are three hyperplanes and they intersect at a point in $\mathbb{CP}_3$. Summarizing the reduced correlator (7.7.99) takes the form

$$\begin{aligned} &\langle \Theta_a^{(0)}(x_1)\, \Theta_b^{(0)}(x_2)\, \Theta_c^{(0)}(x_3) \rangle_{(0,\{k\})} = \\ &\Sigma_{r=1}^{n\{k\}} \# \left( C^{U_{\{k\},r} \circ \mathbf{g}} \cap C_4^a \right) \cdot \# \left( C^{U_{\{k\},r} \circ \mathbf{g}} \cap C_4^b \right) \cdot \# \left( C^{U_{\{k\},r} \circ \mathbf{g}} \cap C_4^c \right) = \\ &\Sigma_{r=1}^{n\{k\}} k_a\, k_b\, k_c = \\ &n_{\{k\}}\, k_a\, k_b\, k_c \end{aligned} \tag{7.7.103}$$

where we have utilized Eq.(7.7.95). In particular, if we consider a case with just one Kähler class deformation, namely $h^{(1,1)} = 1$, we have

$$\langle\, \Theta^{(0)}(x_1)\, \Theta^{(0)}(x_2)\, \Theta^{(0)}(x_3)\, \rangle_{(0,\{k\})} = n_k\, k^3 \tag{7.7.104}$$

Recalling Eq.(7.7.31) we are tempted to conclude that

$$W_{abc}(t) = d_{abc} + \sum_{\{k_a\}} \prod_{a=1}^{h^{(1,1)}} q_a^{k_a}\, n_{\{k_a\}}\, k_a\, k_b\, k_c \tag{7.7.105}$$

Actually Eq.(7.7.105) is not really correct because it overlooks the *multiple coverings of a rational curve*. What do we mean by this? Consider a rational map $Q$ of $\mathbb{CP}'_1$ into $\mathbb{CP}_1$ of degree $m$:

$$Q : \mathbb{CP}'_1 \longrightarrow \mathbb{CP}_1 \tag{7.7.106}$$

Naming $X, Y$ the homogeneous coordinates of the $\mathbb{CP}_1$ playing the role of *target-manifold* and $x, y$ the homogeneous coordinates of the $\mathbb{CP}'_1$ playing the role of world manifold, the map (7.7.106) is described by setting

$$X = Q_X(x,y) \qquad Y = Q_Y(x,y) \tag{7.7.107}$$

where both $Q_X(x,y)$ and $Q_Y(x,y)$ are homogeneous polynomials of degree $m$. The genus zero sphere $\mathbb{CP}_1$ is covered $m$-times by this map, namely every point in the target manifold has $m$ pre-images in the world manifold. Consider now a rational curve $\mathcal{C}^U$ of degree $\{k\}$ in the Calabi–Yau 3-fold $\mathcal{M}_3$. By definition $\mathcal{C}^U$ is a 2-cycle with homology class as in Eq.(7.7.94) that is obtained as the holomorphic image of a $\mathbb{CP}_1$ manifold via the map (7.7.101). Regarding the $\mathbb{CP}_1$ manifold in Eq.(7.7.101) as the target $\mathbb{CP}_1$ of Eq. (7.7.106), it is our privilege to consider the same submanifold $\mathcal{C}^U$ as the holomorphic image of the other $\mathbb{CP}'_1$ that plays the role of world manifold in Eq.(7.7.106). Namely we can set

$$\mathcal{C}^U = \left[U_{\{k\};r} \circ \mathbf{g}(\alpha,\beta,\gamma,\delta) \circ Q\right] (\mathbb{CP}'_1) \tag{7.7.108}$$

The map

$$\mathcal{U}_m \stackrel{\text{def}}{=} U_{\{k\};r} \circ \mathbf{g}(\alpha,\beta,\gamma,\delta) \circ Q : \mathbb{CP}_1 \longrightarrow \mathcal{M}_3 \tag{7.7.109}$$

is an $m$-folded multiple cover of the rational curve $\mathcal{C}^U$ of degree $\{k\}$. In the sum over instantons implied by the functional integral we are supposed to sum over all rational curves and for each of them over all their $m$-folded multiple covers. Indeed the map (7.7.109) is a holomorphic map satisfying the instantonic condition as much as the map

$$\mathcal{U} = U_{\{k\};r} \circ \mathbf{g}(\alpha,\beta,\gamma,\delta) : \mathbb{CP}_1 \longrightarrow \mathcal{M}_3 \tag{7.7.110}$$

The evaluation of the reduced correlator for the case of the map (7.7.109) leads to the same result as for the case of the map (7.7.110):

$$\begin{aligned} &\langle\, \Theta^{(0)}_a\,(x_1)\; \Theta^{(0)}_b\,(x_2)\; \Theta^{(0)}_c\,(x_3)\,\rangle_{(0,\{k\})} = \\ &\textstyle\sum_{r=1}^{n_{\{k\}}} \#\left(C^{U_{\{k\},r}\circ g\circ Q} \cap C^a_4\right) \cdot \#\left(C^{U_{\{k\},r}\circ g\circ Q} \cap C^b_4\right) \cdot \#\left(C^{U_{\{k\},r}\circ g\circ Q} \cap C^c_4\right) = \\ &\textstyle\sum_{r=1}^{n_{\{k\}}} k_a\, k_b\, k_c = \\ &n_{\{k\}}\, k_a\, k_b\, k_c \end{aligned} \tag{7.7.111}$$

since the submanifolds $C^{U_{\{k\},r}\circ g\circ Q}$ and $C^{U_{\{k\},r}\circ g}$ are, by definition, identical and, as such, have the same intersections with the 4-cycles $C^a_4$. The only difference between simple and multiple covers occurs in the evaluation of the prefactors

$$\exp\left[-2t\, S_{class}\right] = \exp\left[2\pi i t \int_{\Sigma_0} X^\star(K)\right] \tag{7.7.112}$$

If for the embedding map $X : \Sigma_0 \longrightarrow \mathcal{M}_3$ we choose the simple cover $\mathcal{U}$, as defined in Eq.(7.7.110), we obtain

$$\begin{aligned} \exp\left[-2t\, S_{class}\right] &= \exp\left[2\pi i t \int_{\Sigma_0} \mathcal{U}^\star(K)\right] \\ &= \exp\left[2\pi i \sum_{a=1}^{h^{(1,1)}} t^a \int_{\mathbb{CP}_1} \mathcal{U}^\star\left(\omega^{(1,1)}_a\right)\right] \\ &= \exp\left[2\pi i \sum_{a=1}^{h^{(1,1)}} t^a\, k_a\right] \\ &= \prod_{a=1}^{h^{(1,1)}} q^{k_a} \end{aligned} \tag{7.7.113}$$

If for $X : \Sigma_0 \longrightarrow \mathcal{M}_3$ we choose the $m$-folded multiple cover $\mathcal{U}_m$ as defined in Eq.(7.7.109), the evaluation of the prefactor yields

$$\begin{aligned} \exp\left[-2t\, S_{class}\right] &= \exp\left[2\pi i t \int_{\Sigma_0} \mathcal{U}^\star(K)\right] \\ &= \exp\left[2\pi i \sum_{a=1}^{h^{(1,1)}} t^a \int_{\mathbb{CP}_1} \mathcal{U}^\star_m\left(\omega^{(1,1)}_a\right)\right] \\ &= \exp\left[2\pi i \sum_{a=1}^{h^{(1,1)}} t^a\, m\, k_a\right] = \left(\prod_{a=1}^{h^{(1,1)}} q^{k_a}\right)^m \end{aligned} \tag{7.7.114}$$

Taking the multiple covers into account the exact formula for the topological correalator (7.7.96), rather than (7.7.105), becomes

$$W_{abc}(t) = \langle\, \Theta^{(0)}_a\,(x_1)\; \Theta^{(0)}_b\,(x_2)\; \Theta^{(0)}_c\,(x_3)\,\rangle$$

$$= d_{abc} + \sum_{\{k_a\}} \sum_{m=1}^{\infty} n_{\{k_a\}} k_a k_b k_c \left( \prod_{a=1}^{h^{(1,1)}} q_a^{k_a} \right)^m$$

$$= d_{abc} + \sum_{\{k_a\}} n_{\{k_a\}} k_a k_b k_c \frac{\prod_{a=1}^{h^{(1,1)}} q_a^{k_a}}{1 - \prod_{a=1}^{h^{(1,1)}} q_a^{k_a}} \qquad (7.7.115)$$

In particular, in the case of a one-dimensional Dolbeault group $H^{(1,1)}$ we have

$$W(t) = d_{111} + \sum_{k=1}^{\infty} \frac{n_k k^3 q^k}{1 - q^k} \qquad (7.7.116)$$

Furthermore, if the Calabi–Yau 3-fold with a one-dimensional Kähler class deformation space is the quintic 3-fold in $\mathbb{CP}_4$, then the intersection number $d_{111}$ is normalized as follows:

$$d_{111} = \int_{\mathbf{CP}_4[5]} K \wedge K \wedge K = 5 \qquad (7.7.117)$$

the Kähler 2-form $K$ on the quintic 3-fold being the pull-back of the Kähler 2-form on $\mathbb{CP}_4$. Equation (7.7.116), with the normalization (7.7.117), will be the starting point in Chapter 8 for the prediction of the number of rational curves $n_k$ through the use of mirror symmetry.

### 7.7.3 The B-twist case and the Hodge structure deformations

In the two previous subsections we studied the correlators of the topological sigma model as it was defined in section 7.4. These correlators depend on a set of topological coupling constants that are identified with the Kähler class moduli of the target manifold $\mathcal{M}_K$. This happens because the BRST non-trivial part of the action (the classical action) is just the pull-back of the Kähler 2-form (see Eq.(7.4.3)). On the other hand we saw that the topological sigma model defined in section 7.4 is nothing but the A-twist of the N=2 sigma model described in section 5.5.6. Hence in view of the fact that any N=2 field theory admits both an A- and a B-twist the natural question that arises is: *What is the B-twist of the N=2 $\sigma$-model?*. As we are going to see, the B-twist selects the complex structure moduli as coupling parameters and the elements of the *Hodge ring* (see section 5.9) as topological observables, so that the B-twisted $\sigma$-model is essentially equivalent to the topological Landau–Ginzburg model with superpotential equal to the the polynomial constraint $\mathcal{W}(X)$ which defines the target space $\mathcal{M}_K$ as a projective (or weighted projective) hypersurface. This is quite natural from the point of view of the interpolating two-phase gauge theory. Indeed, as we already saw, if we perform the twist directly on the gauge theory, the B-twist gives rise to a gauge-coupled Landau–Ginzburg theory. It is anyhow quite instructive to perform the B-twist on the sigma model and see how the Hodge structure deformations come into play.

Therefore let us consider the action (5.5.52), the supersymmetry transformation rules (5.5.54) and the R-symmetry charges (5.5.57). Using the B-twist rules of Eqs. (7.5.3),

(7.5.4) and (7.5.6) we conclude that $X^i$ maintains zero spin and zero ghost number after the twist, the fermions $\psi^{i^*}$, $\widetilde{\psi}^{i^*}$ acquire spin 0 and ghost number +1, while $\psi^i$ and $\widetilde{\psi}^i$ acquire ghost number −1 and spin −1 and +1 respectively. Hence in full analogy with what we did in Eq.(7.6.29) for the B-twist of the Landau–Ginzburg model, we introduce the new variables

$$\begin{aligned} C^{i^*} &= \psi^{i^*} + \widetilde{\psi}^{i^*} \\ \overline{C}^i &= \left(\overline{C}^i_+ \, e^+ + \overline{C}^i_- \, e^-\right) \\ &= \left(\psi^i \, e^+ + \widetilde{\psi}^i \, e^-\right) \\ \theta_i &= g_{ij^*} \left(\psi^{i^*} - \widetilde{\psi}^{i^*}\right) \end{aligned} \qquad (7.7.118)$$

where the only difference with Eq. (7.6.29) is that we lowered the index of $\theta^{i^*}$ by means of the Kähler metric $g_{ij^*}$. This is done for the following reason. As we already noted in Eq. (5.5.55), the only formal difference between the supersymmetry algebra of the Landau–Ginzburg model and of the sigma model is the replacement of the term proportional to the derivatives of the superpotential with a bilinear fermion term proportional to the Christoffel symbols. In the B-twist the BRST variation of $\theta^{i^*}$, as defined in Eq.(7.6.29), is just proportional to such a term (see Eq.(7.6.32)). This is just fine for the Landau–Ginzburg theory. Indeed $\eta^{i^* j} \, \partial_j \mathcal{W} = 0$ is one of the two gauge-fixings and the interpretation of $\theta^{i^*}$ is that of an anti-ghost. In the B-twisted $\sigma$-model, where a gauge-fixing $\Gamma^{i^*}_{j^* k^*} \, \widetilde{\psi}^{j^*} \, \psi^{k^*} = 0$ would have no meaning and where $\theta$ has ghost number +1, the natural variable is $\theta_i$ as defined in Eq.(7.7.118). Indeed a simple calculation shows that, after lowering its index with the metric, the BRST variation of $\theta_i$ is just zero. In terms of the new variables (7.7.118) the BRST algebra derived from Eqs. (5.5.54) with the lower choice of sign in Eq. (7.5.4) is given by

$$\begin{aligned} s\, X^i &= 0 \\ s\, X^{i^*} &= C^{i^*} \\ s\, C^{i^*} &= 0 \\ s\, \theta_i &= 0 \\ s\overline{C}^i &= -\frac{i}{2} \, dX^i \end{aligned} \qquad (7.7.119)$$

which should be compared with Eqs. (7.6.32). As we see, the instantons are given by the constant maps ($dX^i = 0$) from the Riemann surface $\Sigma_g$ to the target manifold $\mathcal{M}_K$, just as in the topological Landau–Ginzburg model they were the constant maps from the Riemann surface to the critical points of the superpotential. However, the role of the anti-ghost $\theta^{i^*}$, whose variation defined the critical points as a gauge-fixing, is now superfluous and this is the reason why we just have $s\, \theta_i = 0$. In terms of the new variables (7.7.118), the N=2 sigma model action (5.5.52) can be rewritten as

$$S_{B\, topol\, sigma} = \Omega\, [R] + K^{(kin)} \qquad (7.7.120)$$

where

$$\Omega[R] = \int \left( -2\, R_{ij^*kl^*}\, \overline{C}^i \wedge \overline{C}^k\, C^{j^*}\, g^{l^*m}\, \theta_m + 2i\, \overline{C}^i \wedge \nabla\theta_i \right) \tag{7.7.121}$$

is BRST-closed but not exact, while

$$\begin{aligned} K^{(Kin)} &= \int \left( dX^i \wedge \left[dX^{j^*}\right]_* - 2i\, \overline{C}^i \wedge \left[\nabla C^{j^*}\right]_* \right) g_{ij^*} \\ &= s \int \Psi^{(Kin)} = s \int 2i\overline{C}^i \wedge \left[dX^{j^*}\right]_* g_{ij^*} \end{aligned} \tag{7.7.122}$$

is the BRST-exact part corresponding to the implementation of the gauge-fixing $dX^j = 0$ of constant maps. Equations (7.7.120), (7.7.121) and (7.7.122) should be compared with the similar equations (7.6.31), (7.6.34) and (7.6.35) describing the action of the B-twisted Landau–Ginzburg model. As one notices, the BRST non-trivial part $\Omega$, which in the Landau–Ginzburg model is a bilinear in the anti-ghosts with the hessian matrix as a coefficient:

$$\mathcal{H}_{ij} \stackrel{\text{def}}{=} \partial_i \partial_j \mathcal{W} \tag{7.7.123}$$

is given here by a quadrilinear in the anti-ghosts and ghosts with coefficients the Riemann curvature tensor: in addition in both cases $\Omega$ contains a ghost kinetic term $\overline{C}^i \wedge \nabla\theta_i$. The other difference between the two theories occurs in the BRST-trivial sector. In the Landau–Ginzburg model we have two independent gauge-fixings, one selecting the constant maps $dX^i = 0$, the second constraining $X^i = X_0^i$ to be one of the critical points of the superpotential where $\partial_i \mathcal{W} = 0$. As already anticipated in the B-twisted sigma model only the first gauge–fixing occurs, the second having no meaning.

As usual the exponential of minus the action (7.7.120) provides the measure for the path integral calculation of the topological correlators:

$$c_{B_1 \ldots B_N}(t) = \int \mathcal{D}X\, \mathcal{D}\overline{C}\, \mathcal{D}C^*\, \mathcal{D}\theta\, \exp\left[-t\left(\Omega[R] + K^{(Kin)}\right)\right] \prod_{i=1}^{N} \Theta^{(0)}_{B_i}(x_i) \tag{7.7.124}$$

The question is: *What are the topological observables $\Theta^{(0)}_{B_i}(x_i)$ in this case?* To answer this question we have to derive the solution of the descent equations (7.6.17) appropriate to the BRST algebra (7.7.119). To this end consider, just as in Eq. (5.9.39), the $\wedge^q T^{(1,0)}\mathcal{M}_K$-twisted Dolbeault complex:

$$\cdots \xrightarrow{\bar\partial} \Omega^{(0,p)} \otimes \wedge^q T^{(1,0)}\mathcal{M}_K \xrightarrow{\bar\partial} \Omega^{(0,p+1)} \otimes \wedge^q T^{(1,0)}\mathcal{M}_K \xrightarrow{\bar\partial} \cdots \tag{7.7.125}$$

where we have replaced the Calabi–Yau $n$-fold $\mathcal{M}_n$ of that discussion with the target space $\mathcal{M}_K$ of the B-twisted N=2 sigma model. Later we shall see that for a sound interpretation of the B-twisting it is indeed convenient that $c_1(\mathcal{M}_K) = 0$, namely that the target manifold be a Calabi–Yau $n$-fold. Such a condition is not, as we have seen, necessary in the A-twisting. Consider next a section $V^{(0,p)}_{(q,0)} \in \Omega^{(0,p)} \otimes \wedge^q T^{(1,0)}\mathcal{M}_K$ which, according to Eq.(5.9.40), is a $(0,p)$-form with values in a section of $\wedge^q T^{(1,0)}\mathcal{M}_K$, namely

$$V = dX^{i_1^\star} \wedge \cdots \wedge dX^{i_p^\star} V^{j_1 \ldots j_q}_{i_1^\star \ldots i_p^\star} \frac{\partial}{\partial X_{j_1}} \wedge \cdots \wedge \frac{\partial}{\partial X_{j_q}} \tag{7.7.126}$$

To any such section we can associate a 0-form operator:

$$\mathcal{O}\left[V^{(0,p)}_{(q,0)}\right] \stackrel{\text{def}}{=} C^{i_1^\star} \cdots C^{i_p^\star} V^{j_1 \ldots j_q}_{i_1^\star \ldots i_p^\star} \theta_{j_1} \cdots \theta_{j_q} \tag{7.7.127}$$

and using the BRST algebra (7.7.119) we obtain

$$s\,\mathcal{O}\left[V^{(0,p)}_{(q,0)}\right] = \mathcal{O}\left[\overline{\partial}\, V^{(0,p)}_{(q,0)}\right] \tag{7.7.128}$$

It follows that $\mathcal{O}\left[V^{(0,p)}_{(q,0)}\right]$ is BRST-closed if $\overline{\partial}\, V^{(0,p)}_{(q,0)} = 0$, namely if $V^{(0,p)}_{(q,0)} = Z^{(0,p)}_{(q,0)} \in H^{(p)}_{\overline{\partial}}\left(\mathcal{M}_K, \wedge^q T^{(1,0)}\mathcal{M}_K\right)$ is a representative of some element in the $p$-th cohomology group of the $\wedge^q T^{(1,0)}\mathcal{M}_K$-twisted Dolbeault complex. This discussion shows that, in the B-twisted sigma model, we have an injective map

$$H^\star_{twisted\ Dolbeault} \stackrel{\text{def}}{=} \oplus_{p,q}\, H^{(p)}_{\overline{\partial}}\left(\mathcal{M}_K, \wedge^q T^{(1,0)}\mathcal{M}_K\right) \longrightarrow \mathcal{H}^\star_{BRST} \tag{7.7.129}$$

between the twisted Dolbeault cohomology ring $H^\star_{twisted\ Dolbeault}$ defined by Eq. (7.7.129) and the BRST-cohomology ring. The product in this ring is the double wedge product as defined in Eq. (5.9.42). Equation (7.7.129) is to be contrasted with Eq.(7.7.10) applying to the case of the A-twist. An important difference between the two cases is that in the A-twist the map Eq.(7.7.10) is an isomophism only if we restrict our attention to the $k = 0$ instanton sector: if, on the other hand, we consider also the contribution of multi-winding instantons ($k > 0$), then the law product in the $\mathcal{H}^\star_{BRST}$ ring (expressed by the correlation functions (7.7.21)) is *quantum-corrected* with respect to the law product in the Dolbeault cohomology ring $H^\star_{Dolbeault}$, expressed by the ordinary wedge product of the closed forms corresponding, via Eqs. (7.4.14), to each 0-form operator $\Theta^{(0)}_{A_i}(x_i)$ . In the B-twist case, instead, the map (7.7.129) is a true isomorphism. This is essentially due to the fact that in the B-twisted model, the instantons are constant maps and the calculation of correlation functions reduces, as we are going to see, to *classical intersection integrals* on the target manifold. In terms of elements $Z^{(0,p)}_{(q,0)} \in H^{(p)}_{\overline{\partial}}\left(\mathcal{M}_K, \wedge^q T^{(1,0)}\mathcal{M}_K\right)$ of the twisted Dolbeault cohomology groups we can now write the solution of the descent equations (7.6.17) as follows:

$$\begin{aligned}
\Theta^{(0)}\left[Z^{(0,p)}_{(q,0)}\right] &= \mathcal{O}\left[Z^{(0,p)}_{(q,0)}\right] = C^{i_1^\star} \cdots C^{i_p^\star} Z^{j_1 \ldots j_q}_{i_1^\star \ldots i_p^\star} \theta_{j_1} \cdots \theta_{j_q} \\
\Theta^{(1)}\left[Z^{(0,p)}_{(q,0)}\right] &= p\, dX^{i_1^\star} C^{i_2^\star} \cdots C^{i_p^\star} Z^{j_1 \ldots j_q}_{i_1^\star \ldots i_p^\star} \theta_{j_1} \cdots \theta_{j_q} \\
\Theta^{(2)}\left[Z^{(0,p)}_{(q,0)}\right] &= p(p-1)\, dX^{i_1^\star} \wedge dX^{i_2^\star} C^{i_3^\star} \cdots C^{i_p^\star} Z^{j_1 \ldots j_q}_{i_1^\star \ldots i_p^\star} \theta_{j_1} \cdots \theta_{j_q}
\end{aligned} \tag{7.7.130}$$

The ghost number of the local observables $\Theta^{(0)}\left[Z^{(0,p)}_{(q,0)}\right]$ is immediately seen to be

$$\#gh\,\Theta^{(0)}\left[Z^{(0,p)}_{(q,0)}\right] = p + q \tag{7.7.131}$$

so that the selection rule for the correlation functions (7.7.124) is

$$\sum_{i=1}^{N}(p_i + q_i) = \Delta U \tag{7.7.132}$$

The calculation of the integrated anomaly $\Delta U$ cannot be directly performed by use of the Hirzebruch–Riemann–Roch theorem as for the A-twisted case if we do not take particular care. Indeed there is an important difference that appears if we compare the kinetic terms of the fermions in the two cases. Recalling Eq. (7.7.34), in the A-twist case we have

$$\text{fermion kinetic terms} = 2\,\overline{c}^{l^\star}_{+}\,g_{l^\star i}\,\nabla_{-}c^{i} + 2\,\overline{c}^{l}_{-}\,g_{li^\star}\,\nabla_{+}c^{i^\star} \tag{7.7.133}$$

so that the anti-ghost anti-holomorphic 1-forms $\overline{c}^{l}_{-}$ are complex conjugates of the anti-ghost holomorphic 1-forms $\overline{c}^{l^\star}_{+}$ while, at the same time, the ghost holomorphic 0-forms $c^i$ are complex conjugates of the ghost anti-holomorphic 0-forms $c^{i^\star}$. It follows that

$$c^{i} \in \Omega^{(0,0)}(\Sigma_g)\otimes\mathcal{E}_U \tag{7.7.134}$$
$$\overline{c}^{l}_{-} \in \Omega^{(0,1)}(\Sigma_g)\otimes\mathcal{E}_U \tag{7.7.135}$$

are two sections in the $\mathcal{E}_U$-twisted Dolbeault complex:

$$\Omega^{(0,0)}\otimes\mathcal{E}_U\xrightarrow{\overline{\partial}}\Omega^{(0,1)}\otimes\mathcal{E}_U\xrightarrow{\overline{\partial}}0 \tag{7.7.136}$$

while

$$c^{i^\star} \in \Omega^{(0,0)}(\Sigma_g)\otimes\overline{\mathcal{E}}_U \tag{7.7.137}$$
$$\overline{c}^{l^\star}_{+} \in \Omega^{(1,0)}(\Sigma_g)\otimes\overline{\mathcal{E}}_U \tag{7.7.138}$$

are two sections in the conjugate complex:

$$\Omega^{(0,0)}\otimes\overline{\mathcal{E}}_U\xrightarrow{\partial}\Omega^{(1,0)}\otimes\overline{\mathcal{E}}_U\xrightarrow{\partial}0 \tag{7.7.139}$$

and the kinetic term is of the form

$$\langle\overline{c}^{i}|\nabla_{-}|c^{i}\rangle + \text{complex conjugate} \tag{7.7.140}$$

Hence the functional integration on ghost and anti-ghosts produces, in the A-twisted model, a fermion determinant that is real (compare with Eqs. (7.7.61)). On the other hand in the B-twisted case, recalling Eqs. (7.7.121) and (7.7.122) we have

$$\text{fermion kinetic terms} = 2i\,\overline{C}^{l}_{+}\,\nabla_{-}\left(\theta_l + g_{li^\star}\,C^{i^\star}\right) - 2i\overline{C}^{l}_{-}\,g_{li^\star}\,\nabla_{+}\left(\theta_l - g_{li^\star}\,C^{i^\star}\right) \tag{7.7.141}$$

Denoting by $X^\star T^{(1,0)}\mathcal{M}_K \xrightarrow{\pi} \Sigma_g$ the pull-back on $\Sigma_g$ through the map $X : \Sigma_g \longrightarrow \mathcal{M}_K$ of the holomorphic tangent bundle $T^{(1,0)}\mathcal{M}_K \xrightarrow{\pi} \mathcal{M}_K$ we see that the anti-ghost anti-holomorphic 1-form is a section of the $X^\star T^{(1,0)}\mathcal{M}_K$-twisted bundle of $(0,1)$-forms:

$$\overline{C}^l_- \in \Omega^{(0,1)} \otimes X^\star T^{(1,0)}\mathcal{M}_K \tag{7.7.142}$$

This is the counterpart of Eq. (7.7.134). However, the holomorphic anti-ghost 1-form $\overline{C}^l_+$ is not the complex conjugate of $\overline{C}^l_-$, since we have

$$\overline{C}^l_+ \in \Omega^{(1,0)} \otimes X^\star T^{(1,0)}\mathcal{M}_K \tag{7.7.143}$$

which is not the counterpart of Eq. (7.7.137). Similarly the 0-form

$$g^{m^\star l}\left(\theta_l + g_{li^\star} C^{i^\star}\right) \in \Omega^{(0,0)} \otimes X^\star T^{(0,1)}\mathcal{M}_K \tag{7.7.144}$$

is not the complex conjugate of

$$g^{m^\star l}\left(\theta_l - g_{li^\star} C^{i^\star}\right) \in \Omega^{(0,0)} \otimes X^\star T^{(0,1)}\mathcal{M}_K \tag{7.7.145}$$

As a consequence the fermionic kinetic terms (7.7.141) are not of the form (7.7.140) and the fermion determinant is intrinsically complex in the B-twisted case. Indeed it has the following structure:

$$\text{fermion determinant} = \left[\det\left(\nabla_-\right)^{i^\star}_{\ j^\star}\right] \cdot \left[\det\left(\nabla_+\right)^{i^\star}_{\ j^\star}\right] \tag{7.7.146}$$

where the determinant operation is both in the functional sense and in the matrix sense, the operators $\left(\nabla_\pm\right)^{i^\star}_{\ j^\star}$ being matrix-valued differential operators. From the functional point of view $\det\left(\nabla_\pm\right)$ are complex conjugate sections of a line bundle over the moduli space of Riemann surfaces and therefore their product is a globally defined function: the problem comes from the matrix point of view since both $\left[\det\left(\nabla_-\right)^{i^\star}_{\ j^\star}\right]$ and $\left[\det\left(\nabla_+\right)^{i^\star}_{\ j^\star}\right]$ are sections of the determinant line bundle $\wedge^n T^{(0,1)}\mathcal{M}_K$, $n = \dim_{\mathbb{C}} \mathcal{M}_K$ being the complex dimension of the target space. The only possibility for the fermion determinant (7.7.146) to be a globally defined function on the bosonic integration manifold is that the determinant line bundle $\wedge^n T^{(0,1)}\mathcal{M}_K \xrightarrow{\pi} \mathcal{M}_K$ be a trivial bundle, so that we can choose a global trivialization. This occurs if the first Chern class of $\wedge^n T^{(0,1)}\mathcal{M}_K$ vanishes:

$$c_1\left(\wedge^n T^{(0,1)}\mathcal{M}_K\right) \stackrel{\text{def}}{=} c_1\left(\mathcal{M}_K\right) = 0 \tag{7.7.147}$$

Equation (7.7.147) shows that the condition for the existence of the fermion determinant in the B-twisted N=2 sigma model is just the Calabi–Yau condition $c_1\left(\mathcal{M}_K\right) = 0$ on the target manifold. Indeed, by definition, a section of the determinant line bundle $s(X) = \xi^{i_1^\star} \cdots \xi^{i_n^\star} \varepsilon_{i_1^\star \ldots i_n^\star}$ is the exterior product of $n$ sections of $T^{(0,1)}\mathcal{M}_K$ and its covariant derivative is

$$\nabla s = \overline{\partial} s + \Gamma^{i^\star}_{\ i^\star} s \tag{7.7.148}$$

where ${\Gamma^{i^\star}}_{i^\star}$ is the trace of the connection on $T^{(0,1)}\mathcal{M}_K$. Hence the curvature of the determinant line bundle is the trace of the curvature on $T^{(0,1)}\mathcal{M}_K$ and this shows that the identification in the first equality of (7.7.147) is correct.

*Hence from now on we assume that $\mathcal{M}_K$ is a Calabi–Yau n-fold $\mathcal{M}_n$.*

With this simplification we can also calculate the ghost number anomaly, obtaining the same result as in Eq.(7.7.57):

$$\Delta U_{B\text{-}model} = 2n\,(1-g) \tag{7.7.149}$$

Indeed under the condition that the determinant line bundle $\wedge^n T^{(0,1)}\mathcal{M}_K$ be trivial we can treat the anti-ghosts $\overline{C}^l_\pm$ and the ghosts $g^{m^\star l}\left(\theta_l \pm g_{li^\star} C^{i^\star}\right)$ as if they were $b$–$c$, $\bar{b}$–$\bar{c}$ systems of conformal weight $\lambda = 1$, their index simply enumerating $n$ identical copies. Hence by use of the standard Riemann–Roch theorem on the Riemann surface $\Sigma_g$ we obtain the result (7.7.149). Combining Eq. (7.7.132) with Eq. (7.7.149), the final selection rule for the topological correlators of a B-twisted sigma model on a Calabi–Yau $n$-fold is

$$\sum_{i=1}^{N}(p_i + q_i) = 2n\,(1-g) \tag{7.7.150}$$

Let us now consider, under the condition that Eq.(7.7.150) be satisfied the explicit calculation of the correlators (7.7.124), now more explicitly rewritten as:

$$\begin{aligned}
&c_{Z_1\ldots Z_N}(t) = \\
&\langle \prod_{i=1}^{N} \Theta^{(0)}\left[Z^{(0,p_i)}_{(q_i,0)}\right](x_i)\rangle = \\
&\int \mathcal{D}X\,\mathcal{D}\overline{C}\,\mathcal{D}C^\star\,\mathcal{D}\theta\,\exp\left[-t\left(\Omega\left[R\right] + K^{(Kin)}\right)\right]\prod_{i=1}^{N}\Theta^{(0)}\left[Z^{(0,p_i)}_{(q_i,0)}\right](x_i)
\end{aligned} \tag{7.7.151}$$

To begin with we can easily determine the dependence of $c_{Z_1\ldots Z_N}(t)$ on the coupling parameter $t$. It suffices to note that the BRST non-trivial part of the action (7.7.121) is linear in the ghost field $\theta_m$ while the BRST trivial part (7.7.122) does not depend on such a field. Hence by a field redefinition $\theta_m \longrightarrow \theta'_m = t\,\theta_m$ we can reabsorb the $t$-dependence of the BRST non-trivial part of the action. After such a redefinition the path integral becomes $t$-independent since the parameter $t$ occurs only through BRST-exact terms. However, rescaling the ghost field $\theta_m$ we gain a power of $t$ due to the $\theta_m$-factors coming from the topological observables. In conclusion we can write

$$\begin{aligned}
&c_{Z_1\ldots Z_N}(t) = \\
&\langle \prod_{i=1}^{N} \Theta^{(0)}\left[Z^{(0,p_i)}_{(q_i,0)}\right](x_i)\rangle = \\
&t^{-\left(1+\sum_{i=1}^{N} q_i\right)}\langle \prod_{i=1}^{N} \Theta^{(0)}\left[Z^{(0,p_i)}_{(q_i,0)}\right](x_i)\rangle_0
\end{aligned} \tag{7.7.152}$$

where the reduced correlator is defined by the following path integral:

$$\langle \prod_{i=1}^{N} \Theta^{(0)} \left[ Z^{(0,p_i)}_{(q_i,0)} \right] (x_i) \rangle_0 =$$
$$\int \mathcal{D}X \, \mathcal{D}\overline{C} \, \mathcal{D}C^* \, \mathcal{D}\theta \, \exp\left[ -\Omega[R] - t\, K^{(Kin)} \right] \prod_{i=1}^{N} \Theta^{(0)} \left[ Z^{(0,p_i)}_{(q_i,0)} \right] (x_i) \tag{7.7.153}$$

Equations (7.7.152) and (7.7.153) should be compared with their analogues for the A-twisted model, namely Eqs. (7.7.31), (7.7.32) and (7.7.33). While in the A-model we have a transcendental dependence on the Kähler structure parameters $t_a$ due to the sum over the instanton sectors, in the B-model the $t$-dependence is much simpler and just power-like. Let us now come to the evaluation of the reduced correlator (7.7.153). Since $\langle \prod_{i=1}^{N} \Theta^{(0)} \left[ Z^{(0,p_i)}_{(q_i,0)} \right] (x_i) \rangle_0$ is $t$-independent we can evaluate the path integral in the limit $\mathrm{Re}\, t \longrightarrow \infty$. Recalling Eq. (7.7.122) the BRST-trivial part of the action, proportional to $t$, has the structure

$$K^{(Kin)} = || \, dX^i \, ||^2 + \text{fermionic terms} \tag{7.7.154}$$

so that in the limit $\mathrm{Re}\, t \longrightarrow \infty$ the constant maps $dX^i = 0$ dominate the functional integral. As we already know the functional determinants due to the integration over the bosonic and fermionic *non-zero-modes* cancel against each other because of the original N=2 supersymmetry of the untwisted model. This was verified in the A-twisted case and such analysis will not be repeated here. So we just focus on the integration over the *zero-modes*. If $\omega^l \; \{l = 1 \dots g\}$ is a basis of holomorphic 1-forms on the Riemann surface, we can write the following *zero-mode* expansion for all the fields appearing in the path integral:

$$\begin{aligned} \overline{C}^i &= \xi^i_l \left[ \mathrm{Im}\Omega^{-1/2} \right]^l_{\ k} \omega^k + \overline{\xi}^i_{l^*} \left[ \mathrm{Im}\Omega^{-1/2} \right]^{l^*}_{\ k^*} \overline{\omega}^{k^*} + \delta \overline{C}^i \\ \theta_m &= \theta^{(0)}_m + \delta\theta_m \\ C^{i^*} &= C^{i^*}_{(0)} + \delta C^{i^*} \\ X^{i^*} &= X^{i^*}_{(0)} + \delta X^{i^*} \end{aligned} \tag{7.7.155}$$

where $X^{i^*}_{(0)}$ are constant bosonic parameters, while $\xi^i_l$, $\overline{\xi}^i_{l^*}$, $\theta^{(0)}_m$, $C^{i^*}_{(0)}$ are constant anti-commuting parameters: $\mathrm{Im}\Omega$ denotes the imaginary part of the period matrix of the Riemann surface $\Sigma_g$ and it is introduced for convenience of normalization. The integration measure on the zero-modes has therefore the following form:

$$\int \mathcal{D}X \, \mathcal{D}\overline{C} \, \mathcal{D}C^* \, \mathcal{D}\theta \; \overset{\text{on zero-modes}}{\longrightarrow}$$
$$\int_{\mathcal{M}_n} \prod_{i=1}^{n} dX^i_{(0)} \, dX^{i^*}_{(0)} \int \prod_{i=1}^{n} d\theta^{(0)}_i \, dC^{i^*}_{(0)} \prod_{l=1}^{g} d\xi^i_l \, d\overline{\xi}^i_{l^*} \tag{7.7.156}$$

and, when restricted to the zero modes, the action (7.7.120) receives non-vanishing contributions only from $\Omega\,[R]$. Using the identity

$$\int_{\Sigma_g} \omega^l \wedge \overline{\omega}^{k^\star} = [\,\mathrm{Im}\Omega\,]^{lk^\star} \tag{7.7.157}$$

we get

$$S_{B\,topol\,sigma} \stackrel{\text{on zero-modes}}{\longrightarrow} -4 \sum_{r=1}^{g} R_{ij^\star kl^\star}\, g^{l^\star m}\, \xi^i_r\, \overline{\xi}^k_{r^\star}\, C^{j^\star}_{(0)}\, \theta^{(0)}_m \tag{7.7.158}$$

Expanding in power series the exponential $\exp\left[-S_{B\,topol\,sigma}\right]$ we obtain terms that are monomials in the fermionic variables and the only contributing ones are those that saturate the Berezin integral. In genus $g \geq 2$ there is in general an obstruction to getting non-vanishing contributions to any correlation function of the observables we are considering. This is evident from the selection rule (7.7.150): for $g \geq 2$ the left hand side of (7.7.150) is negative while the right hand side is strictly positive. In genus one the selection rule allows only the vacuum expectation value of the identity, namely the partition function, to be non zero and from Eq.(7.7.158) we see that the result is just proportional to the integral over $\mathcal{M}_n$ of the top Chern class $c_n\left(\mathcal{M}_n\right) = \det \mathcal{R}_i{}^m$, the symbol $\mathcal{R}_i{}^m$ denoting the matrix valued curvature 2-form. Hence in genus one we have

$$\langle\, \mathbb{1}\, \rangle_{g=1} = const \int_{\mathcal{M}_n} \det \mathcal{R} \tag{7.7.159}$$

This being clarified, the interesting results come from genus zero where, due to the absence of anti-ghosts zero-modes, the action (7.7.158) is just zero. In this case the integration on the fermion zero-modes $dC^{i^\star}_{(0)}$ simply replaces the very same fields with the coordinate differentials $dX^{i^\star}$ while the integration on the zero-modes $\theta^{(0)}_m$ produces an overall $\varepsilon_{m_1\ldots m_n}$. Indeed the zero-mode fermion integration yields a selection rule that is finer than the anomaly saturation (7.7.150) at genus zero, namely

$$\begin{aligned} \sum_{i=1}^{N} p_i &= n \\ \sum_{i=1}^{N} q_i &= n \end{aligned} \tag{7.7.160}$$

Using this selection rule and performing the fermion zero-mode integration we finally get

$$\langle \prod_{i=1}^{N} \Theta^{(0)}\left[Z^{(0,p_i)}_{(q_i,0)}\right](x_i)\rangle_0 = \int_{\mathcal{M}_n} \Omega^{(n,0)} \wedge Z^{(0,p_1)}_{(q_1,0)} \wedge \cdots \wedge Z^{(0,p_N)}_{(q_N,0)} \tag{7.7.161}$$

the wedge product utilized in the above formula being that defined in Eq. (5.9.42) and $\Omega^{(n,0)}$ being the unique $(n,0)$-form of $\mathcal{M}_n$. Indeed by means of the wedge product appearing under integration in Eq.(7.7.161) we realize a map:

$$\otimes_{i=1}^{N} H^{(p_i)}_{\overline{\partial}}\left(\mathcal{M}_n\,,\wedge^{q_i} T^{(1,0)}\mathcal{M}_n\right) \longrightarrow H^{(n)}_{\overline{\partial}}\left(\mathcal{M}_K\,,\wedge^{n} T^{(1,0)}\mathcal{M}_n\right) \tag{7.7.162}$$

The essential thing is that under the Calabi–Yau condition that we have assumed, the cohomology group $H^{(n)}_{\bar\partial}\left(\mathcal{M}_K\,,\wedge^n T^{(1,0)}\mathcal{M}_n\right)$ is one-dimensional, being, due to Eq. (5.9.38), isomorphic to $H^{(0,n)}\left(\mathcal{M}_n\right)$. Hence, if the selection rule (7.7.160) is satisfied, the wedge product in Eq. (7.7.161) produces the unique top form $Z^{(0,n)}_{(n,0)}$ and all that remains is to integrate it on the $n$-fold.

In particular we can restrict our attention to the observables of type $\Theta^{(0)}\left[Z^{(0,k)}_{(k,0)}\right]$. Recalling Eqs. (5.9.38), (5.9.43), (5.9.44) and (5.9.45), if the $n$-fold $\mathcal{M}_n$ is the vanishing locus of a (weighted) homogeneous polynomial $\mathcal{W}(X)$, then these observables are in one-to-one correspondence with elements of the chiral ring $\frac{\mathbf{C}[X]}{\partial\mathcal{W}(X)}$. Furthermore the wedge product just realizes the isomorphism with the chiral ring and with the Hodge ring so that we can write

$$
\begin{aligned}
\langle\prod_{i=1}^{N}\Theta^{(0)}\left[Z^{(0,k_i)}_{(k_i,0)}\right](x_i)\rangle_0 &= \\
&= \int_{\mathcal{M}_n}\Omega^{(n,0)}\wedge Z^{(0,k_1)}_{(k_1,0)}\wedge\cdots\wedge Z^{(0,k_N)}_{(k_N,0)} \\
&= \int_{\mathcal{M}_n}\Omega^{(n,0)}\wedge\omega^{(n-k_1,k_1)}\star\cdots\star\omega^{(n-k_N,k_N)} \\
&= H^{-1}\cdots P_{k_1|\nu}\cdot\ldots\cdot P_{k_N|\nu}
\end{aligned}
\qquad (7.7.163)
$$

where $\star$ is the product in the Hodge ring $Hod^\star\left(\mathcal{M}_n\right)$, defined by Eq. (5.9.44), $\cdot$ is the product in the chiral polynomial ring $\mathcal{R}\left(\mathcal{W}\right)$, $\Omega^{(n,0)}$ is the unique holomorphic $(n,0)$-form and $H=\det\partial_\Lambda\partial_\Sigma\mathcal{W}$ is the unique top grade polynomial of the chiral ring.

This chain of isomorphisms is of the utmost interest. Indeed it implies that the correlation functions of the B-twisted sigma model can be equivalently calculated as correlation functions of the corresponding topological Landau–Ginzburg model. Indeed, as we are going to see, the last line of Eq.(7.7.163) is, essentially, the final recipe for the calculation of the correlators in this type of models. It is also evident from all the above discussion that the calculation of B-twisted correlators is immensely simpler, being a classical algebraic problem, than the calculation of the A-twisted correlators. In the last chapter we shall address the problem of mirror maps that take advantage of such a situation. A mirror pair of Calabi–Yau manifolds $\mathcal{M}_n$ and $M\mathcal{M}_n$ is such that the A-twisted sigma model on the first is equivalent to the B-twisted sigma model on the second and vice versa. When we have such a mirror pair we can solve the usually formidable problem of the instanton sum involved in the A-twisting by calculating B-twisted correlators on the mirror manifold. Mirror symmetry and its uses are addressed in the final chapter.

## 7.8 Topological Conformal Field Theories

In this section we study a particular class of two-dimensional topological field theories: those obtained by twisting N=2 superconformal theories [137].

As we anticipated in section 7.2, the hallmark of a topological field theory is the independence of the correlation functions from the manifold metric. In general this means that the stress–energy tensor of the theory is a BRST commutator. In a two-dimensional quantum field theory we can also require conformal invariance, to study a particular subclass of topological field theories. The additional property of a general topological conformal field theory (TCFT) is the requirement that the energy–momentum tensor should be traceless.

As for all the N=2 cases, we can start from an N=2 superconformal theory and define a suitable twist procedure, following the main lines presented in section 7.5, to get a topological conformal theory. This can be done either at the lagrangian level, or at the abstract algebraic level.

In general the natural framework to implement the twist is at the lagrangian level. This means that we have an N=2 superconformal field theory, with a suitable set of Noether currents which close an N=2 superconformal algebra. Defining the twist in the lagrangian implies a change in the currents which define the particular topological conformal algebra associated with the twisted lagrangian.

However, we can also study, from an abstract algebraic point of view, the class of topological conformal models without reference to a particular lagrangian. If we work at the algebraic level the twist procedure is particularly simple, and it is suggested by the OPE's (5.2.1). Indeed, if we interpret one of the (holomorphic) supersymmetry currents as the BRST current, from the structure of the simple pole in the OPE $G^+G^-$ of Eq. (5.2.1), we immediately recognize that the new stress–energy tensor has the form

$$\widehat{T}_{\pm}(z) = T(z) \pm \frac{1}{2}\partial J(z) \tag{7.8.1}$$

where the $\pm$ sign depends on what we have chosen as BRST current, i.e. $G^+(z)$ or $G^-(z)$. This procedure changes the spin of the supercurrents: one becomes a spin 1 field (the BRST current) and the other a spin 2 field. The complete form of the topological conformal algebra obtained by twisting an N=2 superconformal theory is

$$\begin{aligned} \widehat{T}_{\pm}(z) &= T(z) \pm \frac{1}{2}\partial J(z) \\ \widehat{J}_{\pm}(z) &= \pm J(z) \\ Q_{\pm}(z) &= G^{\pm}(z) \\ G(z) &= G^{\mp}(z) \end{aligned} \tag{7.8.2}$$

To fix ideas let us specialize to the + case in (7.8.2) (we drop for convenience the "+" superscript). The fundamental OPE's are

$$\begin{aligned} \widehat{T}(z)\,\widehat{T}(w) &= \frac{2\,\widehat{T}(w)}{(z-w)^2} + \frac{\partial\widehat{T}(w)}{(z-w)} + reg \\ \widehat{T}(z)\,G(w) &= \frac{2}{(z-w)^2}\,G(w) + \frac{\partial G(w)}{(z-w)} + reg \end{aligned}$$

$$\begin{aligned}
\hat{T}(z)\,Q(w) &= \frac{1}{(z-w)^2}Q(w) + \frac{\partial Q(w)}{(z-w)} + reg \\
\hat{T}(z)\,J(w) &= -\frac{c}{3}\frac{1}{(z-w)^3} + \frac{1}{(z-w)^2}J(w) + \frac{1}{(z-w)}\partial J(w) \\
Q(z)\,G(w) &= \frac{2}{3}c\frac{1}{(z-w)^3} + 2\frac{J(w)}{(z-w)^2} + 2\frac{\hat{T}(z)}{(z-w)} + reg \\
J(z)\,G(w) &= -\frac{G(w)}{(z-w)} + reg \\
J(z)\,Q(w) &= \frac{Q(w)}{(z-w)} + reg \\
J(z)\,J(w) &= \frac{c}{3}\frac{1}{(z-w)^2} + reg
\end{aligned} \tag{7.8.3}$$

In this case $Q(z)$ is interpreted as a BRST current and the cohomology classes of the BRST charge

$$Q^{BRST} = \oint dz\; Q_{\pm}(z) \tag{7.8.4}$$

are identified with the physical fields of the topological theory.

The combined presence of conformal invariance and topological symmetry implies that the the topological algebra is generated by the holomorphic sector (7.8.3) plus its anti-holomorphic counterpart. This implies that the generator $Q_{BRST}$ can be decomposed into a holomorphic plus an anti-holomorphic component defined by the sum of (7.8.4) and $\overline{Q} = \oint \overline{Q}_{\pm}(\overline{z})$.

Notice that the central extension of the Virasoro subalgebra in Eq. (7.8.3) vanishes, while the $U(1)$ current becomes anomalous, with a central extension $\frac{c}{3}$. The $U(1)$ charge $J$ of the superconformal algebra now becomes the ghost current of the topological algebra, and its integrated anomaly, given by $\frac{c}{3}$, is related as usual to the formal dimension of the moduli space. Since

$$\hat{T}(z) = \{Q, G(z)\} = \oint_z dz Q(w)G(z) \tag{7.8.5}$$

the theory is by definition a topological one.

The physical operators spanning the cohomology of the BRST operator are easily determined. If we recall the "Hodge" decomposition that we have introduced in chapter 5, Eq. (5.2.25), we can rewrite it, taking into account the new spin content of the supercurrents, as

$$|\Phi\rangle = |\phi\rangle + Q_0|\chi\rangle \tag{7.8.6}$$

where $\Phi$ is a generic primary field and $\phi$ is a chiral primary field of the superconformal theory. Equivalently, to select a unique representative we can impose the condition

$$\oint Q(z)\phi = 0 \rightarrow Q_0|\phi\rangle = 0, \tag{7.8.7}$$

Any chiral primary operator $\phi$ is a BRST physical observable. As explained in Chapter 5, for unitary theories there are only a finite number of such fields, and their $U(1)$ charges are positive and bounded by $\frac{c}{3} \equiv d$:

$$J_0|\phi_i\rangle = q_i|\phi_i\rangle \quad 0 \leq q_i \leq d \tag{7.8.8}$$

This means that we have an isomorphism between the chiral ring of the N=2 superconformal theory and the cohomology ring of the corresponding topological theory. To each chiral primary field, by applying supersymmetry tranformations $G(z)$, we associate its superpartner, or "1-form":

$$\phi^{(1,0)}(z,\bar{z}) = \int dw G(w)\phi(z,\bar{z}) \tag{7.8.9}$$

The BRST variation of $\phi^{(1,0)}$ is a total derivative, as can be easily checked using Eqs. (6.4.32)

$$\oint dw Q(w)\phi^{(1,0)}(z,\bar{z}) = \partial_z\phi \tag{7.8.10}$$

This means that the integrated operator

$$\oint dw \phi^{(1,0)}(w,\bar{w}) \tag{7.8.11}$$

is another physical invariant observable. Of course we can also construct the anti-holomorphic counterpart of $\phi^{(1)}$, given by

$$\bar{\phi}^{(0,1)} = \int d\bar{w}\overline{G}(\bar{w})\phi(z,\bar{z}) \tag{7.8.12}$$

and finally the non-trivial BRST "2-form"

$$\phi^{(2)} \equiv \phi^{(1,1)} = \oint_w \oint_{\bar{w}} dz d\bar{z} G(z)\overline{G}(\bar{z})\phi(w,\bar{w}) \tag{7.8.13}$$

Descent equations are explicitly solved by

$$\begin{aligned}
\Theta^{(0)} &= \phi \\
\Theta^{(1)} &= \phi^{(1,0)}dz + \phi^{(0,1)}d\bar{z} \\
\Theta^{(2)} &= \phi^2 dz \wedge d\bar{z}
\end{aligned} \tag{7.8.14}$$

This construction shows that to each chiral primary field we can associate a (chiral $n = 2$) superfield

$$A(z,\bar{z},\theta^-,\bar{\theta}^-) = \phi^{(0)} + \theta^-\phi^{(1,0)} + \bar{\theta}^-\bar{\phi}^{(0,1)} + \theta\bar{\theta}\phi^{(2)} \tag{7.8.15}$$

which contains the "0-form" (first component) $\phi^{(0)} \equiv \phi$, as well as the "1-forms" - $(\phi^{(1,0)}, \phi^{(0,1)})$ and "2-form" components $\phi^{(2)}$ of the physical field. As we have shown, these components are obtained by acting with the superconformal generators $G$ and $\overline{G}$.

As can be easily verified using $\hat{T}$ in (7.8.3) the physical operators $\phi_i = \phi_i^{(0)}$ acquire, after the twist, conformal dimensions $(h, \bar{h}) = (0,0)$. Their two point functions define the metric in the topological Hilbert space

$$\langle \phi_i \phi_j \rangle = \eta_{ij} \tag{7.8.16}$$

and their OPE's are nonsingular:

$$\phi_i \phi_j = c_{ij}^k \phi_k \tag{7.8.17}$$

where $c_{ij}^k = \eta^{lk} c_{lij}$ are numerical coefficients related to the three-point functions $c_{ijk}$ via the metric $\eta_{ij}$ (compare with Eq. (7.7.7)).

A generical topological correlation function of physical operators

$$\langle \phi_i \cdots \phi_k \rangle \tag{7.8.18}$$

has to satisfy, at genus zero, the ghost number selection rule

$$\sum_i q_i = q_{top} = \frac{c}{3} \tag{7.8.19}$$

If we consider minimal models with $c = \frac{3k}{k+2}$, the metric $\eta_{ij}$ has a peculiar expression:

$$\langle \phi_i \phi_j \rangle = \eta_{ij} = \delta_{i+j,k} \tag{7.8.20}$$

where we label the chiral ring $\{\phi_i\}$ with $i = 0, \ldots, k$, $\phi_k$ being the top chiral operator with charge $q = \frac{k}{k+2}$. Equation (7.8.20) is true also for a general tensor product of minimal models.

Let us now suppose that our topological field theory can be described by an action $S$. Consider the following family of actions:

$$S(t) = S(0) - \sum t_n \int \phi_n^{(2)} \tag{7.8.21}$$

They are obtained by deforming the action $S = S(0)$, describing the original topological theory, with the operators $\int \phi_i^{(2)}$, corresponding to the coupling constants $t_i$. From the point of view of the N=2 theory these perturbations respect N=2 supersymmetry, being associated with the highest component of a chiral N=2 supermultiplet. This means that in the topological theory they respect the nilpotency of the BRST charge $Q$ and therefore preserve the topological properties of the theory, such as the metric independence and the factorization Eq. (7.7.6). In correspondence with Eq. (7.8.21) we can write the perturbed correlation functions:

$$c_{ijk}(t) = \langle \phi_i \phi_j \phi_k \exp\left(\sum_n t_n \int \phi_n^{(2)}\right) \rangle \tag{7.8.22}$$

In the case of topological conformal field theories, by exploiting the conformal invariance of the $t_n = 0$ point, we can show that the coefficients $c_{ijk}(t)$ satisfy an important integrability condition. Namely

$$\frac{\partial c_{ijk}(t)}{\partial t_l} = \frac{\partial c_{ijl}(t)}{\partial t_k} \tag{7.8.23}$$

This equation shows that one can integrate the $c_{ijk}(t)$, and obtain a single function $F(t)$ that satisfies

$$c_{ijk}(t) = \frac{\partial^3 F(t)}{\partial t_i \partial t_j \partial t_k} \tag{7.8.24}$$

Moreover we can show that the metric $\eta_{ij} = c_{0ij}$ is in fact independent of the couplings $t_i$. Indeed

$$c_{ij0}(t) \equiv \langle \phi_i \phi_j \mathbb{1} \rangle(t) = \langle \phi_i \phi_j \rangle(t) = \eta_{ij}(0) \tag{7.8.25}$$

the index 0 being associated with the identity operator $\mathbb{1}$.

Looking at Eq. (7.8.25) we see that the set of coupling constants $t_i$ form a distinguished basis in the space of couplings. They are called flat coordinates because the metric (7.8.25) is a constant flat metric. They correspond to deformations of a topological theory around the conformal point, and the proof of (7.8.25) indeed makes use of the Ward identities of the topological conformal theory.

Given a generic topological field theory with a non-trivial infrared fixed point, we can define in general the perturbation around the fixed point by taking arbitrary directions $s_1, \ldots, s_n$ in the coupling constant space. In the $s$ coordinates the metric is not in general a constant metric and in this case we must replace in Eq. (7.8.24) the ordinary derivatives by covariant ones defined by the metric $\eta_{ij}$. This is very reminiscent of the fact that in special geometry, one can write (7.8.24) only in "special" coordinates, while in a general coördinate system we must use covariant derivatives. It is obvious that there should be a relation between the flat geometry of topological theories and special geometry. This relation will be analyzed in Chapter 8, using Picard–Fuchs equations.

To prove (7.8.24), (7.8.25) we follow [118, 116] and we use Ward identities associated with the spin 2 field $G(z)$. On the sphere we can write

$$0 = \langle \oint_c \zeta(w) G(w) \phi_1(z_1) \cdots \phi_n(z_n) \rangle = \sum_{i=1}^{n} \zeta(z_i) \langle \phi_1(z_1) \cdots \phi_i^{(1,0)}(z_i) \cdots \phi_n(z_n) \rangle \tag{7.8.26}$$

where $\zeta(w)$ is a globally defined vector field, so that on the sphere it is necessarily of the form

$$\zeta(w) = aw^2 + bw + c \tag{7.8.27}$$

and the closed curve $c$ encircles the insertion operator points $z_1 \cdots z_n$. If we now apply the right-moving supercurrent $\overline{G}(\overline{z})$ and we choose a suitable $\zeta(w)$ vanishing at $z_1, z_2$, for the three-point function we get

$$\langle \phi_1 \phi_2 \phi_3^{(2)} \rangle = 0 \tag{7.8.28}$$

This means that, to first order, $c_{ij0} = \eta_{ij}$. Moreover, using the definition (7.8.13) of $\phi^{(2)}$, we have

$$\oint \zeta(w)G(w)\phi_n^2(z) = 0 \tag{7.8.29}$$

so that (7.8.28) remains unchanged also to higher orders in the $t$ expansion. This concludes the proof of (7.8.25). Now we look at the four point function and we choose $\zeta(w) = (w - z_1)(w - z_2)$, so that

$$\langle \phi_1\phi_2\phi_3\phi_4^{(2)}\rangle = |\frac{\zeta(z_3)}{\zeta(z_4)}|^2 \langle \phi_1\phi_2\phi_3^{(2)}\phi_4\rangle \tag{7.8.30}$$

If we set

$$\langle \phi_1\phi_2\phi_3\phi_4^2\rangle = |\frac{\partial\chi}{\partial z_4}|^2 G_{123,4}(\chi) \tag{7.8.31}$$

where

$$\chi = \frac{(z_1 - z_3)(z_2 - z_4)}{(z_1 - z_2)(z_3 - z_4)} \tag{7.8.32}$$

is the anharmonic ratio, we find

$$G_{123,4} = G_{124,3} \tag{7.8.33}$$

or, more generally,

$$G_{ijk,m} = G_{ijm,k} \tag{7.8.34}$$

so that

$$\partial_m c_{ijk}(t=0) = \partial_k c_{ijm}(t=0) \tag{7.8.35}$$

To prove Eq. (7.8.33) we have used the identity

$$\zeta(z_3)\frac{\partial\chi}{\partial z_3} + \zeta(z_4)\frac{\partial\chi}{\partial z_4} = 0 \tag{7.8.36}$$

Finally, using again Eq. (7.8.29) we conclude that (7.8.35) is also true for $t \neq 0$. Equation (7.8.24) and (7.8.25) are very powerful and state that the marginal and relevant perturbations around a conformal field theory define a peculiar flat geometry. Due to the link between superconformal field theories and Landau–Ginzburg models, it is natural to ask whether we can estabish an effective Landau–Ginzburg lagrangian formulation of topological conformal field theories. In particular we are interested in giving an exlicit expression for the topological correlators, via a path integral argument. The natural candidate lagrangian to perform such a calculation is the topological Landau–Ginzburg lagrangian (7.6.30).

As for the case of N=2 Landau–Ginzburg model, which defines at its infrared fixed point an N=2 conformal field theory, the lagrangian (7.6.30) flows at the large scale limit to a topological conformal field theory. The same ADE classification that holds for N=2 Landau–Ginzburg and N=2 superconformal models is also valid for topological Landau–Ginzburg models and topological conformal field theories. The purpose of the following section is to give a path integral formula for the correlation functions in the topological Landau–Ginzburg model, following arguments similar to those utilized in the case of the B-twisted sigma model. Moreover we study perturbed correlations functions and we analyze a general procedure to find flat coordinates in Landau–Ginzburg models. This is the best way to analyze the general properties of TCFT, and to find a closed formula for their correlation functions.

## 7.9 Correlators of the Topological Landau–Ginzburg Model

As we saw in Chapter 5, in the Landau–Ginzburg formulation of (2,2)-supersymmetric models, the superconformal theory is viewed as the infrared fixed point of the two-dimensional N=2 Wess–Zumino model with a polynomial superpotential $\mathcal{W}(X)$. When $\mathcal{W}(X)$ is an analytic quasi-homogeneous function we can assign a well-defined $U(1)$ charge (the R charge) to these fields as shown in equations (5.5.43). After the twist the R charge counts the ghost number of the fields, and, as usual, the integrated anomaly of the ghost current gives the formal dimension of the moduli space. In the case of topological Landau–Ginzburg models we have

$$\int \partial_\mu J^\mu = q_{top}(1-g) \tag{7.9.1}$$

where $q_{top}$ is the R charge of the top chiral field in the local ring $\mathcal{R}_\mathcal{W} = \frac{\mathbb{C}(X)}{d\mathcal{W}}$. When we choose a minimal model with potential in the $A_k$ series, we have precisely $q_{top} = \frac{k}{k+2}$ which, as expected, coincides with the top $U(1)$ charge of the corresponding conformal model.

Any element in $\mathcal{R}_\mathcal{W}$ of the untwisted Landau–Ginzburg model is associated with a chiral primary field in the corresponding (2,2) superconformal theory. The same is true in the topological case: the physical operator of a TCFT corresponds to elements in the local ring. This means that in studying the descent equations associated with BRST transformations (7.6.32), we should find a BRST cohomology that selects the elements of $\mathcal{R}_\mathcal{W}$ as representatives of the classes. This is explicitly proved in Eq. (7.9.2). As was shown before, once given the chiral ring of the N=2 superconformal theory (i.e. the topological sector of the model), we can consider the perturbed three point functions $c_{ijk}(t)$ as well as the metric $\eta_{ij}(t) = \eta_{ij}(0)$, which in the coupling space defines a *flat constant geometry*.

So we are allowed to consider in the same fashion also a perturbed correlation function of the Landau–Ginzburg model just as in Eq. (7.8.22). In this case, however, the natural couplings associated with the deformations are not necessarily the flat ones, as we shall see in the following.

Let us recall the action (7.6.30) with the transformation rules (7.6.32). The descent equations associated with (7.6.32) are easily solved by the following positions:

$$\Theta^{(0)} = P(X)$$

$$\begin{aligned}\Theta^{(1)} &= 2i\partial_i P\overline{C}^i \\ \Theta^{(2)} &= \left[-2\partial_i\partial_j P\,\overline{C}^i\wedge\overline{C}^j + 4\partial_k P\,\partial_{l^*}\overline{\mathcal{W}}\,\eta^{kl^*}e^+\wedge e^-\right]\end{aligned} \tag{7.9.2}$$

where $P(X)$ is a polynomial corresponding to some non-trivial element of the local ring determined by the superpotential $\mathcal{W}(X)$. Indeed, if $P(X)$ is proportional to the vanishing relations (i.e. if $P(X) = \sum_i p^i(X)\frac{\partial \mathcal{W}}{\partial X_i}$), then using the BRST transformations (7.6.32) (7.168), one can see that $P(X) = s\,K$ and so $\Theta^{(0)}$ would be exact. (For the proof it suffices to set $K = \frac{1}{2}p^i(X)\theta^{j^*}\eta_{ij^*}$.)

It is interesting to observe that under the deformation

$$\mathcal{W} \longrightarrow \mathcal{W} - \frac{1}{2}s_P\,P(X) \tag{7.9.3}$$

where $P(X)$ is some element of the local ring and $t_P$ is the corresponding coupling constant, the topological Landau–Ginzburg action changes as follows (to first order in $s_p$):

$$\int d^2z\,\mathcal{L} \longrightarrow \int d^2z\,\mathcal{L} - s_P\int\Theta^{(2)} - \overline{s}_P\int\overline{\Theta}^{(2)} \tag{7.9.4}$$

where $\mathcal{L}$ is given in (7.6.30), $\Theta^{(2)}$ in (7.9.2) and $\overline{\Theta}^{(2)}$ is given by

$$\overline{\Theta}^{(2)} = -2C^{i^*}\theta^{i^*}\partial_{i^*}\partial_{j^*}\overline{P}e^+\wedge e^- - 4\partial_{i^*}\overline{P}\eta^{j^*k}\partial_k\mathcal{W}e^+\wedge e^- \tag{7.9.5}$$

This means that there is a precise relation between deformations of the potential by elements of the chiral ring and topological perturbations of the lagrangian. Notice that the 2-form $\overline{\Theta}^{(2)}$, corresponding to a deformation of $\overline{\mathcal{W}}(\overline{X})$, is BRST-exact, and so adding or not its integral to the action is completely irrelevant. In fact, using the BRST transformations (7.6.32), one can check that

$$\overline{\Theta}_P^{(2)} = s\left(-2\,\partial_{j^*}\overline{P}\,\theta^{j^*}e^+\wedge e^-\right) \tag{7.9.6}$$

The solution of the descent equations shows that the cohomology of the BRST operator, i.e. the Hilbert space structure of the topological Landau-Ginzburg theory, is completely determined by the local ring of the potential $\mathcal{W}(X)$. We have shown that we can add a suitable deformation to the potential such as $\sum_p s_pP(X)$ with $P(X)$ belonging to the local ring, to obtain a new topological theory. So we can relax the condition of quasi-homogenity of the potential and we assume that $\mathcal{W}$ is deformed so that there are no degenerate critical points and $H = \det\,\partial_i\partial_j\mathcal{W} \neq 0$. We have denoted the coupling constant related to the deformation operators by "$s_P$" because they do not coincide with the flat coordinates. This is so because in the Landau–Ginzburg formulation of topological conformal theories the relevant and marginal deformations of the superpotential do not correspond directly to deformations around the conformal point; rather they are related to the latter by the solution of a uniformization problem, which in general involves higher transcendental functions. This is easily undestood with the following qualitative considerations [116]. The N=2 Landau–Ginzburg action can be put into correspondence with an N=2 superconformal theory only at its infrared fixed point.

When we turn on a coupling $\delta s_i$ corresponding to one of the operators $P_i(X_j)$, we change the potential by

$$\delta\mathcal{W} = \delta s_j P_j \tag{7.9.7}$$

This modifies also the chiral ring, so that the superpotential $\mathcal{W}$ will change to the next order. This reflects the fact that the metric in the coupling constant space $\{s\}$ is not a constant.

As we pointed out in Chapter 5, the critical points of the deformed potential are in one-to-one correspondence with the elements of the local ring. This means that it is possible to give a description of the Hilbert space structure in terms of the minima $d\mathcal{W} = 0$. Let us now give a path integral formula that indeed shows how the Hilbert space is spanned by the critical points of $\mathcal{W}$ [247].

Our purpose is to calculate the correlator:

$$\langle \prod_i \Theta_i^0(X) e^{-S_{Top\,LG}} \rangle = \int \mathcal{D}X\mathcal{D}\overline{\mathcal{C}}\mathcal{D}\mathcal{C}^*\mathcal{D}\theta^* e^{-S_{Top\,LG}} \prod_i \Theta_i^0(X) \tag{7.9.8}$$

where the topological Landau–Ginzburg action is defined in (7.6.30). We are particularly interested in genus zero correlations functions, but we perform the actual calculation for any value of $g$. For $g = 0$, selection rule (7.9.1) states that the non-vanishing contributions to (7.9.8) arise when the product of the $\Theta^{(0)}$ is proportional to the top chiral field $H$ (plus possible contributions proportional to the vanishing relation) so that

$$\sum_i q_i = q_{top} = \frac{c}{3} \tag{7.9.9}$$

For a generic topological theory, correlation functions do not depend on the world-sheet metric $g$. So we can freely rescale $g \to \lambda^2 g$ and take $\lambda \to \infty$, and see what happens to the path integral. Since $\Omega(\mathcal{W})$ does not depend on the metric $g$, while $K_{kin}$ and $K_{\mathcal{W}}$ do, the path integral measure has the following form:

$$\exp\left[-\lambda^2(||dX^i||^2 + |\partial_i\mathcal{W}|^2) + fermions\right] \tag{7.9.10}$$

so that, as anticipated in section 7.6.3, the configurations that dominate the path-integral are the instanton configurations, defined by the following conditions:

$$\begin{aligned} dX^i &= 0 \\ \partial_i\mathcal{W} &= 0 \end{aligned} \tag{7.9.11}$$

These instantons are constant maps from the the Riemann surface $\Sigma$ to the critical points of the superpotential $\mathcal{W}$, namely to a discrete and finite set of points. To go on with the functional integral we observe that, as in the sigma model case, the non-zero-mode contributions of the fermions and of the bosons cancel against each other and we are left with the integrations over the zero-modes. So we can write

$$\int \mathcal{D}X\mathcal{D}\overline{C}\mathcal{D}C^*\mathcal{D}\theta^* = \int \mathcal{D}X_0\mathcal{D}\overline{C}_0\mathcal{D}C_0^*\mathcal{D}\theta_0^* \tag{7.9.12}$$

Now the number of zero-modes associated with $\overline{C}^i, \theta_i, C^{i^*}$ is easily computed. Since the instantons are constant maps, the zero-modes associated with $\theta^*$ and $C^*$ are constant anti-commuting parameters, while those related to $\overline{C}$ are 1-forms, as in (7.7.155). This means that, for a genus $g$ Riemann surface, we have $g$ zero-modes of $\overline{C}^i$ and one zero-mode associated with the fields $\theta, C^{i^*}$ respectively.

In the limit $\lambda^2 \to \infty$ the bosonic sector gets contribution from the term $|\partial_i\mathcal{W}|^2$ in the action. Indeed the bosonic zero-mode fields have to be expanded around the critical points $X_0 \in [d\mathcal{W} = 0]$ of the potential, so that repalcing $\partial_i\mathcal{W}$ by its linear term in these fields (which is correct in the large $\lambda^2$ limit) we get

$$\begin{aligned}&\int \mathcal{D}\phi_0 \prod_i P_i(X) exp[-\lambda^2 \sum_{X_0\in[d\mathcal{W}=0]} (-8|\partial_i\partial_j W X_0^j|^2 e^+ e^- \\ &+4\overline{C}_0^i \wedge \overline{C}_0^j \partial_i\partial_j W + 4C^{i^*}\theta^{j^*}\partial_{i^*}\partial_{j^*}\overline{W})]\end{aligned} \tag{7.9.13}$$

where $\mathcal{D}\phi_0$ is a shorthand notation for the zero-mode measure. Evaluating the path integral (7.9.13) we can conclude that the correlators have the following form:

$$\langle P_1(X)\cdots P_l(X)\rangle = h\langle\langle P_1(X)\cdots P_l(X)\rangle\rangle \tag{7.9.14}$$

where $h$ is some so far undetermined normalization constant and where we have defined

$$\langle\langle P_1(X)\cdots P_l(X)\rangle\rangle \equiv \sum_{X_0\in[d\mathcal{W}=0]} P_1(X_0)\cdots P_l(X_0)H^{g-1}(X_0) \tag{7.9.15}$$

For the case of genus zero, (7.9.14) and (7.9.15) coincide with Eq. (7.7.163), showing the claimed equivalence of the B-twisted sigma model with the Landau–Ginzburg model. Notice that the factor $H^{g-1}$ comes from the product of the bosonic determinant $(H\overline{H})^{-1}$ with the fermionic one $H^g\overline{H}$. Therefore any correlation of our two-dimensional theory reduces to a zero-dimensional correlation function, where the sum is performed over the critical points of $\mathcal{W}$. Performing explicit calculations with formula (7.9.15) is in general quite difficult, but we can rephrase it in a simpler way. To fix ideas we consider a potential depending on one variable $X$; we get

$$\begin{aligned}\langle\langle \prod_i P_i(X)\rangle\rangle &= \sum_{d\mathcal{W}=0} \frac{[\prod_i P_i(X)(\partial^2\mathcal{W})^g]}{\partial^2\mathcal{W}} = \\ \oint\left(\frac{\prod_i P_i(X)(H)^g}{\partial\mathcal{W}(X)}\right) &= \text{res}\left(\frac{\prod_i P_i(X)(H)^g}{\partial\mathcal{W}(X)}\right)\end{aligned} \tag{7.9.16}$$

where, to calculate the residue in the complex variable $X$, we take a contour at large radius. For many variables this generalizes to

$$\langle\langle \prod_i P_i(X_j) \rangle\rangle = \oint \frac{dX_1 \cdots dX_n}{\partial_1 \mathcal{W} \cdots \partial_n \mathcal{W}} \prod_i P_i(X_j) H^g \tag{7.9.17}$$

For genus $g = 0$ formula (7.9.17) defines the "residue pairing formula".

### 7.9.1 Applications of the residue pairing formula

Let us consider formulas (7.9.16) and (7.9.17), with $g = 0$. For a given quasi-homogeneous potential $\mathcal{W}(X_i)$ there exists (up to an overall normalization) a unique chiral field $H = \det \partial_i \partial_j W$ of maximal $U(1)$ charge $\frac{c}{3}$. For this field we easily find, using (7.9.15),

$$\langle\langle H \rangle\rangle = \text{res}_{g=0}(H) \equiv \text{res}_{\mathcal{W}}(H) = \mu \tag{7.9.18}$$

where $\mu$ is the critical index of $\mathcal{W}$, namely the dimension of the chiral ring $\mathcal{R}_{\mathcal{W}}$. In the following, to simplify our computation we normalize the constant $\mu$ to the value $\mu = 1$. If we take a generic monomial $\phi_i(X)$ in the fields $X$ we find that

$$\langle\langle \phi(X) \rangle\rangle = \text{res}\,(\phi_i(X)) = 0 \quad \text{if} \quad q_{\phi_i} < \frac{c}{3} \tag{7.9.19}$$

This formula can be easily understood if we consider the selection rule (7.9.1). However, Eq. (7.9.19) is also true when we consider the perturbed potential:

$$\mathcal{W}(X_i, s_j) = \mathcal{W}(X_i) - s_i \phi^i(X_j) \tag{7.9.20}$$

where $\phi^i(X_j) \in \mathcal{R}_{\mathcal{W}}$ and have charges less than or equal to 1 (i.e. we consider only relevant or marginal perturbations). Formula (7.9.19) can be proved using (7.9.17) and rescaling the fields as $X_i \to \lambda^{q_i} X_i$ (where $q_i$ is the $U(1)$ charge defined by the unperturbed potential). If the fields $\phi_i$ have charge less than 1, then taking the limit $\lambda \to \infty$ we obtain a leading contribution which is entirely governed by the quasi-homogeneous part of $\mathcal{W}(X)$. Moreover the integral in (7.9.17) gets a prefactor:

$$\lambda^{q_\phi - \sum_i (1-2q_i)} = \lambda^{q_\phi - c/3} \to 0 \tag{7.9.21}$$

and this proves formula (7.9.19).

Given a generic monomial $g(X_i)$ we can decompose it as follows:

$$g(X_i) = f(s_i) H + G^i \partial_i \mathcal{W} + (q < \frac{c}{3} \text{ fields }) \tag{7.9.22}$$

This can be understood in the following way: in the unperturbed case if $q_{g(X)} > c/3$, $g(X_i)$ trivially vanishes (it contains a factor proportional to the vanishing relation $\partial_i \mathcal{W} = 0$). In the perturbed case, as we explicitly show in some of the examples discussed below, this is no longer true, but we can prove, by repeated applications of the vanishing relations, that $g(X_i)$ is equivalent to a field of lower charge. In this way we can inductively find a representative of any field which involves only states with charge less than or equal to $c/3$. Thus, when we consider the expectation value of $g(X_i)$, we find

$$\langle g(X_i)\rangle_{g=0} = f(s_i)\langle H(s_i)\rangle \tag{7.9.23}$$

where $\langle H(s_i)\rangle$ is the expectation value of the hessian in the perturbed case. If we consider the chiral field $\phi_i \in \mathcal{R}_{\mathcal{W}}$ used to perturb the potential, we can contruct the metric $g_{ij}$ as

$$g_{ij}(s) = \langle \phi_i \phi_j\rangle(s) \tag{7.9.24}$$

which depends non-trivially on the coupling constants $s_i$. The metric $g_{ij}$ is a flat metric but in general it is not constant, as in the topological conformal case. The flat coordinates $t_i$ are related to the $s_i$ ones by a coordinate tranformation which transforms $g_{ij}$ into the constant metric $\eta_{ij}$ discussed in section 7.8. The explicit form of this transformation is in general easily found when all the couplings are relevant, but in the presence of marginal perturbations it involves a uniformization problem which is deeply related to the geometry of the moduli space (see Chapter 8 on Picard–Fuchs equations). In the following we give some explicit examples of how to compute the metric for a perturbed Landau–Ginzburg potential and how to find the coordinate transformation which relates the $s$ coordinates to the flat coordinates $t$. We essentially follow the method proposed in [218, 203, 250], which consists in calculating all the correlations in the non-flat frame $s_i$ and then finding the coordinate transformation $s = s(t)$ such that $g_{ij}(t) = \eta_{ij} =$ constant.

This method is successful only for theories with $\frac{c}{3} \leq 1$. For theories with $\frac{c}{3} > 1$ the search for flat coordinates is more involved and requires the analysis of Picard–Fuchs equations. In Chapter 8 we shall understand this difference in terms of invariants of the Picard–Fuchs differential equation. In the language of that chapter the case $\frac{c}{3} = 1$ corresponds to a situation where the only invariant is $W_2 = 0$ and the flat coordinate $t_{flat}$ coincides with the schwarzian coordinate $t_{schw}$ satisfying the equation

$$\{t_{schw}, s\} = \frac{1}{5}w_2(s) \tag{7.9.25}$$

where

$$\{f(x), x\} = \frac{f'''}{f'} - \frac{3}{2}(\frac{f''}{f'})^2 \tag{7.9.26}$$

is the schwarzian derivative. For $\frac{c}{3} > 1$, in particular for $\frac{c}{3} = 3$, the Picard–Fuchs differential equation admits more invariants, notably $W_4 = 0$, and the schwarzian coordinates do not coincide with the flat coordinates. The method we present amounts essentially to the search for schwarzian coordinates.

We begin with a $\frac{c}{3} \leq 1$ example, where the flat coordinates can be determined algebraically, since no marginal deformations occur. We consider the $k = 3$ minimal model:

$$\mathcal{W} = \frac{1}{5}X^5 \tag{7.9.27}$$

The chiral ring for the unperturbed potential is given by

$$\{1, X, X^2, X^3\} = \{\phi_0, \phi_1, \phi_2, \phi_3\} \tag{7.9.28}$$

The unperturbed metric is

$$g_{ij} = \langle \phi_i \phi_j \rangle = \delta_{i+j,3} \tag{7.9.29}$$

Then we consider the perturbed potential

$$W = \frac{1}{5}X^5 - s_0 - s_1 X - s_2 X^2 - s_3 X^3 \tag{7.9.30}$$

with the vanishing relation

$$X^4 = s_1 + 2s_2 X + 3s_3 X^2 \tag{7.9.31}$$

By using (7.9.31) (eventually iterating the procedure), we get the following expression for the metric:

$$g_{ij}(s) = \begin{pmatrix} 0 & 0 & 0 & 1 \\ 0 & 0 & 1 & 0 \\ 0 & 1 & 0 & 3s_3 \\ 1 & 0 & 3s_3 & 2s_2 \end{pmatrix} \tag{7.9.32}$$

It is immediate to verify that (7.9.32) defines a flat metric, as it should. However, as anticipated, in the $s_i$ coordinates that metric is not constant. The coordinate transformation which yields a constant metric is easily found following two steps. We first make an ansatz of a general transformation $s = s(t)$ which is compatible with the $U(1)$ charge assignements of the couplings. This means: $\{s_0, s_1, s_2, s_3\}$, as well as the corresponding $t_i$, have $U(1)$ charges $\{1, 4/5, 3/5, 2/5\}$. So we can guess that, for example, $s_0 = a_1 t_0 + a_2 t_2 t_3$ (up to an overall constant) and similarly for the other cases. Then we require that such a transformation should take the metric $g_{ij}(s)$ into the canonical flat one $\eta_{ij}$. After a simple calculation we find

$$\begin{aligned} s_0 &= t_0 - t_2 t_3 \\ s_1 &= t_1 - t_3^2 \\ s_2 &= t_2 \\ s_3 &= t_3 \end{aligned} \tag{7.9.33}$$

If we consider a Landau–Ginzburg potential which has marginal deformations, the problem of finding flat coordinates is, as we anticipated, more involved. In the case where all the couplings are relevant the entries of the metric are polynomial functions of the couplings. Moreover the transformation $s_i = s_i(t_j)$ is also of the polynomial type, as required by the $U(1)$ charge conservation. In the case where marginal deformations are present we have in general non-polynomial entries in the metric as well as in the transformation $s(t)$, so that the expectation value of the hessian $H$ depends non-trivially on the marginal deformations. In this case we can use the a priori information that the metric $g_{ij}(s) = \langle\phi_i\phi_j\rangle(s)$ should be a flat metric in the coupling constant space (*information obtained from the properties valid at the conformal point, as described in section 7.8*) to determine the $s$-dependence of the so-far-undetermined normalization factor appearing in Eq. (7.9.14). We shall see this in the explicit examples we analyze. For the reasons already discussed we consider two $\frac{c}{3} = 1$ cases where the method we shall outline works. Geometrically the examples presented correspond to two different algebraic realizations of the only existing Calabi–Yau 1-fold, namely the complex torus $T$. This simply means that $T$ is the vanishing locus in a weighted projective space of the considered superpotential $\mathcal{W}(X, s)$. Hence the marginal coupling parameter $s$ we shall be discussing is just a coordinate for the torus moduli space.

The cases where $\frac{c}{3} > 1$ are postponed to Chapter 8, since their analysis can be addressed only through the use of Picard–Fuchs equations.

We begin with the example of the $c = 3$ model defined by the potential

$$\mathcal{W} = \frac{X^4}{4} + \frac{Y^4}{4} - sX^2Y^2 \tag{7.9.34}$$

The field $X^2Y^2$ is a marginal perturbation of the unperturbed potential $\mathcal{W}_0 = \frac{X^4}{4} + \frac{Y^4}{4}$. To perform the computation we switch on all the relevant couplings associated with the chiral ring of $\mathcal{W}_0$, and we write down the fully perturbed potential:

$$\mathcal{W} = \frac{X^4}{4} + \frac{Y^4}{4} - s_0 - s_1X - s_2Y - s_3X^2 - s_4XY - s_5XY^2 - s_6X^2Y - s_7Y^2 - sX^2Y^2 \tag{7.9.35}$$

where, as in (7.9.34), by $s$ we denote the coupling constant associated with $\phi_8 = X^2Y^2$. We then compute all the metric elements in the nine-dimensional space $s_0, \ldots s_7, s$. This in general involves lengthy calculations but in this particular case there is a short cut that provides a significant simplification of the procedure. Indeed, in this case, it suffices to look at a subring of the chiral ring.

Let us consider the action of the $\mathbb{Z}_4$ symmetry group of the potential (7.9.34) generated by

$$\mathbf{g} = (\beta, \beta^3) \qquad \beta = e^{\frac{\pi i}{2}} \tag{7.9.36}$$

with the first entry of $\mathbf{g}$ multiplying the X coordinate and the second the Y coordinate. The only elements in the chiral ring that are left invariant by this symmetry are $\phi_0 = \mathbb{1}$, $\phi_4 = XY$ and $\phi_8 = \phi_{top} = X^2Y^2$. Our purpose is to compute the metric and the Riemann tensor for the reduced perturbed potential:

$$W(\mu, s) = \frac{X^4}{4} + \frac{Y^4}{4} - \mu\phi_4 - s\phi_8 \ . \tag{7.9.37}$$

with $\mu = s_4$. Since we are interested only in the $s$ dependence of the potential (7.9.34), we keep only linear and quadratic terms in the relevant coupling $\mu$, since we send it to zero after the Riemann tensor has been computed. The vanishing relations associated with (7.9.37) are

$$\begin{aligned} X^3 &= \mu Y + 2sXY^2 \\ Y^3 &= \mu X + 2sX^2Y \end{aligned} \tag{7.9.38}$$

Using (7.9.34) it is immediate to find the explicit expression of the hessian, namely

$$H(X,Y,s) = \det \partial_i\partial_j W = (1-4s^2)X^2Y^2 \tag{7.9.39}$$

Hence we have

$$\langle H(X,Y,s)\rangle(s) = h(s)\langle\langle H(X,Y,s)\rangle\rangle = \frac{1}{1-4s^2}\langle X^2Y^2\rangle(s) \tag{7.9.40}$$

where we have taken into account the fact that the normalization $h(s)$ of the path integral can be chosen differently for different potentials and correspondingly is some unknown function $h(s)$ of the marginal parameters. From the property of the residue pairing formula (7.9.17) $\langle\langle H(X,Y,s)\rangle\rangle = \text{const} \equiv 1$, so that we get

$$\langle X^2Y^2\rangle(s) = \frac{h(s)}{1-4s^2} \tag{7.9.41}$$

Using this normalization we can compute the metric $g_{ij}(s)$. As is easy to foresee, the information that this metric should be a flat metric provides a differential equation to be satisfied by $h(s)$.

Formula (7.9.40) gives the $g_{08}$ component of the metric, since

$$g_{08} = \langle\phi_0\phi_8\rangle = \langle 1\ \phi_8\rangle = \langle X^2Y^2\rangle(s) \tag{7.9.42}$$

For the other components we have to use repeatedly the vanishing relations (7.9.38). As an example we write explicitly $g_{48}$. The operator product gives

$$\phi_4\phi_8 = X^3Y^3 = (\mu Y + 2sXY^2)(\mu X + 2sX^2Y) \tag{7.9.43}$$

so that

$$(1-4s^2)X^3Y^3 = \mu^2XY + 4sX^2Y^2 \tag{7.9.44}$$

In the vacuum expectation value only the second term survives and

$$g_{48} = \frac{4\mu s}{1-4s^2}f(s) \tag{7.9.45}$$

where we have set $f(s) = \frac{h(s)}{1-4s^2}$. Our final result is expressed by the following three-by-three matrix:

$$g_{ij}(s) = f(s)\begin{pmatrix} 0 & 0 & 1 \\ 0 & 1 & \frac{4\mu s}{1-4s^2} \\ 1 & \frac{4\mu s}{1-4s^2} & \frac{\mu^2}{1-4s^2} + \frac{16s^2\mu^2}{(1-4s^2)^2} \end{pmatrix} \tag{7.9.46}$$

If we compute the $R^4_{884}$ component of the Riemann tensor and send the relevant coupling $\mu$ to zero, we find

$$R^4_{884} = 2(1-4s^2)f\frac{d^2f}{ds^2} - s(1-4s^2)(\frac{df}{ds})^2 - 64s^2f^2 - 12f^2 \tag{7.9.47}$$

Clearly if the metric $g_{ij}(s)$ is flat all the Riemann tensors components vanish, in particular $R^4_{884}$. Hence the unknown prefactor $h(s)$ must satisfy the differential equation that follows from $R^4_{884} = 0$, namely

$$2(1-4s^2)h\frac{d^2h}{ds^2} - 3(1-4s^2)(\frac{dh}{ds})^2 - 16sh\frac{dh}{ds} + 4h^2 = 0 \tag{7.9.48}$$

Next it is almost immediate to verify that once (7.9.48) is fulfilled, the Riemann tensor vanishes identically. Replacing $h = y^{-2}$ in Eq. (7.9.48) we obtain

$$(1-4s^2)\frac{d^2y(s)}{ds^2} - 8s\frac{dy(s)}{ds} - y(s) = 0 \tag{7.9.49}$$

In the Picard–Fuchs approach discussed in Chapter 8, Eq. (7.9.49) will come out to be exactly the differential equation satisfied by the period

$$Y = \int_\gamma \Omega^{(1,0)}(s) \tag{7.9.50}$$

of the holomorphic 1-form on the torus. Such an identification will not be verified for this specific model, but it will be verified in Chapter 8 for the analogue of (7.9.37) pertaining to the next considered model (see below). Equation (7.9.49), upon the change of variable $z = 4s^2$, becomes a hypergeometric equation of parameters $a = \frac{1}{4}$, $b = \frac{1}{4}$ and $c = \frac{1}{2}$. Hence the solution for the $h(s)$ factor can be explicitly given:

$$h(s) = \left(c_1F[\frac{1}{4},\frac{1}{4},\frac{1}{2},4s^2] + c_2F[\frac{3}{4},\frac{3}{4},\frac{3}{2},4s^2]\right)^{-2} \tag{7.9.51}$$

where $c_1, c_2$ are some constants.

To find flat coordinates (those in which the metric is constant), we should write the most general coordinate transformation $s_i = s_i(t_j)$ compatible with the $U(1)$ symmetry and discrete symmetries of the potential and insert it into

$$\eta_{ij} = \langle\phi_i\phi_j\rangle(t) = \frac{\partial s_l}{\partial t_i}\frac{\partial s_m}{\partial t_j}\langle\phi_l\phi_m\rangle(s) = \frac{\partial s_l}{\partial t_i}\frac{\partial s_m}{\partial t_j}g_{lm}(s(t)) \tag{7.9.52}$$

This means that one can write, for a general operator in the chiral ring,

$$\phi_i(t) = \frac{\partial s_k}{\partial t_i}\phi_k(s) = -\frac{\partial s_k}{\partial t_i}\frac{\partial W}{\partial s_k} = -\frac{\partial W}{\partial t_i} \tag{7.9.53}$$

In particular for the top chiral field, we can write

$$\phi_8(t) = \frac{\partial s}{\partial t} X^2 Y^2 + \text{fields of lower charges} \tag{7.9.54}$$

so that, by requiring that $\langle \phi_0 \, \phi_8 \rangle(t) = 1$ we get

$$\langle \phi_0 \phi_8 \rangle(s) = \langle \mathbb{1} X^2 Y^2 \rangle(s) = \frac{\partial t}{\partial s} \langle \phi_8 \rangle(t) = \frac{\partial t}{\partial s} \langle \phi_0 \phi_8 \rangle(t) = \frac{\partial t}{\partial s} \tag{7.9.55}$$

Finally, using (7.9.41) we obtain

$$t(s) = \int ds' \frac{h(s')}{1 - 4(s')^2} \tag{7.9.56}$$

Inserting (7.9.56) into (7.9.48) we get a new differential equation, namely the schwarzian equation

$$\{t, s\} = \frac{8s^2 + 6}{(1 - 4s^2)^2} \tag{7.9.57}$$

which gives a relation between the flat coordinate $t$ and the $s$ coordinate.

We turn now to the next $c = 3$ example defined by the cubic potential

$$\mathcal{W} = \frac{1}{3}(X^3 + Y^3 + Z^3) - sXYZ \tag{7.9.58}$$

The starting point is again the perturbed potential

$$\begin{aligned} \mathcal{W} &= \frac{1}{3}(X^3 + Y^3 + Z^3) - s_0 - s_1 X - s_2 Y - s_3 Z \\ &- s_4 XY - s_5 XZ - s_6 YZ - sXYZ \end{aligned} \tag{7.9.59}$$

This time the calculation is a little more involved but it can be performed along the same lines as those previously presented. In this case we have

$$H(s) = 8(1 - s^3) XYZ \tag{7.9.60}$$

$$\langle XYZ \rangle(s) = \frac{1}{8} \frac{1}{1 - s^3} h(s) \tag{7.9.61}$$

All the non-vanishing components of the Riemann tensor are proportional to the left hand side of the following equation:

$$(1 - s^3) y''(s) - 3s^2 y'(s) - s y(s) = 0 \tag{7.9.62}$$

where we have set $h = y^{-2}$. Notice that the differential Eq. (7.9.62) is the same as the Picard–Fuchs Eq. (8.2.27), as we anticipated. Hence $1/h^2$ has the interpretation of a period. Moreover, by imposing $\langle \phi_0 \, \phi_8 \rangle(t) = 1$ we get

$$t(s) = \int ds' \frac{h(s')}{8(1 - (s')^2)} \tag{7.9.63}$$

so the insertion of Eq. (7.9.63) into (7.9.62) yields

$$\{t, s\} = \frac{1}{2}\frac{8+s^3}{(1-s^3)^2}s \tag{7.9.64}$$

which is precisely the schwarzian equation we shall obtain in Chapter 8 (Eq. (8.2.50)) (using $\{s,t\} = -(s')^2\{t,s\}$).

## 7.10 Topological Observables in the Two-Phase Theory

After our analysis of the topological correlators for the A-twisted and B-twisted sigma model and of the topological Landau–Ginzburg model, we briefly resume our analysis of the two-phase N=2 gauge theory which interpolates between the corresponding N=2 field theories. This theory was described in section 5.5 and its A- and B-twists were discussed in section 7.6. We showed that in the first case we obtain a BRST algebra similar to that of the A-twisted sigma model, while in the second case the resulting BRST algebra is analogous to that of the topological Landau–Ginzburg theory. Such a structure of the twisted models matches with the two phase structure of the untwisted N=2 theory. In this section we consider in greater detail the interplay between the phase structure and the instantons selected by the topological twists. We focus on the case where the superpotential of the matter system, composed of $n+3$ chiral multiplets $X^0$, $X^i$ $\{i = 1, \ldots, n+2\}$, has the form (5.5.58), namely

$$W(X) \;=\; X^0\,\mathcal{W}(X^i) \tag{7.10.1}$$

This matter system is coupled to a single $U(1)$ gauge group, with the following charges: $-d < 0$ for $X^0$, $q^i > 0$ for the $n+2$ fields $X^i$. The reduced superpotential $\mathcal{W}(X^i)$ is a quasi-homogeneous functions of degree $d$ with weights $q^i$:

$$\forall \lambda \in \mathbb{C} \quad \mathcal{W}\left(e^{q^i\lambda} X^i\right) \;=\; e^{d\lambda}\,\mathcal{W}\left(X^i\right) \tag{7.10.2}$$

For the gauge multiplet we make the choice of the linear superpotential (5.5.7), (5.5.8):

$$\mathcal{U} \;=\; -\frac{it}{4}M \qquad t \;=\; ir + \frac{\theta}{2\pi} \tag{7.10.3}$$

Correspondingly the bosonic lagrangian takes the form

$$\begin{aligned}\mathcal{L}_{Bose} \;=\; & \frac{1}{2}\mathcal{F}^2 - 4(\partial_+ M^*\partial_- M + \partial_- M^*\partial_+ M) \\ & + (\nabla_+ X^{i^*}\nabla_- X^i + \nabla_- X^{i^*}\nabla_+ X^i) + \frac{\theta}{2\pi}\mathcal{F} \\ & - U(\mathcal{P}, X, M)\end{aligned} \tag{7.10.4}$$

where $U(\mathcal{P},X,M)$ denotes the scalar potential:

$$
\begin{aligned}
U(\mathcal{P},X,M) &= -2\mathcal{P}^2 + 2r\mathcal{P} + 8\sum_{I=0}^{n+2} \partial_I W \partial_{I^*} W^* \\
&\quad + 8M^*M\left(d^2|X^0|^2 + \sum_{i=1}^{n+2}(q^i)^2 X^{i^*}X^i\right) - 2\mathcal{P}\,\mathcal{D}\left(X,\overline{X}\right)
\end{aligned}
\tag{7.10.5}
$$

and

$$
\mathcal{D}\left(X,\overline{X}\right) \stackrel{\text{def}}{=} \left(\sum_{i=1}^{n+2} q^i X^{i^*}X^i - d|X^0|^2\right) \tag{7.10.6}
$$

is the momentum map function discussed both in Chapter 3 and in section 5.5. In other words, $\mathcal{D}\left(X,\overline{X}\right)$ is the Killing vector prepotential for the isometry gauged by the $U(1)$ group. Eliminating the auxiliary field $\mathcal{P}$ through its own field equation:

$$
\mathcal{P} = \frac{1}{2}\left(r - \mathcal{D}\left(X,\overline{X}\right)\right) \tag{7.10.7}
$$

the scalar potential takes the form (5.5.59), which we repeat here for the reader's convenience:

$$
\begin{aligned}
U &= \frac{1}{2}\left(r + d|X^0|^2 - \sum_i q^i|X^i|^2\right)^2 + 8|\mathcal{W}(X^i)|^2 + 8|X^0|^2|\partial_i \mathcal{W}|^2 \\
&\quad + 8|M|^2\left(d^2|X^0|^2 + \sum_{i=1}^{n+2}(q^i)^2|X^i|^2\right)
\end{aligned}
\tag{7.10.8}
$$

The extrema of (7.10.8) were discussed in subsection 5.5.7 where the two-phase structure was pointed out.

- $r > 0$. In this case the classical vacua span a manifold $\mathcal{M}_{vacua}$ corresponding to the following algebraic locus:

$$
\begin{aligned}
M &= X^0 = 0 \\
\hat{\mathcal{D}}\left(X,\overline{X}\right) &\stackrel{\text{def}}{=} \sum_i q^i|X^i|^2 = r \\
e^{iq^i\theta}X^i &\sim X^i \quad \forall\theta \in [0, 2\pi] \\
\mathcal{W}\left(X^i\right) &= 0
\end{aligned}
\tag{7.10.9}
$$

The first of Eqs. (7.10.9) is obvious. The second and the third define the weighted projective space $W\mathbb{CP}_{n+1}(q_1,\ldots,q_{n+2})$, while the last defines the complex $n$-fold $\mathcal{M}_{vacua}$ as the zero locus, in such a space, of the degree $d$ polynomial $\mathcal{W}(X^i)$. In particular, recalling the results of section 5.7, if

$$
d = \sum_{i=1}^{n+2} q^i \tag{7.10.10}
$$

the vacua manifold $\mathcal{M}_{vacua}$ is a Calabi–Yau $n$-fold. In this phase the fields $M$, $X^0$ and the component of $X^i$ normal to the locus (7.10.9) are massive, while the tangent ones are massless. Also the gauge field is massive. Hence the effective lagrangian of the massless modes is an N=2 sigma model with target manifold identical to the manifold of classical vacua (7.10.9):

$$\mathcal{M}_{target} = \mathcal{M}_{vacua} \tag{7.10.11}$$

- $r < 0$. In this case the classical vacuum is a point:

$$\begin{aligned} M &= X^i = 0 \quad i = 1, \ldots, n+2 \\ X^0 &= \sqrt{\frac{-r}{d}} \end{aligned} \tag{7.10.12}$$

In this phase, the gauge field, $M$ and $X^0$ are massive, while all the $X^i$ are massless. The effective lagrangian of the massless modes corresponds to an N=2 rigid Landau–Ginzburg theory with superpotential $\mathcal{W}(X^i)$.

Let us then reconsider the topological twists of this theory.

- *A-twist.* In this case the instanton conditions are given by the specialization of Eqs. (7.6.26) to the case of the superpotential (7.10.1), namely

$$\begin{aligned} \mathcal{W}^\star\left(\overline{X}\right) &= 0 \\ \overline{X}^{0\star}\,\partial_{i^\star}\mathcal{W}^\star\left(\overline{X}\right) &= 0 \\ \nabla_- X^0 &= 0 \quad \nabla_+ \overline{X}^{0\star} = 0 \\ \nabla_- X^i &= 0 \quad \nabla_+ \overline{X}^{i\star} = 0 \\ \frac{i}{2\pi}\mathcal{F} &= -\frac{1}{\pi}\mathcal{P} = \frac{1}{2\pi}\left(\mathcal{D}(X,\overline{X}) - r\right) \\ \frac{i}{2\pi}\int \mathcal{F} &= N \in \mathbb{Z} \end{aligned} \tag{7.10.13}$$

where by means of the last equation we have specified the Chern class of the gauge connection. In this twist the field $M$ has the interpretation of a ghost just as the fermion fields, so that the no-ghost part of the lagrangian $\mathcal{L}_{no\text{-}ghost}$ coincides with the bosonic lagrangian (7.10.4) with the field $M$ deleted. The value of the no-ghost action on an instanton configuration (7.10.13) was calculated in section 7.6 (see Eqs. (7.6.27), (7.6.28)) and we obtained

$$\int \mathcal{L}_{no-ghost}\, d^2z = -2\pi i t N \tag{7.10.14}$$

In particular on an instanton we have

$$\begin{aligned} \int \mathcal{L}_0 d^2 z &= 2\pi r N \\ \mathcal{L}_0 &= \frac{1}{2}\mathcal{F}^2 + \left(\nabla_+ X^{i^\star}\nabla_- X^i + \nabla_- X^{i^\star}\nabla_+ X^i\right) - U\left(\mathcal{P}, X, M\right) \end{aligned} \tag{7.10.15}$$

where the lagrangian $\mathcal{L}_0$ defined above is, after Wick rotation to the Euclidean region, negative definite. Indeed, as is evident from Eq.(7.10.8), the potential $U(X,\overline{X})$ is positive definite while the kinetic terms are negative definite. It follows that there is a correlation between the sign of the instanton number and the sign of the parameter $r$:

$$2\pi r N < 0 \Longrightarrow \begin{cases} r < 0 & N < 0 \\ r < 0 & N > 0 \end{cases} \tag{7.10.16}$$

Hence in the two phases we have either *instantons* or *anti-instantons*. This has far-reaching consequences. As recalled by Witten [262], a very general theorem states that line bundles of negative degree have no holomorphic sections. Hence the two instanton equations

$$\begin{aligned} \nabla_- X^0 &= 0 \quad \nabla_+ \overline{X}^{0\star} = 0 \\ \nabla_- X^i &= 0 \quad \nabla_+ \overline{X}^{i\star} = 0 \end{aligned} \tag{7.10.17}$$

do not admit simultaneous solutions, since the field $X^0$ and $X^i$ have $U(1)$ charges of opposite sign. Which can be non-zero depends on the sign of the Chern class $N$ and hence on the sign of $r$. We have

$$\nabla_- X^I = 0 \longrightarrow X^I = 0 \quad \text{unless} \quad \text{sign}(q^I) = -\text{sign}(N) \tag{7.10.18}$$

so that in the two phases the instanton configuration reduces to

$$\begin{aligned} r > 0 \quad & \begin{cases} \overline{X}^{0\star} = 0 \\ \mathcal{W}^\star\left(\overline{X}\right) = 0 \\ \nabla_- X^i = 0 \quad \nabla_+ \overline{X}^{i\star} = 0 \\ \frac{i}{2\pi}\mathcal{F} = \frac{1}{2\pi}\left(\hat{\mathcal{D}}(X,\overline{X}) - r\right) \\ \frac{i}{2\pi}\int \mathcal{F} = N \in \mathbb{Z}_- \end{cases} \\ r < 0 \quad & \begin{cases} \overline{X}^{i\star} = 0 \\ \nabla_- X^0 = 0 \quad \nabla_+ \overline{X}^{0\star} = 0 \\ \frac{i}{2\pi}\mathcal{F} = \frac{1}{2\pi}\left(-d|X^0|^2 - r\right) \\ \frac{i}{2\pi}\int \mathcal{F} = N \in \mathbb{Z}_+ \end{cases} \end{aligned} \tag{7.10.19}$$

As we see, also at the instanton level the two-phase structure of the theory becomes manifest.

• $r < 0$. This region corresponds to the Landau–Ginzburg phase: an A-twisted Landau–Ginzburg model is essentially an empty theory, so that quite little is expected to emerge from the A-twisted gauge theory in the $r$ ????? 0 regime. Indeed Eq.(7.10.19) shows that in this regime the fields $X^i$ play no role, the effective physical system being reduced to the abelian gauge field plus the massive scalar field $X^0$ related by the equations

$$\nabla_- X^0 = 0 \quad \nabla_+ \overline{X}^{0\star} = 0$$

$$\frac{i}{2\pi}\mathcal{F} = \frac{1}{2\pi}\left(-d|X^0|^2 - r\right) \tag{7.10.20}$$

which, as Witten remarked, are the equations of a Nielsen–Olesen abelian vortex line.

• $r > 0$. This region corresponds to the sigma model phase and the A-twisted gauge theory is expected to reproduce the essential features of the topological sigma model on the $n$-fold defined by Eqs. (7.10.9). This is indeed the case although there are some subtle differences. First let us discuss the topological observables. There are two kinds of them: those associated with the gauge sector and those associated with the matter sector. We begin with the first. The corresponding descent equations were discussed in section 7.6 and we found that a solution of Eqs. (7.6.17) is associated with each abelian factor in the gauge group and it is given by Eq.(7.6.18). Hence a set of topological deformations of the action are proportional, in the A-model, to the $r$ parameters of the N=2 Fayet–Iliopoulos terms. In the specific case under discussion there is just one $U(1)$ group and correspondingly just one parameter $r$. Our analysis of the A-twisted topological sigma model has revealed that the topological coupling constants appearing in the action have the interpretation of Kähler class moduli. It follows that the parameter $r$ should be interpreted as a deformation parameter of the Kähler class in the effective sigma model. To show this, we recall that the effective sigma model target space $\mathcal{M}_{target}$, namely the locus (7.10.9), is a hypersurface ($\mathcal{W}(X^i) = O$) in the Kähler quotient $\mathcal{D}^{-1}(r)/\mathcal{G}$ of flat space with respect the holomorphic action of the gauge group $\mathcal{G}$. Hence the Kähler 2-form $K_{target}$ of $\mathcal{M}_{target}$ is the pull-back of the Kähler 2-form $K_{quotient}$ of the Kähler quotient. The deformations of $K_{target}$ are simply induced by the deformations of $K_{quotient}$. To see that $r$ is a deformation parameter for $K_{quotient}$ it suffices to recall the way the Kähler potential of the quotient manifold $\mathcal{D}^{-1}(r)/\mathcal{G}$ is determined. Let $\mathcal{K}_0 = \sum_{I=0}^{n+2} \overline{X}^I X^I$ be the Kähler potential of flat space and $\mathcal{D}(X,\overline{X})$ be the momentum map. By definition both $\mathcal{K}_0$ and $\mathcal{D}$ are invariant under the action of the isometry group $\mathcal{G}$ but not under the action of its complexification $\mathcal{G}^c$. On the other hand the superpotential derivatives $\partial_I W(X)$ are invariant not only under $\mathcal{G}$, but also under $\mathcal{G}^c$. Furthermore one shows that the wanted hypersurface in the quotient manifold $\mathcal{D}^{-1}(r)/\mathcal{G}$ is the same thing as the quotient $\frac{\partial_I W(X)=0}{\mathcal{G}^c}$ of the holomorphic hypersurface $\partial_I W(X) = 0$ in the whole $\mathbb{C}^{n+3}$ modded by the action of $\mathcal{G}^c$. If we name $e^V \in \mathcal{G}^c = U(1)^c = \mathbb{C}$ an element of this complexified group such that

$$\mathcal{D}\left(e^V X, e^{-V}\overline{X}\right) = r \tag{7.10.21}$$

is a true equation on the hypersurface $\frac{\partial_I W(X)=0}{\mathcal{G}^c}$, then the Kähler potential of the Kähler quotient manifold $\mathcal{D}^{-1}(r)/\mathcal{G}$ is

$$\mathcal{K} = \mathcal{K}_0(e^V X e^{-V}\overline{X}) + r\,V \tag{7.10.22}$$

Consequently a variation of the $r$ parameters uniquely affects the Kähler potential, the quotient $\frac{\partial_I W(X)=0}{\mathcal{G}^c}$, as an analytic manifold, being insensitive to such a variation. Summarizing the A-model of the two-phase N=2 gauge theory is, as expected, a cohomological theory in the moduli space of Kähler class deformations and $t = \frac{\theta}{2\pi} + ri$ is a modulus parameter for these deformations.

It is then worth discussing the general form of the observables in a topological theory described by the BRST algebra (7.6.22) and coupled to a topological gauge theory (7.3.2). As we have seen in section 7.4 in a topological $\sigma$-model the observables, namely the solutions of the descent equations (7.6.17), are in correspondence with the cohomology classes of the target manifold. If $\omega^{(n)} = \omega_{i_1,\dots,i_n}(X)\, dX^{i_1} \wedge \cdots \wedge dX^{i_n}$ is a closed $n$-form $d\omega^{(n)} = 0$, we promote it to a ghost form $\widehat{\omega}^{(n)}$ by substituting $d \longrightarrow d + s$, $dX^i \longrightarrow dX^i + c^i$; then by expanding this ghost form into addends of definite ghost number $\widehat{\omega}^{(n)} = \sum_{g=0}^{n} \widehat{\omega}^{(n-g)}_{(g)}$ we solve the descent equations by setting $\Theta^{(2)} = \widehat{\omega}^{(2)}_{(n-2)}$, $\Theta^{(1)} = \widehat{\omega}^{(1)}_{(n-1)}$, $\Theta^{(0)} = \widehat{\omega}^{(0)}_{(n)}$. (Note that in this discussion, for simplicity, we do not distinguish holomorphic and anti-holomorphic indices.) In a similar way in the topological model described by Eqs. (7.6.22), (7.3.2), the solutions of the descent equations are in correspondence with the anti-symmetric constant tensors $a_{i_1,\dots,i_n}$ which are invariant under the action of the gauge group, namely which satisfy the condition

$$a_{p,[i_2,\dots,i_n}\, q^{p}_{i_1]} = 0 \tag{7.10.23}$$

Indeed, setting

$$\begin{aligned} \widehat{\nabla} &= \widehat{d} - i\, q \widehat{\mathcal{A}} \\ &= (d + s) - i\, q\, (A + c^g) \\ &= \nabla_{(1,0)} + \nabla_{(0,1)} \\ &= (d - i\, \mathcal{A} q) + (s - i\, c^g\, q) \end{aligned} \tag{7.10.24}$$

we obtain $\widehat{\nabla}^2 = -i\,\widehat{F}\, q$ and to every invariant anti-symmetric tensor we can associate the $\widehat{d}$-closed ghost form $\widehat{\omega} = a_{i_1,\dots,i_n} \widehat{\nabla X^{i_1}} \cdot\cdots\cdot \widehat{\nabla X^{i_n}}$. Expanding it in definite ghost number parts, the solution of the descent equations is obtained in the same way as in the $\sigma$-model case.

When we follow the procedure of section 5.5 and we reproduce the N=2 $\sigma$-model by integrating out the gauge field, the topological observables discussed above and related to the anti-symmetric gauge-invariant tensors become representatives of the cohomology classes of the target manifold. This is essentially a field theory reconstruction of the Griffith residue mapping.

It may then seem that the A-twisted N=2 gauge theory in the $r$ ????? 0 phase is fully equivalent to the A-twisted topological sigma model on the target manifold (7.10.9). As we have already anticipated, this conclusion is not completely true, because there are still some subtle differences. These occur in the definition of the instantons. The first of Eqs. (7.10.19), which defines the instantons in the matter-coupled gauge theory, is weaker than the definition of instantons in the effective sigma model. As a consequence all the sigma model instantons contribute to the sum defining topological correlators in the A-twisted gauge theory, but this latter includes additional *singular instantons* that are absent in the topological sigma model. Let us see how this occurs. We focus on the equations

$$\begin{aligned} \mathcal{W}^*\left(\overline{X}\right) &= 0 \\ \nabla_- X^i &= 0 \qquad \nabla_+ \overline{X}^{i^*} = 0 \\ \frac{i}{2\pi}\int \mathcal{F} &= N \in \mathbb{Z}_- \\ \frac{i}{2\pi}\mathcal{F} &= \frac{1}{2\pi}\left(\widehat{\mathcal{D}}(X,\overline{X}) - r\right) \end{aligned} \tag{7.10.25}$$

and we observe that, by definition, they are invariant under the gauge transformation:

$$X^i \rightarrow e^{iq^i\theta} X^i \qquad \mathcal{A} \rightarrow \mathcal{A} - \mathrm{i}\, d\theta \qquad \theta \in \mathbb{R} \tag{7.10.26}$$

The first three of Eqs. (7.10.25), however, are invariant under the larger group of transformations where the parameter $\theta$ is complexified:

$$X^i \rightarrow e^{iq^i\theta^c} X^i \qquad \mathcal{A} \rightarrow \mathcal{A} - \mathrm{i}\, d\theta^c \qquad \theta^c \in \mathbb{C} \tag{7.10.27}$$

The fourth and last of Eqs. (7.10.25) is not invariant under (7.10.27) and can just be seen as a condition that fixes the complex gauge invariance of the first three equations. In other words, the space of solutions of the first three equations, up to complex gauge transformations (7.10.27), is the same as the space of the set of four equations, up to a real gauge transformation (7.10.26). This is the field theory analogue of the equivalence between the algebro-geometric quotient and the Kähler quotient, namely the fact that the hypersurface $\partial_I W(X) = 0$ in the quotient manifold $\mathcal{D}^{-1}(r)/\mathcal{G}$ is the same thing as the quotient $\frac{\partial_I W(X)=0}{\mathcal{G}^c}$ of the holomorphic hypersurface $\partial_I W(X) = 0$ in the whole $\mathbb{C}^{n+3}$, modded by the action of the complexified group $\mathcal{G}^c$. In view of this we can simply study the first three of Eqs. (7.10.25), up to complex gauge transformations. As we did in section 7.7, we consider the case of a genus zero world-sheet: $\Sigma_0 \sim \mathbb{CP}_1$. Let $u, v$ be homogeneous coordinates on the world-sheet. In an instanton configuration the fields $X^i$ are holomorphic sections of a line bundle of degree $k = -N$ on $\mathbb{CP}_1$, namely they are homogeneous polynomials of degree $k$ in $u, v$:

$$X^i(u,v) = X^i_k u^k + X^i_{k-1} u^{k-1} v + \ldots X^i_0 v^k \tag{7.10.28}$$

The overall scaling

$$X^i(u,v) \rightarrow t^{q_i} X^i(u,v) \qquad t \in \mathbb{C} \quad t \neq 0 \tag{7.10.29}$$

corresponds to the complex gauge transformation (7.10.27) with constant gauge parameter. The moduli space of the gauge theory instantons (7.10.25) is therefore the space of polynomials (7.10.28) satisfying identically the relation

$$\mathcal{W}\left(X^1(u,v), X^2(u,v), \ldots, X^{n+2}(u,v)\right) = 0 \tag{7.10.30}$$

modulo the identification (7.10.29). Let us compare this moduli space with the moduli space of the corresponding sigma model. This is the moduli space of the degree $k$ holomorphic maps $X : \mathbb{CP}_1 \longrightarrow \mathcal{M}_{target}$, $\mathcal{M}_{target}$ being the hypersurface $\mathcal{W}(X) = 0$ in $W\mathbb{CP}_{n+1;q_1,\dots,q_{n+2}}$. Also in this case the homogeneous coordinates $X^i$ of $W\mathbb{CP}_{n+1;q_1,\dots,q_{n+2}}$ are homogeneous polynomials in the $u, v$ coordinates of $\mathbb{CP}_1$, and also in this case they must satisfy the constraint (7.10.30). The difference, however, is that, being homogeneous coordinates, they can never vanish simultaneously. Hence the admitted polynomials satisfy the additional conditions that they should have no common zeros. This shows that all sigma model instantons are covered by the instanton equations (7.10.25). In addition one has the singular *instantons* that correspond to polynomials with common zeros. The instanton sum in the gauge theory must also include these objects. Their effect has not yet been fully analysed in the existing literature.

• B-twist. Let us now turn our attention to the B-model, which describes a topological gauge-coupled Landau–Ginzburg theory. Here the topological observables are in correspondence with the symmetric invariant tensors, rather than with the anti-symmetric ones. To see it we recall the solutions of the descent equations in the case of the topological rigid Landau–Ginzburg model where the topological observables are in correspondence with the elements of the local polynomial ring of the superpotential $\mathcal{W}(\mathcal{X})$:

$$\mathcal{R}_{\mathcal{W}} = \frac{\mathbb{C}[X^i]}{\partial \mathcal{W}(X^i)} \tag{7.10.31}$$

Indeed, let $\mathcal{P}(X) \in \mathcal{R}_{\mathcal{W}}$ be some non-trivial polynomial of this local ring; a solution of the descent equations (7.6.17) was given in Eq. (7.9.2), namely

$$\begin{aligned}
\Theta^{(0)}_{\mathcal{P}} &= \mathcal{P}(X) \\
\Theta^{(1)}_{\mathcal{P}} &= 2i\partial_i \mathcal{P}\,\overline{C}^i \\
\Theta^{(2)}_{\mathcal{P}} &= -2\,\partial_i\partial_j \mathcal{P}\,\overline{C}^i \wedge \overline{C}^j + 4\left[\partial_k \mathcal{P}\,\partial_{l^*}\overline{\mathcal{W}}\,\eta^{kl^*}\right] e^+ \wedge e^-
\end{aligned} \tag{7.10.32}$$

The reason why $\mathcal{P}(X)$ has to be a non-trivial element of the local ring was simple. If $\mathcal{P}(X)$ were proportional to the vanishing relations (i.e. if $\mathcal{P}(X) = \sum_i p^I(X)\frac{\partial W}{\partial X_I}$), then using the BRST transformations (7.6.32), one could see that $\mathcal{P}(X) = s\,K$ and so $\Theta^{(0)}_{P}$ would be exact. In the case where the Landau–Ginzburg theory is gauged-coupled and the BRST transformations are given by Eqs. (7.6.36), the solution of the descent equations has the same form as in Eq.(7.10.32), upon a substitution of the polynomial $\mathcal{P}(X) \in \mathcal{R}_{\mathcal{W}}$ with a polynomial

$$P(X^I) \in \mathcal{R}_W = \mathcal{R}_W = \frac{\mathbb{C}\left[X^I\right]}{\partial W(X^I)} \tag{7.10.33}$$

in the local ring of the full superpotential (7.10.1). In addition, however, the polynomial $P(X^I)$ must be gauge-invariant. This is guaranteed, if the polynomial is quasi-homogeneous of degree zero in $X^0, X^i$, namely if it is of the form

$$P(X) = \left(X^0\right)^{\nu}\,\mathcal{P}\left(X^i\right) \tag{7.10.34}$$

where $\mathcal{P}(X^i)$ is any quasi-homogeneous polynomial of degree $\nu$ in $X^i$ corresponding to some non-trivial element of the local ring of $\mathcal{W}(X)$:

$$\mathcal{P}(X^i) \in \mathcal{R}_{\mathcal{W}} = \frac{\mathbb{C}[X^i]}{\partial \mathcal{W}} \tag{7.10.35}$$

Hence the space of physical observables reduces to the chiral ring (7.10.31) of the superpotential $\mathcal{W}(X)$ which defines the corresponding rigid Landau–Ginzburg model. At the level of the B-twist, the Landau–Ginzburg model and the N=2 matter-coupled gauge theory seem to be fully equivalent.

## 7.11 Bibliographical Note

*Topological field theories of the cohomological type were introduced by Witten in 1988 in [259] and [260]. As he explicitly states in the original papers, the motivation was that of recasting in a field-theoretical path integral framework the problem of calculating Donaldson invariants of 4-manifolds [127] and Gromov invariants of holomorphic maps, the corresponding field theory models being topological Yang–Mills theory and the topologcal sigma model. Witten obtained both theories by means of a twisting procedure on the corresponding N=2 theories. For this reason he did not identify the nilpotent transformations as BRST transformations; rather, he called them supersymmetries. The clarification of the BRST structure underlying topological field theories is due to Baulieu and Singer [33, 32].*

*After these seminal papers a vast literature on topological field–theories has been produced: General references are [47, 116, 127, 21].*

*Specifically:*

- *The geometrical approach to BRST symmetry [34],[35] is due to Bonora, Pasti and Tonin [52] (1982) and to Baulieu and Bellon [31] (1987). The similarity between the BRST formalism for supergravity developed by Baulieu and Bellon and the rheonomy approach is striking. Yet these authors did not formulate the correspondence principle between classical and quantum rheonomic parametrizations explained in this chapter. This is due to Anselmi and Fre' (1992) [8].*

- *The appropriate generalization of Witten's twist procedure to general N=2 theories in four dimensions, including also supergravity and the hypermultiplets is due Anselmi and Fre' [8, 7, 10].*

- *For the correlators of the topological sigma model, basic references are Witten's paper [261], the more mathematically oriented paper by Aspinwall and Morrison [16] and Witten's essay [263]. The topological sigma model was introduced by Witten in the original paper [260] and was formulated à la BRST by Baulieu and Singer in [32].*

- *The two-phase gauge theory that interpolates between sigma models and Landau–Ginzburg models was introduced by Witten in [262] and was further analysed in [45] by Billo' and Fre' who extended the idea to N=4 theories and hyper-Kähler quotients.*

- *The formulation of A- and B-twists for topological sigma models is due to Witten [263]. The formal definition of A and B twists for a general N=2, D=2 theory is due to Billó and Fré [45].*

- *Topological conformal field theories were introduced by Eguchi and Yang in 1990 [137], who defined the purely algebraic rules of the twist. These theories were developed by Dijkgraaf, Verlinde and Verlinde in [118]. In section 7.8 we have mainly followed this paper and [116]. On 2D topological models we also recall [215, 117, 204, 54].*

- *Topological Landau–Ginzburg models were introduced by Vafa in [247] (1991). In section 7.9 we follow this paper and [118, 250, 151, 129, 48, 218, 203]. We refer also to [86, 247].*

- *Some topics introduced in section 7.7, such as the concept of rational curves in relation with topological field theories will be resumed in Chapter 8, so that we refer to the Bibliographical Note 8.6 at the end of that chapter for a more extensive bibliography. Here we mention in particular [16, 17, 68, 69].*

# Chapter 8

# PICARD–FUCHS EQUATIONS AND MIRROR MAPS

## 8.1 Introduction to Mirror Symmetry

The lesson we have learned from the previous chapter is that to each Calabi–Yau $n$-fold $\mathcal{M}_n$ we can associate two topological field-theories, the *A-twisted* and the *B-twisted sigma-models*. Both are significant since they single out complementary geometrical properties of the target manifold that have a physical interpretation in string compactifications and in other applications. The A-twisted theory deals with Kähler structure deformations, while the second deals with complex structure deformations. The difference is that the correlators of the B-twisted model are *easy to calculate* (easy in a sense that we are going discuss below), while the correlators of the A-twisted model are usually *not accessible to direct calculation.* In particular in the case of Calabi–Yau 3-folds the calculation of A-twisted correlators requires the knowledge of the number $n_{\{k\}}$ of rational curves of degree $\{k\}$ embedded in the 3-fold, knowledge that is not provided by the mathematical literature. Hence it would be extremely useful and deeply significant if the following rather miraculous thing happened.

**Conjecture:** *The Calabi–Yau n-folds occur in* **mirror pairs** $\mathcal{M}_n$ *and* $M\mathcal{M}_n$*, such that:*

$$\begin{aligned} \text{A-model}\,[\mathcal{M}_n] &\approx \text{B-model}\,[M\mathcal{M}_n] \\ \text{A-model}\,[M\mathcal{M}_n] &\approx \text{B-model}\,[\mathcal{M}_n] \end{aligned} \tag{8.1.1}$$

Indeed, if the conjecture were true, we could calculate both the A-twisted and the B-twisted correlators relative to the same manifold $\mathcal{M}_n$ using for the *difficult* A-correlators on $\mathcal{M}_n$ the *"easy"* B-correlators on $M\mathcal{M}_n$. Here *difficult* means the solution of a transcendental problem, namely the summation of a series, whose general term is unknown, while *easy* means the solution of an algebraic problem, namely the evaluation of the product (7.7.163) in the underlying chiral ring. Is the conjecture so unmotivated as it might seem at first sight? After some thinking it is not. Let us first see what it implies on the Hodge numbers of the two spaces forming a mirror pair. Since the A-model has to be interchanged with the B-model the underlying *ring structures* have to be interchanged as well. As many times emphasized the A-twisted sigma model, by means of its three-point function, defines a *quantum deformation* of the *classical Dolbeault ring*, while the B-twisted sigma model, also by means of its three-point function defines a ring isomorphic to the *Hodge ring* (see section 7.7 and section 5.9). If these two graded rings must get interchanged, then at least the dimensions of their subspaces of definite grade must agree. This means that the vertical central row and the horizontal central row of the Hodge diamond must be exchanged in the case of two mirror manifolds. In particular we must have

$$h^{(p,p)}(\mathcal{M}_n) = h^{(n-p,p)}(M\mathcal{M}_n) \qquad \forall p = 0,1,\ldots,n \tag{8.1.2}$$

This symmetry can be generalized to the whole diamond by setting:

$$h^{(p,q)}(\mathcal{M}_n) = h^{(n-p,q)}(M\mathcal{M}_n) \qquad \forall p,q = 0,1,\ldots,n \tag{8.1.3}$$

Hence we assume Eq.(8.1.3) as a first necessary but not yet sufficient condition relating mirror manifolds. The first reason why the conjecture of mirror symmetry is not unmotivated is that the classification of Calabi–Yau $n$-folds constructed as complete intersection in projective or weighted projective spaces (*CICY manifolds*), or as quotients of these latter with respect to discrete groups, has revealed a pattern where the symmetry (8.1.3) has a tendency to occur [71]. Another, deeper reason is based on conformal field theories [180]. From the abstract point of view each Calabi–Yau $n$-fold defines a (2,2) superconformal field theory with central charge $c = 3n$ and the harmonic differential forms $\omega^{(n-p,p)}$ are in one-to-one correspondence with the primary chiral fields of the form

$$\psi\begin{pmatrix} \frac{p}{2} & \frac{p}{2} \\ p & p \end{pmatrix} \tag{8.1.4}$$

while the harmonic $\omega^{(p,p)}$ are in correspondence with the primary chiral fields of the form

$$\psi\begin{pmatrix} \frac{p}{2} & \frac{p}{2} \\ -p & p \end{pmatrix} \tag{8.1.5}$$

But reversing all the $U(1)$ charges in the left or right-moving sector is certainly a symmetry of a (2,2) superconformal field theory that leaves all its correlation functions unchanged. Indeed such a sign reversal is achieved by a spectral flow transformation in one of the two sectors that maps the $(chiral, chiral)$ ring into the $(chiral, antichiral)$ ring. Hence the same abstract superconformal field theory must describe the N=2 sigma model on two different Calabi–Yau manifolds whose Hodge diamonds are related by Eq.(8.1.2), which generalizes to Eq.(8.1.3), and the corresponding A-twisted and B-twisted models must be related as in the above conjecture. Indeed if we perform the topological twists at the level of the superconformal model the exchange of the A- and B-twists is nothing but the exchange of the $(chiral, chiral)$ with the $(chiral, antichiral)$ ring.

Given the evidence that *mirror pairs* should exist we consider the situation for some specific values of $n$. We are eventually mostly interested in 3-folds, but to better grasp the implications of the concept it is convenient to look at higher dimensional examples. As a first example we consider the case of a general Calabi–Yau 4-fold $\mathcal{M}_4$. Taking into account the symmetries

$$h^{(p,q)} = h^{(q,p)} \qquad h^{(p,0)} = h^{(0,4-p)} \qquad h^{(p,q)} = h^{(4-p,4-q)} \tag{8.1.6}$$

and using

$$h^{(0,0)} = h^{(4,0)} = 1 \qquad h^{(1,0)} = 1 \tag{8.1.7}$$

we conclude that the most general Hodge diamond for such a type of manifold is parametrized by five integer numbers:

$$a = h^{(2,0)} \quad x = h^{(2,1)} \quad A = h^{(1,1)} \quad B = h^{(3,1)} \quad C = h^{(2,2)} \tag{8.1.8}$$

and has the following form

$$\begin{array}{ccccccccc} & & & & 1 & & & & \\ & & & 0 & & 0 & & & \\ & & a & & A & & a & & \\ & 0 & & x & & x & & 0 & \\ 1 & & B & & C & & B & & 1 \\ & 0 & & x & & x & & 0 & \\ & & a & & A & & a & & \\ & & & 0 & & 0 & & & \\ & & & & 1 & & & & \end{array} \tag{8.1.9}$$

The parameters $A$ and $B$ correspond to *the numbers of Kähler structure and complex structure deformations, respectively.* The Euler character, therefore, is

$$\chi_{Euler}(\mathcal{M}_4) = 4(a - x + 1) + 2(A + B) + C \tag{8.1.10}$$

The mirror partner $M\mathcal{M}_4$ of a given $\mathcal{M}_4$ is a 4-fold with Hodge numbers interchanged according to the rule

$$h^{(p,q)} \leftrightarrow h^{(4-p,q)} \tag{8.1.11}$$

and therefore possessing the following Hodge diamond:

$$\begin{array}{ccccccccc} & & & & 1 & & & & \\ & & & 0 & & 0 & & & \\ & & a & & B & & a & & \\ & 0 & & x & & x & & 0 & \\ 1 & & A & & C & & A & & 1 \\ & 0 & & x & & x & & 0 & \\ & & a & & B & & a & & \\ & & & 0 & & 0 & & & \\ & & & & 1 & & & & \end{array} \tag{8.1.12}$$

which leads to an identical value for the Euler character:

$$\chi_{Euler}\left(M\mathcal{M}_4\right) = 4(a - x + 1) + 2(A + B) + C = \chi_{Euler}\left(\mathcal{M}_4\right) \tag{8.1.13}$$

This is a general feature for Calabi–Yau manifolds of even complex dimension. Their mirror partners have the same Euler characteristic. Mirror pairs of odd-dimensional Calabi–Yau manifolds have instead opposite Euler numbers. As one sees from Eq.(8.1.12), what has happened in the Hodge–diamond of the mirror partner $M\mathcal{M}_4$ is that the central vertical row has been interchanged with the central horizontal row. This is the basic defining feature of mirror symmetry since it corresponds to an exchange of the ring structures respectively associated with the A-twisted and B-twisted sigma models. Indeed on the vertical central line of the diamond (8.1.9) we have the dimensions of the cohomology groups $H^{(p,p)}$ $(p = 0, 1, 2, 3, 4)$ which form a subring:

$$H^{\star}_{A\text{-}class}(\mathcal{M}_4) = \oplus_{p=0}^{4} H^{(p,p)} \subset H^{\star}_{Dbt}(\mathcal{M}_4) \tag{8.1.14}$$

of the classical Dolbeault cohomology ring. Using the three-point correlators of the A-twisted topological sigma model as structure functions, the ring $H^{\star}_{A\text{-}class}(\mathcal{M}_4)$ becomes a *quantum-deformed ring* $H^{\star}_{A\text{-}quantum}(\mathcal{M}_4)$ (see section 7.7). On the other hand, on the horizontal central line of the diamond (8.1.19) we have the dimensions of the cohomology groups $H^{(4-k,k)}$ $(k = 0, 1, 2, 3, 4)$ which, under the $\star$-product defined in Eq.s (5.9.43), form the Hodge ring:

$$Hod^{\star}\left(\mathcal{M}_4\right) = \oplus_{k=0}^{4} H^{(4-k,k)} \tag{8.1.15}$$

We have already stressed this point. As a further example we consider the case of a generic Calabi–Yau 5-fold $\mathcal{M}_5$. Taking into account the symmetries

$$h^{(p,q)} = h^{(q,p)} \qquad h^{(p,0)} = h^{(0,5-p)} \qquad h^{(p,q)} = h^{(5-p,5-q)} \tag{8.1.16}$$

and using

$$h^{(0,0)} = h^{(5,0)} = 1 \qquad h^{(1,0)} = 1 \tag{8.1.17}$$

the most general Hodge diamond is parametrized by seven independent Hodge numbers:

$$\begin{array}{lllll} h^{(1,1)} = A_1 & h^{(2,2)} = A_2 & \\ h^{(4,1)} = B_1 & h^{(3,2)} = B_2 & \\ h^{(2,0)} = a & h^{(2,1)} = & h^{(3,1)} = y \end{array} \tag{8.1.18}$$

and has the following structure:

$$
\begin{array}{ccccccccccc}
 & & & & & 1 & & & & & \\
 & & & & 0 & & 0 & & & & \\
 & & & a & & A_1 & & a & & & \\
 & & a & & x & & x & & a & & \\
 & 0 & & y & & A_2 & & y & & 0 & \\
1 & & B_1 & & B_2 & & B_2 & & B_1 & & 1 \\
 & 0 & & y & & A_2 & & y & & 0 & \\
 & & a & & x & & x & & a & & \\
 & & & a & & A_1 & & a & & & \\
 & & & & 0 & & 0 & & & & \\
 & & & & & 1 & & & & &
\end{array}
\tag{8.1.19}
$$

The Euler characteristic is immediately calculated and takes the form

$$
\chi_{Euler} \;=\; 2\,(A_1 + A_2 - B_1 - B_2) \,+\, 4\,(x - y) \tag{8.1.20}
$$

As in the previous case, on the vertical central line of the diamond (8.1.19) we have the dimensions of the cohomology groups $H^{(p,p)}$ $(p = 0, 1, 2, 3, 4, 5)$ which form the subring:

$$
H^{\star}_{A\text{-}class}(\mathcal{M}_5) \;=\; \oplus_{p=0}^{5} H^{(p,p)} \subset H^{\star}_{Dbt}(\mathcal{M}_5) \tag{8.1.21}
$$

of the classical Dolbeault cohomology ring that upon use of the three-point correlators of the A-twisted topological sigma model as structure functions, becomes the *quantum deformed ring* $H^{\star}_{A\text{-}quantum}(\mathcal{M}_5)$ (see section 7.7). Also as before, on the horizontal central line of the diamond (8.1.19) we have the dimensions of the cohomology groups $H^{(5-k,k)}$ $(k = 0, 1, 2, 3, 4, 5)$ forming the Hodge ring:

$$
Hod^{\star}\,(\mathcal{M}_5) \;=\; \oplus_{k=0}^{5} H^{(5-k,k)} \tag{8.1.22}
$$

under the $\star$-product. As we know from the results of section (5.9), in those cases where $\mathcal{M}_n$ $(n = 1, 2, 3, 4, 5, \ldots)$ is the vanishing locus of a quasi-homogeneous polynomial $\mathcal{W}(X)$, the Hodge ring $Hod^{\star}\,(\mathcal{M}_n)$ is isomorphic with the integer charge subring of the chiral ring $\mathcal{R}_{\mathcal{W}}$. By definition the mirror partner of the 5-fold $\mathcal{M}_5$ is a new Calabi–Yau 5-fold $\widetilde{\mathcal{M}}_5$ characterized by the following Hodge diamond:

$$
\begin{array}{ccccccccccc}
 & & & & & 1 & & & & & \\
 & & & & 0 & & 0 & & & & \\
 & & & a & & B_1 & & a & & & \\
 & & a & & y & & y & & a & & \\
 & 0 & & x & & B_2 & & x & & 0 & \\
1 & & A_1 & & A_2 & & A_2 & & A_1 & & 1 \\
 & 0 & & x & & B_2 & & x & & 0 & \\
 & & a & & y & & y & & a & & \\
 & & & a & & B_1 & & a & & & \\
 & & & & 0 & & 0 & & & & \\
 & & & & & 1 & & & & &
\end{array}
\tag{8.1.23}
$$

where the Hodge numbers have been permuted according to the rule

$$h^{(p,q)} \leftrightarrow h^{(5-p,q)} \tag{8.1.24}$$

Consequently the Euler number of the mirror manifold is related to that of its partner by

$$\chi\left(M\mathcal{M}_5\right) = -\chi\left(\mathcal{M}_5\right) \tag{8.1.25}$$

As already stated, this is a general property of odd-dimensional Calabi–Yau manifolds. The mirror partner $M\mathcal{M}_{2\nu+1}$ has an Euler number of reversed sign. Furthermore, in the Hodg diamond of the mirror 5-fold the vertical and horizontal central rows have got interchanged. This should not be a mere numerical coincidence, rather, it should be the outcome of an isomorphism of rings.

**Definition 8.1.1** *By definition of mirror symmetry, the manifold $M\mathcal{M}_n$ is the mirror partner of $\mathcal{M}_n$ iff:*

*i)*

$$h^{(p,q)}\left(\mathcal{M}_n\right) = h^{(n-p,q)}\left(M\mathcal{M}_n\right) \qquad \forall\, p,q = 0,1,\ldots,n \tag{8.1.26}$$

*ii)*

$$Hod^{\star}\left(M\mathcal{M}_n\right) \approx H^{\star}_{A\text{-}quantum}\left(\mathcal{M}_n\right) \qquad Hod^{\star}\left(\mathcal{M}_n\right) \approx H^{\star}_{A\text{-}quantum}\left(M\mathcal{M}_n\right) \tag{8.1.27}$$

### 8.1.1 The mirror quintic

Having properly defined mirror pairs we consider now the case that will mostly concern us in the sequel, namely that of the mirror quintic. We begin with the family of Calabi–Yau $(p-2)$-folds $\mathbb{CP}_{p-1}[p]$, defined as the vanishing locus in $\mathbb{CP}_{p-1}$ of some homogeneous polynomial $\mathcal{W}(X)$ of degree

$$\deg\mathcal{W}(X) = p \tag{8.1.28}$$

Namely we set

$$\mathbb{CP}_{p-1}[p] \stackrel{\text{def}}{=} \left\{\{X^{\Lambda} \quad (\Lambda = 1,\ldots,p)\} \in \mathbb{CP}_{p-1} \,|\, \mathcal{W}\left(X^{\Lambda}\right) = 0\right\} \tag{8.1.29}$$

By construction the first Chern class of these $(p-2)$-folds vanishes, while their $(p-2)$-th Chern class is easily calculated from the formula (see Chapter 2)

$$c_{total}\left(\mathbb{CP}_{p-1}[p]\right) = \frac{(1+K)^p}{(1+pK)} \tag{8.1.30}$$

where $K$ is the Kähler 2-form of the ambient space $\mathbb{CP}_{p-1}$. We immediately find

$$\begin{aligned} c_{p-2}\left(\mathbb{CP}_{p-1}[p]\right) &= f_p \wedge^{p-2} K \\ f_p &= \frac{1}{p^2}\left[(1-p)^p + p^2 - 1\right] \end{aligned} \tag{8.1.31}$$

so that the Euler characteristic of these manifolds is

$$\chi\left(\mathbb{CP}_{p-1}[p]\right) = \int_{\mathbb{CP}_{p-1}[p]} c_p\left(\mathbb{CP}_{p-1}[p]\right) = f_p \int \Lambda^{p-2} K$$
$$= p\, f_p = \frac{1}{p}\left[(1-p)^p + p^2 - 1\right] \tag{8.1.32}$$

For the first values of $p$ we get

$$\begin{array}{rclrcl} \chi\left(\mathbb{CP}_2[3]\right) & = & 0 & \chi\left(\mathbb{CP}_3[4]\right) & = & 24 \\ \chi\left(\mathbb{CP}_4[5]\right) & = & -200 & \chi\left(\mathbb{CP}_5[6]\right) & = & 2610 \\ \chi\left(\mathbb{CP}_6[7]\right) & = & -39984 & \dots & = & \dots \end{array} \tag{8.1.33}$$

There is an immediate understanding of the Euler number for the first three cases. The cubic 1-fold $\mathbb{CP}_2[3]$ is a torus so that its Euler number vanishes. The quartic 2-fold $\mathbb{CP}_3[4]$ is a $K_3$ surface, and the result $\chi = 24$ is consistent with the Hodge diamond

$$\begin{array}{ccccc} & & 1 & & \\ & 0 & & 0 & \\ 1 & & 20 & & 1 \\ & 0 & & 0 & \\ & & 1 & & \end{array} \tag{8.1.34}$$

Indeed the quartic polynomial has 19 parameters that lead to 19 algebraic complex structure deformations. To these one has to add a non-algebraic deformation corresponding to the (1,1)-form in the same class as the Kähler 2-form and the value $h^{(1,1)} = 20$ is explained. Finally the value $-200$ obtained for the Euler character of the quintic 3-fold $\mathbb{CP}_4[5]$, is consistent with the Hodge diamond

$$\begin{array}{ccccccc} & & & 1 & & & \\ & & 0 & & 0 & & \\ & 0 & & 1 & & 0 & \\ 1 & & 101 & & 101 & & 1 \\ & 0 & & 1 & & 0 & \\ & & 0 & & 0 & & \\ & & & 1 & & & \end{array} \tag{8.1.35}$$

whose structure was derived in Chapter 4, section 4.5. We briefly recall the derivation. There is just one Kähler class (that inherited from the ambient $\mathbb{CP}_4$ space) and that explains $h^{(1,1)} = 1$, while the quintic polynomial admits 101 deformations, yielding such a value for $h^{(2,1)}$. The counting 101 is performed in the following way. The most general homogeneous polynomial of degree five in five unknowns:

$$P_{1|5}(X) = d_{\Lambda_1 \dots \Lambda_5} X^{\Lambda_1} \dots X^{\Lambda^5} \qquad d_{\Lambda_1 \dots \Lambda_5} X^{\Lambda_1} = \text{symm. tensor} \tag{8.1.36}$$

has $\frac{5\cdot 6\cdot 7\cdot 8\cdot 9}{5!} = 126$ parameters. Subtracting the 25 degrees of freedom of the linear reparametrization group $GL(5,\mathbb{C})$, which is equivalent to taking into account the vanishing relations $\partial_\Lambda W \approx 0$, we get the result $h^{(2,1)} = 101$. The subsequent Euler numbers in Eq.(8.1.32) and Eq.(8.1.33) have also an immediate interpretation. For instance if we consider the next case of the sextic 4-fold $\mathbb{CP}_5[6]$, and we recall the structure (8.1.9) of its Hodge diamond, we conclude that the result for the Euler number is consistent with the diamond

$$\begin{array}{ccccccccc} &&&&1&&&& \\ &&&0&&0&&& \\ &&0&&1&&0&& \\ &0&&0&&0&&0& \\ 1&&426&&1752&&426&&1 \\ &0&&0&&0&&0& \\ &&0&&1&&0&& \\ &&&0&&0&&& \\ &&&&1&&&& \end{array} \tag{8.1.37}$$

corresponding to

$$a = x = 0 \quad A = 1 \quad B = 426 \quad C = 1752 \tag{8.1.38}$$

Indeed

$$\chi \;=\; 2+2+1+426+1752+426+2 \;=\; 2610 \tag{8.1.39}$$

On the other hand the numbers appearing on the central horizontal row of (8.1.37) correspond to the counting of polynomials in six unknowns, respectively of degree 6, 12, 18 and 24, modulo the vanishing relations $(X^1)^5 = \ldots (X^6)^5 = 0$. This counting is easily performed and gives the shown result. Generically for all these $(p-2)$-folds, the Hodge diamond has the structure

$$\begin{array}{ccccccccccccc} &&&&&&1&&&&&& \\ &&&&&0&&0&&&&& \\ &&&&0&&1&&0&&&& \\ &&&0&&0&&0&&0&&& \\ &&0&&\cdots&&\cdots&&\cdots&&0&& \\ &\cdots&&\cdots&&\cdots&&\cdots&&\cdots&&\cdots& \\ 1&B_1&&B_2&&\cdots&&\cdots&&B_2&&B_1&1 \\ &\cdots&&\cdots&&\cdots&&\cdots&&\cdots&&\cdots& \\ &&0&&\cdots&&\cdots&&\cdots&&0&& \\ &&&0&&0&&0&&0&&& \\ &&&&0&&1&&0&&&& \\ &&&&&0&&0&&&&& \\ &&&&&&1&&&&&& \end{array} \tag{8.1.40}$$

where

$$\begin{aligned} B_k &= \dim \mathcal{R}_{k|p}(\mathcal{W}) \qquad k = 1, 2, \ldots, p-2 \\ B_0 &= B_{p-2} = 1 \\ B_k &= B_{p-2-k} \end{aligned} \tag{8.1.41}$$

are the dimensions of the spaces of non-trivial polynomials of degree $k|p$. The Euler character, calculated in Eq.(8.1.32) has therefore the interpretation

$$\begin{aligned}\chi\,(\mathbb{CP}_{p-1}[p]) &= \frac{1}{p}\left[(1-p)^p + p^2 - 1\right] \\ &= \begin{cases} 6 + \sum_{k=1}^{p-3} B_k & \text{if } p = \text{even} \\ 2 - \sum_{k=1}^{p-3} B_k & \text{if } p = \text{odd} \end{cases}\end{aligned} \tag{8.1.42}$$

According to the hypothesis of mirror symmetry the mirror manifold $M\mathbb{CP}_{p-1}[p]$ of $\mathbb{CP}_{p-1}[p]$, if it exists, must be a Calabi–Yau 3-fold with the Hodge-diamond

$$\begin{array}{ccccccccccc} & & & & & 1 & & & & & \\ & & & & 0 & & 0 & & & & \\ & & & 0 & & B_1 & & 0 & & & \\ & & 0 & & 0 & & 0 & & 0 & & \\ & 0 & & 0 & & B_2 & & 0 & & 0 & \\ & \cdots & \cdots & \cdots & \cdots & & \cdots & \cdots & \cdots & & \\ 1 \quad 1 & & 0 & & \cdots & & \cdots & & 0 & & 1 \quad 1 \\ & \cdots & \cdots & \cdots & \cdots & & \cdots & \cdots & \cdots & & \\ & 0 & & 0 & & B_2 & & 0 & & 0 & \\ & & 0 & & 0 & & 0 & & 0 & & \\ & & & 0 & & B_1 & & 0 & & & \\ & & & & 0 & & 0 & & & & \\ & & & & & 1 & & & & & \end{array} \tag{8.1.43}$$

which for the particular case of the mirror quintic $M\mathbb{CP}_4[5]$, means

$$\begin{array}{ccccccc} & & & 1 & & & \\ & & 0 & & 0 & & \\ & 0 & & 101 & & 0 & \\ 1 & & 1 & & 1 & & 1 \\ & 0 & & 101 & & 0 & \\ & & 0 & & 0 & & \\ & & & 1 & & & \end{array} \tag{8.1.44}$$

The Euler character of these mirror manifolds is therefore given by

$$\chi\,(M\mathbb{CP}_{p-1}[p]) = \begin{cases} 6 + \sum_{k=1}^{p-3} B_k = \chi\,(\mathbb{CP}_{p-1}[p]) & \text{if } p = \text{even} \\ -2 + \sum_{k=1}^{p-3} B_k = -\chi\,(\mathbb{CP}_{p-1}[p]) & \text{if } p = \text{odd} \end{cases} \tag{8.1.45}$$

so that for the first few cases we get

$$\begin{array}{lcllcl} \chi\,(M\mathbb{CP}_2[3]) & = & 0 & \chi\,(M\mathbb{CP}_3[4]) & = & 24 \\ \chi\,(M\mathbb{CP}_4[5]) & = & 200 & \chi\,(M\mathbb{CP}_5[6]) & = & 2610 \\ \chi\,(M\mathbb{CP}_6[7]) & = & 39984 & \ldots & = & \ldots \end{array} \tag{8.1.46}$$

In recent years, after mirror symmetry has been introduced by physicists, mathematicians have devised various strategies to construct mirror pairs of Calabi–Yau $n$-folds. The involved techniques are mostly based on toric geometry. As already pointed out, we do not want to enter this rather difficult chapter of new mathematics. Indeed our aim is more centred on the illustration of general ideas than on the actual construction and classification of examples. Hence we confine ourselves to presenting the definition of the mirror quintic 3-fold and showing that it is appropriate.

To this end we go once more back to a general case and we consider the class of Calabi–Yau $(p-2)$-folds $M\mathbb{CP}_{p-1}[p]$ defined as follows. Take the the ambient space $\mathbb{CP}_{p-1}$ and, among all the possible $\mathbb{CP}_{p-1}[p]$ manifolds, consider the one-parameter subclass $\widetilde{M\mathbb{CP}}_{p-1}[p,\psi] \subset \mathbb{CP}_{p-1}[p]$ of those that are the vanishing locus:

$$\widetilde{M\mathbb{CP}}_{p-1}[p,\psi] = \left\{X^\Lambda \in \mathbb{CP}_{p-1} \,|\, \mathcal{W}(\psi;X) = 0\right\} \tag{8.1.47}$$

of the specific degree $p$ polynomial of Eq.(5.9.5), namely

$$\mathcal{W}(\psi;X) = \frac{1}{p}\left(\sum_{\Lambda=1}^{p} X_\Lambda^p\right) - \psi \prod_{\Lambda=1}^{p} X_\Lambda \tag{8.1.48}$$

Consider now, in $\mathbb{CP}_{p-1}$, the action of the discrete group

$$H \sim (\mathbb{Z}_p)^{p-2} \tag{8.1.49}$$

constructed in the following way. Name, as usual, $\alpha_p = \exp[\frac{2\pi i}{p}]$ the $p$-th root of the identity and consider the following diagonal $p \times p$ matrices:

$$\begin{aligned} H_1 &= \operatorname{diag}\left(\alpha_p, 1, \ldots, (\alpha_p)^{p-1}\right) \\ H_2 &= \operatorname{diag}\left(1, \alpha_p, 1, \ldots, (\alpha_p)^{p-1}\right) \\ \ldots &= \ldots \\ H_{p-1} &= \operatorname{diag}\left(1, \ldots, \alpha_p, (\alpha_p)^{p-1}\right) \end{aligned} \tag{8.1.50}$$

These matrices are the generators of the $H$ group of Eq. (8.1.49) and act on the $p$-vector of homogeneous coordinates $X^\Lambda$ in an obvious way. For instance, we have

$$H_1 : \begin{pmatrix} X^1 \\ X^2 \\ \ldots \\ \ldots \\ X^p \end{pmatrix} \rightarrow \begin{pmatrix} \alpha_p X^1 \\ X^2 \\ \ldots \\ \ldots \\ (\alpha_p)^{p-1} X^p \end{pmatrix} \tag{8.1.51}$$

The group generated by these matrices is $(\mathbb{Z}_p)^{p-2}$ as stated in Eq.(8.1.49) and not $(\mathbb{Z}_p)^{p-1}$, as it would naively seem from the fact that we have introduced $p-1$ generators. The reason is simple. The action of $\prod_{i=1}^{p-1} H_i$ on the homogeneous coordinates is just an overall multiplication by a single phase factor. In $\mathbb{CP}_{p-1}$ this operation is the identity operation and hence

$$\prod_{i=1}^{p-1} H_i \sim \mathbb{1} \tag{8.1.52}$$

so that only $p-2$ of the $p-1$ generators (8.1.50) are truly independent. Given this discrete group one realizes that the polynomial (8.1.48) is invariant under $H$ and that it is actually the most general homogeneous order $p$ polynomial with such an invariance property. Consequently the manifolds $\widetilde{M\mathbb{CP}}_{p-1}[p,\psi]$, defined by Eq.(8.1.47) are invariant under the action of $H$. We can then define the new one-parameter class of $(p-2)$-folds given by the quotient:

$$M\mathbb{CP}_{p-1}[p,\psi] \stackrel{\text{def}}{=} \frac{\widetilde{M\mathbb{CP}}_{p-1}[p,\psi]}{H} \tag{8.1.53}$$

If the action of $H$ on $\widetilde{M\mathbb{CP}}_{p-1}[p,\psi]$ were free the quotient manifold $M\mathbb{CP}_{p-1}[p,\psi]$ would be smooth and its Euler characteristic would just be given by that of $\widetilde{M\mathbb{CP}}_{p-1}[p,\psi]$, which is identical with $\chi\left(\mathbb{CP}_{p-1}[p]\right)$, divided by the order of the group $H$

$$|H| \;=\; p^{p-2} \tag{8.1.54}$$

Actually the action of $H$ on $\widetilde{M\mathbb{CP}}_{p-1}[p,\psi]$ admits fixed points and fixed submanifolds of higher dimensions so that the quotient (8.1.53) is a singular variety, the singularities occurring at the fixed points and surfaces. These quotient singularities can be repaired by the standard algebraic-geometric procedure of *blow-up* and the resulting smooth manifold is what we properly name $M\mathbb{CP}_{p-1}[p,\psi]$. We claim that $M\mathbb{CP}_4[5,\psi]$ is the looked for mirror quintic, with Hodge diamond as in Eq. (8.1.44). We motivate this by calculating the Euler character of $M\mathbb{CP}_4[5,\psi]$ and showing that it is $\chi_{Euler}\left(M\mathbb{CP}_4[5,\psi]\right) = 200$. It must, however, be recalled that this is a necessary but not sufficient condition. The true proof that $M\mathbb{CP}_4[5,\psi]$ is the mirror partner of $\mathbb{CP}_4[5]$ is given by verifying the isomorphism of the corresponding quantum deformed Dolbeault and Hodge rings. This will be done by computing the Yukawa couplings of $(2,1)$-forms in $M\mathbb{CP}_4[5,\psi]$ and reinterpreting them as Yukawa couplings of $(1,1)$-forms in $\mathbb{CP}_4[5]$. This reinterpretation yields a prediction for the number of rational curves of degree $k$ embedded in $\mathbb{CP}_4[5]$ and the first few predictions agree with results known in mathematics. If this explicit verification were essential for deciding which manifold is the mirror of which partner manifold, then the utility of mirror symmetry would be very much limited. Indeed its value resides precisely in the possibility of studying the structure of the *quantum-deformed Dolbeault ring* $H^{\star}_{A\text{-}quantum}\left(\mathcal{M}_n\right)$ (which usually corresponds to an unsolvable problem) by identifying it with the *Hodge ring* $Hod^{\star}\left(M\mathcal{M}_n\right)$, whose features are accessible through Picard–Fuchs equations. Hence to know *a priori* that certain manifolds constitute *mirror pairs* is a vital information. It is in this direction that mathematicians have been very active in recent years [41, 42, 40, 17] and have found various *a priori* patterns for the construction of mirror pairs. The quotienting by discrete subgroups, as we do in the case of the manifolds $\widetilde{M\mathbb{CP}}_{p-1}[p,\psi]$, has provided one of these patterns, actually the first to be discovered, in chronological order [180, 18]. This is to say that there are more general arguments to support the conjecture that $M\mathbb{CP}_4[5,\psi]$ is the mirror partner of $\mathbb{CP}_4[5]$ than the mere numerical coincidence $\chi_{Euler}\left(M\mathbb{CP}_4[5,\psi]\right) = 200$, yet we confine ourselves to showing this fact.

Given the action (8.1.51) of the $H$-group generators (8.1.50), it is almost immediate to see that the covering manifold $\widetilde{M\mathbb{CP}}_4[5,\psi]$ admits, under $H$ the following ten fixed curves:

$$\mathcal{FC}_{\Lambda_1\Lambda_2\Lambda_3} \stackrel{\text{def}}{=} \left\{X^5_{\Lambda_1} + X^5_{\Lambda_2} + X^5_{\Lambda_3} = 0 \,|\, X_{\Lambda_4} = X_{\Lambda_5} = 0\right\} \\ (\Lambda_1,\Lambda_2,\Lambda_3,\Lambda_4,\Lambda_5 \text{ all distinct}) \tag{8.1.55}$$

Indeed, each of the submanifolds $\mathcal{FC}_{\Lambda_1\Lambda_2\Lambda_3} \subset \widetilde{M\mathbb{CP}}_4[5,\psi]$ is mapped into itself by the whole action of $H$. In addition each of the ten fixed curves $\mathcal{FC}_{\Lambda_1\Lambda_2\Lambda_3}$ is made up of points that are fixed under the action of a subgroup $\mathbb{Z}_5 \subset H$. For instance all the points of $\mathcal{FC}_{123}$ are fixed under the action of the $\mathbb{Z}_5$ generated by $H_4$ as defined in (8.1.50). The fixed curves $\mathcal{FC}_{\Lambda_1\Lambda_2\Lambda_3}$ meet in 50 points that, under the action of the group $H$, form 10 fixed orbits $\mathcal{FP}_{\Lambda_1\Lambda_2}$ of five points each. The orbits $\mathcal{FP}_{\Lambda_1\Lambda_2}$ are defined as the following loci:

$$\mathcal{FC}_{\Lambda_1\Lambda_2} \stackrel{\text{def}}{=} \left\{X^5_{\Lambda_1} + X^5_{\Lambda_2} = 0 \,|\, X_{\Lambda_3} = X_{\Lambda_4} = X_{\Lambda_5} = 0\right\} \\ (\Lambda_1,\Lambda_2,\Lambda_3,\Lambda_4,\Lambda_5 \text{ all distinct}) \tag{8.1.56}$$

The fixed curves $\mathcal{FC}_{\Lambda_1\Lambda_2\Lambda_3}$ are projective algebraic 1-folds of type $\mathbb{CP}_2[5]$, while the fixed orbits $\mathcal{FP}_{\Lambda_1\Lambda_2}$ are projective algebraic 0-folds of type $\mathbb{CP}_1[5]$. The Euler characteristic of these $p$-folds is immediately calculated with the techniques of section 4.5. We have

$$\chi_{Euler}(\mathbb{CP}_1[5]) = -10 \qquad \chi_{Euler}(\mathbb{CP}_1[5]) = 5 \tag{8.1.57}$$

Now each of the fixed curves $\mathcal{FC}_{\Lambda_1\Lambda_2\Lambda_3} \sim \mathbb{CP}_2[5]$ contains three of the fixed orbits $\mathcal{FP}_{\Lambda_1\Lambda_2} \sim \mathbb{CP}_1[5]$. For instance $\mathcal{FC}_{123}$ contains $\mathcal{FP}_{12}$, $\mathcal{FP}_{13}$ and $\mathcal{FP}_{23}$. Hence we can define the open manifolds $\widehat{\mathcal{FC}}_{\Lambda_1\Lambda_2\Lambda_3}$ obtained by subtracting from $\mathcal{FC}_{\Lambda_1\Lambda_2\Lambda_3}$ the three fixed orbits of points and using (8.1.57) we have:

$$\chi_{Euler}\left(\widehat{\mathcal{FC}}_{\Lambda_1\Lambda_2\Lambda_3}\right) = -10 - 3 \times 5 = -25 \tag{8.1.58}$$

Then we define the open manifold:

$$\widehat{\widetilde{M\mathbb{CP}}}_4[5] = \widetilde{M\mathbb{CP}}_4[5] - \cup \widehat{\mathcal{FC}}_{\Lambda_1\Lambda_2\Lambda_3} - \cup \mathcal{FP}_{\Lambda_1\Lambda_2} \tag{8.1.59}$$

obtained by subtracting from the covering manifold all the fixed points and surfaces. We immediately find

$$\begin{aligned} \chi_{Euler}\left(\widehat{\widetilde{M\mathbb{CP}}}_4[5]\right) &= \chi_{Euler}\left(\widetilde{M\mathbb{CP}}_4[5]\right) - 10 \times \chi_{Euler}\left(\widehat{\mathcal{FC}}_{\Lambda_1\Lambda_2\Lambda_3}\right) \\ &\quad - 10 \times \chi_{Euler}\left(\mathcal{FP}_{\Lambda_1\Lambda_2}\right) \\ &= -200 - 10 \times (-25) - 10 \times 5 \\ &= 0 \end{aligned} \tag{8.1.60}$$

Then we define the true mirror manifold as

$$
\begin{aligned}
M\mathbb{CP}_4[5] = & \frac{\widehat{M\mathbb{CP}_4[5]}}{H \sim \mathbb{Z}_5 \otimes \mathbb{Z}_5 \otimes Z_5} \\
& + 10 \times \frac{[\mathbb{CP}_2[5] - 3 \times \mathbb{CP}_1[5]] \otimes \mathrm{Lens}_5}{H' \sim \mathbb{Z}_5 \otimes \mathbb{Z}_5} \\
& + 50 \times \mathrm{Lens}_5
\end{aligned} \tag{8.1.61}
$$

where the fixed curves and fixed points have been replaced with their smooth equivalents (i.e. singularity resolutions). For the fixed points the smooth equivalents are given by copies of the three-dimensional analogue of the Eguchi–Hanson space with a $\mathbb{Z}_5$ Lens boundary. We have named these non-compact spaces $\mathrm{Lens}_n$ and we refer the reader to the literature for further information on their construction and properties [131, 190, 166]. In particular for $n = 2$, the $\mathbb{Z}_2$ Lens spaces are the Kähler manifolds with Calabi metrics constructed in chapter 5 as Kähler quotients of flat space (see section 5.5.7). What matters here is that the Euler number of these manifolds is

$$
\chi_{Euler}\left(\mathrm{Lens}_5\right) = 5 \tag{8.1.62}
$$

For the fixed curves the smooth analogue is the a bundle over the fixed curve with the Lens space as a fiber

$$
\frac{[\mathbb{CP}_2[5] - 3 \times \mathbb{CP}_1[5]] \otimes \mathrm{Lens}_5}{H' \sim \mathbb{Z}_5 \otimes \mathbb{Z}_5} \tag{8.1.63}
$$

In Eq.s(8.1.63) and (8.1.61) we have been careful to mode out the fixed curve by the subgroup of $H$ that acts freely on it, namely $H' \sim \mathbb{Z}_5 \otimes \mathbb{Z}_5$. Now using all this information we can calculate the Euler number of the mirror manifold $M\mathbb{CP}_4[5]$. We have

$$
\begin{aligned}
\chi_{Euler}\left(M\mathbb{CP}_4[5]\right) = & \frac{\chi_{Euler}\left(\widehat{M\mathbb{CP}_4[5]}\right)}{125} \\
& + 10 \times \frac{\chi_{Euler}\left(\mathbb{CP}_2[5] - 3 \times \mathbb{CP}_1[5]\right) \cdot \chi_{Euler}\left(\mathrm{Lens}_5\right)}{25} \\
& + 50 \times \chi_{Euler}\left(\mathrm{Lens}_5\right) \\
= & \frac{0}{125} + 10 \times \frac{-25 \times 5}{25} + 50 \times 5 \\
= & 200
\end{aligned} \tag{8.1.64}
$$

## 8.1.2 The issue of flat coordinates and Picard–Fuchs equation

Given the existence of *mirror pairs* we must now understand how easy is the *easy calculation* of B-correlators. Product multiplication of polynomials in the chiral ring is indeed a very easy operation and if we have a parametrized family of hypersurfaces $\mathcal{W}(X, \psi) = 0$ or, equivalently, a topological Landau–Ginzburg model with superpotential $\mathcal{W}(X, \psi)$, the calculation of B-correlators as functions of the coupling parameters $\psi^\alpha$ appearing in the superpotential is almost immediate. In view of mirror symmetry these correlators are also correlators of the A-twisted sigma model and so the programme we started with seems to be an accomplished fact. The complication we still have to face is one of coordinate patches. In order to make a direct comparison between B-twisted correlators on $M\mathcal{M}_n$ and A-twisted correlators on $\mathcal{M}$ we must be sure that we use the same coordinate system for the moduli spaces or topological coupling constant space in the two cases. The natural coordinate system of the B-model, defined by the parameters $\psi^\alpha$ appearing in the superpotential $\mathcal{W}(X;\psi)$ is not necessarily the natural coordinate system for Kähler class deformations defined in the A-model and given, for instance, by the parameters $t^a$ which appear in the decomposition of the Kähler 2-form along a basis of $(1,1)$-forms. This problem of comparison is solved by choosing for both models the privileged coordinate patch that every topological field theory intrinsically defines: that of *flat coordinates.* We recall from section 7.8 that flat coordinates $t^a$ are those for which the topological metric $\eta_{ij}$, namely the topological 2-point function, is a constant and the three-point functions can be written as the third derivative of a *generating* or *free energy* function $\mathcal{F}(t)$:

$$\langle \mathcal{O}_i \mathcal{O}_j \mathcal{O}_k \rangle = \frac{\partial^3 \mathcal{F}}{\partial t_i \partial t_j \partial t_k} \tag{8.1.65}$$

Hence the basic problem is that of calculating

$$t^a = t^a(\psi^\alpha) \tag{8.1.66}$$

the coordinate transformation from the Landau–Ginzburg coordinates $\psi^\alpha$ to the flat coordinates $t^a$. This problem is solved by studying a system of differential equations, the *Picard–Fuchs equations* satisfied by the $t^a$ as functions of the $\psi^\alpha$. This system of differential equations can be directly deduced from the structure of the superpotential $\mathcal{W}(X;\psi)$, the catch of the method being the Griffiths residue map explained in section 5.9. Indeed what happens is that the flat coordinates $t^a$ are nothing but the periods of the harmonic $(n-k,k)$-forms along a basis of $n$-cycles. This statement can be motivated by considering the case of Calabi–Yau 3-folds. There the topological coupling constant space of marginal deformations is the moduli space of either Kähler class or complex structure deformations. Both are *special Kähler manifolds* and the three-point correlator coincides with the Yukawa couplings $W_{\alpha\beta\gamma}$ or $W_{\mu\nu\rho}$. These three-point functions can be expressed as third derivatives of a generating function only in the *special coordinate system.* Hence the special coordinates of special geometry and the flat coordinates of topological field theory must be the very same thing, at least for marginal deformations. Now considering the derivation of special geometry of $(2,1)$-forms given in section 6.2 we see that special coordinates $X^\Lambda, \partial_\Sigma F(X)$ are precisely the periods of the holomorphic 3-form $\Omega^{(3,0)}$ along a basis of 3-cycles. This shows what we just claimed.

With this basic motivation in mind we now turn to the derivation of the Picard–Fuchs equations satisfied by the periods of harmonic $(n-k,k)$-forms.

## 8.2 Picard–Fuchs Equations for the Period Matrix

In this section our aim is to show that the periods of the holomorphic $n$-form $\Omega^{(n,0)}$ on a parametrized family of Calabi–Yau $n$-folds satisfy a set of higher order linear differential equations with respect to the moduli parameters of the complex structure. By periods we mean the integrals of $\Omega^{(n,0)}$ along a basis $A_i$ ($i = 1, \ldots, b^n$) of homology $n$-cycles and the corresponding differential identities are what go under the name of *Picard–Fuchs equations*. The main reason of interest for this set of differential equations is that they provide a tool to compute the periods and thus they allow the explicit evaluation of the B-twisted sigma model correlators in terms of "flat coordinates". Via mirror symmetry this algorithm provides also a solution of the much more difficult A-twisted sigma model for the corresponding mirror manifold.

In a subsequent section we shall explore the relation between Picard–Fuchs equations and the structure of special geometry. Indeed we shall be able to show that a set of holomorphic differential identities analogous to the Picard–Fuchs equations associated with Calabi–Yau 3-folds holds true for any $m$-dimensional special Kählerian manifold, independently of its interpretation as the moduli space of any Calabi–Yau 3-fold. In this section, on the other hand, we take a geometrical attitude and we show that a holomorphic set of Picard–Fuchs equations exists for the period matrix of any Calabi–Yau $n$-fold, although the moduli space of the complex structures is special Kählerian only in the case $n = 3$. Hence the Picard–Fuchs equations have two different origins that become equivalent in the case of such special manifolds that can be interpreted as the complex structure moduli space of a Calabi–Yau 3-fold.

In the present geometrical approach to Picard–Fuchs equations the starting point is provided by the Griffiths residue mapping discussed in section 5.9. We consider the *Hodge filtration* (5.9.24), (5.9.25) of the *n-th de Rham cohomology group* (5.9.26) and, at the same time, its isomorphism with the filtration (5.9.20) of the space $\mathcal{H}_n$ of *rational meromorphic* $(n+1)$-forms on $\mathbb{CP}_{n+1}$ which admit the Calabi–Yau $n$-fold $\mathcal{M}_n$ as the polar locus. In this description $\mathcal{M}_n$ is the vanishing locus:

$$\mathcal{W}(X;\psi) = 0 \tag{8.2.1}$$

of a homogeneous polynomial of degree $\nu = n+2$ (see Eq.(5.9.3)), depending on a set of parameters $\psi^\alpha$, and the rational $(n+1)$-forms are represented as in Eq.(5.9.9), the numerator being a polynomial of degree (5.9.11), (5.9.12). The spaces $\mathcal{H}_k$ contain all those rational $(n+1)$-forms, the effective order of whose pole at $\mathcal{W} = 0$ is not stronger than $k$. The isomorphism between the two filtrations is realized by the Griffiths residue map expressed by Eq.s (5.9.33), (5.9.34) and (5.9.35).

The derivation of Picard–Fuchs equations, just like the discussion of the Griffiths residue map, could be performed for $n$-folds that are complete intersection of quasi-homogeneous polynomial constraints in a product of weighted projective spaces. Yet, to avoid a clumsy notation and for the sake of simplicity we follow the policy of section 5.9 and we restrict our attention to the case of a single homogeneous polynomial constraint in ordinary projective space. Furthermore, as we already said, we assume the Calabi–Yau condition $\nu = n+2$.

Equation (5.9.35) is of particular interest to us at this junction. Let us consider a homology basis $\{A_i\}$ $\left(i = 1,\ldots,b^n = \sum_{k=0}^{n} h^{(n-k,k)}\right)$ of $n$-cycles on the $n$-fold $\mathcal{W} = 0$. To each of these cycles we can associate an $(n+1)$-dimensional tube in $\mathbb{CP}_{n+1}$ that winds around the $n$-fold. Such a tube can be constructed by attaching, to each point $p \in A_i$ of a homology $n$-cycle a circle $c_p \subset \{\mathbb{CP}_{n+1} - \mathcal{M}_n\}$ in the external space that winds around that point. The union of all such circles $\mathcal{C}_i = \cup_{p\in A_i} c_p$ is an $(n+1)$-dimensional subspace $\mathcal{C}_i \subset \{\mathbb{CP}_{n+1} - \mathcal{M}_n\}$ erected over the homology cycle $A_i \subset \mathcal{M}_n$ and winding around the polar locus $\mathcal{W} = 0$. Making use of Eq. (5.9.35) we can write

$$\begin{aligned}\int_{\mathcal{C}_i} \frac{P_{k|\nu}(X)}{(\mathcal{W}(X))^{k+1}}\,\omega &= \int_{\mathcal{C}_i} \Omega^{(n+1)}_{(k)} = \\ \int_{\mathcal{C}_i} T\left(\Omega^{(n+1)}_{(k)}\right) &= \int_{A_i}\int_{c_p} T\left(\Omega^{(n+1)}_{(k)}\right) = \\ \int_{A_i}\int_{c_p} \frac{\gamma}{\mathcal{W}} \wedge d\mathcal{W} &= \int_{A_i} \gamma \end{aligned} \tag{8.2.2}$$

The main point in the above chain of equations is the equality between the first line and the last. By construction $\gamma \in \mathcal{F}_{(n-k,k)}$ is a harmonic form of type $(n,0)+(n-1,1)+\ldots+(n-k,k)$ on the Calabi-Yau $n$-fold. Its periods $\int_{A_i}\gamma$ along a basis of homology $n$-cycles can be expressed as integrals of the corresponding rational meromorphic form on the tubes $\mathcal{C}_i$. As we know from section 5.9, the polynomials

$$P^{\alpha}_{k|\nu}(X) \qquad \left(\alpha = 1\ldots \dim\mathcal{R}_{k|\nu}\right) \tag{8.2.3}$$

are in one-to-one correspondence with the elements of degree $k|\nu$ in the chiral ring $\mathcal{R}(\mathcal{W})$ (see Eq. (5.9.23) ) and, via the Griffiths residue map, with the elements of $H^{(n-k,k)}$. Furthermore, their product in $\mathcal{R}(\mathcal{W})$ is isomorphic to the $\star$-product of the corresponding elements of $\mathcal{F}_{(n-k,k)}$. Relying on these facts, for each homology cycle $A_i$, we can construct a column vector

$$\Pi^i = \begin{pmatrix} \Pi^i_0 \\ \Pi^i_1 \\ \cdots \\ \cdots \\ \cdots \\ \Pi^i_n \end{pmatrix} = \begin{pmatrix} \int_{\mathcal{C}_i} \frac{1}{(W(X))^1}\,\omega \\ \int_{\mathcal{C}_i} \frac{P^{\alpha}_{1|\nu}(X)}{(W(X))^2}\,\omega \\ \cdots \\ \cdots \\ \cdots \\ \int_{\mathcal{C}_i} \frac{P_{n|\nu}(X)}{(W(X))^{n+1}}\,\omega \end{pmatrix} \tag{8.2.4}$$

whose elements, in view of Eq.(8.2.2), can be interpreted as the periods of a basis of $(n-k,k)$ harmonic forms along the homology cycle $A_i$:

$$\Pi_{k\,\alpha}^{\;\;i}(\psi) = \int_{A_i} \gamma^{(n-k,k)}_{\alpha} \tag{8.2.5}$$

In Eq. (8.2.6) we have made explicit the dependence of such periods on the parameters $\psi^\alpha$ that appear in the polynomial constraint (8.2.1). Indeed, by construction, the elements of the period vector are holomorphic (or meromorphic) functions of such parameters. If we consider a complete basis of homology cycles, then the set of period vectors $\mathbf{\Pi}^i$ constitute an $m \times m$ period matrix $\Pi_{j=k,\alpha}{}^i$, where

$$m = 2 + \sum_{k=1}^{n-1} h^{(n-k,k)} \tag{8.2.6}$$

In Eq.s (8.2.4) and (8.2.6) we have already taken into account the Calabi–Yau condition that guarantees $h^{(n,0)} = h^{(0,n)} = 1$.

The Picard–Fuchs equations are partial differential equations of order $n+1$ satisfied by the periods $\Pi_0{}^i = \int_{C_i} \Omega^{(n,0)}$ of the harmonic $(n,0)$-form, namely by the first entry of the vectors $\mathbf{\Pi}^i$. These higher order differential equations are an immediate consequence of a system of first order differential equations satisfied by the period vectors $\mathbf{\Pi}^i$:

$$\left[ \mathbb{1} \frac{\partial}{\partial \psi^\alpha} - \mathbf{A}_\alpha(\psi) \right] \cdot \mathbf{\Pi}^i = 0, \tag{8.2.7}$$

where $\mathbf{A}_\alpha(\psi)$ is a certain matrix of $\psi$-dependent coefficents that has the following structure:

$$\mathbf{A}_\alpha(\psi) = \begin{pmatrix} * & * & 0 & \dots & \dots & 0 \\ * & * & * & 0 & \dots & 0 \\ \dots & \dots & \dots & \dots & \dots & \dots \\ * & * & * & * & * & * \\ * & * & * & * & * & * \end{pmatrix} \tag{8.2.8}$$

The algorithm needed to work out Eq.s (8.2.7) is very simple: one considers the definition of the periods given by the right hand side of Eq.(8.2.4) and takes the derivative $\frac{\partial}{\partial \psi^\alpha}$ under the integral. Next one makes a systematic use of the identity (5.9.15) to reduce all polynomials to elements of the chiral ring. Let us explain the procedure in general terms. By naive derivation we obtain

$$\frac{\partial}{\partial \psi^\alpha} \Pi^i_{k,\sigma} = \int_{C_i} \left[ \frac{\frac{\partial}{\partial \psi^\alpha} P^\sigma_{k|\nu}(X)}{(\mathcal{W}(X))^{k+1}} - (k+1) \frac{P^\sigma_{k|\nu}(X) \frac{\partial}{\partial \psi^\alpha} \mathcal{W}}{\mathcal{W}^{k+2}} \right] \omega \tag{8.2.9}$$

As far as the first term on the right hand side of Eq.(8.2.9) is concerned, it is the integral of a rational form whose pole at $\mathcal{W} = 0$ is not stronger than $k+1$. The derivative $\frac{\partial}{\partial \psi^\alpha} P^\sigma_{k|\nu}(X)$ is a polynomial of the same degree as $P^\sigma_{k|\nu}(X)$ and in general it is the sum of a non-trivial element of the chiral ring plus a trivial part:

$$\frac{\partial}{\partial \psi^\alpha} P^\sigma_{k|\nu}(X) = t_{\alpha|\sigma}{}^\lambda P^\lambda_{k|\nu}(X) + Y^\Lambda_{\alpha|\sigma} \partial_\Lambda \mathcal{W} \tag{8.2.10}$$

where $t_{\alpha|\sigma}{}^\lambda$ is a coefficient that depends on the parameters $\psi^\alpha$ but not on the coordinates $X^\Lambda$. Substituting Eq.(8.2.10) into the first term appearing on the r.h.s. of (8.2.9), we obtain a first contribution to the derivative $\frac{\partial}{\partial \psi^\alpha} \Pi^i_{k,\sigma}$ of the period that is proportional to the same period

$$t_{\alpha|\sigma}{}^{\lambda}\,\Pi^i_{k,\lambda} \tag{8.2.11}$$

The second contribution, coming from the trivial part $Y^\Lambda_{\alpha|\sigma}\,\partial_\Lambda \mathcal{W}$ can be reduced using Eq.(5.9.15). Indeed, under integration on cycles, we can discard exact forms so that, by partial integration, we obtain a contribution to the derivative $\frac{\partial}{\partial \psi^\alpha} \Pi^i_{k,\sigma}$ of the form

$$\frac{1}{k}\int_{C_i} \frac{\partial_\Lambda Y^\Lambda_{\alpha|\sigma}}{\mathcal{W}^k}\,\omega \tag{8.2.12}$$

Such a term has a pole at $\mathcal{W}=0$ not stronger than $k$ so that it is a linear combination of the periods $\Pi^i_{j,\sigma}$ from $j=0$ up to $j=k-1$. Indeed the same argument used before can be utilized for $\partial_\Lambda Y^\Lambda_{\alpha|\sigma}$: it is a polynomial of degree $(k-1)|\nu$ and, in full generality, it is composed of a non-trivial plus a trivial part. We can now apply the same kind of manipulation to the second term on the right hand side of Eq.(8.2.9). At $\mathcal{W}=0$ it has at most a pole of order $k+2$, so that it contributes a linear combination of the periods $\Pi^i_{j,\sigma}$ from $j=0$ up to $j=k+1$.

This discussion has proved that the periods do indeed satisfy a first order linear differential system with the structure (8.2.8). It has also outlined the algorithm by means of which the matrix $\mathbf{A}_\alpha(\psi)$ can be explicitly computed in terms of the polynomial (8.2.1) and of its associated chiral ring. Next we illustrate the procedure by means of a specific example.

### 8.2.1 Picard–Fuchs equations for the cubic torus

We consider the case $p=3$ in the one-modulus family of Calabi-Yau $(p-2)$-folds $M\mathbb{CP}_{p-1}[p]$ defined as the vanishing loci of the superpotential (5.9.5). In this case, for notational convenience we rename the three homogeneous coordinates of $\mathbb{CP}_2$ as $x, y, z$ and we get

$$\mathcal{W}(x,y,z;\psi) = \frac{1}{3}\left(x^3+y^3+z^3\right) - \psi\, x\, y\, z \tag{8.2.13}$$

The cubic 1-fold $M\mathbb{CP}_2[3]$ is a torus and the coefficient $\psi$ appearing in (8.2.13) is a coordinate parametrizing the torus moduli space $\mathcal{M}^{(torus)}_\tau$, defined by

$$\begin{aligned} \mathcal{M}^{(torus)}_\tau &= \frac{\mathbb{H}}{PSL(2,\mathbb{Z})} & \mathbb{H} &= \{\tau \in \mathbb{C}\,|\,\mathrm{Im}\,\tau > 0\} \\ \forall\gamma &= \pm\begin{pmatrix} a & b \\ c & d\end{pmatrix} \in PSL(2,\mathbb{Z}) & \gamma\cdot\tau &= \frac{a\tau+b}{c\tau+d} \end{aligned} \tag{8.2.14}$$

The relation between the modulus $\tau$ and $\psi$ is highly complicated and it will be uncovered by the study of the Picard–Fuchs equation; furthermore, in a later section, by studying the monodromy properties of the same equation we shall illustrate how the modular group

$PSL(2,\mathbb{Z})$ emerges analytically. Here we begin by observing that the Hodge diamond of the torus is given by

$$\begin{array}{ccc} & h^{(0,0)} & \\ h^{(1,0)} & & h^{(0,1)} \\ & h^{(1,1)} & \end{array} = \begin{array}{ccc} & 1 & \\ 1 & & 1 \\ & 1 & \end{array} \quad (8.2.15)$$

so that, recalling Eq.(5.9.21), for the $U(1)$ integer-charged sectors of the chiral ring we expect the following dimensions:

$$\dim \mathcal{R}_{0|3} = 1 \qquad \dim \mathcal{R}_{1|3} = 1 \quad (8.2.16)$$

Indeed the vanishing relations associated with the superpotential (8.2.13) are

$$\begin{aligned} x^2 &= \psi\, y\, z + \partial_x \mathcal{W} \\ y^2 &= \psi\, x\, z + \partial_y \mathcal{W} \\ z^2 &= \psi\, x\, y + \partial_z \mathcal{W} \end{aligned} \quad (8.2.17)$$

and the full chiral ring, which is composed of the eight elements

$$\begin{aligned} P_{0|3} &= 1 \\ P^x_{\frac{1}{3}|3} &= x \quad P^y_{\frac{1}{3}|3} = y \quad P^z_{\frac{1}{3}|3} = z \\ P^x_{\frac{2}{3}|3} &= y\, z \quad P^y_{\frac{2}{3}|3} = x\, z \quad P^z_{\frac{2}{3}|3} = x\, y \\ P_{1|3} &= x\, y\, z \end{aligned} \quad (8.2.18)$$

just contains two integer-charged polynomials: the identity $P_{0|3} = 1$ and the top-element $P_{1|3} = x\,y\,z$, which, in this case, is also a marginal operator, namely it has the same degree as $\mathcal{W}$. The first Betti number is $b^2 = h^{(1,0)} + h^{(0,1)} = 2$ and a basis of homology 1-cycles is composed of the well known $a$-cycle and $b$-cycle. Utilizing the Griffiths residue map the periods can be represented in the following way:

$$\begin{aligned} \mathbf{\Pi}^{(a)} &= \begin{pmatrix} \Pi^{(a)}_1 \\ \Pi^{(a)}_2 \end{pmatrix} = \begin{pmatrix} \int_{\mathcal{A}} \frac{1}{\mathcal{W}}\, \omega \\ \int_{\mathcal{A}} \frac{x\,y\,z}{\mathcal{W}^2}\, \omega \end{pmatrix} \\ \mathbf{\Pi}^{(b)} &= \begin{pmatrix} \Pi^{(b)}_1 \\ \Pi^{(b)}_2 \end{pmatrix} = \begin{pmatrix} \int_{\mathcal{B}} \frac{1}{\mathcal{W}}\, \omega \\ \int_{\mathcal{B}} \frac{x\,y\,z}{\mathcal{W}^2}\, \omega \end{pmatrix} \end{aligned} \quad (8.2.19)$$

where the explicit form of the volume element (5.9.10) is

$$\omega = 2\,(\, x\, dy \wedge dz + y\, dz \wedge dx + z dx \wedge dy\,) \quad (8.2.20)$$

Taking the derivative of the period vector we obtain

$$\frac{\partial}{\partial \psi} \Pi_1 = \int \frac{x\,y\,z}{\mathcal{W}^2} = \Pi_2 \quad \frac{\partial}{\partial \psi} \Pi_2 = 2 \int \frac{x^2\, y^2\, z^2}{\mathcal{W}^3} \quad (8.2.21)$$

To obtain the first order differential system we need to reduce the polynomial $x^2 y^2 z^2$. We have

$$\begin{aligned} x^2 y^2 z^2 &= x^2 z^2 (\psi x z + \partial_y \mathcal{W}) = \psi x^3 z^3 + x^2 z^2 \partial_y \mathcal{W} \\ &= x^2 z^2 \partial_y \mathcal{W} + \psi z^3 x (\psi y z + \partial_x \mathcal{W}) \\ &= \psi^2 x y z^4 + \psi x z^3 \partial_x \mathcal{W} + x^2 z^2 \partial_y \mathcal{W} \\ &= \psi^2 x y z^2 (\psi x y + \partial_z \mathcal{W}) + \psi x z^3 \partial_x \mathcal{W} + x^2 z^2 \partial_y \mathcal{W} \\ &= \psi^3 x^2 y^2 z^2 + \left( \psi^2 x y z^2 \partial_z \mathcal{W} + \psi x z^3 \partial_x \mathcal{W} + x^2 z^2 \partial_y \mathcal{W} \right) \end{aligned} \quad (8.2.22)$$

so that we can write

$$\begin{aligned} \left(1 - \psi^3\right) x^2 y^2 z^2 &= Y^\Lambda \partial_\Lambda \mathcal{W} \\ Y^\Lambda &= \left( \psi x z^3 , x^2 z^2 , \psi^2 x y z^2 \right) \end{aligned} \quad (8.2.23)$$

Utilizing Eq.(8.2.23) in Eq.(5.9.15) we obtain

$$\begin{aligned} 2 \int \frac{x^2 y^2 z^2}{\mathcal{W}^3} &= \frac{1}{1-\psi^3} \int \frac{\partial_\Lambda Y^\Lambda}{\mathcal{W}^2} \omega \\ &= \frac{\psi}{1-\psi^3} \int \frac{z^3}{\mathcal{W}^2} \omega + \frac{2\psi^2}{1-\psi^3} \int \frac{x y z}{\mathcal{W}^2} \omega \end{aligned} \quad (8.2.24)$$

and with a further reduction of $z^3 = \psi x y z + z \partial_z \mathcal{W}$ that yields

$$\begin{aligned} \int \frac{z^3}{\mathcal{W}^2} \omega &= \psi \int \frac{x y z}{\mathcal{W}^2} \omega + \int \frac{z \partial_z \mathcal{W}}{\mathcal{W}^2} \omega \\ &= \psi \int \frac{x y z}{\mathcal{W}^2} \omega + \int \frac{1}{\mathcal{W}} \omega \\ &= \psi \Pi_2 + \Pi_1 \end{aligned} \quad (8.2.25)$$

we finally obtain

$$\frac{\partial}{\partial \psi} \begin{pmatrix} \Pi_1^{(a)} \\ \Pi_2^{(a)} \end{pmatrix} = \begin{pmatrix} 0 & 1 \\ \frac{\psi}{1-\psi^3} & \frac{3\psi^3}{1-\psi^3} \end{pmatrix} \begin{pmatrix} \Pi_1^{(a)} \\ \Pi_2^{(a)} \end{pmatrix} \quad (8.2.26)$$

which is the explicit form of Eq.(8.2.7) for the cubic torus case. As a consequence of Eq.(8.2.26) the period $\Pi_1$ satisfies a second order differential equation:

$$\left(1 - \psi^3\right) \frac{\partial^2 \Pi}{\partial \psi^2} - 3\psi^2 \frac{\partial \Pi}{\partial \psi} - \psi \Pi = 0 \quad (8.2.27)$$

which, as we will later show, is equivalent to a hypergeometric equation. Notice that Eq. (8.2.27) is the same as Eq. (7.9.62), which establishes the flatness of the metric in the coupling constant space.

### 8.2.2 Picard–Fuchs equation for the one-modulus $\mathbb{MCP}_{p-1}[p]$ hypersurfaces, and its singularity structure

Utilizing the above technique we can generalize our previous result to the case of a generic value of $p$ in Eq.(5.9.5). Setting

$$n = p - 2 \tag{8.2.28}$$

we define the period matrix:

$$\Pi^i_j = \int_{A_i} \Omega^{(n-j,j)} \tag{8.2.29}$$

whose rows are the periods of the harmonic $(n-j,j)$-forms ($j = 0,\dots,n$) along a symplectic basis of $n$-cycles $A_i$ ($i = 0,..,p-1$). The columns of the period matrix satisfy a first order linear system of type (8.2.7), whose explicit structure is worked out just in the same way as in the particular $p = 3$ case. The system has a natural Drinf'eld-Sokholov form [128]. We recall its expression, which will provide the starting point for our analysis of the duality and monodromy groups. We have

$$(1-\psi^p)\frac{\partial}{\partial\psi}\begin{pmatrix}\Pi^0_i\\ \Pi^1_i\\ \dots\\ \dots\\ \dots\\ \Pi^n_i\end{pmatrix} = \tag{8.2.30}$$

$$= \begin{pmatrix} 0 & 1 & 0 & \dots & 0 & 0 \\ 0 & 0 & 1 & \dots & 0 & 0 \\ \dots & \dots & \dots & \dots & \dots & \dots \\ \dots & \dots & \dots & \dots & \dots & \dots \\ 0 & 0 & 0 & \dots & 0 & 1 \\ \psi\, b_1^{(p)} & \psi^2\, b_2^{(p)} & \psi^3\, b_3^{(p)} & \dots & \psi^{p-2}\, b_{p-2}^{(p)} & \psi^{p-1}\, b_{p-1}^{(p)} \end{pmatrix}\begin{pmatrix}\Pi^0_i\\ \Pi^1_i\\ \dots\\ \dots\\ \dots\\ \Pi^n_i\end{pmatrix} \tag{8.2.31}$$

where the coefficients $b^p_l$ are determined by the following recursion relation:

$$\begin{aligned} b_l^{(p)} &= l\, b_l^{(p-1)} + b_{l-1}^{(p-1)} \qquad (l = 1,\dots,p-2) \\ b_1^{(p)} &= 1 \quad b_{p-1}^{(p)} = \frac{1}{2}p(p-1) \end{aligned} \tag{8.2.32}$$

From Eq.s (8.2.30), by iterative substitution, one very easily obtains the Picard–Fuchs equation satisfied by the periods $\Pi^i_0$ of the holomorphic $n$-form. Naming these periods generically $\Pi$, we get

$$(1-\psi^p)\frac{\partial^{p-1}}{\partial\psi^{p-1}}\Pi - \sum_{l=1}^{p-1} b_l^{(p)}\psi^l\frac{\partial^{l-1}}{\partial\psi^{l-1}}\Pi = 0 \tag{8.2.33}$$

Given a complete basis of $p-1$ linearly independent solutions of the above $(p-1)$-th order equation, one can always choose a homology basis for the $(p-1)$-cycles such that the given solutions correspond to the actual periods of $\Omega^{n,0}$. In particular, studying the action of the monodromy and duality group on the solutions of (8.2.33) we shall be able to find bases where this action is represented by integer symplectic matrices, namely by elements of $Sp(p-1,\mathbb{Z})$ in the $p = odd$ case and by integer orthogonal matrices (elements of $SO(p-1,\mathbb{Z})$ ) in the $p = even$ case. Such solution bases correspond to canonical homology bases, where the intersection matrix of the $(p-1)$-cycles is the standard symplectic metric $\mathbf{C} = \begin{pmatrix} 0 & \mathbb{1} \\ -\mathbb{1} & 0 \end{pmatrix}$ or the standard orthogonal metric $\mathbf{O} = \begin{pmatrix} 1 & 0 & 0 \\ 0 & 0 & \mathbb{1} \\ 0 & \mathbb{1} & 0 \end{pmatrix}$, depending on the parity of $p$. Furthermore, when the periods $\Pi_i^0$ $(i = 1,...,p-1)$ of the form $\Omega^{(n,0)}$ are known, those of the other forms $\Omega^{n-j,j}$ are immediately retrieved by setting

$$\Pi_i^j = \frac{\partial^j}{\partial\psi^j}\Pi_i^0 \tag{8.2.34}$$

This is just a trivial consequence of the general properties of the linear systems of first order differential equations. The two cases we are mostly interested in are $p = 3$ and $p = 5$ corresponding, respectively, to the cubic torus, as we have already pointed out and to the mirror quintic. In these examples the Picard–Fuchs equation (8.2.33) takes the form (8.2.27) for the torus and the form

$$\left(1-\psi^5\right)\frac{\partial^4\Pi}{\partial\psi^4} - 10\psi^4\frac{\partial^3\Pi}{\partial\psi^3} - 25\psi^3\frac{\partial^2\Pi}{\partial\psi^2} - 15\psi^2\frac{\partial\Pi}{\partial\psi} - \psi\Pi = 0 \tag{8.2.35}$$

for the mirror quintic. In our analysis of the duality and monodromy groups, we shall occasionaly make use of (8.2.27) or (8.2.35), but most of our results will hold for the general $p$-case. Indeed we can proceed a long way without specializing the number $p$. The regularity in $p$ can be made manifest, by performing the following coordinate transformation:

$$\frac{1}{z} = \psi^p \tag{8.2.36}$$

and by rescaling the period:

$$\overline{\Pi} = z^{1/p}\,\Pi \tag{8.2.37}$$

Upon these substitutions the Picard–Fuchs equation (8.2.33) can be rewritten in the following form:

$$\left[\left(z\frac{d}{dz}\right)^{p-1} - z\left(z\frac{d}{dz}+\frac{1}{p}\right)\left(z\frac{d}{dz}+\frac{2}{p}\right)\cdots\left(z\frac{d}{dz}+\frac{p-1}{p}\right)\right]\overline{\Pi} = 0 \tag{8.2.38}$$

which, in the $z$-variable, is a generalized hypergeometric equation of parameters

$$\left(\frac{1}{p},\frac{2}{p},\ldots,\frac{p-1}{p};1,\ldots,1\right). \tag{8.2.39}$$

The realization of this fact provides an immediate tool to obtain, in terms of a generalized hypergeometric series, a solution of the original equation that is regular in a neighbourhood of $\psi = \infty$. We name such a solution $\Pi^0(\psi)$ and we have:

$$\begin{aligned} \Pi^0(\psi) &= \frac{1}{\psi}\overline{\Pi}^0(\psi) \\ &= \frac{1}{\psi}\,{}_{(p-1)}F_{(p-2)}\left[\frac{1}{p},\frac{2}{p},\ldots,\frac{p-1}{p};1,\ldots,1;\frac{1}{\psi^p}\right] \end{aligned} \tag{8.2.40}$$

The explicit expression of this particular solution will be the starting point for our calculation of the monodromy and duality groups. In order to provide the tools for such an analysis we pause for a moment to discuss the singularity structure and the indicial equations of the Picard–Fuchs equation.

It is just evident from its very form that the $(p-1)$-th order differential equation (8.2.33) has $p+1$ fuchsian singular points located in

$$\begin{aligned} \psi &= (\alpha_p)^r \quad (r = 0,1,\ldots,p-1) \\ \psi &= \infty \end{aligned} \tag{8.2.41}$$

where

$$\alpha_p = \exp[2\pi i/p] \tag{8.2.42}$$

is a $p$-th root of the unity. Indeed each of these points is a simple pole for the equation coefficients. It is also a simple matter to write down the indicial equation corresponding to each of these singular points. From the general theory of differential equations with fuchsian singularities [149, 139], we know that the indices of the solutions, in the neighbourhood of each singular point $\psi_0$, are the roots of the following order $p-1$ algebraic equation:

$$r(r-1)(r-2)\ldots\ldots(r-p+2) + \sum_{l=0}^{p-2} A_0^{(l)}\, r(r-1)\ldots\ldots(r-l+1) = 0 \tag{8.2.43}$$

where

$$A_0^{(l)} = \lim_{\psi\to\psi_0} (\psi-\psi_0)^{p-1-l}\, A^l(\psi) \tag{8.2.44}$$

$A^l(\psi)$ being the coefficients of the equation. In our case we have

$$A^l(\psi) = \frac{\psi^{l+1}}{1-\psi^p}\left(-b^{(p)}_{(l+1)}\right) \tag{8.2.45}$$

so that, for the singular points $\psi = \alpha_p^r$, the only non vanishing coefficient $A_0^{(l)}$ is

$$A_0^{(p-2)} = -\frac{1}{p}\left(-b^{(p)}_{p-1}\right) = \frac{1}{2}(p-1) \tag{8.2.46}$$

the last identity being a consequence of Eq.s (8.2.32). Using this result we conclude that in each singular point $\psi = \alpha_p^r$ the indices are the same and are given by the roots of the indicial equation:

$$r\,(r-1)\,(r-2)\ldots(r-p+3)\left(r-\frac{p-3}{2}\right) = 0 \tag{8.2.47}$$

A similar result can be obtained for the indicial equation in the neighbourhood of $\psi = \infty$. From the above result we realize the different structure of the space of solution in the $p = odd$ and $p = even$ cases. Indeed for $p = odd$ in each of the singular points $\psi = \alpha_p^r$ the indicial equation has integer-valued roots and one of them is degenerate. This guarantees that in the neighbourhood of each of these points there is just one regular solution while the others have a logarithmic singularity. No polydromy of finite order occurs, since there are no fractional indices. In the $p = even$ case, on the other hand, there are no repeated roots of the indicial equation, but one of them is fractional. Hence, in this case we do get solutions with a polydromy of finite order.

### 8.2.3 Perspective

Let us summarize the results of the above discussion. We have seen how the periods of the holomorphic $n$-form on a parametrized family of Calabi–Yau $n$-folds can be calculated, as functions of the moduli parameters, by solving a linear partial differential equation of order $2+2m$ ($m$ being defined in Eq.(8.2.6)). The $2+2m$ linearly independent solutions of this equation can be made to correspond to the periods along a basis of homology $n$-cycles and one can arrange things in such a way that the solutions correspond to an integral homology basis. The solution of this geometrical problem has a correspondence with the solution of the problem of finding "flat coordinates" in topological field theory. This can be seen quite clearly by comparing the Picard–Fuchs equations for the cubic torus and the equation for the flattening factor defined by

$$\phi_8 = \frac{d\psi}{d\tau}\,x\,y\,z \tag{8.2.48}$$

which is found in the analysis of the topological Landau–Ginzburg model with unperturbed superpotential equal to[1]

$$\mathcal{W}_0 = \frac{1}{3}\left(x^3 + y^3 + z^3\right) \tag{8.2.49}$$

Recalling Eq.(7.9.64) the relation between the flat coordinate $\tau$ and the parameter $\psi$ appearing in the perturbed superpotential is given by the schwarzian differential equation:

$$\{\psi;\tau\} = \frac{1}{2}\frac{(8+\psi^3)}{(1-\psi^3)^2}\,\psi\left(\frac{d\psi}{d\tau}\right)^2 \tag{8.2.50}$$

[1]For notational convenience we rename here the coordinates $s, t$ appearing in section 7.9.1, by $\psi, \tau$. This is done to avoid any confusion in the formal definition of these coordinates. Here $\psi$ parametrizes a hypersurface and $\tau$ will be defined as the ratio of the periods. Howewer, the relation between $\psi, \tau$ is the same as the one between $t$ and $s$ and we can identify them.

It is easy to check that Eq.(8.2.50) is the non-linear equation satisfied by the ratio

$$\tau = \frac{\Pi_1^{(a)}}{\Pi_1^{(b)}} \tag{8.2.51}$$

of any two independent solutions of the Picard–Fuchs equation (8.2.27).

In more general terms it can be shown that the connection matrix $\mathbf{A}_\alpha(\psi)$ appearing in the first order Picard–Fuchs differential system is the sum of the flat topological connection plus a term containing the structure constants of the chiral ring. We do not show this in the general case but, in the next section, we prove it for the case of the Picard–Fuchs equations associated with Calabi–Yau 3-folds. Indeed for $n = 3$ the relation between the problem of calculating periods and the problem of finding flat coordinates becomes particularly evident. Recalling Eq.s (6.2.21) the periods of $\Omega^{(3,0)}$ can be identified with a system of special coordinates on the complex structure moduli space, by setting

$$t^a = \frac{X^a}{X^0} = \frac{\int_{A_a} \Omega^{(3,0)}}{\int_{A_0} \Omega^{(3,0)}} \tag{8.2.52}$$

and in these coordinates the three-point correlation function is the third derivative of a generating function:

$$W_{abc} = \frac{\partial^3 \mathcal{F}}{\partial t_a \partial t_b \partial t_c} \tag{8.2.53}$$

Hence the *special coordinates* of special geometry are the *flat coordinates* of topological field theory and they can be calculated solving the Picard–Fuchs differential system. It follows that it is worthwhile to discuss the direct relation between Picard–Fuchs equations and special geometry. This is the subject of the next section.

## 8.3 Picard–Fuchs Equations and Special Geometry

### 8.3.1 Introduction and summary

Summarizing our previous analysis, the problem of determining flat coordinates in the marginal parameter space of a B-twisted sigma model on a Calabi–Yau $n$-fold $\mathcal{M}_n$ (or, equivalently, of a B-twisted Landau–Ginzburg model), reduces to the solution of a system of partial linear differential equations, the *Picard-Fuchs equations*, whose order equals $n+1$. This result, as we have seen, is a consequence of the Griffiths residue mapping and of the representation of harmonic $(n-k, k)$-forms as rational meromorphic $(n+1)$-forms in the ambient space where the $n$-fold is defined as a hypersurface. In the case $n = 3$, where the marginal deformation parameters, associated with the $(2,1)$-forms and interpreted as complex structure moduli, span a *special Kähler manifold*, the holomorphic Picard–Fuchs equations can also be directly derived from special Kählerian geometry. This view-point has some advantages since it allows one to write Picard–Fuchs equations also for systems that have not a geometrical interpretation as moduli spaces of Calabi–Yau 3-folds and to perform a deeper analysis of their formal structure. In this section we focus on this viewpoint. More precisely let us start from the fundamental identity characterizing special Kählerian manifolds (3.3.16), which we repeat here for convenience in a slightly modified notation[2]:

$$\begin{aligned} R^{\delta}_{\alpha\overline{\beta}\gamma} &= g_{\alpha\overline{\beta}}\delta^{\delta}_{\gamma} + g_{\gamma\overline{\beta}}\delta^{\delta}_{\alpha} - C_{\alpha\gamma\epsilon}g^{\epsilon\overline{\epsilon}}C_{\overline{\beta}\overline{\delta}\overline{\epsilon}}g^{\delta\overline{\delta}} \\ C_{\alpha\beta\gamma} &= e^{\mathcal{K}}W_{\alpha\beta\gamma}(z) \end{aligned} \tag{8.3.1}$$

In the above relation, $g_{\alpha\overline{\beta}}(z,\overline{z}) = \partial_{\alpha}\partial_{\overline{\beta}}\mathcal{K}(z,\overline{z})$ is the Kähler metric of the special Kähler manifold $SK_n$, $R^{\delta}_{\alpha\overline{\beta}\gamma}$ is its Riemann tensor and $W_{\alpha\beta\gamma} \in \Gamma\left(SK_n\,,\,\otimes^3_{symm}\,T^{(1,0)^*}SK_n \otimes \mathcal{L}\right)$ is a holomorphic section of the third symmetric power of the canonical bundle $T^{(1,0)^*}SK_n$, tensored with the line bundle $\mathcal{L} \xrightarrow{\pi} SK$, whose first Chern class equals the Kähler class. When the special Kähler manifold is interpreted as the moduli space of either the Kähler class or complex structure deformations in a string compactification, the physical interpretation of $W_{\alpha\beta\gamma}$ is that of *Yukawa coupling* of either the $\overline{27}$ or 27 families of fermions. In both cases the Yukawa coupling is a three point topological correlator of marginal deformations. In the case of Kähler structure moduli, it is a three-point correlator of the A-twisted sigma model; namely, recalling Eq.s (7.7.25) and (7.7.31) we have

$$\begin{aligned} W_{\alpha\beta\gamma} &= \langle\, \Theta^{(0)}\left[\omega^{(1,1)}_{\alpha}\right]\, \Theta^{(0)}\left[\omega^{(1,1)}_{\beta}\right]\, \Theta^{(0)}\left[\omega^{(1,1)}_{\gamma}\right] \rangle \\ &= \sum_{\{k_A\}} \prod_{a=1}^{h^{(1,1)}} q_a^{k_a}\, \langle\, \Theta^{(0)}\left[\omega^{(1,1)}_{\alpha}\right]\, \Theta^{(0)}\left[\omega^{(1,1)}_{\beta}\right]\, \Theta^{(0)}\left[\omega^{(1,1)}_{\gamma}\right] \rangle_{(0,\{k_A\})} \end{aligned} \tag{8.3.2}$$

In the second case, the section $W_{\alpha\beta\gamma}$ is a three-point correlator of the B-twisted topological sigma model. Indeed, recalling Eq. (7.7.163), we can write

$$\begin{aligned} W_{\alpha\beta\gamma} &= \int_{\mathcal{M}_3} \Omega^{(3,0)} \wedge \omega^{(2,1)}_{\alpha} \star \omega^{(2,1)}_{\beta} \star \omega^{(2,1)}_{\gamma} \\ &= H^{-1} \cdot P^{\alpha}_{1|\nu} \cdot P^{\beta}_{1|\nu} \cdot P^{\gamma}_{1|\nu} \\ &= \int_{\mathcal{M}_3} \Omega^{(3,0)} \frac{\partial^3}{\partial z^{\alpha}\partial z^{\beta}\partial z^{\gamma}} \Omega^{(3,0)} \end{aligned} \tag{8.3.3}$$

As many times observed, the direct calculation of $W_{\alpha\beta\gamma}$ as a function of the Kähler class moduli $t_a$ appearing in Eq. (8.3.2) through the combination $q_a = \exp[2\pi i t_a]$, is a rather formidable problem. On the other hand, in the second case, $W_{\alpha\beta\gamma}$ is naturally and simply evaluated as a function of the parameters $\psi^{\alpha}$ which appear in the polynomial constraint $\mathcal{W}(\psi\,,X)$, defining the $n$-fold $\mathcal{M}_n$ as a vanishing locus, or, more generally, which parametrize the unique holomorphic 3-form $\Omega^{(3,0)}(\psi)$. As we already pointed out, the main advantage of mirror symmetry is that we can evaluate the Yukawa couplings

[2] **From now on we will not pay any attention to the the indices labelling (2,1) or (1,1) moduli. We will denote them generically by $\alpha$ ($\alpha = 1,\ldots \mathrm{dimSK}_n$).**

of Kähler classes as Yukawa couplings of complex structures of the mirror manifold, provided this latter is identifiable. However in both cases the moduli parameters fill a special Kähler manifold and, independently of the geometric interpretation of either Eq. (8.3.2) or (8.3.3), the relation between Picard–Fuchs and special geometry emerges when we look at Eq.(8.3.1) as to a system af covariant and non-holomorphic differential equations for the Kähler potential $\mathcal{K}$ in terms of the Yukawa couplings $W_{\alpha\beta\gamma}$ regarded as independent *data*. We assume that these data are given in some natural coordinate system $\psi^\alpha$ which can have a geometrical origin as in the interpretation of Eq.(8.3.3) or not, but which, in general, is not a *special coordinate system* for the special manifold $SK$. Indeed, viewed in this way, the structure of special geometry emerges as the general solution of the constraint (8.3.1) for given Yukawa couplings. This solution can be expressed in terms of $n+1$ holomorphic sections $X^\Lambda(z), \Lambda = 0, 1, \ldots, n$ which obey $\partial_{\overline{\alpha}} X^\Lambda = 0$:

$$\mathcal{K} = -\ln i(X^\Lambda \overline{F}_\Lambda - \overline{X}^\Lambda F_\Lambda) \tag{8.3.4}$$

where

$$W_{\alpha\beta\gamma} = \partial_\alpha X^\Lambda \partial_\beta X^\Gamma \partial_\gamma X^\Delta F_{\Lambda\Gamma\Delta} \tag{8.3.5}$$

$F_\Lambda(X) = \frac{\partial F(X)}{\partial X^\Lambda}$, and $F(X)$ is a homogeneous function of $X$ of degree two. We see that all information about $\mathcal{K}$ and $W_{\alpha\beta\gamma}$ are encoded in the holomorphic objects $X^\Lambda(z)$, $F_\Lambda(z)$ and their complex conjugates.

In order to make contact with the Picard–Fuchs differential equations, one has just to observe that Eq.(8.3.1) is entirely equivalent to the following system of non-holomorphic first order equations;

$$\begin{aligned} D_\alpha V &\stackrel{\text{def}}{=} U_\alpha \\ D_\alpha U_\beta &= -iC_{\alpha\beta\gamma} g^{\gamma\overline{\delta}} \overline{U}_{\overline{\delta}} \\ D_\alpha \overline{U}_{\overline{\beta}} &= g_{\alpha\overline{\beta}} \overline{V} \\ D_\alpha \overline{V} &= 0 \end{aligned} \tag{8.3.6}$$

where $V(z) = \left(X^0, X^\alpha(z), \frac{\partial}{\partial X^\alpha} F, -\frac{\partial}{\partial X^0} F\right)$ $(\alpha = 1, \ldots, n)$ and $D_\alpha$ is the Kähler and reparametrization-covariant derivative, namely the covariant derivative on the bundle $TSK \otimes \mathcal{L}$. The derivation of Eq.s (8.3.6) will be performed in a later subsection: at this introductory level we just note that, by successively inserting these equations into each other one can reduce the first order system to a fourth order linear partial differential equation

$$D_\alpha D_\beta (C^{-1\hat{\gamma}})^{\rho\sigma} D_{\hat{\gamma}} D_\sigma V = 0 \tag{8.3.7}$$

(assuming for the moment that the matrix $(C_\alpha)_{\beta\gamma}$ is invertible). Here, $\hat{\gamma}$ is not summed over (in contrast to $\sigma$). Equation (8.3.7) is actually holomorphic, although its building blocks are not, and thus it is the analogue of the Picard–Fuchs equations in special geometry. Its solution determines $X^\Lambda$ and $F_\Lambda$ and thus, via Eq.(8.3.4), also $\mathcal{K}$, in terms of any given Yukawa coupling $W_{\alpha\beta\gamma}$ (see Eq. (8.3.5)). The existence of the covariant holomorphic differential equation (8.3.7) is intimately connected with the fact that the Christoffel as well as the Kähler connection on $SK$ naturally split into the sum of two terms. One of them is non-holomorphic and transforms as a tensor whereas the other term is holomorphic and transforms like a connection. Furthermore, the holomorphic pieces of these connections are flat and vanish in "special coordinates":

$$t^a(z) = \frac{X^a(z)}{X^0(z)} \qquad (a = 1, \ldots, n) \tag{8.3.8}$$

The holomorphic part of the connection, once reduced to the marginal deformations, is just the flat connection on the parameter space of the underlying B-twisted topological sigma model (or topological Landau–Ginzburg model): the special coordinates are, in the language of topological field theory the *flat coordinates*. To prove these facts we will start by showing that Eq. (8.3.7) in one dimension is not the most general linear fourth order differential equation but rather it is characterized by the vanishing of one of the "invariants" of the equation, namely $w_3 = 0$. The other invariant $w_4$ measures the deviation from Yukawa couplings that are covariantly constant with respect to the holomorphic connection. Covariantly constant Yukawa couplings have a special meaning in the case where they are interpreted as in Eq.(8.3.2), namely as A-twisted correlators: they correspond to the first term in the instanton sum $\{k_A = 0\}$, which is a good approximation in the large radius limit, where the Kaluza–Klein result holds true:

$$W_{\alpha\beta\gamma} = \int_{\mathcal{M}_n} \omega_\alpha^{(1,1)} \wedge \omega_\beta^{(1,1)} \wedge \omega_\gamma^{(1,1)} \tag{8.3.9}$$

Hence the invariant $w_4$ measures the deviation from the large radius limit and in some sense represents the effect of the sum over instantons.

Every $N$-th order differential equation is equivalent to a first order matrix equation of the form $(\partial - \mathbf{A})\mathbf{V} = 0$, where the first row of $\mathbf{V}$ is the solution vector $V$ in (8.3.7). We shall prove that $w_3 = 0$ translates into the statement that the gauge potential $\mathbf{A}$ of this matrix equation takes values in the $sp(4)$ Lie algebra. From the point of view of Calabi–Yau manifolds, the number 4 is understood as the dimension of the third cohomology group $H^3_{DR} = H^{(3,0)} \oplus H^{(2,1)} \oplus H^{(1,2)} \oplus H^{(0,3)}$ in the case of one complex structure modulus, namely $\dim_{\mathbb{C}} H^{(2,1)} = 1$. The symplectic group has then an explanation as the group that leaves the homology 3-cycle intersection matrix invariant.

The statement that the gauge potential $\mathbf{A}$ of the matrix equation takes values in the $sp(4)$ Lie algebra is what admits a nice generalization to the case of an $n$-dimensional moduli space. It suffices to require $\mathbf{A} \in \mathbf{sp}(2n+2)$. The symplectic group $Sp(2n+2)$ can again be interpreted geometrically in terms of homology 3-cycles for a 3-fold with $h^{(2,1)} = n$ or more generally in terms of the flat symplectic holomorphic bundle characterizing a generic special manifold. Indeed we shall derive the Picard–Fuchs equations of special geometry for an arbitrary number of moduli. They are a direct consequence of (8.3.5), and are most easily displayed as $n$ coupled first order holomorphic matrix equations

$$(\partial_\alpha - \mathbf{A}_\alpha)\mathbf{V} = 0 \tag{8.3.10}$$

where $\mathbf{A}$ takes values in $sp(2n+2)$. As just pointed out this is the higher dimensional analogue of the vanishing of $w_3$ in one dimension. $\mathbf{A}_\alpha$ is the sum of a matrix $\mathbf{\Gamma}_\alpha$ which contains the flat connection alluded to above plus the structure constants $\mathbb{C}_\alpha$ of a $2n+2$ dimensional chiral ring $\mathcal{R}^{(3)}$. The structure constants are given in terms of the Yukawa couplings $W_{\alpha\beta\gamma}$ and furthermore satisfy $[\mathbb{C}_\alpha, \mathbb{C}_\beta] = 0$, $\mathbb{C}^4 = 0$. Since the connection is symplectic, $\mathbf{V}$ can always be taken as an element of $Sp(2n+2)$. This means that the symplectic structure of special geometry can ultimately be traced back to the identity (8.3.7).

Next we shall display the relationship between Eq.s (8.3.10) and (8.3.6). Equation (8.3.6) can also be written as a first order matrix equation $(\partial_\alpha - \mathcal{A}_\alpha)\mathbf{U} = 0$, although with a non-holomorphic connection $\mathcal{A}$. Equation (8.3.10) corresponds to a gauge where $\overline{\mathcal{A}} = 0, \mathcal{A} = \mathbf{A}, \overline{\partial}\mathbf{A} = 0$.

Having analysed the structure and the meaning of Picard–Fuchs equations from the special geometry point of view, their explicit evaluation for the periods of specific Calabi–Yau $n$-folds will be addressed in subsequent sections.

### 8.3.2 Differential equations and $W$-generators

In this subsection we pause for a moment and recall some general properties of higher order linear differential equations and of their associated first-order system. To begin with we concentrate on the *one-modulus case* where the partial differential equation is replaced by an ordinary differential equation. As anticipated in the introductory section, this will prove advantageous for the study of the general situation.

Thus, let us first briefly review some facts about linear fourth order differential equations. Their general form reads

$$\sum_{n=0}^{4} a_n(z)\,\partial_z^n V = 0 \tag{8.3.11}$$

where the coefficients $a_n$ obey well-defined transformation laws in order to make Eq. (8.3.11) covariant under coordinate changes $z \to \tilde{z}(z)$, $\partial \to \xi^{-1}\partial$, $\xi \equiv \partial\tilde{z}/\partial z$. Not all of the $a_n$ are relevant. First, one can scale out $a_4$, and furthermore drop the coefficient proportional to $a_3$ by means of the redefinition $V \to V e^{-1/4\int \frac{a_3(u)}{a_4(u)}du}$. This puts the differential equation into the form

$$\mathcal{D}V \equiv (\partial^4 + c_2\partial^2 + c_1\partial + c_0)V = 0 \tag{8.3.12}$$

where the new coefficients $c_n$ are combinations of the $a_n$ and their derivatives. In this basis $V$ transforms as a $-3/2$ differential, but the transformation properties of the $c_n$ are not very illuminating. However, one can find combinations of the $c_n$'s and their derivatives which transform like tensors (with the exception of $w_2 \equiv c_2$, which transforms as a connection):

$$w_2 = c_2$$

$$
\begin{aligned}
w_3 &= c_1 - c_2' \\
w_4 &= c_0 - \tfrac{1}{2}c_1' + \tfrac{1}{5}c_2'' - \tfrac{9}{100}c_2^2
\end{aligned} \tag{8.3.13}
$$

A straightforward computation shows

$$
\begin{aligned}
\tilde{w}_2 &= \xi^{-2}[w_2 - 5\{\tilde{z}; z\}] \\
\tilde{w}_3 &= \xi^{-3}w_3 \\
\tilde{w}_4 &= \xi^{-4}w_4
\end{aligned} \tag{8.3.14}
$$

where $\{\tilde{z}; z\} = (\frac{\partial^2\xi}{\xi} - \frac{3}{2}(\frac{\partial\xi}{\xi})^2)$ is the schwarzian derivative. Actually $w_2, w_3, w_4$ form a classical $W_4$-algebra, a fact, however, that we will not make any use of. Using (8.3.12) and (8.3.13) one finds

$$
\mathcal{D}V = [\partial^4 + w_2\partial^2 + (w_3 + w_2')\partial + \tfrac{3}{10}w_2'' + \tfrac{9}{100}{w_2}^2 + \tfrac{1}{2}w_3' + w_4]\,V \tag{8.3.15}
$$

The advantage of rewriting a differential equation in terms of $W$-generators is that this is a convenient way to display the particular properties of the equation in a reparametrization-covariant way. From Eq.(8.3.14) we learn that there is always a coordinate system in which $w_2 = 0$ holds. On the other hand, $w_3$ and $w_4$ do characterize the fourth order differential operator $\mathcal{D}$ in any coordinate frame.

Let us return to special geometry: according to the discussion of the previous section we expect that there is a holomorphic fourth order differential equation that expresses the constraint of special geometry and thus is equivalent to Eq.(8.3.1). This equation is the one-dimensional version of Eq. (8.3.7), whose proof is postponed to later subsections and reads

$$
DD\,W^{-1}\,DDV = 0 \tag{8.3.16}
$$

where $D$ is the Kähler and reparametrization-covariant derivative, and $W$ is the one-dimensional Yukawa coupling. In special coordinates, this equation becomes very simple:

$$
\partial^2\,W^{-1}\,\partial^2\,V = 0 \tag{8.3.17}
$$

Equation (8.3.16) can be written in the form (8.3.11) and one finds that the coefficients are not arbitrary but are related as follows: $a_3 = 2\partial a_4$, $a_4 = W^{-1}, a_1 = \partial a_2 - \frac{1}{2}\partial^2 a_3$. The coefficients $a_2$ and $a_0$ are complicated functions of $W$ and the connections. The above relations translate into the invariant statement

$$
w_3 \equiv 0 \tag{8.3.18}
$$

Furthermore, the other $W$-generators are given (in special coordinates) by

$$
\begin{aligned}
w_2 &= \tfrac{1}{2W^2}(4WW'' - 5W'^2) \\
w_4 &= \tfrac{1}{100W^4}(175W'^4 - 280WW'^2W'' + 49W^2W''^2 + 70W^2W'W''' \\
&\qquad - 10W^3W'''')
\end{aligned} \tag{8.3.19}
$$

Thus, all special geometries in one dimension lead to a fourth order linear differential equation that is characterized by $w_3 = 0$. This is in close relation with the fact that the solution vector $V$ does not consist of four cómpletely independent elements, but rather has a restricted structure. More precisely, by construction four linear independent solutions are given by the components of a vector $V$, which are related to the section of the symplectic $Sp(4)$ bundle $(X^\Lambda(z), F_\Lambda(z))$ via

$$V = \left(X^0, X^1, F_1, -F_0\right) \qquad \Lambda = 0, 1 \tag{8.3.20}$$

with

$$F_\Lambda(z) \;=\; \frac{\partial}{\partial X^\Lambda(z)} F(z) \tag{8.3.21}$$

where $F$ is a homogeneous function of $X$ of degree 2. However the reverse statement is not true: $w_3 = 0$ does not imply that the solution $V$ can always be written in the form (8.3.20). The proof of this statement is explicitly given in [88].

Note that the property (8.3.21) does not uniquely fix $V$. It is implicit in the notion of special Kähler geometry that for symplectic rotations of $V$,

$$\widetilde{V} = V \cdot M \qquad M \in Sp(4) \tag{8.3.22}$$

one has $\widetilde{F}_A = (\partial\widetilde{F}/\partial\widetilde{X}^A)$, where $\widetilde{F}$ is again a homogeneous function of degree 2. Thus, the elements of $V$ are defined only up to this kind of transformations. Of course, generic linear combinations of the four solutions are still solutions of (8.3.16), but for these the special structure of the solutions (which reflects $w_3 = 0$) is not manifest.

Symplectic transformations belonging to $Sp(2n+2, \mathbb{R})$ have a particular meaning in special geometry. They represent changes of special coordinate bases and are exactly those transformations which leave $\mathcal{K}$ form-invariant and consequently do not change any physical quantity. We show in the following sections how this symplectic structure of special geometry is encoded in the differential equations.

We can similarly discuss the properties of $\mathcal{D}$ when in addition we also have

$$w_4 \;=\; 0 \tag{8.3.23}$$

From (8.3.19) it is clear that this applies in particular if $W = const.$ However, $w_4(W) = 0$ is a non-trivial differential equation that possesses other solutions than $W = const.$ One might thus ask about the significance of general solutions of $w_4(W) = 0$ with non-constant Yukawa couplings. *Indeed one can show that the general solution of $w_4(W) = 0$ corresponds to covariantly constant Yukawa couplings.*

If $w_4 = 0$, Eq. (8.3.15) simplifies to

$$\mathcal{D}V = \left(\partial^4 + w_2\partial^2 + w_2'\partial + \tfrac{3}{10}w_2'' + \tfrac{9}{100}w_2^2\right) V \tag{8.3.24}$$

and the four independent solutions are given by $V = \{\theta_1^3, \theta_1^2\theta_2, \theta_1\theta_2^2, \theta_2^3\}$. Here, $\theta_{1,2}$ are the independent solutions of the second order equation,

$$(\partial^2 + \tfrac{1}{10}w_2)\theta_{1,2} = 0 \tag{8.3.25}$$

From the solution one can easily construct the most general symplectic vector $V$ compatible with the position (8.3.20), i.e.

$$\begin{aligned} X^0 &= \theta_1^3 \\ F_0 &= -\tfrac{1}{6}\frac{(X^1)^3}{(X^0)^2} + 2c_{00}X^0 + 2c_{01}X^1 \\ X^1 &= \theta_1^2\theta_2 \\ F_1 &= \tfrac{1}{2}\frac{(X^1)^2}{X^0} + 2c_{01}X^0 + 2c_{11}X^1 \end{aligned} \tag{8.3.26}$$

where $c_{AB}$ are arbitrary constants. Using the homogeneity property $X^A F_A = 2F$ or integrating $F_A$ we find

$$F = \tfrac{1}{6}\frac{(X^1(z))^3}{X^0(z)} + c_{AB}X^A X^B \tag{8.3.27}$$

From this $F$ using (8.3.5) we can compute the Yukawa coupling and find that it is covariantly constant: $\widehat{D}W = 0$. For $c_{AB} = 0$ (8.3.27) is the $F$-function corresponding to the homogeneous moduli space $SU(1,1)/U(1)$ (which satisfies the stronger constraint $DW = 0$). Moreover, it follows from the inhomogeneous transformation behaviour (8.3.14) of $w_2$ that one can always find a "schwarzian" coordinate where $w_2$ vanishes, by solving a schwarzian differential equation, $\{t; z\} = \frac{1}{5}w_2(t)$. Then one has $\theta_1 = 1, \theta_2 = t$ and thus, with $c_{AB} = 0$, one can write the solution vector as follows:

$$V = (1, t, \tfrac{1}{2}d_0 t^2, \tfrac{1}{6}d_0 t^3) \qquad F = \tfrac{1}{6}d_0 t^3 \qquad W = \partial^3 F = d_0 \tag{8.3.28}$$

where $d_0$ is a constant, which in the last section of the present chapter, will be interpreted as the order zero approximation (without instantonic contribution) of the Yukawa couplings. It is clear that $t = X^1/X^0$ is precisely the special coordinate of Eq.(8.3.8) (note that the coincidence of special coordinates with schwarzian coordinates holds only if $w_4 = 0$). There is an analogous group action that preserves the relationship among the solutions of (8.3.24). This group is just the invariance group of the schwarzian derivative, which is $SL(2,\mathbb{R})$: $\theta' = \frac{a\theta+b}{c\theta+d}$ , $ad - bc = 1$. The action on the solutions of (8.3.24) is easily found through the mapping $V = \theta^3$:

$$M = \begin{pmatrix} a^3 & a^2c & ac^2/2 & c^3/6 \\ 3a^2b & 2abc + a^2d & bc^2/2 + acd & c^2d/2 \\ 6ab^2 & 2b^2c + 4abd & 2bcd + ad^2 & cd^2 \\ 6b^3 & 6b^2d & 3bd^2 & d^3 \end{pmatrix} \tag{8.3.29}$$

which is part of $Sp(4,\mathbb{R})$ . Thus, the specific structure of the solutions is unique up to such $SL(2)$ transformations.

Summarizing, the above means that if $w_4 = 0$, the situation for generic $w_2$ is reparametrization-equivalent to $w_2 = 0$, in which case the solutions are given by (8.3.28). This corresponds to a cubic $F$-function and to constant Yukawa coupling $W$. In general coordinates where $w_2$ does not vanish, $W$ is not constant (but covariantly constant with respect to the holomorphic connection).

Thus, for covariantly constant Yukawa couplings the differential equation is essentially reduced to the differential equation of a torus. This is similar to the situation for the $K_3$ surface where the only non-trivial $W$-generator is $w_2$. These cases shall be addressed in a later section where we study the Picard–Fuchs equations associated with the Calabi–Yau $(p-2)$-folds described by the vanishing of the superpotential (5.9.5). The possibility of having non-trivial Yukawa couplings, or $w_4 \neq 0$, is the new ingredient in special geometry. It reflects the possibility of having instanton corrections to the Yukawa coupling $W$. Specifically, it is easy to see from (8.3.17) that in special coordinates the solutions have the general structure

$$V = (1, t, \tfrac{1}{2}d_0 t^2 + \mathcal{O}(t^3), \tfrac{1}{6}d_0 t^3 + \mathcal{O}(t^4)) \tag{8.3.30}$$

where the higher order "instanton" terms arise from a non-trivial $w_4$. Thus, the invariant $w_4$ measures the deviation from $W = const$, which is the large radius limit of the Calabi–Yau moduli space. One can actually check that the contribution of a given rational curve of degree $k$ to the Yukawa couplings corresponds to a "$w_4$-surface", i.e. to a covariantly constant $w_4$ generator. That is, from (8.3.19) one finds that in special coordinates $w_4(W = e^{kt}) = (const)k^4$ (see [88, 90]).

We now turn to another way of understanding the significance of $w_3 = 0$. This allows us to introduce some concepts which nicely generalize to multi-dimensional moduli spaces.

### 8.3.3 Associated first order linear systems

In this section we consider the properties of the first order linear differential system associated with the fourth order ordinary differential equation (8.3.11). In our analysis we mainly follow the mathematical work of Drinf'eld and Sokholov on N-th order differential equations [128], to which we refer the reader for more details, although we try to keep our presentation self-contained. Additional literature that the interested reader might wish to look at in connection with the present topic is an article by Di Francesco et al. [114] and also an article by Balog et al. [27, 28]. Furthermore, it is proper to stress that here, as elsewhere in the present section, our presentation follows closely a review article by Ceresole et al. [89].

The starting point observation is that any linear fourth order differential equation (8.3.11) is, for a suitable choice of the matrix $\mathbf{A}$, equivalent to a first order matrix equation

$$\left[\mathbb{1}\partial - \mathbf{A}\right] \cdot \mathbf{V} = 0 \tag{8.3.31}$$

where $\mathbf{V}$ is a $4 \times 4$ matrix whose first row is $V$, namely the vector of four independent solutions of the differential equation (8.3.11). A matrix of the form

$$\mathbf{A} = \begin{pmatrix} * & 1 & 0 & 0 \\ * & * & 1 & 0 \\ * & * & * & 1 \\ * & * & * & * \end{pmatrix}, \tag{8.3.32}$$

corresponds to a fourth order operator $\mathcal{D}$ with $a_4 = 1$ whereas $tr\mathbf{A} = 0$ leads to $a_3 = 0$. Hence from our previous discussion of the general form of a fourth order differential equation, we conclude that the most general case is associated with a linear system where the connection matrix $\mathbf{A} \in sl(4)$ is in the special linear Lie algebra and has the form (8.3.32) However the corresponding fourth order differential operator $\mathcal{D}$ is left invariant by local gauge transformations acting as $\mathbf{V} \to S^{-1} \cdot \mathbf{V}$ and $\mathbf{A} \to S^{-1}\mathbf{A}S - S^{-1}\partial S$, if $S$ has the form

$$S = \begin{pmatrix} 1 & 0 & 0 & 0 \\ * & 1 & 0 & 0 \\ * & * & 1 & 0 \\ * & * & * & 1 \end{pmatrix} \in N \subset SL(4) \tag{8.3.33}$$

This is just the usual matrix of lower triangular transformations generated by a nilpotent subalgebra of $sl(4)$. The top row of $\mathbf{V}$ corresponds to a highest weight of the $sl(4)$ Lie algebra and thus it is also $N$-invariant (the other rows of $\mathbf{V}$ are gauge dependent). That is, the solutions of (8.3.15) are completely invariant under the local transformations (8.3.33).

Note also that the more general gauge transformations belonging to a Borel subgroup $B$ of $SL(4)$, where

$$S = \begin{pmatrix} * & 0 & 0 & 0 \\ * & * & 0 & 0 \\ * & * & * & 0 \\ * & * & * & * \end{pmatrix} \in B \tag{8.3.34}$$

do not leave $\mathcal{D}$ invariant but induce $a_3 \neq 0$ and $a_4 \neq 1$. However, this just corresponds to a rescaling of the solution $V \to f(z)V$ (and corresponds to an irrelevant Kähler transformation in this context).

Using the gauge freedom we have discussed, we can put the connection in the form

$$\mathbf{A} = \mathbf{A}_w \equiv \begin{pmatrix} 0 & 1 & 0 & 0 \\ -\frac{3}{10}w_2 & 0 & 1 & 0 \\ -\frac{1}{2}w_3 & -\frac{4}{10}w_2 & 0 & 1 \\ -w_4 & -\frac{1}{2}w_3 & -\frac{3}{10}w_2 & 0 \end{pmatrix} \in sl(4,R) \tag{8.3.35}$$

To understand this form, we have to recall some mathematical results on the relation between the so-called *Kostant group* [205] and the theory of $W$-algebras. We begin with the Kostant group. Let $\mathbb{G}$ be a semi-simple Lie algebra with commutation relations written in a canonical Weyl basis:

$$\begin{aligned} [H^i, H^j] &= 0 \\ [H^i, E^\alpha] &= \alpha^i E^\alpha \\ [E^\alpha, E^\beta] &= \begin{cases} N(\alpha,\beta)\, E^\alpha & \text{if } \alpha+\beta \text{ is a root} \\ 0 & \text{if } \alpha+\beta \text{ is not a root} \end{cases} \\ [E^\alpha, E^{-\alpha}] &= \alpha \cdot H \end{aligned} \tag{8.3.36}$$

$\alpha^i$ denoting the components of the roots in the chosen basis $\{H^i\}$ for the Cartan subalgebra $\mathcal{H} \subset \mathbb{G}$. Choosing an arbitrary linear combination $J_- = \sum_{\substack{simple \\ roots\ \alpha}} b_\alpha E_\alpha$ of the step generators associated with the simple roots, provided all coefficients $b_\alpha$ are non-zero, Kostant [205] has shown that one can always find another linear combination $J_+ = \sum_{\substack{simple \\ roots\ \alpha}} c_\alpha(b_\alpha) E_{-\alpha}$ with appropriate coefficients $c_\alpha(b_\alpha)$ such that the operators

$$J_- = \sum_{\substack{simple \\ roots\ \alpha}} b_\alpha E_\alpha \quad J_+ = \sum_{\substack{simple \\ roots\ \alpha}} c_\alpha(b_\alpha) E_{-\alpha} \quad J_0 = \rho_G \cdot H \tag{8.3.37}$$

close the standard $SL(2)$ Lie algebra

$$\begin{aligned} [J_0, J_\pm] &= \pm J_\pm \\ [J_+, J_-] &= -2i\, J_0 \end{aligned} \tag{8.3.38}$$

$\rho_G = \frac{1}{2}\sum_{\alpha>0}\alpha$ being the Weyl vector. Such a "principally embedded" $SL(2) \subset \mathbb{G}$ subalgebra is the *Kostant subalgebra* $Kost \subset \mathbb{G}$ and it generates the Kostant subgroup of the corresponding Lie group. An intriguing property of $Kost$ is that the adjoint of $G$ decomposes under $Kost$ in a very specific manner, namely

$$adj(G) \to \bigoplus r_j \tag{8.3.39}$$

where $r_j$ are representations of $SL(2)$ labelled by spin $j$, and the values of $j$ that appear on the r.h.s. are equal to the exponents of $G$. By definition the exponents of a semi-simple Lie algebra are just the degrees, as polynomial in the generators, of the independent Casimirs of $\mathbb{G}$ minus one. Indeed the Casimirs are just a set of homogeneous invariant polynomials $P^{(j+1)} = d_{A_1 \ldots A_{j+1}} J^{A_1} \ldots J^{A_{j+1}}$ generating the ring of invariant polynomials on the Lie algebra $\mathbb{G}$. For instance in the case of the $SL(N)$ Lie algebra the exponents are equal to $1, 2, \ldots, N-1$. Each of the Casimirs is in one-to-one correspondence with the additional $W$-generators $W_{j+1}(z) = d_{A_1 \ldots A_{j+1}} J^{A_1}(z) \ldots J^{A_{j+1}} + \cdots$ one can add to the stress–energy tensor $T(z) = \frac{1}{k+C_V} d_{AB}\, J^A(z)\, J^B(z)$ when the Sugawara construction based on the Kač-Mody algebra of $\mathbb{G}$ is extended by means of higher order terms to form a so-called $W$-algebra. In view of these observations, one easily sees that the decomposition (8.3.39) corresponds to writing the connection (8.3.35) in terms of $W$-generators of the corresponding $W$-algebra. More precisely, the result shown by Di Francesco Itzykson and Zuber in [114] and discussed also by the authors of [27, 28], is that for an $N$-th order equation obtained from a first order system where the connection matrix is in $\mathbb{G} = SL(N)$, the connection matrix (8.3.35) itself can be rewritten as follows:

$$\mathbf{A}_w = J_- - \sum_{m=1}^{N-1} w_{m+1} (J_+)^m \tag{8.3.40}$$

in terms of the $W$-generators, which are nothing but the invariants (8.3.13) of the higher order differential equation. In the above equation $J_\pm$ are the step generators of the

Kostant group defined in Eq.(8.3.37). The identification (8.3.40) is done up to an irrelevant normalization of the $w_n$ invariants. Summarizing, the mathematical result we have been describing is the following. Each $N$-th order linear $\mathcal{D}$ differential operator $D$ can be described in terms of a set of invariants that transform tensorially under coordinate change as $W$-algebra generators do. At the same time $\mathcal{D}$ is uniquely determined by a first linear system (8.3.31) where the $SL(N)$ Lie-algebra-valued connection $\mathbf{A}$ can be written in terms of the invariants according to Eq. (8.3.40).

The case N=4 is that relevant to our analysis case and the choice (8.3.35) for $\mathbf{A}$ corresponds to the following emebedding of the Kostant group (8.3.37): $b_1 = b_2 = b_3 = 1$, and $c_1 = c_3 = 3/10, c_2 = 4/10$. With these conventions the decomposition (8.3.39) of the adjoint of $SL(4)$ is given by $j = 1,2,3$, which corresponds to $w_2, w_3$ and $w_4$. We noticed above that special geometry implies $w_3 \equiv 0$ and this means that $\mathbf{A}_w$ belongs to a Lie algebra that decomposes as $j = 1,2$ under $Kost$. It follows that this Lie algebra is $sp(4)$. Indeed, recalling that the algebra $sp(n)$ is spanned by matrices $A$ which satisfy $A\,Q + Q\,A^T = 0$, we can immediately see from (8.3.35) that

$$\mathbf{A}_w \in sp(4) \qquad \longleftrightarrow \qquad w_3 \equiv 0 \tag{8.3.41}$$

Above, the symplectic metric $Q$ is given by

$$Q = \begin{pmatrix} & & & 1 \\ & & -\mathbb{1}_n & \\ & \mathbb{1}_n & & \\ -1 & & & \end{pmatrix} \tag{8.3.42}$$

We chose the gauge in (8.3.35) precisely in such a way such that the symplectic structure be eventually manifest. General gauge transformations conjugate the embedding of $sp(4)$ in $sl(4)$, and in general gauges the fact that $\mathbf{A}_w \in sp(4)$ is not obvious. The invariant way to express this fact is to state that $w_3 = 0$ in the gauge invariant scalar equation.

Similarly, if in addition $w_4 = 0$ (which corresponds to a covariantly constant Yukawa coupling), $\mathbf{A}_w$ further reduces to an $SL(2)$ connection. This $SL(2)$ is identical to the Kostant $SL(2)$ subgroup, $Kost$, since according to (8.3.40) the entries labelled by $w_2$ and 1 in (8.3.35) are directly given by the $Kost$ generators $J_+$ and $J_-$. It consists precisely of the transformations (8.3.29) which preserve the non-trivial relationship between the solutions.

### 8.3.4 The flat holomorphic connection of special Kahler manifolds

As we anticipated in the introductory subsection, special Kähler manifolds are equipped with a *canonical flat holomorphic connection* [143]. The geometry defined by such a connection is the natural *flat holomorphic geometry* on the parameter space of the corresponding topological field theory (the B-twisted sigma model). Extracting such a flat connection from the general structure of a special Kählerian manifold is the first step in order to derive the holomorphic Picard–Fuchs equations associated with any special manifold. In the present subsection we do such a step.

We recall that on a Hodge–Kähler manifold $\mathcal{M}$ of complex dimension $\dim_{\mathbb{C}} = n$ the Kählerian metric (see Eq.(2.6.12)) is given by

$$g_{\alpha\overline{\beta}}(z,\overline{z}) = \partial_\alpha \partial_{\overline{\beta}} \mathcal{K}(z,\overline{z}) \tag{8.3.43}$$

where the Kähler potential $\mathcal{K}(z,\overline{z})$ is the logarithm of the fiber metric on a line bundle $\mathcal{L} \xrightarrow{\pi} \mathcal{M}$, whose first Chern class equals the Kähler class of the metric ( see Eq.(2.8.3)). In local terms, according to the discussion of section 3.2.1, this is described by introducing a connection 1-form $\mathcal{Q}$ on the bundle $\mathcal{L}$ defined by

$$\mathcal{Q} = -\frac{i}{2}(\partial_\alpha \mathcal{K} dz^\alpha - \partial_{\overline{\alpha}} \mathcal{K} d\overline{z}^{\overline{\alpha}}) \tag{8.3.44}$$

(see Eq.(3.2.11). Under Kähler transformations, namely the gauge transformations of the $\mathcal{L}$-bundle:

$$\mathcal{K} \longrightarrow \mathcal{K} + f(z) + \overline{f}(\overline{z}) \tag{8.3.45}$$

the metric $g_{\alpha\overline{\beta}}$ is invariant, while $\mathcal{Q}$ transforms as a connection should:

$$\mathcal{Q} \longrightarrow \mathcal{Q} + d(Im f)\,. \tag{8.3.46}$$

The form $\mathcal{Q}$ was named the Kähler connection in section 3.2.1. The Kähler 2–form

$$K = \frac{i}{2\pi} g_{\alpha\overline{\beta}} dz^\alpha \wedge d\overline{z}^{\overline{\beta}} \qquad dK = 0 \tag{8.3.47}$$

is the curvature of the Kähler connection divided by $2\pi$:

$$\frac{1}{2\pi} d\mathcal{Q} = \mathcal{K} \tag{8.3.48}$$

and this is what guarantees the fulfilment of the global equation (2.8.3).

A section $\psi(z,\overline{z})$ of $\mathcal{L}^p \otimes \overline{\mathcal{L}}^{\overline{p}}$ (to which we often refer, with a little abuse of language, as to a section of $\mathcal{L} \otimes \overline{\mathcal{L}}$ with "Kähler weights" $(p,\overline{p})$) is defined by the transformation laws

$$\psi(z,\overline{z}) \longrightarrow \psi(z,\overline{z}) \;\; e^{-\frac{p}{2}f} \;\; e^{-\frac{\overline{p}}{2}\overline{f}} \tag{8.3.49}$$

Accordingly, we define $\mathcal{L} \otimes \overline{\mathcal{L}}$ covariant derivatives by

$$\begin{aligned} D_\alpha \psi &= (\partial_\alpha + \tfrac{p}{2}\partial_\alpha \mathcal{K})\psi \\ D_{\overline{\alpha}} \psi &= (\partial_{\overline{\alpha}} + \tfrac{\overline{p}}{2}\partial_{\overline{\alpha}} \mathcal{K})\psi \end{aligned} \tag{8.3.50}$$

This is just an extension of the formulae (3.3.8), (3.3.9) (3.3.10). There we considered $\mathcal{Q}$ as a connection on the $U(1)$-bundle obtained by imposing the transformation rule

$$\Phi \longrightarrow \Phi \exp[-i\,p\,\mathrm{Im} f\,] \tag{8.3.51}$$

Here we look at $\mathcal{Q}$ as a connection on the complex line bundle $\mathcal{L}$. As we see, a section of the $U(1)$-bundle is a section of $\mathcal{L}\otimes\overline{\mathcal{L}}$, with weights $(p=p\overline{p}=-p)$. A covariantly holomorphic section, satisfying $D_{\overline{\alpha}}\psi=0$, is related to a purely holomorphic field $\widetilde{\psi}$ by

$$\widetilde{\psi}=e^{\frac{\overline{p}}{2}K}\psi \tag{8.3.52}$$

$\widetilde{\psi}$ has weight $(p-\overline{p},0)$ and satisfies $\partial_{\overline{\alpha}}\widetilde{\psi}=0$. The Levi–Civita connections and their curvatures are defined as in Eq.s (2.5.18) and (2.5.19), namely

$$\Gamma^{\alpha}{}_{\beta\gamma}=-g^{\alpha\overline{\delta}}\partial_{\beta}g_{\gamma\overline{\delta}} \qquad R^{\alpha}_{\beta\overline{\gamma}\delta}=\partial_{\overline{\gamma}}\Gamma^{\alpha}_{\beta\delta} \tag{8.3.53}$$

(Analogous formulas hold for the barred quantities $\Gamma^{\overline{\alpha}}_{\overline{\beta\gamma}}$ and $R^{\overline{\alpha}}_{\overline{\beta}\gamma\overline{\delta}}$.) Thus for a vector $\phi_{\alpha}$ of weight $(p,\overline{p})$ the covariant derivatives read

$$\begin{aligned} D_{\alpha}\phi_{\beta} &= (\partial_{\alpha}+\tfrac{p}{2}\partial_{\alpha}K)\phi_{\beta}+\Gamma^{\gamma}_{\alpha\beta}\phi_{\gamma} \\ D_{\overline{\alpha}}\phi_{\beta} &= (\partial_{\overline{\alpha}}+\tfrac{\overline{p}}{2}\partial_{\overline{\alpha}}K)\phi_{\beta} \end{aligned} \tag{8.3.54}$$

Let us now come to Hodge–Kahler manifolds that are also special Kähler manifolds. In this case, recalling the discussion of section 3.3, we have a finer structure expressed by the flat $Sp(2+2n)$-bundle. If special geometry is derived from the coupling of $n$-vector multiplets to N=2 supergravity, as we did in section 3.3, then the $Sp(2+2n)$-structure arises in the following way. One needs a set of $n+1$ sections $L^{\Lambda}(z,\overline{z})\in\Gamma\left(\mathcal{L}\otimes\overline{\mathcal{L}},\mathcal{M}\right)$ of weight $(1,-1)$ (together with their complex conjugates $\overline{L}^{\Lambda}(z,\overline{z})$ of weights $(-1,1)$) satisfying

$$\begin{aligned} D_{\overline{\alpha}}L^{\Lambda} &= 0 \\ D_{\alpha}\overline{L}^{\Lambda} &= 0 \\ D_{\alpha}L^{\Lambda} &= f^{\Lambda}_{\alpha} \\ D_{\overline{\alpha}}\overline{L}^{\Lambda} &= \overline{f}^{\Lambda}_{\overline{\alpha}} \\ D_{\alpha}\overline{f}^{\Lambda}_{\overline{\beta}} &= g_{\alpha\overline{\beta}}\overline{L}^{\Lambda} \\ D_{\overline{\alpha}}f^{\Lambda}_{\beta} &= g_{\alpha\overline{\beta}}L^{\Lambda} \end{aligned} \tag{8.3.55}$$

and a section $C_{\alpha\beta\gamma}\in\Gamma\left(\otimes^{3}_{symm}\mathcal{L},\mathcal{M}\right)$ of weights $(2,-2)$ and completely symmetric in its indices, which together with its complex conjugate $C_{\overline{\alpha\beta\gamma}}$ of weights $(-2,2)$ satisfies

$$\begin{aligned} C_{\overline{\alpha}\beta\gamma} &= C_{\alpha\overline{\beta\gamma}}=0 \\ D_{\alpha}f^{A}_{\beta} &= -iC_{\alpha\beta\gamma}g^{\gamma\overline{\delta}}\overline{f}^{A}_{\overline{\delta}} \\ D_{\overline{\alpha}}\overline{f}^{A}_{\overline{\beta}} &= -iC_{\overline{\alpha\beta\gamma}}g^{\overline{\gamma}\delta}f^{A}_{\delta} \\ D_{\overline{\alpha}}C_{\beta\gamma\delta} &= D_{[\alpha}C_{\beta]\gamma\delta}=0 \\ D_{\alpha}C_{\overline{\beta\gamma\delta}} &= D_{[\overline{\alpha}}C_{\overline{\beta}]\overline{\gamma\delta}}=0 \end{aligned} \tag{8.3.56}$$

Note that the last set of equations in (8.3.55) is just the integrability condition of the second set. Furthermore as an integrability condition of Eq.s (8.3.56) one finds the constraint

$$R_{\overline{\alpha}\beta\overline{\gamma}\delta} = g_{\overline{\alpha}\beta}g_{\delta\overline{\gamma}} + g_{\overline{\alpha}\delta}g_{\beta\overline{\gamma}} - C_{\beta\delta\mu}g^{\mu\overline{\mu}}C_{\overline{\mu}\overline{\alpha}\overline{\gamma}} \tag{8.3.57}$$

which is with one index lowered the fundamental characterizing identity (8.3.1) of special manifolds. From Eq. (8.3.56) we also learn that $C_{\alpha\beta\gamma}$ obeys

$$C_{\alpha\beta\gamma} = D_\alpha D_\beta D_\gamma S \tag{8.3.58}$$

where $S$ has weight $(2,-2)$.

As explained in section 3.3, a special Kähler manifold can be equivalently defined by introducing a three-index symmetric tensor $C_{\alpha\beta\gamma}$ on a Kähler–Hodge manifold with the properties (8.3.56) furthermore restricting the curvature by the constraint (8.3.57). The existence of the sections $L^\Lambda$ and their properties then follow.

The Kähler potential itself is most easily expressed in terms of the holomorphic sections. By using (8.3.52) one defines $X^\Lambda(z)$ and $W_{\alpha\beta\gamma}(z)$ of Kähler weights $(2,0)$ and $(4,0)$ respectively:

$$\begin{aligned}
X^\Lambda(z) &= e^{-\frac{\mathcal{K}}{2}}L^\Lambda(z,\overline{z}) \\
\partial_{\overline{\alpha}}X^\Lambda &= 0 \\
W_{\alpha\beta\gamma}(z) &= e^{-\mathcal{K}}C_{\alpha\beta\gamma}(z,\overline{z}) \\
\partial_{\overline{\alpha}}W_{\beta\gamma\delta} &= 0
\end{aligned} \tag{8.3.59}$$

Then one introduces the functional $F(X^\Lambda)$, which is holomorphic and homogeneous of degree 2 in the $X^\Lambda$:

$$2F = X^\Lambda F_\Lambda(X) \qquad F_\Lambda \equiv \frac{\partial}{\partial X^\Lambda}F \tag{8.3.60}$$

and in terms of $X^\Lambda$ and $F_\Lambda$, we recall the expression for the Kähler potential which solves the constraints (8.3.56) and (8.3.57), namely

$$\mathcal{K}(z,\overline{z}) = -\ln iY \qquad Y = -\overline{X}^\Lambda N_{\Lambda\Gamma}X^\Gamma = X^\Lambda\overline{F}_\Lambda - \overline{X}^\Lambda F_\Lambda \tag{8.3.61}$$

where

$$N_{\Lambda\Sigma} = F_{\Lambda\Sigma}(X) - \overline{F}_{\Lambda\Sigma}(\overline{X}) \qquad F_{\Lambda\Sigma} = \partial_\Lambda\partial_\Sigma F \tag{8.3.62}$$

Furthermore, $C_{\alpha\beta\gamma}$ is given by

$$\begin{aligned}
C_{\alpha\beta\gamma} &= D_\alpha D_\beta D_\gamma S = e^{\mathcal{K}}\partial_\alpha X^\Lambda\partial_\beta X^\Sigma\partial_\gamma X^\Gamma F_{\Lambda\Sigma\Gamma} \\
S &= -\frac{1}{2}e^{\mathcal{K}}X^\Lambda N_{\Lambda\Sigma}X^\Lambda
\end{aligned} \tag{8.3.63}$$

From Eqs. (8.3.55), (8.3.56) and (8.3.59)–(8.3.63) it is straightforward to verify that $X^\Lambda$ and $F_\Lambda$ satisfy the same set of constraints. In this way one retrieves the symplectic structure discussed in section 3.3, introducing the $(2n+2)$-dimensional row vectors

$$V = (X^0, X^\alpha, F_\alpha, -F_0) \qquad (\alpha = 1,\ldots,n) \tag{8.3.64}$$

which constitute a section of the flat $Sp(2+2n)$-bundle. All these formulae have been recalled from section 3.3, in order to note that, by using (8.3.59)-(8.3.63), we are able to rewrite the identities (8.3.55) and (8.3.56) as it follows

$$\begin{aligned} D_\alpha V &= U_\alpha \\ D_\alpha U_\beta &= -iC_{\alpha\beta\gamma}g^{\gamma\bar\delta}\overline{U}_{\bar\delta} \\ D_\alpha \overline{U}_{\bar\beta} &= g_{\alpha\bar\beta}\overline{V} \\ D_\alpha \overline{V} &= 0 \end{aligned} \tag{8.3.65}$$

which are just the set of equations (8.3.6) mentioned in the introductory subsection. Similarly, one derives the constraints including the anti-holomorphic derivative $D_{\bar\alpha}$.

As noticed in section (3.3), the Kähler potential can be expressed in terms of $V$ and $V^\dagger$ as follows:

$$\mathcal{K} = -\ln\left(V(-iQ)V^\dagger\right) \tag{8.3.66}$$

which makes its $Sp(2n+2,\mathbb{R})$ symmetry manifest. Above, $Q$ is the symplectic metric which satisfies $Q^2=-1\,, Q=-Q^T$. The convention that we use for the symplectic matrix in this chapter is expressed by Eq. (8.3.42). Note that the vector $V$ in (8.3.64) is symplectic with respect to this metric.

After all these preliminaries we can finally come to the main point and show that the connection on $\mathcal{L}\otimes T^{(1,0)}\mathcal{M}$, when $\mathcal{M}=SK$ is a special manifold, naturally decomposes into a flat holomorphic part and a non-holomorphic part. This property can be shown by introducing the *special coordinates*

$$t^a(z) = \frac{X^a}{X^0} \tag{8.3.67}$$

In terms of $t^a$ and $X^0$ one finds

$$\begin{aligned} \mathcal{K}_\alpha(z,\bar z) &= \widehat{K}_\alpha(z) + \Delta_\alpha(z,\bar z) \\ -\Gamma^\gamma_{\alpha\beta}(z,\bar z) &= \widehat{\Gamma}^\gamma_{\alpha\beta}(z) + T^\gamma_{\alpha\beta}(z,\bar z) \end{aligned} \tag{8.3.68}$$

where

$$\begin{aligned} \Delta_\alpha(z,\bar z) &= e^a_\alpha(z)K_a(z,\bar z) \equiv e^a_\alpha(z)\frac{\partial}{\partial t^a}K(t(z),\bar t(\bar z)) \\ \widehat{K}_\alpha(z) &= -\partial_\alpha \ln X^0(z) \\ e^a_\alpha(z) &= \partial_\alpha t^a(z) \\ T^\gamma_{\alpha\beta}(z,\bar z) &= e^a_\alpha e^b_\beta \partial_b g_{a\bar d} g^{-1\bar d c} e_c^{-1\gamma} \\ \widehat{\Gamma}^\gamma_{\alpha\beta}(z) &= (\partial_\beta e^a_\alpha)e_a^{-1\gamma} \end{aligned} \tag{8.3.69}$$

The holomorphic objects $\widehat{K}_\alpha$ and $\widehat{\Gamma}^\alpha_{\beta\gamma}$ transform as connections under Kähler and holomorphic reparametrizations respectively; moreover $T^\alpha_{\beta\gamma}$ is a tensor under holomorphic diffeomorphisms and $\mathcal{K}_\alpha$ is Kähker-invariant. As a consequence one can define holomorphic covariant derivatives in analogy with (8.3.54) by

$$\widehat{D}_\alpha \phi_\beta = (\partial_\alpha + \tfrac{p}{2}\partial_\alpha \widehat{K})\phi_\beta - \widehat{\Gamma}^\gamma_{\alpha\beta}\phi_\gamma \tag{8.3.70}$$

As we are going to see, the covariant Picard–Fuchs equations associated with special geometry precisely use this holomorphic derivative.

Moreover, one can verify that $\widehat{\Gamma}$ is a flat connection, i.e. it satisfies

$$\widehat{R}^\gamma_{\delta\alpha\beta} \equiv \partial_\delta \widehat{\Gamma}^\gamma_{\alpha\beta} - \partial_\alpha \widehat{\Gamma}^\gamma_{\delta\beta} + \widehat{\Gamma}^\mu_{\alpha\beta}\widehat{\Gamma}^\gamma_{\mu\delta} - \widehat{\Gamma}^\mu_{\delta\beta}\widehat{\Gamma}^\gamma_{\mu\alpha} = 0 \tag{8.3.71}$$

The holomorphic metric for which $\widehat{\Gamma}$ is a connection reads

$$\widehat{g}_{\alpha\beta} = e^a_\alpha e^b_\beta \eta_{ab} \tag{8.3.72}$$

where $\eta_{ab}$ is a constant (invertible) symmetric matrix. (Note that $\widehat{g}_{\alpha\beta}$ has two holomorphic indices in contrast to the Kähler metric $g_{\alpha\overline{\beta}}$.) We can think of $\widehat{g}_{\alpha\beta}$ as the topological field theory metric given by the correlation function of a marginal operator with its complement to the top element of the underlying chiral ring.

The flat coordinates of this flat metric are exactly the "special coordinates" $t^a = z^\alpha$. In these coordinates we find

$$e^a_\alpha = \delta^a_\alpha \qquad \widehat{\Gamma}^\delta_{\alpha\beta} = 0 \qquad \widehat{g}_{\alpha\beta} = \eta_{\alpha\beta} \tag{8.3.73}$$

(The gauge choice $X^0 = 1$ implies $\widehat{K}_\alpha = 0$.)

In terms of $t^a$ one defines the Kähler invariant function

$$\mathcal{F}(t^a) \equiv (X^0)^{-2} F(X^A) \tag{8.3.74}$$

The Kähler potential can then be expressed as (up to Kähler freedom)

$$\mathcal{K} = -\ln i[2(\mathcal{F} - \overline{\mathcal{F}}) - (\mathcal{F}_a + \overline{\mathcal{F}}_a)(t^a - \overline{t}^a)] \tag{8.3.75}$$

The special coordinates $t^a$ play the double role of flat coordinates for the holomorphic geometry with flat connection $\widehat{\Gamma}$ and of "free-falling frame" coordinates for (non-holomorphic) special geometry. The analogous of local Lorentz transformations in the free falling frame is given in our case by the symplectic transformations that relate equivalent patches of special coordinates.

### 8.3.5 Holomorphic Picard–Fuchs equations for n-dimensional special manifolds

In this section we generalize the previous analysis to the case of an $n$-dimensional special manifold and we discuss the corresponding Picard–Fuchs differential equations. The basic identities of special geometry are given by the system (8.3.6). As we already mentioned in the introductory subsection, if we assume that $(C_\alpha)_{\beta\gamma}$ is invertible then these identities are equivalent to

$$D_\alpha D_\beta (C^{-1\hat{\gamma}})^{\rho\sigma} D_{\hat{\gamma}} D_\sigma V = 0 \tag{8.3.76}$$

where $\hat{\gamma}$ is not summed over.

Since the solution vector $V$ is holomorphic, we expect that in (8.3.76) the non-holomorphic pieces coming from the connections contained in $D$ should cancel, so that (8.3.76) is actually a purely holomorphic identity. We prove below that this is indeed the case by showing that $V$ satisfies manifestly holomorphic identities that are equivalent to (8.3.76). These equations contain only the holomorphic connections $\widehat{\Gamma}$ and $\partial\widehat{K}$ defined in the previous subsection.

Let us choose special coordinates $t^a = X^a/X^0$ and the Kähler gauge $X^0 = 1$, and consider the following set of equations:

$$\begin{aligned} \partial_a V &= V_a \\ \partial_a V_b &= W_{abc} V^c \\ \partial_a V^b &= \delta_a^b V^0 \\ \partial_a V^0 &= 0 \end{aligned} \tag{8.3.77}$$

where $(V, V_a, V^a, V^0)$ are all holomorphic and $W_{abc}$ are the Yukawa couplings in special coordinates. The last two equations of (8.3.77) give

$$\begin{aligned} V^0 &\equiv (\ 0 \quad 0 \quad 0 \quad 1\ ) \\ V^a &\equiv (\ 0 \quad 0 \quad \delta_b^a \quad t^a\ ) \end{aligned} \tag{8.3.78}$$

while the first two are solved by setting

$$\begin{aligned} V &\equiv (1, \quad t^a, \quad \partial_a \mathcal{F}, \quad t^a \partial_a \mathcal{F} - 2\mathcal{F}) \\ V_a &\equiv (0, \quad \delta_a^b, \quad \partial_a \partial_b \mathcal{F}, \quad t^b \partial_a \partial_b \mathcal{F} - \partial_a \mathcal{F}) \end{aligned} \tag{8.3.79}$$

The holomorphic function $\mathcal{F}$ is defined in Eq. (8.3.74) and satisfies (in special coordinates)

$$\partial_a \partial_b \partial_c \mathcal{F} = W_{abc} \tag{8.3.80}$$

This identity is the only non-trivial input in solving the differential equations. The system (8.3.77) can also be written in matrix form,

$$\begin{aligned} (\, \mathbb{1}\partial_a - \mathbf{C}_a \,) \mathbf{V} &= 0 \\ \mathbf{C}_a &= \begin{pmatrix} 0 & \delta_a^c & 0 & 0 \\ 0 & 0 & W_{abc} & 0 \\ 0 & 0 & 0 & \delta_a^b \\ 0 & 0 & 0 & 0 \end{pmatrix} \end{aligned} \tag{8.3.81}$$

and from the above transcription we see that this system is solved by the columns of the following $[(2n+2)\times(2n+2)]$-dimensional matrix:

$$\mathbf{V} = \begin{pmatrix} V \\ V_b \\ V^b \\ V^0 \end{pmatrix} = \begin{pmatrix} 1 & t^a & \partial_a\mathcal{F} & t^a\partial_a\mathcal{F} - 2\mathcal{F} \\ 0 & \delta^a_b & \partial_a\partial_b\mathcal{F} & t^a\partial_a\partial_b\mathcal{F} - \partial_b\mathcal{F} \\ 0 & 0 & \delta^b_a & t^b \\ 0 & 0 & 0 & 1 \end{pmatrix} \tag{8.3.82}$$

From Eqs. (8.3.77), (8.3.59) we can infer the transformation properties of $\mathbf{V}$ under coordinate and Kähler transformation and thus it is straightforward to write down the covariant and holomorphic version of Eqs. (8.3.77):

$$\begin{aligned} \widehat{D}_\alpha V &= V_\alpha \\ \widehat{D}_\alpha V_\beta &= W_{\alpha\beta\gamma}V^\gamma \\ \widehat{D}_\alpha V^\beta &= \delta^\beta_\alpha V^0 \\ \widehat{D}_\alpha V^0 &= 0 \end{aligned} \tag{8.3.83}$$

where $\widehat{D}$ was defined in Eq.(8.3.70) and contains the holomorphic part of the connections given in Eq.(8.3.69). This system can also be written in the form

$$(\mathbb{1}\partial_\alpha - \mathbf{A}_\alpha)\,\mathbf{V} \equiv 0 \qquad \mathbf{V} = \begin{pmatrix} V \\ V_\beta \\ V^\beta \\ V^0 \end{pmatrix} \tag{8.3.84}$$

which contains the holomorphic "connection"

$$\mathbf{A}_\alpha = \begin{pmatrix} -\partial_\alpha\widehat{K} & \delta^\gamma_\alpha & 0 & 0 \\ 0 & (\widehat{\Gamma}_\alpha - \partial_\alpha\widehat{K}\mathbb{1})^\gamma_\beta & (W_\alpha)_{\gamma\beta} & 0 \\ 0 & 0 & (\partial_\alpha\widehat{K}\mathbb{1} - \widehat{\Gamma}_\alpha)^\beta_\gamma & \delta^\beta_\alpha \\ 0 & 0 & 0 & \partial_\alpha\widehat{K} \end{pmatrix} \tag{8.3.85}$$

The general solution of (8.3.84) is just the covariant version of Eq. (8.3.82) and thus corresponds to the columns of the matrix

$$\mathbf{V} = \begin{pmatrix} X^0 & X^a & X^0\, e^\alpha_a\partial_\alpha\mathcal{F} & X^a\, e^\alpha_a\partial_\alpha\mathcal{F} - 2\mathcal{F}X^0 \\ 0 & X^0\, e^a_\beta & X^0 e^\alpha_a\, \widehat{D}_\alpha\, \partial_\beta\mathcal{F} & X^a\, e^\alpha_a\, \widehat{D}_\alpha\, \partial_\beta\mathcal{F} - X^0\, \partial_\beta\mathcal{F} \\ 0 & 0 & (X^0)^{-1}\, e^\alpha_a & (X^0)^{-2}\, X^a\, e^\alpha_a \\ 0 & 0 & 0 & (X^0)^{-1} \end{pmatrix} \tag{8.3.86}$$

Here $e^a_\alpha = \partial_\alpha t^a(z)$, which satisfies $\widehat{D}_\beta e^a_\alpha = 0$. Furthermore, in arbitrary coordinates $\mathcal{F}$ is Kähler invariant and obeys

$$\widehat{D}_\alpha\widehat{D}_\beta\widehat{D}_\gamma\mathcal{F} = (X^0)^{-2}W_{\alpha\beta\gamma} \tag{8.3.87}$$

The system (8.3.83) implies the following manifestly holomorphic equation for $V$:

$$\widehat{D}_\alpha \widehat{D}_\beta (W^{-1})^{\widehat{\gamma}\rho\sigma} \widehat{D}_{\widehat{\gamma}} \widehat{D}_\sigma V = 0 \qquad (8.3.88)$$

Using Eq. (8.3.74) one checks that the first row of (8.3.86) indeed coincides with $V \equiv (X^0, X^\alpha, \partial_\alpha F, -\partial_0 F)$. We conclude, therefore, that Eq. (8.3.88) is the same as Eq. (8.3.76), except that it is written in a manifestly holomorphic way.

Just as it happens in the case of one variable, the correspondence between Eq. (8.3.88) and the linear system (8.3.84) is not unique. Indeed, (8.3.88) is invariant under gauge transformations (up to Kähler transformations) acting on **V** and **A** via

$$S = \begin{pmatrix} *_{1\times 1} & \mathbf{0} & \mathbf{0} & \mathbf{0} \\ * & *_{n\times n} & \mathbf{0} & \mathbf{0} \\ * & * & *_{n\times n} & \mathbf{0} \\ * & * & * & *_{1\times 1} \end{pmatrix} \in B \qquad (8.3.89)$$

which belong to a Borel subgroup $B$ of $SL(2n+2,\mathbb{C})$.

It is easy to check that for one variable, the connection **A** in (8.3.85) can be gauge transformed to the form (8.3.35), which displays the $W$-generators. More precisely, under a symplectic transformation

$$S = \mathrm{diag}(W^{-1/2}, W^{-1/2}, W^{1/2}, W^{1/2}) \qquad (8.3.90)$$

the connection **A** takes the form

$$\mathbf{A} = \begin{pmatrix} -\partial\widetilde{K} & 1 & 0 & 0 \\ 0 & \widehat{\Gamma} - \partial\widetilde{K} & 1 & 0 \\ 0 & 0 & -\widehat{\Gamma} + \partial\widetilde{K} & 1 \\ 0 & 0 & 0 & \partial\widetilde{K} \end{pmatrix} \qquad (8.3.91)$$

where $\widetilde{K} = \widehat{K} + \frac{1}{2}\ln W = -\ln(X^0 W^{-1/2})$. To bring further $\widetilde{K}$ to the gauge (8.3.35) one obviously needs an additional $Sp(4)$ transformation that belongs to the nilpotent subgroup $N$. This transition from (8.3.91) to (8.3.35) is nothing but a Miura transformation (for the definition of such a transformation see the original paper by Drinf'eld and Sokholov [128]).

We saw above that the Picard–Fuchs equations for one variable can be invariantly characterized by the vanishing of some classical $W$-generators. The vanishing of $w_3$ was related to $\mathbf{A}_w \in sp(4)$. For many variables, we do not know how to characterize the differential equation (8.3.88) in terms of covariant quantities like $w_n$. But in analogy to the one-variable equation, we expect that the statement corresponding to $w_3 = 0$ is just the condition of symplecticity on the connection matrix appearing in in (8.3.85), namely

$$\mathbf{A}_\alpha \in sp(2n+2) \qquad (8.3.92)$$

Indeed, the gauge in which we wrote (8.3.85) is manifestly symplectic: one easily verifies that $Q\mathbf{A} = (Q\mathbf{A})^T$, where $Q$ is the symplectic metric given in (8.3.42).

More generally, each possible choice of $W_{\alpha\beta\gamma}$ leads to some specific subalgebra $g \subset sp(2n+2)$ in which the set of connections actually takes values. This subalgebra is an invariant characterization of the multi-variable differential equation. This is just a generalization of what we saw in the one-dimensional case, where the additional vanishing of $w_4$ was equivalent to that $\mathbf{A}_w \in sl(2)$). Obviously, for large $n$, there exist a large number of distinct possible subgroups. (Note that in general, it is not very easy to determine $g$, as the embedding in $sp(2n+2)$ is gauge-dependent and thus not always obvious. For the many variable case one is missing a gauge invariant criterion analogous to the vanishing of certain $W$-generators that applies to the one-dimensional case.) Accordingly the solution vectors can be viewed as representations of $Sp(2n+2)$ (or of some subgroup therein). The set of solution vectors, when it is written as a matrix $\mathbf{V}$, can always be chosen in a such a way that this matrix becomes a group element, by multiplying $\mathbf{V}$ with an appropriate constant matrix from the right. One can easily check that our choice of the solution matrix (8.3.86) is indeed symplectic with respect to the metric (8.3.42). In this way, one can regard $\mathbf{V}$ as a vielbein $V_{\hat{\alpha}}^{\hat{A}}$ with a well-defined symplectic action on both indices ($\hat{A}, \hat{\alpha} = 1, \ldots, 2n+2$). Under coordinate and Kähler transformations $z \to \tilde{z}(z)$, $\mathcal{K} \to \mathcal{K} + f(z) + \overline{f}(\overline{z})$, the matrix $\mathbf{V}$ transforms as follows:

$$V_{\hat{\beta}}{}^{\hat{A}}(\tilde{z}) = S^{-1}{}_{\hat{\beta}}^{\hat{\alpha}}(z) V_{\hat{\alpha}}{}^{\hat{B}}(z) M_{\hat{B}}{}^{\hat{A}} \tag{8.3.93}$$

where $S$ is the symplectic block diagonal matrix

$$S = \begin{pmatrix} e^{-f} & 0 & 0 & 0 \\ 0 & e^{-f}\xi^{-1} & 0 & 0 \\ 0 & 0 & e^{f}\xi & 0 \\ 0 & 0 & 0 & e^{f} \end{pmatrix} (\in B \tag{8.3.94}$$

with $\xi \equiv \xi_{\beta}^{\alpha} = \partial\tilde{z}^{\alpha}/\partial z^{\beta}$. Furthermore, $M$ is a constant matrix that can always be taken as an element of $Sp(2n+2)$. One easily infers form (8.3.85) that these transformations are nothing but gauge transformations of the holomorphic connections $\partial\widehat{K}$ and $\widehat{\Gamma}$:

$$\begin{aligned} \partial_{\alpha}\widehat{K} &\longrightarrow \partial_{\alpha}\widehat{K} + \partial_{\alpha} f \\ \widehat{\Gamma}_{\alpha} &\longrightarrow \xi^{-1}\widehat{\Gamma}_{\alpha}\xi + \partial_{\alpha}\ln\,\xi \end{aligned} \tag{8.3.95}$$

This point of view allows one to understand how global $Sp(2n+2)$ transformations acting on the index $\hat{A}$ induce local frame rotations acting on the index $\hat{\alpha}$: the local rotations are induced by the requirement that $\mathbf{A}_{\alpha}$ should stay in the gauge (8.3.85). More explicitly, symplectic transformations, $\widetilde{\mathbf{V}} = \mathbf{V} \cdot M$, which act in particular on the symplectic section as

$$\begin{aligned} (\widetilde{X}^{A}, \widetilde{F}_{A}(\widetilde{X}^{A})) &= (X^{A}, F_{A}(X^{A})) \cdot M \\ M &= \begin{pmatrix} A & C \\ B & D \end{pmatrix} \in Sp(2n+2) \end{aligned} \tag{8.3.96}$$

induce the following reparametrizations of special coordinates:

$$\tilde{t}^a = \frac{A^a_B X^B + B^{aB} F_B}{A^0_B X^B + B^{0B} F_B}(t) \tag{8.3.97}$$

These reparametrizations induce local, compensating gauge transformations (8.3.95) with $f = \mathrm{Tr}(\ln \xi)$ and $\xi = \partial\tilde{t}^a/\partial t^b$.

Note that the transformations (8.3.94) belong to the part of the Borel gauge group (8.3.89) that is not fixed by the gauge choice (8.3.85); that is, they lie in (the complexification of) the maximal compact subgroup $U(n) \times U(1)$ of $Sp(2n+2)$. This implies that the group element $\mathbf{V}$ can be thought of as an element of $G/H$, where $G \subset Sp(2n+2)$ and $H \subset U(n) \times U(1)$. More specifically, one can decompose

$$\mathbf{A}_\alpha = \mathbb{\Gamma}_\alpha + \mathbb{C}_\alpha \tag{8.3.98}$$

where the diagonal part, $\mathbb{\Gamma}_\alpha$, consists of the connections $\hat{\Gamma}$ and $\partial\widehat{K}$ (which are flattened by special coordinates $t^a = X^a, X^0 = 1$). Furthermore, $\mathbb{C}_\alpha$ is the covariant version of (8.3.81) and generates an abelian, $n$-dimensional subalgebra of $sp(2n+2)$ that is nilpotent of order three: $\mathbb{C}_\alpha\mathbb{C}_\beta\mathbb{C}_\gamma\mathbb{C}_\delta = 0$. Thus, $G$ is determined by the subalgebra of $sp(2n+2)$ in which $\mathbb{C}_\alpha$ takes values, and $H$ is determined by the subgroup of $U(n) \times U(1)$ which is gauged by $\mathbb{\Gamma}_\alpha$ .

More precisely, $\mathbf{V}$ is an element of $G^c/B$ (which is, essentially, isomorphic to $G/H$), where $G^c$ is the complexification of $G$ and $B$ the Borel subgroup (8.3.89), which contains the complexification of $H$. In the generic case, $G/H = Sp(2n+2)/U(n) \times U(1)$, but the moduli space in which $\mathbf{V}(z)$ actually takes values is usually a complicated subvariety of this space. However, there are special cases where $G$ and $H$ are effectively smaller subgroups; some examples are the theories with cubic $F$-function where the moduli spaces are directly given by $G/H$. For instance for $n = 1$, the generic moduli space is some complicated one-dimensional submanifold of $\frac{Sp(4,\mathbb{R})}{U(1)^2}$ whose complex dimension is four. But for constant coupling $W$ (and for $c_{AB} = 0$ in (8.3.27)), the moduli space in which $\mathbf{V}$ takes values is just the homogeneous one-dimensional submanifold $G/H = \frac{SL(2,\mathbb{R})}{U(1)}$.

### 8.3.6 The non-holomorphic Picard–Fuchs equations of special manifolds

In this subsection we establish the relationship between the first order systems (8.3.6) and (8.3.83). Let us first note that the gauge group (8.3.89) can also be extended to non-holomorphic gauge transformations $\mathcal{S} = \mathcal{S}(z, \bar{z})$, which leave $V$ invariant. The point is that eqs. (8.3.6) and (8.3.83) are precisely related by such a non-holomorphic gauge transformation. That is, the non-holomorphicity of the supergravity equations (8.3.1) and (8.3.6) is a gauge artifact, corresponding to the fact that all quantities in special geometry are determined entirely in terms of holomorphic quantities.

More specifically, one can rewrite the non-holomorphic system (8.3.6) in first order form

$$\mathcal{D}_\alpha \mathbf{U} \equiv (\mathbb{1}\partial_\alpha - \mathcal{A}_\alpha)\mathbf{U} = 0 \tag{8.3.99}$$

where $\mathbf{U} = (V, U_\alpha, \overline{U}_{\overline{\alpha}}, \overline{V})^T$ and

$$\mathcal{A}_\alpha = \begin{pmatrix} -\partial_\alpha K & \delta_\alpha^\beta & 0 & 0 \\ 0 & -\delta_\gamma^\beta \partial_\alpha K + \Gamma_{\gamma\alpha}^\beta & -iC_{\alpha\beta\gamma} g^{\gamma\overline{\gamma}} & 0 \\ 0 & 0 & 0 & g_{\alpha\overline{\beta}} \\ 0 & 0 & 0 & 0 \end{pmatrix} \tag{8.3.100}$$

In addition $\mathbf{U}$ also satisfies

$$\mathcal{D}_{\overline{\alpha}} \mathbf{U} \equiv (\mathbb{1}\partial_{\overline{\alpha}} - \mathcal{A}_{\overline{\alpha}})\mathbf{U} = 0 \tag{8.3.101}$$

where

$$\mathcal{A}_{\overline{\alpha}} = \begin{pmatrix} 0 & 0 & 0 & 0 \\ g_{\overline{\alpha}\beta} & 0 & 0 & 0 \\ 0 & iC_{\overline{\alpha}\overline{\beta}\overline{\gamma}} g^{\overline{\gamma}\gamma} & -\delta_{\overline{\gamma}}^{\overline{\beta}} \partial_{\overline{\alpha}} K + \Gamma_{\overline{\gamma}\overline{\alpha}}^{\overline{\beta}} & 0 \\ 0 & 0 & \delta_{\overline{\alpha}}^{\overline{\beta}} & -\partial_{\overline{\alpha}} K \end{pmatrix} \tag{8.3.102}$$

It is easy to verify that as a consequence of (8.3.1) the connections $\mathcal{A}_\alpha$ and $\mathcal{A}_{\overline{\alpha}}$ have vanishing curvature:

$$[\mathcal{D}_\alpha, \mathcal{D}_\beta] = [\mathcal{D}_{\overline{\alpha}}, \mathcal{D}_{\overline{\beta}}] = [\mathcal{D}_\alpha, \mathcal{D}_{\overline{\beta}}] = 0 \tag{8.3.103}$$

It follows that via non-holomorphic transformations $\mathcal{S}(z, \overline{z})$, which leave $V$ invariant ($\mathcal{S}V = V$), one can gauge away $\mathcal{A}_{\overline{\alpha}}$ and make $\mathcal{A}_\alpha$ purely holomorphic. As a consequence of Eq. (8.3.103) one can go to a gauge where

$$\mathcal{A}_{\overline{\alpha}} = \mathcal{S}\partial_{\overline{\alpha}}\mathcal{S}^{-1} \tag{8.3.104}$$

This implies

$$\partial_{\overline{\alpha}}(\mathcal{S}\mathbf{U}) = 0 \quad \text{and} \quad \partial_{\overline{\alpha}}[\mathcal{S}\mathcal{A}_\alpha\mathcal{S}^{-1} - \mathcal{S}\partial_\alpha\mathcal{S}^{-1}] = 0 \tag{8.3.105}$$

so that the non-holomorphic system (8.3.99) becomes the holomorphic system (8.3.84) with

$$\mathbf{A} = \mathcal{S}\mathcal{A}_\alpha\mathcal{S}^{-1} - \mathcal{S}\partial_\alpha\mathcal{S}^{-1} \qquad \mathbf{V} = \mathcal{S}\,\mathbf{U} \tag{8.3.106}$$

which displays a residual gauge symmetry of holomorphic transformations. Of course, one could also have chosen gauge transformations $\overline{\mathcal{S}}$ that leave the lowest row of $\mathbf{V}$, i.e. $\overline{V}$, invariant; in that instance one would have produced a purely anti-holomorphic connection $A_{\overline{\alpha}}$. The point is that there is no invariant subspace with respect to both $\mathcal{S}$ and $\overline{\mathcal{S}}$, so that the connection cannot be completely flattened.

## 8.4 Monodromy and Duality Groups

### 8.4.1 Introduction

So far we have discussed the differential Picard–Fuchs identities satisfied by the periods of the holomorphic $n$-form as a tool to determine local geometrical properties of the complex structure moduli space of a Calabi–Yau $n$-fold. Via mirror symmetry this provides information also on the Kähler class moduli space of the corresponding mirror manifold. Experience with the moduli spaces of Riemann surfaces, however, shows that the global properties of these manifolds are even more important than their local geometrical properties. The moduli space has the generic structure

$$\mathcal{M}_{moduli} = \frac{Teich}{\Gamma_{modular}} \tag{8.4.1}$$

where *Teich* denotes the usually non-compact Teichmuller space, namely the universal covering manifold of the complex structure deformation space and the discrete modular group $\Gamma_{modular}$ is the mapping class group, namely the group of diffeomorphisms of the $n$ fold that are not connected to the identity, modulo the contractible ones. Knowledge of the modular group of the world sheet is of the utmost relevance in string theory since string amplitudes are given by integrals on moduli space and the correct structure of the integrands is determined by imposing modular invariance with respect to $\Gamma_{modular}$. The amplitude integrands are therefore *modular forms* of the discrete, but usually infinite group $\Gamma_{modular}$. A similar role is played by the modular group of the internal target manifold. The effective supergravity lagrangian depends on the moduli fields and on their local geometry through the Kähler metric and the Yukawa couplings related to each other by the identities of special geometry. We have seen how the prepotential $F(X)$ can in principle be calculated once the periods of the holomorphic 3-form are known. Non-perturbative string effects might create an additional *superpotential* interaction for the moduli fields: what is guaranteed is that the invariance of the effective lagrangian with respect to the discrete modular group $\Gamma_{modular}$ cannot be broken at any level. Hence the superpotential must be a *modular form* and any possible additional coupling generated at the non-perturbative level must be $\Gamma_{modular}$-invariant. From these considerations it follows that the determination of the modular group is quite essential also in this context. One of the nice features of the Picard–Fuchs equations is that they provide a tool to study such a question of global geometry through the study of their monodromy group.

Following the policy of the previous chapters, we choose to illustrate the general ideas on the geometry of moduli spaces by utilizing, as a case study, the example of the $M\mathbb{CP}_{p-1}[p]$ hypersurfaces. As already explained, these hypersurfaces are defined as the vanishing locus in $\mathbb{CP}_{p-1}$ of the following Landau–Ginzburg potential (see Eq. (8.1.48)):

$$\mathcal{W}(\psi; X) = \frac{1}{p}\left(\sum_{\Lambda=1}^{p} X_\Lambda^p\right) - \psi \prod_{i=1}^{p} X_\Lambda \tag{8.4.2}$$

further modded by the action of a discrete group $H \sim (\mathbb{Z}_p)^{p-2}$, defined in Eq. (8.1.50). Hence in the present section we analyse the monodromy group $\mathcal{M}on$ of the Picard–Fuchs equation associated with these surfaces, its relation with both the complete modular group $\Gamma$ and with the group of duality transformations $\Gamma_{\mathcal{W}}$. The notion of $\Gamma_{\mathcal{W}}$ will be discussed in the present section and will play a central role in what follows.

## 8.4.2 The duality group $\Gamma_{\mathcal{W}}$ of $\mathbb{M}\mathbb{CP}_{p-1}[p]$ hypersurfaces

To obtain the structure of the monodromy and duality groups associated with the Picard–Fuchs equation (8.2.33), we follow a strategy that has been introduced in the seminal paper by Candelas et al. [69, 68] on the mirror quintic. This method relies precisely on the existence of the logarithmic singularity we have mentioned while discussing the structure of (8.2.33) and it works best in the $p = odd$ case.

By definition, the monodromy group $\mathcal{M}on$ is the discrete group generated by the transvections $\gamma_A$ around all singular points $(A = (\alpha_p)^0, (\alpha_p)^1, \dots, (\alpha_p)^{p-1}, \infty)$. Given a basis $\{\Pi_0(\psi), \Pi_1(\psi), \dots, \Pi_{p-2}(\psi)\}$ of solutions of the differential equation (8.2.33) [3] one considers the analytic continuation of each $\Pi_i(\psi)$ along a closed path $\gamma_A$ encircling the singular point $\psi = \psi_A$. Name $\gamma_A\, \Pi_i(\psi)$ the result of this analytic continuation: it is still a solution of the same differential equation and hence it can be expressed as a linear combination of the solutions $\{\Pi_0(\psi), \Pi_1(\psi), \dots, \Pi_{p-2}\}$, namely

$$\gamma_A\, \Pi_i(\psi) \;=\; (T_A)_i^{\ j}\, \Pi_j(\psi) \tag{8.4.3}$$

This defines a $(p-1) \times (p-1)$ matrix $T_A$ representing the monodromy generator $\gamma_A$. Since the product $\gamma \;=\; \gamma_\infty\, \gamma_{p-1}\, \gamma_{p-3}\, \dots, \gamma_0$ is homotopically equivalent to a path that encircles no singular points, it follows that the monodromy generators satisfy the following basic relation:

$$T_\infty\, T_{p-1}\, T_{p-2}\, \dots, T_0 \;=\; \mathbb{1} \tag{8.4.4}$$

In principle we should find the explicit form of all the $T_A$ by considering the analytic continuation around all the singular points, however, taking into account the additional group $\Gamma_{\mathcal{W}}$ of duality transformations that preserve the Landau–Ginzburg superpotential (8.4.2), our task is greatly simplified and reduces to calculating the monodromy transvection around a single singular point, say $\psi \;=\; \psi_0 \;=\; 1$. Let us see why. Since the Calabi–Yau $n$-fold we are discussing is the vanishing locus, in $\mathbb{CP}_{p-1}$, of the quasi-homogeneous polynomial $\mathcal{W}(\,\psi\,;\,X\,)$, then, by definition, the duality symmetries of $\mathcal{W}$ are those transformations of the modulus parameter $\psi$ that can be reabsorbed by a unitary linear transformation of the projective coordinates $X_\Lambda$ $(\Lambda = 1, \dots, p)$. Indeeed if it happens that, for a certain transformation of the $\psi$-parameter:

$$\psi' \;=\; f\,(\psi) \tag{8.4.5}$$

[3]Note that differently from previous notation the lower index of the periods now denotes the cycle along which it is calculated. Indeed, from now on, we refer only to periods of the holomorphic $n$-form $\Omega^{(n,0)}$, so that the second index of the period matrix, distinguishing the harmonic forms, becomes superfluous.

we can find a unitary $p \times p$ matrix $U(f)_\Lambda{}^\Sigma$, such that the following equation holds true:

$$\mathcal{W}(\, f\,(\psi);\, X\,) \;=\; \mu(f)\,\mathcal{W}(\,\psi,\, U(f)\,X\,) \tag{8.4.6}$$

where $\mu(f)$ is some complex number and $U(f)\,X$ denotes the action of the matrix $U(f)$ on the vector $X$, then the vanishing loci of $\mathcal{W}(\, f\,(\psi);\, X\,)$ and of $\mathcal{W}(\,\psi;\, X\,)$ are just the same hypersurface in $\mathbb{CP}_{p-1}$. This means that we must identify all the points of the $\psi$-complex plane that are related by transformations $f(\psi)$ with the above property. It is also clear that these transformations $f$ form a group, namely $\Gamma_{\mathcal{W}}$. Because of the same argument, any element $g \in \Gamma_{\mathcal{W}}$ must be a symmetry of the Picard–Fuchs equation (8.2.33); in particular the image through $g$ of a singular point must be another singular point. The property that singular points fill up $\Gamma_{\mathcal{W}}$ orbits is very handy in the construction of the monodromy generators. Indeed if we construct a solution basis that has known and simple transformation properties under $\Gamma_{\mathcal{W}}$ then, once we have obtained the transvection generator $T_0$, we can retrieve all the others by a simple conjugation with $\Gamma_{\mathcal{W}}$.

For this reason we begin with an investigation of $\Gamma_{\mathcal{W}}$. To this end we have to be careful with another important subtlety. As already pointed out, the $n$-fold $\mathcal{M}_n$ one is concerned with is not exactly the vanishing locus $\mathcal{W} = 0$, rather, it is an orbifold thereof, namely

$$\mathcal{M}_n \;=\; (\,\mathcal{W} \,=\, 0\,)\;/\,H \tag{8.4.7}$$

where $H$ is a discrete group acting freely on $\mathcal{W} = 0$ (except possibly at some points of the $\psi$-plane) and exactly preserving its form. In the particular case of the $M\mathbb{CP}_{p-1}[p]$-hypersurfaces, the group $H = (\mathbb{Z}_p)^{p-2}$ was defined in Eq. (8.1.50). The quotient by means of $H$ produces a new manifold (the mirror quintic, in the $p = 5$ case) and it is what forces $M\mathbb{CP}_{p-1}$ to have only one modulus. Indeed, with respect to this symmetry, the only marginal deformation of the Fermat curve $\mathcal{W}_F\,(\,X\,) \;=\; 0$ which is invariant is that provided by the operator $\prod_{i=1}^p X_i$. Taking $H$ into account, the elements of the duality group $\Gamma_{\mathcal{W}}$ are further constrained by the consistency requirement:

$$\Gamma_{\mathcal{W}}\, H\, \Gamma_{\mathcal{W}}^{-1} \subset H \tag{8.4.8}$$

In the generic $p$-case the duality group is abelian ($\Gamma_{\mathcal{W}} = \mathbb{Z}_p$) and it just contains the transformations

$$\psi' \,=\, f_j\,(\psi\,) \,=\, (\alpha_p)^j\,\psi \qquad (\,j = 0, 1, \ldots, p-1\,) \tag{8.4.9}$$

These transformations are compensated by the following $U(f_j)$ matrices:

$$U(f_j) \;=\; \mathrm{diag}\left((\alpha_p)^{p-j}\,,\,1\,,\,\ldots\,,1\right) \tag{8.4.10}$$

as one can immediately check by looking at Eq. (8.4.2). These matrices satisfy Eq. (8.4.8) and are therefore good symmetries. Actually there are many other choices of the matrices $U(f_j)$ that could compensate the transformations (8.4.9) but they are equivalent under $H$, and hence they have the same action on $\mathcal{M}_n$. We name $A$ the generator of this duality $\mathbb{Z}_p$ group. Its action on $\psi$ is given by the case $j = 1$ of Eq. (8.4.9). On the solutions of the Picard–Fuchs equation (8.2.33), which, as expected, are indeed invariant against this transformation, the operation $A$ is represented by a suitable $(p-1) \times (p-1)$ matrix $\mathcal{A}$ possessing the property

$$\mathcal{A}^p = 1 \tag{8.4.11}$$

Furthermore, under the action of this duality group, the singular points $\psi_A$ of the Picard–Fuchs equation form two orbits: the first is composed by the p-th roots of the identity $\psi_i \, (i = 0, 1, \ldots, p-1)$, rotated one into the other by $A$; the second contains only $\psi_\infty$, which is $A$-invariant. This fact explains why the indicial equation (8.2.43) is identical in all the finite singular points. Relying on this, if we obtain the explicit form $T_0$ of the transvection around the singular point $\psi = 1$, we can write

$$T_j = \mathcal{A}^{-j} \, T_0 \, \mathcal{A}^j \qquad (j = 1, 2, \ldots, p-1) \tag{8.4.12}$$

which, combined with (8.4.4), also yields

$$T_\infty = (\mathcal{A} T_0)^{-p} \tag{8.4.13}$$

The strategy introduced by Candelas et al. consists in choosing a solution basis where $\mathcal{A}$ is known a priori and then devising an analytic tool for the evaluation of $T_0$. The method works for any value of $p$. However the case $p = 3$ is exceptional since there the duality group is enlarged by an additional generator and becomes non-abelian. Consider the unitary $3 \times 3$ matrix:

$$S = -\frac{i}{\sqrt{3}} \begin{pmatrix} 1 & 1 & 1 \\ 1 & \alpha_3 & \alpha_3^2 \\ 1 & \alpha_3^2 & \alpha_3 \end{pmatrix} \tag{8.4.14}$$

and fix $p = 3$ in Eq. (8.4.2); a straightforward calculation reveals that

$$\mathcal{W}(\psi\,;\, SX) = \mu_S(\psi)\, \mathcal{W}(\mathcal{S}\psi\,;\, X) \tag{8.4.15}$$

where

$$\begin{aligned} \mu_S(\psi) &= \frac{i}{\sqrt{3}} (1 - \psi) \\ \mathcal{S}\psi &= \frac{\psi + 2}{\psi - 1} \end{aligned} \tag{8.4.16}$$

This shows that the transformation $\mathcal{S}$ defined by Eq. (8.4.16) is another generator of $\Gamma_W$, in addition to the transformation $\mathcal{A}\psi = \alpha_3 \psi$. A direct check reveals that these generators satisfy

$$\mathcal{A}^3 = \mathcal{S}^2 = (\mathcal{A}\mathcal{S})^3 = \mathbb{1} \tag{8.4.17}$$

These relations correspond to the abstract presentation of the tetrahedral group $\mathcal{T}$, whose order is $|\mathcal{T}| = 12$. Hence in the $p = 3$ case we have a non-abelian duality group $\Gamma_W = \mathcal{T}$. This fact has far reaching consequences, both for the case of the torus and for the general lesson one is taught by this case. Let us examine what these consequences are through the following sequence of arguments.

*i)* It is well known that the tetrahedral group can be identified with the following quotient:

$$\mathcal{T} = \Gamma / \Gamma\,(3) = PSL(2, \mathbb{Z}_3) \tag{8.4.18}$$

where $\Gamma$ is the modular group, namely the set of $2 \times 2$ integer valued, unimodular matrices, (up to an overall sign):

$$\begin{aligned} \gamma \in \Gamma \longrightarrow \gamma &= \pm \begin{pmatrix} a & b \\ c & d \end{pmatrix} \\ a, b, c, d \in \mathbb{Z} \qquad ad - bc &= 1 \\ \Gamma &= PSL(2,\mathbb{Z}) = \frac{SL(2,\mathbb{Z})}{\begin{pmatrix} \pm 1 & 0 \\ 0 & \pm 1 \end{pmatrix}} \end{aligned} \tag{8.4.19}$$

$\Gamma(3) \subset \Gamma$ is the subgroup of the modular group formed by those integer-valued, unimodular matrices that are equivalent to the identity modulu 3:

$$\begin{aligned} \gamma \in \Gamma(3) \longrightarrow \gamma &= \pm \begin{pmatrix} a & b \\ c & d \end{pmatrix} \\ a, b, c, d \in \mathbb{Z} \qquad ad - bc &= 1 \\ a = \pm 1 + 3k \quad k \in \mathbb{Z}\,;\; b &= 3k\,;\; k \in \mathbb{Z} \\ c = 3k\,;\; k \in \mathbb{Z} \quad d &= \pm 1 + 3k \quad k \in \mathbb{Z} \end{aligned} \tag{8.4.20}$$

and $PSL(2, \mathbb{Z}_3)$ is the modular group constructed on the field $\mathbb{Z}_3$ of integer numbers modulus 3.

*ii)* Now $\Gamma$ is named the *modular group* precisely because it is well known to be the complete symmetry group of the moduli space of the torus. A topological torus $S_1 \otimes S_1$ is described by two real coordinates $\xi_1, \xi_2$ defined modulus 1, $\xi_i \approx \xi_i + 1$. A complex torus $T_\tau = \frac{\mathbb{C}}{\Lambda_\tau}$ can be defined as the complex plane modded by the action of a lattice group, namely by identifying the complex numbers $z$, through the equivalence relation $z \approx m + n\tau$ ($m, n \in \mathbb{Z}$). In this definition $\tau$ is a complex number with positive imaginary part ($Im\tau > 0$) that parametrizes the possible complex structures of the topological torus and therefore provides a coordinatization of its moduli space. The way the complex structures are parametrized by $\tau$ can be made explicit by setting

$$z = \frac{1}{2\pi}\,(\xi_1 + \tau\,\xi_2) \tag{8.4.21}$$

If we deform $\tau$ by an infinitesimal transformation $\tau \longrightarrow \tau + \delta\tau$, there is no way of accounting for this deformation by means of a diffeomorphism $\xi_i = \xi_i^{'}(\xi)$ contractible to the identity. There are, however, two global diffeomorphisms, corresponding to the two

Dehn twists along the $a$ and $b$ cycles

$$\begin{aligned} D_a &: \begin{cases} \xi_1 \longrightarrow \xi_1 + \xi_2 \\ \xi_2 \longrightarrow \xi_2 \end{cases} \\ D_b &: \begin{cases} \xi_1 \longrightarrow \xi_1 \\ \xi_2 \longrightarrow \xi_1 + \xi_2 \end{cases} \end{aligned} \tag{8.4.22}$$

which can be reabsorbed by the following two transformations of the parameter $\tau$:

$$\begin{aligned} D_a &: \tau \longrightarrow \tau + 1 \\ D_b &: \tau \longrightarrow \frac{\tau}{\tau + 1} \end{aligned} \tag{8.4.23}$$

Regarded as elements of $SL(2, \mathbf{R})$ acting projectively on the complex upper plane:

$$\begin{pmatrix} a & b \\ c & d \end{pmatrix} \in SL(2, \mathbf{R}) : \tau \longrightarrow \frac{a\tau + b}{c\tau + d} \tag{8.4.24}$$

the two transformations (8.4.23) correspond to the two matrices:

$$\begin{aligned} D_a &\approx T = \begin{pmatrix} 1 & 1 \\ 0 & 1 \end{pmatrix} \\ D_b &\approx TST = \begin{pmatrix} 1 & 0 \\ 1 & 1 \end{pmatrix} \end{aligned} \tag{8.4.25}$$

which generate the modular group of Eq. (8.4.19). Hence two tori whose moduli $\tau_1$ and $\tau_2$ are related by a modular transformation $\tau_1 = \gamma \tau_2 > (\gamma \in \Gamma)$ are holomorphic equivalent and have to be identified. The true moduli space of the torus is the quotient $\mathbf{H}/\Gamma$, $\mathbf{H}$ being the upper complex plane.

From a different point of view the parameter $\tau$ coincides with the ratio $\frac{\Pi^0_b}{\Pi^0_a}$ of the two periods of the torus holomorphic 1-form, according to the definition of Eq.(8.2.29). Indeed, with respect to the complex coordinate of Eq.(8.4.21), the holomorphic 1-form is $\Omega^{(1)} = dz$ and we get

$$\begin{aligned} \Pi^0_a &= \int_a dz = \int_0^{2\pi} \frac{d\xi_1}{2\pi} = 1 \\ \Pi^0_b &= \int_b dz = \int_0^{2\pi} \tau \frac{d\xi_2}{2\pi} = \tau \end{aligned} \tag{8.4.26}$$

The transformations (8.4.23) are the effect on the ratio $\tau = \frac{\Pi^0_b}{\Pi^0_a}$ of the action of the diffeomorphisms (8.4.22) on the homology basis $(a, b)$. The diffeomorphisms (8.4.23) generate the *mapping class group*

$$\mathcal{M}_{class} = \frac{\text{Diff}}{\text{Diff}_0} \tag{8.4.27}$$

namely the group of all diffeomorphisms modulo the subgroup of dieffeomorphisms connected to the identity.

*iii)* Considering now the general case of any Calabi–Yau $n$-fold, in particular the $M\mathbb{CP}_{p-1}[p]$ hypersurfaces of Eq.(8.4.2), we can define the *modular group* $\Gamma$ *as the homomorphic image on a canonical homology basis of $n$ – cycles of the mapping class group (8.4.27).* Since the intersection matrix is preserved by the modular group, we have

$$\begin{aligned} \Gamma &\subset Sp\left(2 + 2\sum_{j=1}^{\nu} h^{(n-j,j)}\,,\,\mathbb{Z}\right) \quad \text{if } n = 2\nu + 1 \\ \Gamma &\subset O\left(h^{(\nu,\nu)} + 2\sum_{j=1}^{\nu-1} h^{(n-j,j)}\,,\,\mathbb{Z}\right) \quad \text{if } n = 2\nu \end{aligned} \tag{8.4.28}$$

and in the $M\mathbb{CP}_{p-1}[p]$ case Eq.s (8.4.28) specialize to

$$\begin{aligned} \Gamma &\subset Sp(p-1\,,\mathbb{Z}) \ \text{ if } p = 2\nu + 1 \\ \Gamma &\subset O(p-1\,,\mathbb{Z}) \ \text{ if } p = 2\nu \end{aligned} \tag{8.4.29}$$

This is the same condition respected by the elements of the monodromy group elements. Indeed the transvections around the singular points of the Picard–Fuchs equation must correspond to some global reparametrization of the surface $\mathcal{W}(X, \psi) = 0$. Hence we infer that

$$\mathcal{M}on \subset \Gamma \tag{8.4.30}$$

Furthermore, reconsidering, in the case of the torus, Eq.s (8.4.17) and (8.4.18), we can also conjecture that

$$\Gamma_{\mathcal{W}} = \frac{\Gamma}{\mathcal{M}on} \tag{8.4.31}$$

This conjecture can indeed be verified in the case of the torus, where $\Gamma$ is known a priori, by calculating $\mathcal{M}on$ and showing that $\mathcal{M}on \approx \Gamma(3)$. This verification [242], which we shall present in a moment, encourages us to assume that Eq. (8.4.31) is true in the general case. This assumption is very convenient because it provides an algorithm to reconstruct the unknown $\Gamma$ by calculating $\mathcal{M}on$ with the analytic methods we already referred to and by adjoining to it the $\mathcal{A}$ matrix generating the duality group, whose action on the solutions of the Picard–Fuchs equations will be known a priori.

### 8.4.3 Monodromy group of the cubic torus

We focus on the second order Picard–Fuchs equation (8.2.27): its monodromy group can be derived as follows. As already pointed out, we recall that it is sufficient to compute the transvection $T_0$ around the singular point $\psi = 1$. In fact, applying to this particular case the general formula (8.4.11), the effect of a closed loop around $\psi = \alpha$ and $\psi = \alpha^2$

($\alpha = e^{2\pi i/3}$) can be computed from the matrix $T_0$ by conjugation with $\mathcal{A}$, where $\mathcal{A}$ represents the operation $\psi \to \alpha\psi$:

$$\begin{aligned} T_1 &= \mathcal{A}^{-1} T_0 \mathcal{A} \\ T_2 &= \mathcal{A}^{-2} T_0 \mathcal{A}^2 \end{aligned} \tag{8.4.32}$$

Furthermore a closed loop which encloses all the singular points, including $\infty$, is contractible and therefore, according to the general equation (8.4.3), we can write

$$T_\infty T_2 T_1 T_0 = 1 \quad \to \quad T_\infty = (T_2 T_1 T_0)^{-1} \tag{8.4.33}$$

To compute $T_0$ it is convenient to perform the substitution $z = \psi^3$ in the differential equation (8.2.27). We obtain

$$\left\{9z(1-z)\frac{d^2}{dz^2} + (6-15z)\frac{d}{dz} - 1\right\}\omega = 0 \tag{8.4.34}$$

This is a hypergeometric equation of parameters $a = b = 1/3, c = 2/3$ and therefore a set of independent solutions around $z \equiv \psi^3 = 0$ is given by

$$\begin{cases} U_1 = \frac{\Gamma^2(1/3)}{\Gamma(2/3)} F(1/3, 1/3, 2/3; \psi^3) \\ U_2 = \frac{\Gamma^2(2/3)}{\Gamma(4/3)} \psi F(2/3, 2/3, 4/3; \psi^3) \end{cases} \tag{8.4.35}$$

where $F(a, b, c; z)$ is the hypergeometric function. These two solutions can be analytically continued around $\psi^3 = 1$ by means of classical formulae [139]: one finds

$$\begin{aligned} U_1 &= -\log(1-z)F(1/3, 1/3, 1; 1-\psi^3) + B_1(1-\psi^3) \\ U_2 &= -\log(1-z)F(1/3, 1/3, 1; 1-\psi^3) + B_2(1-\psi^3) \end{aligned} \tag{8.4.36}$$

where $B_1$ and $B_2$ are regular series around $\psi^3 = 1$. (The appearance of the logarithmic factor in (8.4.36) is traceable to the equality of the roots of the indicial equation around $z \equiv \psi^3 = 1$.) Hence the transvection along a closed loop around $\psi = 1$ yields

$$\begin{pmatrix} U_1 \\ U_2 \end{pmatrix} \to \begin{pmatrix} U_1' \\ U_2' \end{pmatrix} = \begin{pmatrix} U_1 \\ U_2 \end{pmatrix} - 2\pi i \; F(\frac{1}{3}, \frac{1}{3}, 1; 1-\psi^3)\begin{pmatrix} 1 \\ 1 \end{pmatrix} \tag{8.4.37}$$

The Kummer relations [139] among hypergeometric functions enable us to re-express $F(\frac{1}{3}, \frac{1}{3}, 1; 1-\psi^3)$ in terms of the original basis $(U_1, U_2)$ of solutions analytic in the regular point $\psi = 0$:

$$F\left(\frac{1}{3}, \frac{1}{3}, 1; 1-z\right) = \frac{\Gamma(\frac{1}{3})}{\Gamma^2(\frac{2}{3})} F\left(\frac{1}{3}, \frac{1}{3}, \frac{2}{3}; z\right) + \frac{\Gamma(-\frac{1}{3})}{\Gamma^2(\frac{1}{3})} F\left(\frac{2}{3}, \frac{2}{3}, \frac{4}{3}; z\right) \tag{8.4.38}$$

Therefore, using the relation $\Gamma(z)\Gamma(1-z) = \frac{\pi}{\sin \pi z}$ and $2\,\mathrm{Sin}(\frac{\Pi}{3}) = -\mathrm{tg}(\frac{\Pi}{3}) = -\sqrt{3}$, one obtains

$$\begin{pmatrix} U_1' \\ U_2' \end{pmatrix} = \begin{pmatrix} 1 + i\, tg\frac{2\pi}{3} & -i\, tg\frac{2\pi}{3} \\ i\, tg\frac{2\pi}{3} & 1 - i\, tg\frac{2\pi}{3} \end{pmatrix} \begin{pmatrix} U_1 \\ U_2 \end{pmatrix} \tag{8.4.39}$$

i.e. the monodromy matrix around $\psi = 1$ is

$$T_0 = \begin{pmatrix} 1 - i\sqrt{3} & i\sqrt{3} \\ -i\sqrt{3} & 1 + i\sqrt{3} \end{pmatrix} \tag{8.4.40}$$

To find $T_1, T_2$ we need to represent $\mathcal{A} : \psi \to \alpha\psi$ on $U_1, U_2$. From (8.2.27) and (8.4.35) we see that under $\psi \to \alpha\psi$ the differential operator is invariant while

$$\begin{pmatrix} U_1 \\ U_2 \end{pmatrix} \to \begin{pmatrix} 1 & 0 \\ 0 & \alpha \end{pmatrix} \begin{pmatrix} U_1 \\ U_2 \end{pmatrix} \tag{8.4.41}$$

Since we are interested in the projective representation of the monodromy group we may rescale our basis in such a way that $\det\mathcal{A} = 1$ (note that $T_0$ already satisfies $\det T_0 = 1$). Hence we have

$$\mathcal{A} = \begin{pmatrix} \alpha^{-1/2} & 0 \\ 0 & \alpha^{1/2} \end{pmatrix} = \frac{1}{2} \begin{pmatrix} 1 - i\sqrt{3} & 0 \\ 0 & 1 - i\sqrt{3} \end{pmatrix} \tag{8.4.42}$$

and from (8.4.32)

$$T_1 = \begin{pmatrix} 1 - i\sqrt{3} & -\frac{3}{2} - \frac{i\sqrt{3}}{2} \\ -\frac{3}{2} + \frac{i\sqrt{3}}{2} & 1 + i\sqrt{3} \end{pmatrix} \qquad T_2 = \begin{pmatrix} 1 - i\sqrt{3} & \frac{3}{2} - \frac{i\sqrt{3}}{2} \\ \frac{3}{2} + \frac{i\sqrt{3}}{2} & 1 + i\sqrt{3} \end{pmatrix} \tag{8.4.43}$$

In this way the monodromy group $\mathcal{M}on$ has been determined. In view of our previous discussion (see Eq.s (8.4.29)) we know that it should be a subgroup of $Sp(2,\mathbb{Z}) \approx SL(2,\mathbb{Z})$, so that we should be able to find a change of basis on the periods $U_i$ such that the entries of the generators $T_0, T_1, T_2$ are integer numbers. Actually if our conjecture is correct we should find a basis where $\mathcal{M}on \approx \Gamma(3)$. The basis $(\mathcal{F}_1, \mathcal{F}_2)$ where this isomorphism becomes indeed manifest is obtained by means of the following linear transformation:

$$\begin{pmatrix} \mathcal{F}_1 \\ \mathcal{F}_2 \end{pmatrix} = M \begin{pmatrix} U_1 \\ U_2 \end{pmatrix} \stackrel{\text{def}}{=} \frac{1}{\pi} \begin{pmatrix} \frac{3i}{2} - \frac{3\sqrt{3}}{2} & -3i \\ \frac{3i}{2} - \frac{3}{2\sqrt{3}} & -\frac{3i}{2} + \frac{\sqrt{3}}{2} \end{pmatrix} \begin{pmatrix} U_1 \\ U_2 \end{pmatrix} \tag{8.4.44}$$

The transformed $\mathcal{M}on$ generators $\widehat{T}_i = M\, T_i\, M^{-1}$ take the following expressions

$$\widehat{T}_0 = \begin{pmatrix} 1 & 3 \\ 0 & 1 \end{pmatrix} \qquad \widehat{T}_1 = \begin{pmatrix} -5 & 12 \\ -3 & 7 \end{pmatrix} \qquad \widehat{T}_2 = \begin{pmatrix} -2 & 3 \\ -3 & 4 \end{pmatrix} \tag{8.4.45}$$

$$\widehat{T}_\infty \equiv (\widehat{T}_2 \widehat{T}_1 \widehat{T}_0)^{-1} = \begin{pmatrix} 1 & 0 \\ -3 & 1 \end{pmatrix} \tag{8.4.46}$$

and are all, manifestly, elements of $\Gamma(3)$ according to the definition (8.4.20). Actually, by comparing with classical last century books (see for example [201]) we see that $T_0, T_1, T_2, T_\infty$ are the classical four generators of $\Gamma(3)$ satisfying the relation (8.4.12). As already remarked in the previous section, the transformation $\mathcal{A} : \psi \to \alpha\psi$ is obviously an invariance of the hypersurface equation $\mathcal{W}(X) = 0$, since it satisfies Eq. (8.4.6), and as a consequence it is an invariance of the differential operator (8.4.41). It is an element of the modular group but not of the monodromy group. In the basis $(\mathcal{F}_1, \mathcal{F}_2)$, the generator $\mathcal{A}$ takes the following form:

$$\widehat{\mathcal{A}} = M \mathcal{A} M^{-1} = \begin{pmatrix} -1 & 3 \\ -1 & 2 \end{pmatrix} \tag{8.4.47}$$

As one sees, $\widehat{\mathcal{A}} \in PSL(2, \mathbb{Z})$ but $\widehat{\mathcal{A}} \notin \Gamma(3)$. This is precisely what we expected. Indeed as stated in Eq. (8.4.18), the tetrahedral group $\mathcal{T}$, isomorphic to the duality group $\Gamma_W$, is also isomorphic to the quotient $PSL(2, \mathbb{Z})/\Gamma(3)$: hence the two operations $\mathcal{A}$ and $\mathcal{S}$, in the basis $\mathcal{F}_1$, $\mathcal{F}_2$ of solutions of the Picard–Fuchs equation, must be represented by integer-valued unimodular matrices that are not equivalent to the identity modulus three. The explicit form of $\mathcal{A}$ we were able to calculate. To find the form of $\mathcal{S}$ we should be able to re-express $\mathcal{F}_i\left(\frac{\psi+2}{\psi-1}\right)$ in terms of $\mathcal{F}_i(\psi)$: there are no known autotransformations of the hypergeometric functions that do this for us and hence we should resort to explicit numerical calculations: we do not feel it necessary to do it for the purpose of our discussion. Indeed by means of our present calculations we have already shown that the monodromy group of the cubic torus is $\mathcal{M}on = \Gamma(3)$, so that the conjectured relation (8.4.31) is indeed verified. Other examples can be provided and it appears that (8.4.31) can be assumed as a true statement in all relevant cases. Hence we can now make a return to the general $M\mathbb{CP}_{p-1}[p]$ case and use our general strategy to calculate the full modular group: it suffices to calculate the monodromy group in a solution bases where the action of the generator $\mathcal{A}$ of the $\mathbb{Z}_p$ duality group is well defined. This is possible by means of Barne's integral transform.

### 8.4.4 Barne's integral transform and the calculation of the monodromy matrix $\mathbf{T_0}$

Our starting point for the explicit calculation of the matrix $T_0$ representing the transvection around the singular point $\psi = 0$ is given by Barne's integral transform of the solution (8.2.40) which is regular in the neighbourhood of the point $\psi = \infty$. By means of such a transformation we shall obtain an analytic continuation of this particular solution to a neighbourhood of the point $\psi = 1$. Then, by means of the replacement $\psi \longrightarrow \alpha^j \psi$, we shall obtain a basis of solutions with known transformation properties under the generator $A$ of the duality group $\Gamma_W$. In this basis the transvection matrix $T_0$ will be explicitly evaluated. Let us then describe the analytic continuation of $\Pi_0(\psi)$. Recalling the definition of the generalized hypergeometric function, we have

$$\begin{aligned} \Pi_0^{(p)}(\psi) &= \frac{1}{\psi}\overline{\Pi}_0^{(p)}(\psi) \\ \overline{\Pi}_0^{(p)}(\psi) &= {}_{(p-1)}F_{(p-2)}\left[\frac{1}{p}, \frac{2}{p}, \ldots, \frac{p-1}{p}; 1, \ldots, 1; \frac{1}{\psi^p}\right] \\ &= \frac{1}{\prod_{k=1}^{p-1} \Gamma\left(\frac{k}{p}\right)} \sum_{n=0}^{\infty} \frac{\prod_{k=1}^{p-1} \Gamma\left(n + \frac{k}{p}\right)}{[\Gamma(n+1)]^p\, \Gamma(n+1)} \psi^{-pn} \end{aligned} \tag{8.4.48}$$

Using the standard product formulae:

$$\begin{aligned} \Gamma(n\,x) &= (2\pi)^{\frac{1-n}{2}}\, n^{n\,x-\frac{1}{2}} \prod_{k=0}^{n-1} \Gamma\left(x+\frac{k}{n}\right) \\ \prod_{k=1}^{n-1} \Gamma\left(\frac{k}{n}\right) \Gamma\left(1-\frac{k}{n}\right) &= \frac{(2\pi)^{n-1}}{n} \end{aligned} \tag{8.4.49}$$

we obtain

$$\overline{\Pi}_0^{(p)}(\psi) = p \sum_{n=0}^{\infty} \frac{\Gamma(pn)}{[\Gamma(n+1)]^{p-1}\,\Gamma(n)}\,(p\,\psi)^{-pn} \tag{8.4.50}$$

The series in Eq. (8.4.50) can be given the following integral representation which holds true for $|\psi| > 1$:

$$\overline{\Pi}_0^{(p)}(\psi) = \frac{1}{2\pi i} \int_{-i\infty}^{+i\infty} ds\, \frac{\Gamma(-s)\,\Gamma(ps+1)}{[\Gamma(s+1)]^{p-1}}\, e^{i\pi s}\,(p\psi)^{-ps} \tag{8.4.51}$$

Indeed, to calculate the integral (8.4.51) for $|\psi| > 1$, we can close the contour in the right half of the complex plane, in this way enclosing the poles at

$$s = 1, 2, \ldots, n, \ldots \tag{8.4.52}$$

Since

$$\operatorname*{Res}_{s=n}\ \Gamma(-s) = -\frac{(-1)^n}{n!} \tag{8.4.53}$$

Eq.(8.4.50) follows. The integral representation allows the analytic continuation to $|\psi| < 1$: indeed, in this range of the variable, we can calculate the integral (8.4.51) by closing the contour in the left half of the complex plane, containing the poles at

$$ps + 1 = -1, -2, \ldots, -n, \ldots \tag{8.4.54}$$

By means of the change of variable $t = ps + 1$ we find

$$\begin{aligned} p\,\overline{\Pi}_0^{(p)}(\psi) &= \frac{1}{2\pi i} \int_{-i\infty}^{+i\infty} dt\, \frac{\Gamma\left(-\frac{t-1}{p}\right)\Gamma(t)}{\left[\Gamma\left(\frac{t+p-1}{p}\right)\right]^{p-1}}\, e^{i\pi\frac{t-1}{p}}\,(p\psi)^{-t-1} \\ &= \sum_{m+1}^{\infty} \frac{\Gamma\left(\frac{m}{p}\right)\Gamma\left(1-\frac{m}{p}\right)}{\Gamma(m)\left[\Gamma\left(1-\frac{m}{p}\right)\right]^{p}}\, \alpha^{m\frac{p-1}{2}}\,(p\,\psi)^{m} \end{aligned} \tag{8.4.55}$$

where we have multiplied and divided by $\Gamma\left(1-\frac{m}{p}\right)$. Using the standard relation

$$\Gamma\left(\frac{m}{p}\right)\Gamma\left(1-\frac{m}{p}\right) = \frac{1}{\sin\left(\pi\frac{m}{p}\right)} \tag{8.4.56}$$

and
$$\left[\sin\left(\pi\frac{m}{p}\right)\right]^{p-1} = \alpha^{-m\frac{p-1}{p}}\left(\frac{\alpha^m-1}{2i}\right)^{p-1} \qquad (8.4.57)$$

we obtain

$$\begin{aligned}\Pi_0^{(p)}(\psi) &= \frac{1}{p\psi}\widehat{\Pi}_0^{(p)}(\psi)\\ \widehat{\Pi}_0^{(p)}(\psi) &\stackrel{\text{def}}{=} \frac{1}{(2\pi i)^{p-1}}\sum_{m=0}^{\infty}\frac{\Gamma\left(\frac{m}{p}\right)^p}{\Gamma(m)}\left(\alpha^m-1\right)^{p-1}(p\psi)^m\end{aligned} \qquad (8.4.58)$$

The function $\Pi_0^{(p)}(\psi)$ being the analytic continuation of a solution is also, by construction, a solution of the Picard–Fuchs differential equation (8.2.33). Because of its expression as a power series in $\psi$ it is a solution that is regular in the neighborhood of the regular point $\psi = 0$. Inspection of Eq. (8.2.33) shows that if $f(\psi)$ is a solution, $f_j(\psi) = f(\alpha^j\psi)$ are also solutions for all values $j = 1, \ldots, p-1$. Hence we can introduce the following set of $p$ solutions of the differential equation:

$$\Pi_j^{(p)}(\psi) = \Pi_0^{(p)}(\alpha^j\psi) \qquad j = 0, 1, \ldots, p-1 \qquad (8.4.59)$$

Clearly, since the equation is linear of degree $p-1$ only $p-1$ of the above $p$ solutions can be independent. Indeed there is just one linear relation, namely

$$\sum_{j=0}^{p-1}\alpha^j\,\Pi_j^{(p)} = 0 \qquad (8.4.60)$$

which immediately follows from the definitions (8.4.58) and from the identity

$$\sum_{j=0}^{p-1}\widehat{\Pi}_j^{(p)} = 0 \qquad (8.4.61)$$

To prove the truth of (8.4.61) it suffices to note that in the series $\widehat{\Pi}_j^{(p)}$ powers of the form $\psi^{p\nu}$ ($\nu \in \mathbb{Z}_+$) are absent, since $m = p\nu$ is a zero of the coefficient $\frac{1}{\Gamma\left(1-\frac{m}{p}\right)}$. The only powers that have non-vanishing coefficients are of the form $\psi^{p\nu+k}$ ($k = 1, 2, \ldots, p-1$). For such terms we have

$$\sum_{j=0}^{p}\left(\alpha^j\psi\right)^{p\nu+k} = \psi^{p\nu+k}\sum_{j=0}^{p}\alpha^{jk} = 0 \qquad (8.4.62)$$

so that Eq.(8.4.61) follows. In view of (8.4.60), as a basis of $p-1$ independent solutions we take $\Pi_j^{(p)}$ for $j = 0, 1, \ldots, p-2$. This basis is very convenient because here the action of the duality group generator $A$ is known by construction. Since we are interested only in the projective realizations of the modular group, factors common to the whole set of solutions of a basis can be disregarded. This means that rather than the action of $A$ on $\Pi_j^{(p)}$ we can just consider its action on the reduced basis $\widehat{\Pi}_j^{(p)}$. We immediately get

$$\mathcal{A}\begin{pmatrix}\widehat{\Pi}_0^{(p)}\\ \widehat{\Pi}_1^{(p)}\\ \dots\\ \dots\\ \widehat{\Pi}_{p-2}^{(p)}\end{pmatrix} = \begin{pmatrix}0 & 1 & 0 & 0 & \dots & 0\\ 0 & 0 & 1 & 0 & \dots & 0\\ \dots & \dots & \dots & \dots & \dots & \dots\\ \dots & \dots & \dots & \dots & \dots & \dots\\ 0 & 0 & 0 & 0 & 0 & 1\\ -1 & -1 & -1 & -1 & -1 & -1\end{pmatrix}\begin{pmatrix}\widehat{\Pi}_0^{(p)}\\ \widehat{\Pi}_1^{(p)}\\ \dots\\ \dots\\ \widehat{\Pi}_{p-2}^{(p)}\end{pmatrix} \tag{8.4.63}$$

In general the solutions $\Pi_j^{(p)}(\psi)$ and hence the power series $\widehat{\Pi}_1^{(p)}(\psi)$ have a singular behaviour at the point $\psi = 1$. In the $p = odd$ case, on which from now on we fix our attention, we can write

$$\widehat{\Pi}_j^{(p)}(\psi) = \frac{1}{2\pi \mathrm{i}} c_j \ln(\psi - 1)\, g(\psi) + f_j(\psi) \tag{8.4.64}$$

where $g(\psi)$ and $f_j(\psi)$ are regular functions at $\psi = 1$. This is so because the indicial equation (8.2.47) implies that in the $p = odd$ case at $\psi = 1$ there are $\frac{p-1}{2}$ regular solutions and $\frac{p-1}{2}$ solutions with a logarithmic singularity. This point will be further discussed later on: here, assuming (8.4.64) we derive its consequences.

The explicit calculation of the coefficients $c_j$ will be performed in a moment. Prior to that we want to show that the knowledege of this $p-1$-tuplet of complex numbers $c_j$ suffices to determine the monodromy generator $T_0$ and hence the complete structure of the modular group for all odd values of $p$. To this end we begin by noting that for $x \in \mathbb{R}\ x > 1$ we can write the discontinuities through the real axis as follows:

$$\widehat{\Pi}_j^{(p)}(x - \mathrm{i}\varepsilon) - \widehat{\Pi}_j^{(p)}(x + \mathrm{i}\varepsilon) = c_j\, g(x) \tag{8.4.65}$$

On the other hand, recalling the original expression (8.4.50) of $\widehat{\Pi}_0^{(p)}(\psi)$, which holds true for $|\psi| > 1$, we obtain the identification of values

$$\widehat{\Pi}_0^{(p)}(x) = \widehat{\Pi}_0^{(p)}(\alpha x) \qquad \forall x \in \mathbb{R}\ x > 1 \tag{8.4.66}$$

By definition we have

$$\widehat{\Pi}_1^{(p)}(x - \mathrm{i}\varepsilon) = \widehat{\Pi}_0^{(p)}(\alpha(x - \mathrm{i}\varepsilon)) \tag{8.4.67}$$

but since

$$\arg[\alpha \mathrm{i}\varepsilon] = \frac{2\pi}{p} - \varepsilon' > 0 \tag{8.4.68}$$

we can write

$$\widehat{\Pi}_1^{(p)}(x - \mathrm{i}\varepsilon) = \widehat{\Pi}_0^{(p)}(\alpha x + \mathrm{i}\varepsilon'') = \widehat{\Pi}_0^{(p)}(x + \mathrm{i}\varepsilon''') \tag{8.4.69}$$

where in the last equality we have made use of (8.4.66). Combining (8.4.65) with (8.4.69) we finally get

$$c_1\, g(\psi) \;=\; \widehat{\Pi}_0^{(p)}(\psi) \;-\; \widehat{\Pi}_1^{(p)}(\psi) \tag{8.4.70}$$

This equation combined with (8.4.64) allows us to express the monodromy generator $T_0$ in terms of the coefficients $c_j$. Under the transformation

$$T_0 \;:\; (\psi - 1) \longrightarrow e^{2\pi i}\,(\psi - 1) \tag{8.4.71}$$

we have

$$T_0 \;:\; \widehat{\Pi}_j^{(p)}(\psi) \;\longrightarrow\; c_j\, g(\psi) + \widehat{\Pi}_j^{(p)}(\psi) = \frac{c_j}{c_1}\left[\widehat{\Pi}_0^{(p)}(\psi) - \widehat{\Pi}_1^{(p)}(\psi)\right] + \widehat{\Pi}_j^{(p)}(\psi) \tag{8.4.72}$$

and we can write:

$$T_0 \;=\; \begin{pmatrix} 1+\frac{c_0}{c_1} & -\frac{c_0}{c_1} & 0 & 0 & \dots & 0 & 0 \\ 1 & 0 & 0 & 0 & \dots & 0 & 0 \\ \frac{c_2}{c_1} & -\frac{c_2}{c_1} & 1 & 0 & \dots & \dots & 0 \\ \frac{c_3}{c_1} & -\frac{c_3}{c_1} & 0 & 1 & 0 & \dots & 0 \\ \dots & \dots & \dots & \dots & \dots & \dots & \dots \\ \frac{c_{p-2}}{c_1} & -\frac{c_{p-2}}{c_1} & 0 & \dots & \dots & 0 & 1 \end{pmatrix} \tag{8.4.73}$$

This form can be further reduced using the identity

$$c_0 \;=\; c_1 \tag{8.4.74}$$

which leads to the following general form for the two generators of the modular group:

$$\mathcal{A} = \begin{pmatrix} 0 & 1 & 0 & 0 & \dots & 0 \\ 0 & 0 & 1 & 0 & \dots & 0 \\ \dots & \dots & \dots & \dots & \dots & \dots \\ \dots & \dots & \dots & \dots & \dots & \dots \\ 0 & 0 & 0 & 0 & 0 & 1 \\ -1 & -1 & -1 & -1 & -1 & -1 \end{pmatrix} \tag{8.4.75}$$

$$T_0 \;=\; \begin{pmatrix} 2 & -1 & 0 & 0 & \dots & 0 & 0 \\ 1 & 0 & 0 & 0 & \dots & 0 & 0 \\ \frac{c_2}{c_1} & -\frac{c_2}{c_1} & 1 & 0 & \dots & \dots & 0 \\ \frac{c_3}{c_1} & -\frac{c_3}{c_1} & 0 & 1 & 0 & \dots & 0 \\ \dots & \dots & \dots & \dots & \dots & \dots & \dots \\ \frac{c_{p-2}}{c_1} & -\frac{c_{p-2}}{c_1} & 0 & \dots & \dots & 0 & 1 \end{pmatrix} \tag{8.4.76}$$

The identity (8.4.74) is an immediate consequence of Eq.(8.4.70). Indeed by definition $g(\psi)$ is a solution of the differential equation that is regular at $\psi = 1$ so that

$$T_0\, g(\psi) \;=\; g(\psi) \tag{8.4.77}$$

Consistency of (8.4.77) with (8.4.70) and (8.4.72) implies (8.4.74).

So we are left with the problem of determining the coefficient vector $c_j$. To do this, always restricting our attention to the $p = odd$ case, we recall the structure of the indicial equation (8.2.47). Because of its structure the Picard–Fuchs differential equation admits $\frac{p-1}{2}$ solutions that are regular in the neighbourhood of $\psi = 1$ and correspond to the indices

$$r = \frac{p-3}{2} + k \qquad k = 0, 1, \ldots \frac{p-3}{2} \tag{8.4.78}$$

These solutions have the power series structure

$$\begin{aligned}
\chi_0(\psi) &= \sum_{m=\frac{p-3}{2}}^{\infty} a_m^{(o)} \psi^m \\
\chi_1(\psi) &= \sum_{m=\frac{p-3}{2}+1}^{\infty} a_m^{(1)} \psi^m \\
\ldots & \ldots \ldots \\
\ldots & \ldots \ldots \\
\chi_k(\psi) &= \sum_{m=\frac{p-3}{2}+k}^{\infty} a_m^{(k)} \psi^m \\
\ldots & \ldots \ldots \\
\ldots & \ldots \ldots \\
\chi_{\frac{p-3}{2}}(\psi) &= \sum_{m=p-3}^{\infty} a_m^{(\frac{p-3}{2})} \psi^m
\end{aligned} \tag{8.4.79}$$

and form half of a basis of $p-1$ independent solutions. The remaining $\frac{p-1}{2}$ solutions have a logarithmic singularity at $\psi = 1$ and take the form

$$\chi'_k = \ln(\psi - 1)\, \chi_0(\psi) + r_k(\psi) \qquad k = 0, 1, \ldots, \frac{p-3}{2} \tag{8.4.80}$$

where $r_k(\psi)$ are power series in $(\psi - 1)$. Since the union of the $\chi_k$ and $\chi'_k$ is a solution basis, it follows that the function $g(\psi)$ defined by Eq.(8.4.64) is just proportional to $\chi_0(\psi)$ and to all effects it can be identified with it. This fact implies that all the reduced solutions $\hat{\Pi}_j^{(p)}(\psi)$ in the vicinity of $\psi = 1$ behave as follows:

$$\begin{aligned}
\hat{\Pi}_j^{(p)}(\psi) \underset{\psi \to 1}{\approx} & \frac{c_j}{2\pi i} \ln(\psi - 1) \left[ \mu (\psi - 1)^{\frac{p-3}{2}} + \mathcal{O}\left( (\psi - 1)^{\frac{p-3}{2}+1} \right) \right] \\
& + const + \mathcal{O}((\psi - 1))
\end{aligned} \tag{8.4.81}$$

where $\mu$ is some fixed, non vanishing constant. From (8.4.81) we easily determine the singular part in the derivative of order $\frac{p-3}{2}$ of the reduced solutions. We have

$$\frac{d^{\frac{p-3}{2}} \hat{\Pi}_j^{(p)}}{d\psi^{\frac{p-3}{2}}} \underset{\psi \to 1}{\approx} \frac{p-3}{2} \frac{c_j}{2\pi i} \mu \ln(\psi - 1) + \text{reg. terms} \tag{8.4.82}$$

Hence we can evaluate the coefficients $c_j$ by separating the singular asymptotic behaviour of $\frac{d^{\frac{p-3}{2}}\widehat{\Pi}_j^{(p)}}{d\psi^{\frac{p-3}{2}}}$. Starting from Eq.(8.4.58) and taking the derivative term by term of the power series we get

$$\frac{d^{\frac{p-3}{2}}\widehat{\Pi}_j^{(p)}}{d\psi^{\frac{p-3}{2}}} = \frac{1}{(2\pi i)^{p-1}} \sum_{m=\frac{p-3}{2}}^{\infty} \frac{\Gamma\left(\frac{m}{p}\right)^p}{\Gamma(m)} \alpha^{jm} (\alpha^m - 1)^{p-1} \times$$
$$pm(m-1)\dots\left(m - \frac{p-3}{2} + 1\right)(p\psi)^{m-\frac{p-3}{2}} \tag{8.4.83}$$

Since we are interested in the asymptotic behaviour for $\psi \to 1$, what matters are the large orders of the power series expansion, namely the behaviour of the coefficients for $m \to \infty$. Correspondingly we can approximate

$$\frac{d^{\frac{p-3}{2}}\widehat{\Pi}_j^{(p)}}{d\psi^{\frac{p-3}{2}}} \underset{\psi\to 1}{\approx} \frac{1}{(2\pi i)^{p-1}} \sum_m^{\infty} \alpha^{jm} (\alpha^m - 1)^{p-1}\, t(p,m)\, \psi^m \tag{8.4.84}$$

where we have used

$$\psi^{m-\frac{p-3}{2}} \underset{m\to\infty}{\approx} \psi^m \tag{8.4.85}$$

and relying on

$$m(m-1)\dots\left(m - \frac{p-3}{2} + 1\right) \underset{m\to\infty}{\approx} m^{\frac{p-3}{2}}$$
$$p^{m-\frac{p-3}{2}} \underset{m\to\infty}{\approx} p^m \tag{8.4.86}$$

we have defined

$$t(p,m) = m^{\frac{p-3}{2}}\, p^m\, \frac{\Gamma\left(\frac{m}{p}\right)^p}{\Gamma(m)} \tag{8.4.87}$$

Using the asymptotic expansion of the gamma function:

$$\Gamma(z) = (2\pi)^{\frac{1}{2}}\, z^{z-\frac{1}{2}}\, e^{-z} \left\{1 + \frac{1}{12z} + \mathcal{O}(z^{-2})\right\} \tag{8.4.88}$$

we have

$$t(p,m) \underset{m\to\infty}{\approx} (2\pi)^{\frac{p-1}{2}} \frac{1}{m} p^{\frac{p}{2}} \left(1 + \frac{p^2-1}{12m}\right) \tag{8.4.89}$$

so that we can write

$$\frac{d^{\frac{p-3}{2}}\widehat{\Pi}_j^{(p)}}{d\psi^{\frac{p-3}{2}}} \underset{\psi\to 1}{\approx} (-1)^{\frac{p-1}{2}} p^{\frac{p}{2}} \sum_m^{\infty} \alpha^{jm} (\alpha^m - 1)^{p-1} \frac{\psi^m}{m} \left(1 + \frac{p^2-1}{12m}\right) \tag{8.4.90}$$

Setting

$$m = pN + l \qquad l = 0, 1, \dots, p-1 \tag{8.4.91}$$

and using

$$\begin{aligned} \psi^{pN+l} & \underset{N\to\infty}{\approx} \psi^{pN} \\ \frac{1}{(pN+l)^{\frac{p-3}{2}}} & \underset{N\to\infty}{\approx} \frac{1}{(pN)^{\frac{p-3}{2}}} \end{aligned} \tag{8.4.92}$$

we get

$$\begin{aligned} \frac{d^{\frac{p-3}{2}}\widehat{\Pi}_j^{(p)}}{d\psi^{\frac{p-3}{2}}} & \underset{\psi\to 1}{\approx} \frac{(-1)^{\frac{p-1}{2}}}{p} p^{\frac{p}{2}} S_j^{(p)} \sum_N^\infty \frac{(p\psi)^N}{N} \\ & \underset{\psi\to 1}{\approx} -\frac{(-1)^{\frac{p-1}{2}}}{p} p^{\frac{p}{2}} S_j^{(p)} \ln\left(1-\psi^p\right) \\ & \underset{\psi\to 1}{\approx} -\frac{(-1)^{\frac{p-1}{2}}}{p} p^{\frac{p}{2}} S_j^{(p)} \ln\left[(1-\psi)\left(\psi+\psi^2+\ldots+\psi^{p-1}\right)\right] \\ & \underset{\psi\to 1}{\approx} -\frac{(-1)^{\frac{p-1}{2}}}{p} p^{\frac{p}{2}} S_j^{(p)} \ln\left(1-\psi\right) \\ & \underset{\psi\to 1}{\approx} -\frac{(-1)^{\frac{p-1}{2}}}{p} p^{\frac{p}{2}} S_j^{(p)} \ln\left(\psi-1\right) \end{aligned} \tag{8.4.93}$$

where

$$S_j^{(p)} \stackrel{\text{def}}{=} \sum_{l=0}^{p-1} \alpha^{jl}\left(\alpha^l-1\right)^{p-1} \tag{8.4.94}$$

Comparing Eq.(8.4.93) with Eq.(8.4.82), we conclude that

$$\frac{p-3}{2}\frac{c_j}{2\pi i}\mu = -\frac{(-1)^{\frac{p-1}{2}}}{p} p^{\frac{p}{2}} S_j^{(p)} \tag{8.4.95}$$

so that

$$\frac{c_j}{c_1} = \frac{S_j^{(p)}}{S_1^{(p)}} \tag{8.4.96}$$

Recalling Eq. (8.4.76) the above ratios are all what is needed to determine the transvection matrix $T_0$. Hence we have reduced the solution of our problem to the evaluation of the numbers defined in Eq.(8.4.94). Expanding the binomial we get

$$S_j^{(p)} = \sum_{k=0}^{p-1}\binom{p-1}{k}(-1)^{p-1-k}\sum_{l=0}^{p-1}\alpha^{kl} \tag{8.4.97}$$

Since, by definition, $\alpha = \exp\left[\frac{2\pi i}{p}\right]$ is a $p$-th root of the unity, it follows that

$$\sum_{l=0}^{p-1}\alpha^{kl} = \begin{cases} p & k = 0 \bmod p \\ 0 & k \neq 0 \bmod p \end{cases} \tag{8.4.98}$$

so that

$$\begin{aligned} S_0^{(p)} &= p \\ S_j^{(p)} &= p \binom{p-1}{p-j} (-1)^{p-j} \quad j = 1, 2, \ldots, p-1 \\ \frac{c_j}{c_1} &= \frac{S_j^{(p)}}{S_1^{(p)}} = \binom{p-1}{p-j} (-1)^{p-j} \quad j = 2, \ldots, p-1 \end{aligned} \tag{8.4.99}$$

In conclusion, if one combines Eq.(8.4.99) with Eq.(8.4.79) the monodromy generator that represents the transvection of the solutions around the singular point $\psi = 1$, for the $p = odd$ case, can be written as follows:

$$T_0 = \begin{pmatrix} 2 & -1 & 0 & 0 & \ldots & 0 & 0 \\ 1 & 0 & 0 & 0 & \ldots & 0 & 0 \\ \binom{p-1}{p-2}(-1)^{p-2} & -\binom{p-1}{p-2}(-1)^{p-2} & 1 & 0 & \ldots & \ldots & 0 \\ \binom{p-1}{p-3}(-1)^{p-3} & -\binom{p-1}{p-3}(-1)^{p-3} & 0 & 1 & 0 & \ldots & 0 \\ \ldots & \ldots & \ldots & \ldots & \ldots & \ldots & \ldots \\ \ldots & \ldots & \ldots & \ldots & \ldots & \ldots & \ldots \\ \ldots & \ldots & \ldots & \ldots & \ldots & \ldots & \ldots \\ \binom{p-1}{2} & -\binom{p-1}{2} & 0 & \ldots & \ldots & 0 & 1 \end{pmatrix} \tag{8.4.100}$$

As one sees, because of the properties of the binomial coefficients, the matrix $T_0$ is always integer valued. In particular, in the case $p = 5$ of the mirror quintic we have

$$\mathcal{A} \begin{pmatrix} \widehat{\Pi}_0^{(5)} \\ \widehat{\Pi}_1^{(5)} \\ \widehat{\Pi}_2^{(5)} \\ \widehat{\Pi}_3^{(5)} \end{pmatrix} = \begin{pmatrix} 0 & 1 & 0 & 0 \\ 0 & 0 & 1 & 0 \\ 0 & 0 & 0 & 1 \\ -1 & -1 & -1 & -1 \end{pmatrix} \begin{pmatrix} \widehat{\Pi}_0^{(5)} \\ \widehat{\Pi}_1^{(5)} \\ \widehat{\Pi}_2^{(5)} \\ \widehat{\Pi}_3^{(5)} \end{pmatrix} \tag{8.4.101}$$

$$T_0 \begin{pmatrix} \widehat{\Pi}_0^{(5)} \\ \widehat{\Pi}_1^{(5)} \\ \widehat{\Pi}_2^{(5)} \\ \widehat{\Pi}_3^{(5)} \end{pmatrix} = \begin{pmatrix} 2 & -1 & 0 & 0 \\ 1 & 0 & 0 & 0 \\ -4 & 4 & 1 & 0 \\ 6 & -6 & 0 & 1 \end{pmatrix} \begin{pmatrix} \widehat{\Pi}_0^{(5)} \\ \widehat{\Pi}_1^{(5)} \\ \widehat{\Pi}_2^{(5)} \\ \widehat{\Pi}_3^{(5)} \end{pmatrix} \tag{8.4.102}$$

From our knowledge of the relation between modular group transformations and changes of homology cycle bases it follows that the generators should not only be integer valued but also symplectic. Hence we should be able to find a solution basis where both $\mathcal{A}$ and $T_0$ are elements of the group $Sp(p-1, \mathbb{Z})$. Candelas et al. [68, 69] have shown that this is indeed true in the mirror quintic case. First, using as a solution basis

$\widehat{\Pi}^{(5)}_2, \widehat{\Pi}^{(5)}_1, \widehat{\Pi}^{(5)}_0, \widehat{\Pi}^{(5)}_4 = -\widehat{\Pi}^{(5)}_0 - \widehat{\Pi}^{(5)}_1 - \widehat{\Pi}^{(5)}_2 - \widehat{\Pi}^{(5)}_3$ Eq.s (8.4.101) and (8.4.102) are rewritten as

$$\mathcal{A}\begin{pmatrix} \widehat{\Pi}^{(5)}_2 \\ \widehat{\Pi}^{(5)}_1 \\ \widehat{\Pi}^{(5)}_0 \\ \widehat{\Pi}^{(5)}_4 \end{pmatrix} = \begin{pmatrix} -1 & -1 & -1 & -1 \\ 1 & 0 & 0 & 0 \\ 0 & 1 & 0 & 0 \\ 0 & 0 & 1 & 0 \end{pmatrix} \begin{pmatrix} \widehat{\Pi}^{(5)}_2 \\ \widehat{\Pi}^{(5)}_1 \\ \widehat{\Pi}^{(5)}_0 \\ \widehat{\Pi}^{(5)}_4 \end{pmatrix} \tag{8.4.103}$$

$$T_0\begin{pmatrix} \widehat{\Pi}^{(5)}_2 \\ \widehat{\Pi}^{(5)}_1 \\ \widehat{\Pi}^{(5)}_0 \\ \widehat{\Pi}^{(5)}_4 \end{pmatrix} = \begin{pmatrix} 1 & 4 & -4 & 0 \\ 0 & 0 & 1 & 0 \\ 0 & -1 & 2 & 0 \\ 6 & 4 & -4 & 1 \end{pmatrix} \begin{pmatrix} \widehat{\Pi}^{(5)}_2 \\ \widehat{\Pi}^{(5)}_1 \\ \widehat{\Pi}^{(5)}_0 \\ \widehat{\Pi}^{(5)}_4 \end{pmatrix} \tag{8.4.104}$$

Then one tries to determine a further change of basis:

$$\begin{pmatrix} \widehat{\Pi}^{(5)}_2 \\ \widehat{\Pi}^{(5)}_1 \\ \widehat{\Pi}^{(5)}_0 \\ \widehat{\Pi}^{(5)}_4 \end{pmatrix}' = \mathcal{M}\begin{pmatrix} \widehat{\Pi}^{(5)}_2 \\ \widehat{\Pi}^{(5)}_1 \\ \widehat{\Pi}^{(5)}_0 \\ \widehat{\Pi}^{(5)}_4 \end{pmatrix} \tag{8.4.105}$$

such that the conjugated generators

$$\mathcal{A}' = \mathcal{M}\mathcal{A}\mathcal{M}^{-1} \quad T_0' = \mathcal{M}T_0\mathcal{M}^{-1} \tag{8.4.106}$$

are both integer valued and symplectic. The matrix $\mathcal{M}$ turns out to be unique up to multiplication by elements of the group $Sp(4,\mathbb{Z})$. We have

$$\mathcal{M} = \begin{pmatrix} -\frac{3}{5} & -\frac{1}{5} & \frac{21}{5} & \frac{8}{5} \\ 0 & 0 & -1 & 0 \\ -1 & 0 & 8 & 3 \\ 0 & 1 & -1 & 0 \end{pmatrix} \tag{8.4.107}$$

$$\mathcal{A}' = \begin{pmatrix} -9 & -3 & 5 & 3 \\ 0 & 1 & 0 & -1 \\ -20 & -5 & 11 & 5 \\ -15 & 5 & 8 & -4 \end{pmatrix} \tag{8.4.108}$$

$$T_0' = \begin{pmatrix} 1 & 0 & 0 & 0 \\ 0 & 1 & 0 & 1 \\ 0 & 0 & 1 & 0 \\ 0 & 0 & 0 & 1 \end{pmatrix} \tag{8.4.109}$$

Using Eq.s (8.4.12) and (8.4.13) one can now evaluate the other generators of the monodromy group that are all symplectic and integer-valued. Hence both the full modular group $\Gamma_{modular}$ and the monodromy group $\mathcal{M}on$ are subgroups of $Sp(p-1,\mathbb{Z})$, the explicit embedding having been exhibited for the $p=5$ case. The difference between the general case and the exceptional $p=3$ case is that $\Gamma_{modular} \subset Sp(p-1,\mathbb{Z})$ is a proper subgroup of the integer symplectic group.

## 8.5 The Mirror Map and the Sum over Instantons

We come now to the main point of the whole programme undertaken in the present chapter. This is the calculation of correlators of the A-twisted sigma model on a manifold $\mathcal{M}$ by means of their identification with homologous correlators of the B-twisted sigma model on a mirror manifold $M(\mathcal{M})$. Such a procedure is named the *mirror map*. The example we consider in order to illustrate the general ideas, by working out all the minor details in a specific case, is provided by the mirror pair

$$\mathcal{M} = \mathbb{CP}_4[5] \iff M(\mathcal{M}) = M\mathbb{CP}_4[5] \tag{8.5.1}$$

$M\mathbb{CP}_4[5]$ is the case $p=5$ of the general family of Calabi–Yau $(p-3)$-folds we have been considering in the previous sections. The A-twisted correlator we focus on is the three-point function

$$\begin{aligned} c_{VVV}(t) &= \langle \Theta^{(0)}_V(x_1)\,\Theta^{(0)}_V(x_1)\,\Theta^{(0)}_V(x_1)\rangle \\ &= \int \mathcal{D}X\,\mathcal{D}c\,\mathcal{D}\overline{c}\,\exp\left[-2t\,S_{quantum}\right]\prod_i^3 \Theta^{(0)}_V(x_i) \\ &= \sum_{k=0}^{\infty} q^k \langle \prod_i^3 \Theta^{(0)}_V(x_i)\rangle_{(0,k)} \end{aligned} \tag{8.5.2}$$

where the harmonic (1,1)-form $V^{(1,1)}$ is a generator of the Dolbeault group $H^{(1,1)}$ which, for the quintic 3-fold $\mathbb{CP}_4[5]$, is one-dimensional, $q = \exp\left[2\pi i t\right]$ and the Kähler 2-form $K$, determining the value of the classical action, is cohomologous to $V^{(1,1)}$

$$[K] = \left[V^{(1,1)}\right] \tag{8.5.3}$$

According to the discussion of section 7.7.2, the value of the three-point correlator (8.5.2) is given by

$$c_{VVV}(t) = W^{(0)}_{ttt} + \sum_{k=1}^{\infty} \frac{n_k\,k^3\,q^k}{1-q} \tag{8.5.4}$$

where

$$W^{(0)}_{ttt} = \int_{\mathbb{CP}_4[5]} V^{(1,1)} \wedge V^{(1,1)} \wedge V^{(1,1)} \tag{8.5.5}$$

is the contribution of $k=0$ instantons, namely of constant maps, and $n_k \in \mathbb{Z}_+$ is the number of degree $k$ instantons, namely of degree $k$ holomorphic maps:

$$\begin{aligned} U^{(k)}_a &: \Sigma_0 \longrightarrow \mathbb{CP}_4[5] \qquad a = 1,\ldots,n_k \\ \int_{\Sigma_0} U^{(k)\star}_a\,[K] &= k \end{aligned} \tag{8.5.6}$$

from a genus zero world-sheet to $\mathbb{CP}_4[5]$.

As we know from the theory developed in Chapters 4 and 6, the deformations of the Kähler class and of the complex structures of a given Calabi–Yau 3-fold $\mathcal{M}$ fill two separate *special Kähler manifolds* named $SK_{KC}(\mathcal{M})$ and $SK_{CS}(\mathcal{M})$, respectively. This result was derived both from macroscopic Kaluza–Klein arguments (see Chapter 6, sections 6.2 and 6.3) and from microscopic N=2 superconformal arguments (see Chapter 6, section 6.4). Due to the constraints of special geometry, the metric $g_{\alpha\beta}$ of these moduli spaces is derivable from the holomorphic prepotential $F(X)$, which, on the other hand, is related by Eq.(8.3.5) to the section $W_{\alpha\beta\gamma}(z,\overline{z})$ which appears in the fundamental identity (8.3.1). As already remarked, the physical meaning of $W_{\alpha\beta\gamma}(z,\overline{z})$ is that of Yukawa coupling. In the effective supergravity lagrangian $W_{\alpha\beta\gamma}(z,\overline{z})$ of the Kähler class deformations appears as a coefficient in the Yukawa coupling for the 27-families, while $W_{\alpha\beta\gamma}(z,\overline{z})$ of the complex structure deformations appears as a coefficient in the Yukawa coupling of the $\overline{27}$-families. The microscopic N=2 argument identifies the Yukawa couplings with an A-twisted three-point correlator in the case of Kähler class deformations and with a B-twisted three-point correlator in the case of complex structure deformations. This identification is encoded in the two equations (8.3.2) and (8.3.3). The main difference between the two cases is that in the complex structure case (B-twist) the result (8.3.3) provided by the microscopic arguments coincides with the result obtained from Kaluza–Klein arguments (see Chapter 4, specifically section 4.4.3) while in the Kähler class case (A-twist) the microscopic non constant result (8.3.2) differs from the constant Kaluza–Klein result:

$$W_{\alpha\beta\gamma} = \int_{\mathcal{M}} \omega^{(1,1)}_{\alpha} \wedge \omega^{(1,1)}_{\beta} \wedge \omega^{(1,1)}_{\gamma} \tag{8.5.7}$$

because of the sum over instantons. Furthermore, while the evaluation of the sum over instantons is a difficult task, Eq. (8.3.3) encodes a fairly manageable algorithm. Hence it becomes quite worthwhile to evaluate $W^{(KC)}_{\alpha\beta\gamma}[\mathcal{M}]$ by identifying it with $W^{(CS)}_{\alpha\beta\gamma}[M(\mathcal{M})]$. This is what we do in this section for the case of the mirror pair (8.5.1).

In view of our discussion we have

$$c_{VVV}(t) \stackrel{\text{def}}{=} W^{(KC)}_{ttt}[\mathbb{CP}_4[5]](t) \stackrel{\text{mirror map}}{=} W^{(SC)}_{ttt}[M\mathbb{CP}_4[5]](t) \stackrel{\text{def}}{=} W_{ttt}(t) \tag{8.5.8}$$

where $c_{VVV}(t)$ is defined in Eq.(8.5.4), while $W_{ttt}(t)$ is to be evaluated with the knowledge we have accumulated in the previous sections.

The first thing to observe is that $c_{VVV}(t)$ is a Yukawa coupling presented in a special coordinate frame where

$$c_{VVV}(t) \stackrel{\text{def}}{=} W^{(KC)}_{ttt}[\mathbb{CP}_4[5]](t) = \frac{\partial^3 \mathcal{F}(t)}{\partial t^3} \tag{8.5.9}$$

Hence the calculation we are going to present is based on two steps:

i) *The evaluation of the B-twisted three-point correlator*

$$W_{\psi\psi\psi} = \int_{M\mathbf{CP}_4[5]} \Omega^{(3,0)} \wedge \frac{\partial^3}{\partial\psi^3}\Omega^{(3,0)} \tag{8.5.10}$$

*in the natural coordinate basis provided by the deformation of the defining polynomial $\mathcal{W}(X,\psi)$ as given in Eq.(5.9.5).*

*ii) The transformation of $W_{\psi\psi\psi}$ to a suitable flat coordinate basis.*

Let us begin with the second point. We must recall that $W_{\alpha\beta\gamma}$ transforms as a section of the bundle:

$$\otimes^3_{symm} \; T^{(1,0)\star}SK \otimes \mathcal{L}^2 \tag{8.5.11}$$

where $T^{(1,0)\star}SK$ is the holomorphic cotangent bundle to the special Kählerian moduli space $SK$, while $\mathcal{L} \xrightarrow{\pi} SK$ is the line bundle whose first Chern class equals the Kähler class of $SK$. Hence we have to take care of two transition functions when we go from one local chart to another one. The transition functions of the bundle $\otimes^3_{symm} T^{(1,0)\star}SK$ are quite familiar. If $\psi'$ and $\psi$ are two coordinate frames for $SK$, then we have $t_{\psi'\psi} = \left(\frac{d\psi}{d\psi'}\right)^3$. Hence we can write

$$W_{\psi'\psi'\psi'} = (f(\psi))^2 \left(\frac{d\psi}{d\psi'}\right)^3 W_{\psi\psi\psi} \tag{8.5.12}$$

where $f(\psi)$ is the transition function of the bundle $\mathcal{L}$. This is easy to determine in terms of the data of special geometry. Recalling the expression (8.3.4) for the Kähler potential which leads to

$$\mathcal{K} = \mathcal{K}(t,\bar{t}) - \ln\left(|X^0|^2\right) \tag{8.5.13}$$

where $\mathcal{K}(t,\bar{t})$ is the value of $\mathcal{K}$ in special coordinates, we conclude that

$$f(\psi) = \frac{X^{0\prime}(\psi)}{X^0(\psi)} \tag{8.5.14}$$

In particular in the flat coordinate system $t$ we can impose the standard *gauge* $X^{0\prime} = 1$, so that

$$W_{ttt} = \left(\frac{d\psi}{dt}\right)^3 W_{\psi\psi\psi} \frac{1}{(X^0(\psi))^2} \tag{8.5.15}$$

In a later subsection, starting from Eq.(8.5.15) we shall develop a general strategy for the evaluation of the coefficients appearing in the instantonic expansion of $W_{ttt}$. In the next subsection we just evaluate $W_{\psi\psi\psi}$.

### 8.5.1 Yukawa coupling as the fusion coefficient of the chiral ring

We start from Eq.s (8.3.3) and (7.7.163). As already remarked, the B-twisted sigma model correlator is nothing but the homologous correlator in the corresponding topological Landau–Ginzburg model, the star product of the Hodge ring being replaced by the ordinary product of the corresponding polynomial chiral ring $\frac{\mathbb{C}[X]}{\partial\mathcal{W}(X;\psi)}$. In the case of a general Calabi–Yau 3-fold described by a polynomial constraint $\mathcal{W}(X;\psi) = 0$ with many complex structure deformations, namely many parameters $\psi^\alpha$, consider the integer charge subring $\mathcal{R}^{Hod}(\mathcal{W}) \subset \mathcal{R}(\mathcal{W})$ of the chiral ring defined as in Eq. (5.9.45), namely

$$\mathcal{R}^{Hod}(\mathcal{W}) = \sum_{k=0}^{3} \mathcal{R}_{k|\nu}(\mathcal{W}) \tag{8.5.16}$$

The linear spaces $\mathcal{R}_{k|\nu}(\mathcal{W})$ contain the non-trivial polynomials $P_{k|\nu}(X)$ of degree $k|\nu$ as defined in Eq.(5.9.12) and are isomorphic to the Dolbeault cohomology groups $H^{(3-k,k)}$. The Calabi–Yau condition implies

$$\begin{array}{rclrcl} \dim \mathcal{R}_{0|\nu}(\mathcal{W}) &=& 1 & \dim \mathcal{R}_{3|\nu}(\mathcal{W}) &=& 1 \\ \dim \mathcal{R}_{1|\nu}(\mathcal{W}) &=& m & \dim \mathcal{R}_{2|\nu}(\mathcal{W}) &=& m \\ \dim_{\mathbb{C}} SK^{(CS)} &\stackrel{\text{def}}{=}& m & & & \end{array} \tag{8.5.17}$$

where $SK^{(CS)}$ is the special Kählerian moduli space of complex structure deformations. Let $P_{1|\nu;\alpha}(X)$ $(\alpha = 1, \ldots, m)$ be a basis of $\mathcal{R}_{1|\nu}(\mathcal{W})$ and $P_{3|\nu}(X) \in \mathcal{R}_{3|\nu}(\mathcal{W})$ be a choice of the top element of the chiral ring, which is unique up to multiplication by a constant $f(\psi)$ depending on the moduli. Given these data, we can choose a dual basis $P^{\alpha}_{2|\nu}(X)$ for $\mathcal{R}_{2|\nu}(\mathcal{W})$ such that

$$P_{1|\nu;\alpha}(X) \cdot P^{\beta}_{2|\nu}(X) = \delta^{\beta}_{\alpha} P_{3|\nu}(X) \mod \partial\mathcal{W} \tag{8.5.18}$$

Then we can introduce the structure constants $\mathcal{C}_{\alpha\beta\gamma}(\psi)$ of the Hodge ring, $\mathcal{R}^{Hod}(\mathcal{W})$, by setting

$$P_{1|\nu;\alpha}(X)\, P_{1|\nu;\beta}(X) = \mathcal{C}_{\alpha\beta\gamma}(\psi)\, P^{\gamma}_{2|\nu}(X) \mod \partial\mathcal{W} \tag{8.5.19}$$

Recalling Eq.(7.9.15) we can now easily evaluate the following three-point correlator of the topological Landau–Ginzburg model based on the superpotential $\mathcal{W}(X;\psi)$:

$$\begin{aligned} \langle P_{1|\nu;\alpha}(X)\, P_{1|\nu;\beta}(X)\, P_{1|\nu;\alpha}(X) \rangle_{(top\ L.G.)} &= \sum_{d\mathcal{W}=0} P_{1|\nu;\alpha}(X)\, P_{1|\nu;\beta}(X)\, P_{1|\nu;\alpha}(X)\, H^{-1}(X) \\ &= \mathcal{C}_{\alpha\beta\gamma}(\psi) \sum_{d\mathcal{W}=0} \frac{P_{3|\nu}(X)}{H(X)} \end{aligned} \tag{8.5.20}$$

where $H(X) = \det \partial_\Lambda \partial_\Sigma \mathcal{W}$ is the hessian polynomial. Since the hessian has maximal grade it is necessarily proportional to $P_{3|\nu}(X)$:

$$H(X) = \frac{1}{f_H(\psi)} P_{3|\nu}(X) \tag{8.5.21}$$

where $f_H(\psi)$ is a holomorphic function of the moduli parameters defined by the above relation. Using this information in (8.5.20) we obtain

$$\langle P_{1|\nu;\alpha}(X)\, P_{1|\nu;\beta}(X)\, P_{1|\nu;\alpha}(X) \rangle_{(top\ L.G.)} = \text{const} \times f_H(\psi)\, \mathcal{C}_{\alpha\beta\gamma}(\psi) \tag{8.5.22}$$

On the other hand, in view of the Griffiths residue map and of the triple identification encoded in Eq.(8.3.3) we also have

$$\begin{aligned} W_{\alpha\beta\gamma} &= \int_{\mathcal{M}} \Omega^{(3,0)} \wedge \frac{\partial^3}{\partial\psi^\alpha \partial\psi^\beta \partial\psi^\gamma} \Omega^{(3,0)} \\ &= \langle P_{1|\nu;\alpha}(X)\, P_{1|\nu;\beta}(X)\, P_{1|\nu;\alpha}(X) \rangle_{(top\ L.G.)} \\ &= \text{const} \times f_H(\psi)\, \mathcal{C}_{\alpha\beta\gamma}(\psi) \end{aligned} \tag{8.5.23}$$

Recalling the structure of the differential Picard–Fuchs system derived from the Griffiths residue map (see Eq.s (8.2.4), (8.2.7) with $n = 3$) and comparing it with the structure of the same system derived from special geometry (see Eq.(8.3.83)), we conclude that the period vector

$$\mathbf{\Pi}^i = \begin{pmatrix} \Pi_0^i \\ \Pi_1^i \\ \Pi_2^i \\ \Pi_3^i \end{pmatrix} = \begin{pmatrix} \int_{C_i} \frac{1}{(W(X))^1}\, \omega \\ \int_{C_i} \frac{P_{1|\nu;\alpha}(X)}{(W(X))^2}\, \omega \\ \int_{C_i} \frac{P^\beta_{2|\nu}(X)}{(W(X))^3}\, \omega \\ \int_{C_i} \frac{P_{3|\nu}(X)}{(W(X))^4}\, \omega \end{pmatrix} \tag{8.5.24}$$

satisfies a differential system:

$$\left[ \mathbb{1} \frac{\partial}{\partial\psi^\alpha} - \mathbf{A}_\alpha(\psi) \right] \cdot \mathbf{\Pi}^i = 0, \tag{8.5.25}$$

where the holomorphic connection $\mathbf{A}_\alpha(\psi)$ has the structure (8.3.98), namely $\mathbf{A}_\alpha = \mathbf{\Gamma}_\alpha + \mathbf{C}_\alpha$ , with

$$\mathbf{\Gamma}_\alpha = \begin{pmatrix} * & 0 & 0 & 0 \\ * & * & 0 & 0 \\ * & * & * & 0 \\ * & * & * & * \end{pmatrix} \qquad \mathbf{C}_\alpha = f_H(\psi) \begin{pmatrix} 0 & 1 & 0 & 0 \\ 0 & 0 & \mathcal{C}_{\alpha\beta\gamma}(\psi) & 0 \\ 0 & 0 & 0 & 1 \\ 0 & 0 & 0 & 0 \end{pmatrix} \tag{8.5.26}$$

the matrix $\mathbf{\Gamma}_\alpha$ being the flat connection of topological field theory while $\mathbf{C}_\alpha$ contains all the information on the fusion rule of the Hodge ring.

This being the general set-up we can apply it to our one-modulus case. The data in the $M\mathbb{CP}_4[5]$ case are

$$\begin{aligned} \mathcal{W}(X,\psi) &= \frac{1}{5} \sum_{\Lambda=1}^{5} X_\Lambda^5 - \psi \prod_{\Lambda=1}^{5} X_\Lambda \\ P_{1|5}(X) &= \prod_{\Lambda=1}^{5} X_\Lambda \\ P_{2|5}(X) &= \prod_{\Lambda=1}^{5} X_\Lambda^2 \\ P_{3|5}(X) &= \prod_{\Lambda=1}^{5} X_\Lambda^3 \end{aligned}$$

$$H(X) = \left(1 - \psi^5\right) \prod_{\Lambda=1}^{5} X_\Lambda^3 \tag{8.5.27}$$

the last line being derived by a straightforward evaluation of the determinant of the $5 \times 5$ hessian matrix, modulo the vanishing relations. From (8.5.27) we immediately obtain

$$\mathcal{C} = 1 \qquad f_H(\psi) = \frac{1}{(1 - \psi^5)} \tag{8.5.28}$$

and using (8.5.23) we conclude

$$W_{\psi\psi\psi} = \text{const} \times \frac{1}{(1 - \psi^5)} \tag{8.5.29}$$

The result (8.5.29) can be checked also at the level of the Picard–Fuchs differential system. First, recalling Eq.(8.2.30), we see that, in the case of the mirror quintic, the Picard–Fuchs differential system has indeed the form (8.5.25) and (8.5.26), with

$$\mathbb{\Gamma}_\alpha = \frac{1}{1 - \psi^5}\begin{pmatrix} 0 & 0 & 0 & 0 \\ 0 & 0 & 0 & 0 \\ 0 & 0 & 0 & 0 \\ \psi & 15\psi^2 & 25\psi^3 & 10\psi^4 \end{pmatrix} \qquad \mathbb{C}_\alpha = \frac{1}{1 - \psi^5}\begin{pmatrix} 0 & 1 & 0 & 0 \\ 0 & 0 & 1 & 0 \\ 0 & 0 & 0 & 1 \\ 0 & 0 & 0 & 0 \end{pmatrix} \tag{8.5.30}$$

This confirms the result (8.5.28). Furthermore, recalling Eq.(8.3.16) and comparing it with Eq.(8.3.11) and Eq.(8.2.35) we see that

$$W_{\psi\psi\psi} = a_4(\psi) = \frac{1}{1 - \psi^5} \qquad a_3(\psi) = -10 \quad \psi^4 = -2\frac{\frac{dW_{\psi\psi\psi}}{\delta\psi}}{W^2_{\psi\psi\psi}} \tag{8.5.31}$$

which are indeed consistent with the result (8.5.29).

### 8.5.2 General strategy for the evaluation of the Yukawa coupling of the mirror quintic

From the above discussion we have come to the conclusion that, in flat coordinates, the Yukawa coupling for the mirror quintic has the form

$$\begin{aligned} W_{ttt} &= \left(\frac{d\psi}{dt}\right)^3 W_{\psi\psi\psi} \frac{1}{(X^0(\psi))^2} \\ &= \left(\frac{d\psi}{dt}\right)^3 \frac{\psi^2}{(1 - z)} \frac{1}{h^2(z)} \end{aligned} \tag{8.5.32}$$

where $z = \psi^p = \psi^5$, and

$$X^0(\psi) = -\frac{1}{\psi} h(z) \tag{8.5.33}$$

is a period of the holomorphic 3-form that we interpret as the auxiliary homogeneous coordinate of special geometry. The structure (8.5.34) is a consequence of the following assumptions that we make at this point and that we shall verify in a later subection.

*There is a basis for the periods*

$$\psi \begin{pmatrix} X^0 \\ X^1 \\ \partial_0 F \\ \partial_1 F \end{pmatrix} = \mathcal{S} \begin{pmatrix} \widehat{\Pi}_2^{(p)} \\ \widehat{\Pi}_1^{(p)} \\ \widehat{\Pi}_0^{(p)} \\ \widehat{\Pi}_4^{(p)} \end{pmatrix} \tag{8.5.34}$$

*given by a suitable matrix $\mathcal{S}$, where:*

*A) The action of the modular group is given by symplectic integer-valued matrices $\Gamma \subset Sp(4,\mathbb{Z})$.*

*B) The period playing the role of the auxiliary homogeneous variable $X^0$ is holomorphic at large complex structures $\psi \to \infty$ and free from logarithmic singularities. Hence it has the structure (8.5.33), where $h(z)$ is a power series in $\frac{1}{z}$.*

*C) There is an element of the modular group, specifically $(\mathcal{AT})^{-1}$, whose action on the special coordinate*

$$t \stackrel{\text{def}}{=} \frac{X^1}{X^0} \tag{8.5.35}$$

*is a translation by a unit:*

$$(\mathcal{AT})^{-1} \; : \; t \longrightarrow t+1 \tag{8.5.36}$$

*D) The expression of the special coordinate $t$ in terms of the deformation parameter $\psi$ is*

$$t = -\frac{1}{2\pi i}\ln(z) + \frac{1}{2\pi i}\ln(c) - \frac{k(z)}{h(z)} \tag{8.5.37}$$

*where $c$ is some constant, and $k(z)$ is another power series in $\frac{1}{z}$ just as $h(z)$.*

Let us now consider the meaning of these assumptions and their consequences. We focus in particular on assumption C). A symplectic integer valued matrix $(\mathcal{AT})^{-1}$ which realizes the translation (8.5.36) has necessarily the form

$$(\mathcal{AT})^{-1} = \begin{pmatrix} 1 & 0 & 0 & 0 \\ 1 & 1 & 0 & 0 \\ a & b & 1 & -1 \\ c & d & 0 & 1 \end{pmatrix} \qquad a,b,c,d \in \mathbb{Z} \tag{8.5.38}$$

Recalling the homogeneity property of the generating function $F(X)$:

$$X^A\,\partial_A F(X) = 2\,F(X) \tag{8.5.39}$$

after the transformation generated by $(\mathcal{A}\mathcal{T})^{-1}$ we obtain

$$\begin{aligned} 2\widetilde{F}\left(\widetilde{X}\right) &= \left(\widetilde{X}^0, \widetilde{X}^1\right)\begin{pmatrix}\widetilde{F}_0\\ \widetilde{F}_1\end{pmatrix} \\ &= \left(X^0, X^0+X^1\right)\begin{pmatrix} aX^0+bX^1+F_0-F_1 \\ cX^0+dX^1+F_1\end{pmatrix} \\ &= 2F(X) + Q(X) \end{aligned} \tag{8.5.40}$$

where

$$Q(X) = (a+c)(X^0)^2 + (b+c+d)X^0X^1 + d(X^1)^2 \tag{8.5.41}$$

is a quadratic homogeneous polynomial. Since the modular group is an isometry group for the special Kählerian structure of the moduli space, it follows that

$$\widetilde{F}(X) = F(X) \tag{8.5.42}$$

So, dividing by $(\widetilde{X}^0)^2 = (X^0)^2$ to obtain the inhomogeneous holomorphic prepotential $\mathcal{F}(t) = (X^0)^{-2}F(X)$, we get

$$\mathcal{F}(t+1) = \mathcal{F}(t) + \mathcal{Q}(t) \tag{8.5.43}$$

where

$$\mathcal{Q}(t) = \frac{1}{2}\left((a+c) + (b+c+d)\,t + d\,t^2\right) \tag{8.5.44}$$

Since in special coordinates the Yukawa coupling is the third derivative of the inhomogeneous generating function:

$$W_{ttt} = \frac{\partial^3\mathcal{F}}{\partial t^3} \tag{8.5.45}$$

Eq.(8.5.43) implies that $W_{ttt}(t)$ is a periodic function of the special coordinate

$$W_{ttt}(t+1) = W_{ttt}(t) \tag{8.5.46}$$

and as such it can be expanded as follows:

$$W_{ttt}(t) = \sum_{N=0}^{\infty} d_N\, q^N \tag{8.5.47}$$

where

$$q \stackrel{\text{def}}{=} \exp\left[2\pi i t\right] \tag{8.5.48}$$

Comparing Eq.(8.5.47) with Eq.(7.7.116), we realize that, in the flat coordinate $t$, singled out by our assumptions A)–D), the Yukawa coupling, namely the correlator of three marginal operators in the B-twisted sigma model on $M\mathbb{CP}_4[5]$, has the same form as the correlator of three marginal operators in the A-twisted sigma model on some other Calabi–Yau 3-fold $M^{-1}(M\mathbb{CP}_4[5])$. In view of the mirror symmetry hypothesis, the B twisted correlator (8.5.47) should not only have the same form, but actually it should be equal to the homologous A-twisted correlator on a mirror manifold. In our example the mirror manifold is the ordinary quintic $M^{-1}(M\mathbb{CP}_4[5]) = \mathbb{CP}_4[5]$. For the mirror symmetry hypothesis to be verified, it is necessary that the coefficients $d_N$ should be such that

$$\begin{aligned}\sum_{N=0}^{\infty} d_N\, q^N &= 5 + \sum_{k=1}^{\infty} \frac{n_k\, k^3\, q^k}{1-q} \\ &= 5 + n_1\, q + (8n_2 + n_1)\, q^2 + (27n_3 + 8n_2 + n_1)\, q^3 \\ &\quad + (48n_4 + 27n_3 + 8n_2 + n1)\, q^4 + \dots \end{aligned} \tag{8.5.49}$$

where the numbers $n_k$, according to Eq.(7.7.116), are the numbers of rational curves of degree $k$ embedded in the ordinary quintic $\mathbb{CP}_4[5]$. By definition, $n_k$ are integer numbers and so have to be the $d_N$. Indeed, if we assume mirror symmetry and we compute the coefficients $d_N$, then we can predict the numbers $n_k$, by setting

$$\begin{aligned} n_1 &= d_1 \quad n_2 = \frac{1}{8}(d_2 - d_1) \\ n_3 &= \frac{1}{27}(d_3 - d_2) \qquad n_4 = \frac{1}{48}(d_4 - d_3) \\ n_5 &= \frac{1}{125}(d_5 - d_4) \quad \dots\dots \\ \dots & \quad \dots \quad \dots \quad \dots \\ n_k &= \frac{1}{k^3}(d_k - d_{k-1}) \quad \dots \quad \dots \\ \dots & \quad \dots \quad \dots \quad \dots \end{aligned} \tag{8.5.50}$$

Clearly the fact that the $d_N \in \mathbb{Z}_+$ be positive integers and that all the above ratios of integers be also positive integers $n_k = \frac{1}{k^3}(d_k - d_{k-1}) \in \mathbb{Z}_+$ is, if verified, an overwhelming evidence of the mirror symmetry hypothesis.

We now outline, starting from the assumptions A)–D) the algorithm that allows the calculations of the $d_N$ coefficients. In two subsequent subsections we verify A)–D) and we explicitly implement the algorithm reporting the predictions obtained by Candelas et al. [68] for the numbers $n_k$. As we shall have the opportunity to remark further on, such predictions have been confirmed by mathematicians.

Using

$$\frac{d\psi}{dz} = \frac{z^{1/p}}{pz} = \frac{z^{\frac{1}{5}}}{5z} \tag{8.5.51}$$

and substituting the result into Eq.(8.5.32), we obtain

$$W_{ttt} = \left(\frac{dz}{dt}\right)^3 \frac{z}{125\, z^3\, (1-z)\, [h(z)]^2} \tag{8.5.52}$$

On the other hand, recalling Eq.(8.5.37) we have

$$q = e^{2\pi i t} = \frac{c}{z} \exp\left[-\frac{k(z)}{h(z)}\right] \tag{8.5.53}$$

and we get

$$\frac{dq}{dz} = -\frac{c}{z^2}\,\delta(z)\,\exp\left[-\frac{k(z)}{h(z)}\right] \tag{8.5.54}$$

where

$$\delta(z) \stackrel{\text{def}}{=} 1 + z\frac{d}{dz}\left(\frac{k(z)}{h(z)}\right) \tag{8.5.55}$$

Then we can write

$$\frac{dz}{dt} = \frac{\frac{dq}{dt}}{\frac{dq}{dz}} = -\frac{2\pi i z}{\delta(z)} \tag{8.5.56}$$

so that

$$\begin{aligned} W_{ttt} &= c'Y(z) \\ Y(z) &\stackrel{\text{def}}{=} \frac{z}{[\delta(z)]^3\,(1-z)\,[h(z)]^2} \\ c' &= \left(\frac{-2\pi i}{5}\right)^3 \end{aligned} \tag{8.5.57}$$

The coefficient $c'$ is given in Eq.(8.5.57) but it can be readjusted to any desirable value by a change of the absolute normalization of the three point function $W_{\psi\psi\psi}$. For this reason we leave it unspecified. By means of Eq.(8.5.57) the Yukawa coupling is explicitly given as a function of $z$. On the other hand from our assumptions we know that it should be a power series in the variable $q$. Using

$$\frac{d}{dq} = \frac{dz}{dq}\frac{d}{dz} = -\frac{z^2}{c\,\delta(z)}\exp\left[\frac{k(z}{h(z)}\right]\frac{d}{dz} \tag{8.5.58}$$

we obtain an explicit formula for the coefficients $d_N$:

$$d_N = \frac{(-)^N c'}{c^N N!}\lim_{z\to\infty}\left(\frac{z^2}{\delta(z)}\exp\left[\frac{k(z}{h(z)}\right]\frac{d}{dz}\right)^N Y(z) \tag{8.5.59}$$

### 8.5.3 Logarithmic behaviour of the solutions in the neighbourhood of $\psi = \infty$

As we just saw, in order to calculate the Yukawa coupling, we need to know the behaviour of the periods at large complex structures, namely in the limit $\psi \longrightarrow \infty$. In particular, in order to verify assumption D) introduced in the above subsection we need to single out their logarithmic behaviour in the neighbourhood of the singular point $\psi = \infty$. Hence, in the present subsection we present a further analytic continuation of the solution basis $\widehat{\Pi}_j^{(p)}$ that extends them to the region $|\psi| >> 1$ and singles out their logarithmic behaviour at $\psi = \infty$.

The starting point is given by Eq.s (8.4.58) and (8.4.59), which yield

$$\widehat{\Pi}_j^{(p)}(\psi) \stackrel{\text{def}}{=} \frac{1}{(2\pi \mathrm{i})^{p-1}} \sum_{m=0}^{\infty} \frac{\Gamma\left(\frac{m}{p}\right)^p}{\Gamma(m)} \left(\alpha^m - 1\right)^{p-1} \alpha^{mj} \left(p\psi\right)^m \tag{8.5.60}$$

On the other hand, if we introduce the power series

$$\begin{aligned}\Delta_k^{(p)}(\psi) &\stackrel{\text{def}}{=} \sum_{n=0}^{\infty} \frac{\left[\Gamma\left(n+\frac{k}{p}\right)\right]^p}{\Gamma(pn+k)} (p\psi)^{pn+k} \\ &= \frac{\left[\Gamma\left(\frac{k}{p}\right)\right]^p}{\Gamma(k)} (p\psi)^p \; {}_pF_{p-1}\left(\frac{k}{p},\frac{k}{p},\ldots,\frac{k}{p},1;\frac{k+1}{p},\frac{k+2}{p},\ldots,\frac{k+p-1}{p};\psi^p\right)\end{aligned} \tag{8.5.61}$$

we get a basis of $p-1$ independent solutions of the differential equation that are all regular in the neighbourhood of the regular point $\psi = 0$. (Just like the $\widehat{\Pi}_j^{(p)}(\psi)$, the $\Delta_k^{(p)}$ are solutions only up to a common factor $\frac{1}{\psi}$. This can be checked directly, or deduced from the fact that the Picard–Fuchs equation (8.2.33), upon the substitution $z = \psi^p$, becomes a generalized hypergeometric equation in this new variable.) In order to discover the relation between the two bases it suffices to set $m = pn + k$ $(k = 0, 1, \ldots, p-1)$ in Eq.(8.5.60). We immediately obtain

$$\widehat{\Pi}_j^{(p)}(\psi) = \frac{1}{(2\pi \mathrm{i})^{p-1}} \sum_{k=0}^{p-1} \alpha^{kj} \left(\alpha^k - 1\right)^{p-1} \Delta_j^{(p)} \tag{8.5.62}$$

To obtain an analytic continuation of the $\Delta_k^{(p)}(\psi)$ power series to values $|\psi| >> 1$ we resort to the following integral representation:

$$\Delta_k^{(p)}(\psi) = -\int_{-\mathrm{i}\infty}^{+\mathrm{i}\infty} \frac{ds}{e^{2\pi \mathrm{i} s} - 1} \frac{\left[\Gamma\left(s+\frac{k}{p}\right)\right]^p}{\Gamma(ps+k)} (p\psi)^{ps+k} \tag{8.5.63}$$

For $|\psi| < 1$ we can close the integration contour in the right half plane enclosing the poles at $s = N \in \mathbb{Z}_+$. The result reproduces the original definition (8.5.61). For $|\psi| > 1$, instead, the integration contour can be closed in the left half plane which encircles the poles at

$$s = -N - \frac{k}{p} \quad N \in \mathbb{Z}_+ \tag{8.5.64}$$

due to the gamma function in the numerator. Since these poles are also poles for the gamma function in the denominator, their order is $p-1$. In this way we obtain

$$\Delta_k^{(p)}(\psi) = \sum_{N=0}^{\infty} \mathrm{Res}_{\left(s=-N-\frac{k}{p}\right)} \frac{\left[\Gamma\left(s+\frac{k}{p}\right)\right]^p}{\Gamma(ps+k)} (p\psi)^{ps+k} \tag{8.5.65}$$

To evaluate the residues we consider the behaviour of the gamma function ratio in the vicinity of the pole. For $s = -N - \frac{k}{p} + \varepsilon$ we get

$$\frac{\left[\Gamma\left(s+\frac{k}{p}\right)\right]^p}{\Gamma(ps+k)} \underset{s=-N-\frac{k}{p}+\varepsilon}{\approx} \sum_{r=0}^{p-2} \frac{c_r(p,k,N)}{\varepsilon^{p-1-r}} + \text{reg} \tag{8.5.66}$$

where the coefficients $c_r(p,k,N)$ , being a combination of derivatives of gamma functions, can be expressed in terms of digammas $\Psi(v)$ for suitable values of the argument $v$. We do not dwell on their explicit form, which the reader can easily work out by himself. At the same time we have

$$(p\psi)^{ps+k} \underset{s=-N-\frac{k}{p}+\varepsilon}{\approx} \sum_{x=0}^{p-2} \varepsilon^x \frac{(\ln(p\psi))^x}{x!} + \mathcal{O}\left(\varepsilon^{p-1}\right) \tag{8.5.67}$$

so that

$$\text{Res}_{\left(s=-N-\frac{k}{p}\right)} \frac{\left[\Gamma\left(s+\frac{k}{p}\right)\right]^p}{\Gamma(ps+k)} (p\psi)^{ps+k} = \sum_{r=0}^{p-2} \frac{(\ln(p\psi))^r}{(r+1)!} c_{p-2-r}(p,k,N) \frac{1}{(p\psi)^N} \tag{8.5.68}$$

Using these results in Eq.s (8.5.62) and (8.5.65), we obtain

$$\widehat{\Pi}_j^{(p)}(\psi) = \sum_{r=0}^{p-2} (\ln(p\psi))^r \sum_{N=0}^{\infty} b_{jrN}(p) \frac{(pN)!}{(N!)^p (p\psi)^{pN}} \tag{8.5.69}$$

where the coefficients $b_{jrN}$ are given by

$$b_{jrN} = \sum_{k=0}^{p-1} \alpha^{kj} \left(\alpha^k - 1\right)^{p-1} \frac{(N!)^p}{(pN)!\,(r+1)!} c_{p-r-2}(p,k,N) \tag{8.5.70}$$

In this way we have obtained expressions for the periods valid at large complex structures that single out a logarithmic singularity times a regular series in $\frac{1}{z} = \frac{1}{\psi^p}$. In the next section we come back to the case of the mirror quintic and we use the above results to evaluate the instanton expansion of the Yukawa coupling.

### 8.5.4 The instanton expansion of the Yukawa coupling and the prediction of the number of rational curves on the quintic 3-fold

After these preliminaries we can explicitly verify the assumptions A)–D), evaluate the coefficients $d_N$ using formula (8.5.59) and verify mirror symmetry, predicting the numbers $n_k$ from Eq.(8.5.50). The first step is given by the identification of the matrix $\mathcal{S}$ whose existence has been advocated in Eq.(8.5.34). It is given by

$$\mathcal{S} = \mathbb{C}_s \cdot \mathcal{M} = \begin{pmatrix} 0 & 0 & -1 & 0 \\ -\frac{1}{5} & -\frac{2}{5} & \frac{2}{5} & \frac{1}{5} \\ 0 & 1 & -1 & 0 \\ -\frac{3}{5} & -\frac{1}{5} & \frac{21}{5} & \frac{8}{5} \end{pmatrix} \tag{8.5.71}$$

$$\mathbb{C}_s = \begin{pmatrix} 0 & 1 & 0 & 0 \\ 2 & 0 & -1 & 0 \\ 0 & 0 & 0 & 1 \\ 1 & 0 & 0 & 0 \end{pmatrix} \tag{8.5.72}$$

where $\mathcal{M}$ is the matrix given by Eq.(8.4.107) which determines the transition to a symplectic basis, while $\mathbb{C}_s \in Sp(4, \mathbb{Z})$ is a further change of basis from one symplectic basis to another one. Performing the conjugation of the modular group generators (8.4.108) and (8.4.109) with $\mathbb{C}_s$:

$$\mathcal{A}'' = \mathbb{C}_s \mathcal{A}' \mathbb{C}_S^{-1} \qquad \mathcal{T}'' = \mathbb{C}_s \mathcal{T}' \mathbb{C}_S^{-1} \tag{8.5.73}$$

we obtain

$$\mathcal{A}'' = \begin{pmatrix} 1 & 0 & -1 & 0 \\ -1 & 1 & 1 & 0 \\ 5 & -8 & -4 & 1 \\ -3 & -5 & 3 & 1 \end{pmatrix} \tag{8.5.74}$$

$$\mathcal{T}'' = \begin{pmatrix} 1 & 0 & 1 & 0 \\ 0 & 1 & 0 & 0 \\ 0 & 0 & 1 & 0 \\ 0 & 0 & 0 & 1 \end{pmatrix} \tag{8.5.75}$$

$$(\mathcal{A}''\mathcal{T}'')^{-1} = \begin{pmatrix} 1 & 0 & 0 & 0 \\ 1 & 1 & 0 & 0 \\ -5 & 3 & 1 & -1 \\ 8 & 5 & 0 & 1 \end{pmatrix} \tag{8.5.76}$$

and comparing Eq.(8.5.76) with Eq.(8.5.38) we see that we have indeed found a basis where assumption C) is verified. Assumption A) is verified by construction since the change of basis has been effected by an element $\mathbb{C}_s$ of the integer symplectic group $Sp(4, \mathbb{Z})$. At the same time looking at the explicit form of the matrix $\mathcal{S}$ as given in Eq.(8.5.71), we see that:

$$\begin{aligned} X^0 &= -\frac{1}{\psi}\widehat{\Pi}_0^{(5)} \\ X^1 &= \frac{2}{5}\left(\widehat{\Pi}_0^{(5)} - \widehat{\Pi}_1^{(5)}\right) - \frac{1}{5}\left(\widehat{\Pi}_2^{(5)} - \widehat{\Pi}_4^{(5)}\right) \\ t &= \frac{2\left(\widehat{\Pi}_1^{(5)} - \widehat{\Pi}_0^{(5)}\right) + \left(\widehat{\Pi}_2^{(5)} - \widehat{\Pi}_4^{(5)}\right)}{5\,\widehat{\Pi}_0^{(5)}} \end{aligned} \tag{8.5.77}$$

Hence recalling the definition (8.4.48) of the period $\Pi_0^{(5)} = \frac{const}{\psi}\widehat{\Pi}_0^{(5)}(\psi)$ we see that assumption B) is also verified. Indeed $\Pi_0^{(5)}$ was precisely chosen as the solution of the Picard–Fuchs differential equation that is regular at $\psi = \infty$. It remains for us to check the validity of the assumption D). Starting from the last of Eq.s (8.5.77) and plugging in the results (8.5.69) and (8.5.70) obtained in the previous subsection, we obtain

$$t = -\frac{5}{2\pi i}\ln(5\psi) - \frac{1}{\widehat{\Pi}_0^{(5)}(\psi)}\sum_{m=1}^{\infty}\frac{5m!}{m!\,(5\psi)^{5m}}\left[\Psi(1+5m) - \Psi(1+5m)\right] \quad (8.5.78)$$

so that, comparing with Eq.(8.5.37), we see that assumption D) is also verified, with

$$\begin{aligned} k(z) &= \sum_{m=1}^{\infty}\frac{5m!}{m!\,(5\psi)^{5m}}\left[\Psi(1+5m) - \Psi(1+5m)\right] \\ h(z) &= \widehat{\Pi}_0^{(5)} \\ c &= \frac{1}{125} \end{aligned} \quad (8.5.79)$$

All the ingredients of the algorithm presented in the previous subsection being now use-ready, we can apply Eq.s (8.5.59) and (8.5.50) to evaluate the $d_N$ and $n_k$ coefficients. The result for the first ten values obtained by Candelas et al. in [69, 68] is reported below. Quite remarkably all the numbers calculated in this way are positive integers as implied by mirror symmetry. They predict the number of rational curves of degree $k$ embedded in the quintic 3-fold.

| *Degree* <br> $k$ | *Number of rational curves* <br> $n_k$ |
|---|---|
| 1 | 2875 |
| 2 | 6 09250 |
| 3 | 3172 06375 |
| 4 | 24 24675 30000 |
| 5 | 22930 58888 87652 |
| 6 | 248 24974 21180 22000 |
| 7 | 2 95091 05057 08456 59250 |
| 8 | 3756 32160 93747 66035 50000 |
| 9 | 50 38405 10416 98524 36451 06250 |
| 10 | 70428 81649 78454 68611 34882 49750 |

### 8.5.5 The special Kählerian metric of the moduli space of Kähler class deformations for the quintic 3-fold

We can now use our information on the instanton sum to determine the inhomogeneous prepotential $\mathcal{F}(t)$, the Kähler potential:

$$\mathcal{K}(t,\bar{t}) \;=\; -\ln\left[i\,Y(t,\bar{t})\right]$$

$$Y(t,\bar{t}) = 2\left(\mathcal{F}(t) - \overline{\mathcal{F}}(\bar{t})\right) - \left(\mathcal{F}'(t) + \overline{\mathcal{F}}'(\bar{t})\right)(t - \bar{t}) \qquad (8.5.80)$$

and the metric of the moduli space of Kähler class deformations:

$$g_{t\bar{t}} = \frac{\partial}{\partial t}\frac{\partial}{\partial \bar{t}}\mathcal{K}(t,\bar{t}) \qquad (8.5.81)$$

Starting from Eq.(8.5.47) and recalling Eq.(8.5.45), by means of a triple integration, we obtain

$$\begin{aligned} \mathcal{F}(t) &= \mathcal{F}_\infty(t) + \Delta\mathcal{F}(t) \\ \mathcal{F}_\infty(t) &= \frac{1}{3!}\,d_0\,t^3 + \mathcal{Q}(t) \\ \mathcal{Q}(t) &= \frac{1}{2}A_2\,t^2 + A_1\,t + i\,A_0 \\ \Delta\mathcal{F}(t) &= -i\sum_{N=1}^{\infty}\frac{d_N}{(2\pi N)^3}\,e^{2\pi i N t} \end{aligned} \qquad (8.5.82)$$

where $A_2, A_1$ and $A_0$ are three arbitrary complex constants introduced by the triple integration. With the additional information provided by our explicit evaluation of the periods, we can determine also the numerical value of these constants. Indeed the period vector associated with a prepotential $\mathcal{F}(t)$ has the general form

$$\frac{1}{X^0}\begin{pmatrix} X^0 \\ X^1 \\ F_0 \\ F_1 \end{pmatrix} = \begin{pmatrix} 1 \\ t \\ 2\mathcal{F}(t) - t\,\mathcal{F}'(t) \\ \mathcal{F}'(t) \end{pmatrix} \qquad (8.5.83)$$

and in the limit $\mathrm{Im}\,t \to \infty$ we can approximate $\mathcal{F}(t)$ with the its cubic polynomial part $\mathcal{F}_\infty(t)$. Indeed $\mathcal{F}_\infty(t)$ is the prepotential of the metric in the large complex structure limit that, via mirror map, coincides with the large radius limit on the mirror manifold $\mathbb{CP}_4[5]$, while $\Delta\mathcal{F}(t)$ contains all the instanton corrections. In this approximation we have

$$\frac{1}{X^0}\begin{pmatrix} X^0 \\ X^1 \\ F_0 \\ F_1 \end{pmatrix} \underset{\mathrm{Im}\,t\to\infty}{\approx} \begin{pmatrix} 1 \\ t \\ -\frac{1}{3!}\,d_0\,t^3 + A_1\,t + 2\mathrm{i}\,A_0 \\ \frac{1}{2}\,d_0\,t^2 + A_2\,t + A_1 \end{pmatrix} \qquad (8.5.84)$$

On the other hand, using the explicit form of the matrix $\mathcal{S}$ (8.5.71), together with the expansions (8.5.69) and the explicit form (8.5.70) of the coefficients $b_{jrN}$ we obtain the asymptotic expansion

$$\frac{1}{X^0}\begin{pmatrix} X^0 \\ X^1 \\ F_0 \\ F_1 \end{pmatrix} \underset{\mathrm{Im}\,t\to\infty}{\approx} \begin{pmatrix} 1 \\ t \\ -\frac{5}{6}\,t^3 + -\frac{25}{12}\,t + 2\mathrm{i}\,\frac{25}{\pi^3}\,\zeta(3) \\ \frac{5}{2}\,t^2 + \frac{11}{2}\,t - \frac{25}{12} \end{pmatrix} \qquad (8.5.85)$$

which is consistent with (8.5.84) and implies

$$d_0 = 5 \quad A_2 = \frac{11}{2} \quad A_1 = -\frac{25}{12} \quad A_0 = \frac{25}{\pi^3}\zeta(3) \tag{8.5.86}$$

The special geometry determined by the cubic prepotential $\mathcal{F}_\infty(t)$ is the geometry of Kähler class deformations as we determined it in Chapter 6 through the Kaluza–Klein approximation. In section 6.3, we found

$$\mathcal{F}_{KK}(t) = \frac{1}{3!} W^{(0)}_{ttt}\, t^3 + \mathcal{Q}(t) \tag{8.5.87}$$

where

$$W^{(0)}_{ttt} = \int_{\mathcal{M}_3} \omega^{(1,1)} \wedge \omega^{(1,1)} \wedge \omega^{(1,1)} \tag{8.5.88}$$

is the intersection integral of three times the generator $V^{(1,1)}$ of the one dimensional Dolbeault group $H^{(1,1)}$, the cohomology class of the Kähler 2-form being $[K] = \mathrm{Imt}\left[\omega^{(1,1)}\right]$. The intersection integral (8.5.88) is also the first contribution (that of the constant maps) to the three-point correlator in the A-twisted topological sigma model: this is the reason for its name. The quadratic polynomial $\mathcal{Q}(t)$ is undetermined in the Kaluza–Klein approximation and uninfluential on the Kähler metric, as long as the coefficients of all the terms are real. As is evident from Eq.(8.5.86), this is what happens, in our case, for the quadratic and the linear terms that have real coefficients, but not for the constant term, which is purely imaginary. Hence we can write

$$\mathcal{F}_{KK}(t) = \mathcal{F}_\infty(t) - \mathrm{i}\, A_0 \quad \Longrightarrow \quad d_0 = W^{(0)}_{ttt} = 5 \tag{8.5.89}$$

and recalling Eq.(8.5.80) we can set

$$\begin{aligned} Y(t,\bar{t}) &= Y_\infty(t,\bar{t}) + \Delta Y(t,\bar{t}) \\ Y_\infty(t,\bar{t}) &= Y_{KK}(t,\bar{t}) + 2\,\mathrm{i}\, A_0 \\ Y_{KK}(t,\bar{t}) &= -\frac{1}{3!}\, d_0\, (t-\bar{t})^3 \\ \Delta Y(t,\bar{t}) &= 2\,\mathrm{i} \sum_{N=1}^{\infty} \frac{d_N}{(2\pi N)^3} \left[\pi N \mathrm{Imt} + 1\right] \left(q^N + \bar{q}^N\right) \end{aligned} \tag{8.5.90}$$

If we calculate the Kähler metric, using the definition (8.5.81), namely

$$g_{t\bar{t}} = -\frac{1}{Y(t,\bar{t})^2} \left(Y(t,\bar{t})\, \partial_t \partial_{\bar{t}} Y(t,\bar{t}) - \partial_t Y(t,\bar{t})\, \partial_{\bar{t}} Y(t,\bar{t})\right) \tag{8.5.91}$$

we obtain the exact result by plugging in the expressions given in Eq.(8.5.90). We can also utilize (8.5.91) to obtain the asymptotic behaviour of the metric at large complex structures (or large radius, via the mirror map) by replacing the exact $Y(t,\bar{t})$ with either $Y_{KK}(t,\bar{t})$ or $Y_{\infty}(t,\bar{t})$. We find

$$\begin{aligned} g_{t\bar{t}}^{(KK)} &= \frac{-3}{(t-\bar{t})^2} \\ g_{t\bar{t}}^{(\infty)} &= \frac{-3}{(t-\bar{t})^2} + -144\mathrm{i}\,\frac{A_0}{d_0}\frac{1}{(t-\bar{t})^5} + \mathcal{O}\left((t-\bar{t})^{-8}\right) \end{aligned} \tag{8.5.92}$$

As we see, the metric $g_{t\bar{t}}^{(KK)}$ obtained in the Kaluza–Klein approximation $g_{t\bar{t}}^{(KK)}$ is that of the homogeneous space $SU(1,1)/U(1)$. The quadratic terms appearing in the prepotential $\mathcal{F}_{\infty}$ of the metric $g_{t\bar{t}}^{(\infty)}$ make to differ from the $SU(1,1)$-symmetric metric $g_{t\bar{t}}^{(KK)}$, but the difference tends to zero as $\frac{1}{(t-\bar{t})^5}$ in the limit $\mathrm{Im}\,t \to \infty$. Hence the leading approximation in the large radius limit is indeed the metric $g_{t\bar{t}}^{(KK)}$ obtained from the Kaluza-Klein approach, as we have been claiming all the time.

### 8.5.6 Summary and conclusion

Summarizing, we have calculated the exact metric for the moduli space of Kähler class deformations in the case of the quintic Calabi–Yau 3-fold $\mathbb{CP}_4[5]$. Using the information that this Kählerian geometry is of the special type, we can determine it starting only from holomorphic data, namely from the prepotential $\mathcal{F}(t)$. This latter is obtained by means of a triple integration, once the Yukawa coupling $W_{ttt}(t)$ is given as a function of the flat coordinate $t$. From the point of view of topological field theory $W_{ttt}(t)$ is a three-point correlator of the A-twisted topological sigma model on the manifold $\mathcal{M} = \mathbb{CP}_4[5]$. As such the determination of its exact value involves the knowledge of the number $n_k$ of rational curves of degree $k$ embedded in $\mathcal{M}$. This requires the solution of a rather formidable mathematical problem. However, the hypothesis of mirror symmetry implies that this three-point correlator in the A-twisted sigma model on $\mathcal{M} = \mathbb{CP}_4[5]$ is equal to the homologous three-point correlator in the B-twisted sigma model on the mirror manifold $M^{-1}[\mathcal{M}] = M\mathbb{CP}_4[5]$. The calculation of three-point correlators in the B-twisted sigma model is rather straightforward and is fully determined by the structure constants of the chiral ring $\frac{\mathbb{C}[X]}{\partial W}(X,\psi)$ associated with the defining polynomial constraint $\mathcal{W}(X,\psi) = 0$. In this way, however, one obtains $W_{\psi\psi\psi}(\psi)$, namely the three-point function in the natural coordinate $\psi$ which is not the flat one $t$. What remains to be done is to find the relation $t = t(\psi)$, between the flat coordinate and the natural one. This is done by means of the Picard–Fuchs differential equation. It suffices to recognize that the column vector $(X^0, X^a, F_0, F_a)$ is the vector of periods of the holomorphic 3-form $\Omega^{(3,0)}$ along a basis of homology 3-cycles. Then using the Griffiths residue map representation of these periods, we are able to derive a linear system of first order differential equations they satisfy in terms of the natural variable $\psi$. The explicit solution of this differential problem provides the desired relation $t = t(\psi)$ and concludes the programme. As a by-product one is able to predict the number of rational curves embedded in the quintic 3-fold. The logic of this derivation can be applied to the manifolds with many moduli, provided one knows their mirror manifold and is able to solve the Picard–Fuchs differential system of equations.

## 8.6 Bibliographical Note

- *The idea of mirror symmetry was introduced in 1990 by Greene and Plesser in [180] and by Aspinwall, Lutken and Ross in [18]. A strong evidence for this conjecture had been provided by the plot of $h^{(1,1)}$ and $h^{(2,1)}$ numbers given by Candelas, Lynker and Schimmrigk in [71]. Other references on mirror manifolds are [41] and [42], [40, 16, 17, 192], [58].*

- *The idea of the mirror map and the first example of the use of Picard–Fuchs equations in such a context are contained in the two seminal 1991 papers by Candelas, de la Ossa, Green and Parkes [69, 68]. Additional references on mirror maps are [67, 39], [193, 148]. The formulation of mirror symmetry in terms of the exchange of A and B-twisted topological models is due to Witten [263].*

- *The mathematical theory of Picard–Fuchs equations had been developed in the 1960s by Katz [193], by Griffiths [182] and by Atiyah, Bott and Gäarding [148]. The general method for the derivation of these equations was explained in a language accessible to physicists by Cadavid and Ferrara in [59], who followed Katz's work, and by Lerche, Smit and Warner [213] whose work is based more on Griffiths' papers. In this chapter we have mainly followed the approach of [213].*

- *The relation between special geometry and Picard–Fuchs equations illustrated in section 8.3 is a result due to Ceresole, D'Auria, Ferrara, Lerche and Louis [88] (1993). The flat holomorphic connection implied by special geometry had been found the year before, by Ferrara and Louis in [143]. Altogether for the topics dealt with in section 8.3 a convenient set of references is [143, 88, 59, 90, 89], [114, 149, 229, 230, 199], [20, 27, 28, 128].*

- *Duality symmetries in string theory have been studied by many authors. A short list of references on such a topic is provided by: [172, 120, 207, 212, 130], [170, 171, 169, 144, 220], [235, 251, 145, 174] and [58, 180, 147] [211, 90, 148]. The notion of duality group of the Landau–Ginzburg superpotential as we presented it in section 8.4 is due to Lerche, Smit and Warner [213].*

- *The derivation of the monodromy group for the quintic 3-fold is also due to the seminal papers of Candelas et al. [69, 68]. Other examples of monodromy groups for one-modulus cases were worked out by several authors. The first derivation of the monodromy group in a two-moduli example is due to Ceresole, D'Auria and Regge [90], who found an extremely interesting connection with the braid group.*

*Altogether for the topics discussed in section 8.4 a convenient set of references is [68, 174], [213, 180, 58].*

- *In the discussion of the mirror map presented in section 8.5 we have mostly followed [68, 16].*

# Chapter 9

# FAREWELL

*The patient reader who has gone through the above pages, if he shares the feelings of the authors, has had an opportunity to admire the remarkable conceptual machinery, developed in recent years, for the calculation of Yukawa couplings in Calabi–Yau compactifications of string theory. From the physical point of view, it is clear that these couplings have a deep meaning, provided the whole scenario of superstring is correct. In principle these couplings contain the basic information on the low energy effective gauge theory of quarks and leptons. In the authors' opinion, however, the conceptual machinery presented in this book and the*

$$\text{Physics} \bigcap \text{Mathematics}$$

*ideas originated in this environment, in particular the whole programme of*

Topological Field Theories

*are much more important than their actual applications to string theory. Indeed it is the hope of establishing new methods in non-perturbative field theory which should be considered the main motivation of the whole set of results presented here. Actually it now appears that the most promising applications of topological field theories might be in the context of four-dimensional field theories, including gravity and Yang–Mills theory (see for example [237, 266, 267, 264, 87, 13, 202, 9]). In this book we have restricted our attention to two dimensions, where these ideas have been developed in a most exhaustive way. The future is probably four-dimensional but the two-dimensional lesson is, in our opinion, essential.*

# Bibliography

[1] M. Ademollo, L. Brink, A. D' Adda, R. D'Auria, E. Napolitano, S. Sciuto, E. del Giudice, P. Di Vecchia, S. Ferrara, F. Gliozzi, R. Musto, and R. Pettorino. "Supersymmetric strings and color confinement". *Phys. Lett.*, 62B:105, (1976).

[2] M. Ademollo, L. Brink, A. D' Adda, R. D'Auria, E. Napolitano, S. Sciuto, E. del Giudice, P. Di Vecchia, S. Ferrara, F. Gliozzi, R. Musto, R. Pettorino, and J.H. Schwarz. "Dual string with U(1) color symmetry". *Nucl. Phys.*, B111:77, (1976).

[3] M. Ademollo, L. Brink, A. D' Adda, R. D'Auria, E. Napolitano, S. Sciuto, E. del Giudice, P. Di Vecchia, S. Ferrara, F. Gliozzi, R. Musto, and R. Pettorino. "Dual string models with non abelian colour and flavour symmetries". *Nucl. Phys.*, B114:297, (1976).

[4] L. Alvarez-Gaume'. "Supersymmetry and the Atiyah–Singer index theorem". *Comm. Math. Phys.*, 90:161, (1983).

[5] L. Alvarez-Gaume' and E. Witten. "Gravitational anomalies". *Nucl. Phys.*, B324:268, (1983).

[6] D. Anselmi, M. Billo', P. Fre', L. Girardello, and A. Zaffaroni. "ALE manifolds and conformal field theoris". *Int. J. Mod. Phys.*, A9:3007, (1994).

[7] D. Anselmi and P. Fre'. "Topological twist in four dimensions, R duality and hyperinstantons". *Nucl. Phys.*, B404:288, (1993).

[8] D. Anselmi and P. Fre'. "Twisted N=2 supergravity as topological gravity in four dimensions". *Nucl. Phys.*, B392:401, (1993).

[9] D. Anselmi and P. Fre'. "Gauged hyperinstantons and monopole equations". *hep-th*, 9411205, (1994).

[10] D. Anselmi and P. Fre'. "Topological sigma models in four dimensions and tri–holo morphic maps". *Nucl. Phys.*, B416:255, (1994).

[11] D. Anselmi, P. Fre', L. Girardello, and P. Soriani. "Constrained topological field theory". *Phys. Lett.*, 335B:416, (1994).

[12] D. Anselmi, P. Fre', L. Girardello, and P. Soriani. "Constrained topological gravity from twisted N=2 Liouville theory". *Nucl. Phys.*, B427:351, (1994).

[13] P. Argyres and A. Faraggi. "The vacuum structure and spectrum of N=2 supersymmetric SU(N) gauge theory". *hep-th*, 9411057, (1994).

[14] V.I. Arnold, S.M. Gusein-Zade, and A.N. Varchenko. *Singularities of Differentiable Maps*. Birkhäuser, 1975.

[15] M. Artin. "On isolated rational singularities". *Am. J. Math.*, 88:129, (1966).

[16] P. Aspinwall and D. Morrison. "Topological field theories and rational curves". *Comm. Math. Phys.*, 151:245, (1993).

[17] P.S. Aspinwall and C.A. Lutken. "Quantum algebraic geometry of superstring compactifications". *Nucl. Phys.*, B355:482, (1991).

[18] P.S. Aspinwall, C.A. Lutken, and G.G. Ross. "Construction and couplings of mirror manifolds". *Phys. Lett.*, 241B:373, (1990).

[19] J.J Atick, A. Dhar, and B. Ratra. "Superstring propagation in curved superspace in the presence of background super Yang Mills fields". *Phys. Lett.*, 169B:54, (1986).

[20] M.F. Atiyah, R. Bott, and L. Gäarding. "Lacunas of four hyperbolic differential operators with constant coefficients: II ". *Acta Math.*, 131:145, (1973).

[21] M.F. Atiyah and L. Jeffrey. "Topological lagrangians and cohomology". *J. Diff. Geom.*, 7:119, (1990).

[22] M.F. Atiyah and G.B. Segal. "Index of elliptic operators II". *Ann. Math.*, 87:531, (1968).

[23] M.F. Atiyah and I.M. Singer. "Index of elliptic operators I-III". *Ann. Math.*, 87:485, (1968).

[24] M.F. Atiyah and I.M. Singer. "Index of elliptic operators IV-V". *Ann. Math.*, 93:119, (1971).

[25] J. Bagger and E. Witten. "Quantization of Newton's constant in certain supergravity theories". *Phys. Lett.*, 115B:202, (1982).

[26] J. Bagger and E. Witten. "Coupling the gauge invariant supersymmetric sigma model to supergravity". *Nucl. Phys.*, B211:302, (1983).

[27] J. Balog, L. Feher, L. O'Raifeartaigh, P. Forgacs, and A. Wipf. "Kac-Moody realization of W-algebras". *Phys. Lett.*, 244B:435, (1990).

[28] J. Balog, L. Feher, L. O'Raifeartaigh, P. Forgacs, and A. Wipf. "Toda theory and W-algebra from a gauged WZWN point of view". *Ann. Phys.*, 203:194, (1990).

[29] T. Banks, L. Dixon, D. Friedan, and S. Shenker. "Constraints on string vacua with space–time supersymmetry". *Nucl. Phys.*, B307:93, (1988).

[30] T. Banks, L. Dixon, D. Friedan, and S. Shenker. "Phenomenology and conformal field theory or can string predict the weak mixing angle". *Nucl. Phys.*, B299:613, (1988).

[31] L. Baulieu and M. Bellon. "BRS symmetry of supergravity and its projection to component formalism". *Nucl. Phys.*, B294:279, (1987).

[32] L. Baulieu and I. M. Singer. "The topological sigma model". *Comm. Math. Phys.*, 125:227, (1989).

[33] L. Baulieu and I.M. Singer. "Topological Yang–Mills theory". *Nucl. Phys. B (proc. suppl.)*, 5B:12, (1988).

[34] C. Becchi, A. Rouet, and R. Stora. "The abelian Higgs–Kibble model, unitarity of the S-operator". *Phys. Lett.*, 52B:344, (1974).

[35] C. Becchi, A. Rouet, and R. Stora. "Renormalization of gauge theories". *Ann. Phys.*, 98:287, (1976).

[36] C.M. Becchi, R. Collina, and C. Imbimbo. "A Functional and Lagrangian formulation of two-dimensional topological gravity". In *Trieste Spring Workshop on String Theory*, (1994).

[37] S. Bellucci, D. A. Depireux, and S.J. Gates. "Consistent and universal inclusion of the Lorentz Chern–Simons form in D=10, N=1 supergravity theories". *Phys. Lett.*, 238B:315, (1990).

[38] S. Bellucci and S.J. Gates. "D=10 , N=10 supespace supergravity and the Lorentz Chern–Simons term". *Phys. Lett.*, 208B:456, (1988).

[39] P. Berglund, P. Candelas, X. C. de la Ossa, A. Font, T. Hübsch, D. Jancic, and F. Quevedo. "Periods for Calabi–Yau and Landau–Ginzburg vacua". *Nucl. Phys.*, B419:352, (1994).

[40] P. Berglund, E. Derrick, T. Hübsh, and D. Jancic. "On periods for string compactifications". *Nucl. Phys.*, B420:268, (1994).

[41] P. Berglund and T. Hübsh. "A generalized construction of mirror manifolds". *Nucl. Phys.*, B393:377, (1993).

[42] P. Berglund and S. Katz. "Mirror symmetry for hypersurfaces in weighted projective space and topological couplings". *Nucl. Phys.*, B420:289, (1994).

[43] E. Bergshoeff, M. de Roo, B. de Wit, and P. van Nieuwenhuizen. "Ten dimensional Maxwell Einstein supergravity, its currents and the issue of its auxiliary fields". *Nucl. Phys.*, B195:97, (1982).

[44] E. Bergshoeff, A. Salam, and E. Sezgin. "Supersymmetric $R^2$ actions, conformal invariance and the Lorentz Chern–Simons term in 6 and 10 dimensions". *Nucl. Phys.*, B279:659, (1987).

[45] M. Billo' and P. Fre'. "N=4 versus N=2 phases, hyper-Kähler quotients and the 2d topological twist". *Class. Quantum Grav.*, 11:785, (1994).

[46] M. Billo', P. Fre', L. Girardello, and A. Zaffaroni. "Gravitational instantons in heterotic string theory: the H–map and the moduli deformations of (4,4) superconformal theories". *Int. J. Mod. Phys.*, A8:2351, (1993).

[47] D. Birmingham, M. Blau, and M. Rakowski. "Topological field theories". *Phys. Rep.*, 209:129, (1991).

[48] B. Block and A. Varchenko. "Topological conformal field theories and the flat coordinates". *Int. J. Mod. Phys.*, A7:1467, (1992).

[49] L. Bonora, M. Bregola, R. D'Auria, P. Fre', K. Lechner, P. Pasti, I. Pesando, M. Raciti, F. Riva, M. Tonin, and D. Zanon. "Some remarks on the supersymmetrization of the Lorentz Chern–Simons form in D=10, N=1 supergravity theories". *Phys. Lett.*, 277B:306, (1992).

[50] L. Bonora, M. Bregola, K. Lechner, P. Pasti, and M. Tonin. "Anomaly-free supergravity and super Yang–Mills theories in ten dimensions". *Nucl. Phys.*, B296:877, (1988).

[51] L. Bonora, M. Bregola, K. Lechner, P. Pasti, and M. Tonin. "A discussion of the constraints in N=1 SUGRA + YM theory in 10D". *Int. J. Mod. Phys.*, A5:461, (1990).

[52] L. Bonora, P. Pasti, and M. Tonin. "Superspace approach to quantum gauge theories". *Ann. Phys.*, 144:15, (1982).

[53] L. Bonora, P. Pasti, and M. Tonin. "Supespace formulation of 10D supergravity and super Yang Mills theory à la Green–Schwarz". *Ann. Phys.*, 175:112, (1987).

[54] L. Bonora and C. S. Xiong. "Matrix models without scaling limits". *Int. J. Mod. Phys.*, A8:2973, (1993).

[55] W. Boucher, D. Friedan, and A. Kent. "Determinant formulae and unitarity for the N=2 superconformal algebras, or exact results on string compactification". *Phys. Lett.*, B315:193, (1989).

[56] E. Brieskorn. "Singular elements of semisimple algebraic groups". In *Proc. Int. Congr. Math.*, page 279, Nice, (1970).

[57] R. Bryant and P. Griffiths. "Some observations on the infinitesimal period relations for the regular threefolds with trivial canonical bundle". In *Arithmetic and Geometry, papers dedicated to I. R. Shafarevith*, page 77. Birkhauser, (1983).

[58] A. Cadavid, M. Bodner, and S. Ferrara. "Calabi–Yau supermoduli space, field strength duality and mirror manifolds". *Phys. Lett.*, 247B:25, (1991).

[59] A. Cadavid and S. Ferrara. "Picard–Fuchs equations and the moduli space of supercoformal field theories". *Phys. Lett.*, 267B:193, (1991).

[60] E. Calabi. "Metriques Kähleriennes et fibres holomorphes". *Ann. Scie. Ec. Norm. Sup.*, 12:269, (1979).

[61] E. Calabi. "On Kähler manifolds with vanishing canonical class". In *Algebraic Geometry and Topology : A Symposium in Honor of S. Lefschetz*, page 78. Princeton University Press, (1995).

[62] C. G. Callan, D. Friedan, E. Martinec, and M. Perry. "Strings in background fields". *Nucl. Phys.*, B262:593, (1985).

[63] P. Candelas. "Yukawa couplings between (2,1) forms". *Nucl. Phys.*, B298:458, (1988).

[64] P. Candelas, A.M. Dale, C.A. Lutken, and R. Schimmrigk. "Complete intersection Calabi–Yau manifolds". *Nucl. Phys.*, B298:493, (1988).

[65] P. Candelas and X. de la Ossa. "Comments on conifolds". *Nucl. Phys.*, B342:246, (1990).

[66] P. Candelas and X. C. de la Ossa. "Moduli space of Calabi–Yau manifolds". *Nucl. Phys.*, B355:455, (1991).

[67] P. Candelas, X. C. de la Ossa, A. Font, S. Katz, and D. Morrison. "Mirror symmetry for two-parameter models". *Nucl. Phys.*, B416:481, (1994).

[68] P. Candelas, X. C. de la Ossa, P. S. Green, and L. Parkes. "A pair of Calabi–Yau manifolds as an exatly soluble superconformal theory". *Nucl. Phys.*, B359:21, (1991).

[69] P. Candelas, X. C. de la Ossa, P. S. Green, and L. Parkes. "An exactly soluble superconformal theory from a mirror pair of Calabi–Yau manifolds". *Phys. Lett.*, 258B:118, (1991).

[70] P. Candelas, C.T. Horowitz, A. Strominger, and E. Witten. "Vacuum configurations for superstrings". *Nucl. Phys.*, B258:46, (1985).

[71] P. Candelas, M. Lynker, and R. Schimmrigk. "Calabi–Yau manifolds in weighted $P_4$". *Nucl. Phys.*, B341:383, (1990).

[72] A. Cappelli, C. Itzykson, and J.B. Zuber. "Modular invariant partition functions in two dimensions". *Nucl. Phys.*, B280:445, (1987).

[73] L. Castellani. "Supergrav: a Reduce package for Bose–Fermi exterior calculus and the construction of supergravity actions". *Int. J. Mod. Phys.*, A3:1435, (1988).

[74] L. Castellani, R. D'Auria, and S. Ferrara. "Special geometry without special coordinates". *Class. Quantum Grav.*, 7:1767, (1990).

[75] L. Castellani, R. D'Auria, and S. Ferrara. "Special Kähler geometry: an intrinsic formulation from N=2 spacetime supersymmetry ". *Phys. Lett.*, 241B:57, (1990).

[76] L. Castellani, R. D'Auria, and P. Fre. *Supergravity and Superstrings: A Geometric Perspective.* World Scientific, 1991.

[77] S. Cecotti. "Homogeneous Kähler manifolds and T-algebras in N=2 supergravity and superstrings". *Comm. Math. Phys.*, 124:23, (1989).

[78] S. Cecotti. "N=2 supergravity, type II superstrings and algebraic geometry". *Comm. Math. Phys.*, 131:517, (1990).

[79] S. Cecotti. "Geometry of N=2 Landau–Ginzburg families". *Nucl. Phys.*, B355:754, (1991).

[80] S. Cecotti. "N=2 Landau-Ginzburg versus Calabi–Yau sigma models: non-perturbative aspects". *Int. J. Mod. Phys.*, A6:1749, (1991).

[81] S. Cecotti, S. Ferrara, and L. Girardello. "A topological formula for the Kähler potential of 4D, N=1,2 strings and its implications for the moduli problem". *Phys. Lett.*, 213B:443, (1988).

[82] S. Cecotti, S. Ferrara, and L. Girardello. "Geometry of type II superstrings and the moduli of superconformal field theories". *Int. J. Mod. Phys.*, A4:2475, (1989).

[83] S. Cecotti, S. Ferrara, and M. Villasante. "Linear multiplets and super Chern–Simons forms in 4d supergravity". *Int. J. Mod. Phys.*, A2:1839, (1987).

[84] S. Cecotti, L. Girardello, and A. Pasquinucci. "Non-perturbative aspects and exact results for the N= Landau-Ginzbug models". *Nucl. Phys.*, B328:701, (1989).

[85] S. Cecotti, L. Girardello, and A. Pasquinucci. "Singularity theory and N=2 supersymmetry". *Int. J. Mod. Phys.*, A6:2427, (1991).

[86] S. Cecotti and C. Vafa. "Topological anti-topological fusions". *Nucl. Phys.*, B367:359, (1991).

[87] A. Ceresole, R. D'Auria, and S. Ferrara. "On the geometry of moduli space of vacua in N=2 supersymmetric Yang-Mills theory". *Phys. Lett.*, 339B:71, (1994).

[88] A. Ceresole, R. D'Auria, S. Ferrara, W. Lerche, and J. Louis. "Picard Fuchs equations and special geometry". *Int. J. Mod. Phys.*, 8:79, (1993).

[89] A. Ceresole, R. D'Auria, S. Ferrara, W. Lerche, J. Louis, and T. Regge. "Picard-Fuchs equations, special geometry and target space duality". In *Essays on Mirror Manifolds.* S.T. Yau International Press, 1994.

[90] A. Ceresole, R. D'Auria, and T. Regge. "Duality group for Calabi-Yau two moduli space". *Nucl. Phys.*, B414:517, (1994).

[91] G.F. Chapline and N. Manton. "Unification of Yang-Mills theory and supergravity in ten dimensions". *Phys. Lett.*, 120B:105, (1983).

[92] S.S. Chern. *Complex Manifolds Without Potential Theory.* Springer-Verlag, 1979.

[93] E. Cremmer, S. Ferrara, L. Girardello, and A. Van Proeyen. "Yang-Mills theories with local supersymmetry: lagrangian, transformation laws and super-Higgs effect". *Nucl. Phys.*, B212:413, (1983).

[94] E. Cremmer, B. Julia, J. Scherk, S. Ferrara, L. Girardello, and P. van Nieuwenhuizen. "Super Higgs effect in supergravity with general scalar interactions". *Phys. Lett.*, 79B:231, (1978).

[95] E. Cremmer, B. Julia, J. Scherk, S. Ferrara, L. Girardello, and P. van Nieuwenhuizen. "Spontaneous symmetry breaking and Higgs effect in supergravity without a cosmological constant". *Nucl. Phys.*, B147:105, (1979).

[96] E. Cremmer, C. Kounnas, A. Van Proeyen, J.P. Derendinger, S. Ferrara, B. de Wit, and L. Girardello. "Vector multiplets coupled to N=2 supergravity: superHiggs effect, flat potentials and geometric structures". *Nucl. Phys.*, B250:385, (1985).

[97] E. Cremmer and A. Van Proeyen. "Classifications of Kähler manifolds in N=2 vector multiplet supergravity couplings". *Class. Quantum Grav.*, 2:445, (1985).

[98] A. D'Adda, R. D'Auria, P. Fre', and T. Regge. "Geometric formulation of supergravity theories on orthosymplectic supergroup manifolds". *Nuovo Cimento*, 3:1, (1979).

[99] R. D'Auria, S. Ferrara, and P. Fre'. "Special and quaternionic isometries: general couplings in N=2 supergravity and the scalar potential". *Nucl. Phys.*, B359:705, (1991).

[100] R. D'Auria and P. Fre'. "About bosonic rheonomic symmetry and the generation of a spin 1 field in D=5 supergravity". *Nucl. Phys.*, B173:454, (1980).

[101] R. D'Auria and P. Fre'. "Duality in superspace and Anomaly-free supergravity: some remarks". *Mod. Phys. Lett.*, A3:673, (1988).

[102] R. D'Auria and P. Fre'. "Minimal 10D anomaly-free supergravity and the effective superstring theory". *Phys. Lett.*, 200B:63, (1988).

[103] R. D'Auria, P. Fre', and A. da Silva. "Geometric structure of N=1, D=10 and N=4, D=4 super Yang Mills theory". *Nucl. Phys.*, B196:205, (1982).

[104] R. D'Auria, P. Fre', G. de Matteis, and I. Pesando. "Superspace constraints and Chern–Simons cohomology in D=4 superstring effective theories". *Int. J. Mod. Phys.*, A4:3577, (1989).

[105] R. D'Auria, P. Fre', M. Raciti, and F. Riva. "Anomaly-free supergravity in D=10: I ) the Bianchi identities and the bosonic lagrangian". *Int. J. Mod. Phys.*, A3:953, (1988).

[106] R. D'Auria, P. Fre', and T. Regge. "Graded Lie algebra cohomology and supergravity". *Rivista del Nuovo Cimento*, 3:12, (1980).

[107] R. D'Auria, P. Fre', and T. Regge. "Group manifold approach to gravity and supergravity theories". In *Supergravity 81*, Trieste, 1981. ICTP, S. Ferrara et al. editors.

[108] B. de Wit, P. G. Lauwers, and A. Van Proeyen. "Lagrangians of N=2 supergravity". *Nucl. Phys.*, B255:569, (1985).

[109] B. de Wit, R. Philippe, S.Q. Su, and A. Van Proeyen. "Gauge and matter fields coupled to N=2 supergravity". *Phys. Lett.*, 134B:37, (1984).

[110] B. de Wit and A. Van Proeyen. "Potentials and supersymmetries of general gauged N=2 supergravity Yang–Mills models". *Nucl. Phys.*, B245:89, (1984).

[111] B. de Wit and A. Van Proeyen. "Potentials and symmetries of general gauged N=2 supergravity-Yang–Mills models". *Nucl. Phys.*, B245:89, (1984).

[112] B. de Wit and A. Van Proeyen. "Special geometry, cubic polynomials and homogeneous quaternionic spaces". *Comm. Math. Phys.*, 149:307, (1992).

[113] S. Deser and B. Zumino. "Consistent supergravity". *Phys. Lett.*, 62B:335, (1976).

[114] P. Di Francesco, C. Itzykson, and J. B. Zuber. "Classical W-algebras". *Comm. Math. Phys.*, 140:543, (1991).

[115] P. Di Vecchia, J.L. Petersen, and H.B. Zheng. "Explicit construction of unitary representations of the N=2 superconformal algebra". *Phys. Lett.*, 174B:280, (1986).

[116] R. Dijkgraaf, E. Verlinde, and H. Verlinde. "Notes on topological string theory and 2d quantum gravity". In *Spring School on Strings and Quantum Gravity*, Trieste, (1990). ICTP.

[117] R. Dijkgraaf, E. Verlinde, and H. Verlinde. "Loop equations and Virasoro constraints in non-perturbative two-dimensional gravity". *Nucl. Phys.*, B348:435, (1991).

[118] R. Dijkgraaf, E. Verlinde, and H. Verlinde. "Topological strings in $D < 1$". *Nucl. Phys.*, B352:59, (1991).

[119] R. Dijkgraaf and E. Witten. "Mean field theory, topological field theory and multimatrix models". *Nucl. Phys.*, B342:486, (1990).

[120] M. Dine, P. Huet, and N. Seiberg. "Large and small radius in string theory". *Nucl. Phys.*, B322:301, (1989).

[121] M. Dine, I. Ichinoise, and N. Seiberg. "F terms and D terms in string theory". *Nucl. Phys.*, B293:253, (1987).

[122] M. Dine, R. Rohm, N. Seiberg, and E. Witten. "Gluino condensation in superstring models". *Phys. Lett.*, 156B:55, (1985).

[123] M. Dine, N. Seiberg, X.G. Wen, and E. Witten. "Non-perturbative effects on the string world sheet: I". *Nucl. Phys.*, B278:769, (1986).

[124] M. Dine, N. Seiberg, X.G. Wen, and E. Witten. "Non-perturbative effects on the string world sheet: II". *Nucl. Phys.*, B289:319, (1987).

[125] J. Distler and B. Greene. "Some exact results on the superpotential from Calabi–Yau compactifications". *Nucl. Phys.*, B309:295, (1988).

[126] L. J. Dixon, V. Kaplunowski, and J. Louis. "On the effective field theories describing (2,2) vacua of the heterotic string". *Nucl. Phys.*, 329:27, (1990).

[127] S.K. Donaldson. "An application of gauge theories to the topology of four manifolds". *J. Diff. Geom.*, 18:269, (1983).

[128] Drinfel'd and V. G. Sokolov. "Hamiltonian reduction". *Jour. Sov. Math.*, 30:1975, (1985).

[129] B. Dubrovin. "Integrable systems and classification of two-dimensional topological field theories". In *Integrable systems*, page 313, Marseille, (1991).

[130] M. Duff. "Duality rotations in string theory". *Nucl. Phys.*, B335:610, (1990).

[131] T. Eguchi, P. Gilkey, and A. Hanson. "Gravitation, gauge theories and differential geometry". *Phys. Rep.*, 66:213, (1980).

[132] T. Eguchi and A.J. Hanson. "Self-dual solutions to Euclidean gravity". *Ann. Phys.*, 120:82, (1979).

[133] T. Eguchi, H. Ooguri, A. Taormina, and S.K. Yang. "Superconformal algebras and string compactifications on manifolds of SU(N) holonomy". *Nucl. Phys.*, B315:193, (1989).

[134] T. Eguchi and A. Taormina. "Unitary representations of the N=4 superconformal algebra". *Phys. Lett.*, 196B:75, (1986).

[135] T. Eguchi and A. Taormina. "Character formulas for the N=4 superconformal algebra". *Phys. Lett.*, 200B:315, (1988).

[136] T. Eguchi and A. Taormina. "On the unitary representations of the N=2 and N=4 superconformal algebras". *Phys. Lett.*, 210B:125, (1988).

[137] T. Eguchi and S. K. Yang. "N=2 superconformal models as toppological field theories". *Mod. Phys. Lett.*, A5:1693, (1990).

[138] F. Englert, H. Nicolai, and A. Schellekens. "Superstrings from 26 dimensions". *Nucl. Phys.*, B274:315, (1986).

[139] A. Erdelyi, F. Oberhettinger, W. Magnus, and F.G. Tricomi. *Higher Transcendental Functions.* MacGraw–Hill, 1953.

[140] P. Fayet and J. Iliopulos. "Spontaneously broken supergauge symmetries and Goldstone spinors". *Phys. Lett.*, 51B:46, (1974).

[141] S. Ferrara, P. Fre', , and M. Porrati. "First order higher curvature supergravities and effective theories of strings". *Ann. Phys.*, 175:112, (1987).

[142] S. Ferrara and P. Fre'. "Type II superstrings on twisted group manifolds and their heterotic counterparts". *Int. J. Mod. Phys.*, A5:989, (1990).

[143] S. Ferrara and J. Louis. "Flat holomorphic connections and Picard–Fuchs identities from N=2 supergravity". *Phys. Lett.*, B278:240, (1992).

[144] S. Ferrara, D. Lüst, A. Shapere, and S. Theisen. "Target space modular invariance and low energy couplings in orbifold compactifications". *Phys. Lett.*, 233B:147, (1989).

[145] S. Ferrara, D. Lüst, and S. Theisen. "Duality transformations for blownup orbifolds". *Phys. Lett.*, 242B:39, (1990).

[146] S. Ferrara and A. Strominger. "N=2 spacetime supersymmetry and Calabi–Yau moduli spaces". In *Strings 89*, College Station, 1989. Texas A. and M., World Scientific, Singapore.

[147] S. Ferrara and S. Theisen. "Moduli spaces, effective actions and duality symmetry in string compactification". In *Third Ellenic Summer School*, Corfu', (1989).

[148] A. Font. "Periods and duality symmetries in Calabi–Yau compactifications". *Nucl. Phys.*, B391:358, (1993).

[149] A. Forsyth. *Theory of Differential Equations.* Dover Publications, 1959.

[150] E.S. Fradkin and A.A. Tseytlin. "Quantum string theory effective action". *Nucl. Phys.*, B261:1, (1985).

[151] P. Fre', L. Girardello, A. Lerda, and P; Soriani. "Topological first order systems with Landau–Ginzburg interactions". *Nucl. Phys.*, B387:333, (1992).

[152] P. Fre', F. Gliozzi, R. Monteiro, and A. Piras. "A moduli-dependent lagrangian for (2,2) theories on Calabi–Yau n-folds". *Class. Quantum Grav.*, 8:1455, (1991).

[153] P. Fre' and P. Soriani. "Symplectic embeddings, special Kähler geometry and automorphic functions: the case of SK(n+1)". *Nucl. Phys.*, B371:659, (1992).

[154] D.Z. Freedman, S. Ferrara, and P. van Nieuwenhuizen. "Progress toward a theory of supergravity". *Phys. Rev.*, D13:3214, (1976).

[155] D. Friedan, E. Martinec, and S. Schenker. "Conformal invariance, supersymmetry and string theory". *Nucl. Phys.*, B271:93, (1986).

[156] D. Friedan, E. Martinec, and S.H. Shenker. "Conformal invariance, supersymmetry and string theory". *Nucl. Phys.*, B271:93, (1986).

[157] D. Friedan and P. Windey. "Supersymmetric derivation of the Atiyah–Singer index and the chiral anomaly". *Nucl. Phys.*, B235:395, (1984).

[158] K. Galicki. "A generalization of the momentum mapping construction for quaternionic Kähler manifolds". *Comm. Math. Phys.*, 108:117, (1987).

[159] S.J. Gates, M. Grisaru, M. Rocek, and W. Siegel. *Superspace, or 1001 Lessons in Supersymmetry.* Benjamin Cummings, 1983.

[160] S.J. Gates and H. Nishino. "Manifestly supersymmetric $O(\alpha')$ superstring corrections in new D=10, N=1 supergravity Yang–Mills theory". *Phys. Lett.*, 173B:52, (1986).

[161] D. Gepner. "Exactly solvable string theory on manifolds of SU(N) holonomy". *Phys. Lett.*, 199B:380, (1987).

[162] D. Gepner. "Space–time supersymmetry in compactified string theory and superconformal models". *Nucl. Phys.*, B296:757, (1988).

[163] D. Gepner. "Field identifications in coset conformal field theories". *Phys. Lett.*, 222B:207, (1989).

[164] D. Gepner and Z. Qiu. "Modular invariant partition functions for parafermionic theories". *Nucl. Phys.*, B285:423, (1987).

[165] D. Gepner and E. Witten. "String theory on group manifolds". *Nucl. Phys.*, B278:493, (1986).

[166] G.W. Gibbons and S. Hawking. "Gravitational multi-instantons". *Phys. Lett.*, 78B:430, (1978).

[167] P.B. Gilkey. *Invariance Theory, the Heat Equation and the Atiyah–Singer Index Theorem.* Publish or Perish, Wilmingthon Delaware, 1984.

[168] P. Ginsparg. "Applied conformal field theory". In *Les Houches Summer School*, (1988).

[169] A. Giveon, N. Malkin, and E. Rabinovici. "On discrete symmetries and fundamental domains of target space". *Phys. Lett.*, 238B:57, (1990).

[170] A. Giveon and M. Porrati. "A completely duality invariant effective action of N=4 heterotic strings". *Phys. Lett.*, B246:54, (1990).

[171] A. Giveon and M. Porrati. "Duality invariant string algebra and D=4 effective actions". *Nucl. Phys.*, B355:422, (1991).

[172] A. Giveon, E. Rabinovici, and G. Veneziano. "Duality in string background space". *Nucl. Phys.*, B322:167, (1989).

[173] A. Giveon and D. J. Smit. "Exact Yukawa couplings from topological Landau Ginzburg models ". *Mod. Phys. Lett.*, A6:2211, (1991).

[174] A. Giveon and D.J. Smit. "Symmetries on the moduli space of (2,2) superstring vacua". *Nucl. Phys.*, B349:168, (1991).

[175] F. Gliozzi, D. Olive, and J. Scherk. "Supersymmetry, supergravity theories and the dual spinor spinor model". *Nucl. Phys.*, B122:253, (1977).

[176] P. Goddard, A. Kent, and D. Olive. "Virasoro algebras and coset space models". *Phys. Lett.*, 152B:88, (1985).

[177] P. Goddard and D. Olive. "Kac–Moody and Virasoro algebras in relation to quantum physics". *Int. J. Mod. Phys.*, A1:303, (1986).

[178] M. Green and J.H. Schwarz. "Anomaly cancellation in supersymmetric D=10 gauge theory and superstring theory". *Phys. Lett.*, 149B:117, (1984).

[179] M.B. Green, J.H. Schwarz, and E. Witten. *Superstring theory.* Cambridge U.P., 1986.

[180] B. Greene and M. Plesser. "Duality in Calabi–Yau moduli space". *Nucl. Phys.*, B338:15, (1990).

[181] B. Greene, C. Vafa, and N.P. Warner. "Calabi–Yau manifolds and renormalization group flows". *Nucl. Phys.*, B324:371, (1989).

[182] P. Griffiths. "On the periods of certain rational integrals: I and II". *Ann. Math.*, 90:460, (1969).

[183] P. Griffiths and J. Harris. *Principles of Algebraic Geometry.* A. Wiley, New York, 1978.

[184] M.T. Grisaru, A.M. Van de Ven, and D. Zanon. "Four-loop beta functions for the N=1 and N=2 supersymmetric non linear sigma model in two dimensions". *Phys. Lett.*, 172B:423, (1986).

[185] D. J. Gross and J. Sloan. "The quartic effective action for the heterotic string". *Nucl. Phys.*, B291:41, (1987).

[186] D.J. Gross, J.A. Harvey, E. Martinec, and R. Rohm. "Heterotic string theory I: the free heterotic string". *Nucl. Phys.*, B256:253, (1985).

[187] D.J. Gross, J.A. Harvey, E. Martinec, and R. Rohm. "Heterotic string theory II: the interacting heterotic string". *Nucl. Phys.*, B267:75, (1986).

[188] D.J. Gross and E. Witten. "Superstring modifications of Einstein equation". *Nucl. Phys.*, B277:1, (1986).

[189] S.K. Hau, J.K. Kim, I.G. Koh, and Y. Tanii. "Supersymmetrization of N=1 ten dimensional supergravity with Lorentz Chern–Simons term". *Phys. Rev.*, D34:533, (1986).

[190] S. Hawking and C. Pope. "Symmetry breaking by instantons in supergravity". *Nucl. Phys.*, B146:381, (1978).

[191] N.J. Hitchin, A. Karlhede, U. Lindström, and M. Rocek. "Hyper-Kähler metrics and supersymmetry". *Comm. Math. Phys.*, 108:535, (1987).

[192] S. Hosono, A. Klemm, and S. Theisen. "Lectures on Mirror Symmetry". In *Third Baltic Student Seminar*, (1993).

[193] S. Hosono, A. Klemm, S. Theisen, and S. T. Yau. "Mirror symmetry, mirror map and applications to Calabi–Yau hypersurfaces". *preprint*, HUTMP-93:0801, (1993).

[194] P. Howe and P. West. "N=2 superconformal models, Landau–Ginzburg Hamiltonians and the epsilon expansion". *Phys. Lett.*, 223B:377, (1989).

[195] J. Iliopulos and B. Zumino. "Broken supergauge symmetry and renormalization". *Nucl. Phys.*, B76:310, (1974).

[196] K. Intrilligator and C. Vafa. "Landau–Ginzburg orbifolds". *Nucl. Phys.*, B339:95, (1990).

[197] C. Itzkison and J. Drouffe. *Statistical Field Theory.* Cambridge University Press, 1991.

[198] D. Kastor, E. Martinec, and S. Schenker. "RG flow in N=1 discrete series". *Nucl. Phys.*, B316:590, (1989).

[199] N. Katz. "On the differential equations satisfied by period matrices". *Publ. Math. I.H.E.S.*, 35, (1968).

[200] Y. Kazama and H. Suzuki. "New N=2 superconformal field theories and superstring compactification". *Nucl. Phys.*, B321:232, (1989).

[201] R. Klein and F. Fricke. *Vorlesungen über die Theorie der elliptishen Modulfunktionen.* B.G. Teubner, 1890.

[202] A. Klemm, W. Lerche, S. Yankielowicz, and S. Theisen. "Simple singularities and N=2 supersymmetric Yang–Mills theory". *hep-th*, 9411048, (1994).

[203] A. Klemm, M.G. Schmidt, and S. Theisen. "Correlation functions for topological Landau-Ginzburg models with $c \leq 3$". *Int. J. Mod. Phys.*, 7:6215, (1992).

[204] M. Kontsevich. "Intersection theory on the moduli space of curves and the matrix Airy function". *Comm. Math. Phys.*, 147:1, (1992).

[205] B. Kostant. "Betti numbers of Lie groups". *Am. J. Math.*, 81:973, (1959).

[206] P.B. Kronheimer. "The construction of ALE spaces as c-Kähler quotients". *J. Diff. Geom.*, 29:665, (1989).

[207] J. Lauer, J. Maas, and H.P. Nilles. "Twisted sector representations of discrete background symmetries for two-dimensional orbifolds". *Nucl. Phys.*, B351:353, (1991).

[208] M. Le Bellac. *Quantum and Statistical Field Theory.* Oxford Science Publications, 1990.

[209] K. Lechner, P. Pasti, and M. Tonin. "Anomaly-free sugra and the R superstring term". *Mod. Phys. Lett.*, A2:929, (1987).

[210] W. Lerche, D. Lüst, and A. N. Schellekens. "Chiral four-dimensional heterotic strings from self-dual lattices". *Nucl. Phys.*, B287:477, (1987).

[211] W. Lerche, D. Lüst, and N.P. Warner. "Duality symmetries in N=2 Landau–Ginzburg models". *Phys. Lett.*, 231B:417, (1989).

[212] W. Lerche, D. Lüst, and N.P. Warner. "Duality symmetries in N=2 Landau–Ginzburg models". *Phys. Lett.*, 231B:418, (1989).

[213] W. Lerche, D. Smit, and N. Warner. "Differential equations for periods and flat coordinates in two-dimensional topological matter fields". *Nucl. Phys.*, B372:87, (1992).

[214] W. Lerche, C. Vafa, and N.P. Warner. "Chiral rings in N=2 superconformal theories" . *Nucl. Phys.*, B324:427, (1989).

[215] K. Li. "Recursion relations in topological gravity with minimal matter". *Nucl. Phys.*, B354:711, (1991).

[216] C.A. Lutken and G.C. Ross. "Taxonomy of heterotic superconformal field theories". *Phys. Lett.*, 213B:152, (1988).

[217] M. Lynker and R. Schimmrigk. "Heterotic string compactification on N=2 superconformal theories with c=9". *Phys. Lett.*, 208B:216, (1988).

[218] Z. Maassarani. "On the solution of topological Landau–Ginzburg models with c=3". *Phys. Lett.*, 273B:457, (1991).

[219] E. Martinec. "Algebraic geometry and effective lagrangians". *Phys. Lett.*, 217B:431, (1989).

[220] J. Molera and B. Ovrut. "Sigma model duality and duality transformations in string theories". *Phys. Rev.*, D40:1150, (1989).

[221] V. P. Nair, A. Shapere, A. Strominger, and F. Wilczec. "Compactification of the twisted heterotic string". *Nucl. Phys.*, B287:402, (1987).

[222] M. Nakahara. *Geometry, Topology and Physics.* Adam Hilger, Bristol, 1990.

[223] S. Nam. "The Kac formula for the N=1 and the N=2 superconformal algebras". *Phys. Lett.*, 172B:323, (1986).

[224] C. Nash. *Topology and Geometry for Physicists.* Academic Press, 1983.

[225] C. Nash. *Differential Topology and Quantum Field Theory.* Academic Press, 1990.

[226] Y. Ne'eman and T. Regge. "Gravity and supergravity as gauge theories on a group manifold". *Phys. Lett.*, 74B:54, (1978).

[227] I. Pesando. "The equation of motion of 10D Anomaly-free supergravity". *Phys. Lett.*, 272B:45, (1991).

[228] I. Pesando. "Completion of the ten dimensional Anomaly-free supergravity program: the field equation". *Class. Quantum Grav.*, 9:828, (1992).

[229] G. Ponzano, T. Regge, E.R. Speer, and M.J. Westwater. "The monodromy rings of a class of self-energy graphs". *Comm. Math. Phys.*, 15:83, (1969).

[230] G. Ponzano, T. Regge, E.R. Speer, and M.J. Westwater. "The monodromy rings of one loops Feynmann integrals". *Comm. Math. Phys.*, 18:1, (1970).

[231] M. Raciti, F. Riva, and D. Zanon. "Perturbative approach to D=10 superspace supergravity with Lorentz Chern–Simons form". *Phys. Lett.*, 227B:118, (1989).

[232] T. Regge. "The fundamental group of Poincare' and the analytic properties of Feynmann relativistic amplitudes". In *Nobel Symposium Series VII*, 1968.

[233] L. Romans and N.P. Warner. "Some supersymmetric counterparts of the Lorentz Chern–Simons term". *Nucl. Phys.*, B273:320, (1986).

[234] A. S. Schwarz. "The partition function of a degenerate quadratic function and the Ray–Singer invariants". *Lett. Math. Phys.*, 2:247, (1978).

[235] J. H. Schwarz. "Can string theory overcome deep problems in quantum gravity". *Phys. Lett.*, 272B:239, (1991).

[236] N. Seiberg. "Observations on the moduli space of superconformal theories". *Nucl. Phys.*, B303:286, (1988).

[237] N. Seiberg. "Supersymmetry and non-perturbative beta functions". *Phys. Lett.*, 206:75, (1988).

[238] N. Seiberg and A. Schwimmer. "Comments on N=2,3,4 superconformal algebras in two dimensions". *Phys. Lett.*, 184B:191, (1987).

[239] A. Sen. "(2,0) supersymmetry and space–time supersymmetry in the Heterotic string theory". *Nucl. Phys.*, B278:289, (1986).

[240] A. Sen. "Heterotic string theory on Calabi–Yau manifolds in the Green–Schwarz formalism". *Nucl. Phys.*, B284:423, (1987).

[241] P. Slodowy. *Simple Singularities and Simple Algebraic Groups.* Lect. Notes in Math. 815, Springer Verlag, 1980.

[242] P. Soriani. *"Aspects of special Kähler geometry and moduli space theory in string compactification and (2,2) superconformal models"*. PhD thesis, SISSA/ISAS, 1992.

[243] A. Strominger. "Yukawa couplings in superstring compactifications ". *Phys. Rev. Lett.*, 55:2547, (1985).

[244] A. Strominger. "Special geometry". *Comm. Math. Phys.*, 133:163, (1990).

[245] C. Vafa. "Quantum symmetries of string vacua". *Mod. Phys. Lett.*, A4:1615, (1989).

[246] C. Vafa. "String vacua and orbifoldized Landau Ginzburg models". *Mod. Phys. Lett.*, A4:1169, (1989).

[247] C. Vafa. "Topological Landau Ginzburg models". *Mod. Phys. Lett.*, A6:337, (1991).

[248] C. Vafa and N. P. Warner. "Catastrophes and the classification of conformal theories". *Phys. Lett.*, 218B:51, (1989).

[249] P. van Nieuwenhuizen. "Supergravity". *Phys. Rep.*, 68:189, (1981).

[250] E. Verlinde and N. P. Warner. "Topological Landau–Ginzburg matter at c=3". *Phys. Lett.*, 269B:96, (1991).

[251] M. Villasante. "Duality invariance in four-dimensional N=2 supergravity". *Phys. Rev.*, D45:1831, (1992).

[252] M.A. Virasoro. "Subsidiary conditions and ghosts in dual resonance models". *Phys. Rev.*, D1:2933, (1970).

[253] R. S. Ward and R.O. Wells. *Twistor Geometry and Field Theory.* Cambridge University Press, 1990.

[254] N.P. Warner. "Lectures on N=2 superconformal theories and singularity theory". In *Superstrings 89*, page 197, Trieste, 1989. ICTP, M. Green et al. editors.

[255] R.O. Wells. *Differential Analysis on Complex Manifolds.* Springer Verlag, 1980.

[256] J. Wess and J. Bagger. *Supersymmetry and Supergravity.* Princeton University Press, 1983.

[257] J. Wess and B. Zumino. "A Lagrangian model invariant under supergauge transformations". *Phys. Lett.*, 49B:52, (1974).

[258] E. Witten. "New issues in manifolds with SU(3) holonomy". *Nucl. Phys.*, B268:79, (1986).

[259] E. Witten. "Topological quantum field theories". *Comm. Math. Phys.*, 117:353, (1988).

[260] E. Witten. "Topological sigma models". *Comm. Math. Phys.*, 118:411, (1988).

[261] E. Witten. "On the structure of the topological phase of two dimensional gravity". *Nucl. Phys.*, B340:281, (1990).

[262] E. Witten. "Phases of N=2 theories in two dimensions". *Nucl. Phys.*, B403:159, (1993).

[263] E. Witten. "Mirror symmetry and topological field theory". In *Essays on mirror manifolds.* S.T. Yau International Press, 1994.

[264] E. Witten. "Monopoles and four manifolds". *hep-th*, 9411102, (1994).

[265] E. Witten. "On the Landau–Ginzburg description of N=2 minimal models". *Int. J. Mod. Phys.*, A9:4783, (1994).

[266] E. Witten and N. Seiberg. "Electric–magnetic duality, monopole condensation and confinement in N=2 supersymmetric Yang–Mills theory". *Nucl. Phys.*, B426:19, (1994).

[267] E. Witten and N. Seiberg. "Monopoles, duality and chiral supersymmetry breaking in N=2 QCD". *Nucl. Phys.*, B431:484, (1994).

[268] S.T. Yau. "Calabi's conjecture and some new results in algebraic geometry". *Proc. Natl. Acad. Sci.*, 74:1798, (1977).

[269] A. B. Zamolodchikov. "Renormalization group and perturbation theory about fixed points in two-dimensional field theory". *JEPT*, 46:1090, (1987).

[270] A.B. Zamolodchikov. "Conformal symmetry and multicritical points in two-dimensional quantum field theory". *Sov. J. Nucl. Phys.*, 44:529, (1986).

[271] P. Zoglin. "Heterotic string compactifications using N=2 minimal field theories: quotient models". *Phys. Lett.*, 218B:444, (1989).

[272] B. Zumino. "Supersymmetry and Kähler geometry". *Phys. Lett.*, 87B:203, (1979).